Ergebnisse der Mathematik und ihrer Grenzgebiete. 3. Folge / A Series of Modern Surveys in Mathematics

Volume 60

Ergebnisse der Mathematik und ihrer Grenzgebiete, now in its third sequence, aims to provide summary reports, on a high level, on important topics of mathematical research. Each book is designed as a reliable reference covering a significant area of advanced mathematics, spelling out related open questions, and incorporating a comprehensive, up-to-date bibliography.

More information about this series at http://www.springer.com/series/728

Michel Talagrand

Upper and Lower Bounds for Stochastic Processes

Decomposition Theorems

Second Edition

Springer

Michel Talagrand
Paris, France

ISSN 0071-1136 ISSN 2197-5655 (electronic)
Ergebnisse der Mathematik und ihrer Grenzgebiete. 3. Folge / A Series of Modern Surveys in Mathematics
ISBN 978-3-030-82594-2 ISBN 978-3-030-82595-9 (eBook)
https://doi.org/10.1007/978-3-030-82595-9

Mathematics Subject Classification: 60G17, 60G15, 60E07, 46B09

This Springer imprint is published by the registered company Springer Nature Switzerland AG.
The registered company address is: Gewerbestrasse 11, 6330 Cham, Switzerland

Dedicated to the memory of Xavier Fernique.

Xavier Fernique (1934–2020), by Pierre Fernique

Preface

This book had a previous edition [132]. The changes between the two editions are not only cosmetic or pedagogical, and the degree of improvement in the mathematics themselves is almost embarrassing at times. Besides significant simplifications in the arguments, several of the main conjectures of [132] have been solved and a new direction came to fruition. It would have been more appropriate to publish this text as a brand new book, but the improvements occurred gradually and the bureaucratic constraints of the editor did not allow a change at a late stage without further delay and uncertainty.

We first explain in broad terms the contents of this book, and then we detail some of the changes from [132].

What is the maximum level a certain river is likely to reach over the next 25 years? What is the likely magnitude of the strongest earthquake to occur during the life of a planned nuclear plant? These fundamental practical questions have motivated (arguably also fundamental) mathematics, some of which are the object of this book. The value X_t of the quantity of interest at time t is modeled by a random variable. What can be said about the maximum value of X_t over a certain range of t? How can we guarantee that, with probability close to one, this maximum will not exceed a given threshold?

A collection of random variables $(X_t)_{t \in T}$, where t belongs to a certain index set T, is called a stochastic process, and the topic of this book is the study of the suprema of certain stochastic processes, and more precisely the search for upper and lower bounds for these suprema. The keyword of the book is

INEQUALITIES.

The "classical theory of processes" deals mostly with the case where T is a subset of the real line or of $\mathbb{R}^n$. We do not focus on that case, and the book does not really expand on the most basic and robust results which are important in this situation. Our most important index sets are "high-dimensional": the large sets of data which are currently the focus of so much attention consist of data which usually depend on many parameters. Our specific goal is to demonstrate the impact and the range of modern abstract methods, in particular through their treatment of several classical questions which are not accessible to "classical methods."

Andrey Kolmogorov invented the most important idea to bound stochastic processes: chaining. This wonderfully efficient method answers with little effort a number of basic questions but fails to provide a complete understanding, even in natural situations. This is best discussed in the case of Gaussian processes, where the family $(X_t)_{t\in T}$ consists of centered jointly Gaussian random variables (r.v.s). These are arguably the most important of all. A Gaussian process defines in a canonical manner a distance d on its index set T by the formula

$$d(s,t) = (\mathsf{E}(X_s - X_t)^2)^{1/2} . \tag{0.1}$$

Probably the single most important conceptual progress about Gaussian processes was the gradual realization that the metric space (T, d) is the key object to understand them, even if T happens to be an interval of the real line. This led Richard Dudley to develop in 1967 an abstract version of Kolmgorov's chaining argument adapted to this situation. The resulting very efficient bound for Gaussian processes is unfortunately not always tight. Roughly speaking, "there sometimes remains a parasitic logarithmic factor in the estimates".

The discovery around 1985 (by Xavier Fernique and the author) of a precise (and in a sense, *exact*) relationship between the "size" of a Gaussian process and the "size" of this metric space provided the missing understanding in the case of these processes. Attempts to extend this result to other processes spanned a body of work that forms the core of this book.

A significant part of the book is devoted to situations where skill is required to "remove the last parasitic logarithm in the estimates". These situations occur with unexpected frequency in all kinds of problems. A particularly striking example is as follows. Consider n^2 independent uniform random points $(X_i)_{i\le n^2}$, which are uniformly distributed in the unit square $[0,1]^2$. How far is a typical sample from being very uniformly spread on the unit square? To measure this we construct a one-to-one map π from $\{1,\ldots,n^2\}$ to the vertices $v_1,\ldots,v_{n^2}$ of a uniform $n \times n$ grid in the unit square. If we try to minimize the *average* distance between X_i and $v_{\pi(i)}$, we can do as well as about $\sqrt{\log n}/n$ but no better. If we try to minimize the *maximum* distance between X_i and $v_{\pi(i)}$, we can do as well as about $(\log n)^{3/4}/n$ but no better. The factor $1/n$ is just due to scaling, but the fractional powers of $\log n$ require a surprising amount of work.

The book is largely self-contained, but it mostly deals with rather subtle questions such as the previous one. It also devotes considerable energy to the problem of finding *lower* bounds for certain processes, a topic far more difficult and less developed than the search for upper bounds. Even though some of the main ideas of at least Chap. 2 could (and should!) be taught at an elementary level, this is an advanced text.

This book is in a sense a continuation of the monograph [53], or at least of part of it. I made no attempt to cover again all the relevant material of [53], but familiarity with [53] is certainly not a prerequisite and maybe not even helpful. The way certain results are presented there is arguably obsolete, and, more importantly, many of the problems considered in [53] (in particular, limit theorems) are no longer the focus of much interest.

One of my main goals is to communicate as much as possible of my experience from working on stochastic processes, and I have covered most of my results in this area. A number of these results were proved many years ago. I still like them, but some seem to be waiting for their first reader. The odds of these results meeting this first reader while staying buried in the original papers seemed nil, but they might increase in the present book form. In order to present a somewhat coherent body of work, I have also included rather recent results by others in the same general direction.[1] I find these results deep and very beautiful. They are sometimes difficult to access for the non-specialist. Explaining them here in a unified and often simplified presentation could serve a useful purpose. Still, the choice of topics is highly personal and does not represent a systematic effort to cover all the important directions. I can only hope that the book contains enough state-of-art knowledge about sufficiently many fundamental questions to be useful.

Let me now try to outline the progress since the previous edition.[2] While attempting to explain better my results to others, I ended up understanding them much better myself. The material of the previous edition was reduced by about 100 pages due to better proofs.[3] More importantly, reexamination of the material resulted in new methods, and a new direction came to fruition, that of

DECOMPOSITION THEOREMS.

The basic idea is that there are two fundamentally different ways to control the size of a sum $\sum_{i\le N} X_i$. One may take advantage of cancellations between terms, or one may bound the sum by the sum of the absolute values. One may also *interpolate* between the two methods, which in that case means writing a decomposition $X_i = X_i' + X_i''$ and controlling the size of the sum $\sum_{i\le N} X_i'$ by taking advantage of the cancellations between terms, but controlling the sum $\sum_{i\le N} X_i''$ by the sum of the absolute values. The same schoolboy idea, in the setting of stochastic processes, is that a process can be bounded on the one hand using chaining, and on the other hand can often be bounded by cruder methods, involving replacing certain sums by the sums of the absolute values. The amazing fact is that many processes can be controlled by interpolating between these two methods, that is can be decomposed into the sum of two pieces, each of which can be controlled by one of these methods.

[1] With one single exception I did not include results by others proved after the first edition of this book.

[2] A precise comparison between the two editions may be found in Appendix G.

[3] A limited quantity of material of secondary importance was also removed. The current edition is not shorter than the previous one because many details have been added, as well as an entire chapter on the new results, and a sketch of proof for many exercises.

Such is the nature of the landmark result of Bednorz and Latała [16], the proof of the Bernoulli conjecture, which is the towering result of this book. Several conjectures of [132] in the same general directions have been solved[4] concerning in particular empirical processes and random series of functions.

Despite the considerable progress represented by the solution of these conjectures, a number of seemingly important questions remain open, and one of my main goals is to popularize these. Opinions differ as to what constitutes a really important problem, but I like those I explain here because they deal with fundamental structures. These problems might be challenging. At least, I tried my best to make progress on them, but they have seen little progress and received little attention.

I would like to express my infinite gratitude to Shahar Mendelson. While he was donating his time to help another of my projects, it became clear through our interactions that, while I had produced great efforts toward the quality of the mathematics contained in my books, I certainly had not put enough efforts into the exposition of this material. I concluded that there should be real room for improvement in the text of [132], and this is why I started to revise it, and this led to the major advances presented here.

While preparing the current text I have been helped by a number of people. I would like to thank some of them here (and to apologize to all those whom I do not mention). Ramon van Handel suggested a few almost embarrassing simplifications. Hengrui Luo and Zhenyuan Zhang suggested literately hundreds of improvements, and Rafał Meller's comments had a great impact too. Further luck had it that, almost at the last minute, my text attracted the attention of Kevin Tanguy whose efforts resulted in a higher level of detail and a gentler pace of exposition. In particular, his and Zhang's efforts gave me the energy to make a fresh attempt at explaining and detailing the proof of the Bernoulli conjecture obtained by Bednorz and Latała in [16]. This proof is the most stupendously beautiful piece of mathematics I have met in my entire life. I wish the power of this result and the beauty of this proof become better understood.

I dedicate this work to the memory of Xavier Fernique. Fernique was a deeply original thinker. His groundbreaking contributions to the theory of Gaussian processes were viewed as exotic by mainstream probabilists, and he never got the recognition he deserved. I owe a great debt to Fernique: it is his work on Gaussian processes which made my own work possible, first on Gaussian processes, and then on all the situations beyond this case. This work occupied many of my most fruitful years. A large part of it is presented in the present volume. It would not have existed without Fernique's breakthroughs.

Paris, France Michel Talagrand

[4] After another crucial contribution of Bednorz and Martynek [18].

Contents

Part I The Generic Chaining

Chapter 1
What Is This Book About?

This short chapter describes the philosophy underlying this book and some of its highlights. This description, often using words rather than formulas, is necessarily imprecise and is only intended to provide some insight into our point of view.

1.1 Philosophy

The practitioner of stochastic processes is likely to be struggling at any given time with his favorite model of the moment, a model which typically involves a rich and complicated structure. There is a near infinite supply of such models. The importance with which we view any one of them is likely to vary over time.

The first advice I received from my advisor Gustave Choquet was as follows: always consider a problem under the minimum structure in which it makes sense. This advice has literally shaped my mathematical life. It will probably be as fruitful in the future as it has been in the past. By following it, one is naturally led to study problems with a kind of minimal and intrinsic structure. Not so many structures are really basic, and one may hope that these will remain of interest for a very long time. This book is devoted to the study of such structures which arise when one tries to estimate the suprema of stochastic processes.

The feeling, real or imaginary, that one is studying objects of intrinsic importance is enjoyable, but the success of the approach of studying "minimal structures" has ultimately to be judged by its results. As we shall demonstrate, the tools arising from this approach provide the final words on a number of classical problems.

M. Talagrand, *Upper and Lower Bounds for Stochastic Processes*, Ergebnisse der Mathematik und ihrer Grenzgebiete. 3. Folge / A Series of Modern Surveys in Mathematics 60, https://doi.org/10.1007/978-3-030-82595-9_1

1.2 What Is Chaining?

A stochastic process is a collection of random variables (r.v.s) $(X_t)_{t\in T}$ indexed by a set T. To study it, Kolmogorov invented chaining, the main tool of this book. The fundamental idea of chaining is to replace the index set T by a sequence of finite approximations T_n and to study the r.v.s X_t through successive approximations $X_{\pi_n(t)}$ where $\pi_n(t) \in T_n$. The first approximation consists of a single point t_0 so $T_0 = \{t_0\}$. The fundamental relation is then

$$X_t - X_{t_0} = \sum_{n\geq 1}(X_{\pi_n(t)} - X_{\pi_{n-1}(t)}) \ . \tag{1.1}$$

When T is finite, the only case we really need, the sum on the right is finite. This relation gives its name to the method: the chain of approximations $\pi_n(t)$ links t_0 and t. To control the differences $X_t - X_{t_0}$, it suffices then to control all the differences $|X_{\pi_n(t)} - X_{\pi_{n-1}(t)}|$.

1.3 The Kolmogorov Conditions

Kolmogorov stated the "Kolmogorov conditions", which robustly ensure the good behavior of a stochastic process indexed by a subset of $\mathbb{R}^m$. These conditions are studied in any advanced probability course. If you have taken such a course, this section will refresh your memory about these conditions, and the next few sections will present the natural generalization of the chaining method in an abstract metric space, as it was understood in, say, 1970. Learning in detail about these historical developments now makes sense only if you have already heard of them, because the modern chaining method, which is presented in Chap. 2, is in a sense *far simpler* than the classical method. For this reason, the material up to Sect. 1.4 included is directed toward a reader who is already fluent in probability theory. If, on the other hand, you have never heard of these things and if you find this material too difficult, you should start directly with Chap. 2, which *is written at a far greater level of detail and assumes minimal familiarity with even basic probability theory.*

We say that a process $(X_t)_{t\in T}$, where $T = [0,1]^m$, satisfies the Kolmogorov conditions if

$$\forall s,t \in [0,1]^m \ , \ \mathsf{E}|X_s - X_t|^p \leq d(s,t)^\alpha \ . \tag{1.2}$$

where $d(s,t)$ denotes the Euclidean distance and $p > 0, \alpha > m$. Here E denotes mathematical expectation. In our notation, the operator E applies to whatever expression is placed behind it, so that $\mathsf{E}|Y|^p$ stands for $\mathsf{E}(|Y|^p)$ and not for $(\mathsf{E}|Y|)^p$. *This convention is in force throughout the book.*

Let us apply the idea of chaining to processes satisfying the Kolmogorov conditions. The most obvious candidate for the approximating set T_n is the set G_n of points x in $[0,1[^m$ such that the coordinates of $2^n x$ are positive integers.[1] Thus, $\operatorname{card} G_n = 2^{nm}$. It is completely natural to choose $\pi_n(u) \in G_n$ as close to u as possible, so that $d(u, \pi_n(u)) \le \sqrt{m}2^{-n}$ and $d(\pi_n(u), \pi_{n-1}(u)) \le d(\pi_n(u), u) + d(u, \pi_{n-1}(u)) \le 3\sqrt{m}2^{-n}$.

For $n \ge 1$, let us then define

$$U_n = \{(s,t) \; ; \; s \in G_n \; , \; t \in G_n \; , \; d(s,t) \le 3\sqrt{m}2^{-n}\} \; . \tag{1.3}$$

Given $s = (s_1, \ldots, s_m) \in G_n$, the number of points $t = (t_1, \ldots, t_m) \in G_n$ with $d(s,t) \le 3\sqrt{m}2^{-n}$ is bounded independently of s and n because $|t_i - s_i| \le d(s,t)$ for each $i \le m$, so that we have the crucial property

$$\operatorname{card} U_n \le K(m) 2^{nm} \; , \tag{1.4}$$

where $K(m)$ denotes a number depending only on m, which need not be the same on each occurrence. Consider then the r.v.

$$Y_n = \max\{|X_s - X_t| \; ; \; (s,t) \in U_n\} \; , \tag{1.5}$$

so that (and since $G_{n-1} \subset G_n$) for each u,

$$|X_{\pi_n(u)} - X_{\pi_{n-1}(u)}| \le Y_n \; . \tag{1.6}$$

To avoid having to explain what is "a version of the process" and since we care only about inequalities, we will consider only the r.v.s X_t for $t \in G =: \bigcup_{n \ge 0} G_n$. We first claim that

$$\sup_{s,t \in G \, ; \, d(s,t) \le 2^{-k}} |X_s - X_t| \le 3 \sum_{n \ge k} Y_n \; . \tag{1.7}$$

To prove this, consider $n \ge k$ such that $s, t \in G_n$, so that $s = \pi_n(s)$ and $t = \pi_n(t)$. Assuming $d(s,t) \le 2^{-k}$, we have

$$d(\pi_k(s), \pi_k(t)) \le d(s, \pi_k(s)) + d(s,t) + d(t, \pi_k(t)) \le 3\sqrt{m}2^{-k} \; ,$$

so that $(\pi_k(s), \pi_k(t)) \in U_k$ and thus

$$|X_{\pi_k(s)} - X_{\pi_k(t)}| \le Y_k \; .$$

[1] There is no other reason for using the points x in $[0,1[^m$ such that the coordinates of $2^n x$ are positive integers rather than the points x in $[0,1]^m$ with the same property than the fact that there are 2^{nm} such points rather than the typographically unpleasant number $(2^n + 1)^m$.

Next, for $u \in \{s,t\}$,

$$X_u - X_{\pi_k(u)} = X_{\pi_n(u)} - X_{\pi_k(u)} = \sum_{k \le \ell < n} X_{\pi_{\ell+1}(u)} - X_{\pi_\ell(u)} ,$$

and since $|X_{\pi_{\ell+1}(u)} - X_{\pi_\ell(u)}| \le Y_{\ell+1}$, we obtain $|X_u - X_{\pi_k(u)}| \le \sum_{\ell \ge k} Y_{\ell+1}$. To obtain (1.7), we then use the previous inequalities and the identity

$$X_s - X_t = X_s - X_{\pi_k(s)} + X_{\pi_k(s)} - X_{\pi_k(t)} + X_{\pi_k(t)} - X_t .$$

Let us now draw some consequences of (1.7). For a finite family of numbers $V_i \ge 0$, we have

$$(\max_i V_i)^p \le \sum_i V_i^p , \tag{1.8}$$

and thus

$$\mathsf{E} Y_n^p \le \mathsf{E} \sum_{(s,t)\in U_n} |X_s - X_t|^p \le K(m,\alpha)2^{n(m-\alpha)} ,$$

since $\mathsf{E}|X_s - X_t|^p \le K(m,\alpha)2^{-n\alpha}$ for $(s,t) \in U_n$ by (1.2) and using (1.4). To proceed, one needs to distinguish whether or not $p \ge 1$. For specificity, we assume $p \ge 1$. Since, as we just proved, $\|Y_n\|_p := (\mathsf{E}|Y_n|^p)^{1/p} \le K(m,p,\alpha)2^{n(m-\alpha)/p}$, the triangle inequality in L^p yields[2]

$$\Big\| \sum_{n \ge k} Y_n \Big\|_p \le \sum_{n \ge k} K(m,p,\alpha)2^{n(m-\alpha)/p} \le K(m,p,\alpha)2^{k(m-\alpha)/p} . \tag{1.9}$$

Combining with (1.7), we then obtain

$$\Big\| \sup_{s,t \in G; d(s,t) \le 2^{-k}} |X_s - X_t| \Big\|_p \le K(m,p,\alpha)2^{k(m-\alpha)/p} , \tag{1.10}$$

a sharp inequality from which it is then simple to prove (with some loss of sharpness) results such as the fact that for $0 < \beta < \alpha - m$, one has

$$\mathsf{E} \sup_{s,t \in G} \frac{|X_s - X_t|^p}{d(s,t)^\beta} < \infty . \tag{1.11}$$

[2] There of course the two occurrences of the constant $K(m,p,\alpha)$ are not the same.

Exercise 1.3.1 Prove (1.11). Hint: Prove that

$$\sum_{k\geq 0} \mathsf{E} \sup_{s,t\in G; d(s,t)\leq 2^{-k}} 2^{k\beta}|X_s - X_t|^p < \infty . \tag{1.12}$$

Thus, chaining not only proves that the process $(X_t)_{t\in T}$ has a continuous version; it also provides the very good estimate (1.10). One reason for which everything is so easy in this case is that the size of the terms $X_{\pi_{n+1}(u)} - X_{\pi_n(u)}$ decreases like a geometric series.

Let us then pause for a moment and reflect on what we have been doing.

- The Euclidean metric structure of T is not really intrinsic to the problem. Far more intrinsic is the (quasi) distance on T given by $\delta(s,t) = \|X_s - X_t\|_p$. The condition (1.2), which we may now write as $\delta(s,t) \leq d(s,t)^{\alpha/p}$, simply enforces a kind of "smallness condition" on the metric space (T, δ).
- The use of the bound (1.6) is rather pessimistic, as it bounds each of the increments along the chain by the worst possible case among each increment.

These two remarks contain in germ much of the future progress we will make. Following the first remark, we will learn, starting with the next section, to look at problems in a more intrinsic manner. And our sharp chaining methods will avoid the crude bound of each increment by the worst possible case.

There are many variations on the previous ideas. The next two exercises explore one.

Exercise 1.3.2 Consider a convex function $\varphi \geq 0$ with $\varphi(0) = 0$. Prove that for r.v.s $V_i \geq 0$ one has

$$\mathsf{E} \max_i V_i \leq \varphi^{-1}\Big(\sum_i \mathsf{E}\varphi(V_i)\Big) . \tag{1.13}$$

Exercise 1.3.3 Consider the function φ as above, and consider positive numbers c_n, d_n. Assume that the process $(X_t)_{t\in T}$ satisfies

$$\forall n \geq 0 \ , \ \forall s,t \in T \ , \ d(s,t) \leq 3\sqrt{m}2^{-n} \Rightarrow \mathsf{E}\varphi\Big(\frac{|X_s - X_t|}{c_n}\Big) \leq d_n . \tag{1.14}$$

Prove that

$$\mathsf{E} \sup_{s,t\in G, d(s,t)\leq 2^{-k}} |X_s - X_t| \leq 3\sum_{n\geq k} c_n \varphi^{-1}(K(m)2^{nm} d_n) . \tag{1.15}$$

The series in (1.15) has no reason to converge like a geometric series, so we already are being more sophisticated than in the case of the Kolmogorov conditions.[3]

1.4 Chaining in a Metric Space: Dudley's Bound

Suppose now that we want to study the uniform convergence on [0, 1] of a random Fourier series $X_t = \sum_{k\geq 1} a_k g_k \cos(2\pi i k t)$ where a_k are numbers and (g_k) are independent standard Gaussian r.v.s. The Euclidean structure of [0, 1] is not intrinsic to the problem. Far more relevant is the distance d given by

$$d(s,t)^2 = \mathsf{E}(X_s - X_t)^2 = \sum_k a_k^2(\cos(2i\pi ks) - \cos(2i\pi kt))^2 . \tag{1.16}$$

This simple idea took a very long time to emerge. Once one thinks about the distance d, then in turn the fact that the index set T is [0, 1] is no longer very relevant because this particular structure does not connect very well with the distance d. One is then led to consider Gaussian processes indexed by an abstract set T.[4] We say that $(X_t)_{t\in T}$ is a *Gaussian process* when the family $(X_t)_{t\in T}$ is jointly Gaussian and centered.[5] Then, just as in (1.16), the process induces a canonical distance d on T given by $d(s,t) = (\mathsf{E}(X_s - X_t)^2)^{1/2}$. We will express that Gaussian r.v.s have small tails by the inequality

$$\forall s,t \in T\ ,\ \mathsf{E}\varphi\Big(\frac{|X_s - X_t|}{d(s,t)}\Big) \leq 1\ , \tag{1.17}$$

where $\varphi(x) = \exp(x^2/4) - 1$. This inequality holds because if g is a standard Gaussian r.v., then $\mathsf{E}\exp(g^2/4) \leq 2$.[6]

To perform chaining for such a process, in the absence of further structure on our metric space (T,d), how do we choose the approximating sets T_n? Thinking back to the Kolmogorov conditions, it is very natural to introduce the following definition:

---

[3] In the left-hand side of (1.15), we would like to do better than controlling the expectation, but one really needs some regularity of the function φ for this. It suffices here to say that when $\varphi(x) = |x|^p$ for $p \geq 1$, we may replace the expectation by the norm of L^p, proceeding exactly as we did in the case of the Kolmogorov conditions.

[4] Let us stress the point. Even though the index set is a subset of $\mathbb{R}^m$, we have *no chance* to really understand the process unless we forget this irrelevant structure.

[5] Centered means that $\mathsf{E}X_t = 0$ for each t.

[6] Starting with the next chapter, we will control the r.v.s $|X_s - X_t|$ through their tail properties, and (1.17) is just another way to present the same situation.

Definition 1.4.1 For $\epsilon > 0$, the covering number $N(T, d, \epsilon)$ of a metric space (T, d) is the smallest integer N such that T can be covered by N balls of radius ϵ.[7]

Equivalently, $N(T, d, \epsilon)$ is the smallest number N such that there exists a set $V \subset T$ with card $V \le N$ and such that each point of T is within distance ϵ of V.

Let us denote by $\Delta(T) = \sup_{s,t \in T} d(s,t)$ the diameter of T and observe that $N(T, d, \Delta(T)) = 1$. We construct our approximating sets T_n as follows: Consider the largest integer n_0 with $\Delta(T) \le 2^{-n_0}$. For $n \ge n_0$, consider a set $T_n \subset T$ with card $T_n = N(T, d, 2^{-n})$ such that each point of T is within distance 2^{-n} of a point of T_n.[8] In particular T_0 consists of a single point.

We then perform the chaining as in the case of the Kolmogorov conditions, using for $\pi_n(t)$ a point in T_n with $d(t, \pi_n(t)) \le 2^{-n}$. Consider

$$U_n = \{(s,t) \ ; \ s, t \in T_n \ , \ d(s,t) \le 3 \cdot 2^{-n}\} \ ,$$

so that

$$\operatorname{card} U_n \le (\operatorname{card} T_n)^2 = N(T, d, 2^{-n})^2 \ .$$

This crude bound is hard to improve in general and should be compared to (1.4). We now apply (1.13) to the r.v.s $V_i = |X_s - X_t|/(3 \cdot 2^{-n})$ for $i = (s,t) \in U_n$. Since $\mathsf{E}\varphi(V_i) \le 1$, we obtain that the r.v.

$$Y_n = \max\{|X_s - X_t| \ ; \ (s,t) \in U_n\}$$

satisfies

$$\mathsf{E} Y_n \le 3 \cdot 2^{-n} \varphi^{-1}(N(T, d, 2^{-n})^2) \ ,$$

and exactly as in the case of the Kolmogorov conditions, we obtain

$$\mathsf{E} \sup_{d(s,t) \le 2^{-k}} |X_s - X_t| \le L \sum_{n \ge k} 2^{-n} \varphi^{-1}(N(T, d, 2^{-n})^2) \ ,$$

where L is a number (which may change between occurrences). We delay the exercise of writing this inequality in integral form as

$$\mathsf{E} \sup_{d(s,t) \le \delta} |X_s - X_t| \le L \int_0^{\delta} \varphi^{-1}(N(T, d, \epsilon)^2) d\epsilon \ . \tag{1.18}$$

[7] Here our balls are closed balls. One could also use open balls in this definition. There seems to be no universal agreement about this. For our purpose, it makes no difference whatsoever.

[8] We do not require that $T_n \subset T_{n+1}$. In Sect. 1.3, it does happen that $G_n \subset G_{n+1}$, but this was not really used in the arguments.

In the case of the function $\varphi(x) = \exp(x^2/4) - 1$, so that $\varphi^{-1}(x) = 2\sqrt{\log(1+x)}$, inequality (1.18) is easily shown to be equivalent to the following more elegant formulation:

Theorem 1.4.2 (Dudley's Bound) *If $(X_t)_{t\in T}$ is a Gaussian process with natural distance d, then*

$$\mathsf{E} \sup_{d(s,t)\le\delta} |X_s - X_t| \le L \int_0^\delta \sqrt{\log N(T,d,\epsilon)}\, d\epsilon \;. \tag{1.19}$$

This *very general* inequality is *by far the most useful result* on continuity of Gaussian processes.

Exercise 1.4.3 Prove that the previous bound gives the correct uniform modulus of continuity for Brownian motion on [0, 1]: for $\delta \le 1$,

$$\mathsf{E} \sup_{|s-t|\le\delta} |B_s - B_t| \le L\sqrt{\delta \log(2/\delta)} \;.$$

The message of Chap. 2 is simple:

- However useful, Dudley's bound is not optimal in a number of fundamentally important situations.
- It requires no more work to obtain a better bound which is optimal in every situation.

1.5 Overall Plan of the Book

A specific feature of the index set $T = [0,1]^m$ (provided with the Euclidean distance) occurring in the Kolmogorov conditions is that it is really "m-dimensional" and "the same around each point". This is not the case for index sets which occur in a great many natural situations. If one had to summarize in one sentence the content of the upper bounds presented in this book, it would be that they develop methods which are optimal even when this feature does not occur.

The main tools are built in Parts I and II. Part I is devoted to the most important situation we consider in the book, the study of Gaussian processes, and we learn the basic concepts on how to measure the "size" of a metric space. The effectiveness of the corresponding tools is then demonstrated by proving classical results on matchings.

The goal of Part II is to extend the results of the Gaussian case to other more general processes. This program of building the proper tools to go beyond the Gaussian case was started by the author soon after he obtained his results on Gaussian processes (which are presented in Chap. 2). It is a significant endeavor which requires a number of new concepts. The most important of these is the idea

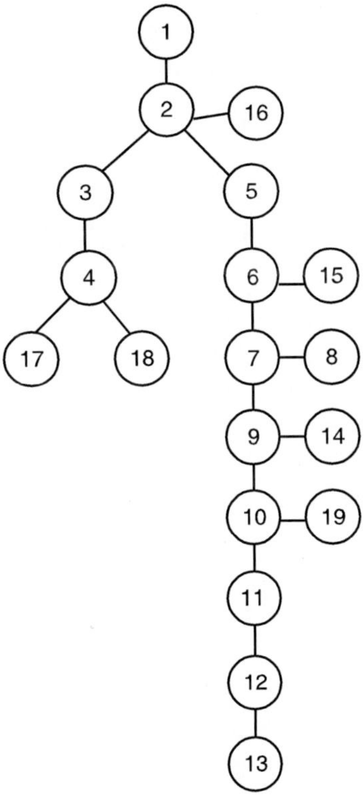

Fig. 1.1 Dependence chart between chapters. Very marginal dependence is not indicated

of families of distances. We can no longer entirely describe the situation using a single distance on the index set (as is the case for Gaussian processes). In some sense, *this program has been completed.* Most of the results which were dreamed by the author[9] between 1985 and 1990 are now proved in Chap. 11.

Part III explores situations which belong to the same circle of ideas but in diverse directions. The dependence chart between the chapters is given in Fig. 1.1.

1.6 Does This Book Contain any Ideas?

At this stage, it is not really possible to precisely describe any of the new ideas which will be presented, but if the following statements are not crystal clear to you, you may have something to learn from this book:

Idea 1 It is possible to organize chaining optimally using increasing sequences of partitions.

[9] Including some which sounded like crazily optimistic conjectures!

Idea 2 There is an automatic device to construct such sequences of partitions, using "functionals", quantities which measure the size of the subsets of the index set. This yields a complete understanding of boundedness of Gaussian processes.

Idea 3 Ellipsoids are much smaller than one would think, because they (and, more generally, sufficiently convex bodies) are thin around the edges. This explains the funny fractional powers of logarithms in certain matching theorems.

Idea 4 One may witness that a metric space is large by the fact that it contains large trees or equivalently that it supports an extremely scattered probability measure.

Idea 5 Consider a set T on which you are given a distance d and a random distance d_ω such that, given $s, t \in T$, it is rare that the distance $d_\omega(s, t)$ is much smaller than $d(s, t)$. Then if in the appropriate sense (T, d) is large, it must be the case that (T, d_ω) is typically large. This principle enormously constrains the structure of many bounded processes built on random series.

Idea 6 There are different ways a random series might converge. It might converge because chaining witnesses that there is cancellation between terms, or it might converge because the sum of the absolute values of its terms already converges. Many processes built on random series can be split in two parts, each one converging according to one of the previous phenomena.

The book contains many more ideas, but you will have to read more to discover them.

1.7 Overview by Chapters

1.7.1 Gaussian Processes and the Generic Chaining

This subsection gives an overview of Chap. 2. More generally, Sect. 1.7.n gives the overview for Chapter $n + 1$.

The most important question considered in this book is the boundedness of Gaussian processes. The key object is the metric space (T, d) where T is the index set and d the intrinsic distance (0.1). As investigated in Sect. 2.11, this metric space is far from being arbitrary: it is isometric to a subset of a Hilbert space. It is, however, a deadly trap to try to use this specific property of the metric space (T, d). The proper approach is to just think of it as a general metric space.

After reviewing some elementary facts, in Sect. 2.4, we explain the basic idea of the "generic chaining", one of the key ideas of this work. Chaining is a succession of steps that provide successive approximations of the index space (T, d). In the Kolmogorov chaining, for each n, the difference between the n-th and the $(n+1)$-th approximation of the process, which we call here "the variation of the process during the n-th chaining step", is "controlled uniformly over all possible chains". Generic chaining allows that the variation of the process during the n-th chaining step "may depend on which chain we follow". Once the argument is properly organized, it is not any more complicated than the classical argument. It is in fact exactly the same. Yet, while Dudley's classical bound is not always sharp, the bound obtained through the generic chaining is optimal. Entropy numbers are reviewed in Sect. 2.5.

It is technically convenient to formulate the generic chaining bound using special sequences of partitions of the metric space (T, d), that we shall call *admissible sequences* throughout the book. The key to make the generic chaining bound useful is then to be able to construct admissible sequences. These admissible sequences measure an aspect of the "size" of the metric space and are introduced in Sect. 2.7. In Sect. 2.8, we introduce another method to measure the "size" of the metric space, through the behavior of certain "functionals", which are simply numbers attached to each subset of the entire space. The fundamental fact is that the two measures of the size of the metric space one obtains either through admissible sequences or through functionals are equivalent in full generality. This is proved in Sect. 2.8 for the easy part (that the admissible sequence approach provides a larger measure of size than the functional approach) and in Sect. 2.9 for the converse. This converse is, in effect, an algorithm to construct sequences of partitions in a metric space given a functional. Functionals are of considerable use throughout the book.

In Sect. 2.10, we prove that the generic bound can be reversed for Gaussian processes, therefore providing a characterization of their sample-boundedness. Generic chaining entirely explains the size of Gaussian processes, and the dream of Sect. 2.12 is that a similar situation will occur for many processes.

In Sect. 2.11, we explain why a Gaussian process in a sense *is* nothing but a subset of Hilbert space. Remarkably, a number of basic questions remain unanswered, such as how to relate through geometry the size of a subset of Hilbert space seen as a Gaussian process with the corresponding size of its convex hull.

Dudley's bound fails to explain the size of the Gaussian processes indexed by ellipsoids in Hilbert space. This is investigated in Sect. 2.13. Ellipsoids will play a basic role in Chap. 4.

1.7.2 Trees and Other Measures of Size

We describe different notions of trees and show how one can measure the "size" of a metric space by the size of the largest trees it contains, in a way which is equivalent to the measures of size introduced in Chap. 2. This idea played an important part in the history of Gaussian processes. Its appeal is mostly that trees are easy to visualize. Building a large tree in a metric space is an efficient method to bound its size from below. We then learn a method of Fernique to measure the size of a metric space through certain properties of the probability measures on it. It will be amenable to vast generalizations.

1.7.3 Matching Theorems

Chapter 4 makes the point that the generic chaining (or some equivalent form of it) is already required to really understand the irregularities occurring in the distribution of N points $(X_i)_{i \le N}$ independently and uniformly distributed in the unit square. These irregularities are measured by the "cost" of pairing (=matching) these points with N fixed points that are very uniformly spread, for various notions of cost.

These optimal results involve mysterious powers of $\log N$. We are able to trace them back to the geometry of ellipsoids in Hilbert space, so we start the chapter with an investigation of these ellipsoids in Sect. 4.1. The philosophy of the main result, the ellipsoid theorem, is that an ellipsoid is in some sense somewhat smaller than it appears at first. This is due to convexity: an ellipsoid gets "thinner" when one gets away from its center. The ellipsoid theorem is a special case of a more general result (with the same proof) about the structure of sufficiently convex bodies, one that will have important applications in Chap. 19.

In Sect. 4.3, we provide general background on matchings. In Sect. 4.5, we investigate the case where the cost of a matching is measured by the average distance between paired points. We prove the result of Ajtai, Komlós and Tusnády that the expected cost of an optimal matching is at most $L\sqrt{\log N}/\sqrt{N}$ where L is a number. The factor $1/\sqrt{N}$ is simply a scaling factor, but the fractional power of log is optimal as shown in Sect. 4.6. In Sect. 4.7, we investigate the case where the cost of a matching is measured instead by the maximal distance between paired points. We prove the theorem of Leighton and Shor that the expected cost of a matching is at most $L(\log N)^{3/4}/\sqrt{N}$, and the power of log is shown to be optimal in Sect. 4.8.

With the exception of Sect. 4.1, the results of Chap. 4 are not connected to any subsequent material before Chap. 17.

1.7.4 Warming Up with *p*-Stable Processes

With this chapter, we start the program of vastly extending the results of Chap. 2 concerning Gaussian processes. We outline several of the fruitful methods on the class of p-stable processes, based on their property of being conditionally Gaussian.

1.7.5 Bernoulli Processes

Random signs are obviously important r.v.s and occur frequently in connection with "symmetrization procedures", a very useful tool. In a Bernoulli process, the individual random variables X_t are linear combinations of independent random signs. Each Bernoulli process is associated with a Gaussian process in a canonical manner, when one replaces the random signs by independent standard Gaussian r.v.s. The Bernoulli process has better tails than the corresponding Gaussian process (it is "sub-Gaussian") and is bounded whenever the corresponding Gaussian process is bounded. There is, however, a completely different reason for which a Bernoulli process might be bounded, namely, that the sum of the absolute values of the coefficients of the random signs remain bounded independently of the index t. A natural question is then to decide whether these two extreme situations are the only fundamental reasons why a Bernoulli process can be bounded, in the sense that a suitable "mixture" of them occurs in every bounded Bernoulli process. This was the "Bernoulli conjecture" (to be stated formally on page 179), which has been so brilliantly solved by W. Bednorz and R. Latała.

It is a long road to the solution of the Bernoulli conjecture, and we start to build the main tools bearing on Bernoulli processes. A linear combination of independent random signs looks like a Gaussian r.v. when the coefficients of the random signs are small. We can expect that a Bernoulli process will look like a Gaussian process when these coefficients are suitably small. This is a fundamental idea: the key to understanding Bernoulli processes is to reduce to situations where these coefficients are small.

The Bernoulli conjecture, on which the author worked so many years, greatly influenced the way he looked at various processes. In the case of empirical processes, this is explained in Sect. 6.8.

1.7.6 Random Fourier Series and Trigonometric Sums

The basic example of a random Fourier series is

$$X_t = \sum_{k \geq 1} \xi_k \exp(2\pi i k t) , \tag{1.20}$$

where $i^2 = -1$, where $t \in [0, 1]$ and the r.v.s ξ_k are independent symmetric. In this chapter, we provide a final answer to the question of the convergence of such series.

The fundamental case where $\xi_k = a_k g_k$ for numbers a_k and independent Gaussian r.v.s (g_k) is of great historical importance. There is, however, another motivation for the study of such series. The generic chaining and related methods are well adapted to the case of a "nonhomogeneous index space". The study of certain of the processes we will consider in the next chapters is already subtle even in the absence of the extra difficulty due to this lack of homogeneity. The setting of random Fourier series allows us to put aside the issue of lack of homogeneity and to concentrate on the other difficulties and played a great part in the development of the theory. It provides an ideal setting to understand a basic fact: many processes can be exactly controlled, not by using one or two distances, but by using an entire family of distances. This concept of "family of distances" will play a major role later. It is also while analyzing the lower bounds discovered in the setting of random Fourier series that the author discovered the method which allows to extend these bounds to general random series as explained in Chap. 11. In this chapter, we also meet our first "decomposition theorem": there are two distinct reasons which explain the size of a random trigonometric sum. First, there can be a lot of cancellation between the terms. Second, it may happen that the sum of the absolute values of the terms is already small. We show that every random trigonometric sum is the sum of two such pieces, one of each type.

1.7.7 Partition Scheme for Families of Distances

Once one has survived the initial surprise of the new idea that many processes are naturally associated with an entire family of distances, it is very pleasant to realize that the tools of Sect. 2.9 can be extended to this setting with essentially the same proof. This is the purpose of Sect. 8.1.

In Sect. 8.3, we apply these tools to the situation of "canonical processes" where the r.v.s X_t are linear combinations of independent copies of symmetric r.v.s with density proportional to $\exp(-|x|^\alpha)$ where $\alpha \geq 1$ (and to considerably more general situations as discovered by R. Latała). In these situations, the size of the process can be completely described from the geometry of the index space, a far-reaching extension of the Gaussian case.

1.7.8 Peaky Parts of Functions

We learn how to measure the size of sets of functions on a measured space using an appropriate family of distances. We show that when we control this size, for each function of the set, we can distinguish its "peaky part" in a coherent way over the whole set of functions which then has in a sense a simple structure, as it is built from simpler pieces.

1.7.9 Proof of the Bernoulli Conjecture

Having learned how to manipulate "families of distances", we are now better prepared to prove the Bernoulli conjecture. This is the (overwhelmingly important) Latała-Bednorz theorem. The challenging proof occupies most of Chap. 10.[10] In the last section, we investigate how to get lower bounds on Bernoulli processes using "witnessing measures".

[10] It is a good research program to discover a more intuitive approach to this result.

1.7.10 Random Series of Functions

For a large class of random series of functions, we prove in full generality that chaining explains all the part of the boundedness of these processes created by cancellations, in the spirit of the Bernoulli conjecture. This covers the cases both of empirical processes and of the closely related class of selector processes. Our main tool is to reduce to processes which are conditionally Bernoulli processes and to use the Latała-Bednorz theorem and its consequences.

1.7.11 Infinitely Divisible Processes

The infinitely divisible processes we study are indexed by a general set and are to Lévy processes what a general Gaussian process (index by an arbitrary index set) is to Brownian motion (a Gaussian process indexed by $\mathbb{R}$ with stationary increments). We extend to these processes our results on random series of functions: chaining explains all the part of the boundedness of these processes which is due to cancellations. The results are described in complete detail with all definitions in Sect. 12.3.

1.7.12 Unfulfilled Dreams

Having proved in several general settings that "chaining explains all the part of the boundedness which is due to cancellation", we concentrate on the problem of describing the "part of the boundedness which owes nothing to cancellation". We propose sweeping conjectures. The underlying hope behind these conjectures is that, ultimately, a bound for a selector process always arises from the use of the "union bound" $\mathsf{P}(\cup_n A_n) \le \sum_n \mathsf{P}(A_n)$ in a simple situation, the use of basic principles such as linearity and positivity, or combinations of these.

1.7.13 Empirical Processes

We focus on a special yet fundamental topic: the control of the supremum of the empirical process over a class of functions.

We demonstrate again the power of the chaining scheme of Sect. 9.4 by providing a sharper version of Ossiander's bracketing theorem with a very simple proof. We then illustrate various techniques by presenting proofs of two deep recent results.

1.7.14 Gaussian Chaos

Our satisfactory understanding of the properties of Gaussian processes should bring information about processes that are, in various senses, related to Gaussian processes. Such is the case of an order 2 Gaussian chaos (which is essentially a family of second-degree polynomials of Gaussian random variables). It seems at present a hopelessly difficult task to give lower and upper bounds of the same order for these processes, but in Sect. 15.1, we obtain a number of results in this direction. Chaos processes are very instructive because there exist other methods than chaining to control their size (a situation which we do not expect to occur for processes defined as sums of a random series).

In Sect. 15.2, we study the tails of a single multiple-order Gaussian chaos and present (yet another) deep result of R. Latała which provides a rather complete description of the size of these tails.

1.7.15 Convergence of Orthogonal Series: Majorizing Measures

The old problem of characterizing the sequences (a_m) such that for each orthonormal sequence (φ_m) the series $\sum_{m\geq 1} a_m \varphi_m$ converges a.s. was solved by A. Paszkiewicz. Using a more abstract point of view, we present a very much simplified proof of his results (due essentially to W. Bednorz). This leads us to the question of discussing when a certain condition on the "increments" of a process implies its boundedness. When the increment condition is of "polynomial type", this is more difficult than in the case of Gaussian processes and requires the notion of "majorizing measure". We present several elegant results of this theory, in their seemingly final forms recently obtained by W. Bednorz.

1.7.16 Shor's Matching Theorem

This chapter continues Chap. 4. We prove a deep improvement of the Ajtai-Komlós-Tusnády theorem due to P. Shor. Unfortunately, due mostly to our lack of geometrical understanding, the best conceivable matching theorem, which would encompass this result as well as those of Chap. 4, and much more, remains as a challenging problem, "the ultimate matching conjecture" (a conjecture which is solved in the next chapter in dimension ≥ 3).

1.7.17 The Ultimate Matching Theorem in Dimension Three

In this case, which is easier than the case of dimension two (but still apparently rather non-trivial), we are able to obtain the seemingly final result about matchings, a strong version of "the ultimate matching conjecture". There are no more fractional powers of $\log N$ here, but in a random sample of N points uniformly distributed in $[0, 1]^3$, local irregularities occur at all scales between $N^{-1/3}$ and $(\log N)^{1/3} N^{-1/3}$, and our result can be seen as a precise global description of these irregularities.

1.7.18 Applications to Banach Space Theory

Chapter 19 gives applications to Banach space theory. As interest in this theory has decreased in recent years, we have not reproduced many of the results of [132], and we urge the interested reader to consult this earlier edition. We have kept only the results which make direct use of results presented elsewhere in the book (rather than including results based on the methods of the book). In Sect. 19.1.2, we study the cotype of operators from ℓ_N^∞ into a Banach space. In Sect. 19.1.3, we prove a comparison principle between Rademacher (=Bernoulli) and Gaussian averages of vectors in a finite-dimensional Banach space, and we use it to compute the Rademacher cotype-2 of a finite-dimensional space using only a few vectors. In Sect. 19.2.1 we discover how to classify the elements of the unit ball of L^1 "according to the size of the level sets". In Sect. 19.2.3 we explain, given a 1-unconditional sequence $(e_i)_{i \le N}$ in a Banach space E how to "compute" the quantity $\mathsf{E}\| \sum_i g_i e_i \|$ when g_i are independent Gaussian r.v.s, a further variation on the fundamental theme of the interplay between the L^1, L^2 and L^∞ norms. In Sect. 19.3.1 we study the norm of the restriction of an operator from ℓ_N^q to the subspace generated by a randomly chosen small proportion of the coordinate vectors, and in Sect. 19.3.2 we use these results to deduce the celebrated results of J. Bourgain on the Λ_p problem. Recent results of Gilles Pisier on Sidon sets conclude this chapter in Sect. 19.4.

Part I
The Generic Chaining

Chapter 2
Gaussian Processes and the Generic Chaining

2.1 Overview

The overview of this chapter is given in Chap. 1, Sect. 1.7.1. More generally, Sect. 1.7.n is the overview of Chapter $n + 1$.

2.2 Measuring the Size of the Supremum

In this section, we consider a metric space (T, d) and a process $(X_t)_{t\in T}$. Unless explicitly specified otherwise (and even when we forget to repeat it), we will *always* assume that the process is centered, i.e.,

$$\forall t \in T\,,\ \ \mathsf{E}X_t = 0\,. \tag{2.1}$$

We will measure the "size of the process $(X_t)_{t\in T}$" by the quantity $\mathsf{E}\sup_{t\in T} X_t$. Why this quantity is a good measure of the "size of the process" is explained in Lemma 2.2.1.

When T is uncountable, it is not obvious what the quantity $\mathsf{E}\sup_{t\in T} X_t$ means.[1] We *define* it by the following formula

$$\mathsf{E}\sup_{t\in T} X_t = \sup\left\{\mathsf{E}\sup_{t\in F} X_t\ ;\ F\subset T\ ,\ F \text{ finite}\right\}\,, \tag{2.2}$$

where the right-hand side makes sense as soon as each r.v. X_t is integrable. This will be the case in almost all the situations considered in this book.

[1] Such questions are treated in detail, for example, in [53] pages 42–43.

M. Talagrand, *Upper and Lower Bounds for Stochastic Processes*, Ergebnisse der Mathematik und ihrer Grenzgebiete. 3. Folge / A Series of Modern Surveys in Mathematics 60, https://doi.org/10.1007/978-3-030-82595-9_2

Let us say that a process $(X_t)_{t\in T}$ is *symmetric* if it has the same law as the process $(-X_t)_{t\in T}$. Almost all the processes we shall consider are symmetric (although this hypothesis is not necessary for some of our results). The following lemma justifies using the quantity $\mathsf{E}\sup_t X_t$ to measure "the size of a symmetric process":

Lemma 2.2.1 *If the process $(X_t)_{t\in T}$ is symmetric, then*

$$\mathsf{E}\sup_{s,t\in T}|X_s - X_t| = 2\mathsf{E}\sup_{t\in T} X_t\ .$$

Proof We note that

$$\sup_{s,t\in T}|X_s - X_t| = \sup_{s,t\in T}(X_s - X_t) = \sup_{s\in T} X_s + \sup_{t\in T}(-X_t)\,,$$

and we take expectations.[2] □

Exercise 2.2.2 Consider a symmetric process $(X_t)_{t\in T}$. Given any t_0 in T, prove that

$$\mathsf{E}\sup_{t\in T}|X_t| \le 2\mathsf{E}\sup_{t\in T} X_t + \mathsf{E}|X_{t_0}| \le 3\mathsf{E}\sup_{t\in T}|X_t|\ . \tag{2.3}$$

The previous exercise is easy, but this need not be always the case. The author has never taught this material in a classroom, and cannot really evaluate the level of difficulty of the exercises for a beginner. So please do not feel discouraged if most of the exercises feel like research problems.[3] A sketch of a solution is provided for almost every exercise. For the exercises which are too difficult, understanding this very concise sketch is in itself a good exercise. Just try to peek at the solution one line at a time.

In this book, we often state inequalities about the supremum of a symmetric process using the quantity $\mathsf{E}\sup_{t\in T} X_t$ simply because this quantity looks typographically more elegant than the equivalent[4] quantity $\mathsf{E}\sup_{s,t\in T}|X_s - X_t|$. It is good to remember that when $X_{t_0} = 0$ for some $t_0 \in T$, (2.3) shows that there is not so much difference between $\mathsf{E}\sup_{t\in T} X_t$ and $\mathsf{E}\sup_{t\in T}|X_t|$.

We actually often need to control the tails of the r.v. $\sup_{s,t\in T}|X_s - X_t|$, not only its first moment. Emphasis is given to the first moment because this is the difficult

[2] To be really rigorous, we should first consider the case where T is finite and then appeal to (2.2), but it is better to skip this kind of tedious detail.

[3] I had feedback from talented readers who felt that way. Consequently, I did not shy away to state as "exercises" rather non-trivial material complementing the text while being fully aware that one has to have achieved a rather complete understanding of the concepts as well as a mastery of the techniques to solve them.

[4] Equivalent does not mean equal; we have been dropping a factor 2 here. Generally speaking, the methods of this book are not appropriate to find sharp numerical constants, and all the crucial inequalities are "sharp within a multiplicative constant".

part, and once this is achieved, control of higher moments is often provided by the same arguments.

2.3 The Union Bound and Other Basic Facts

From now on, we assume that the process $(X_t)_{t\in T}$ satisfies the increment condition:

$$\forall u > 0\ ,\ \mathsf{P}(|X_s - X_t| \geq u) \leq 2\exp\Big(-\frac{u^2}{2d(s,t)^2}\Big)\ , \tag{2.4}$$

where d is a distance on T. In particular this is the case when $(X_t)_{t\in T}$ is a Gaussian process and $d(s,t)^2 = \mathsf{E}(X_s - X_t)^2$. Our goal is to find bounds on $\mathsf{E}\sup_{t\in T} X_t$ depending on the structure of the metric space (T,d). We will assume that T is finite, which, as shown by (2.2), does not decrease generality.

Given any t_0 in T, the centering hypothesis (2.1) implies

$$\mathsf{E}\sup_{t\in T} X_t = \mathsf{E}\sup_{t\in T}(X_t - X_{t_0})\ . \tag{2.5}$$

The latter form has the advantage that we now seek estimates for the expectation of the nonnegative r.v. $Y = \sup_{t\in T}(X_t - X_{t_0})$. For such a variable, we have the formula

$$\mathsf{E}Y = \int_0^\infty \mathsf{P}(Y \geq u)\,du\ . \tag{2.6}$$

Let us note that since the function $u \mapsto \mathsf{P}(Y \geq u)$ is non-increasing, for any $u > 0$, we have the following:

$$\mathsf{E}Y \geq u\mathsf{P}(Y \geq u)\ . \tag{2.7}$$

In particular $\mathsf{P}(Y \geq u) \leq \mathsf{E}Y/u$, a very important fact known as Markov's inequality. Arguments such as the following one will be of constant use:

Exercise 2.3.1 Consider a r.v. $Y \geq 0$ and $a > 0$. Prove that $\mathsf{P}(Y \leq a\mathsf{E}Y) \geq 1 - 1/a$.

Let us stress a consequence of Markov's inequality: when Y is a kind a random error, of very small expectation, $\mathsf{E}Y = b^2$ where b is small. Then most of time Y is small: $\mathsf{P}(Y \leq b) \geq 1 - b$.

According to (2.6), it is natural to look for bounds of

$$\mathsf{P}\Big(\sup_{t\in T}(X_t - X_{t_0}) \geq u\Big)\ . \tag{2.8}$$

The first bound that comes to mind is the "union bound"

$$\mathsf{P}\Big(\sup_{t\in T}(X_t - X_{t_0}) \ge u\Big) \le \sum_{t\in T} \mathsf{P}(X_t - X_{t_0} \ge u) \,. \tag{2.9}$$

It seems worthwhile to immediately draw some consequences from this bound and to discuss at leisure a number of other simple, yet fundamental facts. This will take a bit over three pages, after which we will come back to the main story of bounding Y. Throughout this work, $\Delta(T)$ denotes the diameter of T,

$$\Delta(T) = \sup_{t_1,t_2\in T} d(t_1,t_2) \,. \tag{2.10}$$

When we need to make clear which distance we use in the definition of the diameter, we will write $\Delta(T,d)$ rather than $\Delta(T)$. Consequently (2.4) and (2.9) imply

$$\mathsf{P}\Big(\sup_{t\in T}(X_t - X_{t_0}) \ge u\Big) \le 2\,\mathrm{card}\,T \exp\Big(-\frac{u^2}{2\Delta(T)^2}\Big) \,. \tag{2.11}$$

Let us now record a simple yet important computation, which will allow us to use the information (2.11).

Lemma 2.3.2 *Consider a r.v. $Y \ge 0$ which satisfies*

$$\forall u > 0 \,,\ \mathsf{P}(Y \ge u) \le A \exp\Big(-\frac{u^2}{B^2}\Big) \tag{2.12}$$

for certain numbers $A \ge 2$ and $B > 0$. Then

$$\mathsf{E}Y \le LB\sqrt{\log A} \,. \tag{2.13}$$

Here, as in the entire book, L denotes a universal constant.[5] We make the convention that this constant *is not necessarily* the same on each occurrence (even in the same equation). This should be remembered at all times. One of the benefits of the convention (as opposed to writing explicit constants) is to make clear that one is not interested in getting sharp constants. Getting sharp constants might be useful for certain applications, but it is a different game.[6] The convention is very convenient, but one needs to get used to it. Now is the time for this, so we urge the reader to pay the greatest attention to the next exercise.

[5] When meeting an unknown notation such as this previous L, the reader might try to look at the *index*, where some of the most common notation is recorded.

[6] Our methods here are not appropriate for this.

Exercise 2.3.3

(a) Prove that for $x, y \in \mathbb{R}^+$ we have $xy - Lx^3 \le Ly^{3/2}$. (Please understand this statement as follows: given a number L_1, there exist a number L_2 such that for all $x, y \in \mathbb{R}$ we have $xy - L_1 x^{1/3} \le L_2 y^{3/2}$.)
(b) Consider a function $p(u) \le 1$ for $u \ge 0$. Assume that for $u > L$, we have $p(u) \le L\exp(-u^2/L)$. Prove that for all $u > 0$ we have $p(Lu) \le 2\exp(-u^2)$. (Of course, this has to be understood as follows: assume that for a certain number L_1, for $u > L_1$, we have $p(u) \le L_1 \exp(-u^2/L_1)$. Prove that there exist a number L_2 such that for all $u > 0$ we have $p(L_2 u) \le \exp(-u^2)$.)
(c) Consider an integer $N \ge 1$. Prove that

$$N^L \exp(-(\log N)^{3/2}/L) \le L\exp(-(\log N)^{3/2}/L)\ .$$

Proof of Lemma 2.3.2 We use (2.6), and we observe that since $\mathsf{P}(Y \ge u) \le 1$, for any number u_0, we have

$$\begin{aligned}
\mathsf{E}Y = \int_0^\infty \mathsf{P}(Y \ge u)\mathrm{d}u &= \int_0^{u_0} \mathsf{P}(Y \ge u)\mathrm{d}u + \int_{u_0}^\infty \mathsf{P}(Y \ge u)\mathrm{d}u \\
&\le u_0 + \int_{u_0}^\infty A\exp\Big(-\frac{u^2}{B^2}\Big)\mathrm{d}u \\
&\le u_0 + \frac{1}{u_0}\int_{u_0}^\infty uA\exp\Big(-\frac{u^2}{B^2}\Big)\mathrm{d}u \\
&= u_0 + \frac{AB^2}{2u_0}\exp\Big(-\frac{u_0^2}{B^2}\Big)\ . \qquad (2.14)
\end{aligned}$$

The choice of $u_0 = B\sqrt{\log A}$ gives the bound

$$B\sqrt{\log A} + \frac{B}{2\sqrt{\log A}} \le LB\sqrt{\log A}$$

since $A \ge 2$. □

Next, recalling that the process $(X_t)_{t\in T}$ is assumed to satisfy (2.4) throughout the section, we claim that

$$\mathsf{E}\sup_{t\in T} X_t \le L\Delta(T)\sqrt{\log \operatorname{card} T}\ . \qquad (2.15)$$

Indeed, this is obvious if $\operatorname{card} T = 1$. If $\operatorname{card} T \ge 2$, it follows from (2.11) that (2.12) holds for $Y = \sup_{t\in T}(X_t - X_{t_0})$ with $A = 2\operatorname{card} T$ and $B = \Delta(T)$, and the result follows from (2.13) since $\log(2\operatorname{card} T) \le 2\log\operatorname{card} T$ and $\mathsf{E}Y = \mathsf{E}\sup_{t\in T} X_t$.

The following special case is fundamental:

Lemma 2.3.4 *If $(g_k)_{k\ge 1}$ are standard Gaussian r.v.s, then*

$$\mathsf{E}\sup_{k\le N} g_k \le L\sqrt{\log N}\ . \tag{2.16}$$

Exercise 2.3.5

(a) Prove that (2.16) holds for any r.v.s g_k which satisfy

$$\mathsf{P}(g_k \ge u) \le 2\exp\Big(-\frac{u^2}{2}\Big) \tag{2.17}$$

for $u > 0$.

(b) For $N \ge 2$, construct N centered r.v.s $(g_k)_{k\le N}$ satisfying (2.17) and taking only the values $0, \pm\sqrt{\log N}$ and for which $\mathsf{E}\sup_{k\le N} g_k \ge \sqrt{\log N}/L$. (You are not yet asked to make these r.v.s independent.)

(c) After learning (2.18), solve (b) with the further requirement that the r.v.s g_k are independent. If this is too hard, look at Exercise 2.3.7 (b).

This is taking us a bit ahead, but an equally fundamental fact is that when the r.v.s (g_k) are jointly Gaussian and "significantly different from each other", i.e., $\mathsf{E}(g_k - g_\ell)^2 \ge a^2 > 0$ for $k \ne \ell$, the bound (2.16) can be reversed, i.e., $\mathsf{E}\sup_{k\le N} g_k \ge a\sqrt{\log N}/L$, a fact known as Sudakov's minoration. Sudakov's minoration is a non-trivial fact, and to understand it, it should be really useful to solve Exercise 2.3.7. However, before that, let us point out a simple fact, which will be used many times.

Exercise 2.3.6 Consider independent events $(A_k)_{k\ge 1}$. Prove that

$$\mathsf{P}\Big(\bigcup_{k\le N} A_k\Big) \ge 1 - \exp\Big(-\sum_{k\le N}\mathsf{P}(A_k)\Big)\ . \tag{2.18}$$

In words, independent events such that the sum of their probabilities is small are basically disjoint.

Exercise 2.3.7

(a) Consider independent r.v.s $Y_k \ge 0$ and $u > 0$ with

$$\sum_{k\le N}\mathsf{P}(Y_k \ge u) \ge 1\ . \tag{2.19}$$

Prove that

$$\mathsf{E}\sup_{k\le N} Y_k \ge \frac{u}{L}\ .$$

Hint: Use (2.18) to prove that $\mathsf{P}(\sup_{k\le N} Y_k \ge u) \ge 1/L$.

(b) We assume (2.19), but now Y_k need not be ≥ 0. Prove that

$$\mathsf{E}\sup_{k\leq N} Y_k \geq \frac{u}{L} - \mathsf{E}|Y_1| .$$

Hint: Observe that for each event Ω, we have $\mathsf{E}\mathbf{1}_\Omega \sup_k Y_k \geq -\mathsf{E}|Y_1|$.

(c) Prove that if $(g_k)_{k\geq 1}$ are independent standard Gaussian r.v.s, then $\mathsf{E}\sup_{k\leq N} g_k \geq \sqrt{\log N}/L$.

Before we go back to our main story, we consider in detail the consequences of an "exponential decay of tails" such as in (2.12). This is the point of the next exercise.

Exercise 2.3.8

(a) Assume that for a certain $B > 0$, the r.v. $Y \geq 0$ satisfies

$$\forall u > 0\,,\ \mathsf{P}(Y \geq u) \leq 2\exp\Big(-\frac{u}{B}\Big)\,. \tag{2.20}$$

Prove that

$$\mathsf{E}\exp\Big(\frac{Y}{2B}\Big) \leq L\,. \tag{2.21}$$

Prove that for $x, a > 0$ one has $(x/a)^a \leq \exp x$. Use this for $a = p$ and $x = Y/2B$ to deduce from (2.21) that for $p \geq 1$ one has

$$(\mathsf{E}Y^p)^{1/p} \leq LpB\,. \tag{2.22}$$

(b) Assuming now that for a certain $B > 0$ one has

$$\forall u > 0\,,\ \mathsf{P}(Y \geq u) \leq 2\exp\Big(-\frac{u^2}{B^2}\Big)\,, \tag{2.23}$$

prove similarly (or deduce from (a)) that $\mathsf{E}\exp(Y^2/2B^2) \leq L$ and that for $p \geq 1$ one has

$$(\mathsf{E}Y^p)^{1/p} \leq LB\sqrt{p}\,. \tag{2.24}$$

(c) Consider a r.v. $Y \geq 0$ and a number $B > 0$. Assuming that for $p \geq 1$ we have $(\mathsf{E}Y^p)^{1/p} \leq Bp$, prove that for $u > 0$ we have $\mathsf{P}(Y > u) \leq 2\exp(-u/(LB))$. Assuming that for each $p \geq 1$ we have $(\mathsf{E}Y^p)^{1/p} \leq B\sqrt{p}$, prove that for $u > 0$ we have $\mathsf{P}(Y > u) \leq 2\exp(-u^2/(LB^2))$.

In words, (2.22) states that "as p increases, the L^p norm of an exponentially integrable r.v. does not grow faster than p", and (2.24) asserts that if the square of the r.v. is exponentially integrable, then its L^p norm does not grow faster than

$\sqrt{p}$. These two statements are closely related. More generally, it is very classical to relate the size of the tails of a r.v. with the rate of growth of its L^p norm. This is not explicitly used in the sequel, but is good to know as background information. As the following shows, (2.24) provides the correct rate of growth in the case of Gaussian r.v.s.

Exercise 2.3.9 If g is a standard Gaussian r.v., it follows from (2.24) that for $p \geq 1$, one has $(\mathsf{E}|g|^p)^{1/p} \leq L\sqrt{p}$. Prove one has also

$$(\mathsf{E}|g|^p)^{1/p} \geq \frac{\sqrt{p}}{L} . \tag{2.25}$$

One knows how to compute exactly $\mathsf{E}|g|^p$, from which one can deduce (2.25). You are, however, asked to provide a proof in the spirit of this work by deducing (2.25) solely from the information that, say, for $u > 0$, we have (choosing on purpose crude constants) $\mathsf{P}(|g| \geq u) \geq \exp(-u^2/3)/100$.

You will find basically no exact computations in this book. The aim is different. We study quantities which are far too complicated to be computed exactly, and we try to bound them from above and sometimes from below by simpler quantities with as little a gap as possible between the upper and the lower bounds. Ideally the gap is only a (universal) multiplicative constant.

2.4 The Generic Chaining

We go back to our main story. The bound (2.9) (and hence (2.15)) will be effective if the variables $X_t - X_{t_0}$ are rather uncorrelated (and if there are not too many of them). But it will be a disaster if many of the variables $(X_t)_{t\in T}$ are nearly identical. Thus, it seems a good idea to gather those variables X_t which are nearly identical. To do this, we consider a subset T_1 of T, and for t in T, we consider a point $\pi_1(t)$ in T_1, which we think of as a (first) approximation of t. The elements of T which correspond to the same point $\pi_1(t)$ are, at this level of approximation, considered as identical. We then write

$$X_t - X_{t_0} = X_t - X_{\pi_1(t)} + X_{\pi_1(t)} - X_{t_0} . \tag{2.26}$$

The idea is that it will be effective to use (2.9) for the variables $X_{\pi_1(t)} - X_{t_0}$, because there are not too many of them and, if we have done a good job at finding $\pi_1(t)$, they are rather different from each other (at least in some global sense). On the other hand, since $\pi_1(t)$ is an approximation of t, the variables $X_t - X_{\pi_1(t)}$ are "smaller" than the original variables $X_t - X_{t_0}$, so that their supremum should be easier to handle. The procedure will then be iterated.

Let us set up the general procedure. For $n \geq 0$, we consider a subset T_n of T, and for $t \in T$, we consider $\pi_n(t)$ in T_n. (The idea is that the points $\pi_n(t)$ are successive approximations of t.) We assume that T_0 consists of a single element t_0, so that $\pi_0(t) = t_0$ for each t in T. The fundamental relation is

$$X_t - X_{t_0} = \sum_{n\geq 1} \left(X_{\pi_n(t)} - X_{\pi_{n-1}(t)}\right) , \tag{2.27}$$

which holds provided we arrange that $\pi_n(t) = t$ for n large enough, in which case the series is actually a finite sum. Relation (2.27) decomposes the increments of the process $X_t - X_{t_0}$ along the "chain" $(\pi_n(t))_{n\geq 0}$ (and this is why this method is called "chaining").

It will be convenient to control the set T_n through its cardinality with the condition

$$\operatorname{card} T_n \leq N_n \tag{2.28}$$

where

$$N_0 = 1 ; \; N_n = 2^{2^n} \text{ if } n \geq 1 . \tag{2.29}$$

Notation (2.29) will be used throughout the book. It is at this stage that the procedure to control T_n differs from the traditional one, and it is the crucial point of the generic chaining method.

It is good to notice right away that $\sqrt{\log N_n}$ is about $2^{n/2}$, which will explain the ubiquity of this latter quantity. The occurrence of the function $\sqrt{\log x}$ itself is related to the fact that it is the inverse of the function $\exp(x^2)$ and that the function $\exp(-x^2)$ governs the size of the tails of a Gaussian r.v. Let us also observe the fundamental inequality

$$N_n^2 \leq N_{n+1} ,$$

which makes it very convenient to work with this sequence.

Since $\pi_n(t)$ approximates t, it is natural to assume that[7]

$$d(t, \pi_n(t)) = d(t, T_n) := \inf_{s\in T_n} d(t, s) . \tag{2.30}$$

For $u > 0$, (2.4) implies

$$\mathsf{P}\left(|X_{\pi_n(t)} - X_{\pi_{n-1}(t)}| \geq u2^{n/2} d(\pi_n(t), \pi_{n-1}(t))\right) \leq 2\exp(-u^2 2^{n-1}) .$$

---

[7] The notation := below stresses that this is a definition, so that you should not worry that your memory failed and that you did not see this before.

The number of possible pairs $(\pi_n(t), \pi_{n-1}(t))$ is bounded by

$$\operatorname{card} T_n \cdot \operatorname{card} T_{n-1} \leq N_n N_{n-1} \leq N_{n+1} = 2^{2^{n+1}} .$$

We define the (favorable) event $\Omega_{u,n}$ by

$$\forall t\,,\ |X_{\pi_n(t)} - X_{\pi_{n-1}(t)}| \leq u 2^{n/2} d(\pi_n(t), \pi_{n-1}(t)) . \tag{2.31}$$

and we define $\Omega_u = \bigcap_{n \geq 1} \Omega_{u,n}$. Then

$$p(u) := \mathsf{P}(\Omega_u^c) \leq \sum_{n \geq 1} \mathsf{P}(\Omega_{u,n}^c) \leq \sum_{n \geq 1} 2 \cdot 2^{2^{n+1}} \exp(-u^2 2^{n-1}) . \tag{2.32}$$

Here again, at the crucial step, we have used the union bound $\mathsf{P}(\Omega_u^c) \leq \sum_{n \geq 1} \mathsf{P}(\Omega_{u,n}^c)$. When Ω_u occurs, (2.27) yields

$$|X_t - X_{t_0}| \leq u \sum_{n \geq 1} 2^{n/2} d(\pi_n(t), \pi_{n-1}(t)) ,$$

so that

$$\sup_{t \in T} |X_t - X_{t_0}| \leq uS$$

where

$$S := \sup_{t \in T} \sum_{n \geq 1} 2^{n/2} d(\pi_n(t), \pi_{n-1}(t)) .$$

Thus

$$\mathsf{P}\Big(\sup_{t \in T} |X_t - X_{t_0}| > uS \Big) \leq p(u) .$$

For $n \geq 1$ and $u \geq 3$, we have

$$u^2 2^{n-1} \geq \frac{u^2}{2} + u^2 2^{n-2} \geq \frac{u^2}{2} + 2^{n+1} ,$$

from which it follows that

$$p(u) \leq L \exp\Big(-\frac{u^2}{2} \Big) .$$

We observe here that since $p(u) \leq 1$, the previous inequality holds not only for $u \geq 3$ but also for $u > 0$, because $1 \leq \exp(9/2) \exp(-u^2/2)$ for $u \leq 3$. This type of argument (i.e., changing the universal constant in front of the exponential, cf. Exercise (2.3.3)(b)) will be used repeatedly. Therefore,

$$\mathsf{P}\Big(\sup_{t\in T}|X_t - X_{t_0}| > uS\Big) \le L\exp\Big(-\frac{u^2}{2}\Big)\,. \tag{2.33}$$

In particular (2.33) implies

$$\mathsf{E}\sup_{t\in T} X_t \le LS\,.$$

The triangle inequality yields

$$d(\pi_n(t), \pi_{n-1}(t)) \le d(t, \pi_n(t)) + d(t, \pi_{n-1}(t)) = d(t, T_n) + d(t, T_{n-1})\,,$$

so that (making the change of variable $n = n' + 1$ in the second sum below)

$$S \le \sup_{t\in T}\sum_{n\ge 1} 2^{n/2} d(t, T_n) + \sup_{t\in T}\sum_{n\ge 1} 2^{n/2} d(t, T_{n-1}) \le 3\sup_{t\in T}\sum_{n\ge 0} 2^{n/2} d(t, T_n)\,,$$

and we have proved the fundamental bound

$$\mathsf{E}\sup_{t\in T} X_t \le L\sup_{t\in T}\sum_{n\ge 0} 2^{n/2} d(t, T_n)\,. \tag{2.34}$$

Now, how do we construct the sets T_n? It is obvious that we should try to make the right-hand side of (2.34) small, but this is obvious only because we have used an approach which naturally leads to this bound. In the next section, we investigate how this was traditionally done. Before this, we urge the reader to fully understand the next exercise. It will be crucial to understand a typical case where the traditional methods are not effective.

Exercise 2.4.1 Consider a countable metric space, $T = \{t_1, t_2, \dots\}$. Assume that for each $i \ge 2$, we have $d(t_1, t_i) \le 1/\sqrt{\log i}$. Prove that if $T_n = \{t_1, t_2, \dots, t_{N_n}\}$, then for each $t \in T$, we have $\sum_{n\ge 0} 2^{n/2} d(t, T_n) \le L$.

We end this section by reviewing at a high level the scheme of the previous proof (which will be used again and again). The goal is to bound $\mathsf{E}Y$ where Y is a r.v. ≥ 0 (here $Y = \sup_t (X_t - X_{t_0})$.) The method consists of two steps:

- Given a parameter $u \ge 0$, one identifies a "good set" Ω_u, where some undesirable events do not happen. As u becomes large, $\mathsf{P}(\Omega_u^c)$ becomes small.
- When Ω_u occurs, we bound Y, say $Y \le f(u)$ where f is an increasing function on $\mathbb{R}^+$.

One then obtains the bound

$$\mathsf{E}Y = \int_0^\infty \mathsf{P}(Y \geq u)\mathrm{d}u \leq f(0) + \int_{f(0)}^\infty \mathsf{P}(Y \geq u)\mathrm{d}u$$
$$= f(0) + \int_0^\infty f'(u)\mathsf{P}(Y \geq f(u))\mathrm{d}u \;, \qquad (2.35)$$

where we have used a change of variable in the last equality. Now, since $Y \leq f(u)$ on Ω_u, we have $\mathsf{P}(Y \geq f(u)) \leq \mathsf{P}(\Omega_u^c)$ and finally

$$\mathsf{E}Y \leq f(0) + \int_0^\infty f'(u)\mathsf{P}(\Omega_u^c)\mathrm{d}u \;.$$

In practice, we will always have $\mathsf{P}(\Omega_u^c) \leq L\exp(-u/L)$ and $f(u) = A + u^\alpha B$, yielding the bound $\mathsf{E}Y \leq A + K(\alpha)B$.

2.5 Entropy Numbers

For a number of years, chaining was systematically performed (as in Sect. 1.4) by choosing the sets T_n so that $\sup_{t\in T} d(t, T_n)$ is as small as possible for card $T_n \leq N_n$. We define

$$e_n(T) = e_n(T, d) = \inf_{T_n \subset T, \operatorname{card} T_n \leq N_n} \sup_{t\in T} d(t, T_n) \;, \qquad (2.36)$$

where the infimum is taken over all subsets T_n of T with card $T_n \leq N_n$. (Since here T is finite, the infimum is actually a minimum.) We call the numbers $e_n(T)$ the *entropy numbers*.

Let us recall that in a metric space, a (closed) ball is a set of the type $B(t, r) = \{s \in T; d(s, t) \leq r\}$. Balls are basic sets in a metric space and will be of constant use. It should be obvious to reformulate (2.36) as follows: $e_n(T)$ is the infimum of the set of numbers $r \geq 0$ such that T can be covered by $\leq N_n$ balls of radius $\leq r$ (the set T_n in (2.36) being the set of centers of these balls).

Definition (2.36) is not consistent with the conventions of operator theory, which uses e_{2^n} to denote what we call e_n.[8] When T is infinite, the numbers $e_n(T)$ are also defined by (2.36) but are not always finite (e.g., when $T = \mathbb{R}$).

Let us note that since $N_0 = 1$,

$$\frac{\Delta(T)}{2} \leq e_0(T) \leq \Delta(T) \;. \qquad (2.37)$$

[8] We can't help it if operator theory gets it wrong.

Recalling that T is finite, let us then choose for each n a subset T_n of T with $\operatorname{card} T_n \leq N_n$ and $e_n(T) = \sup_{t\in T} d(t, T_n)$. Since $d(t, T_n) \leq e_n(T)$ for each t, (2.34) implies the following:

Proposition 2.5.1 (Dudley's Entropy Bound [29]) *Under the increment condition (2.4), it holds that*

$$\mathsf{E} \sup_{t\in T} X_t \leq L \sum_{n\geq 0} 2^{n/2} e_n(T) . \tag{2.38}$$

We proved this bound only when T is finite, but using (2.2), it also extends to the case where T is infinite, as is shown by the following easy fact:

Lemma 2.5.2 *If U is a subset of T, we have $e_n(U) \leq 2e_n(T)$.*

The point here is that in the definition of $e_n(U)$, we insist that the balls are centered in U, not in T.

Proof Indeed, if $a > e_n(T)$, by definition one can cover T by N_n balls (for the distance d) with radius a, and the intersections of these balls with U are of diameter $\leq 2a$, so U can be covered by N_n balls in U with radius $2a$. □

Exercise 2.5.3 Prove that the factor 2 in the inequality $e_n(U) \leq 2e_n(T)$ cannot be improved even if $n = 0$.

Dudley's entropy bound is usually formulated using the covering numbers of Definition 1.4.1. These relate to the entropy numbers by the formula

$$e_n(T) = \inf\{\epsilon \ ; \ N(T, d, \epsilon) \leq N_n\} .$$

Indeed, it is obvious by definition of $e_n(T)$ that for $\epsilon > e_n(T)$, we have $N(T, d, \epsilon) \leq N_n$ and that if $N(T, d, \epsilon) \leq N_n$, we have $e_n(T) \leq \epsilon$. Consequently,

$$\begin{aligned} \epsilon < e_n(T) &\Rightarrow N(T, d, \epsilon) > N_n \\ &\Rightarrow N(T, d, \epsilon) \geq 1 + N_n . \end{aligned}$$

Therefore,

$$\sqrt{\log(1 + N_n)}(e_n(T) - e_{n+1}(T)) \leq \int_{e_{n+1}(T)}^{e_n(T)} \sqrt{\log N(T, d, \epsilon)}\, d\epsilon .$$

Since $\log(1 + N_n) \geq 2^n \log 2$ for $n \geq 0$, summation over $n \geq 0$ yields

$$\sqrt{\log 2} \sum_{n\geq 0} 2^{n/2}(e_n(T) - e_{n+1}(T)) \leq \int_0^{e_0(T)} \sqrt{\log N(T, d, \epsilon)}\, d\epsilon . \tag{2.39}$$

Now,

$$\sum_{n\ge 0} 2^{n/2}(e_n(T) - e_{n+1}(T)) = \sum_{n\ge 0} 2^{n/2} e_n(T) - \sum_{n\ge 1} 2^{(n-1)/2} e_n(T)$$
$$\ge \left(1 - \frac{1}{\sqrt{2}}\right) \sum_{n\ge 0} 2^{n/2} e_n(T) ,$$

so (2.39) yields

$$\sum_{n\ge 0} 2^{n/2} e_n(T) \le L \int_0^{e_0(T)} \sqrt{\log N(T,d,\epsilon)}\, d\epsilon . \tag{2.40}$$

Hence Dudley's bound now appears in the familiar form

$$\mathsf{E} \sup_{t\in T} X_t \le L \int_0^{\infty} \sqrt{\log N(T,d,\epsilon)}\, d\epsilon . \tag{2.41}$$

Here, since $\log 1 = 0$, the integral takes place in fact over $0 \le \epsilon \le e_0(T)$. The right-hand side is often called *Dudley's entropy integral.*

Exercise 2.5.4 Prove that

$$\int_0^{\infty} \sqrt{\log N(T,d,\epsilon)}\, d\epsilon \le L \sum_{n\ge 0} 2^{n/2} e_n(T) ,$$

showing that (2.38) is not an improvement over (2.41).

Exercise 2.5.5 Assume that for each $0 < \epsilon < A$ and some $\alpha > 0$, we have $\log N(T,d,\epsilon) \le (A/\epsilon)^\alpha$. Prove that $e_n(T) \le K(\alpha) A 2^{-n/\alpha}$.

Here $K(\alpha)$ is a number depending only on α.[9] This and similar notation are used throughout the book. It is understood that such numbers *need not be the same on every occurrence*, and it would help to *remember this at all times*. The difference between the notations K and L is that L is a universal constant, i.e., a number that do not depend on anything, while K might depend on some parameters, such as α here.

When writing a bound such as (2.41), the immediate question is how sharp is it? The word "sharp" is commonly used, even though people do not agree on what it means exactly. Let us say that a bound of the type $A \le LB$ *can be reversed* if it is true that $B \le LA$. We are not concerned with the value of the universal

[9] It just happens that in this particular case $K(\alpha) = 1$ works, but we typically do not care about the precise dependence of $K(\alpha)$ on α.

constants.[10] Inequalities which can be reversed are our best possible goal. Then, in any circumstance, A and B are of the same order.

We give now a simple (and classical) example that illustrates well the difference between Dudley's bound (2.38) and the bound (2.34) and which shows in particular that *Dudley's bound cannot be reversed.* Consider an independent sequence $(g_i)_{i\geq 1}$ of standard Gaussian r.v.s. Set $X_1 = 0$, and for $i \geq 2$, set

$$X_i = \frac{g_i}{\sqrt{\log i}} . \tag{2.42}$$

Consider an integer $s \geq 3$ and the process $(X_i)_{1\leq i\leq N_s}$ so the index set is $T = \{1, 2, \ldots, N_s\}$. The distance d associated with the process, given by $d(i, j)^2 = \mathsf{E}(X_s - X_t)^2$, satisfies for $i, j \geq 2, i \neq j$,

$$\frac{1}{\sqrt{\log(\min(i, j))}} \leq d(i, j) \leq \frac{2}{\sqrt{\log(\min(i, j))}} . \tag{2.43}$$

Consider $1 \leq n \leq s - 1$ and $T_n \subset T$ with $\operatorname{card} T_n = N_n$. There exists $i \leq N_n + 1$ with $i \notin T_n$. Then (2.43) implies that $d(i, j) \geq 2^{-n/2}/L$ for $j \in T_n$. This proves that the balls of radius $2^{-n/2}/L$ centered on T_n do not cover T, so that $e_n(T) \geq 2^{-n/2}/L$. Therefore,

$$\sum_n 2^{n/2} e_n(T) \geq \frac{s - 1}{L} . \tag{2.44}$$

In the reverse direction, since for $i \geq 1$ we have $d(1, i) \leq 1/\sqrt{\log i}$, Exercise 2.4.1 proves that the bound (2.34) is $\leq L$. Thus, the bound (2.38) is worse than the bound (2.34) by a factor about s.

Exercise 2.5.6 Prove that when T is finite, the bound (2.41) cannot be worse than (2.34) by a factor greater than about $\log\log \operatorname{card} T$. This shows that the previous example is in a sense extremal. Hint: Use $2^{n/2} e_n(T) \leq L \sup_{t\in T} \sum_{n\geq 0} 2^{n/2} d(t, T_n)$ and $e_n(T) = 0$ if $N_n \geq \operatorname{card} T$.

How does one estimate covering numbers (or, equivalently, entropy numbers)? Let us first stress a trivial but nonetheless fundamental fact.

Lemma 2.5.7 *Consider a number $\epsilon > 0$ and a subset W of T maximal with respect to the property*

$$s, t \in W \Rightarrow d(s, t) > \epsilon .$$

Then, $N(T, d, \epsilon) \leq \operatorname{card} W$.

[10] Not that these values are unimportant, but our methods are not appropriate for this.

Proof Since W is maximum, the balls of radius a centered at the points of W cover T. □

Exercise 2.5.8 Consider a probability measure μ on T, a number $\epsilon > 0$ and a number a. Let $U = \{t \in T; \mu(B(t,\epsilon)) \geq a\}$. Prove that $N(U,d,2\epsilon) \leq 1/a$.

The next exercise introduces the reader to "volume estimates", a simple yet fundamental method for this purpose. It deserves to be fully understood. If this exercise is too hard, you can find all the details below in the proof of Lemma 2.13.7.

Exercise 2.5.9

(a) If (T,d) is a metric space, define the packing number $N^*(T,d,\epsilon)$ as the largest integer N such that T contains N points with mutual distances $\geq \epsilon$. Prove that $N(T,d,\epsilon) \leq N^*(T,d,\epsilon)$. Prove that if $\epsilon' > 2\epsilon$, then $N^*(T,d,\epsilon') \leq N(T,d,\epsilon)$.

(b) Consider a distance d on $\mathbb{R}^k$ which arises from a norm $\|\cdot\|$, $d(x,y) = \|x - y\|$ and denote by B the unit ball of center 0. Let us denote by $\mathrm{Vol}(A)$ the k-dimensional volume of a subset A of $\mathbb{R}^k$. By comparing volumes, prove that for any subset A of $\mathbb{R}^k$,

$$N(A,d,\epsilon) \geq \frac{\mathrm{Vol}(A)}{\mathrm{Vol}(\epsilon B)} \tag{2.45}$$

and

$$N(A,d,2\epsilon) \leq N^*(A,d,2\epsilon) \leq \frac{\mathrm{Vol}(A+\epsilon B)}{\mathrm{Vol}(\epsilon B)} . \tag{2.46}$$

(c) Conclude that

$$\Big(\frac{1}{\epsilon}\Big)^k \leq N(B,d,\epsilon) \leq \Big(\frac{2+\epsilon}{\epsilon}\Big)^k . \tag{2.47}$$

(d) Use (c) to find estimates of $e_n(B)$ of the correct order for each value of n. Hint: $e_n(B)$ is about $2^{-2^n/k}$. This decreases very fast as n increases. Estimate Dudley's bound for B provided with the distance d.

(e) Prove that if T is a subset of $\mathbb{R}^k$ and if n_0 is any integer, then for $n \geq n_0$, one has $e_{n+1}(T) \leq L2^{-2^n/k}e_{n_0}(T)$. Hint: Cover T by N_{n_0} balls of radius $2e_{n_0}(T)$, and cover each of these by balls of smaller radius using (d).

(f) This part provides a generalization of (2.45) and (2.46) to a more abstract setting but with the same proofs. Consider a metric space (T,d) and a positive measure μ on T such that all balls of a given radius have the same measure, $\mu(B(t,\epsilon)) = \varphi(\epsilon)$ for each $\epsilon > 0$ and each $t \in T$. For a subset A of T and $\epsilon > 0$, let $A_\epsilon = \{t \in T; d(t,A) \leq \epsilon\}$, where $d(t,A) = \inf_{s\in A} d(t,s)$. Prove that

$$\frac{\mu(A)}{\varphi(2\epsilon)} \leq N(A,d,2\epsilon) \leq \frac{\mu(A_\epsilon)}{\varphi(\epsilon)} .$$

There are many simple situations where Dudley's bound is not of the correct order. We gave a first example on page 35. We give such another example in Exercise 2.5.11. There the set T is particularly appealing: it is a simplex in $\mathbb{R}^m$. Yet other examples based on fundamental geometry (ellipsoids in $\mathbb{R}^k$) are explained in Sect. 2.13.

The result of the following exercise is very useful in all kinds of examples.

Exercise 2.5.10 Consider two integers k, m with $k \le m/4$. Assume for simplicity that k is even.

(a) Prove that

$$\sum_{0\le\ell\le k/2} \binom{m}{\ell} \le 2\Big(\frac{2k}{m}\Big)^{k/2}\binom{m}{k} . \tag{2.48}$$

(b) Denote by $\mathcal{I}$ the class of subsets of $\{1, \dots, m\}$ of cardinality k. Prove that you can find in $\mathcal{I}$ a family $\mathcal{F}$ such that for $I, J \in \mathcal{F}$ one has $\operatorname{card}(I\setminus J)\cup(J\setminus I) \ge k/2$ and $\operatorname{card}\mathcal{F} \ge (m/(2k))^{k/2}/2$. Hint: Use (a) and part (f) of Exercise 2.5.9 for μ the counting measure on $\mathcal{I}$. Warning: This is not so easy.

Exercise 2.5.11 Consider an integer m and an i.i.d. standard Gaussian sequence $(g_i)_{i\le m}$. For $t = (t_i)_{i\le m} \in \mathbb{R}^m$, let $X_t = \sum_{i\le m} t_i g_i$. This is called the canonical Gaussian process on $\mathbb{R}^m$. Its associated distance is the Euclidean distance on $\mathbb{R}^m$. It will be much used later. Consider the set

$$T = \Big\{(t_i)_{i\le m} \in \mathbb{R}^m \ ; \ t_i \ge 0 \, , \ \sum_{i\le m} t_i = 1\Big\} , \tag{2.49}$$

the convex hull of the canonical basis. By (2.16), we have $\mathsf{E}\sup_{t\in T} X_t = \mathsf{E}\sup_{i\le m} g_i \le L\sqrt{\log m}$. Prove, however, that the right-hand side of (2.41) is $\ge (\log m)^{3/2}/L$. (Hint: For an integer $k \le m$, consider the subset T_k of T consisting of sequences $t = (t_i)_{i\le m} \in T$ for which $t_i \in \{0, 1/k\}$, so that $t \in T_k$ is determined by the set $I = \{i \le m \ ; \ t_i = 1/k\}$ and $\operatorname{card} I = k$. Using Exercise 2.5.10, prove that $\log N(T_k, d, 1/(L\sqrt{k})) \ge k\log(em/k)/L$ and conclude.[11]) Thus, in this case, Dudley's bound is off by a multiplicative factor of about $\log m$. Exercise 2.7.9 will show that in $\mathbb{R}^m$ the situation cannot be worse than this.

[11] In case you wonder why e occurs in this formula, it is just to take care of the case where k is nearly m. This term is not needed here, but is important in upper bounds of the same nature that we will use below.

2.6 Rolling Up Our Sleeves: Chaining in the Simplex

The bound (2.34) seems to be genuinely better than the bound (2.38) because when going from (2.34) to (2.38) we have used the somewhat brutal inequality:

$$\sup_{t\in T}\sum_{n\geq 0} 2^{n/2} d(t, T_n) \leq \sum_{n\geq 0} 2^{n/2} \sup_{t\in T} d(t, T_n)\,.$$

The method leading to the bound (2.34) is probably the most important idea of this work. The fact that it appears now so naturally does not reflect the history of the subject, but rather that the proper approach is being used. When using this bound, we will choose the sets T_n in order to minimize the right-hand side of (2.34) instead of choosing them as in (2.36). As we will demonstrate later, *this provides essentially the best possible bound for* $\mathsf{E}\sup_{t\in T} X_t$. It is remarkable that *despite the fact that this result holds in complete generality, it is a non-trivial task to find sets* T_n *witnessing this, even in very simple situations.* In the present situation, we perform this task by an explicit construction for the set T of (2.49).

Proposition 2.6.1 *There exist sets* $T_n \subset \mathbb{R}^m$ *with* card $T_n \leq N_n$ *such that*

$$\sup_{t\in T}\sum_{n\geq 0} 2^{n/2} d(t, T_n) \leq L\sqrt{\log m} \quad (= L\mathsf{E}\sup_{t\in T} X_t)\,.$$

Of course here d is the Euclidean distance in $\mathbb{R}^m$. The reader may try to find these sets herself before reading the rest of this section, as there seems to be no better way to get convinced of the depth of the present theory. The sets T_n are not subsets of T. Please figure out by yourself how to correct this.[12]

Lemma 2.6.2 *For each* $t \in T$*, we can find a sequence* $(p(n,t))_{n\geq 0}$ *of integers* $0 \leq p(n,t) \leq 2n$ *with the following properties:*

$$\sum_{n\geq 0} 2^{n-p(n,t)} \leq L\,, \tag{2.50}$$

$$\forall n \geq 0\,,\ p(n+1,t) \leq p(n,t) + 2\,, \tag{2.51}$$

$$\operatorname{card}\left\{i \leq m\,;\ t_i \geq 2^{-p(n,t)}\right\} < 2^n\,. \tag{2.52}$$

Proof There is no loss of generality to assume that the sequence $(t_i)_{i\leq m}$ is non-increasing. We set $t_i = 0$ for $i > m$. Then for any $n \geq 1$ and $2^{n-1} < i \leq 2^n$, we have $t_i \geq t_{2^n}$, so that $2^{n-1} t_{2^n} \leq \sum_{2^{n-1}<i\leq 2^n} t_i$. By summation over $n \geq 1$, we obtain $\sum_{n\geq 1} 2^n t_{2^n} \leq 2$, and thus $\sum_{n\geq 0} 2^n t_{2^n} \leq 3$. For $n \geq 0$, consider the

[12] The argument can be found in Sect. 2.14.

largest integer $q(n,t) \le 2n$ such that $2^{-q(n,t)} > t_{2^n}$. Thus, $2^{-q(n,t)-1} \le t_{2^n}$ when $q < 2n$. In any case, $2^{-q(n,t)} \le 2t_{2^n} + 2^{-2n}$, and thus $\sum_{n\ge 0} 2^{n-q(n,t)} \le L$. Also if $t_i \ge 2^{-q(n,t)} > t_{2^n}$, then $i < 2^n$. In particular $\operatorname{card}\{i \le m; t_i \ge 2^{-q(t,n)}\} < 2^n$. Finally we define

$$p(n,t) = \min\{q(k,t) + 2(n-k); 0 \le k \le n\}.$$

Taking $k = n$ shows that $p(n,t) \le q(n,t) \le 2n$, implying (2.52). If $k \le n$ is such that $p(n,t) = q(k,t) + 2(n-k)$, then $p(n+1,t) \le q(k,t) + 2(n+1-k) = p(n,t) + 2$, proving (2.51). Also, since $2^{n-p(n,t)} \le \sum_{k\le n} 2^{n-2(n-k)-q(k,t)} = \sum_{k\le n} 2^{k-n+(k-q(k,t))}$, we have

$$\sum_{n\ge 0} 2^{n-p(n,t)} \le \sum_{k\ge 0} 2^{k-q(k,t)} \sum_{n\ge k} 2^{k-n} \le L. \qquad \square$$

Given a set $I \subset \{1,\dots,m\}$ and a integer p, we denote by $V_{I,p}$ the set of elements $u = (u_i)_{i\le m} \in \mathbb{R}^m$ such that $u_i = 0$ if $i \notin I$ and $u_i = r_i 2^{-p}$ if $i \in I$, where r_i is an integer $0 \le r_i \le 3$. Then $\operatorname{card} V_{I,p} \le 4^{\operatorname{card} I}$. For $n \ge 1$, we denote by V_n the union of all the sets $V_{I,p}$ for $\operatorname{card} I \le 2^n$ and $0 \le p \le 2n$. Crudely we have $\operatorname{card} V_n \le m^{L2^n}$. We set $V_0 = \{0\}$ and for $n \ge 1$ we denote by U_n the set of all sums $\sum_{0\le k\le n} x_k$ where $x_k \in V_k$. Then $\operatorname{card} U_n \le m^{L2^n}$.[13]

Lemma 2.6.3 *Consider $t \in T$ and the sequence $(p(n,t))_{n\ge 0}$ constructed in Lemma 2.6.2. Then for each n, we can write $t = u(n) + v(n)$ where $u(n) \in U_n$ and where $v(n) = (v(n)_i)_{i\le m}$ satisfies $0 \le v(n)_i \le \min(t_i, 2^{-p(n,t)})$.*

Proof The proof is by induction over n. For $n = 0$, we set $u(0) = 0$, $v(0) = t$. For the induction from n to $n+1$, consider the set $I = \{i \le m; v(n)_i > 2^{-p(n+1,t)}\}$. Since $v(n)_i \le t_i$, it follows from (2.52) that $\operatorname{card} I < 2^{n+1}$. For each $i \in I$, let r_i be the largest integer with $r_i 2^{-p(n+1,t)} < v(n)_i$ so that $v(n)_i - r_i 2^{-p(n+1,t)} \le 2^{-p(n+1,t)}$. Since $v(n)_i \le 2^{-p(n,t)}$ by induction and since $p(n+1,t) \le p(n,t) + 2$ by (2.51), we have $r_i \le 3$. Define $u = (u_i)_{i\le m} \in \mathbb{R}^m$ by $u_i = r_i 2^{-p(n+1,t)}$ if $i \in I$ and $u_i = 0$ otherwise. Then, $u \in V_{I,p(n+1,t)} \subset V_n$. Thus, $t = u(n+1) + v(n+1)$ where $u(n+1) := u(n) + u \in U_{n+1}$ and $v(n+1) := v(n) - u$ satisfies $v(n+1)_i \le \min(t_i, 2^{-p(n+1,t)})$. $\square$

Lemma 2.6.4 *For each $t \in T$, we have $\sum_{n\ge 0} 2^{n/2} d(t, U_n) \le L$.*

Proof Consider the sequence $(v(n))_{n\ge 0}$ constructed in Lemma 2.6.3, so that $d(t, U_n) \le \|v(n)\|_2$ since $t = u(n) + v(n)$. Let $I_n = \{i \le m; t_i \ge 2^{-p(n,t)}\}$ so that by (2.52) we have $\operatorname{card} I_n < 2^n$. For $n \ge 1$, set $J_n = I_n \setminus I_{n-1}$ so that for $i \in J_n$, we have $t_i < 2^{-p(n-1,t)}$. Then, $\|v(n)\|_2^2 = \sum_{i\le m} v(n)_i^2 = \sum_{i\in I_n} v(n)_i^2 + \sum_{k>n} \sum_{i\in J_k} v(n)_i^2$. Since $v(n)_i \le 2^{-p(n,t)}$ and $\operatorname{card} I_n \le 2^n$, the

[13] Controlling the cardinality of U_n is the key point.

first sum is $\le 2^{n-2p(n,t)}$. Since $v(n)_i \le t_i \le 2^{-p(k-1,t)}$ for $i \in J_k$ and card $J_k \le 2^k$, we have $\sum_{i\in J_k} v(n)_i^2 \le 2^{k-2p(k-1,t)}$. Thus, $\|v(n)\|_2^2 \le \sum_{k\ge n} 2^{k-2p(k-1,t)}$ so that $\|v(n)\|_2 \le \sum_{k\ge n} 2^{k/2-p(k-1,t)}$ and

$$\sum_{n\ge1} 2^{n/2}\|v(n)\|_2 \le \sum_{n\ge1}\sum_{k\ge n} 2^{n/2+k/2-p(k-1,t)} = \sum_{k\ge1} 2^{k/2-p(k-1,t)} \sum_{n\le k} 2^{n/2}$$
$$\le L\sum_{k\ge1} 2^{k-p(k-1,t)} \le L\,, \tag{2.53}$$

where we have used (2.50) in the last inequality. □

Proof of Proposition 2.6.1 Consider the smallest integer k_0 with $m \le N_{k_0}$ so that $2^{k_0/2} \le L\sqrt{\log m}$. Observe also that $m^{2^n} \le (2^{2^{k_0}})^{2^n} = 2^{2^{k_0+n}} = N_{k_0+n}$. Thus, card $U_n \le m^{L2^n} \le N_{k_0+n+k_1}$ where k_1 is a universal constant. For $n \ge k_0+k_1+1$, we set $T_n = U_{n-n_0-k_1}$, so that card $T_n \le N_n$. For $n \le k_0+k_1$, we set $T_n = \{0\}$. Finally, given $t \in T$ (and keeping in mind that k_1 is a universal constant), we have

$$\sum_{n\ge0} 2^{n/2} d(t,T_n) \le L2^{k_0/2} + \sum_{n\ge k_0+k_1+1} 2^{n/2} d(t,T_n)$$

and, using Lemma 2.6.4 in the last inequality,

$$\sum_{n\ge k_0+k_1+1} 2^{n/2} d(t,T_n) = \sum_{n\ge k_0+k_1+1} 2^{n/2} d(t,U_{n-k_0-k_1})$$
$$= \sum_{n\ge1} 2^{(n+k_0+k_1)/2} d(t,U_n) \le L2^{k_0/2}\,.$$ □

2.7 Admissible Sequences of Partitions

The idea behind the bound (2.34) admits a technically more convenient formulation.[14]

Definition 2.7.1 Given a set T, an *admissible sequence* is an increasing sequence $(\mathcal{A}_n)_{n\ge0}$ of partitions of T such that card $\mathcal{A}_n \le N_n$, i.e., card $\mathcal{A}_0 = 1$ and card $\mathcal{A}_n \le 2^{2^n}$ for $n \ge 1$.

[14] We will demonstrate why this is the case only later, in Theorem 4.5.13.

By an *increasing* sequence of partitions, we mean that every set of $\mathcal{A}_{n+1}$ is contained in a set of $\mathcal{A}_n$. Admissible sequences of partitions will be constructed recursively, by breaking each element C of $\mathcal{A}_n$ into at most N_n pieces, obtaining then a partition $\mathcal{A}_{n+1}$ of T consisting of at most $N_n^2 \le N_{n+1}$ pieces.

Throughout the book, we denote by $A_n(t)$ the unique element of $\mathcal{A}_n$ which contains t. The double exponential in the definition of N_n (see (2.29)) occurs simply since for our purposes the proper measure of the "size" of a partition $\mathcal{A}$ is $\log \operatorname{card} \mathcal{A}$. This double exponential ensures that "the size of the partition $\mathcal{A}_n$ doubles at every step". This offers a number of technical advantages which will become clear gradually.

Theorem 2.7.2 (The Generic Chaining Bound) *Under the increment condition (2.4) (and if* $\mathsf{E}X_t = 0$ *for each* t*), then for each admissible sequence* $(\mathcal{A}_n)$ *we have*

$$\mathsf{E} \sup_{t\in T} X_t \le L \sup_{t\in T} \sum_{n\ge 0} 2^{n/2} \Delta(A_n(t)) \,. \tag{2.54}$$

Here as always, $\Delta(A_n(t))$ denotes the diameter of $A_n(t)$ for d. One could think that (2.54) could be much worse than (2.34), but it will turn out that this is not the case when the sequence $(\mathcal{A}_n)$ is appropriately chosen.

Proof We may assume T to be finite. We construct a subset T_n of T by taking exactly one point in each set A of $\mathcal{A}_n$. Then for $t \in T$ and $n \ge 0$, we have $d(t, T_n) \le \Delta(A_n(t))$ and the result follows from (2.34). □

Definition 2.7.3 Given $\alpha > 0$ and a metric space (T, d) (that need not be finite), we define

$$\gamma_\alpha(T, d) = \inf \sup_{t\in T} \sum_{n\ge 0} 2^{n/\alpha} \Delta(A_n(t)) \,,$$

where the infimum is taken over all admissible sequences.

It is useful to observe that since $A_0(t) = T$, we have $\gamma_\alpha(T, d) \ge \Delta(T)$. The most important cases by far are $\alpha = 2$ and $\alpha = 1$. For the time being, we need only the case $\alpha = 2$. The case $\alpha = 1$ is first met in Theorem 4.5.13, although more general functionals occur first in Definition 4.5.

Exercise 2.7.4 Prove that if $d \le Bd'$, then $\gamma_2(T, d) \le B\gamma_2(T, d')$.

Exercise 2.7.5 Prove that $\gamma_\alpha(T, d) \le K(\alpha)\Delta(T)(\log \operatorname{card} T)^{1/\alpha}$ when T is finite. Hint: Ensure that $\Delta(A_n(t)) = 0$ if $N_n \ge \operatorname{card} T$.

A large part of our arguments will take place in abstract metric spaces, and this may represent an obstacle to the reader who has never thought about this. Therefore, we cannot recommend too highly the following exercise:

Exercise 2.7.6

(a) Consider a metric space (T, d), and assume that for each $n \geq 0$, you are given a covering $\mathcal{B}_n$ of T with card $\mathcal{B}_n \leq N_n$. Prove that you can construct an admissible sequence $(\mathcal{A}_n)$ of partitions of T with the following property:

$$\forall n \geq 1 \;,\; \forall A \in \mathcal{A}_n \;,\; \exists B \in \mathcal{B}_{n-1} \;,\; A \subset B \;. \tag{2.55}$$

(b) Prove that for any metric space (T, d), we have

$$\gamma_2(T, d) \leq L \sum_{n \geq 0} 2^{n/2} e_n(T) \;. \tag{2.56}$$

The following exercise explains one of the reasons admissible sequences of sets are so convenient: given two such sequences, we can construct a third sequence which merges the good properties of the two sequences.

Exercise 2.7.7 Consider a set T and two admissible sequences $(\mathcal{B}_n)$ and $(\mathcal{C}_n)$. Prove that there is an admissible sequence $(\mathcal{A}_n)$ such that

$$\forall n \geq 1 \;,\; \forall A \in \mathcal{A}_n \;,\; \exists B \in \mathcal{B}_{n-1} \;,\; A \subset B \;,\; \exists C \in \mathcal{C}_{n-1}, A \subset C \;.$$

The following simple property should be clear in the reader's mind:

Exercise 2.7.8

(a) Prove that for $n \geq 0$, we have

$$2^{n/2} e_n(T) \leq L\gamma_2(T, d) \;. \tag{2.57}$$

Hint: Observe that $2^{n/2} \max\{\Delta(A); A \in \mathcal{A}_n\} \leq \sup_{t \in T} \sum_{n \geq 0} 2^{n/2} \Delta(A_n(t))$.

(b) Prove that, equivalently, for $\epsilon > 0$, we have

$$\epsilon\sqrt{\log N(T, d, \epsilon)} \leq L\gamma_2(T, d) \;.$$

The reader should compare (2.57) with (2.56).

Exercise 2.7.9 Use (2.57) and Exercise 2.5.9 (e) to prove that if $T \subset \mathbb{R}^m$, then

$$\sum_{n \geq 0} 2^{n/2} e_n(T) \leq L \log(m + 1)\gamma_2(T, d) \;. \tag{2.58}$$

In words, Dudley's bound is never off by more than a factor of about $\log(m + 1)$ in $\mathbb{R}^m$.[15]

[15] And we have shown in Exercise 2.5.6 that it is never off by a factor more than about log log card T either.

Exercise 2.7.10 Prove that the estimate (2.58) is essentially optimal. Warning: This requires some skill.

Combining Theorem 2.7.2 with Definition 2.7.3 yields the following very important result:

Theorem 2.7.11 *Under (2.4) and (2.1), we have*

$$\mathsf{E} \sup_{t \in T} X_t \le L\gamma_2(T, d) . \tag{2.59}$$

To make (2.59) of interest, we must be able to control $\gamma_2(T, d)$, i.e., we must learn how to construct admissible sequences, a topic we shall first address in Sect. 2.9.

Exercise 2.7.12 When the process $(X_t)_{t \in T}$ satisfies (2.4) but is no longer assumed to be centered, prove that

$$\mathsf{E} \sup_{s,t \in T} |X_s - X_t| \le L\gamma_2(T, d) . \tag{2.60}$$

We now turn to the control of the tails of the process, which will follow by a small variation of the same argument.

Theorem 2.7.13 *Under (2.4) and (2.1), we have*

$$\mathsf{P}\Big(\sup_{s,t \in T} |X_s - X_t| \ge L\gamma_2(T, d) + Lu\Delta(T)\Big) \le L \exp(-u^2) . \tag{2.61}$$

Proof We use the notation of the proof of (2.34). We may assume $u \ge 1$. Let us consider the smallest integer $k \ge 0$ such that $u^2 \le 2^k$ so that $2^k \le 2u^2$. Consider the event Ω_1 defined by

$$\forall t \in T_k \ , \ |X_t - X_{t_0}| \le 4u\Delta(T) , \tag{2.62}$$

so that by the union bound,

$$P(\Omega_1^c) \le 2^{2^k} \cdot 2\exp(-8u^2) \le 2\exp(2u^2 - 8u^2) \le \exp(-u^2) . \tag{2.63}$$

Consider the event Ω_2 given by

$$\forall n \ge k \ , \ \forall t \in T \ , \ |X_{\pi_n(t)} - X_{\pi_{n+1}(t)}| \le 2^{n/2+2}\Delta(A_n(t)) , \tag{2.64}$$

so that by the union bound again,

$$\mathsf{P}(\Omega_2^c) \le \sum_{n \ge k} 2^{2^{n+1}} \cdot 2\exp(-2^{n+3}) \le \sum_{n \ge k} 2\exp(-2^{n+2}) \le 4\exp(-u^2) , \tag{2.65}$$

using, for example, in the last inequality that $2^{n+2} \geq 2^{k+2} + n - k \geq 4u^2 + n - k$. Consequently, $\mathsf{P}(\Omega_1 \cup \Omega_2) \geq 1 - 5\exp(-u^2)$, and on this event, we have

$$|X_t - X_{\pi_k(t)}| \leq \sum_{n \geq k} 2^{n/2+2} \Delta(A_n(t)) \leq L\gamma_2(T, d) ,$$

so that $|X_t - X_{t_0}| \leq |X_t - X_{\pi_k(t)}| + |X_{\pi_k(t)} - X_{t_0}| \leq L\gamma_2(T, d) + Lu\Delta(T)$. □

Let us note in particular that using (2.24),

$$\Big(\mathsf{E} \sup_{s,t \in T} |X_s - X_t|^p\Big)^{1/p} \leq L\sqrt{p}\gamma_2(T, d) . \tag{2.66}$$

Needless to say that we will look for extensions of Theorem 2.7.13. We can prove right away a particularly elegant result (due independently to R. Latała and S. Mendelson). Let us consider a process $(X_t)_{t \in T}$, which is assumed to be centered but need not to be symmetric. For $n \geq 1$, consider the distance δ_n on T given by $\delta_n(s, t) = \|X_s - X_t\|_{2^n}$. Denote by $\Delta_n(A)$, the diameter of a subset A of T for the distance δ_n.

Theorem 2.7.14 *Consider an admissible sequence $(\mathcal{A}_n)_{n \geq 0}$ of partitions of T. Then,*

$$\mathsf{E} \sup_{s,t \in T} |X_s - X_t| \leq L \sup_{t \in T} \sum_{n \geq 0} \Delta_n(A_n(t)) . \tag{2.67}$$

Moreover, given $u > 0$ and the largest integer k with $2^k \leq u^2$, we have

$$\mathsf{P}\Big(\sup_{s,t \in T} |X_s - X_t| \geq L\Delta_k(T) + \sup_{t \in T} \sum_{n \geq 0} \Delta_n(A_n(t))\Big) \leq L\exp(-u^2) . \tag{2.68}$$

Proof The increment condition (2.4) will be replaced by the following: For a r.v. Y and $p \geq 1$, we have

$$\mathsf{P}(|Y| \geq u) \leq \mathsf{P}(|Y|^p \geq u^p) \leq \Big(\frac{\|Y\|_p}{u}\Big)^p . \tag{2.69}$$

Let us then consider the points $\pi_n(t)$ as usual. For $u \geq 1$, let us consider the event Ω_u defined by[16]

$$\forall n \geq 1 ,\ |X_{\pi_n(t)} - X_{\pi_{n+1}(t)}| \leq u\Delta_n(A_n(t)) , \tag{2.70}$$

[16] We are following here the general method outlined at the end of Sect. 2.4.

so that by the union bound and (2.69) for $u \geq 4$, we have

$$\mathsf{P}(\Omega_u^c) \leq \sum_{n\geq 1}\Big(\frac{2}{u}\Big)^{2^n} \leq \sum_{k\geq 2}\Big(\frac{2}{u}\Big)^k \leq \frac{L}{u^2} . \tag{2.71}$$

On Ω_u, summation of the inequalities (2.70) for $n \geq 1$ yields $\sup_{t\in T}|X_t - X_{\pi_1(t)}| \leq Lu\sum_{n\geq 1}\Delta_n(A_n(t))$. Combining with (2.71), we obtain

$$\mathsf{E}\sup_{t\in T}|X_t - X_{\pi_1(t)}| \leq L\sum_{n\geq 1}\Delta_n(A_n(t)) .$$

Since $\mathsf{E}\sup_{t\in T}|X_{\pi_1(t)} - X_{\pi_0(t)}| \leq L\Delta_0(T)$, we have $\mathsf{E}\sup_{t\in T}|X_t - X_{\pi_0(t)}| \leq L\sum_{n\geq 0}\Delta_n(A_n(t))$, and (2.67) follows. The proof of (2.68) is nearly identical to the proof of (2.61) and is left to the reader. □

2.8 Functionals

Given a metric space (T, d), how do we calculate $\gamma_2(T, d)$? Of course there is no free lunch. The quantity $\gamma_2(T, d)$ reflects a highly non-trivial geometric characteristic of the metric space. This geometry must be understood in order to compute $\gamma_2(T, d)$. There are unsolved problems in this book (such as Conjecture 17.1.4) which boil down to estimating $\gamma_2(T, d)$ for a certain metric space.

In this section, we introduce *functionals*, which are an efficient way to bring up the geometry of a metric space and to build competent admissible sequences, providing upper bounds for $\gamma_2(T, d)$. We will say that a map F is a *functional* on a set T if to each subset H of T it associates a number $F(H) \geq 0$ and if it is increasing, i.e.,

$$H \subset H' \subset T \Rightarrow F(H) \leq F(H') . \tag{2.72}$$

Intuitively a functional is a measure of "size" for the subsets of T. It allows to identify which subsets of T are "large" for our purposes. A first example is given by $F(H) = \Delta(H)$. In the same direction, a fundamental example of a functional is

$$F(H) = \gamma_2(H, d) . \tag{2.73}$$

A second example, equally important, is the quantity

$$F(H) = \mathsf{E}\sup_{t\in H} X_t \tag{2.74}$$

where $(X_t)_{t\in T}$ is a given process indexed by T and satisfying (2.4).

For our purposes, the relevant property of functionals is by no means intuitively obvious yet (but we shall soon see that the functional (2.73) does enjoy this property). Let us first try to explain it in words: if a set is the union of many small pieces far enough from each other, then this set is significantly larger (as measured by the functional) than the *smallest* of its pieces. "Significantly larger" depends on the scale of the pieces and on their number. This property will be called a "growth condition".

Let us address a secondary point before we give definitions. We denote by $B(t, r)$ the ball centered at t of radius r, and we note that

$$\Delta(B(t, r)) \le 2r \ .$$

This factor 2 is a nuisance. It is qualitatively the same to say that a set is contained in a ball of small radius or has small diameter, but quantitatively we have to account for this factor 2. In countless constructions, we will produce sets A which are "small" because they are contained in a ball of small radius r. Either we keep track of this property, which is cumbersome, or we control the size of A through its diameter and we deal with this inelegant factor 2. We have chosen here the second method.[17]

What do we mean by "small pieces far from each other"? There is a scale $a > 0$ at which this happens and a parameter $r \ge 8$ which gives us some room. The pieces are small at that scale: they are contained in balls with radius $2a/r$.[18] The balls are far from each other: any two centers of such balls are at mutual distance $\ge a$. The reason why we require $r \ge 8$ is that we want the following: Two points taken in different balls with radius $2a/r$ whose centers are at distance $\ge a$ cannot be too close to each other. This would not be true for, say, $r = 4$, so we give ourselves some room and take $r \ge 8$. Here is the formal definition.

Definition 2.8.1 Given $a > 0$ and an integer $r \ge 8$, we say that subsets $H_1, \dots, H_m$ of T are (a, r)-separated if

$$\forall \ell \le m \, , \; H_\ell \subset B(t_\ell, 2a/r) \, , \tag{2.75}$$

where the points $t_1, t_2, \dots, t_m$ in T satisfy

$$\forall \ell \, , \, \ell' \le m \, , \, \ell \ne \ell' \Rightarrow a \le d(t_\ell, t_{\ell'}) \le 2ar \ . \tag{2.76}$$

A secondary feature of this definition is that the small pieces H_ℓ are not only well separated (on a scale a), but they are in the "same region of T" (on the larger scale ra). This is the content of the last inequality in condition (2.76).

Exercise 2.8.2 Find interesting examples of metric spaces for which there are no points $t_1, \dots, t_m$ as in (2.76), for all large enough values of m.

---

[17] The opposite choice was made in [132].

[18] This coefficient 2 is motivated by the considerations of the previous paragraph.

Now, what does “the union of the pieces is significantly larger than the *smallest* of these pieces” mean? This is an “additive property”, not a multiplicative one. In this first version of the growth condition, it means that the size of this union is larger than the size of the smallest piece by a quantity $a\sqrt{\log N}$ where N is the number of pieces.[19] Well, sometimes it will only be larger by a quantity of say $a\sqrt{\log N}/100$. This is how the parameter c^* below comes into the picture. One could also multiply the functionals by a suitable constant (i.e., $1/c^*$) to always reduce to the case $c^* = 1$, but this is a matter of taste.

Another feature is that we do not need to consider the case with N pieces for a general value of N, but only for the case where $N = N_n$ for some n. This is because we care about the value of $\log N$ only within, say, a factor of 2, and this is precisely what motivated the definition of N_n. In order to understand the definition below, one should also recall that $\sqrt{\log N_n}$ is about $2^{n/2}$.

Definition 2.8.3 We say that the functional F satisfies the *growth condition* with parameters $r \geq 8$ and $c^* > 0$ if for any integer $n \geq 1$ and any $a > 0$ the following holds true, where $m = N_n$: For each collection of subsets $H_1, \ldots, H_m$ of T that are (a, r)-separated, we have

$$F\Big(\bigcup_{\ell \leq m} H_\ell\Big) \geq c^* a 2^{n/2} + \min_{\ell \leq m} F(H_\ell) \ . \tag{2.77}$$

This definition is motivated by the fundamental fact that when $(X_t)_{t \in T}$ is a Gaussian process, the functional (2.74) satisfies a form of the growth condition (see Proposition 2.10.8).

The following illustrates how we might use the first part of (2.76):

Exercise 2.8.4 Let (T, d) be isometric to a subset of $\mathbb{R}^k$ provided with the distance induced by a norm. Prove that in order to check that a functional satisfies the growth condition of Definition 2.8.3, it suffices to consider the values of n for which $N_{n+1} \leq (1 + 2r)^k$. Hint: It follows from (2.47) that for larger values of n and $m = N_n$, there are no points $t_1, \ldots, t_m$ as in (2.76).

You may find it hard to give simple examples of functionals which satisfy the growth condition (2.77). It will become gradually apparent that this condition imposes strong restrictions on the metric space (T, d) and in particular a control from above of the quantity $\gamma_2(T, d)$. It bears repeating that $\gamma_2(T, d)$ reflects the geometry of the space (T, d). Once this geometry is understood, it is usually possible to guess a good choice for the functional F. Many examples will be given in subsequent chapters.

As we show now, we really have no choice. Functionals with the growth property are intimately connected with the quantity $\gamma_2(T, d)$.

[19] We remind the reader that the function $\sqrt{\log y}$ arises from the fact that it is the inverse of the function $\exp(x^2)$.

Proposition 2.8.5 *Assume $r \geq 16$. Then the functional $F(H) = \gamma_2(H, d)$ satisfies the growth condition with parameters r and $c^* = 1/8$.*

Proof Let $m = N_n$ and consider points $(t_\ell)_{\ell \leq m}$ of T with $d(t_\ell, t_{\ell'}) \geq a$ if $\ell \neq \ell'$. Consider sets $H_\ell \subset B(t_\ell, a/8)$ and the set $H = \bigcup_{\ell \leq m} H_\ell$. We have to prove that

$$\gamma_2(H, d) \geq \frac{1}{8} a 2^{n/2} + \min_{\ell \leq m} \gamma_2(H_\ell, d) . \tag{2.78}$$

Consider an admissible sequence of partitions $(\mathcal{A}_n)$ of H, and consider the set

$$I_n = \{\ell \leq m \; ; \; \exists A \in \mathcal{A}_{n-1} \; ; \; A \subset H_\ell\} .$$

Picking for $\ell \in I_n$ an arbitrary element $A \in \mathcal{A}_{n-1}$ with $A \subset H_\ell$ defines a one-to-one map from I_n to $\mathcal{A}_{n-1}$. Thus, card $I_n \leq$ card $\mathcal{A}_{n-1} \leq N_{n-1} < m = N_n$. Hence, there exists $\ell_0 \notin I_n$. Next, we prove that for $t \in H_{\ell_0}$, we have

$$\Delta(A_{n-1}(t)) \geq \Delta(A_{n-1}(t) \cap H_{\ell_0}) + \frac{1}{4} a . \tag{2.79}$$

Since $\ell_0 \notin I_n$, we have $A_{n-1}(t) \not\subset H_{\ell_0}$, so that since $A_{n-1}(t) \subset H$, the set $A_{n-1}(t)$ must intersect a set $H_\ell \neq H_{\ell_0}$, and consequently it intersects the ball $B(t_\ell, a/8)$. Since $t \in H_{\ell_0}$, we have $d(t, B(t_\ell, a/8)) \geq a/2$. Since $t \in A_{n-1}(t)$, this implies that $\Delta(A_{n-1}(t)) \geq a/2$. This proves (2.79) since $\Delta(A_{n-1}(t) \cap H_{\ell_0}) \leq \Delta(H_{\ell_0}) \leq a/4$.

Now, since for each $k \geq 0$ we have $\Delta(A_k(t)) \geq \Delta(A_k(t) \cap H_{\ell_0})$, we have

$$\begin{aligned}
&\sum_{k \geq 0} 2^{k/2}(\Delta(A_k(t)) - \Delta(A_k(t) \cap H_{\ell_0}))) \\
&\qquad \geq 2^{(n-1)/2}(\Delta(A_{n-1}(t)) - \Delta(A_{n-1}(t) \cap H_{\ell_0})) \\
&\qquad \geq \frac{1}{4} a 2^{(n-1)/2} ,
\end{aligned}$$

where we have used (2.79) in the last inequality, and, consequently,

$$\sum_{k \geq 0} 2^{k/2} \Delta(A_k(t)) \geq \frac{1}{4} a 2^{(n-1)/2} + \sum_{k \geq 0} 2^{k/2} \Delta(A_k(t) \cap H_{\ell_0}) . \tag{2.80}$$

Next, consider the admissible sequence $(\mathcal{A}'_n)$ of H_{ℓ_0} given by $\mathcal{A}'_n = \{A \cap H_{\ell_0}; A \in \mathcal{A}_n\}$. We have by definition

$$\sup_{t \in H_{\ell_0}} \sum_{k \geq 0} 2^{k/2} \Delta(A_k(t) \cap H_{\ell_0}) \geq \gamma_2(H_{\ell_0}, d) .$$

Hence, taking the supremum over t in H_{ℓ_0} in (2.80), we get

$$\sup_{t\in H_{\ell_0}}\sum_{k\geq 0}2^{k/2}\Delta(A_k(t))\geq\frac{1}{4}a2^{(n-1)/2}+\gamma_2(H_{\ell_0},d)\geq\frac{1}{8}a2^{n/2}+\min_{\ell\leq m}\gamma_2(H_\ell,d)\ .$$

Since the admissible sequence $(\mathcal{A}_n)$ is arbitrary, we have proved (2.78). □

2.9 Partitioning Schemes

In this section, we use functionals satisfying the growth condition to construct admissible sequences of partitions. The basic result is as follows:

Theorem 2.9.1 *Assume that there exists on T a functional F which satisfies the growth condition of Definition 2.8.3 with parameters r and c^*. Then*[20]

$$\gamma_2(T,d)\leq\frac{Lr}{c^*}F(T)+Lr\Delta(T)\ . \tag{2.81}$$

This theorem and its generalizations form the backbone of this book. The essence of this theorem is that it produces (by actually constructing them) a sequence of partitions that witnesses the inequality (2.81). For this reason, it could be called "the fundamental partitioning theorem".

Exercise 2.9.2 Consider a metric space T consisting of exactly two points. Prove that the functional given by $F(H)=0$ for each $H\subset T$ satisfies the growth condition of Definition 2.8.3 for $r=8$ and any $c^*>0$. Explain why we cannot replace (2.81) by the inequality $\gamma_2(T,d)\leq LrF(T)/c^*$.

Let us first stress the following trivial fact (connected to Exercise 2.5.9 (a)). It will be used many times. The last statement of (a) is particularly useful.

Lemma 2.9.3

(a) Consider an integer N. If we cannot cover T by at most $N-1$ balls of radius a, then there exist points $(t_\ell)_{\ell\leq N}$ with $d(t_\ell,t_{\ell'})\geq a$ for $\ell\neq\ell'$. In particular if $e_n(T)>a$, we can find points $(t_\ell)_{\ell\leq N_n}$ with $d(t_\ell,t_{\ell'})\geq a$ for $\ell\neq\ell'$.
(b) Assume that any sequence $(t_\ell)_{\ell\leq m}$ with $d(t_\ell,t_{\ell'})\geq a$ for $\ell\neq\ell'$ satisfies $m\leq N$. Then, T can be covered by N balls of radius a.
(c) Consider points $(t_\ell)_{\ell\leq N_n+1}$ such that $d(t_\ell,t_{\ell'})\geq a$ for $\ell\neq\ell'$. Then, $e_n(T)\geq a/2$.

[20] It is certain that as r grows, we must obtain a weaker result. The dependence of the right-hand side of (2.81) on r is not optimal. It may be improved with further work.

Proof

(a) We pick the points t_ℓ recursively with $d(t_\ell, t_{\ell'}) \geq a$ for $\ell' < \ell$. By hypothesis, the balls of radius a centered on the previously constructed points do not cover the space if there are $< N$ of them so that the construction continues until we have constructed N points.
(b) You can either view this as a reformulation of (a) or argue directly that when m is taken as large as possible the balls $B(t_\ell, a)$ cover T.
(c) If T is covered by sets $(B_{\ell'})_{\ell' \leq N_n}$, by the pigeon hole principle, at least two of the points t_ℓ must fall into one of these sets, which therefore cannot be a ball of radius $< a/2$. □

The admissible sequence of partitions witnessing (2.81) will be constructed by recursive application of the following basic principle:

Lemma 2.9.4 *Under the conditions of Theorem 2.9.1, consider $B \subset T$ with $\Delta(B) \leq 2r^{-j}$ for a certain $j \in \mathbb{Z}$, and consider any $n \geq 0$. Let $m = N_n$. Then we can find a partition $(A_\ell)_{\ell \leq m}$ of B into sets which have* either *of the following properties:*

$$\Delta(A_\ell) \leq 2r^{-j-1}\,, \tag{2.82}$$

or else

$$t \in A_\ell \Rightarrow F(B \cap B(t, 2r^{-j-2})) \leq F(B) - c^* 2^{n/2} r^{-j-1}\,. \tag{2.83}$$

In words, the piece of the partitions have two further properties. Either (case (2.82)) we have reduced the bound on their diameter from $2r^{-j}$ for B to $2r^{-j-1}$, or else we have no new information on the diameter, but we have gathered the information (2.83).

Proof Consider the set

$$C = \left\{t \in B\,;\; F(B \cap B(t, 2r^{-j-2})) > F(B) - c^* 2^{n/2} r^{-j-1}\right\}.$$

Consider points $(t_\ell)_{\ell \leq m'}$ in C such that $d(t_\ell, t_{\ell'}) \geq r^{-j-1}$ for $\ell \neq \ell'$. We prove that $m' < m$. For otherwise, using (2.77) for $a = r^{-j-1}$ and for the sets $H_\ell := B \cap B(t_\ell, 2r^{-j-2})$ shows that

$$F(B) \geq F\Big(\bigcup_{\ell \leq m} H_\ell\Big) \geq c^* r^{-j-1} 2^{n/2} + \min_{\ell \leq m} F(H_\ell) > F(B)\,.$$

This contradiction proves that $m' < m$. Consequently, using Lemma 2.9.3 (b) for $N = m - 1$, we may cover C by $m' < m$ balls $(B_\ell)_{\ell \leq m'}$ of radius $\leq r^{-j-1}$. We then set $A_\ell = C \cap (B_\ell \setminus \cup_{\ell' < \ell} B_{\ell'})$ for $\ell \leq m'$, $A_\ell = \emptyset$ for $m' < \ell < m$ and $A_m = B \setminus C$. □

So, in picturesque terms, Lemma 2.9.4 produces many small pieces and (possibly) a large one (on which one has further information).

Before we start the proof of Theorem 2.9.1, we need the following technical fact which will be used many times: The sum of a geometric series is basically of the size of either its first or its last term.

Lemma 2.9.5 *Consider numbers $(a_n)_{n\geq 0}$, $a_n > 0$, and assume $\sup_n a_n < \infty$. Consider $\alpha > 1$, and define*

$$I = \left\{n \geq 0\ ;\ \forall k \geq 0\ , k \neq n\ ,\ a_k < a_n \alpha^{|n-k|}\right\}\ . \tag{2.84}$$

Then $I \neq \emptyset$, and we have

$$\sum_{k\geq 0} a_k \leq \frac{2\alpha}{\alpha - 1} \sum_{n\in I} a_n\ . \tag{2.85}$$

Proof Let us write $k \prec n$ when $a_k \leq a_n \alpha^{-|n-k|}$. This relation is a partial order: if $k \prec n$ and $n \prec p$, then $a_k \leq a_p \alpha^{-|n-k|-|n-p|} \leq a_p \alpha^{-|k-p|}$, so that $k \prec p$. We can then restate the definition of I:

$$I = \left\{n \geq 0\ ;\ \forall k \geq 0\ , n \prec k \Rightarrow n = k\right\}\ .$$

In words, I is the set of elements n of $\mathbb{N}$ that are maximal for the partial order $\prec$.

Next, we prove that for each k in $\mathbb{N}$, there exists $n \in I$ with $k \prec n$. Indeed otherwise we can recursively construct an infinite sequence $n_1 = n \prec n_2 \prec \cdots$, and this is absurd because $a_{n_{\ell+1}} \geq \alpha a_{n_\ell}$ and we assume that the sequence (a_n) is bounded.

Thus, for each k in $\mathbb{N}$, there exists $n \in I$ with $k \prec n$. Then $a_k \leq a_n \alpha^{-|n-k|}$, and therefore

$$\sum_{k\geq 0} a_k \leq \sum_{n\in I} \sum_{k\geq 0} a_n \alpha^{-|k-n|} \leq \frac{2}{1-\alpha^{-1}} \sum_{n\in I} a_n\ . \qquad \square$$

Proof of Theorem 2.9.1 There is no question that this proof is the most demanding up to this point. The result is, however, absolutely central, on its own and also because several of our main results will follow the same overall scheme of proof.

We have to construct an admissible sequence of partitions which witnesses the inequality (2.81). The construction of this sequence is as simple as it could be: we recursively use Lemma 2.9.4. More precisely, we construct an admissible sequence of partitions $\mathcal{A}_n$, and for $A \in \mathcal{A}_n$, we construct an integer $j_n(A) \in \mathbb{Z}$ with

$$\Delta(A) \leq 2r^{-j_n(A)}\ . \tag{2.86}$$

We start with $\mathcal{A}_0 = \{T\}$ and $j_0(T)$ the largest integer $j_0 \in \mathbb{Z}$ with $\Delta(T) \le 2r^{-j_0}$, so that $2r^{-j_0} \le r\Delta(T)$. Having constructed $\mathcal{A}_n$, we construct $\mathcal{A}_{n+1}$ as follows: for each $B \in \mathcal{A}_n$, we use Lemma 2.9.4 with $j = j_n(B)$ to split B into sets $(A_\ell)_{\ell \le N_n}$. If A_ℓ satisfies (2.82), we set $j_{n+1}(A_\ell) = j_n(B) + 1$, and otherwise (since we have no new information on the diameter) we set $j_{n+1}(A_\ell) = j_n(B)$. Thus, in words, $j_{n+1}(A_\ell) = j_n(B) + 1$ if A_ℓ is a small piece of B and $j_{n+1}(A_\ell) = j_n(B)$ if A_ℓ is the large piece of B.

The sequence thus constructed is admissible, since each set B in $\mathcal{A}_n$ is split into at most N_n sets and since $N_n^2 \le N_{n+1}$. We note also by construction that if $B \in \mathcal{A}_n$ and $A \subset B$, $A \in \mathcal{A}_{n+1}$, then

- Either $j_{n+1}(A) = j_n(B) + 1$
- Or else $j_{n+1}(A) = j_n(B)$ and

$$t \in A \Rightarrow F(B \cap B(t, 2r^{-j_{n+1}(A)-2})) \le F(B) - c^* 2^{n/2} r^{-j_{n+1}(A)-1} . \tag{2.87}$$

Now we start the hard part of the proof, proving that the sequence of partitions we just constructed witnesses (2.81). For this, we fix $t \in T$. We want to prove that

$$\sum_{n \ge 0} 2^{n/2} \Delta(A_n(t)) \le \frac{Lr}{c^*} F(T) + Lr\Delta(T) .$$

We set $j(n) = j_n(A_n(t))$, so that $j(n) \le j(n+1) \le j(n) + 1$. We set $a(n) = 2^{n/2} r^{-j(n)}$. Since $2^{n/2}\Delta(A_n(t)) \le 2a(n)$, it suffices to show that

$$\sum_{n \ge 0} a(n) \le \frac{Lr}{c^*} F(T) + Lr\Delta(T) . \tag{2.88}$$

First, we prove a side result, that for $n \ge 0$ we have

$$a(n) \le \frac{Lr}{c^*} F(T) + L\Delta(T) . \tag{2.89}$$

If $n \ge 1$ and $j(n-1) = j(n)$, then using (2.87) for $n-1$ rather than n yields (2.89). Next, if $n \ge 1$ and $j(n-1) = j(n) - 1$, then $a(n) = \sqrt{2} r^{-1} a(n-1) \le a(n-1)$ since $r \ge 8$, and iterating this relation until we reach an integer n' with either $j(n'-1) = j(n')$ or $n' = 0$ proves (2.89) since $a(0) \le L\Delta(T)$.

In particular the sequence $(a(n))$ is bounded. Consider then the set I as provided by Lemma 2.9.5 for $\alpha = \sqrt{2}$ and $a_n = a(n)$, that is,

$$I = \left\{ n \ge 0 \, ; \; \forall k \ge 0 \, , n \ne k \, , \; a(k) < a(n) 2^{|k-n|/2} \right\} .$$

Recalling that $a(0) = r^{-j_0} \leq r\Delta(T)/2$, it suffices to prove that

$$\sum_{n \in I \setminus \{0\}} a(n) \leq \frac{Lr}{c^*} F(T) . \tag{2.90}$$

For $n \in I, n \geq 1$, we have $a(n+1) < \sqrt{2}a(n)$ and $a(n-1) < \sqrt{2}a(n)$. Since $a(n+1) = \sqrt{2}r^{j(n)-j(n+1)}a(n)$, this implies

$$j(n+1) = j(n) + 1 \; ; \; j(n-1) = j(n) . \tag{2.91}$$

Proving (2.90) is the difficult part. Assuming first that I is infinite (the important case), let us enumerate the elements of $I \setminus \{0\}$ as $n_1 < n_2 < \ldots$ so that (2.91) implies

$$j(n_k + 1) = j(n_k) + 1 \; ; \; j(n_k - 1) = j(n_k) . \tag{2.92}$$

In words, n_k is at the end of a sequence of partition steps in which $A_{\ell+1}(t)$ was the large piece of $A_\ell(t)$ and $A_{n_k+1}(t)$ is a small piece of $A_{n_k}(t)$. Let us note that as a consequence of (2.92), we have

$$j_{n_{k+1}} \geq j_{n_k+1} \geq j_{n_k} + 1 .$$

The key to the proof is to show that for $k \geq 1$, we have

$$a(n_k) \leq \frac{Lr}{c^*}(f(n_k - 1) - f(n_{k+2})) \tag{2.93}$$

where $f(n) = F(A_n(t))$. Now the sequence $(f(n))$ is decreasing because $A_n(t) \subset A_{n-1}(t)$ and $f(0) = F(T)$. When $k \geq 2$, then $f(n_k - 1) \leq f(n_{k-1})$, so that (2.93) implies

$$a(n_k) \leq \frac{Lr}{c^*}(f(n_{k-1}) - f(n_{k+2})) . \tag{2.94}$$

Summation of the inequalities (2.94) for $k \geq 2$ then yields

$$\sum_{k \geq 2} a(n_k) \leq \frac{Lr}{c^*} F(T) , \tag{2.95}$$

and combining with (2.93) for $k = 1$ proves (2.90) and concludes the proof of the theorem when I is infinite.

We now prove (2.93). Since $n_k \geq 1$, we may define $n^* := n_k - 1$. By (2.92), we have $j(n_k - 1) = j(n_k)$, i.e., $j(n^*) = j(n^* + 1)$. We may then use (2.87) for $B = A_{n^*}(t)$, $A = A_{n_k}(t) = A_{n^*+1}(t)$ to obtain that

$$F(B \cap B(t, 2r^{-j_{n^*+1}(A)-2})) \leq F(B) - c^* 2^{n^*/2} r^{-j_{n^*+1}(A)-1} .$$

Recalling that $n^* = n_k - 1$, this means

$$F(B \cap B(t, 2r^{-j_{n_k}(A)-2})) \leq F(B) - c^* 2^{(n_k-1)/2} r^{-j_{n_k}(A)-1} , \tag{2.96}$$

so that

$$a(n_k) \leq \frac{Lr}{c^*}\big(F(B) - F(B \cap B(t, 2r^{-j_{n_k}(A)-2}))\big) . \tag{2.97}$$

Furthermore, by (2.92),

$$j(n_{k+2}) \geq j(n_{k+1}) + 1 \geq j(n_k) + 2 . \tag{2.98}$$

Since $j(n_{k+2}) = j_{n_{k+2}}(A_{n_{k+2}}(t))$, (2.86) implies $\Delta(A_{n_{k+2}}(t)) \leq 2r^{-j(n_{k+2})} \leq 2r^{-j(n_k)-2}$ so that

$$A_{n_{k+2}}(t) \subset B \cap B(t, 2r^{-j(n_k)-2}) , \tag{2.99}$$

and thus $f(n_{k+2}) = F(A_{n_{k+2}}(t)) \leq F(B \cap B(t, 2r^{-j_{n_k}(A)-2}))$. Combining with (2.97) and since $F(B) = f(n_k - 1)$, we have proved (2.93).

Assuming now that I is finite, it has a largest element $n_{\bar{k}}$. We use the previous argument to control $a(n_k)$ when $k + 2 \leq \bar{k}$, and for $k = \bar{k} - 1$ and $k = \bar{k}$, we simply use (2.89). □

It is important for the sequel that you fully master the previous argument.

Exercise 2.9.6 We say that a sequence $(F_n)_{n \geq 0}$ of functionals on (T, d) satisfies the growth condition with parameters $r \geq 8$ and $c^* > 0$ if

$$\forall n \geq 0 , \ F_{n+1} \leq F_n$$

and if for any integer $n \geq 0$ and any $a > 0$ the following holds true, where $m = N_n$: For each collection of subsets $H_1, \ldots, H_m$ of T that are (a, r)-separated, we have

$$F_n\Big(\bigcup_{\ell \leq m} H_\ell\Big) \geq c^* a 2^{n/2} + \min_{\ell \leq m} F_{n+1}(H_\ell) . \tag{2.100}$$

Prove that then

$$\gamma_2(T,d) \le \frac{Lr}{c^*} F_0(T) + Lr\Delta(T) . \tag{2.101}$$

Hint: Copy the previous arguments by replacing everywhere $F(A)$ by $F_n(A)$ when $A \in \mathcal{A}_n$.

Proposition 2.9.7 *Consider a metric space (T,d), and for $n \ge 0$, consider subsets T_n of T with* $\operatorname{card} T_0 = 1$ *and* $\operatorname{card} T_n \le N_n$ *for $n \ge 1$. Consider a number S, and let*

$$U = \Big\{ t \in T\, ; \sum_{n\ge 0} 2^{n/2} d(t, T_n) \le S \Big\} .$$

Then $\gamma_2(U,d) \le LS$.

Proof For $H \subset U$, we define $F(H) = \inf \sup_{t\in H} \sum_{n\ge 0} 2^{n/2} d(t, V_n)$ where the infimum is taken over all choices of $V_n \subset T$ with $\operatorname{card} V_n \le N_n$. It is important here not to assume that $V_n \subset H$ to ensure that F is increasing. We then prove that F satisfies the growth condition by an argument very similar to that of Proposition 2.8.5. The proof follows from Theorem 2.9.1 since $\Delta(U,d) \le 2S$, as each point of U is within distance S of the unique point of T_0. □

A slightly different partitioning scheme has recently been discovered by R. van Handel [141], and we describe a variant of it now. We consider a metric space (T,d) and an integer $r \ge 8$. We assume that for $j \in \mathbb{Z}$, we are given a function $s_j(t) \ge 0$ on T.

Theorem 2.9.8 *Assume that the following holds:*

$$\begin{aligned} &\text{For each subset } A \text{ of } T, \text{for each } j \in \mathbb{Z} \text{ with } \Delta(A) \le 2r^{-j} \text{ and for each}\\ &n \ge 1,\ \text{then either } e_n(A) \le r^{-j-1} \text{ or else there exists } t \in A\\ &\text{with } s_j(t) \ge 2^{n/2} r^{-j-1} . \end{aligned} \tag{2.102}$$

Then,

$$\gamma_2(T,d) \le Lr\Big(\Delta(T) + \sup_{t\in T} \sum_{j\in\mathbb{Z}} s_j(t)\Big) . \tag{2.103}$$

We will show later how to construct functions $s_j(t)$ satisfying (2.102) using a functional which satisfies the growth condition.[21]

[21] See [141] for other constructions.

The right-hand side of (2.103) is the supremum over t of a sum of terms. It need not always be the same terms which will contribute the most for different values of t, and the bound is definitely better than if the supremum and the summation were exchanged.

Proof of Theorem 2.9.8 Consider the largest $j_0 \in \mathbb{Z}$ with $\Delta(T) \le 2r^{-j_0}$, so that $2r^{-j_0} \le r\Delta(T)$. We construct by induction an increasing sequence of partitions $\mathcal{A}_n$ with $\operatorname{card} \mathcal{A}_n \le N_n$, and for $A \in \mathcal{A}_n$, we construct an integer $j_n(A) \in \mathbb{Z}$ with $\Delta(A) \le 2r^{-j_n(A)}$. We start with $\mathcal{A}_0 = \mathcal{A}_1 = \{T\}$ and $j_0(T) = j_1(T) = j_0$.

Once $\mathcal{A}_n$ has been constructed ($n \ge 1$), we further split every element $B \in \mathcal{A}_n$. The idea is to first split B into sets which are basically level sets for the function $s_j(t)$ in order to achieve the crucial relation (2.107) and then to further split each of these sets according to its metric entropy. More precisely, we may assume that $S = \sup_{t \in T} \sum_{j \in \mathbb{Z}} s_j(t) < \infty$, for there is nothing to prove otherwise. Let us set $j = j_n(B)$, and define the sets A_k for $1 \le k \le n$ by setting for $k < n$

$$A_k = \{t \in B \; ; \; 2^{-k} S < s_j(t) \le 2^{-k+1} S\} \,, \tag{2.104}$$

and

$$A_n = \{t \in B \; ; \; s_j(t) \le 2^{-n+1} S\} \,. \tag{2.105}$$

The purpose of this construction is to ensure the following:

$$k \le n \; ; \; t, t' \in A_k \Rightarrow s_j(t') \le 2(s_j(t) + 2^{-n} S) \,. \tag{2.106}$$

This is obvious since $s_j(t') \le 2 s_j(t)$ for $k < n$ and $s_j(t') \le 2^{-n+1} S$ if $k = n$. For each set A_k, $k \le n$, we use the following procedure:

- If $e_{n-1}(A_k) \le r^{-j-1}$, we may cover A_k by at most N_{n-1} balls of radius $2r^{-j-1}$, so we may split A_k into N_{n-1} pieces of diameter $\le 4r^{-j-1}$. We decide that each of these pieces A is an element of $\mathcal{A}_{n+1}$, for which we set $j_{n+1}(A) = j + 1$. Thus, $\Delta(A) \le 4r^{-j-1} = 4r^{-j_{n+1}(A)}$.
- Otherwise we decide that $A_k \in \mathcal{A}_{n+1}$ and we set $j_{n+1}(A_k) = j$. Thus, $\Delta(A_k) \le 2r^{-j} = 2r^{-j_{n+1}(A)}$. From (2.102), there exists $t' \in A_k$ for which $s_j(t') \ge 2^{(n-1)/2} r^{-j-1}$. Then by (2.106), we have

$$\forall t \in A_k \; ; \; 2^{(n-1)/2} r^{-j-1} \le 2(s_j(t) + 2^{-n} S) \,. \tag{2.107}$$

In summary, if $B \in \mathcal{A}_n$ and $A \in \mathcal{A}_{n+1}$, $A \subset B$, then

- Either $j_{n+1}(A) = j_n(B) + 1$
- Or else $j_{n+1}(A) = j_n(B)$ and, from (2.107),

$$\forall t \in A \; ; \; 2^{(n-1)/2} r^{-j_{n+1}(A)-1} \le 2(s_{j_n(B)}(t) + 2^{-n} S) \,. \tag{2.108}$$

This completes the construction. Now for $n \geq 1$, we have $n \leq N_{n-1}$ so that $\operatorname{card} \mathcal{A}_{n+1} \leq nN_{n-1}N_n \leq N_{n+1}$ and the sequence $(\mathcal{A}_n)$ is admissible. Next, we fix $t \in T$. We set $j_n = j_n(A_n(t))$, and we observe that by construction $j_n \leq j_{n+1} \leq j_n + 1$. Since $\Delta(A_n(t)) \leq 4r^{-j_n(t)}$, we have $2^{n/2}\Delta(A_n(t)) \leq 4a(n)$ where $a(n) := 2^{n/2}r^{-j_n(t)}$. To complete the argument, we prove that

$$\sum_{n\geq 0} a(n) \leq Lr(S + \Delta(T)) . \tag{2.109}$$

For this, consider the set I provided by Lemma 2.9.5 for $\alpha = \sqrt{2}$, so that since $r^{-j_0} \leq 2r\Delta(T)$ it suffices to prove that

$$\sum_{n\in I\setminus\{0\}} a(n) \leq LrS . \tag{2.110}$$

For $n \in I \setminus \{0\}$, it holds that $j_{n-1} = j_n < j_{n+1}$ (since otherwise this contradicts the definition of I). In particular, the integers j_n for $n \in I$ are all different so that $\sum_{n\geq 0} s_{j_n}(t) \leq S$. Using (2.108) for $n - 1$ instead of n yields $2^{(n-2)/2}r^{-j_n-1} \leq 2(s_{j_{n-1}}(t) + 2^{-n+1}S)$. Since $j_{n-1} = j_n$, we get

$$a(n) \leq Lr(s_{j_n}(t) + 2^{-n}S) ,$$

and summing these relations, we obtain the desired result. □

The following connects Theorems 2.9.1 and 2.9.8:

Proposition 2.9.9 *Assume that the functional F satisfies the growth condition with parameters r and c^*. Then the functions*

$$s_j(t) = \frac{1}{c^*}\big(F(B(t, 2r^{-j+1})) - F(B(t, 2r^{-j-2}))\big)$$

satisfy (2.102).

Proof Consider a subset A of T, $j \in \mathbb{Z}$ with $\Delta(A) \leq 2r^{-j}$ and $n \geq 1$. Let $m = N_n$. If $e_n(A) > r^{-j-1}$, then by Lemma 2.9.3, we may find $(t_\ell)_{\ell\leq m}$ in A with $d(t_\ell, t_{\ell'}) \geq r^{-j-1}$ for $\ell \neq \ell'$. Consider the set $H_\ell = B(t_\ell, 2r^{-j-2})$ so that by (2.77) used for $a = r^{-j-1}$, it holds that

$$F\Big(\bigcup_{\ell\leq m} H_\ell\Big) \geq c^* r^{-j-1}2^{n/2} + \min_{\ell\leq m} F(H_\ell) . \tag{2.111}$$

Let us now consider $\ell_0 \leq m$ such that $F(H_{\ell_0})$ achieves the minimum in the right-hand side, so that $\min_{\ell\leq m} F(H_\ell) = F(B(t_{\ell_0}, 2r^{-j-2}))$. The crude inequality $2r^{-j-2} + 2r^{-j} \leq 2r^{-j+1}$ implies that $H_\ell \subset B(t_{\ell_0}, 2r^{-j+1})$ for each ℓ, so that

$F(\bigcup_{\ell\le m} H_\ell) \le F(B(t_{\ell_0}, 2r^{-j+1}))$. Then (2.111) implies

$$F(B(t_{\ell_0}, 2r^{-j+1})) \ge c^* r^{-j-1} 2^{n/2} + F(B(t_{\ell_0}, 2r^{-j-2}))$$

i.e., $s_j(t_{\ell_0}) \ge 2^{n/2} r^{-j-1}$. □

Despite the fact that the proof of Theorem 2.9.8 is a few lines shorter than the proof of Theorem 2.9.1, in the various generalizations of this principle, we will mostly follow the scheme of proof of Theorem 2.9.1. The reason for this choice is simple: it should help the reader that our various partition theorems follow a common pattern. The most difficult partition theorem we present is Theorem 6.2.8 (the Latała-Bednorz theorem), which is one of the highlights of this work, and it is not clear at this point whether the method of Theorem 2.9.8 can be adapted to the proof of this theorem.

The following simple observation allows us to construct a sequence which is admissible from one which is slightly too large. It will be used several times.

Lemma 2.9.10 *Consider $\alpha > 0$, an integer $\tau \ge 0$, and an increasing sequence of partitions $(\mathcal{B}_n)_{n\ge 0}$ with* $\operatorname{card} \mathcal{B}_n \le N_{n+\tau}$. *Let*

$$S := \sup_{t\in T} \sum_{n\ge 0} 2^{n/\alpha} \Delta(B_n(t)) .$$

Then we can find an admissible sequence of partitions $(\mathcal{A}_n)_{n\ge 0}$ such that

$$\sup_{t\in T} \sum_{n\ge 0} 2^{n/\alpha} \Delta(A_n(t)) \le 2^{\tau/\alpha}(S + K(\alpha)\Delta(T)) . \tag{2.112}$$

Of course (for the last time) here $K(\alpha)$ denotes a number depending on α only (that need not be the same at each occurrence).

Proof We set $\mathcal{A}_n = \{T\}$ if $n < \tau$ and $\mathcal{A}_n = \mathcal{B}_{n-\tau}$ if $n \ge \tau$ so that $\operatorname{card} \mathcal{A}_n \le N_n$ and

$$\sum_{n\ge\tau} 2^{n/\alpha} \Delta(A_n(t)) = 2^{\tau/\alpha} \sum_{n\ge 0} 2^{n/\alpha} \Delta(B_n(t)) .$$

Using the bound $\Delta(A_n(t)) \le \Delta(T)$, we obtain

$$\sum_{n<\tau} 2^{n/\alpha} \Delta(A_n(t)) \le K(\alpha) 2^{\tau/\alpha} \Delta(T) .$$

□

Exercise 2.9.11 Prove that (2.112) might fail if one replaces the right-hand side by $K(\alpha, \tau)S$. Hint: S does not control $\Delta(T)$.

2.10 Gaussian Processes: The Majorizing Measure Theorem

Consider a Gaussian process $(X_t)_{t\in T}$, that is, a jointly Gaussian family of centered r.v.s indexed by T. We provide T with the canonical distance

$$d(s,t) = \left(\mathsf{E}(X_s - X_t)^2\right)^{1/2} . \tag{2.113}$$

Recall the functional γ_2 of Definition 2.7.3.

Theorem 2.10.1 (The Majorizing Measure Theorem) *For a universal constant L, it holds that*

$$\frac{1}{L}\gamma_2(T,d) \le \mathsf{E}\sup_{t\in T} X_t \le L\gamma_2(T,d) . \tag{2.114}$$

The reason for the name is explained in Sect. 3.1. We will meditate on this statement in Sect. 2.12. We will spend much time trying to generalize this theorem to other classes of processes. To link the statements of these generalizations with that of (2.114), it may be good to reformulate the lower bound $\gamma_2(T,d) \le L\mathsf{E}\sup_{t\in T} X_t$ in the following general terms:

The control from above of $\mathsf{E}\sup_{t\in T} X_t$ implies the existence of a

"small" sequence of admissible partitions of T .

The right-hand side inequality in (2.114) is Theorem 2.7.11. To prove the lower bound, we will use Theorem 2.9.1 and the functional

$$F(H) = \mathsf{E}\sup_{t\in H} X_t := \sup_{H^*\subset H, H^* \text{finite}} \mathsf{E}\sup_{t\in H^*} X_t . \tag{2.115}$$

For this, we need to prove that this functional satisfies the growth condition with c^* a universal constant and to bound $\Delta(T)$. We strive to give a proof that relies on general principles and lends itself to generalizations.

Lemma 2.10.2 (Sudakov Minoration) *Assume that*

$$\forall p,q \le m\,,\quad p \ne q \;\Rightarrow d(t_p,t_q) \ge a .$$

Then we have

$$\mathsf{E}\sup_{p\le m} X_{t_p} \ge \frac{a}{L_1}\sqrt{\log m} . \tag{2.116}$$

Here and below $L_1, L_2, \ldots$ are specific universal constants. Their values remain the same, at least within the same section.

The proof of the Sudakov minoration is given just after Lemma 15.2.7.

Exercise 2.10.3 Prove that Lemma 2.10.2 is equivalent to the following statement: If $(X_t)_{t\in T}$ is a Gaussian process and d is the canonical distance, then

$$e_n(T,d) \le L2^{-n/2}\mathsf{E}\sup_{t\in T} X_t \ . \tag{2.117}$$

Compare with Exercise 2.7.8.

To understand the relevance of Sudakov minoration, let us consider the case where $\mathsf{E}X_{t_p}^2 \le 100a^2$ (say) for each p. Then (2.116) means that the bound (2.15) is of the correct order in this situation.

Exercise 2.10.4 Prove (2.116) when the r.v.s X_{t_p} are independent. That is, assume that these variables are Gaussian independent centered. Hint: Use the method of Exercise 2.3.7 (b).

Exercise 2.10.5 A natural approach ("the second moment method") to prove that $\mathsf{P}(\sup_{p\le m} X_{t_p} \ge u)$ is at least $1/L$ for a certain value of u is as follows: consider the r.v. $Y = \sum_p \mathbf{1}_{\{X_{t_p}\ge u\}}$, prove that $\mathsf{E}Y^2 \le L(\mathsf{E}Y)^2$, and then use the Paley-Zygmund inequality (6.15) to prove that $\sup_{p\le m} X_{t_p} \ge a\sqrt{\log m}/L_1$ with probability $\ge 1/L$. Prove that this approach works when the r.v.s X_{t_ℓ} are independent, but find examples showing that this naive approach does not work in general to prove (2.116).

The following is a very important property of Gaussian processes and one of the keys to Theorem 2.10.1. It is a facet of the theory of concentration of measure, a leading idea of modern probability theory. We refer the reader to [52] to learn about this.

Lemma 2.10.6 *Consider a Gaussian process $(X_t)_{t\in U}$, where U is finite, and let $\sigma = \sup_{t\in U}(\mathsf{E}X_t^2)^{1/2}$. Then for $u \ge 0$, we have*

$$\mathsf{P}\Big(\Big|\sup_{t\in U} X_t - \mathsf{E}\sup_{t\in U} X_t\Big| \ge u\Big) \le 2\exp\Big(-\frac{u^2}{2\sigma^2}\Big)\ . \tag{2.118}$$

In words, the size of the fluctuations of $\sup_{t\in U} X_t$ are governed by the size of the individual r.v.s X_t, rather than by the (typically much larger) quantity $\mathsf{E}\sup_{t\in U} X_t$. It is essential that the cardinality of U does not appear in (2.118).

Exercise 2.10.7 Find an example of a Gaussian process for which

$$\mathsf{E}\sup_{t\in T} X_t \gg \sigma = \sup_{t\in T}(\mathsf{E}X_t^2)^{1/2}\ ,$$

whereas the fluctuations of $\sup_{t\in T} X_t$ are of order σ, e.g., the variance of $\sup_t X_t$ is about σ^2. Hint: $T = \{(t_i)_{i\le n}; \sum_{i\le n} t_i^2 \le 1\}$ and $X_t = \sum_{i\le n} t_i g_i$ where g_i are independent standard Gaussian r.v.s

Proposition 2.10.8 *Consider points $(t_\ell)_{\ell\le m}$ in T. Assume that $d(t_\ell, t_{\ell'}) \ge a$ if $\ell \ne \ell'$. Consider $\sigma > 0$ and for $\ell \le m$ a finite set $H_\ell \subset B(t_\ell, \sigma)$. Then if $H = \bigcup_{\ell\le m} H_\ell$, we have*

$$\mathsf{E}\sup_{t\in H} X_t \ge \frac{a}{L_1}\sqrt{\log m} - L_2\sigma\sqrt{\log m} + \min_{\ell\le m}\mathsf{E}\sup_{t\in H_\ell} X_t\ . \tag{2.119}$$

When $\sigma \le a/(2L_1L_2)$, (2.119) implies

$$\mathsf{E}\sup_{t\in H} X_t \ge \frac{a}{2L_1}\sqrt{\log m} + \min_{\ell\le m}\mathsf{E}\sup_{t\in H_\ell} X_t\ , \tag{2.120}$$

which can be seen as a generalization of Sudakov's minoration (2.116) by taking $H_\ell = \{t_\ell\}$. When $m = N_n$, (2.120) proves that the functional $F(H) = \mathsf{E}\sup_{t\in T} X_t$ satisfies the growth condition (2.77).

Proof We can and do assume $m \ge 2$. For $\ell \le m$, we consider the r.v.

$$Y_\ell = \Big(\sup_{t\in H_\ell} X_t\Big) - X_{t_\ell} = \sup_{t\in H_\ell}(X_t - X_{t_\ell})\ .$$

For $t \in H_\ell$, we set $Z_t = X_t - X_{t_\ell}$. Since $H_\ell \subset B(t_\ell, \sigma)$, we have $\mathsf{E}Z_t^2 = d(t, t_\ell)^2 \le \sigma^2$, and for $u \ge 0$, Eq. (2.118) used for the process $(Z_t)_{t\in H_\ell}$ implies

$$\mathsf{P}(|Y_\ell - \mathsf{E}Y_\ell| \ge u) \le 2\exp\Big(-\frac{u^2}{2\sigma^2}\Big)\ . \tag{2.121}$$

Thus, if $V = \max_{\ell\le m}|Y_\ell - \mathsf{E}Y_\ell|$, then combining (2.121) and the union bound, we get

$$\mathsf{P}(V \ge u) \le 2m\exp\Big(-\frac{u^2}{2\sigma^2}\Big)\,, \tag{2.122}$$

and (2.13) implies

$$\mathsf{E}V \le L_2\sigma\sqrt{\log m}\ . \tag{2.123}$$

Now, for each $\ell \le m$,

$$Y_\ell \ge \mathsf{E}Y_\ell - V \ge \min_{\ell\le m}\mathsf{E}Y_\ell - V\ ,$$

and thus

$$\sup_{t \in H_\ell} X_t = Y_\ell + X_{t_\ell} \geq X_{t_\ell} + \min_{\ell \leq m} \mathsf{E} Y_\ell - V$$

so that

$$\sup_{t \in H} X_t \geq \max_{\ell \leq m} X_{t_\ell} + \min_{\ell \leq m} \mathsf{E} Y_\ell - V \ .$$

Taking expectations, we obtain

$$\mathsf{E} \sup_{t \in H} X_t \geq \mathsf{E} \max_{\ell \leq m} X_{t_\ell} + \min_{\ell \leq m} \mathsf{E} Y_\ell - \mathsf{E} V \ ,$$

and we use (2.116) and (2.123). □

Exercise 2.10.9 Prove that (2.120) might fail if one allows $\sigma = a$. Hint: The intersection of the balls $B(t_\ell, a)$ might contain a ball with positive radius.

Exercise 2.10.10 Consider subsets $(H_\ell)_{\ell \leq m}$ of $B(0, a)$ and $H = \cup_{\ell \leq m} H_\ell$. Prove that

$$\mathsf{E} \sup_{t \in H} X_t \leq La\sqrt{\log m} + \max_{\ell \leq m} \mathsf{E} \sup_{t \in H_\ell} X_t \ . \tag{2.124}$$

Try to find improvements on this bound. Hint: Peek at (19.61).

Proof of Theorem 2.10.1 We fix $r = \max(8, 4L_1L_2)$, so that $2a/r \leq a/2L_1L_2$. The growth condition for the functional F of (2.115) follows from (2.120), which implies that (2.77) holds for $c^* = 1/L$. Theorem 2.9.1 implies

$$\gamma_2(T, d) \leq L \mathsf{E} \sup_{t \in T} X_t + L\Delta(T) \ .$$

To control the term $\Delta(T)$, we write that for $t_1, t_2 \in H$,

$$\mathsf{E} \max(X_{t_1}, X_{t_2}) = \mathsf{E} \max(X_{t_1} - X_{t_2}, 0) = \frac{1}{\sqrt{2\pi}} d(t_1, t_2) \ ,$$

so that $\Delta(T) \leq \sqrt{2\pi} \mathsf{E} \sup_{t \in T} X_t$. □

The proof of Theorem 2.10.1 displays an interesting feature. This theorem aims at understanding $\mathsf{E} \sup_{t \in T} X_t$, and for this, we use functionals that are based on precisely this quantity. This is not a circular argument. The content of Theorem 2.10.1 is that there is simply no other way to bound a Gaussian process than to control the quantity $\gamma_2(T, d)$. The miracle of this theorem is that it relates in complete generality two quantities, namely, $\mathsf{E} \sup_{t \in T} X_t$ and $\gamma_2(T, d)$ which are both very hard to estimate. Still, in concrete situations, to estimate these quantities, we must in some way gain understanding of the underlying geometry.

The following is a noteworthy consequence of Theorem 2.10.1:

Theorem 2.10.11 *Consider two processes $(Y_t)_{t\in T}$ and $(X_t)_{t\in T}$ indexed by the same set. Assume that the process $(X_t)_{t\in T}$ is Gaussian and that the process $(Y_t)_{t\in T}$ satisfies the increment condition*

$$\forall u > 0\,,\ \forall s,t \in T\,,\ \mathsf{P}(|Y_s - Y_t| \ge u) \le 2\exp\Big(-\frac{u^2}{d(s,t)^2}\Big)\,, \tag{2.125}$$

where d is the distance (2.113) associated with the process X_t. Then we have

$$\mathsf{E}\sup_{s,t\in T} |Y_s - Y_t| \le L\mathsf{E}\sup_{t\in T} X_t\ . \tag{2.126}$$

Processes satisfying the condition (2.125) are sometimes called sub-Gaussian. We will see many examples later (see (6.2)).

Proof We combine (2.60) with the left-hand side of (2.114). □

Let us also note the following consequence of (2.126) and Lemma 2.2.1:[22]

Corollary 2.10.12 *Consider two Gaussian processes $(X_t)_{t\in T}$ and $(Y_t)_{t\in T}$. Assume that*

$$\forall s,t \in T\,,\ \mathsf{E}(Y_s - Y_t)^2 \le \mathsf{E}(X_s - X_t)^2\ .$$

Then,

$$\mathsf{E}\sup_{t\in T} Y_t \le L\mathsf{E}\sup_{t\in T} X_t\ . \tag{2.127}$$

2.11 Gaussian Processes as Subsets of a Hilbert Space

In this section, we learn to think of a Gaussian process as *a subset of a Hilbert space.* This will reveal our lack of understanding of basic geometric questions.

First, consider a Gaussian process $(Y_t)_{t\in T}$, and assume (the only case which is of interest to us) that there is a countable set $T' \subset T$ which is dense in T. We view each Y_t as a point in the Hilbert space $L^2(\Omega, \mathsf{P})$ where (Ω, P) is the basic probability space. The closed linear span of the r.v.s $(Y_t)_{t\in T}$ in $L^2(\Omega, \mathsf{P})$ is a separable Hilbert space, and the map $t \mapsto Y_t$ is an isometry from (T, d) to its image (by the very

[22] It is known that (2.127) holds with $L = 1$, a result known as Slepian's lemma. Please see the comments at the end of Sect. 2.16.

definition of the distance d). In this manner, we associate a subset of a Hilbert space to each Gaussian process.

Conversely, consider a separable Hilbert space, which we may assume to be $\ell^2 = \ell^2(\mathbb{N})$.[23] Consider an independent sequence $(g_i)_{i\geq 1}$ of standard Gaussian r.v.s. We can then define the Gaussian process $(X_t)_{t\in\ell^2}$, where

$$X_t = \sum_{i\geq 1} t_i g_i \tag{2.128}$$

(the series converges in $L^2(\Omega)$). Thus,

$$\mathsf{E}X_t^2 = \sum_{i\geq 1} t_i^2 = \|t\|^2 \; . \tag{2.129}$$

In this manner, for each subset T of ℓ^2, we can consider the Gaussian process $(X_t)_{t\in T}$. The distance induced on T by the process coincides with the distance of ℓ^2 by (2.129).

A subset T of ℓ^2 will always be provided with the distance induced by ℓ^2, so we may also write $\gamma_2(T)$ rather than $\gamma_2(T,d)$. We denote by $\operatorname{conv} T$ the convex hull of T.

Theorem 2.11.1 *For a subset T of ℓ^2, we have*

$$\gamma_2(\operatorname{conv} T) \leq L\gamma_2(T) \; . \tag{2.130}$$

Of course we also have $\gamma_2(T) \leq \gamma_2(\operatorname{conv} T)$ since $T \subset \operatorname{conv} T$.

Proof To prove (2.130), we observe that since $X_{a_1t_1+a_2t_2} = a_1X_{t_1} + a_2X_{t_2}$, we have

$$\sup_{t\in\operatorname{conv} T} X_t = \sup_{t\in T} X_t \; . \tag{2.131}$$

We then use (2.114) to write

$$\frac{1}{L}\gamma_2(\operatorname{conv} T) \leq \mathsf{E} \sup_{t\in\operatorname{conv} T} X_t = \mathsf{E} \sup_{t\in T} X_t \leq L\gamma_2(T) \; . \qquad \square$$

A basic problem is that it is absolutely not obvious how to construct an admissible sequence of partitions on $\operatorname{conv} T$ witnessing (2.130).

Research Problem 2.11.2 Give a geometrical proof of (2.130).

---

[23] Throughout the book, $\mathbb{N}$ is the set of natural numbers starting at 0, $\mathbb{N} = \{0, 1, \ldots\}$, whereas $\mathbb{N}^* = \mathbb{N} \setminus \{0\}$.

What we mean by geometrical proof is a proof that does not use Gaussian processes but only the geometry of Hilbert space. The difficulty of the problem is that the structure of an admissible sequence which witnesses that $\gamma_2(\operatorname{conv} T) \le L\gamma_2(T)$ must depend on the "geometry" of the set T. A really satisfactory argument would give a proof that holds in Banach spaces more general than Hilbert space, for example, by providing a positive answer to the following, where the concept of q-smooth Banach space is explained in [57]:

Research Problem 2.11.3 Consider a 2-smooth Banach space and the distance d induced by its norm. Is it true that for each subset T of its unit ball, one has $\gamma_2(\operatorname{conv} T, d) \le K\sqrt{\log \operatorname{card} T}$? More generally, is it true that for each finite subset T, one has $\gamma_2(\operatorname{conv} T, d) \le K\gamma_2(T, d)$? (Here K may depend on the Banach space, but not on T.)

Research Problem 2.11.4 Still more generally, is it true that for a finite subset T of a q-smooth Banach space, one has $\gamma_q(\operatorname{conv} T) \le K\gamma_q(T)$?

Even when the Banach space is ℓ^p, I do not know the answer to these problems (unless $p = 2$!). (The Banach space ℓ^p is 2-smooth for $p \ge 2$ and q-smooth for $p < 2$, where $1/p + 1/q = 1$.) One concrete case is when the set T consists of the first N vectors of the unit basis of ℓ^p. It is possible to show in this case that $\gamma_q(\operatorname{conv} T) \le K(p)(\log N)^{1/q}$, where $1/p + 1/q = 1$. We leave this as a challenge to the reader. The proof here is pretty much the same as for the case $p = q = 2$ which was covered in Sect. 2.6.

Exercise 2.11.5 Prove that if $a \ge 2$, we have $\sum_{k\ge1}(k+1)^{-a} \le L2^{-a}$.

We recall the ℓ^2 norm $\|\cdot\|$ of (2.129). Here is a simple fact.

Proposition 2.11.6 *Consider a sequence* $(t_k)_{k\ge1}$ *such that*

$$\forall k \ge 1 \,,\ \|t_k\| \le 1/\sqrt{\log(k+1)} \,.$$

Let $T = \{\pm t_k, k \ge 1\}$. *Then* $\mathsf{E}\sup_{t\in T} X_t \le L$ *and thus* $\mathsf{E}\sup_{t\in\operatorname{conv} T} X_t \le L$ *by (2.131).*

Proof We have

$$\mathsf{P}\Big(\sup_{k\ge1}|X_{t_k}| \ge u\Big) \le \sum_{k\ge1}\mathsf{P}(|X_{t_k}| \ge u) \le \sum_{k\ge1} 2\exp\Big(-\frac{u^2}{2}\log(k+1)\Big) \tag{2.132}$$

since X_{t_k} is Gaussian with $\mathsf{E}X_{t_k}^2 \le 1/\log(k+1)$. For $u \ge 2$, the right-hand side of (2.132) is at most $L\exp(-u^2/L)$ by the result of Exercise 2.11.5, and as usual the conclusion follows from (2.6). □

Exercise 2.11.7 Deduce Proposition 2.11.6 from (2.34). Hint: Use Exercise 2.4.1.

It is particularly frustrating not to be able to solve the following special instance of Problem 2.11.2:

Research Problem 2.11.8 In the setting of Proposition 2.11.6, find a geometrical proof that $\gamma_2(\operatorname{conv} T) \le L$.

The following shows that the situation of Proposition 2.11.6 is in a sense generic:

Theorem 2.11.9 *Consider a countable set $T \subset \ell^2$, with $0 \in T$. Then we can find a sequence (t_k) with*

$$\forall k \ge 1 \;,\; \|t_k\|\sqrt{\log(k+1)} \le L\mathsf{E} \sup_{t \in T} X_t$$

and

$$T \subset \operatorname{conv}(\{t_k \;;\; k \ge 1\}) \;.$$

Furthermore, we may assume that each t_k is a multiple of the difference of two elements of T.[24]

Proof By Theorem 2.10.1, we can find an admissible sequence $(\mathcal{A}_n)$ of T with

$$\forall t \in T \;,\; \sum_{n \ge 0} 2^{n/2} \Delta(A_n(t)) \le L\mathsf{E} \sup_{t \in T} X_t := S \;. \tag{2.133}$$

We construct sets $T_n \subset T$, such that each $A \in \mathcal{A}_n$ contains exactly one element of T_n. We ensure in the construction that $T = \bigcup_{n \ge 0} T_n$ and that $T_0 = \{0\}$. (To do this, we simply enumerate the elements of T as $(v_n)_{n \ge 1}$ with $v_0 = 0$, and we ensure that v_n is in T_n.) For $n \ge 1$, consider the set U_n that consists of all the points

$$2^{-n/2} \frac{t - v}{\|t - v\|}$$

where $t \in T_n$, $v \in T_{n-1}$, and $t \ne v$. Thus, each element of U_n has norm $2^{-n/2}$, and U_n has at most $N_n N_{n-1} \le N_{n+1}$ elements. Let $U = \bigcup_{n \ge 1} U_n$. Then since $\sum_{\ell \le n} N_{\ell+1} \le N_{n+2}$, U contains at most N_{n+2} elements of norm $\ge 2^{-n/2}$. We enumerate U as $\{z_k; k = 1, \ldots\}$ where the sequence $(\|z_k\|)$ is non-increasing, so that $\|z_k\| < 2^{-n/2}$ for $k > N_{n+2}$. Let us now prove that $\|z_k\| \le L/\sqrt{\log(k+1)}$. If $k \le N_2$, this holds because $\|z_k\| \le 1$. Assume then that $k > N_2$, and let $n \ge 0$ be the largest integer with $k > N_{n+2}$. Then by definition of n, we have $k \le N_{n+3}$ and thus $2^{-n/2} \le L/\sqrt{\log k}$. But then $\|z_k\| \le 2^{-n/2} \le L/\sqrt{\log k}$, proving the required inequality.

[24] This information is of secondary importance and will be used only much later.

Consider $t \in T$, so that $t \in T_m$ for some $m \geq 0$. Writing $\pi_n(t)$ for the unique element of $T_n \cap A_n(t)$, since $\pi_0(t) = 0$, we have

$$t = \sum_{1 \leq n \leq m} \pi_n(t) - \pi_{n-1}(t) = \sum_{1 \leq n \leq m} a_n(t) u_n(t) , \tag{2.134}$$

with $a_n(t) = 2^{n/2} \|\pi_n(t) - \pi_{n-1}(t)\|$ and

$$u_n(t) = 2^{-n/2} \frac{\pi_n(t) - \pi_{n-1}(t)}{\|\pi_n(t) - \pi_{n-1}(t)\|} \in U .$$

Since

$$\sum_{1 \leq n \leq m} a_n(t) \leq \sum_{n \geq 1} 2^{n/2} \Delta(A_{n-1}(t)) \leq 2S$$

and since $u_n(t) \in U_n \subset U$, we see from (2.134) that

$$t = \sum_{1 \leq n \leq m} a_n(t) u_n(t) + \Big(2S - \sum_{1 \leq n \leq m} a_n(t)\Big) \times 0 \in 2S \operatorname{conv}(U \cup \{0\}) .$$

Thus, $T \subset 2S \operatorname{conv}(U \cup \{0\}) = \operatorname{conv}(2SU \cup \{0\})$, and it suffices to take $t_k = 2S z_k$. □

Exercise 2.11.10 What is the purpose of the condition $0 \in T$?

Exercise 2.11.11 Prove that if $T \subset \ell^2$ and $0 \in T$, then (even when T is not countable) we can find a sequence (t_k) in ℓ^2, with $\|t_k\| \sqrt{\log(k+1)} \leq L \mathsf{E} \sup_{t \in T} X_t$ for all k and

$$T \subset \overline{\operatorname{conv}}\{t_k \; ; \; k \geq 1\} ,$$

where $\overline{\operatorname{conv}}$ denotes the closed convex hull. (Hint: Do the obvious thing – apply Theorem 2.11.9 to a dense countable subset of T.) Denoting now by $\operatorname{conv}^*(A)$ the set of infinite sums $\sum_i \alpha_i a_i$ where $\sum_i |\alpha_i| = 1$ and $a_i \in A$, prove that one can also achieve

$$T \subset \operatorname{conv}^*\{t_k \; ; \; k \geq 1\} .$$

Exercise 2.11.12 Consider a set $T \subset \ell^2$ with $0 \in T \subset B(0, \delta)$. Prove that we can find a sequence (t_k) in ℓ^2, with the following properties:

$$\forall k \geq 1 \, , \; \|t_k\| \sqrt{\log(k+1)} \leq L \mathsf{E} \sup_{t \in T} X_t \, , \tag{2.135}$$

$$\|t_k\| \leq L\delta \, , \tag{2.136}$$

$$T \subset \overline{\operatorname{conv}}\{t_k \; ; \; k \geq 1\} , \tag{2.137}$$

where $\overline{\text{conv}}$ denotes the closed convex hull. Hint: Copy the proof of Theorem 2.11.9, observing that since $T \subset B(0, \delta)$ one may chose $\mathcal{A}_n = \{T\}$ and $T_n = \{0\}$ for $n \le n_0$, where n_0 is the smallest integer for which $2^{n_0/2} \ge \delta^{-1} \mathsf{E} \sup_{t \in T} X_t$, and thus $U_n = \emptyset$ for $n \le n_0$.

The following problems are closely related to Problem 2.11.2:

Research Problem 2.11.13 Give a geometric proof of the following fact: Given subsets $(T_k)_{k \le N}$ of a Hilbert space and $T = \sum_{k \le N} T_k = \{x_1 + \ldots + x_N; \forall k \le N, x_k \in T_k\}$, prove that $\gamma_2(T) \le L \sum_{k \le N} \gamma_2(T_k)$.

We do not even know how to solve the following special case:

Research Problem 2.11.14 Consider sets $(T_k)_{k \le N}$ in a Hilbert space, and assume that each T_k consists of M vectors of length 1. Let $T = \sum_{k \le N} T_k$. Give a geometrical proof of the fact that $\gamma_2(T) \le LN\sqrt{\log M}$.

The next exercise is inspired by the paper [5] of S Artstein. It is more elaborate and may be omitted on first reading. A Bernoulli r.v. ε is such that $\mathsf{P}(\varepsilon = \pm 1) = 1/2$.[25]

Exercise 2.11.15 Consider a subset $T \subset \mathbb{R}^n$, where $\mathbb{R}^n$ is provided with the Euclidean distance. We assume that for some $\delta > 0$, we have

$$0 \in T \subset B(0, \delta) \ . \tag{2.138}$$

Consider independent Bernoulli r.v.s $(\varepsilon_{i,p})_{i,p \ge 1}$. Given a number $q \le n$, consider the operator $U_q : \mathbb{R}^n \to \mathbb{R}^q$ given by

$$U_q(x) = \Big(\sum_{i \le n} \varepsilon_{i,p} x_i\Big)_{p \le q} \ .$$

(a) Prove that $\|U_q\| \ge \sqrt{n}$.
We want to prove that despite (a), there exist a number L such that if $\mathsf{E} \sup_{t \in T} \sum_{i \le n} g_i t_i \le \delta\sqrt{q}$, then with high probability

$$U_q(T) \subset B(0, L\delta\sqrt{q}) \ , \tag{2.139}$$

whereas from (2.138) we would not expect better than $U_q(T) \subset B(0, \delta\sqrt{n})$.
(b) Use the sub-Gaussian inequality (6.1.1) to prove that if $\|x\| = 1$, then

$$\mathsf{E} \exp\Big(\frac{1}{4}\Big(\sum_{i \le n} \varepsilon_{i,p} x_i\Big)^2\Big) \le L \ . \tag{2.140}$$

[25] One must distinguish Bernoulli r.v.s ε_i from positive numbers ϵ_k!

(c) Use (2.140) and independence to prove that for $x \in \mathbb{R}^n$ and $v \geq 1$,

$$\mathsf{P}(\|U_q(x)\| \geq Lv\sqrt{q}\|x\|) \leq \exp(-v^2 q)\ . \tag{2.141}$$

(d) Use (2.141) to prove that with probability close to 1, for each of the vectors t_k of Exercise 2.11.12, one has $\|U_q(t_k)\| \leq L\delta\sqrt{q}$ and conclude.

We end this section with a discussion of a question which shares some features with Problem 2.11.2, in a sense that it is a property which is obvious, on one hand, but difficult to prove without using the magic of linearity.[26] For $k \leq N$, let us consider Gaussian processes $(X_{k,t})_{t \in T_k}$ with associated distances d_k. On the space $T = \prod_{k \leq N} T_k$, let us consider the distance d given by

$$d((t_k)_{k \leq N}, (t'_k)_{k \leq N}) = \Big(\sum_{k \leq N} d_k(t_k, t'_k)^2\Big)^{1/2}\ . \tag{2.142}$$

Proposition 2.11.16 *We have*

$$\gamma_2(T, d) \leq L \sum_{k \leq N} \gamma_2(T_k, d_k)\ . \tag{2.143}$$

Proof Assuming without loss of generality that the processes $(X_k)_{k \leq N}$ are independent, we can consider the Gaussian process $(X_t)_{t \in T}$ given for $t = (t_k)_{k \leq N}$ by $X_t = \sum_{k \leq N} X_{k,t_k}$. It is obvious that the distance d of (2.143) is associated with this process. It is also obvious that

$$\sup_{t \in T} X_t = \sum_{k \leq N} \sup_{t \in T_k} X_{t_k}\ .$$

Taking expectation and combining with (2.114) conclude the proof. □

The question now is to prove (2.143) without using Gaussian processes, for example, by proving it for any sequence $((T_k, d_k))_{k \leq N}$ of metric spaces. The most interesting part of that project is that it is unexpectedly hard. Is it the sign that we are still missing an important ingredient? In the next exercise, we show how to prove about the simplest possible case of (2.143), which is already pretty challenging.

Exercise 2.11.17 Throughout this exercise, each space T_k consists of M_k points, and the mutual distance of any two different points of T_k is $\epsilon_k > 0$. The goal is to prove the inequality

$$\int \sqrt{\log N(T, d, \epsilon)}\, d\epsilon \leq L \sum_{k \leq N} \epsilon_k \sqrt{\log M_k}\ . \tag{2.144}$$

[26] There are several equally frustrating instances of this situation.

Throughout the exercise, I denotes a subset of $\{1, \ldots, N\}$, and I^c denotes its complement.

(a) Prove that if $\sum_{k\in I} \epsilon_k^2 \le \epsilon^2$, then $N(T, d, \epsilon) \le \prod_{k\in I^c} M_k$.
(b) Show that to prove (2.144), it suffices to prove the following: Consider two sequences $(\epsilon_k)_{k\le N}$ and $(\eta_k)_{k\le N}$. For $\epsilon > 0$, define $S(\epsilon)$ by

$$S(\epsilon)^2 = \inf\Big\{\sum_{k\in I^c} \eta_k^2 \; ; \; \sum_{k\in I} \epsilon_k^2 \le \epsilon^2\Big\} ,$$

where the infimum is over all choices of I. Then

$$\int_0^\infty S(\epsilon)\mathrm{d}\epsilon \le L \sum_{k\le N} \epsilon_k \eta_k . \tag{2.145}$$

(c) To prove (2.145), show that it suffices to prove the following: Consider a function $h > 0$ on a probability space. For $\epsilon > 0$, define $\tilde{S}(\epsilon)$ by

$$\tilde{S}(\epsilon)^2 = \inf\Big\{\int_{A^c} \frac{1}{h}\mathrm{d}\mu \; ; \; \int_A h\mathrm{d}\mu \le \epsilon^2\Big\} ,$$

where the infimum is over all choices of A. Then $\int_0^\infty \tilde{S}(\epsilon)\mathrm{d}\epsilon \le L$. Hint: Reduce to the case where $\sum_{k\le N} \epsilon_k \eta_k = 1$. Use the probability μ on $\{1, \ldots, N\}$ such that $\mu(\{k\}) = \epsilon_k \eta_k$ and the function h given by $h(k) = \epsilon_k/\eta_k$.
(d) Show that it suffices to prove that $\sum_{\ell\in\mathbb{Z}} 2^{-\ell} \tilde{S}(2^{-\ell}) \le L$.
(e) Assuming for simplicity that μ has no atoms,[27] prove the statement given in (d). Hint: For $\ell \in \mathbb{Z}$ and $2^{-2\ell} \le \int h\mathrm{d}\mu$, consider the set A_ℓ of the type $A_\ell = \{h \le t_\ell\}$ where t_ℓ is such that $\int_{A_\ell} h\mathrm{d}\mu = 2^{-2\ell}$, so that $\tilde{S}(2^{-\ell})^2 \le \int_{A_\ell^c} (1/h)\mathrm{d}\mu$. Warning: This is not easy.

Exercise 2.11.18 This exercise continues the previous one. The spaces (T_k, d_k) are now any metric spaces, and the goal is to prove that

$$\int_0^\infty \sqrt{\log N(T, d, \epsilon)}\mathrm{d}\epsilon \le L \sum_{k\le N} \int_0^\infty \sqrt{\log N(T_k, d_k, \epsilon)}\mathrm{d}\epsilon . \tag{2.146}$$

Proving this requires passing the main difficulty of proving (2.143), but to prove (2.143) itself, it will be convenient to use different tools, and that proof is the object of Exercise 3.1.6.

---

[27] I am sure that this is true without this hypothesis, but I did not find the energy to carry out the details.

(a) Show that to prove (2.146), it suffices to prove the following: Consider decreasing functions $f_k : \mathbb{R}^+ \to \mathbb{R}^+$, and for $\epsilon > 0$, define $V(\epsilon)$ by

$$V(\epsilon)^2 = \inf\Big\{ \sum_{k\le N} f_k(\epsilon_k)^2 \; ; \; \sum_{k\le N} \epsilon_k^2 \le \epsilon^2 \Big\} ,$$

where the infimum is taken over all families $(\epsilon_k)_{k\le N}$. Then

$$\int_0^\infty V(\epsilon)\mathrm{d}\epsilon \le L \sum_{k\le N} \int_0^\infty f_k(\epsilon)\mathrm{d}\epsilon \; . \tag{2.147}$$

(b) When each f_k is of the type $f_k = \eta_k \mathbf{1}_{[0,\theta_k[}$, deduce (2.146) from (2.145).
(c) Convince yourself that by approximation, it suffices to consider the case where each f_k is a finite sum $\sum_\ell 2^{-\ell} \mathbf{1}_{[0,\theta_{k\ell}[}$.
(d) In the case (c), prove (2.147) by applying the special case (b) to the family $f_{k,\ell}$ of functions given by $f_{k,\ell} := 2^{-\ell} \mathbf{1}_{[0,\theta_{k,\ell}[}$ for all relevant values of k, ℓ. Hint: This is a bit harder.

2.12 Dreams

We may reformulate the inequality (2.114)

$$\frac{1}{L}\gamma_2(T,d) \le \mathsf{E} \sup_{t\in T} X_t \le L\gamma_2(T,d)$$

of Theorem 2.10.1 by the statement

$$\textit{Chaining suffices to explain the size of a Gaussian process.} \tag{2.148}$$

We simply mean that the "natural" chaining bound for the size of a Gaussian process (i.e., the right-hand side inequality in (2.114)) is of correct order, *provided* one uses the best possible chaining. This is what the left-hand side of (2.114) shows. We may dream of removing the word "Gaussian" in that statement. The desire to achieve this lofty goal in as many situations as possible motivates much of the rest of the book.

Besides the generic chaining, we have found in Theorem 2.11.9 another optimal way to bound Gaussian processes: to put them into the convex hull of a "small" process, that is, to use the inequality

$$\mathsf{E} \sup_{t\in T} X_t \le L \inf\big\{ S \; ; \; T \subset \mathrm{conv}\{t_k, k\ge 1\}, \|t_k\| \le S/\sqrt{\log(k+1)} \big\} \; .$$

Since we do not really understand the geometry of going from a set to its convex hull, it is better (for the time being) to consider this method as somewhat distinct from the generic chaining. Let us try to formulate it in a way which is suitable for generalizations. Given a countable set $\mathcal{V}$ of r.v.s, let us define the (possibly infinite) quantity

$$S(\mathcal{V}) = \inf\Big\{S > 0\ ;\ \int_S^\infty \sum_{V\in\mathcal{V}} \mathsf{P}(|V| > u)du \le S\Big\}\ . \tag{2.149}$$

Lemma 2.12.1 *It holds that*

$$\mathsf{E} \sup_{V\in\operatorname{conv}\mathcal{V}} |V| \le 2S(\mathcal{V})\ . \tag{2.150}$$

Proof We combine (2.6) with the fact that for $S > S(\mathcal{V})$, we have

$$\int_0^\infty \mathsf{P}\Big(\sup_{V\in\operatorname{conv}\mathcal{V}} |V| \ge u\Big)du \le S + \int_S^\infty \sum_{V\in\mathcal{V}} \mathsf{P}(|V| > u)du \le 2S\ . \qquad \square$$

Thus, (2.150) provides a method to bound stochastic processes. This method may look childish, but for Gaussian processes, the following reformulation of Theorem 2.11.9 shows that it is in fact optimal:

Theorem 2.12.2 *Consider a countable set T. Consider a Gaussian process $(X_t)_{t\in T}$, and assume that $X_{t_0} = 0$ for some $t_0 \in T$. Then there exists a countable set $\mathcal{V}$ of Gaussian r.v.s, each of which is a multiple of the difference of two variables X_t, with*

$$\forall t \in T\ ;\ X_t \in \operatorname{conv}\mathcal{V}\ , \tag{2.151}$$

$$S(\mathcal{V}) \le L\mathsf{E}\sup_{t\in T} X_t\ . \tag{2.152}$$

To understand the need of the condition $X_{t_0} = 0$ for some t_0, think of the case where T consists of one single point. The proof of Theorem 2.12.2 is nearly obvious by using (2.132) to bound $S(\mathcal{V})$ for the set $\mathcal{V}$ consisting of the variables X_{t_k} for the sequence (t_k) constructed in Theorem 2.11.9. We may dream of proving statements such as Theorem 2.12.2 for many classes of processes.

Also worthy of detailing is another remarkable geometric consequence of Theorem 2.11.9 in a somewhat different direction. Consider an integer N. Considering i.i.d. standard Gaussian r.v.s, we define as usual the process $X_t = \sum_{i\le N} g_i t_i$. We may view an element t of ℓ^2_N as a function on ℓ^2_N by the canonical duality, and therefore view t as a r.v. on the probability space (ℓ^2_N, μ), where μ is the law of the sequence $(g_i)_{i\le N}$. The processes (X_t) and (t) have the same law; hence, they are really the same object viewed in two different ways. Consider a subset T of ℓ^2_N, and assume that $T \subset \mathrm{conv}\{t_k; k \geq 1\}$. Then for any $v > 0$, we have

$$\Big\{\sup_{t\in T} t > v\Big\} \subset \bigcup_{k\geq 1}\{t_k > v\}\,. \tag{2.153}$$

The sets $\{t_k \geq v\}$ on the right are very simple: they are *half-spaces*. Assume now that for $k \geq 1$ and a certain S, we have $\|t_k\|\sqrt{\log(k+1)} \leq S$. Then for $u \geq 2$

$$\sum_{k\geq 1}\mu(\{t_k \geq Su\}) \leq \sum_{k\geq 1}\exp\Big(-\frac{u^2}{2}\log(k+1)\Big) \leq L\exp(-u^2/L)\,,$$

the very same computation as in (2.132). Theorem 2.11.9 implies that one may find such t_k for $S = L\mathsf{E}\sup_t X_t$. Therefore for $v \geq L\mathsf{E}\sup_t X_t$, the fact that the set in the left-hand side of (2.153) is small (in the sense of probability) may be witnessed by the fact that this set can be covered by a union of *simple sets* (half-spaces), the *sum* of the probabilities of which is small.

We may dream that something similar occurs in many other settings. In Chap. 13, which can be read right now, we will meet such a fundamental setting, which inspired the author's lifetime favorite problem (see Sect. 13.3).

2.13 A First Look at Ellipsoids

We have illustrated the gap between Dudley's bound (2.41) and the sharper bound (2.34), using examples (2.49) and (2.42). These examples might look artificial, but here we demonstrate that the gap between Dudley's bound (2.41) and the generic chaining bound (2.34) already exists for *ellipsoids* in Hilbert space. Truly understanding ellipsoids will be fundamental in several subsequent questions, such as the matching theorems of Chap. 4. A further study of ellipsoids is proposed in Sect. 3.2.

Given a sequence $(a_i)_{i\geq 1}$, $a_i > 0$, we consider the ellipsoid

$$\mathcal{E} = \Big\{t \in \ell^2\,;\ \sum_{i\geq 1}\frac{t_i^2}{a_i^2} \leq 1\Big\}\,. \tag{2.154}$$

Proposition 2.13.1 *When $\sum_{i\geq 1} a_i^2 < \infty$, we have*

$$\frac{1}{L}\Big(\sum_{i\geq 1} a_i^2\Big)^{1/2} \leq \mathsf{E}\sup_{t\in\mathcal{E}} X_t \leq \Big(\sum_{i\geq 1} a_i^2\Big)^{1/2}\,. \tag{2.155}$$

Proof The Cauchy-Schwarz inequality implies

$$Y := \sup_{t\in\mathcal{E}} X_t = \sup_{t\in\mathcal{E}} \sum_{i\geq 1} t_i g_i \leq \Big(\sum_{i\geq 1} a_i^2 g_i^2\Big)^{1/2} . \tag{2.156}$$

Taking $t_i = a_i^2 g_i / (\sum_{j\geq 1} a_j^2 g_j^2)^{1/2}$ yields that actually $Y = (\sum_{i\geq 1} a_i^2 g_i^2)^{1/2}$ and thus $\mathsf{E}Y^2 = \sum_{i\geq 1} a_i^2$. The right-hand side of (2.155) follows from the Cauchy-Schwarz inequality:

$$\mathsf{E}Y \leq (\mathsf{E}Y^2)^{1/2} = \Big(\sum_{i\geq 1} a_i^2\Big)^{1/2} . \tag{2.157}$$

For the left-hand side, let $\sigma = \max_{i\geq 1} |a_i|$. Since $Y = \sup_{t\in\mathcal{E}} X_t \geq |a_i||g_i|$ for any i, we have $\sigma \leq L\mathsf{E}Y$. Also,

$$\mathsf{E}X_t^2 = \sum_i t_i^2 \leq \max_i a_i^2 \sum_j \frac{t_j^2}{a_j^2} \leq \sigma^2 . \tag{2.158}$$

Then (2.118) implies[28]

$$\mathsf{E}(Y - \mathsf{E}Y)^2 \leq L\sigma^2 \leq L(\mathsf{E}Y)^2 ,$$

so that $\sum_{i\geq 1} a_i^2 = \mathsf{E}Y^2 = \mathsf{E}(Y - \mathsf{E}Y)^2 + (\mathsf{E}Y)^2 \leq L(\mathsf{E}Y)^2$. □

As a consequence of Theorem 2.10.1,

$$\gamma_2(\mathcal{E}) \leq L\Big(\sum_{i\geq 1} a_i^2\Big)^{1/2} . \tag{2.159}$$

This statement is purely about the geometry of ellipsoids. The proof we gave was rather indirect, since it involved Gaussian processes. Later on, in Chap. 4, we will learn how to give "purely geometric" proofs of similar statements that will have many consequences.

Let us now assume that the sequence $(a_i)_{i\geq 1}$ is *non-increasing*. Since

$$2^n \leq i \leq 2^{n+1} \Rightarrow a_{2^n} \geq a_i \geq a_{2^{n+1}}$$

we get

$$\sum_{i\geq 1} a_i^2 = \sum_{n\geq 0} \sum_{2^n \leq i < 2^{n+1}} a_i^2 \leq \sum_{n\geq 0} 2^n a_{2^n}^2$$

[28] One may extend (2.118) to the case where U is infinite by a proper definition of $\sup_{t\in U} X_t$.

and

$$\sum_{i\ge 1} a_i^2 \ge \sum_{n\ge 0} 2^n a_{2^{n+1}}^2 = \frac{1}{2}\sum_{n\ge 1} 2^n a_{2^n}^2 \ ,$$

and thus $\sum_{n\ge 0} 2^n a_{2^n}^2 \le 3\sum_{i\ge 1} a_i^2$. So we may rewrite (2.155) as

$$\frac{1}{L}\Big(\sum_{n\ge 0} 2^n a_{2^n}^2\Big)^{1/2} \le \mathsf{E}\sup_{t\in\mathcal{E}} X_t \le \Big(\sum_{n\ge 0} 2^n a_{2^n}^2\Big)^{1/2} . \tag{2.160}$$

Proposition 2.13.1 describes the size of ellipsoids with respect to Gaussian processes. Our next result describes their size with respect to Dudley's entropy bound (2.38).

Proposition 2.13.2 *We have*

$$\frac{1}{L}\sum_{n\ge 0} 2^{n/2} a_{2^n} \le \sum_{n\ge 0} 2^{n/2} e_n(\mathcal{E}) \le L\sum_{n\ge 0} 2^{n/2} a_{2^n} . \tag{2.161}$$

The right-hand sides in (2.160) and (2.161) are distinctly different.[29] Dudley's bound fails to describe the behavior of Gaussian processes on ellipsoids. This is a simple occurrence of a general phenomenon. In some sense, an ellipsoid is smaller than what one would predict just by looking at its entropy numbers $e_n(\mathcal{E})$. This idea will be investigated further in Sect. 4.1.

Exercise 2.13.3 Prove that for an ellipsoid $\mathcal{E}$ of $\mathbb{R}^m$, one has

$$\sum_{n\ge 0} 2^{n/2} e_n(\mathcal{E}) \le L\sqrt{\log(m+1)}\,\gamma_2(\mathcal{E}, d) \ ,$$

and that this estimate is essentially optimal. Compare with (2.58).

The proof of (2.161) hinges on ideas which are at least 50 years old and which relate to the methods of Exercise 2.5.9. The left-hand side is the easier part (it is also the most important for us). It follows from the next lemma, the proof of which is basically a special case of (2.45).

Lemma 2.13.4 *We have* $e_n(\mathcal{E}) \ge \frac{1}{2} a_{2^n}$.

Proof Consider the following ellipsoid in $\mathbb{R}^{2^n}$:

$$\mathcal{E}_n = \Big\{(t_i)_{i\le 2^n} \ ; \ \sum_{i\le 2^n} \frac{t_i^2}{a_i^2} \le 1\Big\} .$$

[29] This difference may seem rather small, but, as we shall see in Chap. 4, there are natural situations where it really matters.

Since $\mathcal{E}_n$ is the image of $\mathcal{E}$ by a contraction[30](namely, the "projection on the first 2^n coordinates"), it holds that $e_n(\mathcal{E}_n) \le e_n(\mathcal{E})$.

Throughout the rest of this section, we denote by B the centered unit Euclidean ball of $\mathbb{R}^{2^n}$ and by Vol the volume in this space. Let us consider a subset T of $\mathcal{E}_n$, with card $T \le 2^{2^n}$ and $\epsilon > 0$; then

$$\mathrm{Vol}\Big(\bigcup_{t\in T}(\epsilon B + t)\Big) \le \sum_{t\in T} \mathrm{Vol}(\epsilon B + t) \le 2^{2^n}\epsilon^{2^n}\mathrm{Vol}B = (2\epsilon)^{2^n}\mathrm{Vol}B\,.$$

Since we have assumed that the sequence (a_i) is non-increasing, we have $a_i \ge a_{2^n}$ for $i \le 2^n$ and thus $a_{2^n}B \subset \mathcal{E}_n$, so that $\mathrm{Vol}\mathcal{E}_n \ge a_{2^n}^{2^n}\mathrm{Vol}B$. Thus, whenever $2\epsilon < a_{2^n}$, we cannot have $\mathcal{E}_n \subset \bigcup_{t\in T}(\epsilon B + t)$, so that $e_n(\mathcal{E}_n) \ge a_{2^n}/2$. □

We now turn to the upper bound, which relies on a special case of (2.46). We keep the notation of the proof of Lemma 2.13.4.

Lemma 2.13.5 *We have*

$$e_{n+3}(\mathcal{E}) \le e_{n+3}(\mathcal{E}_n) + a_{2^n}\,. \tag{2.162}$$

Proof We observe that when $t \in \mathcal{E}$, then, using that $a_i \le a_{2^n}$ for $i > 2^n$ in the last inequality,

$$1 \ge \sum_{i\ge 1}\frac{t_i^2}{a_i^2} \ge \sum_{i>2^n}\frac{t_i^2}{a_i^2} \ge \frac{1}{a_{2^n}^2}\sum_{i>2^n} t_i^2$$

so that $(\sum_{i>2^n} t_i^2)^{1/2} \le a_{2^n}$ and, viewing $\mathcal{E}_n$ as a subset of $\mathcal{E}$, we have $d(t,\mathcal{E}_n) \le a_{2^n}$. Thus, if for $k \ge 1$ we cover $\mathcal{E}_n$ by N_k balls of radius ϵ, the balls with the same centers but radius $\epsilon + a_{2^n}$ cover $\mathcal{E}$. This proves that $e_k(\mathcal{E}_n) \le e_k(\mathcal{E}) + a_{2^n}$ and hence (2.162). □

Lemma 2.13.6 *Let* $\epsilon = \max_{k\le n} a_{2^k}2^{k-n}$. *Consider a subset A of $\mathcal{E}_n$ with the following property:*

$$\textit{Any two points of } A \textit{ are at mutual distance} \ge 2\epsilon\,. \tag{2.163}$$

Then card $A \le N_{n+3}$.

Proof The balls centered at the points of A, with radius ϵ, have disjoint interiors, so that the volume of their union is card A Vol(ϵB), and since these balls are entirely contained in $\mathcal{E}_n + \epsilon B$, we have

$$\mathrm{card}\,A\ \mathrm{Vol}(\epsilon B) \le \mathrm{Vol}(\mathcal{E}_n + \epsilon B)\,. \tag{2.164}$$

---

[30] Generally speaking, a map φ from a metric space (T, d) to a metric space (T', d') is called a contraction if $d'(\varphi(x), \varphi(y)) \le d(x, y)$.

For $t = (t_i)_{i\le 2^n} \in \mathcal{E}_n$, we have $\sum_{i\le 2^n} t_i^2/a_i^2 \le 1$, and for t' in ϵB, we have $\sum_{i\le 2^n} t_i'^2/\epsilon^2 \le 1$. Let $c_i = 2\max(\epsilon, a_i)$. Since

$$\sum_{i\le 2^n} \frac{(t_i + t_i')^2}{c_i^2} \le \sum_{i\le 2^n} \frac{2t_i^2 + 2t_i'^2}{c_i^2} \le \sum_{i\le 2^n} \frac{1}{2}\Big(\frac{t_i^2}{a_i^2} + \frac{t_i'^2}{\epsilon^2}\Big) \le 1 ,$$

we have

$$\mathcal{E}_n + \epsilon B \subset \mathcal{E}^1 := \Big\{t\ ;\ \sum_{i\le 2^n} \frac{t_i^2}{c_i^2} \le 1\Big\} .$$

Therefore

$$\mathrm{Vol}(\mathcal{E}_n + \epsilon B) \le \mathrm{Vol}\,\mathcal{E}^1 = \mathrm{Vol}B \prod_{i\le 2^n} c_i$$

and comparing with (2.164) yields

$$\mathrm{card}\,A \le \prod_{i\le 2^n} \frac{c_i}{\epsilon} = 2^{2^n} \prod_{i\le 2^n} \max\Big(1, \frac{a_i}{\epsilon}\Big) .$$

Next it follows from the choice of ϵ that for any $k \le n$, we have $a_{2^k} 2^{k-n} \le \epsilon$. Then $a_i \le a_{2^k} \le \epsilon 2^{n-k}$ for $2^k < i \le 2^{k+1}$, so that

$$\begin{aligned}\prod_{i\le 2^n} \max\Big(1, \frac{a_i}{\epsilon}\Big) &= \prod_{k\le n-1} \prod_{2^k < i \le 2^{k+1}} \max\Big(1, \frac{a_i}{\epsilon}\Big) \\ &\le \prod_{k\le n-1} (2^{n-k})^{2^k} = 2^{\sum_{k\le n-1}(n-k)2^k} \le 2^{2^{n+2}}\end{aligned}$$

since $\sum_{i\ge 0} i2^{-i} = 4$. Therefore, $\mathrm{card}\,A \le 2^{2^n} \cdot 2^{2^{n+2}} \le N_{n+3}$. □

Lemma 2.13.7 *We have*

$$e_{n+3}(\mathcal{E}_n) \le 2 \max_{k\le n}(a_{2^k} 2^{k-n}) . \tag{2.165}$$

Proof Assume now that A is as large as possible under (2.163). Then the balls centered at points of A and with radius $\le 2\epsilon$ cover $\mathcal{E}_n$, for otherwise we could add a point to A. Since $\mathrm{card}\,A \le N_{n+3}$, we have $e_{n+3}(\mathcal{E}_n) \le 2\epsilon$. □

Combining (2.165) with (2.162), we obtain

Corollary 2.13.8 *We have*

$$e_{n+3}(\mathcal{E}) \le 3 \max_{k\le n}(a_{2^k} 2^{k-n}) \,. \tag{2.166}$$

Proof of Proposition 2.13.2 We have, using (2.166),

$$\sum_{n\ge 3} 2^{n/2} e_n(\mathcal{E}) = \sum_{n\ge 0} 2^{(n+3)/2} e_{n+3}(\mathcal{E}) \le L \sum_{n\ge 0} 2^{n/2}\Big(\sum_{k\le n} 2^{k-n} a_{2^k}\Big)$$
$$= L \sum_{k\ge 0} 2^k a_{2^k} \sum_{n\ge k} 2^{-n/2} \le L \sum_{k\ge 0} 2^{k/2} a_{2^k} \,.$$

Since $\mathcal{E}$ is contained in the ball centered at the origin with radius a_1, we have $e_n(\mathcal{E}) \le a_1$ for each n. The result follows. □

2.14 Rolling Up Our Sleeves: Chaining on Ellipsoids

Let us recall the ellipsoid $\mathcal{E}$ of (2.154). We have proved (2.159) as a consequence of the majorizing measure theorem, Theorem 2.10.1. We will later give a more geometrical proof of this result. In the present section, we demonstrate the hard way that these results are deep, by explicitly constructing a chaining on the ellipsoid $\mathcal{E}$.[31] This is surprisingly non-trivial.[32] Let us assume that the sequence (a_i) is non-increasing, and for $n \ge 0$, let $I_n = \{i ; 2^n \le i < 2^{n+1}\}$ so that card $I_n = 2^n$ and

$$\mathcal{E} \subset \mathcal{E}' = \Big\{t \in \ell^2 \,;\, \sum_{n\ge 0} \frac{1}{a_{2^n}^2} \sum_{i\in I_n} t_i^2 \le 1\Big\} = \Big\{t \in \ell^2 \,;\, \sum_{n\ge 0} \frac{2^n}{c_n} \sum_{i\in I_n} t_i^2 \le 1\Big\} \,, \tag{2.167}$$

where $c_n = 2^n a_{2^n}^2$. Furthermore (as in the previous section), $\sum_{n\ge 0} c_n = \sum_{n\ge 0} 2^n a_{2^n}^2 \le 3 \sum_{i\ge 1} a_i^2$. For such an ellipsoid $\mathcal{E}'$, we will construct sets $U_n \subset \ell^2$ with card $U_n \le N_{n+n_0}$ (where n_0 is a universal constant) such that

$$\forall t \in \mathcal{E}' \,,\; \sum_{n\ge 0} 2^{n/2} d(t, U_n) \le L\Big(\sum_{n\ge 0} c_n\Big)^{1/2} \,. \tag{2.168}$$

[31] There are obvious similarities between this section and Sect. 2.6. It is a good challenge to figure out by yourself how to do the chaining on ellipsoids after having studied Sect. 2.6.

[32] I am grateful to Dali Liu for having suggested to include this section.

Let us now deduce from this result how to perform the chaining on the ellipsoid $\mathcal{E}$ of (2.154). As we have just seen, such an ellipsoid is contained in an ellipsoid $\mathcal{E}'$ of the type (2.167) for which $\sum_{n\geq 0} c_n \leq L\sum_{i\geq 1} a_i^2$. Consider the sets $U_n \subset \ell^2$ as in (2.168). Consider a map $\varphi : \ell^2 \to \mathcal{E}$ such that $d(x, \varphi(x)) \leq 2d(x, \mathcal{E})$, and observe that for $t \in \mathcal{E}$ and $x \in \ell^2$ we have $d(x, \varphi(x)) \leq 2d(x, t)$ so that $d(t, \varphi(x)) \leq d(t, x) + d(x, \varphi(x)) \leq 3d(t, x)$. Consequently, $d(t, \varphi(U_n)) \leq 3d(t, U_n)$. The sets $\varphi(U_n) \subset \mathcal{E}$ satisfy $\operatorname{card}\varphi(U_n) \leq \operatorname{card} U_n \leq N_{n+n_0}$. We define $T_n = \{0\}$ for $n \leq n_0$ and $T_n = U_{n-n_0}$ for $n > n_0$. Thus, $\operatorname{card} T_n \leq N_n$ and (2.168) implies

$$\forall t \in \mathcal{E}\,,\ \sum_{n\geq 0} 2^{n/2} d(t, T_n) \leq L\Big(\sum_i a_i^2\Big)^{1/2}\,.$$

We now prepare for the construction of the sets U_n. There is no loss of generality to assume that $\sum_{n\geq 0} c_n = 1$.

Lemma 2.14.1 *Given $t \in \mathcal{E}'$, we can find a sequence $(p(n,t))_{n\geq 0}$ of integers with the following properties:*

$$\sum_{i\in I_n} t_i^2 \leq 2^{-p(n,t)}\,, \tag{2.169}$$

$$\sum_{n\geq 0} 2^{n/2-p(n,t)/2} \leq L\,, \tag{2.170}$$

$$\forall n \geq 0\,,\ p(n+1,t) \leq p(n,t) + 2\,. \tag{2.171}$$

Proof Define $q(n,t)$ as the largest integer $q \leq 2n$ such that $\sum_{i\in I_n} t_i^2 \leq 2^{-q}$. Let $A = \{n \geq 0; q(n,t) < 2n\}$, so that for $n \in A$ by definition of $q(n,t)$ we have $2^{-q(n,t)} \leq 2\sum_{i\in I_n} t_i^2$. Thus, since $t \in \mathcal{E}'$,

$$\sum_{n\in A} \frac{2^{n-q(n,t)}}{c_n} \leq 2\sum_{n\geq 0} \frac{2^n}{c_n} \sum_{i\in I_n} t_i^2 \leq 2.$$

Since $\sum_n c_n = 1$, then $\sum_{n\in A} 2^{n/2-q(n,t)/2} \leq L$ by the Cauchy-Schwarz inequality. Since $2^{n/2-q(n,t)/2} = 2^{-n/2}$ for $n \notin A$, we have $\sum_{n\geq 0} 2^{n/2-q(n,t)/2} \leq L$. We define now $p(n,t) = \min\{q(k,t) + 2(n-k); 0 \leq k \leq n\}$ so that (2.171) holds. Also, since $2^{-p(n,t)/2} \leq \sum_{k\leq n} 2^{-q(k,t)/2-(n-k)}$, we obtain

$$\sum_{n\geq 0} 2^{n/2-p(n,t)/2} \leq \sum_{n\geq 0}\sum_{k\leq n} 2^{k/2-q(k,t)/2} 2^{-(n-k)/2}$$

$$= \sum_{k\geq 0} 2^{k/2-q(k,t)/2} \sum_{n\geq k} 2^{-(n-k)/2} \leq L\,. \quad \square$$

For each $n \geq 1$ and $p \geq 0$, consider the set $B(n, p) \subset \ell^2$ which consists of the $t = (t_i)_{i\geq 1}$ such that $t_i = 0$ if $i \geq 2^n$ and $\|t\|_2 \leq 2^{-p/2+2}$. This is a ball of dimension $2^n - 1$ and radius $2^{-p/2+2}$. Using (2.47) for $\epsilon = 1/4$, there is a set $V_{n,p} \subset B(n, p)$ with card $V_{n,p} \leq L^{2^n}$ such that every point of $B(n, p)$ is within distance $\leq 2^{-p/2}$ of $V_{n,p}$. We consider the set $V_n = \cup_{0\leq p\leq 2n} V_{n,p}$ so that card $V_{n,p} \leq L^{2^n}$. We set $U_0 = \{0\}$, and we consider the sets U_n consisting of the elements $x_0 + \ldots + x_n$ where $x_k \in V_k$

For $t \in \ell^2$ and $n \geq 0$, we define $t^{(n)} \in \ell^2$ by $t_i^{(n)} = t_i$ if $i < 2^n$ and $t_i^{(n)} = 0$ if $i \geq 2^n$. Note that $t^{(n)} = t$ for $t \in U_n$.

Lemma 2.14.2 *For $t \in \mathcal{E}'$, consider the sequence $(p(n, t))$ of Lemma 2.14.1. Then for each n, we can find $u(n) \in U_n$ such that $d(u(n), t^{(n)}) \leq 2^{-p(n,t)/2}$.*

Proof The proof is by induction over n. For $n = 0$, it suffices to take $u(0) = 0$ since $t^{(0)} = 0$. For the induction step from n to $n + 1$, we have $t^{(n)} = u(n) + v(n)$ where $u(n) \in U_n$ and $\|v(n)\|_2 \leq 2^{-p(n,t)/2}$, so that $t^{(n+1)} = u(n) + v'(n)$ where $v'(n) = v(n) + t^{(n+1)} - t^{(n)}$. By (2.169) $\|t^{(n+1)} - t^{(n)}\|_2 = (\sum_{i\in I_n} t_i^2)^{1/2} \leq 2^{-p(n,t)/2}$. Thus, $\|v'(n)\|_2 \leq 2^{-p(n,t)/2+1} \leq 2^{-p(n+1,t)/2+2}$ where we have used (2.171) in the second inequality. Since $v(n)_i = 0$ for $i \geq 2^n$, we have $v'(n)_i = 0$ for $i \geq 2^{n+1}$ so that $v'(n) \in B(n + 1, p(n + 1, t))$. Thus, there is an element $w \in V_{n+1,p(n+1,t)} \subset V_{n+1}$ for which $\|v'(n) - w\|_2 \leq 2^{-p(n+1,t)/2}$. Setting $u(n+1) := u(n)+w \in U_{n+1}$, we then have $t^{(n+1)} - u(n + 1) = v'(n) - w$. □

Corollary 2.14.3 *For $t \in \mathcal{E}'$ we have $\sum_{n\geq 0} 2^{n/2} d(t, U_n) \leq L$.*

Proof Recalling (2.169), we have

$$\|t - t^{(n)}\|_2^2 = \sum_{k>n} \sum_{i\in I_k} t_i^2 \leq \sum_{k>n} 2^{-p(k,t)}$$

so that $\|t - t^{(n)}\|_2 \leq \sum_{k>n} 2^{-p(k,t)/2}$. Then

$$\sum_{n\geq 0} 2^{n/2} \|t - t^{(n)}\|_2 \leq \sum_{n\geq 0} \sum_{k>n} 2^{-p(k,t)/2} = \sum_{k\geq 1} 2^{-p(k,t)/2} \sum_{0\leq n<k} 2^{n/2}$$

$$\leq L \sum_{k\geq 1} 2^{k/2-p(k,t)/2} \leq L\,,$$

using (2.170) in the last inequality. Since $d(t, U_n) \leq d(t^{(n)}, U_n) + \|t - t^{(n)}\|_2$, the result follows, using also Lemma 2.14.2. □

2.15 Continuity of Gaussian Processes

By far, the most important result concerning continuity of Gaussian processes is Dudley's bound (1.19). However since the finiteness of the right-hand side of (1.19) is not necessary for the Gaussian process to be continuous, there are situations where this bound is not appropriate.[33] In the present section, we show that a suitable form of the generic chaining allows us to capture the exact modulus of continuity of a Gaussian process with respect to its canonical distance in full generality. Not surprisingly, the modulus of continuity is closely related to the rate at which the series $\sum_n 2^{n/2}\Delta(A_n(t))$ converges uniformly on T for a suitable admissible sequence $(\mathcal{A}_n)$. Our first result shows how to obtain a modulus of continuity using the generic chaining.

Lemma 2.15.1 *Consider a metric space (T, d) and a process $(X_t)_{t\in T}$ which satisfies the increment condition (2.4):*

$$\forall u > 0\ ,\ \mathsf{P}(|X_s - X_t| \geq u) \leq 2\exp\left(-\frac{u^2}{2d(s,t)^2}\right) . \tag{2.4}$$

Assume that there exists a sequence (T_n) of subsets of T with card $T_n \leq N_n$ *such that for a certain integer m and a certain number B one has*

$$\sup_{t\in T} \sum_{n\geq m} 2^{n/2} d(t, T_n) \leq B . \tag{2.172}$$

Consider $\delta > 0$. Then, for any $u \geq 1$, with probability $\geq 1 - \exp(-u^2 2^m)$, we have

$$\forall s, t \in T\ ,\ d(s,t) \leq \delta \Rightarrow |X_s - X_t| \leq Lu(2^{m/2}\delta + B) . \tag{2.173}$$

Proof We assume T finite for simplicity. For $n \geq m$ and $t \in T$, denote by $\pi_n(t)$ an element of T_n such that $d(t, \pi_n(t)) = d(t, T_n)$. Consider the event Ω_u defined by[34]

$$\forall n \geq m+1\ ,\ \forall t \in T\ ,\ |X_{\pi_{n-1}(t)} - X_{\pi_n(t)}| \leq Lu2^{n/2} d(\pi_{n-1}(t), \pi_n(t))\ , \tag{2.174}$$

and

$$\forall s', t' \in T_m\ ,\ |X_{s'} - X_{t'}| \leq Lu2^{m/2} d(s', t') . \tag{2.175}$$

[33] In practice, however, as of today, the Gaussian processes for which continuity is important can be handled through Dudley's bound, while for those which cannot be handled through this bound (such as in Chap. 4), it is boundedness which matters. For this reason, the considerations of the present section are of purely theoretical interest and may be skipped at first reading.

[34] We are again following the general method outlined at the end of Sect. 2.4.

Then as in Sect. 2.4, we have $\mathsf{P}(\Omega_u) \geq 1 - \exp(-u^2 2^m)$. Now, when Ω_u occurs, using chaining as usual and (2.174), we get

$$\forall t \in T \; , \; |X_t - X_{\pi_m(t)}| \leq LuB \; . \tag{2.176}$$

Moreover, using (2.172) in the inequality, $d(t, \pi_m(t)) = d(t, T_m) \leq B2^{-m/2}$, so that, using (2.175),

$$\begin{aligned} d(s,t) \leq \delta &\Rightarrow d(\pi_m(s), \pi_m(t)) \leq \delta + 2B2^{-m/2} \\ &\Rightarrow |X_{\pi_m(s)} - X_{\pi_m(t)}| \leq Lu(\delta 2^{m/2} + B) \; . \end{aligned}$$

Combining with (2.175) proves that $|X_s - X_t| \leq Lu(\delta 2^{m/2} + B)$ and completes the proof. □

Exercise 2.15.2 Deduce Dudley's bound (1.19) from Lemma 2.15.1.

We now turn to our main result, which exactly describes the modulus of continuity of a Gaussian process in terms of certain admissible sequences. It implies in particular the remarkable fact (discovered by X. Fernique) that for Gaussian processes the "local modulus of continuity" (as in (2.177)) is also "global".

Theorem 2.15.3 *There exists a constant L^* with the following property. Consider a Gaussian process $(X_t)_{t \in T}$, with canonical associated distance d given by (0.1). Assume that $S = \mathsf{E} \sup_{t \in T} X_t < \infty$. For $k \geq 1$, consider $\delta_k > 0$, and assume that*

$$\forall t \in T \; ; \; \mathsf{E} \sup_{\{s \in T ; d(s,t) \leq \delta_k\}} |X_s - X_t| \leq 2^{-k} S \; . \tag{2.177}$$

Let $n_0 = 0$, and for $k \geq 1$, consider an integer n_k for which

$$L^* S 2^{-n_k/2 - k} \leq \delta_k \; . \tag{2.178}$$

Then we can find an admissible sequence $(\mathcal{A}_n)$ of partitions of T such that

$$\forall k \geq 0 \; ; \; \sup_{t \in T} \sum_{n \geq n_k} 2^{n/2} \Delta(A_n(t)) \leq LS2^{-k} \; . \tag{2.179}$$

Conversely, given integers n_k and an admissible sequence $(\mathcal{A}_n)$ as in (2.179) and defining now $\delta_k^* = S2^{-n_k/2-k}$, with probability $\geq 1 - \exp(-u^2 2^{n_k})$, we have

$$\sup_{\{s,t \in T ; d(s,t) \leq \delta_k^*\}} |X_s - X_t| \leq Lu2^{-k} S \; . \tag{2.180}$$

The abstract formulation here might make it hard at first to feel the power of the statement. The numbers δ_k control the (local) modulus of continuity of the process. The numbers n_k control the uniform convergence (over t) of the series

$\sum_{n\geq 0} 2^{n/2}\Delta(A_n(t))$. They relate to each other by the relation $\delta_k \sim S2^{-n_k/2-k}$. The second part of the theorem asserts that in turn the numbers n_k control the uniform modulus of continuity (2.180).

Proof According to the majorizing measure theorem and specifically (2.114), there exists a constant L^* such that for each subset U of T there exists an admissible sequence $(\mathcal{A}_n)$ of partitions of U such that

$$\forall t \in U\ ,\ \sum_{n\geq 0} 2^{n/2}\Delta(A_n(t)) \leq \frac{L^*}{2}\mathsf{E}\sup_{s\in U} X_s\ . \tag{2.181}$$

Assuming (2.177), by induction over k, we construct an admissible sequence $(\mathcal{A}_n)_{n\leq n_k}$ such that

$$1 \leq p \leq k \Rightarrow \sup_{t\in T} \sum_{n_{p-1}<n\leq n_p} 2^{n/2}\Delta(A_n(t)) \leq L^* S2^{-p}\ . \tag{2.182}$$

For $k = 1$, the existence of the sequence $(\mathcal{A}_n)_{n<n_1}$ follows from the Majorizing Measure Theorem through (2.181) as explained, so we turn to the induction step from k to $k+1$. Using (2.182) for $p = k$, we deduce that for each $t \in T$, $2^{n_k/2}\Delta(A_{n_k}(t)) \leq L^* S2^{-k}$, so that $\Delta(A_{n_k}(t)) \leq L^* S2^{-n_k/2-k} \leq \delta_k$ using (2.178) in the last inequality. Consequently, for any element C of $\mathcal{A}_{n_k}$, we have $\Delta(C) \leq \delta_k$, so that considering any point t of C we have, using (2.177) in the last inequality,

$$\mathsf{E}\sup_{s\in C} X_s = \mathsf{E}\sup_{s\in C}(X_s - X_t) \leq \mathsf{E}\sup_{\{s\in T; d(s,t)\leq\delta_k\}} |X_s - X_t| \leq S2^{-k}\ .$$

Using the majorizing measure theorem, we construct for each $C \in \mathcal{A}_{n_k}$ an admissible sequence $(\mathcal{A}_{C,n})_{n\geq 0}$ of partitions of C for which

$$\forall t \in C\ ,\ \sum_{n\geq 0} 2^{n/2}\Delta(A_{C,n}(t)) \leq L^* S2^{-k-1}\ . \tag{2.183}$$

For $n_k < n \leq n_{k+1}$, we simply define $\mathcal{A}_n$ as the collection of all sets in one of the partitions $A_{C,n-1}$ where $C \in \mathcal{A}_{n_k}$, so that $\operatorname{card}\mathcal{A}_n \leq N_{n-1}\operatorname{card}\mathcal{A}_{n_k} \leq N_{n-1}^2 \leq N_n$. Since for $t \in C$ we have $A_n(t) = A_{C,n-1}(t)$, it follows from (2.183) that for any $C \in \mathcal{A}_{n_k}$, we have

$$\sup_{t\in C} \sum_{n_k<n\leq n_{k+1}} 2^{n/2}\Delta(A_n(t)) \leq \sup_{t\in C}\sum_{n>n_k} 2^{n/2}\Delta(A_{C,n-1}(t)) \leq L^* S2^{-k}\ .$$

This completes the induction and the construction of the sequence $(\mathcal{A}_n)$ since (2.182) implies (2.179).

It remains to prove the "conversely" part. For this for each $n \geq 0$, we simply consider a subset T_n of T such that

$$\forall A \in \mathcal{A}_n \ , \ \operatorname{card}(T_n \cap A) = 1 \ .$$

For each k, we then use Lemma 2.15.1 for $m = n_k$ and $B = S2^{-k}$. □

Key Ideas to Remember[35]

- The generic chaining efficiently organizes the standard chaining argument for processes whose increments have Gaussian-like tails governed by a distance d as in (2.4).
- The generic chaining applied to such processes motivates the introduction of our main measure $\gamma_2(T, d)$ of the size of a metric space (T, d). This measure involves the existence of suitable sequences of partitions.
- The fundamental problem then becomes how to construct such sequences of partitions in a metric space.
- There is a machine (called a partitioning scheme) to construct such sequences of partitions. The input to the machine is a functional, a function of the subsets of our basic metric space, which in a sense is a measure of their size. The existence of such functionals with specific growth properties is intrinsically linked to the existence of such sequences of partitions.
- The majorizing measure theorem is the statement that for a Gaussian process with index set T and canonical distance d the quantity $\mathsf{E} \sup_{t \in T} X_t$ is exactly of order $\gamma_2(T, d)$. The proof relies on a partitioning scheme, used for the functional $F(A) = \mathsf{E} \sup_{t \in A} X_t$. Sudakov minoration and concentration of measure are the main tools to prove that this functional satisfies the required growth condition.
- Gaussian processes can be seen as subsets of a standard Hilbert space, but the geometric understanding that would relate the size of a set with the size of its convex hull is still lacking.
- The traditional way to organize chaining uses entropy numbers. Even for sets as basic as ellipsoids in Hilbert space, entropy numbers provide only a suboptimal description of their size.

2.16 Notes and Comments

I have heard people saying that the problem of characterizing continuity and boundedness of Gaussian processes goes back (at least implicitly) to Kolmogorov. The understanding of Gaussian processes was long delayed by the fact that in

[35] The function of this brief summary is *not* to explain the material again, but is a way for the reader to check that she did understand the main ideas. If any of the points made below is not clear to the reader, she may not be ready to proceed and may want to review the corresponding material.

the most immediate examples the index set is a subset of $\mathbb{R}$ or $\mathbb{R}^n$ and that the temptation to use the special structure of this index set is nearly irresistible. Probably the single most important conceptual progress about Gaussian processes was the realization, in the late 1960s, that the boundedness of a (centered) Gaussian process is determined by the structure of the metric space (T, d), where d is the usual distance $d(s,t) = (\mathsf{E}(X_s - X_t)^2)^{1/2}$. It is difficult now to realize what a tremendous jump in understanding this was, since this seems so obvious *a posteriori*.

In 1967, R. Dudley obtained the inequality (2.38). (As he pointed out, R. Dudley did not state (2.38) though he performed all the essential steps and (2.38) totally deserves to be called Dudley's bound.) A few years later, X. Fernique proved that in the "stationary case", Dudley's inequality can be reversed [32], i.e., he proved in that case the lower bound of Theorem 2.10.1. This historically important result was central to the work of Marcus and Pisier [61, 62] who built on it to solve all the classical problems on random Fourier series. Some of their results will be presented in Chap. 7. Interestingly, now that the right approach has been found, the proof of Fernique's result is not really easier than that of Theorem 2.10.1.

Another major contribution of Fernique (building on earlier ideas of C. Preston) was an improvement of Dudley's bound based on a new tool called majorizing measures (which we will study in Sect. 3.1.3). Fernique conjectured that his bound was essentially optimal. Gilles Pisier suggested in 1983 that I should work on this conjecture. In my first attempt, I proved fast that Fernique's conjecture held in the case where the metric space (T, d) is ultrametric. I learned that Fernique had already done this, so I was discouraged for a while. In the second attempt, I tried to decide whether a majorizing measure existed on ellipsoids. I had the hope that some simple density with respect to the volume measure would work. It was difficult to form any intuition, and I struggled in the dark for months. At some point, I tried a combination of suitable point masses and easily found a direct construction of the majorizing measure on ellipsoids. This made it believable that Fernique's conjecture was true, but I still tried to disprove it. Then I realized that I did not understand why a direct approach using a partitioning scheme should fail, while this understanding should be useful to construct a counterexample. Once I tried this direct approach, it was a matter of 3 days to prove Fernique's conjecture. Gilles Pisier made two comments about this discovery. The first one was "you are lucky", by which he meant that I was lucky that Fernique's conjecture was true, since a counter example would have been of limited interest. I am grateful to this day for his second comment: "I wish I had proved this myself, but I am very glad you did it".

Fernique's concept of majorizing measures is difficult to grasp and was dismissed by the main body of probabilists as a mere curiosity. (I myself found it very difficult to understand.) This could be the main reason why Fernique's pathbreaking work did not receive the recognition it should have. I have tried to repair this and to express my personal admiration by dedicating this book to his memory and by paying homage to his work at numerous places in this book.

In 2000, while discussing one of the open problems of this book with Keith Ball (be he blessed for his interest in it!), I discovered that one could replace majorizing measures by the totally natural variation on the usual chaining arguments that was presented here. That this was not discovered much earlier is a striking illustration of the inefficiency of my brain. For two decades, it looked like majorizing measures would not be of any use anymore, but they now play again a major role again for reasons to be explained in Chap. 5.

In [111], the author presented a particularly simple proof of Theorem 2.10.1 (expressed in terms of majorizing measures since the generic chaining had not been invented yet). It is based on a partition scheme related to the one we use here. The precise relationship is discussed on page 72 of [132].

It is *on purpose* that I did not stress Slepian's lemma, which is the statement that (2.127) holds for $L = 1$. This lemma is very specific to Gaussian processes, and focusing on it seems a good way to guarantee that one will never move beyond these. One notable progress I made was to discover (ages ago) the scheme of proof of Proposition 2.10.8 that dispenses with Slepian's lemma and that we shall use in many situations. Comparison results such as Slepian's lemma are not at the root of results such as the majorizing measure theorem, but rather are (at least qualitatively) a consequence of them as in Corollary 2.10.12. This being said, Slepian's lemma is historically very important as it crystallizes the link between $\mathsf{E} \sup_{t \in T} X_t$ and the structure of the metric space (T, d).

Chapter 3
Trees and Other Measures of Size

In this chapter, we systematically investigate different ways to measure the size of a metric space. One of them, Fernique's functional of Sect. 3.3 will play a major role in the sequel, as it is the form which lends itself to vast generalizations. The concept of a tree presented in Sect. 3.1 is historically important: the author discovered many of the results he presents while thinking in terms of trees. We know now how to present these results and their proofs without ever mentioning trees, arguably in a more elegant fashion, so that trees are not used explicitly elsewhere in this book. However, it might be too early to dismiss this concept, at least as an instrument of discovery.

3.1 Trees

We shall describe different ways to measure the size of a metric space and show that they are all equivalent to the functional $\gamma_2(T, d)$.[1]

In a nutshell, a tree is a certain structure that requires a "lot of space" to be constructed, so that a metric space needs to be large in order to contain large trees. At the simplest level, it already takes some space to construct in a set A sets $B_1, \dots, B_n$ which are appropriately separated from each other. This is even more so if the sets $B_1, \dots, B_n$ are themselves large (e.g., because they contain many sets far from each other). Trees are a proper formulation of the iteration of this idea. The basic use of trees is to measure the size of a metric space by the size of the largest tree (of a certain type) which it contains. Different types of trees yield different measures of size.

[1] It is possible to consider more general notions corresponding to other functionals considered in the book, but for simplicity we consider only the case of γ_2.

M. Talagrand, *Upper and Lower Bounds for Stochastic Processes*, Ergebnisse der Mathematik und ihrer Grenzgebiete. 3. Folge / A Series of Modern Surveys in Mathematics 60, https://doi.org/10.1007/978-3-030-82595-9_3

A *tree* $\mathcal{T}$ of a metric space (T, d) is a *finite* collection of non-empty subsets of T with the following two properties:

$$\text{Given } A, B \text{ in } \mathcal{T}\,,\ \text{ if } A \cap B \neq \emptyset\,,\ \text{ then either } A \subset B \text{ or else } B \subset A\,. \tag{3.1}$$

$$\mathcal{T} \text{ has a largest element}\,. \tag{3.2}$$

The important condition here is (3.1), and (3.2) is just for convenience.

If $A, B \in \mathcal{T}$ and $B \subset A$, $B \neq A$, we say that B is a *child* of A if

$$C \in \mathcal{T}\,,\ B \subset C \subset A \Rightarrow C = B \text{ or } C = A\,. \tag{3.3}$$

We denote by $c(A)$ the number of children of A. Since our trees are finite, some of their sets will have no children. It is convenient to "shrink these sets to a single point", so we will consider only trees with the following property:

$$\text{If } A \in \mathcal{T} \text{ and } c(A) = 0\,,\ \text{then } A \text{ contains exactly one point}\,. \tag{3.4}$$

A fundamental property of trees is as follows: consider trees $\mathcal{T}_1, \ldots, \mathcal{T}_m$, and for $1 \le \ell \le m$, let A_ℓ be the largest element of $\mathcal{T}_\ell$. Assume that the sets A_ℓ are disjoint, and consider a set A with $\bigcup_{\ell \le m} A_\ell \subset A \subset T$. Then the collection of subsets of T consisting of A and of $\bigcup_{\ell \le m} \mathcal{T}_\ell$ is a tree. The proof is straightforward. This fact allows one to construct iteratively more and more complicated (and larger) trees.

An important structure in a tree is a *branch*. A sequence $A_0, A_1, \ldots, A_k$ is a branch if $A_{\ell+1}$ is a child of A_ℓ and if moreover A_0 is the largest element of $\mathcal{T}$ while A_k has no child. Then by (3.4), the set A_k is reduced to a single point t, and $A_0, \ldots, A_k$ are exactly those elements of $\mathcal{T}$ which contain t. So in order to describe the branches of $\mathcal{T}$, it is convenient to introduce the set

$$S_{\mathcal{T}} = \{t \in T\ ;\ \{t\} \in \mathcal{T}\}\,, \tag{3.5}$$

which we call the *support* of $\mathcal{T}$. If a set A in a tree has no child, one may call it a *leaf*. Thus, a leaf of a tree is reduced to one single point, and the support of a tree is the union of its leaves. By considering all the collections $\{A \in \mathcal{T}; t \in A\}$ as t varies in $S_{\mathcal{T}}$, we obtain all the branches of $\mathcal{T}$.

3.1.1 Separated Trees

We now quantify our desired property that the children of a given set should be far from each other in an appropriate sense. A *separated* tree is a tree $\mathcal{T}$ such that to each A in $\mathcal{T}$ with $c(A) \geq 1$ is associated an integer $s(A) \in \mathbb{Z}$ with the following properties: First,

$$\text{If } B_1 \text{ and } B_2 \text{ are distinct children of } A\text{, then } d(B_1, B_2) \geq 4^{-s(A)} . \tag{3.6}$$

Here, $d(B_1, B_2) = \inf\{d(x_1, x_2); x_1 \in B_1, x_2 \in B_2\}$. We observe that in (3.6), we make no restriction on the diameter of the children of A, see Fig 3.1. (Such restrictions will, however, occur in the other notion of tree that we consider later.) Second, to rule out pathologies, we will also make the following purely technical assumption:

$$\text{If } B \text{ is a child of } A\text{, then } s(B) > s(A) . \tag{3.7}$$

An example of separated tree is shown on Fig. 3.1. To measure the size of a separated tree T, we introduce its *depth*, i.e.,

$$\rho(\mathcal{T}) := \inf_{t \in S_\mathcal{T}} \sum_{t \in A \in \mathcal{T}} 4^{-s(A)} \sqrt{\log c(A)} . \tag{3.8}$$

Here and below, we make the convention that the summation does not include the term $A = \{t\}$ (for which $c(A) = 0$). The quantity (3.8) takes into account both the separation between the children of A (through the term $4^{-s(A)}$) and their number

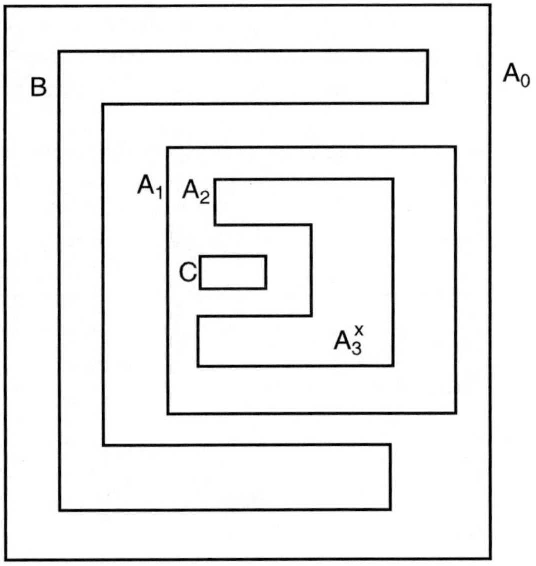

Fig. 3.1 A separated tree. The children of A_0 are A_1 and B. The children of A_1 are A_2 and C. A_0, A_1, A_2, and A_3 form a branch, of which A_3 is a leaf

(through the term $\sqrt{\log c(A)}$). This will be a common feature of all our notions of sizes of trees.

We observe that in (3.8), we have the *infimum* over $t \in S_{\mathcal{T}}$. In words,

A tree is large if it is large along *every* branch.

We can then measure the size of T by

$$\sup\{\rho(\mathcal{T})\ ;\ \mathcal{T} \text{ separated tree} \subset T\}\ . \tag{3.9}$$

3.1.2 Organized Trees

The notion of separated tree we just considered is but one of many possible notions of trees, and it does not seem fundamental. Rather, the quantity (3.9) is used as a convenient intermediate technical step to prove the equivalence of several more important quantities. Let us now consider another notion of trees, which is more restrictive (and apparently much more important). An *organized* tree is a tree $\mathcal{T}$ such that to each $A \in \mathcal{T}$ with $c(A) \geq 1$ is associated an integer $j = j(A) \in \mathbb{Z}$ and points $t_1, \ldots, t_{c(A)}$ with the properties that

$$1 \leq \ell < \ell' \leq c(A) \Rightarrow 4^{-j-1} \leq d(t_\ell, t_{\ell'}) \leq 4^{-j+2} \tag{3.10}$$

and that each ball $B(t_\ell, 4^{-j-2})$ contains exactly one child of A. In some sense, $4^{-j(A)}$ tells you at which scale the children of A live. Please note that it may happen that $4^{-j(A)}$ is much smaller than $\Delta(A)$. An example of organized tree is drawn in Fig. 3.2.

If B_1 and B_2 are distinct children of A in an organized tree, then

$$d(B_1, B_2) \geq 4^{-j(A)-2}\,, \tag{3.11}$$

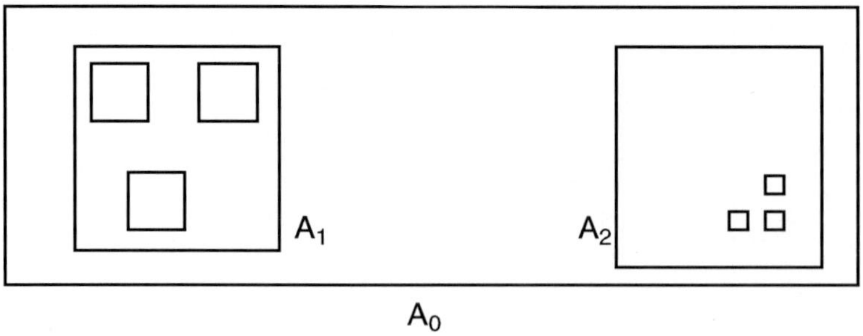

Fig. 3.2 An organized tree. Here $j(A_2) > j(A_1)$

so that an organized tree is also a separated tree, with $s(A) = j(A) + 2$, but the notion of organized tree is more restrictive. (For example, we have no control over the diameter of the children of A in a separated tree.)

We define the depth $\tau(\mathcal{T})$ of an organized tree by

$$\tau(\mathcal{T}) := \inf_{t \in S_{\mathcal{T}}} \sum_{t \in A \in \mathcal{T}} 4^{-j(A)} \sqrt{\log c(A)} . \tag{3.12}$$

Another way to measure the size of T is then

$$\sup\{\tau(\mathcal{T}) \ ; \ \mathcal{T} \text{ organized tree} \subset T\} . \tag{3.13}$$

If we simply view an organized tree $\mathcal{T}$ as a separated tree using (3.11), then $\rho(\mathcal{T}) = \tau(\mathcal{T})/16$ (where $\rho(\mathcal{T})$ is the depth of $\mathcal{T}$ as a separated tree). Thus, we have shown the following:

Proposition 3.1.1 *We have*

$$\sup\{\tau(\mathcal{T}) \ ; \ \mathcal{T} \textit{ organized tree}\} \le 16 \sup\{\rho(\mathcal{T}) \ ; \ \mathcal{T} \textit{ separated tree}\} . \tag{3.14}$$

The next result provides the fundamental connection between trees and the functional γ_2.

Proposition 3.1.2 *We have*

$$\gamma_2(T, d) \le L \sup\{\tau(\mathcal{T}) \ ; \ \mathcal{T} \textit{ organized tree}\} . \tag{3.15}$$

Proof We consider the functional

$$F(A) = \sup\{\tau(\mathcal{T}) \ ; \ \mathcal{T} \subset A \, , \ \mathcal{T} \text{ organized tree}\} ,$$

where we write $\mathcal{T} \subset A$ as a shorthand for "$\forall B \in \mathcal{T}$, $B \subset A$".

Next we prove that this functional satisfies the growth condition (2.77) for $r = 16$ whenever a is of the type 16^{-j}, for $c^* = 1/L$. For this, consider $n \ge 1$ and $m = N_n$. Consider $j \in \mathbb{Z}$ and $t_1, \ldots, t_m \in T$ with

$$1 \le \ell < \ell' \le m \Rightarrow 16^{-j} \le d(t_\ell \, , \, t_{\ell'}) \le 2 \cdot 16^{-j+1} . \tag{3.16}$$

Consider sets $H_\ell \subset B(t_\ell, 2 \cdot 16^{-j-1})$ and $\alpha < \min_{\ell \le m} F(H_\ell)$. Consider, for $\ell \le m$, an organized tree $\mathcal{T}_\ell \subset H_\ell$ with $\tau(\mathcal{T}_\ell) > \alpha$, and denote by A_ℓ its largest element. Next we claim that the tree $\mathcal{T}$ consisting of $C = \bigcup_{\ell \le m} H_\ell$ (its largest element) and the union of the trees $\mathcal{T}_\ell$, $\ell \le m$, is organized, with $j(C) = 2j - 1$ and $A_1, \ldots, A_m$ as children of C (so that $c(C) = m$). To see this, we observe that since $4^{-j(C)-1} = 16^{-j}$ we have $4^{-j(C)-1} \le d(t_\ell, t_{\ell'}) \le 2 \cdot 16^{-j+1} \le 4^{-j(C)+2}$, so that (3.10) holds for C. Furthermore, $A_\ell \subset H_\ell \subset B(t_\ell, 2 \cdot 16^{-j-1}) \subset B(t_\ell, 4^{-j(C)-2})$, so that this ball contains exactly one child of C. Other conditions follow from the fact that the trees $\mathcal{T}_\ell$ are themselves organized. Moreover, $S_{\mathcal{T}} = \bigcup_{\ell \le m} S_{\mathcal{T}_\ell}$.

Consider $t \in S_{\mathcal{T}}$, and let ℓ with $t \in S_{\mathcal{T}_\ell}$. Then $t \in C \in \mathcal{T}$, and also $t \in A \in \mathcal{T}$ whenever $t \in A \in \mathcal{T}_\ell$. Thus, using also in the second line that $j(C) = 2j - 1$ and that $c(C) = m$, we obtain

$$\sum_{t \in A \in \mathcal{T}} 4^{-j(A)} \sqrt{\log c(A)} \geq 4^{-j(C)} \sqrt{\log c(C)} + \sum_{t \in A \in \mathcal{T}_\ell} 4^{-j(A)} \sqrt{\log c(A)}$$

$$\geq 4 \cdot 16^{-j} \sqrt{\log m} + \tau(\mathcal{T}_\ell) \geq \frac{1}{L} 16^{-j} 2^{n/2} + \alpha \,.$$

Since α is arbitrary, we have proved that

$$F\Big(\bigcup_{\ell \leq m} H_\ell\Big) \geq \tau(\mathcal{T}) \geq \frac{1}{L} 16^{-j} 2^{n/2} + \min_{\ell \leq m} F(H_\ell) \,.$$

This completes the proof of the growth condition (2.77).

If one examines the proof of Theorem 2.9.1, one observes that it requires only the growth condition (2.77) to hold true when a is of the type r^{-j}, and we have just proved that this is the case (for $r = 16$), so that from (2.81) we have proved that $\gamma_2(T, d) \leq L(F(T) + \Delta(T))$. It remains only to prove that $\Delta(T) \leq LF(T)$. For this, we simply note that if $s, t \in T$, and j_0 is the largest integer with $4^{-j_0} \geq d(s, t)$, then the tree $\mathcal{T}$ consisting of T, $\{t\}$, $\{s\}$, is organized with $j(T) = j_0$ and $c(T) = 2$, so that $F(T) \geq \tau(\mathcal{T}) \geq 4^{-j_0} \sqrt{\log 2}$ and $4^{-j_0} \leq LF(T)$. □

3.1.3 Majorizing Measures

For a probability measure μ on a metric space (T, d), with countable support,[2] we define for each $t \in T$ the quantity

$$I_\mu(t) := \int_0^\infty \sqrt{\log \frac{1}{\mu(B(t,\epsilon))}} d\epsilon = \int_0^{\Delta(T)} \sqrt{\log \frac{1}{\mu(B(t,\epsilon))}} d\epsilon \,. \tag{3.17}$$

The second equality follows from the fact that $\mu(B(t, \epsilon)) = 1$ when $B(t, \epsilon) = T$, so that then the integrand is 0. It is important to master the mechanism at play in the following elementary exercise:

[2] We assume μ with countable support because we do not need a more general setting. The advantage of this hypothesis is that there are no measurability problems.

Exercise 3.1.3 Consider a number $\Delta > 0$ and non-increasing function $f : [0, \Delta] \to \mathbb{R}^+$. Define $\epsilon_0 = \Delta$ and for $n \geq 1$ define $\epsilon_n = \inf\{\epsilon > 0; f(\epsilon) \leq 2^n\}$. Prove that

$$\frac{1}{2}\sum_{n\geq 1} 2^n \epsilon_n \leq \int_0^{\Delta} f(\epsilon)\mathrm{d}\epsilon \leq 2\sum_{n\geq 0} 2^n \epsilon_n \ .$$

Proposition 3.1.4 *Given a metric space (T, d), we can find on T a probability measure μ, supported by a countable subset of T and such that*

$$\sup_{t\in T} I_\mu(t) = \sup_{t\in T} \int_0^{\infty} \sqrt{\log \frac{1}{\mu(B(t,\epsilon))}}\mathrm{d}\epsilon \leq L\gamma_2(T, d) \ . \tag{3.18}$$

Any probability measure[3] μ on (T, d) is called a *majorizing measure*. The reason for this somewhat unsatisfactory name is that Xavier Fernique proved that for a Gaussian process and a probability measure μ on T, one has

$$\mathsf{E} \sup_{t\in T} X_t \leq L \sup_{t\in T} I_\mu(t) \ , \tag{3.19}$$

so that μ can be used to "majorize" the process $(X_t)_{t\in T}$.[4] This was a major advance over Dudley's bound. The (in)famous theory of majorizing measures used the quantity

$$\inf_{\mu} \sup_{t\in T} I_\mu(t) \tag{3.20}$$

as a measure of the size of the metric space (T, d), where the infimum is over all choices of the probability measure μ. This method is technically quite challenging. We are going to prove that the quantity (3.20) is equivalent to $\gamma_2(T, d)$. A related idea which is still very useful is explained in Sect. 3.3.

Proof Consider an admissible sequence $(\mathcal{A}_n)$ with

$$\forall t \in T \ , \ \sum_{n\geq 0} 2^{n/2}\Delta(A_n(t)) \leq 2\gamma_2(T, d) \ .$$

Let us now pick a point $t_{n,A}$ in each set $A \in \mathcal{A}_n$, for each $n \geq 0$. Since card $\mathcal{A}_n \leq N_n$, for each n, there are at most N_n points of the type $t_{n,A}$. Attributing a mass $1/(2^n N_n)$ to each of them, we obtain a total mass ≤ 1. Thus, there is a probability measure μ on T, supported by a countable set and satisfying $\mu(\{t_{n,A}\}) \geq 1/(2^n N_n)$

[3] To avoid technicalities, one may assume that μ has countable support.

[4] One typically uses the name only when such the right-hand side of (3.19) is usefully small.

for each $n \geq 0$ and each $A \in \mathcal{A}_n$. Then,

$$\forall n \geq 1\,,\ \forall A \in \mathcal{A}_n\,,\ \mu(A) \geq \mu(\{t_{n,A}\}) \geq \frac{1}{2^n N_n} \geq \frac{1}{N_n^2}$$

so that given $t \in T$ and $n \geq 1$,

$$\begin{aligned} \epsilon > \Delta(A_n(t)) \Rightarrow \mu(B(t,\epsilon)) &\geq \frac{1}{N_n^2} \\ \Rightarrow \sqrt{\log \frac{1}{\mu(B(t,\epsilon))}} &\leq 2^{n/2+1}\,. \end{aligned} \tag{3.21}$$

Now, since μ is a probability, $\mu(B(t,\epsilon)) = 1$ for $\epsilon > \Delta(T)$, and then $\log(1/\mu(B(t,\epsilon))) = 0$. Thus

$$\begin{aligned} I_\mu(t) = \int_0^\infty \sqrt{\log \frac{1}{\mu(B(t,\epsilon))}}\, d\epsilon &= \sum_{n \geq 0} \int_{\Delta(A_{n+1}(t))}^{\Delta(A_n(t))} \sqrt{\log \frac{1}{\mu(B(t,\epsilon))}}\, d\epsilon \\ &\leq \sum_{n \geq 1} 2^{(n+1)/2+1} \Delta(A_n(t)) \leq L\gamma_2(T,d) \end{aligned}$$

using (3.21). □

Proposition 3.1.5 *If μ is a probability measure on T (supported by a countable set) and $\mathcal{T}$ is a separated tree on T, then*

$$\rho(\mathcal{T}) \leq L \sup_{t \in T} I_\mu(t)\,.$$

Combining with (3.14), (3.15), and (3.18), this completes the proof that the four "measures of the size of T" considered in this section, namely, (3.9), (3.13), (3.20), and $\gamma_2(T,d)$ are indeed equivalent.

Proof The basic observation is as follows: the sets

$$B(C, 4^{-s(A)-1}) = \{x \in T\ ;\ d(x,C) < 4^{-s(A)-1}\}$$

are disjoint as C varies over the children of A (as follows from (3.6)), so that one of them has measure $\leq c(A)^{-1}$.

We then proceed in the following manner, constructing recursively an appropriate branch of the tree. This is a typical and fundamental way to proceed when working with trees. We start with the largest element A_0 of $\mathcal{T}$. We then select a child A_1 of A_0 with $\mu(B(A_1, 4^{-s(A_0)-1})) \leq 1/c(A_0)$, and a child A_2 of A_1 with $\mu(B(A_2, 4^{-s(A_1)-1})) \leq 1/c(A_1)$, etc., and continue this construction as long as we can. It ends only when we reach a set of $\mathcal{T}$ that has no child and hence by (3.4) is reduced to a single point t which we now fix. For any set A with $t \in A \in \mathcal{T}$, by construction, we have

$$\mu(B(t, 4^{-s(A)-1})) \le \frac{1}{c(A)}$$

so that

$$4^{-s(A)-2}\sqrt{\log c(A)} \le \int_{4^{-s(A)-2}}^{4^{-s(A)-1}} \sqrt{\frac{1}{\log \mu(B(t,\epsilon))}}\,\mathrm{d}\epsilon \;, \tag{3.22}$$

because the integrand is $\ge \sqrt{\log c(A)}$ and the length of the interval of integration is larger than $4^{-s(A)-2}$. By (3.7), the intervals $]4^{-s(A)-2}, 4^{-s(A)-1}[$ are disjoint for different sets A with $t \in A \in \mathcal{T}$, so summation of the inequalities (3.22) yields

$$\frac{1}{16}\rho(\mathcal{T}) \le \sum_{t \in A \in \mathcal{T}} 4^{-s(A)-2}\sqrt{\log c(A)} \le \int_0^\infty \sqrt{\frac{1}{\log \mu(B(t,\epsilon))}}\,\mathrm{d}\epsilon = I_\mu(t) \;. \quad \square$$

In the rest of this chapter, we will implicitly use the previous method of "selecting recursively the branch of the tree we follow" to prove lower bounds without mentioning trees.

We end this section by an exercise completing the proof of (2.143).

Exercise 3.1.6 Consider metric spaces $(T_k, d_k)_{k \le N}$ and probability measures μ_k on T_k. Consider the product probability μ on $T = \prod_{k \le N} T_k$ and the distance (2.142).

(a) Prove that for $t = (t_k)_{k \le N}$, we have

$$I_\mu(t) \le L \sum_{k \le N} I_{\mu_k}(t_k) \;.$$

Hint: Use (2.147).

(b) Complete the proof of (2.143).

3.2 Rolling Up Our Sleeves: Trees in Ellipsoids

It is one thing to have proved abstract results but quite another thing to visualize the combinatorics in concrete situations. Consider an ellipsoid $\mathcal{E}$ as in (2.154), so that according to (2.155), $S := \sqrt{\sum_{i \le N} a_i^2}$ measures its size. Assuming $S < \infty$, the goal of the present section is to construct explicitly an organized tree $\mathcal{T}$ whose depth $\tau(\mathcal{T})$ witnesses the size of the ellipsoid, i.e., $\tau(\mathcal{T}) \ge S/L$. This elaborate exercise will have us confront a number of technical difficulties.

We first reduce to the case where each a_i is of the type 2^{-k} for some $k \in \mathbb{Z}$. Let $\mathbb{N}^* = \mathbb{N} \setminus \{0\}$. For $k \in \mathbb{Z}$, let us set $I_k = \{i \in \mathbb{N}^*; 2^{-k} \le a_i \le 2^{-k-1}\}$, so that the sets I_k cover $\mathbb{N}^*$ and $\sum_k 2^{-2k} \operatorname{card} I_k \ge \sum_{i \ge 1} a_i^2/4 \ge S^2/4$, while at the same time

$$\mathcal{E} \supset \mathcal{E}' = \Big\{ t \in \ell^2 \; ; \; \sum_k 2^{2k} \sum_{i \in I_k} t_i^2 \le 1 \Big\} . \tag{3.23}$$

Thus, we have reduced the problem to considering only ellipsoids of the type $\mathcal{E}'$. Here comes a somewhat unexpected argument. We are going to replace $\mathcal{E}'$ by a set of the type

$$\mathcal{P} = \Big\{ t \in \ell^2 \; ; \; \forall k \, , \; 2^{2k} \sum_{i \in I_k} t_i^2 \le \alpha_k \Big\} \, ,$$

where $\sum_k \alpha_k = 1$ (a condition which ensures that $\mathcal{P} \subset \mathcal{E}'$). At first sight, the set $\mathcal{P}$ looks much smaller than $\mathcal{E}'$, but this is not the case. First, how should we chose α_k to ensure that $\mathcal{P}$ is as large as possible? Considering independent Gaussian r.v.s (g_i), we have $\sup_{t \in \mathcal{P}} \sum_{i \ge 1} t_i g_i = \sum_k \sqrt{\alpha_k} 2^{-k} \sqrt{\sum_{i \in I_k} g_i^2}$, so that since $\mathsf{E}\sqrt{\sum_{i \in I_k} g_i^2}$ is about $\sqrt{\operatorname{card} I_k}$ by (2.155), we obtain

$$\mathsf{E} \sup_{t \in \mathcal{P}} \sum_{i \ge 1} t_i g_i \text{ is about } \sum_k \sqrt{\alpha_k} 2^{-k} \sqrt{\operatorname{card} I_k} \, . \tag{3.24}$$

So to maximize this quantity (and, in a sense, the size of $\mathcal{P}$), it is a good idea to choose $\alpha_k = 2^{-2k} \operatorname{card} I_k / S'^2$ where $S'^2 = \sum_\ell 2^{-2\ell} \operatorname{card} I_\ell$. Then, $\mathcal{P}$ takes the form

$$\mathcal{P} = \Big\{ t \in \ell^2 \; ; \; \forall k \, , \; \sum_{i \in I_k} t_i^2 \le \frac{2^{-4k}}{S'^2} \operatorname{card} I_k \Big\} \, .$$

This set is very simple: geometrically, $\mathcal{P}$ is a product of spheres of dimension $\operatorname{card} I_k$ and radius r_k, where r_k is defined by $r_k^2 = 2^{-4k} \operatorname{card} I_k / S'^2$. It will be very useful to reduce to the case where the radii of these spheres are quite different from each other. This is the purpose of the next lemma.

Lemma 3.2.1 *There is a subset J of $\mathbb{Z}$ with the following two properties:*

$$\sum_{k \in J} 2^{-2k} \operatorname{card} I_k \ge S'^2 / L \, . \tag{3.25}$$

$$k, n \in J \; , \; k < n \Rightarrow r_n \le 2^{-6} r_k \, . \tag{3.26}$$

Proof Since $S < \infty$, the sequence $a_k = 2^{-2k} \operatorname{card} I_k$ is bounded. We apply Lemma 2.9.5 (or more precisely the version of this lemma where the index set is

$\mathbb{Z}$ rather than $\mathbb{N}$) with $\alpha = 2$ to find a finite subset $I \subset \mathbb{Z}$ such that

$$k, n \in I\,,\; k \neq n \Rightarrow a_n < a_k 2^{|n-k|}\,. \tag{3.27}$$

and

$$\sum_{k \in I} a_k \geq S'^2/L\,. \tag{3.28}$$

Consider $k, n \in I$ with $k < n$. Then $a_n < 2^{n-k} a_k$, and recalling the value of a_n, this means that card $I_n \leq 2^{3(n-k)}$ card I_k. Recalling the value of r_k, this means that $r_n^2 \leq r_k^2/2$. Let us now enumerate I as a finite sequence $(n(\ell))_{\ell \leq N}$ in increasing order, so that $r_{n(\ell+1)} \leq r_{n(\ell)}/\sqrt{2}$ and $r_{n(\ell+12)} \leq 2^{-6} r_{n(\ell)}$. For $0 \leq p \leq 11$, consider the set $J_p = \{n(p+12q);\, q \geq 0,\, p+12q \leq N\}$, so that $I = \cup_{0 \leq p \leq 11} J_p$. Consequently, $\sum_{k \in I} a_k = \sum_{0 \leq p \leq 11} \sum_{k \in J_p} a_k$. Thus, using (3.28), there exists some p, $0 \leq p \leq 11$ such that $\sum_{k \in J_p} a_k \geq S'^2/L$, and the set J_p satisfies the desired requirements. □

Consider a set J as constructed in Lemma 3.2.1. We then replace $\mathcal{P}$ by the subset $\mathcal{P}'$ consisting of the points $t \in \mathcal{P}$ such that $t_i = 0$ when $i \in I_k$ and $k \notin J$. We note that $\sum_{k \in J} r_k \sqrt{\operatorname{card} I_k} = \sum_{k \in J} 2^{-2k} \operatorname{card} I_k / S' \geq S'/L$, where the last inequality follows from (3.25).

We have now finished our preliminary reductions. To construct inside any ellipsoid $\mathcal{E}$ an organized tree $\mathcal{T}$ such that its depth $\tau(\mathcal{T})$ witnesses the size of $\mathcal{E}$, it suffices to perform the same task for a set of the type

$$\mathcal{P}' = \Big\{ t \in \ell^2(I^*)\,;\; \forall k \leq N\,,\; \sum_{i \in I_k} t_i^2 \leq r_k^2 \Big\}\,,$$

where N is a given integer, where $(I_k)_{k \leq N}$ are disjoint subsets of $\mathbb{N}^*$ of union I^*, and where $r_{k+1} \leq 2^{-6} r_k$. Just as in (3.24), the size of $\mathcal{P}'$ is about $\sum_{k \leq N} r_k \sqrt{\operatorname{card} I_k}$.

For $k \leq N$, let us consider the sphere

$$\mathbb{S}_k = \Big\{ t \in \ell^2(I^*)\,;\; \sum_{i \in I_k} t_i^2 \leq r_k^2\,,\; i \notin I_k \Rightarrow t_i = 0 \Big\}\,.$$

It follows from the volume argument (2.45) (used for $A = B$ and $\epsilon = 1/2$) that there is a subset U_k of $\mathbb{S}_k$ with

$$\operatorname{card} U_k \geq 2^{\operatorname{card} I_k}\,, \tag{3.29}$$

such that any two distinct points of U_k are at distance $\geq r_k/2$. Given $1 \leq m \leq N$ and for $k \leq m$ given $y_k \in U_k$, $y_k = (y_{k,i})_{i \in I^*}$ consider the set $A = A(y_1, \ldots, y_m) \subset \mathcal{P}'$

defined as

$$A(y_1, \ldots, y_m) = \left\{ t \in \ell^2(I^*) \; ; \; \forall k \le m \; , \; \forall i \in I_k \; , \; t_i = y_{k,i} \right\} .$$

We will show that these sets, together with the set $\mathcal{P}'$, form an organized tree $\mathcal{T}$ (as defined in the previous section). When $m < N$, we have $c(A) = \operatorname{card} U_{m+1}$: the children of A are the sets $A(y_1, \ldots, y_m, y)$ where $y \in U_{m+1}$. When $m = N$, we have $c(A) = 0$ and $A(y_1, \ldots, y_N)$ consists of single point. We now check the condition (3.10). Consider $m < N$. Define $j(A)$ as the smallest integer j such that $4^{-j-1} \le r_{m+1}/2$, so that $r_{m+1} \le 2 \cdot 4^{-j}$. For $y \in U_{m+1}$, consider the unique point $t(y) \in A(y_1, \ldots, y_m, y)$ such that $t(y)_i = 0$ if $i \in I_k$ for $k > m+1$. Then for $y, y' \in U_{m+1}$, $y \ne y'$, the distance $d(t(y), t(y'))$ is the same as the distance between y and y', which are different points of $U_{m+1} \subset \mathbb{S}_{m+1}$. Thus

$$4^{-j(A)-1} \le r_{m+1}/2 \le d(t(y), t(y')) \le 2r_{m+1} \le 4^{-j(A)+1} .$$

Furthermore, recalling that $r_{k+1} \le 2^{-6} r_k$ (so that in particular $\sum_{k \ge m} r_k \le 2r_m$) if $x \in A(y_1, \ldots, y_m, y)$, then

$$\|x - t(y)\| \le \sum_{k \ge m+2} r_k \le 2r_{m+2} \le 2 \cdot 2^{-6} r_{m+1} \le 2^{-6} 4^{-j(A)+1} = 4^{-j(A)-2} ,$$

so that $A(y_1, \ldots, y_m, y) \subset B(t(y), 4^{-j(A)-2})$ as required. Let us now study $\tau(\mathcal{T})$. A branch in $\mathcal{T}$ is defined by a point $t \in S_{\mathcal{T}}$, which is the unique point of a set of the type $A(y_1, \ldots, y_N)$. Let us set $A_0 = \mathcal{P}'$ and $A_m := A(y_1, \ldots, y_m)$ for $1 \le m < N$. Then $t \in A_m$ for $0 \le m < N$. Also, $c(A_m) = \operatorname{card} U_{m+1}$ and $4^{-j(A_m)} \ge r_{m+1}/2$. Thus, using (3.29) in the last equality,

$$\begin{aligned} \sum_{t \in A \in \mathcal{T}} 4^{-j(A)} \sqrt{\log c(A)} &\ge \sum_{0<m<N} 4^{-j(A_m)} \sqrt{\log c(A_m)} \\ &\ge \sum_{0<m<N} r_{m+1} \sqrt{\log \operatorname{card} U_{m+1}} / L \ge \sum_{1 \le k \le N} r_k \sqrt{\operatorname{card} I_k} / L , \end{aligned}$$

and, as we have seen, this last quantity is the size of $\mathcal{P}'$.

3.3 Fernique's Functional

3.3.1 *Fernique's Functional*

In Sect. 3.1, we presented four equivalent methods to measure the size of a metric space. (Besides our usual $\gamma_2(T, d)$, these were the maximum depth of a separated or organized tree contained in T and majorizing measures.) Recalling (3.17), it will turn out that a fifth measure of size, the quantity

$$\mathrm{Fer}(T,d) := \sup_{\mu} \int_T I_\mu(t)\mathrm{d}\mu(t) \tag{3.30}$$

will play an important role. Here, the supremum is over all probability measures on T which are supported by a countable set.

Why is the functional (3.30) important? This is far from obvious. In the context of Gaussian processes, this functional has no special importance, and the related notion of majorizing measures is not particularly useful. We will first understand the usefulness of Fernique's functional in Chap. 5 while studying a class of processes which are conditionally Gaussian. Furthermore, in the following chapters, we will be able to use similar ideas in far more general situations, while the proper generalization of the other functionals has not been found yet. In this section, we will prove that Fernique's functional is equivalent to the functional $\gamma_2(T,d)$:

$$\frac{1}{L}\gamma_2(T,d) \le \mathrm{Fer}(T,d) \le L\gamma_2(T,d)\ . \tag{3.31}$$

We will not give the simplest possible proof of this fact. Rather we will prepare for future work by giving arguments which contain in germ the ideas which will prove fruitful. The ideas of this section will not be critically used before Chap. 11.

A further understanding of Fernique's functional will be reached in Sect. 3.5 where we will basically show the remarkable fact that the supremum in the right-hand side of (3.30) is obtained when μ is the "law of the supremum", i.e., the law of a r.v. such that $X_\tau = \sup_{t\in T} X_t$ (see Theorem 3.5.1).

The right-hand side inequality in (3.31) is the easiest and is a consequence of the following:

Proposition 3.3.1 *Consider a probability measure μ on a metric space (T,d). Then*

$$\int_T I_\mu(t)\mathrm{d}\mu(t) \le L\gamma_2(T,d)\ . \tag{3.32}$$

Proof For each $t \in T$, we define $\epsilon_0(t) = \Delta(T)$, and for $n \ge 1$, we define

$$\epsilon_n(t) = \inf\left\{\epsilon > 0\ ;\ \mu(B(t,\epsilon)) \ge N_{n+1}^{-1}\right\}\ . \tag{3.33}$$

Thus, $\sqrt{\log(1/\mu(B(t,\epsilon)))} \le L2^{n/2}$ for $\epsilon \ge \epsilon_n(t)$ and then

$$I_\mu(t) = \int_0^{\Delta(T)} \sqrt{\frac{1}{\log(\mu(B(t,\epsilon)))}}\mathrm{d}\epsilon = \sum_{n\ge 0}\int_{\epsilon_{n+1}(t)}^{\epsilon_n(t)} \sqrt{\frac{1}{\log(\mu(B(t,\epsilon)))}}\mathrm{d}\epsilon$$
$$\le L\sum_{n\ge 0} 2^{n/2}\epsilon_n(t)\ . \tag{3.34}$$

Consider an admissible sequence $(\mathcal{A}_n)$ of partitions with

$$\forall t \in T \ , \ \sum_{n\geq 0} 2^{n/2}\Delta(A_n(t)) \leq 2\gamma_2(T,d) \ . \tag{3.35}$$

Let us fix $n \geq 0$ and set

$$T_n = \{t \in T \ ; \ \mu(A_n(t)) \geq N_{n+1}^{-1}\}; \ T_n' = T \setminus T_n = \{t \in T \ ; \ \mu(A_n(t)) < N_{n+1}^{-1}\}.$$

Thus, T_n' is the union of the sets $A \in \mathcal{A}_n$ which are of measure $\leq N_{n+1}^{-1}$. Since $\operatorname{card}\mathcal{A}_n \leq N_n$, we have $\mu(T_n') \leq N_n N_{n+1}^{-1} = N_n^{-1}$. Also, by definition of $\epsilon_n(t)$, we have $\epsilon_n(t) \leq \Delta(A_n(t))$ if $t \in T_n$ and $\epsilon_n(t) \leq \Delta(T)$ if $t \in T_n'$. Consequently,

$$\begin{aligned}\int_T 2^{n/2}\epsilon_n(t)\mathrm{d}\mu(t) &= \int_{T_n} 2^{n/2}\epsilon_n(t)\mathrm{d}\mu(t) + \int_{T_n'} 2^{n/2}\epsilon_n(t)\mathrm{d}\mu(t) \\ &\leq \int_{T_n} 2^{n/2}\Delta(A_n(t))\mathrm{d}\mu(t) + L\Delta(T)2^{n/2}N_n^{-1} \ .\end{aligned}$$

Combining with (3.34), we obtain

$$\int_T I_\mu(t)\mathrm{d}\mu(t) \leq L\Delta(T) + L\int_T \sum_{n\geq 1} 2^{n/2}\Delta(A_n(t))\mathrm{d}\mu(t) \ . \tag{3.36}$$

Integrating (3.35) with respect to μ proves that the last term of (3.36) is $\leq K\gamma_2(T,d)$ and concludes the proof since $\Delta(T) \leq L\gamma_2(T,d)$. □

3.3.2 *Fernique's Convexity Argument*

Our approach to the left-hand side of (3.31) is based on the following elementary fact, which is a consequence of the Hahn-Banach theorem:

Lemma 3.3.2 *Consider a number $a > 0$. Consider a set $\mathcal{S}$ of functions on a finite set T. Assume that for each probability measure ν on T, there exists $f \in \mathcal{S}$ such that $\int f\mathrm{d}\nu \leq a$. Then for each $\epsilon > 0$, there is a function f in the convex hull of $\mathcal{S}$ such that $f \leq a + \varepsilon$.*

Proof Denote $\mathcal{S}^+$ the set of functions g such that there exists $f \in \mathcal{S}$ with $f \leq g$. Denote by $\mathcal{C}$ the closed convex hull of $\mathcal{S}^+$. We prove that the constant function a equal to a everywhere belongs to $\mathcal{C}$. We proceed by contradiction. If this is not the case, by the Hahn-Banach theorem, we may separate $\mathcal{C}$ and a. That is, there exists a linear functional φ on the space of functions on T such that $\varphi(f) > \varphi(\mathsf{a})$ for $f \in \mathcal{C}$. Consider then a function g on T with $g \geq 0$. For each $\lambda > 0$ and each

$f \in \mathcal{S}$, we have $f + \lambda g \geq f$ so that $f + \lambda g \in \mathcal{C}$ and hence $\varphi(f) + \lambda\varphi(g) > \varphi(\mathsf{a})$. This proves that $\varphi(g) \geq 0$, i.e., φ is positive. Since T is finite, φ is of the form $\varphi(g) = \sum_{t\in T} \alpha_t g(t)$ for numbers $\alpha_t \geq 0$. Setting $\beta = \sum_{t\in T} \alpha(t)$, consider the probability measure ν on T given by $\nu(\{t\}) = \alpha_t/\beta$ for $t \in T$. Then $\varphi(g) = \beta \int g\mathrm{d}\nu$ for each function g on T. Taking $g = \mathsf{a}$ shows that $\beta a = \varphi(\mathsf{a})$. By hypothesis, there exists $f \in \mathcal{C}$ with $\int f\mathrm{d}\nu \leq a$. Then $\varphi(f) = \beta \int f\mathrm{d}\nu \leq \beta a = \varphi(\mathsf{a})$, a contradiction.

So we have proved that $\mathsf{a} \in \mathcal{C}$, the *closure* of the convex hull of $\mathcal{S}^+$. Consequently, there is one point of this convex hull which is $\leq a + \epsilon$ everywhere. The result follows. □

Of course, the hypothesis that T is finite is inessential; it is just to avoid secondary complications.

Let us give a version of the basic lemma sufficiently general to cover all our needs.

Lemma 3.3.3 *Consider a finite metric space (T, d). Consider a convex function $\Phi :]0, 1] \to \mathbb{R}^+$. Assume that for each probability measure μ on T and a certain number D, one has*

$$\int_T \mathrm{d}\mu(t) \int_0^{\Delta(T)} \Phi(\mu(B(t,\epsilon)))\mathrm{d}\epsilon \leq D \,. \tag{3.37}$$

Then there exists a probability measure μ on T for which

$$\sup_{t\in T} \int_0^{\Delta(T)} \Phi(\mu(B(t,\epsilon)))\mathrm{d}\epsilon \leq 2D \,. \tag{3.38}$$

Proof Let us denote by $\mathcal{M}(T)$ the set of probability measures on T. The class $\mathcal{C}$ of functions f on T that satisfy

$$\exists \mu \in \mathcal{M}(T) \; ; \; \forall t \in T \; , \; f_\mu(t) := \int_0^{\Delta(T)} \Phi(\mu(B(t,\epsilon)))\mathrm{d}\epsilon \leq f(t)$$

is convex. This is immediate to check using the convexity of Φ. For each probability measure ν on T, there exists f in $\mathcal{C}$ with $\int f\mathrm{d}\nu \leq B$: this is true for $f = f_\nu$ by (3.37). Consequently by Lemma 3.3.2, there exists $f \in \mathcal{C}$ such that $f \leq 2B$, which is the content of the lemma. □

Corollary 3.3.4 *Consider a finite metric space (T, d). Assume that for a certain number C and for each probability measure μ on T, we have*

$$\int_T I_\mu(t)\mathrm{d}\mu(t) \leq C \,. \tag{3.39}$$

Then there is probability measure μ on T such that

$$\forall t \in T \; , \; I_\mu(t) \leq 2C + 2\Delta(T) \,. \tag{3.40}$$

Proof Calculus shows that the function $\Phi(x) = \sqrt{\log(e/x)}$ is convex for $x \in]0, 1]$ and $\sqrt{\log(1/x)} \le \Phi(x) \le 1 + \sqrt{\log(1/x)}$. Thus

$$I_\mu(t) \le \int_0^{\Delta(T)} \Phi(\mu(B(t,\epsilon)))\mathrm{d}\epsilon \le \Delta(T) + I_\mu(t) ,$$

so that (3.39) implies that (3.37) holds for $D = C + \Delta(T)$ and (3.40) then follows from (3.38). □

Lemma 3.3.5 *We have* $\Delta(T) \le L\mathrm{Fer}(T, d)$.

Proof Consider $s, u \in T$ with $d(s, u) \ge \Delta(T)/2$ and the probability μ on T such that $\mu(\{s\}) = \mu(\{u\}) = 1/2$. For $\epsilon < \Delta(T)/2 \le d(s, u)$, we have $\mu(B(s, \epsilon)) \le 1/2$, and this implies that $I_\mu(s) \ge \Delta(T)\sqrt{\log 2}/2$. Similarly we have $I_\mu(u) \ge \Delta(T)/L$ so that $\int_T I_\mu(t)\mathrm{d}\mu(t) \ge \Delta(T)/L$. □

Proof of (3.31) When T is Finite Combining Corollary 3.3.4 and Lemma 3.3.5, we obtain that there exist a probability μ on T such that $\sup_{t\in T} I_\mu(t) \le L\,\mathrm{Fer}(T, d)$. On the other hand, we have proved in Sect. 3.1 that $\gamma_2(T, d) \le L \sup_{t\in T} I_\mu(t)$.[5] □

3.3.3 From Majorizing Measures to Sequences of Partitions

In Sect. 3.1, we have proved that given a probability measure μ on a metric space T, we have

$$\gamma_2(T, d) \le L \sup_{t\in T} I_\mu(t) . \tag{3.41}$$

We do not know how to generalize the arguments of the proof to the more general settings we will consider later. We give now a direct proof, following a scheme which we know how to generalize. The contents of this section will not be relevant until Chap. 11. First, we prove that

$$\Delta(T) \le L \sup_{t\in T} I_\mu(t) . \tag{3.42}$$

For this, we consider $s, t \in T$ with $d(s, t) > \Delta(T)/2$ so that since the balls $B(t, \Delta(T)/4)$ and $B(s, \Delta(T)/4)$ are disjoint, one of them, say the first one, has a measure $\le 1/2$. Then $\mu(B(t, \epsilon)) \le 1/2$ for $\epsilon \le \Delta(T)/4$ and thus $I_\mu(t) \ge \sqrt{\log 2}\Delta(T)/4$. We have proved (3.42).

[5] The argument by which we have proved this inequality will not generalize, but fortunately there is another route, which is described in the next section.

We start the main argument. We will construct an admissible sequence $(\mathcal{A}_n)$ of partitions of T which witnesses (3.41). For $A \in \mathcal{A}_n$, we also construct an integer $j_n(A)$ as follows: First, we set $\mathcal{A}_0 = \mathcal{A}_1 = \{T\}$ and $j_0(T) = j_1(T) = j_0$, the largest integer with $\Delta(T) \le 2^{-j_0(T)}$. Next for $n \ge 1$, we require the conditions

$$A \in \mathcal{A}_n \Rightarrow \Delta(A) \le 2^{-j_n(A)+2} , \tag{3.43}$$

$$t \in A \in \mathcal{A}_n \Rightarrow \mu(B(t, 2^{-j_n(A)})) \ge N_{n-1}^{-1} . \tag{3.44}$$

The construction proceeds as follows: Having constructed $\mathcal{A}_n$, we split each element A of $\mathcal{A}_n$ into at most N_n pieces, ensuring that card $\mathcal{A}_{n+1} \le N_n^2 = N_{n+1}$. For this, we set

$$A_0 = \{t \in A; \mu(B(t, 2^{-j_n(A)-1})) \ge 1/N_n\} . \tag{3.45}$$

We claim first that we may cover A_0 by $< N_n$ sets, each of diameter $\le 2^{-j_n(A)+1}$. For this, we consider a subset W of A_0, maximal with respect to the property that any two points of W are at distance $> 2^{-j_n(A)}$. The balls of radius $2^{-j_n(A)-1}$ centered at the points of W are disjoint, and each of them is of measure $> N_n^{-1}$ by (3.45), so that there are $< N_n$ of them. Since W is maximum, the balls of radius $2^{-j_n(A)}$ centered at the points of W cover A_0, and each of them has diameter $\le 2^{-j_n(A)+1}$. Thus, there exists a partition of A_0 in $< N_n$ sets of diameter $\le 2^{-j_n(A)+1}$. The required partition of A consists of these sets B and of $A_1 = A \setminus A_0$. For each set B, we set $j_{n+1}(B) = j_n(A) + 1$, and we set $j_{n+1}(A_1) = j_n(A)$, so that conditions (3.43) and (3.44) hold.

This completes the construction. The important point is that

$$\begin{aligned} B \in \mathcal{A}_{n+1}, B \subset A \in \mathcal{A}_n, j_{n+1}(B) = j_n(A) \Rightarrow \mu(&B(t, 2^{-j_{n+1}(B)})) \\ &= \mu(B(t, 2^{-j_n(A)})) \le N_n^{-1} . \end{aligned} \tag{3.46}$$

This property holds because if $t \in A$ and $\mu(B(t, 2^{-j_n(A)})) > N_n^{-1}$, then $t \in A_0$ and the element B of $\mathcal{A}_{n+1}$ which contains t has been assigned a value $j_{n+1}(B) = j_n(A) + 1$.

To prove (3.41), we will prove that

$$\forall t \in T , \sum_{n \ge 0} 2^{n/2} \Delta(A_n(t)) \le L I_\mu(t) .$$

We fix $t \in T$. We set $j(n) = j_n(A_n(t))$ and $a(n) = 2^{n/2} 2^{-j(n)}$. Using (3.43), it suffices to prove that

$$\sum_{n \ge 0} a(n) \le L I_\mu(t) . \tag{3.47}$$

We consider $\alpha = \sqrt{2}$ and the corresponding set I as in (2.84) (leaving to the reader to prove that the sequence $(a(n))$ is bounded, which will become clear soon). Thus, as in (2.91), we have

$$n \in I, n \geq 1 \Rightarrow j(n+1) = j(n) + 1 \ , \ j(n-1) = j(n) \ . \tag{3.48}$$

We enumerate $I \setminus \{0\}$ as a sequence $(n_k)_{k\geq 0}$ (leaving again to the reader the easier case where I is finite), so that $j(n_{k+1}) \geq j(n_k + 1) = j(n_k) + 1$ and

$$\sum_{n\geq 0} a(n) \leq La(0) + L\sum_{k\geq 1} a(n_k) \ . \tag{3.49}$$

From (3.48), we have $j(n_k - 1) = j(n_k)$. Using (3.46) for $n = n_k - 1$, we obtain

$$\mu(B(t, 2^{-j(n_k)})) < N_{n_k-1}^{-1} \ ,$$

so that $\sqrt{\log(1/\mu(B(t,\epsilon))} \geq 2^{n_k/2}/L$ for $\epsilon < 2^{-j(n_k)}$. Since $j(n_{k+1}) > j(n_k)$, this implies

$$\begin{aligned} a(n_k) = 2^{n_k/2 - j(n_k)} &\leq 2 \cdot 2^{n_k/2}(2^{-j(n_k)} - 2^{j(n_{k+1})}) \\ &\leq L \int_{2^{-j(n_{k+1})}}^{2^{-j(n_k)}} \sqrt{\log \frac{1}{\mu(B(t,\epsilon))}} \mathrm{d}\mu(\epsilon) \ . \end{aligned}$$

Summation of these inequalities and use of (3.42) and (3.49) proves (3.47). □

3.4 Witnessing Measures

Proposition 3.4.1 *([65]) For a metric space (T, d), define*

$$\delta_2(T, d) = \sup_{\mu} \inf_{t\in T} I_\mu(t) \ , \tag{3.50}$$

where the supremum is taken over all probability measures μ on T.[6] *Then*

$$\frac{1}{L}\gamma_2(T, d) \leq \delta_2(T, d) \leq L\gamma_2(T, d) \ . \tag{3.51}$$

[6] Please observe that the order of the infimum and the supremum is not as in (3.20).

It is obvious that $\inf_{t\in T} I_\mu(t) \le \int_T I_\mu(t)\mathrm{d}\mu(t)$ so that $\delta_2(T,d) \le \mathrm{Fer}(T,d)$. Thus, the right-hand side of (3.51) follows from the right-hand side of (3.31), while the left-hand side of (3.31) follows from the left-hand side of (3.51).

The most important consequence of (3.31) is that there exists a probability measure μ on T for which $\inf_{t\in T} I_\mu(t) \ge \gamma_2(T,d)/L$. Such a probability measure "witnesses that the value of $\gamma_2(T,d)$ is large" because of the right-hand side of (3.51). In this spirit, we will call μ a *witnessing measure*, and we define its "size" as the quantity $\inf_{t\in T} I_\mu(t)$.[7] Witnessing measures can be magically convenient. One of the first advances the author made beyond the results of [132] was the realization that witnessing measures yield a proof of Theorem 5.2.1 below an order of magnitude easier than the original proof. This approach is now replaced by the use of Fernique's functional because, unfortunately, we do not know how to extend the idea of witnessing measure to settings where multiple distances will be considered. Finding proper generalizations of Proposition 3.4.1 to more general settings is an attractive research problem (see in particular Problem 10.15.4).

Proof of Proposition 3.4.1 The right-hand side inequality follows from Proposition 3.3.1 and the trivial fact that $\inf_{t\in T} I_\mu(t) \le \int I_\mu(t)\mathrm{d}\mu(t)$. The reader should review the material of Sect. 3.1 to follow the proof of the converse. Recalling (3.5), given an organized tree $\mathcal{T}$, we define a measure μ on T by $\mu(A) = 0$ if $A \cap S_{\mathcal{T}} = \emptyset$ and by

$$t \in S_{\mathcal{T}} \Rightarrow \mu(\{t\}) = \frac{1}{\prod_{t\in A\in\mathcal{T}} c(A)} \,.$$

The intuition is that the mass carried by $A \in \mathcal{T}$ is equally divided between the children of A. Then, $I_\mu(t) = \infty$ if $t \notin S_{\mathcal{T}}$. Consider $t \in A \in \mathcal{T}$ and $j = j(A)$. Then, since $\mathcal{T}$ is an organized tree, $B(t, 4^{-j-2})$ meets only one child of A, so that $\mu(B(t,4^{-j-2})) \le 1/c(A)$. Copying the argument of (3.22) readily implies that $LI_\mu(t) \ge \sum_{t\in A\in\mathcal{T}} 4^{-j(A)}\sqrt{\log c(A)}$ from which the result follows by (3.15). □

Exercise 3.4.2 For a metric space (T,d), define

$$\chi_2(T,d) = \sup_\mu \inf \int \sum_{n\ge 0} 2^{n/2}\Delta(A_n(t))\mathrm{d}\mu(t) \,,$$

where the infimum is taken over all admissible sequences and the supremum over all probability measures. It is obvious that $\chi_2(T,d) \le \gamma_2(T,d)$. Prove that $\gamma_2(T,d) \le L\chi_2(T,d)$. Hint: Prove that the functional $\chi_2(T,d)$ satisfies the appropriate growth condition. Warning: The argument takes about half a page and is fairly non-trivial.

[7] Thus, a probability measure μ on T is both a majorizing and a witnessing measure. It bounds $\gamma_2(T,d)$ from above by $L\sup_{t\in T} I_\mu(t)$ and from below by $\inf_{t\in T} I_\mu(t)/L$. Furthermore, one may find μ such that these two bounds are of the same order.

3.5 An Inequality of Fernique

We end up this chapter with a beautiful inequality of Fernique. It will not be used anywhere else in this work, but is presented to emphasize Fernique's lasting contributions to the theory of Gaussian processes.

Theorem 3.5.1 *Consider a Gaussian process $(X_t)_{t\in T}$. Provide T with the canonical distance associated with this process, and consider a probability measure μ on T. Consider a r.v. τ of law μ. Then for any probability measure ν on T, we have*

$$\mathsf{E}X_\tau \le L\int I_\nu(t)\mathrm{d}\mu(t) + L\Delta(T,d)\ . \tag{3.52}$$

Here of course, X_τ is the r.v. $X_{\tau(\omega)}(\omega)$. We leave some technical details aside, and prove the result only when T is finite. The basic principle is as follows:

Lemma 3.5.2 *Consider a standard Gaussian r.v. g and a set A. Then, $\mathsf{E}\mathbf{1}_A|g| \le L\mathsf{P}(A)\sqrt{\log(2/\mathsf{P}(A))}$.*

Proof We write

$$\mathsf{E}\mathbf{1}_A|g| = \int_0^\infty \mathsf{P}(A\cap\{|g|\ge t\})\mathrm{d}t \le \int_0^\infty \min(\mathsf{P}(A), 2\exp(-t^2/2))\mathrm{d}t\ .$$

Letting $\alpha = \sqrt{2\log(2/\mathsf{P}(A))}$, we split the integral in the regions $t\le\alpha$ and $t>\alpha$. We bound the first part by $\alpha\mathsf{P}(A)$ and the second by

$$\int_\alpha^\infty 2\exp(-t^2/2)\mathrm{d}t \le \frac{1}{\alpha}\int_\alpha^\infty 2t\exp(-t^2/2)\mathrm{d}t = \mathsf{P}(A)/\alpha \le L\mathsf{P}(A)\alpha\ . \qquad \square$$

Corollary 3.5.3 *Consider Gaussian r.v.s $(g_i)_{i\le N}$ with $\mathsf{E}g_i^2 \le a^2$. Consider a r.v. τ valued in $\{1,\dots,N\}$, and let $\alpha_i = \mathsf{P}(\tau = i)$. Then $\mathsf{E}|g_\tau| \le La\sum_{i\le N}\alpha_i\sqrt{\log 2/\alpha_i}$.*

Proof Since $|g_\tau| = \sum_{i\le N}\mathbf{1}_{\{\tau=i\}}|g_i|$, and using Lemma 3.5.2 to obtain $\mathsf{E}\mathbf{1}_{\{\tau=i\}}|g_i| \le La\alpha_i\sqrt{\log 2/\alpha_i}$. $\square$

We also need the following elementary convexity inequality:

Lemma 3.5.4 *Consider numbers $\alpha_i > 0$ with $\sum_{i\le N}\alpha_i \le \alpha \le 1$. Then*

$$\sum_{i\le N}\alpha_i\sqrt{\log(2/\alpha_i)} \le \alpha\sqrt{\log(2N/\alpha)}\ . \tag{3.53}$$

Proof Calculus shows that the function $\varphi(x) = x\sqrt{\log(2/x)}$ is concave increasing for $x\le 1$, so that if $\alpha' = \sum_{i\le N}\alpha_i$, then $N^{-1}\sum_{i\le N}\varphi(\alpha_i) \le \varphi(\alpha'/N) \le \varphi(\alpha/N)$. $\square$

The slightly technical part of the proof of Theorem 3.5.1 is contained in the following:

Lemma 3.5.5 *Consider probability measures μ and ν on a metric space (T,d). Then there exist $n_0 \in \mathbb{Z}$ and a sequence of partitions $(\mathcal{B}_n)_{n\ge n_0}$ (which need not be increasing) of T with the following properties: First, $\mathcal{B}_{n_0}$ contains only one set. Next, the sets of $\mathcal{B}_n$ are of diameter $\le 2^{-n+2}$. Finally,*

$$\sum_{n\ge n_0} 2^{-n} \sum_{A\in\mathcal{B}_n, B\in\mathcal{B}_{n+1}} \mu(A\cap B)\sqrt{\log(2/\mu(A\cap B))}$$
$$\le L\int I_\nu(t)\mathrm{d}\mu(t) + L\Delta(T,d) \ . \tag{3.54}$$

Proof We consider the largest integer n_0 with $2^{-n_0} \ge \Delta(T,d)$. We set $\mathcal{B}_{n_0} = \{T\}$. For $n > n_0$, we proceed as follows: for $k \ge 0$, we set

$$T_{n,k} = \{t \in T \ ; \ 1/N_{k+1} < \nu(B(t,2^{-n})) \le 1/N_k\} \ .$$

The sets $(T_{n,k})_{k\ge 0}$ form a partition of T. Consider a subset V of $T_{n,k}$ such that the points of V are at mutual distance $> 2^{-n+1}$. The balls of radius 2^{-n} centered at the points of V are disjoint, and by definition of $T_{n,k}$, they have a ν-measure $> 1/N_{k+1}$. Thus, card $V \le N_{k+1}$, and according to Lemma 2.9.3, $e_{k+1}(T_{n,k}) \le 2^{-n+1}$, and thus $T_{n,k}$ can be partitioned into at most N_{k+1} sets of diameter $\le 2^{-n+2}$. We construct such a partition $\mathcal{B}_{n,k}$ of $T_{n,k}$ for each k, and we consider the corresponding partition $\mathcal{B}_n$ of T.

We now turn to the proof of (3.54). First we note that $\operatorname{card}\mathcal{B}_{n,k} \le N_{k+1}$ and $\operatorname{card}\mathcal{B}_{n+1,\ell} \le N_{\ell+1}$. Also, $\sum_{A\in\mathcal{B}_{n,k}, B\in\mathcal{B}_{n+1,\ell}} \mu(A\cap B) \le \mu(T_{n,k}\cap T_{n+1,\ell})$. We then use (3.53) to obtain

$$S_{n,k,\ell} := \sum_{A\in\mathcal{B}_{n,k}, B\in\mathcal{B}_{n+1,\ell}} \mu(A\cap B)\sqrt{\log(2/\mu(A\cap B))}$$
$$\le \mu(T_{n,k}\cap T_{n+1,\ell})\sqrt{\log(2N_{k+1}N_{\ell+1}/\mu(T_{n,k}\cap T_{n+1,\ell}))} \ .$$

The left-hand side of (3.54) is $\sum_{n\ge n_0} 2^{-n}\sum_{k,\ell} S_{n,k,\ell}$. We will use the decomposition

$$\sum_{k,\ell} S_{n,k,\ell} = \sum_{(k,\ell)\in I(n)} S_{n,k,\ell} + \sum_{(k,\ell)\in J(n)} S_{n,k,\ell}$$

where $I(n) = \{(k,\ell);\, \mu(T_{n,k}\cap T_{n+1,\ell}) \ge 1/(N_{k+1}N_{\ell+1})\}$ and $J(n) = \{(k,\ell);\, \mu(T_{n,k}\cap T_{n+1,\ell}) < 1/(N_{k+1}N_{\ell+1})\}$. Then

$$\begin{aligned}\sum_{(k,\ell)\in I(n)} S_{n,k,\ell} &\le \sum_{k,\ell}\mu(T_{n,k}\cap T_{n+1,\ell})\sqrt{\log(2N_{k+1}^2N_{\ell+1}^2)}\\ &\le L\sum_{k,\ell}\mu(T_{n,k}\cap T_{n+1,\ell})(2^{k/2}+2^{\ell/2})\\ &\le L\sum_{k}\mu(T_{n,k})2^{k/2}+L\sum_{\ell}\mu(T_{n+1,\ell})2^{\ell/2}\,. \end{aligned}\tag{3.55}$$

Now the definition of $T_{n,k}$ shows that

$$\sum_{k\ge 1}\mu(T_{n,k})2^{k/2}\le\int\sqrt{\log(1/\nu(B(t,2^{-n}))}\mathrm{d}\mu(t)\,, \tag{3.56}$$

and thus

$$\sum_{n\ge n_0}2^{-n}\sum_{k\ge 1}2^{k/2}\mu(T_{n,k})\le L\int I_\nu(t)\mathrm{d}\mu(t)\,.$$

Next, since the function $\varphi(x)=x\sqrt{\log 2N_{k+1}N_{\ell+1}/x}$ increases for $x\le N_{k+1}N_{\ell+1}$, for $(k,\ell)\in J(n)$, we have $\varphi(\mu(T_{n,k}\cap T_{n+1,\ell}))\le\varphi(1/(N_{k+1}N_{\ell+1}))$ so that

$$\begin{aligned}\sum_{(k,\ell)\in J(n)} S_{n,k,\ell} &\le \sum_{(k,\ell)\in J(n)}\varphi(\mu(T_{n,k}\cap T_{n+1,\ell}))\le\sum_{(k,\ell)\in J(n)}\varphi(1/N_{k+1}N_{\ell+1})\\ &\le L\sum_{k,\ell}\frac{2^{k/2}+2^{\ell/2}}{N_{k+1}N_{\ell+1}}\le L\,. \end{aligned}\tag{3.57}$$

Combining these estimates yields the desired inequality:

$$\sum_{n\ge n_0}2^{-n}\sum_{k,\ell\ge 0}S_{n,k,\ell}\le L\int I_\nu(t)\mathrm{d}\mu(t)+L2^{-n_0}\,.$$ □

Proof of Theorem 3.5.1 For $n\ge n_0$ and $A\in\mathcal{B}_n$, we fix an arbitrary point $t_{n,A}\in A$. We lighten notation by writing $t_0=t_{n_0,T}$. We define $\pi_n(t)=t_{n,A}$ for $t\in A\in\mathcal{B}_n$. We write

$$X_\tau-X_{t_0}=\sum_{n\ge n_0}X_{\pi_{n+1}(\tau)}-X_{\pi_n(\tau)}\,, \tag{3.58}$$

so that defining $Y_{n,\tau} := X_{\pi_{n+1}(\tau)} - X_{\pi_n(\tau)}$, we have

$$\mathsf{E}|X_\tau - X_{t_0}| \le \sum_{n \ge n_0} \mathsf{E}|Y_{n,\tau}| \,. \tag{3.59}$$

Given $A \in \mathcal{B}_n$ and $B \in \mathcal{B}_{n+1}$, let us define the variable $Y_{A,B} := X_{t_{n+1,B}} - X_{t_{n,A}}$. The sets $A \cap B$ for $A \in \mathcal{B}_n$ and $B \in \mathcal{B}_{n+1}$ form a partition of T. When $\tau(\omega) \in A \cap B$, we have $\pi_n(\tau) = t_{n,A}$ and $\pi_{n+1}(\tau) = t_{n+1,B}$ so that $Y_{n,\tau} = Y_{A,B}$. The event $\tau(\omega) \in A \cap B$ has probability $\mu(A \cap B)$ since τ has law μ. When $A \cap B \neq \emptyset$, we have $d(t_{n+1,B}, t_{n,A}) \le \Delta(A) + \Delta(B) \le L2^{-n}$, so that $\mathsf{E}Y_{A,B}^2 = d(t_{n+1,B}, t_{n,A})^2 \le L2^{-2n}$. It then follows from Corollary 3.5.3 that

$$\mathsf{E}|Y_{n,\tau}| \le L2^{-n} \sum_{A \in \mathcal{B}_n, B \in \mathcal{B}_{n+1}} \mu(A \cap B)\sqrt{\log(2/\mu(A \cap B))} \,,$$

and summation over n and use of (3.54) finishes the proof. □

It is actually possible to give a complete geometric description of the quantity $\sup_\tau \mathsf{E}X_\tau$ where the supremum is taken over all the τ of given law μ (see [98]).

Key Ideas to Remember

- Trees in a metric space (T, d) are well-separated structures which are easy to visualize and provide a convenient way to measure the size of this metric space, by the size of the largest tree it contains. For suitable classes of trees, this measure of size is equivalent to $\gamma_2(T, d)$.
- One may also measure the size of a metric space by the existence of certain probability measures on this space. Fernique's majorizing measures were used early to control from above the size of a metric space in a way very similar to the functional $\gamma_2(T, d)$, which is, however, far more technically convenient than majorizing measures.
- An offshoot of the idea of majorizing measures, Fernique's functional, is an equivalent way to measure the size of a metric space and will be of fundamental importance in the sequel.
- The size of a metric space (T, d) can also be bounded from below by the existence of well-scattered probability measures on T.

Chapter 4
Matching Theorems

We remind the reader that before attacking any chapter, she should find useful to read the overview of this chapter, which is provided in the appropriate subsection of Chap. 1. Here, this overview should help to understand the overall approach.

4.1 The Ellipsoid Theorem

As pointed out after Proposition 2.13.2, an ellipsoid $\mathcal{E}$ is in some sense quite smaller than what one would predict by looking only at the numbers $e_n(\mathcal{E})$. We will trace the roots of this phenomenon to a simple geometric property, namely, that an ellipsoid is "sufficiently convex", and we will formulate a general version of this principle for sufficiently convex bodies. The case of ellipsoids already suffices to provide tight upper bounds on certain matchings, which is the main goal of the present chapter. The general case is at the root of certain very deep facts of Banach space theory, such as Bourgain's celebrated solution of the Λ_p problem in Sects. 19.3.1 and 19.3.2.

Recall the ellipsoid $\mathcal{E}$ of (2.154), which is defined as the set

$$\mathcal{E} = \left\{ t \in \ell^2 \; ; \; \sum_{i \geq 1} \frac{t_i^2}{a_i^2} \leq 1 \right\} \tag{2.154}$$

and is the unit ball of the norm

$$\|x\|_{\mathcal{E}} := \left(\sum_{i \geq 1} \frac{x_i^2}{a_i^2} \right)^{1/2} . \tag{4.1}$$

M. Talagrand, *Upper and Lower Bounds for Stochastic Processes*, Ergebnisse der Mathematik und ihrer Grenzgebiete. 3. Folge / A Series of Modern Surveys in Mathematics 60, https://doi.org/10.1007/978-3-030-82595-9_4

Lemma 4.1.1 *We have*

$$\|x\|_{\mathcal{E}}\,,\ \|y\|_{\mathcal{E}} \le 1 \Rightarrow \left\|\frac{x+y}{2}\right\|_{\mathcal{E}} \le 1 - \frac{\|x-y\|_{\mathcal{E}}^2}{8}\,. \tag{4.2}$$

Proof The parallelogram identity implies

$$\|x-y\|_{\mathcal{E}}^2 + \|x+y\|_{\mathcal{E}}^2 = 2\|x\|_{\mathcal{E}}^2 + 2\|y\|_{\mathcal{E}}^2 \le 4$$

so that

$$\|x+y\|_{\mathcal{E}}^2 \le 4 - \|x-y\|_{\mathcal{E}}^2$$

and

$$\left\|\frac{x+y}{2}\right\|_{\mathcal{E}} \le \left(1-\frac{1}{4}\|x-y\|_{\mathcal{E}}^2\right)^{1/2} \le 1 - \frac{1}{8}\|x-y\|_{\mathcal{E}}^2\,. \qquad \square$$

Since (4.2) is the only property of ellipsoids we will use, it clarifies matters to state the following definition:

Definition 4.1.2 Consider a number $p \ge 2$. A norm $\|\cdot\|$ in a Banach space is called p-convex if for a certain number $\eta > 0$ we have

$$\|x\|\,,\ \|y\| \le 1 \Rightarrow \left\|\frac{x+y}{2}\right\| \le 1 - \eta\|x-y\|^p\,. \tag{4.3}$$

Saying just that the unit ball of the Banach space is convex implies that for $\|x\|, \|y\| \le 1$, we have $\|(x+y)/2\| \le 1$. Here, (4.3) quantitatively improves on this inequality. Geometrically, it means that the unit ball of the Banach space is "round enough".

Thus, (4.2) implies that the Banach space ℓ^2 provided with the norm $\|\cdot\|_{\mathcal{E}}$ is 2-convex. For $1 < q < \infty$, the classical Banach space L^q is p-convex where $p = \max(2, q)$. The reader is referred to [57] for this result and any other classical facts about Banach spaces. Let us observe that taking $y = -x$ in (4.3), we must have

$$2^p\eta \le 1\,. \tag{4.4}$$

In this section, we shall study the metric space (T, d) where T is the unit ball of a p-convex Banach space B and where d is the distance induced on B by another norm $\|\cdot\|_\sim$. This concerns in particular the case where T is the ellipsoid (2.154) and $\|\cdot\|_\sim$ is the ℓ^2 norm.

Given a metric space (T, d), we consider the functionals

$$\gamma_{\alpha,\beta}(T,d) = \inf\Big(\sup_{t\in T}\sum_{n\geq 0}\big(2^{n/\alpha}\Delta(A_n(t))\big)^{\beta}\Big)^{1/\beta}\,, \tag{4.5}$$

where α and β are positive numbers and where the infimum is over all admissible sequences $(\mathcal{A}_n)$. Thus, with the notation of Definition 2.7.3, we have $\gamma_{\alpha,1}(T,d) = \gamma_\alpha(T,d)$. For matchings, the important functionals are $\gamma_{2,2}(T,d)$ and $\gamma_{1,2}(T,d)$ (but it requires no extra effort to consider the general case). The importance of these functionals is that under certain conditions, they nicely relate to $\gamma_2(T,d)$ through Hölder's inequality. We explain right now how this is done, even though this spoils the surprise of how the terms $\sqrt{\log N}$ occur in Sect. 4.5.

Lemma 4.1.3 *Consider a finite metric space* T*, and assume that* $\operatorname{card} T \leq N_m$*. Then,*

$$\gamma_2(T,d) \leq \sqrt{m}\,\gamma_{2,2}(T,d)\,. \tag{4.6}$$

Proof Since T is finite, there exists[1] an admissible sequence $(\mathcal{A}_n)$ of T for which

$$\forall t\in T\,,\ \sum_{n\geq 0} 2^n\Delta(A_n(t))^2 \leq \gamma_{2,2}(T,d)^2\,. \tag{4.7}$$

Since $\operatorname{card} T \leq N_m$, we may assume that $\mathcal{A}_m$ consists of all the sets $\{t\}$ for $t\in T$. Then, $A_m(t) = \{t\}$ for each t, so that in (4.7) the sum is really over $n \leq m-1$. Since for any numbers $(a_n)_{0\leq n\leq m-1}$ we have $\sum_{0\leq n\leq m-1} a_n \leq \sqrt{m}(\sum_{0\leq n\leq m-1} a_n^2)^{1/2}$ by the Cauchy-Schwarz inequality, it follows that

$$\forall t\in T\,,\ \sum_{n\geq 0} 2^{n/2}\Delta(A_n(t)) \leq \sqrt{m}\,\gamma_{2,2}(T,d)\,. \qquad \square$$

How to relate the functionals $\gamma_{1,2}$ and γ_2 by a similar argument is shown in Lemma 4.7.9.

We may wonder how it is possible, using something as simple as the Cauchy-Schwarz inequality in Lemma 4.1.3, that we can ever get essentially exact results. At a general level, the answer is obvious: it is because we use this inequality in the case of near equality. That this is indeed the case for the ellipsoids of Corollary 4.1.7 is a non-trivial fact about the geometry of these ellipsoids.

[1] Since there are only finitely many admissible sequences, the infimum over these is achieved.

Theorem 4.1.4 *If T is the unit ball of a p-convex Banach space, if η is as in (4.3) and if the distance d on T is induced by another norm, then*

$$\gamma_{\alpha,p}(T,d) \le K(\alpha,p,\eta)\sup_{n\ge 0} 2^{n/\alpha} e_n(T,d) . \tag{4.8}$$

Before we prove this result (in Sect. 4.2), we explore some of its consequences. The following exercise stresses the main point of this theorem:

Exercise 4.1.5 Consider a general metric space (T,d).

(a) Prove that

$$\gamma_{\alpha,p}(T,d) \le K(\alpha)\Big(\sum_{n\ge 0}(2^{n/\alpha}e_n(T,d))^p\Big)^{1/p} , \tag{4.9}$$

and that

$$\sup_{n\ge 0} 2^{n/\alpha} e_n(T,d) \le K(\alpha)\gamma_{\alpha,p}(T,d) . \tag{4.10}$$

(b) Prove that it is essentially impossible in general to improve on (4.9). Hint: You probably want to review Chap. 3 before you try this.

Thus, knowing only the numbers $e_n(T,d)$, we would expect only the general bound (4.9). The content of Theorem 4.1.4 is that the size of T, as measured by the functional $\gamma_{\alpha,p}$, is actually much smaller than that.

Corollary 4.1.6 (The Ellipsoid Theorem) *Consider the ellipsoid $\mathcal{E}$ of (2.154) and $\alpha \ge 1$. Then*[2]

$$\gamma_{\alpha,2}(\mathcal{E}) \le K(\alpha)\sup_{\epsilon>0}\epsilon(\mathrm{card}\{i\ ;\ a_i \ge \epsilon\})^{1/\alpha} . \tag{4.11}$$

Proof Without loss of generality, we may assume that the sequence (a_i) is non-increasing. We apply Theorem 4.1.4 to the case $\|\cdot\| = \|\cdot\|_{\mathcal{E}}$, where d is the distance of ℓ^2, and we get

$$\gamma_{\alpha,2}(\mathcal{E}) \le K(\alpha)\sup_{n\ge 0} 2^{n/\alpha} e_n(\mathcal{E}) .$$

To bound the right-hand side, we write

$$\sup_{n\ge 0} 2^{n/\alpha} e_n(\mathcal{E}) \le 2^{2/\alpha} e_0(\mathcal{E}) + \sup_{n\ge 0} 2^{(n+3)/\alpha} e_{n+3}(\mathcal{E}) .$$

[2] Recalling that a subset of ℓ^2 is always provided with the distance induced by the ℓ^2 norm.

We now proceed as in the proof of Proposition 2.13.2. Using (2.166), we have

$$\begin{aligned}\sup_{n\geq 0} 2^{(n+3)/\alpha} e_{n+3}(\mathcal{E}) &\leq K(\alpha) \sup_{n\geq 0} 2^{n/\alpha} \max_{k\leq n} 2^{k-n} a_{2^k} \\ &= K(\alpha) \sup_{0\leq k\leq n} 2^{k-n(1-1/\alpha)} a_{2^k} \\ &= K(\alpha) \sup_{k\geq 0} 2^{k/\alpha} a_{2^k} , \end{aligned} \tag{4.12}$$

and since $e_0(\mathcal{E}) \leq a_1$, we have proved that $\gamma_{\alpha,2}(\mathcal{E}) \leq K(\alpha) \sup_{n\geq 0} 2^{n/\alpha} a_{2^n}$. Finally, the choice $\epsilon = a_{2^n}$ shows that

$$2^{n/\alpha} a_{2^n} \leq \sup_{\epsilon>0} \epsilon (\mathrm{card}\{i \; ; \; a_i \geq \epsilon\})^{1/\alpha}$$

since $\mathrm{card}\{i; a_i \geq a_{2^n}\} \geq 2^n$ because the sequence (a_i) is non-increasing. □

The restriction $\alpha \geq 1$ is inessential and can be removed by a suitable modification of (2.166). The important cases are $\alpha = 1$ and $\alpha = 2$. We will use the following convenient reformulation:

Corollary 4.1.7 *Consider a countable set J, numbers $(b_i)_{i\in J}$, and the ellipsoid*

$$\mathcal{E} = \left\{ x \in \ell^2(J) \; ; \; \sum_{j\in J} b_j^2 x_j^2 \leq 1 \right\} .$$

Then

$$\gamma_{\alpha,2}(\mathcal{E}) \leq K(\alpha) \sup_{u>0} \frac{1}{u} (\mathrm{card}\{j \in J \; ; \; |b_j| \leq u\})^{1/\alpha} .$$

Proof Without loss of generality, we can assume that $J = \mathbb{N}$. We then set $a_i = 1/b_i$, we apply Corollary 4.1.6, and we set $\epsilon = 1/u$. □

We give right away a striking application of this result. This application is at the root of the results of Sect. 4.7. We denote by λ Lebesgue's measure.

Proposition 4.1.8 *Consider the set $\mathcal{L}$ of functions $f : [0, 1] \to \mathbb{R}$ such that $f(0) = f(1) = 0$, f is continuous on $[0, 1]$, f is differentiable outside a finite set, and $\sup |f'| \leq 1$.*[3] *Then $\gamma_{1,2}(\mathcal{L}, d_2) \leq L$, where $d_2(f, g) = \|f - g\|_2 = \left(\int_{[0,1]} (f - g)^2 \mathrm{d}\lambda\right)^{1/2}$.*

[3] The same result holds for the set $\mathcal{L}'$ of 1-Lipschitz functions f with $f(0) = f(1) = 0$, since $\mathcal{L}$ is dense in $\mathcal{L}'$.

Proof The very beautiful idea (due to Coffman and Shor [26]) is to use the Fourier transform to represent $\mathcal{L}$ as a subset of an ellipsoid. The Fourier coefficients of a function $f \in \mathcal{L}$ are defined for $p \in \mathbb{Z}$ by

$$c_p(f) = \int_0^1 \exp(2\pi i p x) f(x) \mathrm{d}x \; .$$

The key fact is the Plancherel formula,

$$\|f\|_2 = \Big(\sum_{p \in \mathbb{Z}} |c_p(f)|^2\Big)^{1/2} \; , \tag{4.13}$$

which states that the Fourier transform is an isometry from $L^2([0,1])$ into $\ell^2_{\mathbb{C}}(\mathbb{Z})$. Thus, if

$$\mathcal{D} = \{(c_p(f))_{p \in \mathbb{Z}} \; ; \; f \in \mathcal{L}\} \; ,$$

the metric space $(\mathcal{L}, d_2)$ is isometric to a subspace of $(\mathcal{D}, d)$, where d is the distance induced by $\ell^2_{\mathbb{C}}(\mathbb{Z})$. It is then obvious from the definition that $\gamma_{1,2}(\mathcal{L}, d_2) \le \gamma_{1,2}(\mathcal{D}, d)$, so that it suffices to prove that $\gamma_{1,2}(\mathcal{D}, d) < \infty$. By integration by parts and since $f(0) = f(1) = 0$, $c_p(f') = -2\pi i p c_p(f)$, so that, using (4.13) for f', we get

$$\sum_{p \in \mathbb{Z}} p^2 |c_p(f)|^2 \le \sum_{p \in \mathbb{Z}} |c_p(f')|^2 = \|f'\|_2^2 \; .$$

For $f \in \mathcal{L}$, we have $f(0) = 0$ and $|f'| \le 1$ so that $|f| \le 1$ and $|c_0(f)| \le 1$. Thus for $f \in \mathcal{L}$, we have

$$|c_0(f)|^2 + \sum_{p \in \mathbb{Z}} p^2 |c_p(f)|^2 \le 2 \; ,$$

and thus $\mathcal{D}$ is a subset of the complex ellipsoid $\mathcal{E}$ in $\ell^2_{\mathbb{C}}(\mathbb{Z})$ defined by

$$\mathcal{E} := \Big\{(c_p) \in \ell^2_{\mathbb{C}}(\mathbb{Z}) \; ; \; \sum_{p \in \mathbb{Z}} \max(1, p^2)|c_p|^2 \le 2\Big\} \; .$$

Viewing each complex number c_p as a pair (x_p, y_p) of real numbers with $|c_p|^2 = x_p^2 + y_p^2$ yields that $\mathcal{E}$ is (isometric to) the real ellipsoid defined by

$$\sum_{p \in \mathbb{Z}} \max(1, p^2)(x_p^2 + y_p^2) \le 2 \; .$$

We then apply Corollary 4.1.7 as follows: The set J consists of two copies of $\mathbb{Z}$. There is a two-to-one map φ from J to $\mathbb{Z}$ and $b_j = \max(1, |\varphi(j)|)$. Then $\operatorname{card}\{j \in J; |b_j| \le u\} \le Lu$ for $u \ge 1$ and $= 0$ for $u < 1$. □

Exercise 4.1.9

(a) For $k \ge 1$, consider the space $T = \{0, 1\}^{2^k}$. Writing $t = (t_i)_{i \le 2^k}$ a point of T, consider on T the distance $d(t, t') = 2^{-j-1}$, where $j = \min\{i \le 2^k; t_i \ne t'_i\}$. Consider the set $\mathcal{L}$ of 1-Lipschitz functions on (T, d) which are zero at $t = (0, \dots, 0)$. Prove that $\gamma_{1,2}(\mathcal{L}, d_\infty) \le L\sqrt{k}$, where d_∞ denotes the distance induced by the uniform norm. Hint: Use Lemma 4.5.18 to prove that $e_n(\mathcal{L}, d_\infty) \le L2^{-n}$, and conclude using (4.9).

(b) Let μ denote the uniform probability on T and d_2 the distance induced by $L^2(\mu)$. It can be shown that $\gamma_{1,2}(\mathcal{L}, d_2) \ge \sqrt{k}/L$. (This could be challenging even if you master Chap. 3.) Meditate upon the difference with Proposition 4.1.8.

4.2 Partitioning Scheme II

Consider parameters $\alpha, p \ge 1$.

Theorem 4.2.1 *Consider a metric space (T, d) and a number $r \ge 4$. Assume that for $j \in \mathbb{Z}$, we are given functions $s_j \ge 0$ on T with the following property:*

$$\begin{aligned} &\textit{Whenever we consider a subset } A \textit{ of } T \textit{ and } j \in \mathbb{Z} \textit{ with } \Delta(A) \le 2r^{-j},\\ &\textit{then for each } n \ge 1 \textbf{ either } e_n(A) \le r^{-j-1}, \textbf{ or else } \textit{there exists } t \in A\\ &\textit{with } s_j(t) \ge (2^{n/\alpha} r^{-j-1})^p . \end{aligned} \tag{4.14}$$

Then we can find an admissible sequence $(\mathcal{A}_n)$ of partitions such that

$$\forall t \in T\ ;\ \sum_{n \ge 0} (2^{n/\alpha} \Delta(A_n(t)))^p \le K(\alpha, p, r)\Big(\Delta(T, d)^p + \sup_{t \in T} \sum_{j \in \mathbb{Z}} s_j(t)\Big) . \tag{4.15}$$

The proof is identical to that of Theorem 2.9.8 which corresponds to the case $\alpha = 2$ and $p = 1$.

Proof of Theorem 4.1.4 We recall that by hypothesis T is the unit ball for the norm $\|\cdot\|$ of p-convex Banach space (but we study T for the metric d induced by a different norm). For $t \in T$ and $j \in \mathbb{Z}$, we set

$$c_j(t) = \inf\{\|v\|\ ;\ v \in B_d(t, r^{-j}) \cap T\} \le 1 , \tag{4.16}$$

where the index d emphasizes that the ball is for the distance d rather than for the norm. Since T is the unit ball, we have $c_j(t) \le 1$. Let us set

$$D = \sup_{n \ge 0} 2^{n/\alpha} e_n(T, d) \ . \tag{4.17}$$

The proof relies on Theorem 4.2.1 for the functions

$$s_j(t) = K D^p (c_{j+2}(t) - c_{j-1}(t)) \ , \tag{4.18}$$

for a suitable value of K. Since $c_j(t) \le 1$, it is clear that

$$\forall t \in T \ , \ \sum_{j \in \mathbb{Z}} s_j(t) \le 3K D^p \ ,$$

and (using also that $\Delta(T, d) \le 2e_0(T, d)$) the issue is to prove that (4.14) holds for a suitable constant K in (4.18). Consider then a set $A \subset T$ with $\Delta(A) \le 2r^{-j}$, consider $n \ge 1$, and assume that $e_n(A) > a := r^{-j-1}$. The goal is to find $t \in A$ such that $s_j(t) \ge (2^{n/\alpha} r^{-j-1})^p$, i.e.,

$$K D^p (c_{j+2}(t) - c_{j-1}(t)) \ge (2^{n/\alpha} r^{-j-1})^p \ . \tag{4.19}$$

For this, let $m = N_n$. According to Lemma 2.9.3, (a) there exist points $(t_\ell)_{\ell \le m}$ in A, such that $d(t_\ell, t_{\ell'}) \ge a$ whenever $\ell \ne \ell'$. We will show that one of the points t_ℓ satisfies (4.19). Consider $H_\ell = T \cap B_d(t_\ell, a/r) = T \cap B_d(t_\ell, r^{-j-2})$. By definition of $c_{j+1}(t_\ell)$, we have $c_{j+2}(t_\ell) = \inf\{\|v\| \ ; \ v \in H_\ell\}$. The basic idea is that the points of the different sets H_ℓ cannot be too close to each other for the norm of T because there are N_n such sets. So, since the norm is sufficiently convex, we will find a point in the convex hull of these sets with a norm quite smaller than $\max_{\ell \le m} c_{j+2}(t_\ell)$. To implement the idea, consider u' such that

$$2 > u' > \max_{\ell \le m} \inf\{\|v\| \ ; \ v \in H_\ell\} = \max_{\ell \le m} c_{j+2}(t_\ell) \ . \tag{4.20}$$

For $\ell \le m$, consider $v_\ell \in H_\ell$ with $\|v_\ell\| \le u'$. It follows from (4.3) that for $\ell, \ell' \le m$,

$$\left\| \frac{v_\ell + v_{\ell'}}{2u'} \right\| \le 1 - \eta \left\| \frac{v_\ell - v_{\ell'}}{u'} \right\|^p \ . \tag{4.21}$$

Set

$$u = \inf \left\{ \|v\| \ ; \ v \in \operatorname{conv} \bigcup_{\ell \le m} H_\ell \right\} . \tag{4.22}$$

Since $(v_\ell + v_{\ell'})/2 \in \operatorname{conv} \bigcup_{\ell \le m} H_\ell$, by definition of u , we have $u \le \|v_\ell + v'_\ell\|/2$, and (4.21) implies

$$\frac{u}{u'} \le 1 - \eta \Big\| \frac{v_\ell - v_{\ell'}}{u'} \Big\|^p ,$$

so that, using that $u' < 2$ in the second inequality below,

$$\|v_\ell - v_{\ell'}\| \le u' \Big(\frac{u' - u}{\eta u'} \Big)^{1/p} < R := 2 \Big(\frac{u' - u}{\eta} \Big)^{1/p} ,$$

and hence the points $w_\ell := R^{-1}(v_\ell - v_1)$ belong to the unit ball T. Now, since $H_\ell \subset B_d(t_\ell, a/r)$, we have $v_\ell \in B_d(t_\ell, a/r)$. Since $r \ge 4$, we have $d(v_\ell, v_{\ell'}) \ge a/2$ for $\ell \ne \ell'$, and since the distance d arises from a norm, by homogeneity, we have $d(w_\ell, w_{\ell'}) \ge R^{-1}a/2$ for $\ell \ne \ell'$. Then Lemma 2.9.3, (c) implies that $e_{n-1}(T, d) \ge R^{-1}a/4$, so that from (4.17) it holds that $2^{(n-1)/\alpha} R^{-1} a/4 \le D$, and recalling that $R = 2((u' - u)/\eta)^{1/p}$, we obtain

$$(2^{n/\alpha} r^{j-1})^p \le K D^p (u' - u) ,$$

where K depends on α only. Since this holds for any u' as in (4.20), there exists ℓ such that

$$(2^{n/\alpha} r^{j-1})^p \le K D^p (c_{j+2}(t_\ell) - u) . \tag{4.23}$$

Now, by construction, for $\ell' \le m$, we have

$$H_{\ell'} \subset B_d(t_{\ell'}, a/r) = B_d(t_{\ell'}, r^{-j-2}) \subset B_d(t_\ell, r^{-j+1})$$

since $d(t_\ell, t_{\ell'}) \le 2r^{-j}$ as $t_\ell, t_{\ell'} \in A$ and $\Delta(A) \le 2r^{-j}$. Thus $\operatorname{conv} \bigcup_{\ell' \le m} H_{\ell'} \subset B_d(t_\ell, r^{-j+1}) \cap T$, and from (4.16) and (4.22), we have $u \ge c_{j-1}(t_\ell)$, and we have proved (4.19). □

Exercise 4.2.2 Write the previous proof using a certain functional with an appropriate growth condition.

The following generalization of Theorem 4.1.4 yields very precise results when applied to ellipsoids. It will not be used in the sequel, so we refer to [132] for a proof.

Theorem 4.2.3 *Consider* β , β' , $p > 0$ *with*

$$\frac{1}{\beta} = \frac{1}{\beta'} + \frac{1}{p} . \tag{4.24}$$

Then, under the conditions of Theorem 4.1.4, we have

$$\gamma_{\alpha,\beta}(T,d) \le K(p,\eta,\alpha)\Big(\sum_n (2^{n/\alpha} e_n(T,d))^{\beta'}\Big)^{1/\beta'} .$$

Exercise 4.2.4 Use Theorem 4.2.3 to obtain a geometrical proof of (2.159). Hint: Choose $\alpha = 2$, $\beta = 1$, $\beta' = p = 2$ and use (2.166).

4.3 Matchings

The rest of this chapter is devoted to the following problem. Consider N r.v.s $X_1, \ldots, X_N$ independently and uniformly distributed in the unit cube $[0,1]^d$, where $d \ge 1$. Consider a typical realization of these points. How evenly distributed in $[0,1]^d$ are the points $X_1, \ldots, X_N$? To measure this, we will match the points $(X_i)_{i \le N}$ with *nonrandom* "evenly distributed" points $(Y_i)_{i \le N}$, that is, we will find a permutation π of $\{1, \ldots, N\}$ such that the points X_i and $Y_{\pi(i)}$ are "close". There are different ways to measure "closeness". For example, one may wish that the sum of the distances $d(X_i, Y_{\pi(i)})$ be as small as possible (Sect. 4.5), that the maximum distance $d(X_i, Y_{\pi(i)})$ be as small as possible (Sect. 4.7), or one can use more complicated measures of "closeness" (Sect. 17.1).

The case $d = 1$ is by far the simplest. Assuming that the X_i are labeled in a way that $X_1 \le X_2 \le \ldots$ and similarly for the Y_i, one has $\mathsf{E}\sup_{i \le N} |X_i - Y_i| \le L\sqrt{N}$. This is a consequence of the classical inequality (which we will later prove as an exercise):

$$\mathsf{E} \sup_{0 \le t \le 1} |\operatorname{card}\{i \le N \,;\, X_i \le t\} - Nt| \le L\sqrt{N} . \tag{4.25}$$

The case where $d = 2$ is very special and is the object of the present chapter. The case $d \ge 3$ will be studied in Chap. 18. The reader having never thought of the matter might think that the points $X_1, \ldots, X_N$ are very evenly distributed. This is not quite the case; for example, with probability close to one, one is bound to find a little square of area about $N^{-1}\log N$ that contains no point X_i. This is a very local irregularity. In a somewhat informal manner, one can say that this irregularity occurs at scale $\sqrt{\log N}/\sqrt{N}$. This specific irregularity is mentioned just as an easy illustration and plays no part in the considerations of the present chapter. What matters here[4] is that in some sense, there are irregularities at all scales 2^{-k} for $1 \le k \le L^{-1}\log N$ and that these are all of the same order. To see this, let us think that we actually move the points X_i to the points $Y_{\pi(i)}$ in straight lines. In a given small square of side 2^{-k}, there is often an excess of points X_i of order

[4] This is much harder to visualize and is specific to the case $d = 2$.

$\sqrt{N2^{-2k}} = 2^{-k}\sqrt{N}$. When matched these points will *leave* the square and will cross its boundary. The number of points crossing this boundary per unit of length is independent of the scale 2^{-k}. It will also often happen that there is a deficit of points X_i in this square of side 2^{-k}, and in this case, some points X_i will have to cross the boundary to *enter* it. The flows at really different scales should be roughly independent, and there are about $\log N$ such scales, so when we combine what happens at different scales we should get an extra factor $\sqrt{\log N}$ (and not $\log N$). Crossing our fingers, we should believe that about $\sqrt{N \log N}$ points X_i per unit of length cross a typical interval contained in the square, so that the total length of the segments joining the points X_i to the points $Y_{\pi(i)}$ should be of that order.[5] This fact that all scales have the same weight is typical of dimension 2. In dimension 1, it is the large scales that matter most, while in dimension ≥ 3, it is the small ones.

Exercise 4.3.1 Perform this calculation.

One can summarize the situation by saying that

$$\textit{obstacles to matchings at different scales may combine}$$
$$\textit{in dimension 2 but not in dimension} \geq 3\ . \tag{4.26}$$

It is difficult to state a real theorem to this effect, but this is actually seen with great clarity in the proofs. The crucial estimates involve controlling sums, each term of which represents a different scale. In dimension 2, many terms contribute to the final sum (which therefore results in the contribution of many different scales), while in higher dimension, only a few terms contribute. (The case of higher dimension remains non-trivial because *which* terms contribute depend on the value of the parameter.) Of course, these statements are very mysterious at this stage, but we expect that a serious study of the methods involved will gradually bring the reader to share this view.

What does it mean to say that the nonrandom points $(Y_i)_{i \leq N}$ are evenly distributed? When N is a square, $N = n^2$, everybody will agree that the N points $(k/n, \ell/n)$, $1 \leq k\,,\ \ell \leq n$ are evenly distributed, and unless you love details, you are welcomed to stick to this case. More generally, we will say that the nonrandom points $(Y_i)_{i \leq N}$ are *evenly spread* if one can cover $[0,1]^2$ with N rectangles with disjoint interiors, such that each rectangle R has an area $1/N$, contains exactly one point Y_i, and is such that[6] $R \subset B(Y_i, 10/\sqrt{N})$. To construct such points, one may proceed as follows: Consider the largest integer k with $k^2 \leq N$, and observe that $k(k+3) \geq (k+1)^2 \geq N$, so that there exist integers $(n_i)_{i \leq k}$ with $k \leq n_i \leq k+3$ and $\sum_{i \leq k} n_i = N$. Cut the unit square into k vertical strips, in a way that the i-th

[5] As we will see later, we have guessed the correct result.

[6] There is nothing magic about the number 10. Thinks of it as a universal constant. The last thing I want is to figure out the best possible value. That 10 works should be obvious from the following construction.

strip has width n_i/N and to this i-th strip attribute n_i points placed at even intervals $1/n_i$.[7]

The basic tool to construct matchings is the following classical fact. The proof, based on the Hahn-Banach theorem, is given in Sect. B.1.

Proposition 4.3.2 *Consider a matrix* $C = (c_{ij})_{i,j\le N}$. *Let*

$$M(C) = \inf \sum_{i\le N} c_{i\pi(i)} ,$$

where the infimum is over all permutations π *of* $\{1, \dots, N\}$. *Then*

$$M(C) = \sup \sum_{i\le N} (w_i + w'_i) , \tag{4.27}$$

where the supremum is over all families $(w_i)_{i\le N}$, $(w'_i)_{i\le N}$ *that satisfy*

$$\forall i, j \le N , \; w_i + w'_j \le c_{ij} . \tag{4.28}$$

Thus, if c_{ij} is the cost of matching i with j, $M(C)$ is the minimal cost of a matching and is given by the "duality formula" (4.27).

A well-known application of Proposition 4.3.2 is another "duality formula".

Proposition 4.3.3 *Consider points* $(X_i)_{i\le N}$ *and* $(Y_i)_{i\le N}$ *in a metric space* (T, d). *Then*

$$\inf_{\pi} \sum_{i\le N} d(X_i, Y_{\pi(i)}) = \sup_{f\in\mathcal{C}} \sum_{i\le N} (f(X_i) - f(Y_i)) , \tag{4.29}$$

where $\mathcal{C}$ *denotes the class of 1-Lipschitz functions on* (T, d), *i.e., functions* f *for which* $|f(x) - f(y)| \le d(x, y)$.

Proof Given any permutation π and any 1-Lipschitz function f, we have

$$\sum_{i\le N} f(X_i) - f(Y_i) = \sum_{i\le N} (f(X_i) - f(Y_{\pi(i)})) \le \sum_{i\le N} d(X_i, Y_{\pi(i)}) .$$

[7] A more elegant approach dispenses from this slightly awkward construction. It is the concept of "transportation cost". One attributes mass $1/N$ to each point X_i, and one measures the "cost of transporting" the resulting probability measure to the uniform probability on $[0, 1]^2$. In the presentation, one thus replaces the evenly spread points Y_i by a more canonical object, the uniform probability on $[0, 1]^2$. This approach does not make the proofs any easier, so we shall not use it despite its aesthetic appeal.

This proves the inequality $\geq$ in (4.29). To prove the converse, we use (4.27) with $c_{ij} = d(X_i, Y_j)$, so that

$$\inf_{\pi} \sum_{i \leq N} d(X_i, Y_{\pi(i)}) = \sup \sum_{i \leq N} (w_i + w'_i) , \tag{4.30}$$

where the supremum is over all families (w_i) and (w'_i) for which

$$\forall i \, , \, j \leq N \, , \, w_i + w'_j \leq d(X_i, Y_j) . \tag{4.31}$$

Given a family $(w'_i)_{i \leq N}$, consider the function

$$f(x) = \min_{j \leq N} (-w'_j + d(x, Y_j)) . \tag{4.32}$$

It is 1-Lipschitz, since it is the minimum of functions which are themselves 1-Lipschitz. By definition, we have $f(Y_j) \leq -w'_j$, and by (4.31) for $i \leq N$, we have $w_i \leq f(X_i)$, so that

$$\sum_{i \leq N} (w_i + w'_i) \leq \sum_{i \leq N} (f(X_i) - f(Y_i)) . \qquad \square$$

Exercise 4.3.4 Consider a function f which achieves the supremum in the right-hand side of (4.29). Prove that for an optimal matching, we have $f(X_i) - f(Y_{\pi(i)}) = d(X_i, Y_{\pi(i)})$. If you know f, this basically tells you how to find the matching. To find $Y_{\pi(i)}$, move from X_i in the direction of steepest descent of f until you find a points Y_j.

The following is a well-known and rather useful result of combinatorics. We deduce it from Proposition 4.3.2 in Sect. B.1, but other proofs exist, based on different ideas (see, for example, [21] § 2).

Corollary 4.3.5 (Hall's Marriage Lemma) *Assume that to each $i \leq N$, we associate a subset $A(i)$ of $\{1, \ldots, N\}$ and that, for each subset I of $\{1, \ldots, N\}$, we have*

$$\operatorname{card}\Big(\bigcup_{i \in I} A(i)\Big) \geq \operatorname{card} I . \tag{4.33}$$

Then we can find a permutation π of $\{1, \ldots, N\}$ for which

$$\forall i \leq N \, , \, \pi(i) \in A(i) .$$

4.4 Discrepancy Bounds

Generally speaking, the study of expressions of the type

$$\sup_{f \in \mathcal{F}} \Big| \sum_{i \le N} (f(X_i) - \int f \mathrm{d}\mu) \Big| \tag{4.34}$$

for a class of functions $\mathcal{F}$ will be important in the present book, particularly in Chap. 14. A bound on such a quantity is called a *discrepancy bound* because since

$$\Big| \sum_{i \le N} (f(X_i) - \int f \mathrm{d}\mu) \Big| = N \Big| \frac{1}{N} \sum_{i \le N} f(X_i) - \int f \mathrm{d}\mu \Big|$$

it bounds uniformly on $\mathcal{F}$ the "discrepancy" between the true measure $\int f \mathrm{d}\mu$ and the "empirical measure" $N^{-1} \sum_{i \le N} f(X_i)$. Finding such a bound simply requires finding a bound for the supremum of the process $(|Z_f|)_{f \in \mathcal{F}}$, where the (centered) r.v.s Z_f are given by[8]

$$Z_f = \sum_{i \le N} (f(X_i) - \int f \mathrm{d}\mu) \ , \tag{4.35}$$

a topic at the very center of our attention.

A relation between discrepancy bounds and matching theorems can be guessed from Proposition 4.3.3 and will be made explicit in the next section. In this book, *every* matching theorem will be proved through a discrepancy bound.

4.5 The Ajtai-Komlós-Tusnády Matching Theorem

Theorem 4.5.1 ([3]) *If the points $(Y_i)_{i \le N}$ are evenly spread and the points $(X_i)_{i \le N}$ are i.i.d. uniform on $[0, 1]^2$, then (for $N \ge 2$)*

$$\mathsf{E} \inf_{\pi} \sum_{i \le N} d(X_i, Y_{\pi(i)}) \le L\sqrt{N \log N} \ , \tag{4.36}$$

where the infimum is over all permutations of $\{1, \dots, N\}$ and where d is the Euclidean distance.

The term $\sqrt{N}$ is just a scaling effect. There are N terms $d(X_i, Y_{\pi(i)})$, each of which should be about $1/\sqrt{N}$. The non-trivial part of the theorem is the factor

[8] Please remember this notation which is used throughout this chapter.

$\sqrt{\log N}$. In Sect. 4.6, we shall show that (4.36) can be reversed, i.e.,

$$\mathsf{E} \inf_{\pi} \sum_{i \le N} d(X_i, Y_{\pi(i)}) \ge \frac{1}{L} \sqrt{N \log N} \,. \tag{4.37}$$

In order to understand that the bound (4.36) is not trivial, you can study the following greedy matching algorithm which was shown to me by Yash Kanoria:

Exercise 4.5.2 For each $n \ge 0$, consider the partition $\mathcal{H}_n$ of $[0,1]^2$ into 2^{2n} equal squares. Consider the largest integer n_0 with $2^{2n_0} \le N$, and proceed as follows: For each small square in $\mathcal{H}_{n_0}$, match as many as possible of the points X_i with points Y_i in the same square. Remove the points X_i and the points Y_i that you have matched this way. For the remaining points, proceed as follows: In each small square of $\mathcal{H}_{n_0-1}$, match as many of the remaining points X_i to remaining points Y_i inside the same square. Remove all the points X_i and the points Y_i that you have removed at this stage, and continue in this manner. Prove that the expected cost of the matching thus constructed is $\le L\sqrt{N} \log N$.[9]

Let us state the "discrepancy bound" at the root of Theorem 4.5.1. Consider the class $\mathcal{C}$ of 1-Lipschitz functions on $[0,1]^2$, i.e., of functions f that satisfy

$$\forall x, y \in [0,1]^2 \,, \; |f(x) - f(y)| \le d(x, y) \,,$$

where d denotes the Euclidean distance. We denote by λ the uniform measure on $[0,1]^2$.

Theorem 4.5.3 *We have*

$$\mathsf{E} \sup_{f \in \mathcal{C}} \Big| \sum_{i \le N} \Big(f(X_i) - \int f \mathrm{d}\lambda\Big) \Big| \le L \sqrt{N \log N} \,. \tag{4.38}$$

Research Problem 4.5.4 Prove that the following limit

$$\lim_{N \to \infty} \frac{1}{\sqrt{N \log N}} \mathsf{E} \sup_{f \in \mathcal{C}} \Big| \sum_{i \le N} \Big(f(X_i) - \int f \mathrm{d}\lambda\Big) \Big|$$

exists.

At the present time, there does not seem to exist the beginning of a general approach for attacking a problem of this type, and certainly the methods of the present book are not appropriate for this. Quite amazingly, however, the corresponding problem has been solved in the case where the cost of the matching is measured by the square

[9] It can be shown that this bound can be reversed.

of the distance (see [4]). The methods seem rather specific to the case of the square of a distance.

Theorem 4.5.3 is obviously interesting in its own right and proving it is the goal of this section. Before we discuss it, let us put matchings behind us.

Proof of Theorem 4.5.1 We recall (4.29), i.e.,

$$\inf_{\pi} \sum_{i\le N} d(X_i, Y_{\pi(i)}) = \sup_{f\in\mathcal{C}} \sum_{i\le N} (f(X_i) - f(Y_i)) \,, \tag{4.39}$$

and we simply write

$$\sum_{i\le N} (f(X_i) - f(Y_i)) \le \Big|\sum_{i\le N} (f(X_i) - \int f d\lambda)\Big| + \Big|\sum_{i\le N} (f(Y_i) - \int f \mathrm{d}\lambda)\Big| \,. \tag{4.40}$$

Next, we claim that

$$\Big|\sum_{i\le N} (f(Y_i) - \int f \mathrm{d}\lambda)\Big| \le L\sqrt{N} \,. \tag{4.41}$$

We recall that since $(Y_i)_{i\le N}$ are evenly spread, one can cover $[0,1]^2$ with N rectangles R_i with disjoint interiors, such that each rectangle R_i has an area $1/N$ and is such that $Y_i \in R_i \subset B(Y_i, 10/\sqrt{N})$. Consequently, $N \int f \mathrm{d}\lambda = N \sum_{i\le N} \int_{R_i} f \mathrm{d}\lambda$ and

$$\begin{aligned} \Big|\sum_{i\le N} (f(Y_i) - \int f \mathrm{d}\lambda)\Big| &= \Big|\sum_{i\le N} f(Y_i) - N \int f \mathrm{d}\lambda\Big| \\ &\le \sum_{i\le N} \Big|(f(Y_i) - N \int_{R_i} f \mathrm{d}\lambda)\Big| \\ &\le \sum_{i\le N} N\Big|\int_{R_i} (f(Y_i) - f(x))\mathrm{d}\lambda(x)\Big| \,. \end{aligned} \tag{4.42}$$

Since f is 1-Lipschitz and R_i is of diameter $\le L/\sqrt{N}$, we have $|f(Y_i) - f(x)| \le L/\sqrt{N}$ when $x \in R_i$. This proves the claim.

Now, using (4.39) and taking expectation,

$$\begin{aligned} \mathsf{E} \inf_{\pi} \sum_{i\le N} d(X_i, Y_{\pi(i)}) &\le L\sqrt{N} + \mathsf{E} \sup_{f\in\mathcal{C}} \Big|\sum_{i\le N} (f(X_i) - \int f \mathrm{d}\lambda)\Big| \\ &\le L\sqrt{N \log N} \end{aligned}$$

by (4.38). □

4.5.1 The Long and Instructive Way

S. Bobkov and M. Ledoux recently found [19] a magically simple proof of Theorem 4.5.3. We will present it in Sect. 4.5.2. This proof relies on very specific features, and it is unclear as to whether it will apply to other matching theorems. In the present section, we write a far more pedestrian (but far more instructive) proof with the general result Theorem 6.8.3 in mind.

To prove Theorem 4.5.3, the overall strategy is clear. We think of the left-hand side as $\mathsf{E}\sup_{f\in\mathcal{C}}|Z_f|$, where Z_f is the random variable of (4.35). We then find nice tail properties for these r.v.s, and we use the methods of Chap. 2. In the end (and because we are dealing with a deep fact), we shall have to prove some delicate "smallness" property of the class $\mathcal{C}$. This smallness property will ultimately be derived from the ellipsoid theorem. The (very beautiful) strategy for the hard part of the estimates relies on a kind of two-dimensional version of Proposition 4.1.8 and is outlined on page 129.

The class $\mathcal{C}$ of 1-Lipschitz function on the unit square is not small in any sense for the simple reason that it contains all the constant functions. However, the expression $\sum_{i\le N}(f(X_i)-\int f\mathrm{d}\lambda)$ does not change if we replace f by $f+a$ where a is a constant. In particular

$$\sup_{f\in\mathcal{C}}\Big|\sum_{i\le N}(f(X_i)-\int f\mathrm{d}\lambda)\Big| = \sup_{f\in\widehat{\mathcal{C}}}\Big|\sum_{i\le N}(f(X_i)-\int f\mathrm{d}\lambda)\Big|$$

where we define $\widehat{\mathcal{C}}$ as the set of 1-Lipschitz functions on the unit square for which $f(1/2,1/2)=0$.[10] The gain is that we now may hope that $\widehat{\mathcal{C}}$ is small in the appropriate sense. To prove Theorem 4.5.3, we will prove the following:

Theorem 4.5.5 *We have*

$$\mathsf{E}\sup_{f\in\widehat{\mathcal{C}}}\Big|\sum_{i\le N}(f(X_i)-\int f\mathrm{d}\lambda)\Big| \le L\sqrt{N\log N}\,. \tag{4.43}$$

The following fundamental classical result will allow us to control the tails of the r.v. Z_f of (4.35). It will be used many times.

Lemma 4.5.6 (Bernstein's Inequality) *Let $(W_i)_{i\ge 1}$ be independent r.v.s with $\mathsf{E}W_i=0$, and consider a number a with $|W_i|\le a$ for each i. Then, for $v>0$,*

$$\mathsf{P}\Big(\Big|\sum_{i\ge 1}W_i\Big|\ge v\Big)\le 2\exp\Big(-\min\Big(\frac{v^2}{4\sum_{i\ge 1}\mathsf{E}W_i^2},\frac{v}{2a}\Big)\Big)\,. \tag{4.44}$$

[10] There is no real reason other than my own fancy to impose that the functions are zero right in the middle of the square.

Proof For $|x| \le 1$, we have

$$|e^x - 1 - x| \le x^2 \sum_{k \ge 2} \frac{1}{k!} = x^2(e-2) \le x^2$$

and thus, since $\mathsf{E} W_i = 0$, for $a|\lambda| \le 1$, we have

$$|\mathsf{E} \exp \lambda W_i - 1| \le \lambda^2 \mathsf{E} W_i^2 \,.$$

Therefore, $\mathsf{E} \exp \lambda W_i \le 1 + \lambda^2 \mathsf{E} W_i^2 \le \exp \lambda^2 \mathsf{E} W_i^2$, and thus

$$\mathsf{E} \exp \lambda \sum_{i \ge 1} W_i = \prod_{i \ge 1} \mathsf{E} \exp \lambda W_i \le \exp \lambda^2 \sum_{i \ge 1} \mathsf{E} W_i^2 \,.$$

Now, for $0 \le \lambda \le 1/a$, we have

$$\begin{aligned} \mathsf{P}\Big(\sum_{i \ge 1} W_i \ge v\Big) &\le \exp(-\lambda v) \mathsf{E} \exp \lambda \sum_{i \ge 1} W_i \\ &\le \exp\Big(\lambda^2 \sum_{i \ge 1} \mathsf{E} W_i^2 - \lambda v\Big) \,. \end{aligned}$$

If $av \le 2\sum_{i \ge 1} \mathsf{E} W_i^2$, we take $\lambda = v/(2\sum_{i \ge 1} \mathsf{E} W_i^2)$, obtaining a bound $\exp(-v^2/(4\sum_{i \ge 1} \mathsf{E} W_i^2))$. If $av > 2\sum_{i \ge 1} \mathsf{E} W_i^2$, we take $\lambda = 1/a$, and we note that

$$\frac{1}{a^2} \sum_{i \ge 1} \mathsf{E} W_i^2 - \frac{v}{a} \le \frac{av}{2a^2} - \frac{v}{a} = -\frac{v}{2a} \,,$$

so that $\mathsf{P}(\sum_{i \ge 1} W_i \ge v) \le \exp(-\min(v^2/4\sum_{i \ge 1} \mathsf{E} W_i^2, v/2a))$. Changing W_i into $-W_i$ we obtain the same bound for $\mathsf{P}(\sum_{i \ge 1} W_i \le -v)$. □

Corollary 4.5.7 *For each $v > 0$, we have*

$$\mathsf{P}(|Z_f| \ge v) \le 2 \exp\Big(-\min\Big(\frac{v^2}{4N\|f\|_2^2}, \frac{v}{4\|f\|_\infty}\Big)\Big) \,, \tag{4.45}$$

where $\|f\|_p$ denotes the norm of f in $L_p(\lambda)$.

Proof We use Bernstein's inequality with $W_i = f(X_i) - \int f \mathrm{d}\lambda$ if $i \le N$ and $W_i = 0$ if $i > N$. We then observe that $\mathsf{E} W_i^2 \le \mathsf{E} f^2 = \|f\|_2^2$ and $|W_i| \le 2 \sup|f| = 2\|f\|_\infty$. □

Let us then pretend for a while that in (4.45), the bound was instead $2\exp(-v^2/(4N\|f\|_2^2))$. Thus, we would be back to the problem we considered first, bounding the supremum of a stochastic process under the increment condition (2.4), where the distance on $\mathcal{C}$ is given by $d(f_1, f_2) = \sqrt{2N}\|f_1 - f_2\|_2$. The first thing to point out is that Theorem 4.5.3 is a prime example of a natural situation where using covering numbers does not yield the correct result, where we recall that for a metric space (T, d), the covering number $N(T, d, \epsilon)$ denotes the smallest number of balls of radius ϵ that are needed to cover T. This is closely related to the fact that, as explained in Sect. 2.13, covering numbers do not describe well the size of ellipsoids. It is hard to formulate a theorem to the effect that covering numbers do not suffice, but the root of the problem is described in the next exercise, and a more precise version can be found later in Exercise 4.5.20.

Exercise 4.5.8 Prove that for each $0 < \epsilon \le 1$

$$\log N(\widehat{\mathcal{C}}, d_2, \epsilon) \ge \frac{1}{L\epsilon^2}\ , \tag{4.46}$$

where d_2 denotes the distance in $L^2([0,1]^2)$. Hint: Consider an integer $n \ge 0$, and divide $[0,1]^2$ into 2^{2n} equal squares of area 2^{-2n}. For every such square C, consider a number $\epsilon_C = \pm 1$. Consider then the function $f \in \mathcal{C}$ such that for $x \in C$, one has $f(x) = \epsilon_C d(x, B)$, where B denotes the boundary of C. There are $2^{2^{2n}}$ such functions. Prove that by appropriate choices of the signs ϵ_C, one may find at least $\exp(2^{2n}/L)$ functions of this type which are at mutual distance $\ge 2^{-n}/L$.

Since covering numbers do not suffice, we will appeal to the generic chaining, Theorem 2.7.2. As we will show later, in Exercise 4.5.21, we have $\gamma_2(\widehat{\mathcal{C}}, d_2) = \infty$. To overcome this issue, we will replace $\widehat{\mathcal{C}}$ by a sufficiently large finite subset $\mathcal{F} \subset \widehat{\mathcal{C}}$, for which we shall need the crucial estimate $\gamma_2(\mathcal{F}, d_2) \le L\sqrt{\log N}$. This will be done by proving that $\gamma_{2,2}(\widehat{\mathcal{C}}, d_2) < \infty$ where $\gamma_{2,2}$ is the functional of (4.5), so that $\gamma_{2,2}(\mathcal{F}, d_2) < \infty$, and appealing to Lemma 4.1.3.

The main ingredient toward the control of $\gamma_{2,2}(\widehat{\mathcal{C}}, d_2)$ is the following two-dimensional version of Proposition 4.1.8:

Lemma 4.5.9 *Consider the space $\mathcal{C}^*$ of 1-Lipschitz functions on $[0,1]^2$ which are zero on the boundary of $[0,1]^2$. Then $\gamma_{2,2}(\mathcal{C}^*, d_2) < \infty$.*

Proof We represent $\mathcal{C}^*$ as a subset of an ellipsoid using the Fourier transform. The Fourier transform associates with each function f on $L^2([0,1]^2)$ the complex numbers $c_{p,q}(f)$ given by

$$c_{p,q}(f) = \iint_{[0,1]^2} f(x_1, x_2) \exp(2i\pi(px_1 + qx_2))\mathrm{d}x_1\mathrm{d}x_2\ . \tag{4.47}$$

The Plancherel formula

$$\|f\|_2 = \Big(\sum_{p,q\in\mathbb{Z}} |c_{p,q}(f)|^2\Big)^{1/2} \tag{4.48}$$

asserts that Fourier transform is an isometry, so that if

$$\mathcal{D} = \{(c_{p,q}(f))_{p,q\in\mathbb{Z}}\ ;\ f \in \mathcal{C}^*)\}\,,$$

it suffices to show that $\gamma_{2,2}(\mathcal{D}, d) < \infty$ where d is the distance in the complex Hilbert space $\ell^2_{\mathbb{C}}(\mathbb{Z}\times\mathbb{Z})$. Using (4.47) and integration by parts, we get

$$-2i\pi p c_{p,q}(f) = c_{p,q}\Big(\frac{\partial f}{\partial x}\Big)\,.$$

Using (4.48) for $\partial f/\partial x$ and since $\|\partial f/\partial x\|_2 \le 1$, we get $\sum_{p,q\in\mathbb{Z}} p^2|c_{p,q}(f)|^2 \le 1/4\pi^2$. Proceeding similarly for $\partial f/\partial y$, we get

$$\mathcal{D} \subset \mathcal{E} = \Big\{(c_{p,q}) \in \ell^2_{\mathbb{C}}(\mathbb{Z}\times\mathbb{Z})\ ;\ |c_{0,0}| \le 1\,,\ \sum_{p,q\in\mathbb{Z}} (p^2+q^2)|c_{p,q}|^2 \le 1\Big\}\,.$$

We view each complex number $c_{p,q}$ as a pair $(x_{p,q}, y_{p,q})$ of real numbers and $|c_{p,q}|^2 = x^2_{p,q} + y^2_{p,q}$, so that

$$\begin{aligned}\mathcal{E} = \ &\Big\{\big((x_{p,q}), (y_{p,q})\big) \in \ell^2(\mathbb{Z}\times\mathbb{Z}) \times \ell^2(\mathbb{Z}\times\mathbb{Z})\ ;\\ &x^2_{0,0} + y^2_{0,0} \le 1\,,\ \sum_{p,q\in\mathbb{Z}} (p^2+q^2)(x^2_{p,q}+y^2_{p,q}) \le 1\Big\}\,. \end{aligned} \tag{4.49}$$

For $u \ge 1$, we have

$$\operatorname{card}\big\{(p,q) \in \mathbb{Z}\times\mathbb{Z}\ ;\ p^2+q^2 \le u^2\big\} \le (2u+1)^2 \le Lu^2\,.$$

We then deduce from Corollary 4.1.7 that $\gamma_{2,2}(\mathcal{E}, d) < \infty$. □

Proposition 4.5.10 *We have* $\gamma_{2,2}(\widehat{\mathcal{C}}, d_2) < \infty$.

I am grateful to R. van Handel who showed me the following simple arguments, which replaces pages of gritty work in [132]. The basic idea is to deduce this from Lemma 4.5.9, essentially by showing that $\widehat{\mathcal{C}}$ is a Lipschitz image of a subset of $\mathcal{C}^*$ or more exactly of the clone considered in the next lemma.

Lemma 4.5.11 *The set* $C^\sharp$ *of* 1*-Lipschitz functions on* $[-1,2]^2$ *which are zero on the boundary of this set satisfies* $\gamma_{2,2}(C^\sharp, d^\sharp) < \infty$ *where* $d^\sharp$ *is the distance induced by* $L^2([-1,2]^2, d\lambda)$.

Proof This should be obvious form Lemma 4.5.9; we just perform the same construction on two squares of different sizes, $[0,1]^2$ and $[-1,2]^2$. □

Lemma 4.5.12 *Each 1-Lipschitz function* $f \in \widehat{\mathcal{C}}$ *is the restriction to* $[0,1]^2$ *of a function* $f^\sharp$ *of* $C^\sharp$.

Proof A function $f \in \widehat{\mathcal{C}}$ may be extended to a 1-Lipschitz function $\tilde{f}$ on $\mathbb{R}^2$ by the formula $\tilde{f}(y) = \inf_{x \in [0,1]^2} f(x) + d(x, y)$. Since $f(1/2, 1/2) = 0$ by definition of $\widehat{\mathcal{C}}$ and since f is 1-Lipschitz, then $|f(x)| \le 1/\sqrt{2} \le 1$ for $x \in [0, 1]^2$. The function $f^\sharp(y) = \min(\tilde{f}(y), d(y, \mathbb{R}^2 \setminus [-1, 2]^2))$ is 1-Lipschitz. Since each point of $[0, 1]^2$ is at distance ≥ 1 of $\mathbb{R}^2 \setminus [-1, 2]^2$, $f^\sharp$ coincides with f on $[0, 1]^2$, and it is zero on the boundary of $[-1, 2]^2$. □

Proof of Proposition 4.5.10 To each function f of $\mathcal{C}^\sharp$, we associate its restriction $\varphi(f)$ to $[0, 1]^2$. Since the map φ is a contraction, by Lemma 4.5.11, we have $\gamma_{2,2}(\varphi(\mathcal{C}^\sharp)) < \infty$, and by Lemma 4.5.12, we have $\widehat{\mathcal{C}} \subset \varphi(\mathcal{C}^\sharp)$. □

Let us now come back to Earth and deal with the actual bound (4.45). For this, we develop an appropriate version of Theorem 2.7.2. It will be used many times. The ease with which one deals with two distances is remarkable. The proof of the theorem contains a principle which will be used many times: if we have two admissible sequences of partitions such that for each of them, the sets of the partition as small in a certain sense, then we can construct an admissible sequence of partitions whose sets are small in both senses.

Theorem 4.5.13 *Consider a set T provided with two distances d_1 and d_2. Consider a centered process $(X_t)_{t \in T}$ which satisfies*

$$\forall s, t \in T \ , \ \forall u > 0 \, ,$$

$$\mathsf{P}(|X_s - X_t| \ge u) \le 2 \exp\Big(- \min \Big(\frac{u^2}{d_2(s,t)^2}, \frac{u}{d_1(s,t)} \Big) \Big) \, . \tag{4.50}$$

Then

$$\mathsf{E} \sup_{s,t \in T} |X_s - X_t| \le L(\gamma_1(T, d_1) + \gamma_2(T, d_2)) \, . \tag{4.51}$$

This theorem will be applied when d_1 is the ℓ_∞ distance, but it sounds funny, when considering two distances, to call them d_2 and d_∞.

Proof We denote by $\Delta_j(A)$ the diameter of the set A for d_j. We consider an admissible sequence $(\mathcal{B}_n)_{n \ge 0}$ such that[11]

$$\forall t \in T \, , \sum_{n \ge 0} 2^n \Delta_1(B_n(t)) \le 2\gamma_1(T, d_1) \tag{4.52}$$

[11] The factor 2 in the right-hand side below is just in case the infimum over all partitions is not attained.

and an admissible sequence $(\mathcal{C}_n)_{n\ge 0}$ such that

$$\forall t \in T\,,\ \sum_{n\ge 0} 2^{n/2}\Delta_2(C_n(t)) \le 2\gamma_2(T,d_2)\,. \tag{4.53}$$

Here $B_n(t)$ is the unique element of $\mathcal{B}_n$ that contains t (etc.). We define partitions $\mathcal{A}_n$ of T as follows: we set $\mathcal{A}_0 = \{T\}$, and, for $n \ge 1$, we define $\mathcal{A}_n$ as the partition generated by $\mathcal{B}_{n-1}$ and $\mathcal{C}_{n-1}$, i.e., the partition that consists of the sets $B \cap C$ for $B \in \mathcal{B}_{n-1}$ and $C \in \mathcal{C}_{n-1}$. Thus $\operatorname{card}\mathcal{A}_n \le N_{n-1}^2 \le N_n$, and the sequence $(\mathcal{A}_n)$ is admissible. We then choose for each $n \ge 0$ a set T_n such that $\operatorname{card} T_n \le N_n$ which meets all the sets in $\mathcal{A}_n$. It is convenient to reformulate (4.50) as follows: when $u \ge 1$, we have

$$\forall s,t \in T\,,\ \mathsf{P}\big(|X_s - X_t| \ge u^2 d_1(s,t) + u d_2(s,t)\big) \le 2\exp(-u^2)\,.$$

We then copy the proof of (2.34), replacing (2.31) by

$$\forall t\,,\ |X_{\pi_n(t)} - X_{\pi_{n-1}(t)}| \le u^2 2^n d_1(\pi_n(t),\pi_{n-1}(t)) + u 2^{n/2} d_2(\pi_n(t),\pi_{n-1}(t))\,. \ \square$$

Exercise 4.5.14 The purpose of this exercise is to deduce Theorem 4.5.13 from Theorem 2.7.14.

(a) Prove that if for some numbers $A, B > 0$ a r.v. $Y \ge 0$ satisfies

$$\mathsf{P}(Y \ge u) \le 2\exp\Big(-\min\Big(\frac{u^2}{A^2}, \frac{u}{B}\Big)\Big)\,,$$

then for $p \ge 1$, we have $\|Y\|_p \le L(A\sqrt{p} + Bp)$.

(b) We denote by $D_n(A)$ the diameter of a subset A of T for the distance $\delta_n(s,t) = \|X_s - X_t\|_{2^n}$. Prove that under the conditions of Theorem 4.5.13, there exists an admissible sequence of partitions $(\mathcal{A}_n)$ such that

$$\sup_{t\in T}\sum_{n\ge 0} D_n(A_n(t)) \le L(\gamma_1(T,d_1) + \gamma_2(T,d_2))\,. \tag{4.54}$$

Exercise 4.5.15 Consider a space T equipped with two different distances d_1 and d_2. Prove that

$$\gamma_2(T, d_1 + d_2) \le L(\gamma_2(T,d_1) + \gamma_2(T,d_2))\,. \tag{4.55}$$

We can now state a general bound, from which we will deduce Theorem 4.5.3.

Theorem 4.5.16 *Consider a class $\mathcal{F}$ of functions on $[0,1]^2$, and assume that $0 \in \mathcal{F}$. Then*

$$\mathsf{E}\sup_{f\in\mathcal{F}}\Big|\sum_{i\le N}\Big(f(X_i) - \int f\,\mathrm{d}\lambda\Big)\Big| \le L\big(\sqrt{N}\gamma_2(\mathcal{F},d_2) + \gamma_1(\mathcal{F},d_\infty)\big)\,, \tag{4.56}$$

where d_2 and d_∞ are the distances induced on $\mathcal{F}$ by the norms of L^2 and L^∞, respectively.

Proof Combining Corollary 4.5.7 with Theorem 4.5.13, we get, since $0 \in \mathcal{F}$ and $Z_0 = 0$,

$$\mathsf{E} \sup_{f \in \mathcal{F}} |Z_f| \le \mathsf{E} \sup_{f,f' \in \mathcal{F}} |Z_f - Z_{f'}| \le L\big(\gamma_2(\mathcal{F}, 2\sqrt{N} d_2) + \gamma_1(\mathcal{F}, 4d_\infty)\big) . \quad (4.57)$$

Finally, $\gamma_2(\mathcal{F}, 2\sqrt{N}d_2) = 2\sqrt{N}\gamma_2(\mathcal{F}, d_2)$ and $\gamma_1(\mathcal{F}, 4d_\infty) = 4\gamma_1(\mathcal{F}, d_\infty)$. □

Exercise 4.5.17 Try to prove (4.25) now. Hint: Consider $\mathcal{F} = \{\mathbf{1}_{[0,k/N]}; k \le N\}$. Use Exercise 2.7.5 and entropy numbers.

In the situation which interests us, there will plenty of room to control the term $\gamma_1(\mathcal{F}, d_\infty)$, and this term is a lower-order term, which can be considered as a simple nuisance. For this term, entropy numbers suffice. To control these, we first state a general principle, which was already known to Kolmogorov.

Lemma 4.5.18 *Consider a metric space (U, d), and assume that for certain numbers B and $\alpha \ge 1$ and each $0 < \epsilon < B$, we have*

$$N(U, d, \epsilon) \le \Big(\frac{B}{\epsilon}\Big)^\alpha . \quad (4.58)$$

Consider the set $\mathcal{B}$ of 1-Lipschitz functions f on U with $\|f\|_\infty \le B$. Then for each $\epsilon > 0$, we have

$$\log N(\mathcal{B}, d_\infty, \epsilon) \le K(\alpha)\Big(\frac{B}{\epsilon}\Big)^\alpha , \quad (4.59)$$

where $K(\alpha)$ depends only on α. In particular,

$$e_n(\mathcal{B}, d_\infty) \le K(\alpha) B 2^{-n/\alpha} . \quad (4.60)$$

Proof By homogeneity, we may and do assume that $B = 1$. Using (4.58) for $\epsilon = 2^{-n}$, for each $n \ge 0$, consider a set $V_n \subset U$ with card $V_n \le 2^{n\alpha}$ such that any point of U is within distance 2^{-n} of a point of V_n. We define on $\mathcal{B}$ the distance d_n by $d_n(f, g) = \max_{x \in V_n} |f(x) - g(x)|$. We prove first that

$$d_\infty(f, g) \le 2^{-n+1} + d_n(f, g) . \quad (4.61)$$

Indeed, for any $x \in U$, we can find $y \in V_n$ with $d(x, y) \le 2^{-n}$ and then $|f(x) - g(x)| \le 2^{-n+1} + |f(y) - g(y)| \le 2^{-n+1} + d_n(f, g)$.

Denote by $W_n(f, r)$ the ball for d_n of center f and radius r. We claim that

$$W_{n-1}(f, 2^{-n+1}) \subset W_n(f, 2^{-n+3}) . \quad (4.62)$$

Indeed, using (4.61) for $n-1$ rather than n, we see that $d_n(f,g) \le d_\infty(f,g) \le 2^{-n+2} + 2^{-n+1} \le 2^{-n+3}$ for $g \in W_{n-1}(f, 2^{-n+1})$.

Next, we claim that

$$N(W_n(f, 2^{-n+3}), d_n, 2^{-n}) \le L^{\operatorname{card} V_n} . \tag{4.63}$$

Since $d_n(f,g) = \|\varphi_n(f) - \varphi_n(g)\|_\infty$ where $\varphi_n(f) = (f(x))_{x\in V_n}$, we are actually working here in $\mathbb{R}^{\operatorname{card} V_n}$, and (4.63) is a consequence of (2.47): in $\mathbb{R}^{\operatorname{card} V_n}$, we are covering a ball of radius 2^{-n+3} by balls of radius 2^{-n}.

Covering $\mathcal{B}$ by $N(\mathcal{B}, d_{n-1}, 2^{-n+1})$ balls $W_{n-1}(f, 2^{-n+1})$ and hence by $N(\mathcal{B}, d_{n-1}, 2^{-n+1})$ balls $W_n(f, 2^{-n+3})$ and then covering each of these by $N(W_n(f, 2^{-n+3}), d_n, 2^{-n}) \le L^{\operatorname{card} V_n}$ balls for d_n of radius 2^{-n}, we obtain

$$N(\mathcal{B}, d_n, 2^{-n}) \le L^{\operatorname{card} V_n} N(\mathcal{B}, d_{n-1}, 2^{-n+1}) . \tag{4.64}$$

Since $\operatorname{card} V_n = 2^{\alpha n}$, iteration of (4.64) proves that $\log N(\mathcal{B}, d_n, 2^{-n}) \le K2^{\alpha n}$. Finally, (4.61) implies that

$$\log N(\mathcal{B}, d_\infty, 2^{-n+2}) \le \log N(\mathcal{B}, d_{n-1}, 2^{-n-1}) \le K2^{\alpha n}$$

and concludes the proof. □

We apply the previous lemma to $U = [0,1]^2$ which obviously satisfies (4.58) for $\alpha = 2$, so that (4.60) implies that for $n \ge 0$,

$$e_n(\hat{\mathcal{C}}, d_\infty) \le L2^{-n/2} . \tag{4.65}$$

Proposition 4.5.19 *We have*

$$\mathsf{E} \sup_{f\in\widehat{\mathcal{C}}} \Big|\sum_{i\le N} f(X_i) - \int f \mathrm{d}\lambda\Big| \le L\sqrt{N \log N} . \tag{4.66}$$

An interesting feature of this proof is that it does not work to try to use (4.56) directly. Rather we will use (4.56) for an appropriate subset T of $\widehat{\mathcal{C}}$, which can be thought of as the "main part" of $\widehat{\mathcal{C}}$, and for the "rest" of $\widehat{\mathcal{C}}$, we will use other (and much cruder) bounds. This method is not artificial. As we will learn much later, in Theorem 6.8.3, when properly used, it always yields the best possible estimates.

Proof Consider the largest integer m with $2^{-m} \ge 1/N$. By (4.65), we may find a subset T of $\widehat{\mathcal{C}}$ with $\operatorname{card} T \le N_m$ and

$$\forall f \in \widehat{\mathcal{C}},\ d_\infty(f, T) \le L2^{-m/2} \le L/\sqrt{N} .$$

Thus for each $f \in \widehat{\mathcal{C}}$, consider $\tilde{f} \in T$ with $d_\infty(f, \tilde{f}) \le L/\sqrt{N}$. Then

$$|Z_f| \le |Z_{\tilde{f}}| + |Z_f - Z_{\tilde{f}}| = |Z_{\tilde{f}}| + |Z_{f-\tilde{f}}| \le |Z_{\tilde{f}}| + L\sqrt{N} ,$$

where we have used the obvious inequality $|Z_{f-\tilde{f}}| \le 2d_\infty(f, \tilde{f})$. Since $\tilde{f} \in T$, we obtain

$$\mathsf{E} \sup_{f \in \widehat{\mathcal{C}}} |Z_f| \le \mathsf{E} \sup_{f \in T} |Z_f| + L\sqrt{N} . \tag{4.67}$$

To prove (4.66), it suffices to show that

$$\mathsf{E} \sup_{f \in T} |Z_f| \le L\sqrt{N \log N} . \tag{4.68}$$

Proposition 4.5.10 and Lemma 4.1.3 imply $\gamma_2(T, d_2) \le L\sqrt{m} \le L\sqrt{\log N}$. Now, as in (2.56), we have

$$\gamma_1(T, d_\infty) \le L \sum_{n \ge 0} 2^n e_n(T, d_\infty) .$$

Since $e_n(T, d_\infty) = 0$ for $n \ge m$, (4.65) yields $\gamma_1(T, d_\infty) \le L2^{m/2} \le L\sqrt{N}$. Thus (4.68) follows from Theorem 4.5.16 and this completes the proof. □

Exercise 4.5.20 Use Exercise 4.5.8 to prove that Dudley's bound cannot yield better than the estimate $\gamma_2(T, d_2) \le L \log N$.

Exercise 4.5.21 Assuming $\gamma_2(\mathcal{C}^*, d_2) < \infty$, show that the previous arguments prove that

$$\mathsf{E} \sup_{f \in \mathcal{C}^*} \Big| \sum_{i \le N} f(X_i) - \int f \mathrm{d}\lambda \Big| \le L\sqrt{N}(1 + \gamma_2(\mathcal{C}^*, d_2)) .$$

Comparing with (4.78), conclude that $\gamma_2(\mathcal{C}^*, d_2) = \infty$. Convince yourself that the separated trees implicitly constructed in the proof of (4.78) also witness this.

Exercise 4.5.22 Suppose now that you are in dimension $d = 3$. Prove that $\mathsf{E} \sup_{f \in \widehat{\mathcal{C}}} | \sum_{i \le N} f(X_i) - \int f \mathrm{d}\lambda | \le L N^{2/3}$. Hint: According to Lemma 4.5.18, we have $e_n(\widehat{\mathcal{C}}, d_\infty) \le L 2^{-n/3}$. This is the only estimate you need, using the trivial fact that $e_n(\widehat{\mathcal{C}}, d_2) \le e_n(\widehat{\mathcal{C}}, d_\infty)$.

Exercise 4.5.23 Consider the space $T = \{0, 1\}^{\mathbb{N}}$ provided with the distance $d(t, t') = 2^{-j/2}$, where $j = \min\{i \ge 1; t_i \ne t'_i\}$ for $t = (t_i)_{i \ge 1}$. This space somewhat resembles the unit square, in the sense that $N(T, d, \epsilon) \le L\epsilon^{-2}$ for $\epsilon \le 1$. Prove that if $(X_i)_{i \le N}$ are i.i.d. uniformly distributed in T and $(Y_i)_{i \le N}$ are uniformly spread (in a manner which is left to the reader to define precisely), then

$$\mathsf{E} \inf_{\pi} \sum_{i \le N} d(X_i, Y_{\pi(i)}) \le L\sqrt{N} \log N \,, \tag{4.69}$$

where the infimum is over all the permutations of $\{1, \ldots, N\}$. Hint: You can do this from scratch, and for this, covering numbers suffice, e.g., in the form of (4.59). The method of Exercise 4.5.2 also works here. In Exercise 4.6.8, you will be asked to prove that this bound *is of the correct order*.

4.5.2 The Short and Magic Way

We now start studying the Bobkov-Ledoux approach [19] which considerably simplifies previous results such as the following one:

Theorem 4.5.24 ([110]) *Consider the class $\mathcal{C}^*$ of 1-Lipschitz functions on* $[0, 1]^2$ *which are zero on the boundary of* $[0, 1]^2$. *Consider points* $(z_i)_{i \le N}$ *in* $[0, 1]^2$ *and standard independent Gaussian r.v.s* g_i. *Then*

$$\mathsf{E} \sup_{f \in \mathcal{C}^*} \Big| \sum_{i \le N} g_i f(z_i) \Big|^2 \le LN \log N \,. \tag{4.70}$$

It should be obvious from Lemma 4.5.12 that in the previous result, one may replace $\mathcal{C}^*$ by $\widehat{\mathcal{C}}$ of Theorem 4.5.5. The following improves on Theorem 4.5.3:

Corollary 4.5.25 *Consider an independent sequence* $(X_i)_{i \le N}$ *of r.v.s valued in* $[0, 1]^2$. *(It is not assumed that these r.v.s have the same distribution.) Then*

$$\mathsf{E} \sup_{f \in \widehat{\mathcal{C}}} \Big| \sum_{i \le N} (f(X_i) - \mathsf{E} f(X_i)) \Big| \le L\sqrt{N \log N} \,. \tag{4.71}$$

Proof Consider i.i.d. standard Gaussian r.v.s g_i. Taking first expectation in the g_i given the X_i, it follows from Theorem 4.5.24 (or more accurately from the version of this theorem for the class $\widehat{\mathcal{C}}$) that $\mathsf{E} \sup_{f \in \widehat{\mathcal{C}}} |\sum_{i \le N} g_i f(X_i)|^2 \le LN \log N$. The Cauchy-Schwarz inequality yields $\mathsf{E} \sup_{f \in \widehat{\mathcal{C}}} |\sum_{i \le N} g_i f(X_i)| \le L\sqrt{N \log N}$. We will learn later the simple tools which allow to deduce (4.71) from this inequality, in particular the Giné-Zinn inequalities and specifically (11.35) (which has to be combined with (6.6)). □

Let us consider an integer $n \ge \sqrt{N}$ and the set

$$G = \{(k/n, \ell/n) \,;\; 0 \le k, \ell \le n - 1\} \,.$$

Using the fact that the functions $f \in \mathcal{C}^*$ are 1-Lipschitz and replacing each point z_i by the closest point in G (which is at distance $\le \sqrt{2}/n \le L/\sqrt{N}$ of z), the following is obvious:

Lemma 4.5.26 *To prove Theorem 4.5.24, we may assume that each $z_i \in G$.*

Let us define an ellipsoid $\mathcal{E}$ in $\mathbb{R}^N$ as a set $\mathcal{E} = \{\sum_{k\ge1} \alpha_k u_k\}$ where $(u_k)_{k\ge1}$ is a given sequence in $\mathbb{R}^N$ and where $(\alpha_k)_{k\ge1}$ varies over all the possible sequences with $\sum_{k\ge1} \alpha_k^2 \le 1$.[12] For $t \in \mathbb{R}^N$, we write $X_t = \sum_{i\le N} t_i g_i$ as usual.

Lemma 4.5.27 *We have*

$$\mathsf{E} \sup_{t\in\mathcal{E}} |X_t|^2 \le \sum_{k\ge1} \|u_k\|^2 .$$

Proof This is exactly the same argument as to prove (2.157). For $t = \sum_{k\ge1} \alpha_k u_k \in \mathcal{E}$ we have $X_t = \sum_{k\ge1} \alpha_k X_{u_k}$ so that by the Cauchy-Schwarz inequality we have $|X_t|^2 \le \sum_{k\ge1} |X_{u_k}|^2$ and the result follows from taking the supremum in t and expectation since $\mathsf{E}|X_{u_k}|^2 = \|u_k\|^2$. □

To prove Theorem 4.5.24, we will show that the set $\{(f(z_i))_{i\le N};\ f \in \mathcal{C}^*\}$ is a subset of an appropriate ellipsoid.

For this, we identify G with the group $\mathbb{Z}_n \times \mathbb{Z}_n$ where $\mathbb{Z}_n = \mathbb{Z}/n\mathbb{Z}$, with the idea to use Fourier analysis in G, keeping in mind that a function on $[0,1]^2$ which is zero on the boundary of this set will give rise to a function on $\mathbb{Z}_n \times \mathbb{Z}_n$. Consider the elements $\tau_1 = (1,0)$ and $\tau_2 = (0,1)$ of G. For a function $f : G \to \mathbb{R}$, we define the functions $f_1(\tau) = f(\tau+\tau_1) - f(\tau)$ and $f_2(\tau) = f(\tau+\tau_2) - f(\tau)$ and the class $\tilde{\mathcal{C}}$ of functions $G \to \mathbb{R}$ which satisfy

$$\forall \tau \in G\ ,\ |f(\tau)| \le 1\ ;\ \forall \tau \in G\ ;\ |f_1(\tau)| \le 1/n\ ;\ |f_2(\tau)| \le 1/n\ . \tag{4.72}$$

Thus, seeing the functions on $\mathcal{C}^*$ as functions on G, they belong to $\tilde{\mathcal{C}}$. Let us denote by $\widehat{G}$ the set of characters χ on G.[13] The Fourier transform $\hat{f}$ of a function f on G is the function $\hat{f}$ on $\widehat{G}$ given by $\hat{f}(\chi) = (\operatorname{card} G)^{-1} \sum_{\tau\in G} f(\tau)\bar{\chi}(\tau)$ where we recall that $|\chi(\tau)| = 1$. One then has the Fourier expansion

$$f = \sum_{\chi\in\widehat{G}} \hat{f}(\chi)\chi\ , \tag{4.73}$$

[12] The name is justified, a bit of algebra allows one to show that such a set is an ellipsoid in the usual sense, but we do not need that.

[13] A character of a group G is a group homomorphism from G to the unit circle.

and the Plancherel formula

$$\frac{1}{\operatorname{card} G}\sum_{\tau\in G}|f(\tau)|^2=\sum_{\chi\in\widehat{G}}|\hat{f}(\chi)|^2\,. \tag{4.74}$$

The key to the argument is the following:

Proposition 4.5.28 *There exist positive numbers* $(c(\chi))_{\chi\in\widehat{G}}$ *such that*

$$\sum_{\chi\in\widehat{G}}\frac{1}{c(\chi)}\le L\log n \tag{4.75}$$

and

$$\forall f\in\tilde{\mathcal{C}}\,,\ \sum_{\chi\in\widehat{G}}c(\chi)|\hat{f}(\chi)|^2\le 1\,. \tag{4.76}$$

We start the preparations for the proof of Proposition 4.5.28. The following lemma performs integration by parts:

Lemma 4.5.29 *For each function* f *on* G *and every* $\chi\in\widehat{G}$, *we have* $\hat{f}_1(\chi)=(\chi(-\tau_1)-1)\hat{f}(\chi)\,.$

Proof Since $\sum_{\tau\in G}f(\tau+\tau_1)\chi(\tau)=\sum_{\tau\in G}f(\tau)\chi(\tau-\tau_1)$ by change of variable, we have

$$\begin{aligned}(\operatorname{card} G)\hat{f}_1(\chi)&=\sum_{\tau\in G}(f(\tau+\tau_1)-f(\tau))\chi(\tau)=\sum_{\tau\in G}f(\tau)(\chi(\tau-\tau_1)-\chi(\tau))\\&=(\chi(\tau_1)-1)\sum_{\tau\in G}f(\tau)\chi(\tau)=(\chi(\tau_1)-1)(\operatorname{card} G)\hat{f}(\chi)\,,\end{aligned}$$

where we have used in the third equality that $\chi(\tau-\tau_1)=\chi(\tau)\chi(-\tau_1)$. □

Corollary 4.5.30 *For* $f\in\tilde{\mathcal{C}}$, *we have*

$$\sum_{\chi\in\widehat{G}}(|\chi(-\tau_1)-1|^2+|\chi(-\tau_2)-1|^2)|\hat{f}(\chi)|^2\le\frac{2}{n^2}\,. \tag{4.77}$$

Proof Using Lemma 4.5.29 and then the Plancherel formula (4.74) and (4.72), we obtain

$$\sum_{\chi\in\widehat{G}}|\chi(-\tau_1)-1|^2|\hat{f}(\chi)|^2=\sum_{\chi\in\widehat{G}}|\hat{f}_1(\chi)|^2=\frac{1}{\operatorname{card} G}\sum_{\tau\in G}|f_1(\tau)|^2\le\frac{1}{n^2}\,,$$

and we proceed similarly for τ_2. □

Proof of Proposition 4.5.28 For $\chi \in \widehat{G}$, let us set $c(\chi) = 1/2$ if χ is the constant character χ_0 equal to 1, and otherwise

$$c(\chi) = \frac{n^2}{4}\big(|\chi(-\tau_1) - 1|^2 + |\chi(-\tau_2) - 1|^2\big) .$$

Then, since $|\hat{f}(\chi_0)| \le 1$ because $|f(\tau)| \le 1$ for each τ and using (4.77) in the second inequality,

$$\sum_{\chi\in\hat{G}} c(\chi)|\widehat{f}(\chi)|^2 = \frac{1}{2}|\hat{f}(\chi_0)|^2 + \sum_{\chi\in\hat{G},\chi\neq\chi_0} c(\chi)|\hat{f}(\chi)|^2 \le \frac{1}{2} + \frac{n^2}{4}\frac{2}{n^2} \le 1 ,$$

and this proves (4.76). To prove (4.75), we use that $\widehat{G}$ is exactly the set of characters of the type $\chi_{p,q}(a,b) = \exp(2i\pi(ap+bq)/n)$ where $0 \le p,q \le n-1$. Thus $\chi_{p,q}(-\tau_1) = \exp(-2i\pi p/n)$ and$\chi_{p,q}(-\tau_2) = \exp(-2i\pi q/n)$. Now, for $0 \le x \le 1$, we have $|1 - \exp(-2i\pi x)| \ge \min(x, 1-x)$ so that

$$|\chi_{p,q}(-\tau_1) - 1| \ge \frac{1}{Ln}\min(p, n-p) \; ; \; |\chi_{p,q}(-\tau_2) - 1| \ge \frac{1}{Ln}\min(q, n-q) .$$

Thus

$$\sum_{\chi\in\widehat{G}} \frac{1}{c(\chi)} \le \frac{1}{c(\chi_0)} + L\sum \frac{1}{\min(p,n-p)^2 + \min(q,n-q)^2} ,$$

where the sum is over $0 \le p,q \le n-1$ and $(p,q) \ne (0,0)$. Distinguishing whether $p \le n/2$ or $p \ge n/2$ (and similarly for q), we obtain

$$\sum \frac{1}{\min(p,n-p)^2 + \min(q,n-q)^2} \le 4\sum \frac{1}{p^2+q^2} ,$$

where the sum is over the same set and this sum is $\le L\log n$. □

Proof of Theorem 4.5.24 We write (4.73) as $f = \sum_{\chi\in\widehat{G}} \alpha_\chi \chi/\sqrt{c(\chi)}$ where $\alpha_\chi = \sqrt{c(\chi)}\hat{f}(\chi)$ so that $\sum_{\chi\in\widehat{G}} |\alpha_\chi|^2 \le 1$ by (4.76). Now we come back to real numbers by taking the real part of the identity $f = \sum_{\chi\in\widehat{G}} \alpha_\chi\chi/\sqrt{c(\chi)}$. This gives an equality of the type $f = \sum_{\chi\in\widehat{G}} (\alpha'_\chi\chi' + \beta'_\chi\chi'')/\sqrt{c(\chi)}$ where $\sum_{\chi\in\widehat{G}}((\alpha'_\chi)^2 + (\beta'_\chi)^2) \le 1$ and $|\chi'|, |\chi''| \le 1$. That is, the set $\{(f(z_i))_{i\le N};\, f \in \mathcal{C}^*\}$ is a subset of the ellipsoid $\mathcal{E} = \{\sum_{k\ge 1}\alpha_k u_k;\, \sum_k \alpha_k^2 \le 1\}$, where the family (u_k) of points of $\mathbb{R}^N$ consists of the points $(\chi'(z_i)/\sqrt{c(\chi)})_{i\le N}$ and $(\chi''(z_i)/\sqrt{c(\chi)})_{i\le N}$ where χ takes all possible values in $\widehat{G}$. For such a u_k, we have $|u_k(z_i)| \le 1/\sqrt{c(\chi)}$ so that $\|u_k\|^2 = \sum_{i\le N} u_k(z_i)^2 \le N/c(\chi)$, and then by (4.75), we have $\sum_k \|u_k\|^2 \le$

$LN \log n$. Finally we apply Lemma 4.5.27, and we take for n the smallest integer $\geq \sqrt{N}$. □

Exercise 4.5.31 Let us denote by ν the uniform measure on G and by d_ν the distance in the space $L^2(\nu)$. Prove that $\gamma_2(\mathcal{C}^*, d_\nu) \geq \sqrt{\log n}/L$. Warning: This is not so easy, and the solution is not provided. Hint: Make sure you understand the previous chapter, and construct an appropriate tree. The ingredients on how to build that tree are contained in the proof given in the next section, and Sect. 3.2 should also be useful. You may assume that N is a power of 2 to save technical work. Furthermore, you may also look at [132] where trees were explicitly used.

4.6 Lower Bound for the Ajtai-Komlós-Tusnády Theorem

Recall that $\mathcal{C}^*$ denotes the class of 1-Lipschitz functions on the unit square which are zero on the boundary of the square. We shall prove the following, where $(X_i)_{i \leq N}$ are i.i.d. in $[0,1]^2$:

Theorem 4.6.1 *We have*

$$\mathsf{E} \sup_{f \in \mathcal{C}^*} \Big| \sum_{i \leq N} \Big(f(X_i) - \int f \mathrm{d}\lambda \Big) \Big| \geq \frac{1}{L} \sqrt{N \log N} \,. \tag{4.78}$$

Since

$$\sup_{f \in \mathcal{C}^*} \Big| \sum_{i \leq N} \Big(f(X_i) - \int f \mathrm{d}\lambda \Big) \Big| \leq \sup_{f \in \mathcal{C}^*} \Big| \sum_{i \leq N} (f(X_i) - f(Y_i)) \Big| + \sup_{f \in \mathcal{C}^*} \Big| \sum_{i \leq N} \Big(f(Y_i) - \int f \mathrm{d}\lambda \Big) \Big| \,,$$

taking expectation and using (4.41), it follows from (4.78) that if the points Y_i are evenly spread, then (provided $N \geq L$)

$$\mathsf{E} \sup_{f \in \mathcal{C}^*} \Big| \sum_{i \leq N} (f(X_i) - f(Y_i)) \Big| \geq \frac{1}{L} \sqrt{N \log N} \,,$$

and since $\mathcal{C}^* \subset \mathcal{C}$, the duality formula (4.29) implies that the expected cost of matching the points X_i and the points Y_i is at least $\sqrt{N \log N}/L$.

The proof of Theorem 4.6.1 occupies this entire section and starts now. The beautiful argument we present goes back to [3]. We can expect that this proof is non-trivial. To explain why, let us recall the set T used in the proof of Proposition 4.5.19. Analysis of the proof of that proposition leads us to guess that the reason why the bound it provides cannot be improved is that $\gamma_2(T, d_2)$ is actually of order $\sqrt{\log N}$

(and not of a smaller order). So a proof of (4.78) must contain a proof that this is the case. In the previous chapter, we learned a technique to prove such results, the construction of "trees". Not surprisingly, our proof implicitly uses a tree which witnesses just this, somewhat similar to the tree we constructed in Sect. 3.2.[14]

We may assume N large, and we consider a number $r \in \mathbb{N}$ which is a small proportion of $\log N$, say, $r \simeq (\log N)/100$.[15] The structure of the proof is to recursively construct for $k \le r$ certain (random) functions f_k such that for any $q \le r$

$$\sum_{k \le q} f_k \text{ is 1-Lipschitz} \tag{4.79}$$

and for each $k \le r$,

$$\mathsf{E} \sum_{i \le N} \Big(f_k(X_i) - \int f_k \mathrm{d}\lambda \Big) \ge \frac{\sqrt{N}}{L\sqrt{r}} \,. \tag{4.80}$$

The function $f^* = \sum_{k \le r} f_k$ is then 1-Lipschitz and satisfies

$$\mathsf{E} \sum_{i \le N} \Big(f^*(X_i) - \int f^* \mathrm{d}\lambda \Big) \ge \frac{\sqrt{Nr}}{L} \,.$$

This completes the proof of (4.78). The function f_k looks to what happens at scale 2^{-k}, and (4.80) states that each such scale 2^{-k} contributes about equally to the final result.

Following the details of the construction is not that difficult, despite the fact that ensuring (4.79) requires technical work. What is more difficult is to see why one makes such choices as we do. There is no magic there, making the right choices means that we have understood which aspect of the geometric complexity of the class $\mathcal{C}^*$ is relevant here.

The main idea behind the construction of the function f_k is that if we divide $[0, 1]^2$ into little squares of side 2^{-k}, in each of these little squares, there is some irregularity of the distribution of the X_i. The function f_k is a sum of terms, each corresponding to one of the little squares (see (4.90)). It is designed to, in a sense, *add* the irregularities over these different squares.

---

[14] It should also be useful to solve Exercise 4.5.31.

[15] We absolutely need for the proof a number r which is a proportion of $\log N$, and taking a small proportion gives us some room. More specifically, each square of side 2^{-r} will have an area larger than, say, $1/\sqrt{N}$, so that it will typically contain many points X_i, as we will use when we perform normal approximation.

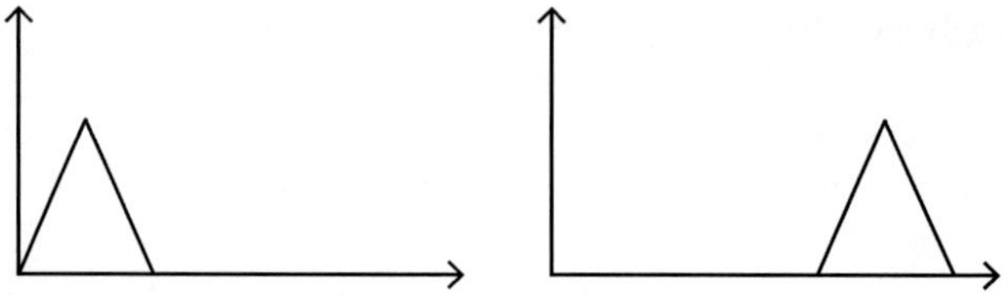

Fig. 4.1 The graphs of $f_{k,1}$ and $f_{k,3}$

The functions f_k will be built out of simple functions which we describe now. For $1 \le k \le r$ and $1 \le \ell \le 2^k$, we consider the function $f'_{k,\ell}$ on $[0, 1]$ defined as follows:

$$f'_{k,\ell}(x) = \begin{cases} 0 & \text{unless } x \in [(\ell-1)2^{-k}, \ell 2^{-k}[\\ 1 & \text{for } x \in [(\ell-1)2^{-k}, (\ell-1/2)2^{-k}[\\ -1 & \text{for } x \in [(\ell-1/2)2^{-k}, \ell 2^{-k}[\;. \end{cases} \tag{4.81}$$

We define (Fig. 4.1)

$$f_{k,\ell}(x) = \int_0^x f'_{k,\ell}(y)\mathrm{d}y \;. \tag{4.82}$$

We now list a few useful properties of these functions. In these formulas, $\|.\|_2$ denotes the norm in $L^2([0, 1])$, etc. The proofs of these assertions are completely straightforward and better left to the reader.

Lemma 4.6.2 *The following holds true:*

$$\textit{The family } (f'_{k,\ell}) \textit{ is orthogonal in } L^2([0,1]) \;. \tag{4.83}$$

$$\|f'_{k,\ell}\|_2^2 = 2^{-k} \;. \tag{4.84}$$

$$\|f'_{k,\ell}\|_1 = 2^{-k} \;. \tag{4.85}$$

$$f_{k,\ell} \textit{ is zero outside }](\ell-1)2^{-k}, \ell 2^{-k}[\;. \tag{4.86}$$

$$\|f_{k,\ell}\|_1 = 2^{-2k-2} \;. \tag{4.87}$$

$$\|f'_{k,\ell}\|_\infty = 1 \;;\; \|f_{k,\ell}\|_\infty = 2^{-k-1} \;. \tag{4.88}$$

$$\|f_{k,\ell}\|_2^2 = \frac{1}{12}2^{-3k} \;. \tag{4.89}$$

The functions f_k will be of the type

$$f_k = \frac{2^{k-5}}{\sqrt{r}} \sum_{\ell,\ell' \le 2^k} z_{k,\ell,\ell'} f_{k,\ell} \otimes f_{k,\ell'} \,, \tag{4.90}$$

where $f_{k,\ell} \otimes f_{k,\ell'}(x,y) = f_{k,\ell}(x) f_{k,\ell'}(y)$ and $z_{k,\ell,\ell'} \in \{0,1,-1\}$. Note that $f_{k,\ell} \otimes f_{k,\ell'}$ is zero outside the little square $[(\ell-1)2^{-k}, \ell 2^{-k}[\times [(\ell'-1)2^{-k}, \ell' 2^{-k}]$ and that these little squares are disjoint as ℓ and ℓ' vary. The term $z_{k,\ell,\ell'} f_{k,\ell} \otimes f_{k,\ell'}$ is designed to take advantage of the irregularity of the distribution of the X_i in the corresponding little square. The problem of course is to choose the numbers $z_{k,\ell,\ell'}$. There are two different ideas here: First, as a technical device to ensure (4.79), $z_{k,\ell,\ell'}$ may be set to zero. This will happen on a few little squares, those where are getting dangerously close to violate this condition (4.79). The second idea is that we will adjust the signs of $z_{k,\ell,\ell'}$ in a way that the contributions of the different little squares add properly (rather than canceling each other).

Let us now explain the scaling term $2^{k-5}/\sqrt{r}$ in (4.90). The coefficient 2^{-5} is just a small numerical constant ensuring that we have enough room. The idea of the term $2^k/\sqrt{r}$ is that the partial derivatives of f_k will be of order $1/\sqrt{r}$. Taking a sum of a most r such terms and taking cancellations in effect will give us partial derivatives which at most of the points are ≤ 1. This is formally expressed in the next lemma. So, such sums are not necessarily 1-Lipschitz, but are pretty close to being so, and some minor tweaking will ensure that they are.

Lemma 4.6.3 *Consider $q \le r$. Consider a function of the type $f = \sum_{k \le q} f_k$, where f_k is given by (4.90) and where $z_{k,\ell,\ell'} \in \{0,1,-1\}$. Then*

$$\left\| \frac{\partial f}{\partial x} \right\|_2 \le 2^{-6} \,. \tag{4.91}$$

Proof First we note that

$$\frac{\partial f}{\partial x}(x,y) = \sum_{k \le q} \frac{2^{k-5}}{\sqrt{r}} \sum_{\ell,\ell' \le 2^k} z_{k,\ell,\ell'} f'_{k,\ell}(x) f_{k,\ell'}(y) \,, \tag{4.92}$$

which we rewrite as

$$\frac{\partial f}{\partial x}(x,y) = \sum_{k \le q} \frac{2^{k-5}}{\sqrt{r}} \sum_{\ell \le 2^k} a_{k,\ell}(y) f'_{k,\ell}(x) \,,$$

where $a_{k,\ell}(y) = \sum_{\ell' \le 2^k} z_{k,\ell,\ell'} f_{k,\ell'}(y)$. Using (4.83), we obtain

$$\int \left(\frac{\partial f}{\partial x}\right)^2 \mathrm{d}x = \sum_{k \le q} \frac{2^{2k-10}}{r} \sum_{\ell \le 2^k} a_{k,\ell}(y)^2 \| f'_{k,\ell} \|_2^2 \,.$$

Since $z_{k,\ell,\ell'}^2 \le 1$ and since the functions $(f_{k,\ell})_{\ell \le 2^k}$ have disjoint support, we have $a_{k,\ell}(y)^2 \le \sum_{\ell' \le 2^k} f_{k,\ell'}(y)^2$, so that

$$\int \Big(\frac{\partial f}{\partial x}\Big)^2 \mathrm{d}x \le \sum_{k \le q} \frac{2^{2k-10}}{r} \sum_{\ell,\ell' \le 2^k} \|f'_{k,\ell}\|_2^2 f_{k,\ell'}(y)^2 .$$

Integrating in y and using (4.84) and (4.89) yield

$$\Big\|\frac{\partial f}{\partial x}\Big\|_2^2 \le \sum_{k \le q} \frac{2^{2k-10}}{r} \frac{2^{-2k}}{12} = \frac{q}{r} \frac{2^{-10}}{12} \le 2^{-12} . \qquad \square$$

Naturally, we have the same bound for $\|\partial f/\partial y\|_2$. These bounds do not imply that f is 1-Lipschitz, but they imply that it is 1-Lipschitz "most of the time".

We construct the functions f_k recursively. Having constructed $f_1, \ldots, f_q$, let $f := \sum_{k \le q} f_k$, and assume that it is 1-Lipschitz. We will construct f_{q+1} of the type (4.90) by choosing the coefficients $z_{q+1,\ell,\ell'}$. Let us say that a square of the type

$$[(\ell - 1)2^{-q}, \ell 2^{-q}[\times[(\ell' - 1)2^{-q}, \ell' 2^{-q}[\tag{4.93}$$

for $1 \le \ell, \ell' \le 2^q$ is a q-*square*. There are 2^{2q} such q-squares.

Definition 4.6.4 We say that a $(q+1)$-square is *dangerous* if it contains a point for which either $|\partial f/\partial x| \ge 1/2$ or $|\partial f/\partial y| \ge 1/2$. We say that it is *safe* if it is not dangerous.

The danger is that on this square (4.93), the function $f + f_{q+1}$ might not be 1-Lipschitz.

Lemma 4.6.5 *At most half of the $(q+1)$-squares are dangerous, so at least half of the $(q+1)$-squares are safe.*

This lemma is a consequence of the fact that "f is 1-Lipschitz most of the time." The proof is a bit technical, so we delay it to the end of the section.

The following is also a bit technical but is certainly expected. It will also be proved later.

Lemma 4.6.6 *If $z_{q+1,\ell,\ell'} = 0$ whenever the corresponding $(q+1)$-square (4.93) is dangerous, then $f + f_{q+1}$ is 1-Lipschitz.*

We now complete the construction of the function f_{q+1}. For a dangerous square, we set $z_{q+1,\ell,\ell'} = 0$. Let us define

$$h_{\ell,\ell'}(x) = f_{q+1,\ell} \otimes f_{q+1,\ell'}(x) - \int f_{q+1,\ell} \otimes f_{q+1,\ell'} \mathrm{d}\lambda .$$

Using (4.89) and (4.87), we obtain

$$\|h_{\ell,\ell'}\|_2^2 \ge 2^{-6q}/L \ . \tag{4.94}$$

Let us define then

$$D_{\ell,\ell'} = \sum_{i\le N} h_{\ell,\ell'}(X_i) \ . \tag{4.95}$$

For a safe square, we choose $z_{q+1,\ell,\ell'} = \pm 1$ such that

$$z_{q+1,\ell,\ell'} D_{\ell,\ell'} = |D_{\ell,\ell'}| \ .$$

Thus, if

$$f_{q+1} = \frac{2^{q-4}}{\sqrt{r}} \sum_{\ell,\ell'\le 2^{q+1}} z_{q+1,\ell,\ell'} f_{q+1,\ell} \otimes f_{q+1,\ell'}$$

we have

$$\sum_{i\le N}\Bigl(f_{q+1}(X_i) - \int f_{q+1}\mathrm{d}\lambda\Bigr) = \frac{2^{q-4}}{\sqrt{r}} \sum_{\text{safe}} |D_{\ell,\ell'}| \ . \tag{4.96}$$

where the sum is over all values of (ℓ, ℓ') such that the corresponding square (4.93) is safe.

We turn to the proof of (4.80) and for this we estimate $\mathsf{E}\sum_{\text{safe}} |D_{\ell,\ell'}|$. An obvious obstacle to perform this estimate is that the r.v.s $D_{\ell,\ell'}$ are not independent of the set of safe squares. But we know that at least half of the squares are safe, so we can bound below $\sum_{\text{safe}} |D_{\ell,\ell'}|$ by the sum of the 2^{2q+1} smallest among the 2^{2q+2} r.v.s $|D_{\ell,\ell'}|$.

Let us estimate $\mathsf{E}D^2_{\ell,\ell'}$. By definition, (4.95) $D_{\ell,\ell'}$ is a sum $\sum_{i\le N} h_{\ell,\ell'}(X_i)$ of independent centered r.v.s so that $\mathsf{E}D^2_{\ell,\ell'} = N\|h_{\ell,\ell'}\|_2^2$, and using (4.94), we obtain an estimate

$$\mathsf{E}D^2_{\ell,\ell'} \ge 2^{-6q} N/L \ . \tag{4.97}$$

Let us then *pretend* for a moment that the r.v.s $D_{\ell,\ell'}$ are Gaussian and independent as ℓ, ℓ' vary. For a Gaussian r.v. g, we have $\mathsf{P}(|g| \ge (\mathsf{E}g^2)^{1/2}/100) \ge 7/8$. Then for each ℓ, ℓ', we have $|D_{\ell,\ell'}| \ge 2^{-3q}\sqrt{N/L}$ with probability $\ge 7/8$. In other words, the r.v. $Y_{\ell,\ell'} = \mathbf{1}_{|D_{\ell,\ell'}|\ge 2^{-3q}\sqrt{N/L}}$ satisfies $\mathsf{E}Y_{\ell,\ell'} \ge 7/8$. Then Bernstein's inequality shows that with overwhelming probability, at least 3/4 of these variables equal 1. For further use let us state the following more general principle: Considering M

independent r.v.s $Z_i \in \{0,1\}$ with $\mathsf{P}(Z_i = 1) = a_i = \mathsf{E}Z_i$, then for $u > 0$, we have

$$\mathsf{P}\Big(\big|\sum_{i\le M}(Z_i - a_i)\big| \ge uM\Big) \le 2\exp(-Mu^2/L)\,, \tag{4.98}$$

and in particular $\mathsf{P}(\sum_{i\le M} Z_i \le \sum_{i\le M} a_i - Mu) \le 2\exp(-Mu^2/L)$.

Thus, $|D_{\ell,\ell'}| \ge 2^{-3q}\sqrt{N}/L$ for at least 3/4 of the squares, so that at least 1/4 of the squares are both safe and satisfy this inequality. Consequently, it follows as desired from (4.96) that (4.80) holds (for $q+1$ rather than q).

It is not exactly true that the r.v.s $D_{\ell,\ell'}$ are independent and Gaussian. Standard techniques exist to take care of this, namely, Poissonization and normal approximation. There is all the room in the world because $r \le (\log N)/100$. As these considerations are not related to the rest of the material of this work, they are better omitted.

We now turn to the proofs of Lemmas 4.6.5 and 4.6.6. The next lemma prepares for these proofs.

Lemma 4.6.7 *Consider $q \le r$ and a function of the type $f = \sum_{k\le q} f_k$, where f_k is given by (4.90) and where $z_{k,\ell,\ell'} \in \{0,1,-1\}$. Then*

$$\Big|\frac{\partial^2 f}{\partial x \partial y}\Big| \le \frac{2^{q-4}}{\sqrt{r}}\,. \tag{4.99}$$

Proof We have

$$\Big|\frac{\partial^2 f}{\partial x \partial y}\Big| = \Big|\sum_{k\le q}\frac{2^{k-5}}{\sqrt{r}}\sum_{\ell,\ell'\le 2^k} z_{k,\ell,\ell'} f'_{k,\ell}\otimes f'_{k,\ell'}\Big| \le \sum_{k\le q}\frac{2^{k-5}}{\sqrt{r}}\sum_{\ell,\ell'\le 2^k}|f'_{k,\ell}\otimes f'_{k,\ell'}|\,.$$

The functions $f'_{k,\ell}\otimes f'_{k,\ell'}$ have disjoint support and by the first part of (4.88) $|f'_{k,\ell}\otimes f'_{k,\ell'}| \le 1$. Also, $\sum_{k\le q} 2^k \le 2^{q+1}$. □

Proof of Lemma 4.6.5 We will observe from the definition that all functions $f'_{k,\ell}$ for $k \le q$ are constant on the intervals $[\ell 2^{-q-1}, (\ell+1)2^{-q-1}[$. Thus according to (4.92), on a $(q+1)$-square, $\partial f/\partial x$ does not depend on x. If (x,y) and (x',y') belong to the same $(q+1)$-square, then

$$\frac{\partial f}{\partial x}(x',y') = \frac{\partial f}{\partial x}(x,y')\,. \tag{4.100}$$

Moreover, $|y-y'| \le 2^{-q-1}$ so that (4.99) implies

$$\Big|\frac{\partial f}{\partial x}(x,y) - \frac{\partial f}{\partial x}(x,y')\Big| \le |y-y'|\frac{2^{q-5}}{\sqrt{r}} \le \frac{2^{-6}}{\sqrt{r}}\,,$$

and combining with (4.100), we obtain

$$\Big|\frac{\partial f}{\partial x}(x,y)-\frac{\partial f}{\partial x}(x',y')\Big|=\Big|\frac{\partial f}{\partial x}(x,y)-\frac{\partial f}{\partial x}(x,y')\Big|\le\frac{2^{-6}}{\sqrt{r}}\,.$$

In particular if a $(q+1)$-square contains a point at which $|\partial f/\partial x|\ge 1/2$, then at each point of this square, we have $|\partial f/\partial x|\ge 1/2-2^{-6}/\sqrt{r}\ge 1/4$. The proportion α of $(q+1)$-squares with this property satisfies $\alpha(1/4)^2\le\|\partial f/\partial x\|_2^2\le 2^{-12}$, where we have used (4.91) in the last inequality. This implies that at most a proportion 2^{-8} of $(q+1)$-squares can contain a point with $|\partial f/\partial x|\ge 1/2$. Repeating the same argument for $\partial f/\partial y$ shows that as desired at most half of $(q+1)$-squares are dangerous. □

Proof of Lemma 4.6.6 To ensure that $g:=f+f_{q+1}$ is 1-Lipschitz, it suffices to ensure that it is 1-Lipschitz on each $(q+1)$-square. When the square is dangerous, $f_{q+1}=0$ on this square by construction, and g is 1-Lipschitz on it because there $g=f$ and f is 1-Lipschitz.

When the square is safe, everywhere on the square we have $|\partial f/\partial x|\le 1/2$ and $|\partial f/\partial y|\le 1/2$. Now the second part of (4.88) implies

$$\Big\|\frac{\partial g}{\partial x}-\frac{\partial f}{\partial x}\Big\|_\infty=\Big\|\frac{1}{\sqrt{r}}2^{q-4}\sum_{\ell,\ell'\le 2^{q+1}}z_{q+1,\ell,\ell'}f'_{q+1,\ell}\otimes f_{q+1,\ell'}\Big\|_\infty\le\frac{1}{2^5\sqrt{r}}$$

and

$$\Big\|\frac{\partial g}{\partial y}-\frac{\partial f}{\partial y}\Big\|_\infty=\Big\|\frac{1}{\sqrt{r}}2^{q-4}\sum_{\ell,\ell'\le 2^{q+1}}z_{q+1,\ell,\ell'}f_{q+1,\ell}\otimes f'_{q+1,\ell'}\Big\|_\infty\le\frac{1}{2^5\sqrt{r}}\,,$$

where we have used that the elements of the sum have disjoint supports. So we are certain that at each point of a safe square we have $|\partial g/\partial x|\le 1/\sqrt{2}$ and $|\partial g/\partial y|\le 1/\sqrt{2}$ and hence that g is 1-Lipschitz on a safe square. □

Exercise 4.6.8 This is a continuation of Exercise 4.5.23. Adapt the method you learned in this section to prove that the bound (4.69) is of the correct order.

4.7 The Leighton-Shor Grid Matching Theorem

Theorem 4.7.1 ([55]) *If the points $(Y_i)_{i\le N}$ are evenly spread and if $(X_i)_{i\le N}$ are i.i.d. uniform over $[0,1]^2$, then (for $N\ge 2$), with probability at least $1-L\exp(-(\log N)^{3/2}/L)$, we have*

$$\inf_\pi\sup_{i\le N}d(X_i,Y_{\pi(i)})\le L\frac{(\log N)^{3/4}}{\sqrt{N}}\,. \tag{4.101}$$

In particular

$$\mathsf{E} \inf_{\pi} \sup_{i \le N} d(X_i, Y_{\pi(i)}) \le L \frac{(\log N)^{3/4}}{\sqrt{N}} . \tag{4.102}$$

To deduce (4.102) from (4.101), one simply uses any matching in the (rare) event that (4.101) fails. We shall prove in Sect. 4.8 that the inequality (4.102) can be reversed. A close cousin of this theorem can be found in Appendix A.

A first simple idea is that to prove Theorem 4.7.1, we do not care about what happens at a scale smaller than $(\log N)^{3/4}/\sqrt{N}$. Therefore, consider the largest integer ℓ_1 with $2^{-\ell_1} \ge (\log N)^{3/4}/\sqrt{N}$ (so that in particular $2^{\ell_1} \le \sqrt{N}$). We divide $[0, 1]^2$ into little squares of side $2^{-\ell_1}$. For each such square, we are interested in how many points (X_i) it contains, but we do not care where these points are located in the square. We shall deduce Theorem 4.7.1 from a discrepancy theorem for a certain class of functions.[16] What we really have in mind is the class of functions which are indicators of a union A of little squares with sides of length $2^{-\ell_1}$ and such that the boundary of A has a given length. It turns out that we shall have to parametrize the boundaries of these sets by curves, so it is convenient to turn things around and to consider the class of sets A that are the interiors of curves of given length.

To make things precise, let us define the *grid* G of $[0, 1]^2$ of mesh width $2^{-\ell_1}$ by

$$G = \left\{(x_1, x_2) \in [0, 1]^2 \; ; \; 2^{\ell_1} x_1 \in \mathbb{N} \textbf{ or } 2^{\ell_1} x_2 \in \mathbb{N}\right\} .$$

A *vertex* of the grid is a point $(x_1, x_2) \in [0, 1]^2$ with $2^{\ell_1} x_1 \in \mathbb{N}$ **and** $2^{\ell_1} x_2 \in \mathbb{N}$. There are $(2^{\ell_1}+1)^2$ vertices. An *edge* of the grid is the segment between two vertices that are at distance $2^{-\ell_1}$ of each other. A *square* of the grid is a square of side $2^{-\ell_1}$ whose edges are edges of the grid. Thus, an edge of the grid is a subset of the grid, but a square of the grid is not a subset of the grid (see Fig. 4.2).

A *curve* is the image of a continuous map $\varphi : [0, 1] \to \mathbb{R}^2$. We say that the curve is a *simple curve* if it is one-to-one on $[0, 1[$. We say that the curve is *traced on G* if $\varphi([0, 1]) \subset G$ and that it is *closed* if $\varphi(0) = \varphi(1)$. If C is a closed simple curve in $\mathbb{R}^2$, the set $\mathbb{R}^2 \setminus C$ has two connected components. One of these is bounded. It is called the interior of C and is denoted by $\overset{o}{C}$.

The proof of Theorem 4.7.1 has a probabilistic part (the hard one) and a deterministic part. The probabilistic part states that with high probability the number of points inside a closed curve differs from its expected value by at most the length of the curve times $L\sqrt{N}(\log N)^{3/4}$. The deterministic part will be given at the end of the section and will show how to deduce Theorem 4.7.1 from Theorem 4.7.2.

[16] This is the case for every matching theorem we prove.

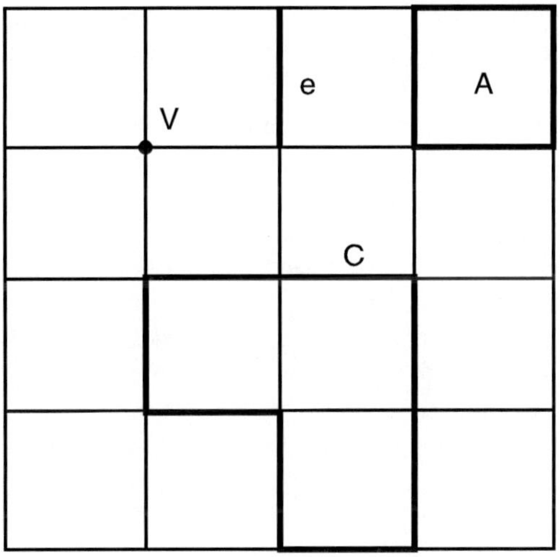

Fig. 4.2 A square A, and edge e, a vertex V, and a simple curve C traced on G

Theorem 4.7.2 *With probability at least* $1 - L\exp(-(\log N)^{3/2}/L)$, *the following occurs: Given any closed simple curve* C *traced on* G, *we have*

$$\Big|\sum_{i\le N}\big(\mathbf{1}_{\overset{o}{C}}(X_i) - \lambda(\overset{o}{C})\big)\Big| \le L\ell(C)\sqrt{N}(\log N)^{3/4}\,, \tag{4.103}$$

where $\lambda(\overset{o}{C})$ *is the area of* $\overset{o}{C}$ *and* $\ell(C)$ *is the length of* C.

We will reduce the proof of this theorem to the following result, which concerns curves of a given length going through a given vertex:

Proposition 4.7.3 *Consider a vertex* τ *of* G *and* $k \in \mathbb{Z}$. *Define* $\mathcal{C}(\tau, k)$ *as the set of closed simple curves traced on* G *that pass through* τ[17] *and have length* $\le 2^k$. *Then, if* $-\ell_1 \le k \le \ell_1 + 2$, *with probability at least* $1 - L\exp(-(\log N)^{3/2}/L)$, *for each* $C \in \mathcal{C}(\tau, k)$, *we have*

$$\Big|\sum_{i\le N}\big(\mathbf{1}_{\overset{o}{C}}(X_i) - \lambda(\overset{o}{C})\big)\Big| \le L2^k\sqrt{N}(\log N)^{3/4}\,. \tag{4.104}$$

It would be easy to control the left-hand side if one considered only curves with a simple pattern, such as boundaries of rectangles. The point, however, is that the curves we consider can be very complicated and the longer we allow them to be, the more so. Before we discuss Proposition 4.7.3 further, we show that it implies Theorem 4.7.2.

[17] That is, τ is an end vertex of an edge which belongs to the curve.

Proof of Theorem 4.7.2 Since there are at most $(2^{\ell_1}+1)^2 \le LN$ choices for the vertex τ, we can assume with probability at least

$$1 - L(2^{\ell_1}+1)^2(2\ell_1+4)\exp(-(\log N)^{3/2}/L) \ge 1 - L'\exp\big(-(\log N)^{3/2}/L'\big) \tag{4.105}$$

that (4.104) occurs for all choices of $C \in \mathcal{C}(\tau,k)$, for any τ and any k with $-\ell_1 \le k \le \ell_1+2$.

Consider a simple curve C traced on G. Bounding the length of C by the total length of the edges of G, we have $2^{-\ell_1} \le \ell(C) \le 2(2^{\ell_1}+1) \le 2^{\ell_1+2}$. Then the smallest integer k for which $\ell(C) \le 2^k$ satisfies $-\ell_1 \le k \le \ell_1+2$. Since $2^k \le 2\ell(C)$, the proof is finished by (4.104). □

Exercise 4.7.4 Prove the second inequality in (4.105) in complete detail.

The main step to prove Proposition 4.7.3 is the following:

Proposition 4.7.5 *Consider a vertex τ of G and $k \in \mathbb{Z}$. Define $\mathcal{C}(\tau,k)$ as in Proposition 4.7.3. Then, if $-\ell_1 \le k \le \ell_1+2$, we have*

$$\mathsf{E} \sup_{C\in\mathcal{C}(\tau,k)} \Big|\sum_{i\le N}\big(\mathbf{1}_{\overset{o}{C}}(X_i) - \lambda(\overset{o}{C})\big)\Big| \le L2^k\sqrt{N}(\log N)^{3/4}\,. \tag{4.106}$$

Proof of Proposition 4.7.3 To prove Proposition 4.7.3, we have to go from the control in expectation provided by (4.106) to the control in probability of (4.104). There is powerful tool to do this: concentration of measure. The function

$$f(x_1,\ldots,x_N) = \sup_{C\in\mathcal{C}(\tau,k)} \Big|\sum_{i\le N}\big(\mathbf{1}_{\overset{o}{C}}(x_i) - \lambda(\overset{o}{C})\big)\Big|$$

of points $x_1,\ldots,x_N \in [0,1]^2$ has the property that changing the value of a given variable x_i can change the value of f by at most one. One of the earliest "concentration of measure" results (for which we refer to [52]) asserts that for such a function, the r.v. $W = f(X_1,\ldots,X_N)$ satisfies a deviation inequality of the form

$$\mathsf{P}(|W - \mathsf{E}W| \ge u) \le 2\exp\Big(-\frac{u^2}{2N}\Big)\,. \tag{4.107}$$

Using (4.106) to control $\mathsf{E}W$ and taking $u = L2^k\sqrt{N}(\log N)^{3/4}$ prove Proposition 4.7.3 in the case $k \ge 0$. A little bit more work is needed when $k < 0$. In that case, a curve of length 2^k is entirely contained in the square V of center τ and side 2^{k+1} and $\mathbf{1}_{\overset{o}{C}}(X_i) = 0$ unless $X_i \in V$. To take advantage of this, we work conditionally on $I = \{i \le N; X_i \in V\}$, and we can then use (4.107) with card I instead of N. This provides the desired inequality when card $I \le L2^{2k}N$. On the other hand, by (4.98) and since $\lambda(V) = 2^{2k+2}$, we have $\mathsf{P}(\operatorname{card} I \geq L2^{2k}N) \leq \exp(-N2^{2k}) \leq L\exp(-(\log N)^{3/2}/L)$ because $k \geq -\ell_1$ and the choice of ℓ_1. □

We start the proof of Proposition 4.7.5. We denote by $\mathcal{F}_k$ the class of functions of the type $\mathbf{1}_{\overset{\circ}{C}}$, where $C \in \mathcal{C}(\tau, k)$ so we can rewrite (4.106) as

$$\mathsf{E} \sup_{f \in \mathcal{F}_k} \Big| \sum_{i \leq N} (f(X_i) - \int f \mathrm{d}\lambda) \Big| \leq L2^k \sqrt{N} (\log N)^{3/4} . \tag{4.108}$$

The key point again is the control on the size of $\mathcal{F}_k$ with respect to the distance of $L^2(\lambda)$. The difficult part of this control is the following:

Proposition 4.7.6 *We have*

$$\gamma_2(\mathcal{F}_k, d_2) \leq L2^k (\log N)^{3/4} . \tag{4.109}$$

Another much easier fact is the following:

Proposition 4.7.7 *We have*

$$\gamma_1(\mathcal{F}_k, d_\infty) \leq L2^k \sqrt{N} . \tag{4.110}$$

Proof of (4.108) and of Proposition 4.7.5 Combine Propositions 4.7.6 and 4.7.7 and Theorem 4.5.16. □

Let us first prove the easy Proposition 4.7.7.

Lemma 4.7.8 *We have* $\operatorname{card} \mathcal{C}(\tau, k) \leq 2^{2^{k+\ell_1+1}} = N_{k+\ell_1+1}$.

Proof A curve $C \in \mathcal{C}(\tau, k)$ consists of at most $2^{k+\ell_1}$ edges of G. If we move through C, at each vertex of G, we have at most *four* choices for the next edge, so $\operatorname{card} \mathcal{C}(\tau, k) \leq 4^{2^{k+\ell_1}} = N_{k+\ell_1+1}$. □

Proof of Proposition 4.7.7 Generally speaking, a set T of cardinality $\leq N_m$ and diameter Δ satisfies $\gamma_1(T, d) \leq L\Delta 2^m$, as is shown by taking $\mathcal{A}_n = \{T\}$ for $n < m$ and $A_m(t) = \{t\}$. We use this for $T = \mathcal{F}_k$, so that $\operatorname{card} T = \operatorname{card} \mathcal{C}(\tau, k) \leq N_{k+\ell_1+1}$ by Lemma 4.7.8 and $2^{k+\ell_1+1} \leq L2^k\sqrt{N}$. □

We now attack the difficult part, the proof of Proposition 4.7.6. The exponent 3/4 occurs through the following general principle, where we recall that if d is a distance, so is $\sqrt{d}$:

Lemma 4.7.9 *Consider a finite metric space* (T, d) *with* $\operatorname{card} T \leq N_m$. *Then*

$$\gamma_2(T, \sqrt{d}) \leq m^{3/4} \gamma_{1,2}(T, d)^{1/2} . \tag{4.111}$$

Proof Since T is finite, there exists an admissible sequence $(\mathcal{A}_n)$ of T such that

$$\forall t \in T\,, \sum_{n\geq 0}(2^n \Delta(A_n(t), d))^2 \leq \gamma_{1,2}(T, d)^2\,. \tag{4.112}$$

Without loss of generality, we can assume that $A_m(t) = \{t\}$ for each t, so that in (4.112) the sum is over $n \leq m-1$. Now

$$\Delta(A, \sqrt{d}) = \Delta(A, d)^{1/2}$$

so that, using Hölder's inequality,

$$\begin{aligned}\sum_{0\leq n\leq m-1} 2^{n/2}\Delta(A_n(t), \sqrt{d}) &= \sum_{0\leq n\leq m-1} (2^n \Delta(A_n(t), d))^{1/2} \\ &\leq m^{3/4}\Big(\sum_{n\geq 0}(2^n \Delta(A_n(t), d))^2\Big)^{1/4} \\ &\leq m^{3/4}\gamma_{1,2}(T, d)^{1/2}\,,\end{aligned}$$

which concludes the proof. □

Let us denote by $A\triangle B$ the symmetric difference $(A \setminus B) \cup (B \setminus A)$ between two sets A and B. On the set of closed simple curves traced on G, we define the distance d_1 by $d_1(C, C') = \lambda(\overset{o}{C} \triangle \overset{o}{C}{}')$ and the distance

$$\delta(C_1, C_2) := \|\mathbf{1}_{\overset{o}{C_1}} - \mathbf{1}_{\overset{o}{C_2}}\|_2 = (\lambda(\overset{o}{C}_1\triangle\overset{o}{C}_2))^{1/2} = \big(d_1(C_1, C_2)\big)^{1/2}\,, \tag{4.113}$$

so that

$$\gamma_2(\mathcal{F}_k, d_2) = \gamma_2(\mathcal{C}(\tau, k), \delta) = \gamma_2(\mathcal{C}(\tau, k), \sqrt{d_1})\,,$$

and using Lemma 4.7.8 and (4.111) for $m := k + \ell_1 + 1$, we obtain

$$\gamma_2(\mathcal{F}_k, d_2) \leq L(\log N)^{3/4}\gamma_{1,2}(\mathcal{C}(\tau, k),\, d_1)^{1/2}\,,$$

because $m \leq L \log N$ for $k \leq \ell_1 + 2$.

Therefore, it remains only to prove the following:

Proposition 4.7.10 *We have*

$$\gamma_{1,2}(\mathcal{C}(\tau, k),\, d_1) \leq L2^{2k}\,. \tag{4.114}$$

The reason why this is true is that the metric space $(\mathcal{L}, d_2)$ of Proposition 4.1.8 satisfies $\gamma_{1,2}(\mathcal{L}, d_2) < \infty$, while $(\mathcal{C}(\tau, k), d_1)$ is a Lipschitz image of a subset of this metric space $(\mathcal{L}, d_2)$. The elementary proof of the following may be found in Sect. B.2.

Lemma 4.7.11 *There exists a map W from a subset T of $\mathcal{L}$ onto $\mathcal{C}(\tau, k)$ which for any $f_0, f_1 \in T$ satisfies*

$$d_1(W(f_0), W(f_1)) \le L2^{2k} \| f_0 - f_1 \|_2 \,. \tag{4.115}$$

To conclude the proof of Proposition 4.7.10, we check that the functionals $\gamma_{\alpha,\beta}$ behave as expected under Lipschitz maps.

Lemma 4.7.12 *Consider two metric spaces (T, d) and (U, d') and a map $f : (T, d) \to (U, d')$ which is onto and satisfies*

$$\forall x, y \in T \,,\; d'(f(x), f(y)) \le A d(x, y)$$

for a certain constant A. Then

$$\gamma_{\alpha,\beta}(U, d') \le K(\alpha, \beta) A \gamma_{\alpha,\beta}(T, d) \,.$$

Proof This is really obvious when f is one-to-one. We reduce to that case by considering a map $\varphi : U \to T$ with $f(\varphi(x)) = x$ and replacing T by $\varphi(U)$. □

It remains to deduce Theorem 4.7.1 from Theorem 4.7.2. The argument is purely deterministic and unrelated to any other material in the present book. The basic idea is very simple, and to keep it simple, we describe it in slightly imprecise terms. Consider a union A of little squares of side length $2^{-\ell_1}$ and the union A' of all the little squares that touch A (see Fig. 4.3).

We want to prove that A' contains as many points Y_i as A contains points X_i, so that by Hall's Marriage Lemma each point X_i can be matched to a point Y_i in the same little square or in a neighbor of it. Since the points Y_i are evenly spread, the number of such points in A' is very nearly $N\lambda(A')$. There may be more than $N\lambda(A)$ points X_i in A, but (4.103) tells us that the excess number of points cannot be more than a proportion of the length ℓ of the boundary of A. The marvelous fact is that we may also expect that $\lambda(A') - \lambda(A)$ is also proportional to ℓ, so that we may hope that the excess number of points X_i in A should not exceed $N(\lambda(A') - \lambda(A))$, proving the result. The proportionality constant is not quite right to make the argument work, but this difficulty is bypassed simply by applying the same argument to a slightly coarser grid.

When one tries to describe precisely what is meant by the previous argument, one has to check a number of details. This elementary task which requires patience is performed in Appendix B.3.

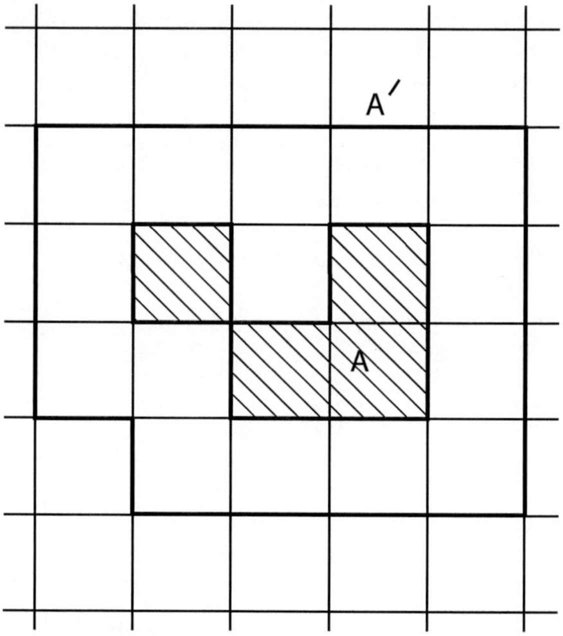

Fig. 4.3 A union A of little squares and the boundary of A'

4.8 Lower Bound for the Leighton-Shor Theorem

Theorem 4.8.1 *If the points* $(X_i)_{i\le N}$ *are i.i.d. uniform over* $[0,1]^2$ *and the points* $(Y_i)_{i\le N}$ *are evenly spread, then*

$$\mathsf{E}\inf_{\pi}\max_{i\le N} d(X_i, Y_{\pi(i)}) \ge \frac{(\log N)^{3/4}}{L\sqrt{N}} . \tag{4.116}$$

We consider the class of functions

$$\mathcal{C} = \Big\{f : [0,1] \to [0,1]\ ;\ f(0) = f(1) = 0\ ;\ \int_0^1 f'^2(x)\mathrm{d}x \le 1\Big\} . \tag{4.117}$$

For $f \in \mathcal{C}$, we consider its subgraph

$$S(f) := \{(x,y) \in [0,1]^2\ ;\ y \le f(x)\} . \tag{4.118}$$

To prove (4.116), the key step will be to show that with high probability we may find $f \in \mathcal{C}$ with

$$\mathrm{card}\{i \le N\ ;\ X_i \in S(f)\} \ge N\lambda(S(f)) + \frac{1}{L}\sqrt{N}(\log N)^{3/4} . \tag{4.119}$$

With a little more work, we could actually prove that we can find such a function f which moreover satisfies $|f'| \le 1$. This extra work is not needed. The key property of f here is that its graph has a bounded length, and this is already implied by the condition $\|f'\|_2 \le 1$, since the length of this graph is $\int_0^1 \sqrt{1 + f'^2(x)}\mathrm{d}x \le 2$.

Lemma 4.8.2 *The set of points within distance $\epsilon > 0$ of the graph of f has an area $\le L\epsilon$. The set of points within distance $\epsilon > 0$ of $S(f)$ has an area $\le \lambda(S(f)) + L\epsilon$.*

Proof The graph of $f \in \mathcal{C}$ has length ≤ 2. One can find a subset of the graph of f of cardinality $\le L/\epsilon$ such that each point of the graph is within distance ϵ of this set.[18] A point within distance ϵ of the graph then belongs to one of L/ϵ balls of radius 2ϵ. This proves the first assertion. The second assertion follows from the fact that a point which is within distance ϵ of $S(f)$ either belongs to $S(f)$ or is within distance ϵ of the graph of f. □

Proof of Theorem 4.8.1 We prove that when there exists a function f satisfying (4.119), then $\inf_\pi \max_{i \le N} d(X_i, Y_{\pi(i)}) \ge (\log N)^{3/4}/L\sqrt{N}$. Let us denote by $S(f)_\epsilon$ the ϵ-neighborhood[19] of $S(f)$ in $[0, 1]^2$. We first observe that for any $f \in \mathcal{C}$, we have

$$\mathrm{card}\{i \le N \; ; \; Y_i \in S(f)_\epsilon\} \le N\lambda(S(f)) + L\epsilon N + L\sqrt{N} \,. \tag{4.120}$$

This is because, by definition of an evenly spread family, each point Y_i belongs to a small rectangle R_i of area $1/N$ and of diameter $\le 10/\sqrt{N}$ and a pessimistic upper bound for the left-hand side of (4.120) is the number of such rectangles that intersect $S(f)_\epsilon$. These rectangles are entirely contained in the set of points within distance $L/\sqrt{N}$ of $S(f)_\epsilon$, i.e., in the set of points within distance $\le \epsilon + L/\sqrt{N}$ of $S(f)$ and by Lemma 4.8.2, this set has area $\le \lambda(S(f)) + L\epsilon + L/\sqrt{N}$, hence the bound (4.120).

Consequently (and since we may assume that N is large enough), (4.119) implies that for $\epsilon = (\log N)^{3/4}/(L\sqrt{N})$, it holds that

$$\mathrm{card}\{i \le N \; ; \; Y_i \in S(f)_\epsilon\} < \mathrm{card}\{i \le N \; ; \; X_i \in S(f)\} \,,$$

and therefore any matching must pair at least one point $X_i \in S(f)$ with a point $Y_j \notin S(f)_\epsilon$, so that $\max_{i \le N} d(X_i, Y_{\pi(i)}) \ge \epsilon$. □

Recalling the functions $f_{k,\ell}$ of (4.82), we consider now an integer $c \ge 2$ which will be determined later. The purpose of c is to give us room. Thus, by (4.87),

$$\int_0^1 f_{ck,\ell}(x)\mathrm{d}x = 2^{-2ck-2} \,. \tag{4.121}$$

[18] This is true for any curve of length 2. If one consider a parameterization $\varphi(t)$ $0 \le t \le 2$ of the curve by arc length, the points $\varphi(k\epsilon)$ for $k \le 2/\epsilon$ have this property.

[19] That is, the set of points within distance $\le \epsilon$ of a point of $S(f)$.

Let us set

$$\tilde{f}_{k,\ell} = \frac{1}{\sqrt{r}} f_{ck,\ell} \ .$$

Consider the functions of the type

$$f = \sum_{k \le r} f_k \text{ with } f_k = \sum_{1 \le \ell \le 2^{ck}} x_{k,\ell} \tilde{f}_{k,\ell} \ , \tag{4.122}$$

where $x_{k,\ell} \in \{0, 1\}$. Then $f(0) = f(1) = 0$.

Lemma 4.8.3 *A function f of the type (4.122) satisfies*

$$\int_0^1 f'(x)^2 \mathrm{d}x \le 1 \ . \tag{4.123}$$

Proof Using (4.83) and (4.84), we obtain

$$\int_0^1 f'(x)^2 \mathrm{d}x = \sum_{k \le r} \sum_{\ell \le 2^{ck}} \frac{x_{k,\ell}^2}{r} \| f_{ck,\ell} \|_2^2 = \sum_{k \le r} \sum_{\ell \le 2^{ck}} \frac{x_{k,\ell}^2}{r} 2^{-ck} \le 1 \ . \qquad \square$$

Consequently, each function of the type (4.122) belongs the class $\mathcal{C}$ of (4.117).

Proof of (4.119) Given N large, we choose r as the largest integer for which $2^{cr} \le N^{1/100}$, so that $r \le \log N / Lc$. The construction of the functions f_k is inductive. Assume that we have already constructed $f_1, \ldots, f_q$, and let $g = \sum_{k \le q} f_k$. For $\ell \le 2^{c(q+1)}$, let us consider the region

$$R_\ell := S(g + \tilde{f}_{q+1,\ell}) \setminus S(g) \ ,$$

so that by (4.121)

$$\lambda(R_\ell) = \frac{2^{-2c(q+1)}}{4\sqrt{r}} \ . \tag{4.124}$$

These regions are disjoint because the functions $\tilde{f}_{q+1,\ell}$ have disjoint support. Furthermore, if we choose $f_{q+1} = \sum_{\ell \le 2^{c(q+1)}} x_{q+1,\ell} \tilde{f}_{q+1,\ell}$ where $x_{q+1,\ell} \in \{0, 1\}$, then we have

$$S(g + f_{q+1}) \setminus S(g) = \bigcup_{\ell \in J} R_\ell \ ,$$

where

$$J = \{\ell \le 2^{c(q+1)} \; ; \; x_{q+1,\ell} = 1\} ,$$

and thus

$$\lambda(S(g + f_{q+1}) \setminus S(g)) = \sum_J \lambda(R_\ell) . \tag{4.125}$$

Since our goal is to construct functions such that there is an excess of points X_i in their subgraph, we do the obvious thing; we take $x_{q+1,\ell} = 1$ if there is an excess of points X_i in R_ℓ, that if

$$\delta_\ell := \mathrm{card}\{i \le N; X_i \in R_\ell\} - N\lambda(R_\ell) \ge 0 , \tag{4.126}$$

and otherwise we set $x_{k+1,\ell} = 0$. We have, recalling (4.125),

$$\begin{aligned} \mathrm{card}\{i \le N \; ; \; X_i \in S(g + f_{q+1}) \setminus S(g)\} &= \sum_J \mathrm{card}\{i \le N \; ; \; X_i \in R_\ell\} \\ &= \sum_J \delta_\ell + N\lambda(S(g + f_{q+1}) \setminus S(g)) . \end{aligned} \tag{4.127}$$

We will show that with high probability, we have $\sum_J \delta_\ell \ge \sqrt{N}/(Lr^{1/4})$. Recalling that $g = \sum_{k \le q} f_k$ and $g + f_{q+1} = \sum_{k \le q+1} f_k$, summation of the inequalities (4.127) over $q < r$ then proves (4.119), where f is the function $\sum_{k \le r} f_k$.

Let us say that the region R_ℓ is *favorable* if

$$\delta_\ell \ge \sqrt{N\lambda(R_\ell)}/L^* = 2^{-c(q+1)}\sqrt{N}/(Lr^{1/4}) ,$$

where the universal constant L^* will be determined later. The idea underlying this definition is that given a subset A of the square, with $1/N \ll \lambda(A) \le 1/2$, the number of points X_i which belong to A has typical fluctuations of order $\sqrt{N\lambda(A)}$. Since $\delta_\ell \ge 0$ for $\ell \in J$ and since by construction $\ell \in J$ when R_ℓ is favorable, we have

$$\sum_J \delta_\ell \ge \mathrm{card}\{\ell; R_\ell \text{ favorable}\} \times 2^{-c(q+1)}\sqrt{N}/(Lr^{1/4}) .$$

To conclude the proof, it then suffices to show that with overwhelming probability at least a fixed proportion of the regions R_ℓ for $\ell \le 2^{c(q+1)}$ are favorable. One has to be cautious that the r.v.s X_i are *not* independent of the function g and of the regions R_ℓ because in particular the construction of g uses the values of the X_i.

One simple way around that difficulty is to proceed as follows: There are at most $\prod_{k\le q} 2^{2^{ck}} \le 2^{2^{cq+1}}$ possibilities for g. To each of these possibilities corresponds a family of $2^{2^{c(q+1)}}$ regions R_ℓ. If we can ensure that with overwhelming probability for each of these families a least a fixed proportion of the R_ℓ are favorable, we are done. Since there are at most $2^{2^{cq+1}}$ families, it suffices to prove that for a given family, this fails with probability $\le 2^{-2^{cq+2}}$. To achieve this, we proceed as follows: by normal approximation of the tails of the binomial law, there exists a constant L^* and a number $N_0 > 0$ such that given any set $A \subset [0,1]^2$ with $1/2 \ge \lambda(A)$ and $N\lambda(A) \ge N_0$, we have

$$\mathsf{P}\big(\mathrm{card}\{i \le N; X_i \in A\} - N\lambda(A) \ge \sqrt{N\lambda(A)}/L^*\big) \ge 1/4 . \tag{4.128}$$

Since c is a universal constant and $2^{rc} \le N^{1/100}$, (4.124) shows that $N\lambda(R_\ell)$ becomes large with N. In particular (4.128) shows that the probability that a given region R_ℓ is favorable is $\ge 1/4$. Now, using Poissonization, we can pretend that these probabilities are independent as ℓ varies. As noted in (4.98), given M independent r.v.s $Z_i \in \{0,1\}$ with $\mathsf{P}(Z_i = 1) \ge 1/4$, then $\mathsf{P}(\sum_{i\le M} Z_i \le M/8) \le \exp(-\beta M)$ for some universal constant β. Since here we have $M = 2^{c(q+1)}$, then $\exp(-\beta M) = \exp(-\beta 2^{c-2} 2^{cq+2})$. This is $\le 2^{-2^{cq+2}}$ as required provided we have chosen c large enough that $\beta 2^{c-2} \ge 1$. □

4.9 For the Expert Only

Having proved both the Ajtai-Komlós-Tusnády and the Leighton-Shor matching theorems, we should not fall under the illusion that we understand everything about matchings. The most important problem left is arguably the ultimate matching conjecture, stated later as Problem 17.1.2. A first step in that direction would be to answer the following question:[20]

Question 4.9.1 Can we find a matching which achieves simultaneously both (4.36) and (4.101)?

The existence of such a matching does not seem to be of any particular importance, but the challenge is that the Ajtai-Komlós-Tusnády (AKT) theorem and the Leighton-Shor matching theorems are proved by rather different routes, and it is far from obvious to find a common proof.

In the rest of the section, we discuss a special matching result. Consider the space $T = \{0,1\}^{\mathbb{N}}$ provided with the distance $d(t,t') = 2^{-j}$, where $j = \min\{i \ge 1; t_i \ne$

[20] The difference between a problem and a question is that a question is permitted to sound less central.

$t_i'\}$ for $t = (t_i)_{i\geq 1}$. This space somewhat resembles the unit interval, in the sense that $N(T, d, \epsilon) \leq L\epsilon^{-1}$ for $\epsilon \leq 1$. The space of Exercise 4.5.23 is essentially the space $T \times T$. The AKT theorem tells us what happens for matchings in $[0, 1]^2$, and Exercise 4.5.23 tells us what happens for matchings in T^2. But what happens in the space $U := [0, 1] \times T$? It does not really matter which specific sensible distance we use on U; let us say that we define $d((x, t), (x', t')) = |x - x'| + d(t, t')$.

Theorem 4.9.2 *The expected cost of the optimal matching of N random i.i.d. uniformly distributed*[21] *points in U with N evenly spread points is exactly of order* $\sqrt{N}(\log N)^{3/4}$.

The appealing part of this special result is of course the fractional power of log. This result is as pretty as almost anything found in this book, but its special nature makes it appropriate to guide the (expert) reader to the proof through exercises.

Let us start with a finite approximation of T. We consider the space $T_m = \{0, 1\}^m$ provided with the distance defined for $t \neq t'$ by $d(t, t') = 2^{-j}$, where $j = \min\{i \geq 1; t_i \neq t_i'\}$ for $t = (t_i)_{i\leq m}$. We set $U_m = [0, 1] \times T_m$, and we denote by θ_m the uniform measure on U_m. Surely the reader who has reached this stage knows how to deduce[22] the upper bound of Theorem 4.9.2 from the following:

Theorem 4.9.3 *The set $\mathcal{L}$ of 1-Lipschitz functions f on U_m which satisfy $|f| \leq 1$ satisfies $\gamma_2(\mathcal{L}, d_2) \leq Lm^{3/4}$.*

Here of course $\mathcal{L}$ is seen as a subset of $L^2(U_m, \theta_m)$. The proof of Theorem 4.9.3 will use an expansion of the elements of $\mathcal{L}$ on a suitable basis. Using the same method as in Lemma 4.5.12, one can assume furthermore that the functions of $\mathcal{L}$ are zero on $\{0\} \times T_m$ and $\{1\} \times T_m$. For $0 \leq n \leq m$, we consider the natural partition $\mathcal{C}_n$ of T_m into 2^n sets obtained by fixing the first n coordinates of $t \in T_m$. Denoting by μ_m the uniform measure on T_m, for $C \in \mathcal{C}_n$, we have $\mu_m(C) = 2^{-n}$. A set $C \in \mathcal{C}_n$ with $n < m$ is the union of two sets C_1 and C_2 in $\mathcal{C}_{n+1}$. We denote by h_C a function on T_n which equals $2^{n/2}$ on one of these sets and $-2^{n/2}$ on the other. Consider also the function $h_\emptyset$ on T_m, constant equal to 1. In this manner, we obtain an orthogonal basis (h_C) of $L^2(T_m, \mu_m)$. For $f \in \mathcal{L}$, we consider the coefficients of f on this basis,

$$a_{p,C}(f) := \int_{U_m} \exp(2ip\pi x) h_C(t) f(x, t) \mathrm{d}x \mathrm{d}\mu_m(t) .$$

There and always, $p \in \mathbb{Z}$, $n \geq 0$, $C \in \mathcal{C}_n$ or $p \in \mathbb{Z}$ and $C = \emptyset$. We will lighten notation by writing simply $\sum_{p,C}$ sums over all possible values of (p, C) as above.

[21] It should be obvious what is meant by "uniform probability on U".

[22] By following the scheme of proof of (4.43).

Exercise 4.9.4

(a) Prove that

$$\sum_{p,C} p^2|a_{p,C}|^2 \le L\,. \tag{4.129}$$

Hint: Just use that $|\partial f/\partial x| \le 1$.

(b) Prove that for each n and each $C \in \mathcal{C}_n$, we have

$$\sum_{p\in\mathbb{Z}} |a_{p,C}|^2 \le L2^{-3n}\,. \tag{4.130}$$

Hint: Prove that $|\int h_C(t)f(x,t)\mathrm{d}\mu_m(t)| \le L2^{-3n/2}$.

We have just shown that $\mathcal{L}$ is isometric to a subset of the set $\mathcal{A}$ of sequences $(a_{p,C})$ which satisfy (4.129) and (4.130).

Exercise 4.9.5 We will now show that $\gamma_2(\mathcal{A}) \le Lm^{3/4}$.

(a) Prove that $\mathcal{A}$ is contained in an ellipsoid of the type

$$\mathcal{E} = \Big\{(a_{c,C})\ ;\ \sum_{p,C}\alpha_{p,C}^2|a_{p,C}|^2 \le 1\Big\}$$

where $\alpha_{p,C}^2 = (p^2 + 2^{2n}/m)/L$ if $C \in \mathcal{C}_n$, $n \ge 0$ and $\alpha_{p,\emptyset}^2 = p^2/L$.

(b) Conclude using (2.155). (The reader must be careful for the unfortunate clash of notation.)

The goal of the next exercise is to prove the lower bound in Theorem 4.9.2. This lower bound is obtained by a non-trivial twist on the proof of the lower bound for the AKT theorem, so you must fully master that argument to have a chance.

Exercise 4.9.6 Let us recall the functions $f_{q,\ell}$ of (4.82) where we take $r \simeq (\log N)/100$. For $n \ge 0$, we still consider the natural partition $\mathcal{C}_n$ of T into 2^n sets obtained by fixing the first n coordinates of $t \in T$. We consider an integer p with $2^{-p} \simeq 1/\sqrt{r}$. For each q, each $\ell \le 2^q$, and each $C \in \mathcal{C}_{q+p}$, we consider the function $f_{q,\ell,C}$ on U given by $f_{q,\ell,C}(x,t) = 2^{-p-20}f_{q,\ell}(x)\mathbf{1}_C(t)$. We consider functions of the type $f_q = \sum_{\ell\le 2^q, C\in\mathcal{C}_{q+p}} z_{q,\ell,C}f_{q,\ell,C}$ where $z_{q,\ell,C} \in \{0,1,-1\}$. Copy the proof of the lower bound of the AKT theorem to prove that with high probability, one can construct these functions such that $\sum_{k\le q} f_k$ is 1-Lipschitz, and for each q, $\sum_{i\le N}(f_q(X_i) - \int f_q \mathrm{d}\theta) \ge \sqrt{N}/(Lr^{1/4})$, where X_i are i.i.d. uniform on U and θ is the uniform probability measure on U. Summation over $q \le r$ yields the desired result.

While making futile attempts in the direction of Theorem 4.9.2 arose further questions which we cannot answer. We describe one of these now. We recall the functionals $\gamma_{\alpha,\beta}$ of (4.5) and the uniform measure μ_m on T_m.

Question 4.9.7 Is it true that for any metric space (T, d) the space U of 1-Lipschitz maps f from T_m to T, provided with the distance D given by $D(f, f')^2 = \int_{T_m} d(f(s), f'(s))^2 \mathrm{d}\mu_m(s)$ satisfies $\gamma_2(U, D) \le Lm^{3/4}\gamma_{1,2}(T)$?

The motivation for this result is that if T is the set of 1-Lipschitz functions on $[0, 1]$, then $\gamma_{1,2}(T) \le L$ (using Fourier transform to compare with an ellipsoids), and with minimal effort, this would provide an alternate and more conceptual proof for the upper bound of Theorem 4.9.2.

Exercise 4.9.8 In the setting of Question 4.9.7, assume that $e_n(T, d) \le 2^{-n}$. Prove that $e_{2n}(U, D) \le L2^{-n}$ (and better if $n > m$). Conclude that $\sum_{n\ge 0} 2^{n/2} e_n(U, D) \le Lm$. Prove that $\gamma_2(U, D) \le Lm\gamma_{1,2}(T)$.

Key Ideas to Remember

- Ellipsoids in a Hilbert space are in a sense smaller than their entropy numbers indicate. This is true more generally for sufficiently convex sets in a Banach space. This phenomenon explains the fractional powers of logarithms occurring in the most famous matching theorems.
- The size of ellipsoids is sometimes accurately described by using proper generalizations $\gamma_{\alpha,\beta}(T, d)$ of the basic functional $\gamma_2(T, d)$.
- Matching theorems are typically proved through a discrepancy bound, which evaluates the supremum of the empirical process over a class $\mathcal{F}$ of functions.
- Bernstein's inequality is a convenient tool to prove discrepancy bounds. It involves the control of $\mathcal{F}$ both for the L^2 and the supremum distance.
- Using two different distances reveals the power of approaching chaining through sequences of partitions.

4.10 Notes and Comments

The original proof of the Leighton-Shor theorem amounts basically to perform by hand a kind of generic chaining in this highly non-trivial case, an incredible tour de force.[23] A first attempt was made in [92] to relate (an important consequence of) the Leighton-Shor theorem to general methods for bounding stochastic processes but runs into technical complications. Coffman and Shor [26] then introduced the use of Fourier transforms and brought to light the role of ellipsoids, after which it became clear that the structure of these ellipsoids plays a central part in these matching results, a point of view systematically expounded in [114].

---

[23] There is a simple explanation as to why this was possible: as you can check through Wikipedia, both authors are geniuses.

Chapter 17 is a continuation of the present chapter. The more difficult material it contains is presented later for fear of scaring readers at this early stage. A notable feature of the result presented there is that ellipsoids do not suffice, a considerable source of complication. The material of Appendix A is closely related to the Leighton-Shor theorem.

The original results of [3] are proved using an interesting technique called the *transportation method*. A version of this method, which avoids many of the technical difficulties of the original approach, is presented in [134]. With the notation of Theorem 4.5.1, it is proved in [134] (a stronger version of the fact) that with probability $\geq 9/10$, one has

$$\inf_{\pi} \frac{1}{N} \sum_{i \leq N} \exp\Big(\frac{N d(X_i, Y_{\pi(i)})^2}{L \log N}\Big) \leq 2 . \tag{4.131}$$

Since $\exp x \geq x$, (4.131) implies that $\sum_{i \leq N} d(X_i, Y_{\pi(i)})^2 \leq L \log N$ and hence using the Cauchy-Schwarz inequality $\sum_{i \leq N} d(X_i, Y_{\pi(i)}) \leq L\sqrt{N \log N}$. Moreover, (4.131) also implies $\max_{i \leq N} d(X_i, Y_{\pi(i)}) \leq L \log N/\sqrt{N}$. This unfortunately fails to bring a positive answer to Question 4.9.1.

For results about matching for unbounded distributions, see the work of J. Yukich [146] as well as the nonstandard results of [133].

Part II
Some Dreams Come True

Chapter 5
Warming Up with p-Stable Processes

Later, in Chap. 11, we will prove far-reaching generalizations of the results of Chap. 2 to many processes which are not close to Gaussian processes. Getting there will require many new ideas, and in this chapter, we will present some of them in a setting which remains close to that of Gaussian processes.

5.1 p-Stable Processes as Conditionally Gaussian Processes

Consider a number $0 < p \le 2$. A r.v. X is called (real, symmetric) p-stable if for each $\lambda \in \mathbb{R}$, we have

$$\mathsf{E} \exp i\lambda X = \exp\Big(-\frac{\sigma^p |\lambda|^p}{2} \Big) , \tag{5.1}$$

where $\sigma = \sigma_p(X) \ge 0$ is called the parameter of X. The name "p-stable" comes from the fact that if $X_1, \ldots, X_m$ are independent and p-stable, then for real numbers α_i, the r.v. $\sum_{j \le m} a_j X_j$ is also p-stable and

$$\sigma_p\Big(\sum_{j \le m} a_j X_j \Big) = \Big(\sum_{j \le m} |a_j|^p \sigma_p(X_j)^p \Big)^{1/p} . \tag{5.2}$$

This is obvious from (5.1).

The reason for the restriction $p \le 2$ is that for $p > 2$, no r.v. satisfies (5.1). The case $p = 2$ is the Gaussian case, which we now understand very well, so from now on we assume $p < 2$. Despite the formal similarity, this is very different from the

M. Talagrand, *Upper and Lower Bounds for Stochastic Processes*, Ergebnisse der Mathematik und ihrer Grenzgebiete. 3. Folge / A Series of Modern Surveys in Mathematics 60, https://doi.org/10.1007/978-3-030-82595-9_5

Gaussian case. It can be shown that

$$\lim_{s\to\infty} s^p \mathsf{P}(|X| \geq s) = c_p \sigma_p(X)^p \tag{5.3}$$

where $c_p > 0$ depends on p only. Thus X does not have moments of order p, but it has moments of order q for $q < p$. We refer the reader to [53] for a proof of this and for general background on p-stable processes.

A process $(X_t)_{t\in T}$ is called p-stable if, for every family $(\alpha_t)_{t\in T}$ for which only finitely many of the numbers α_t are not 0, the r.v. $\sum_t \alpha_t X_t$ is p-stable. We can then define a (quasi-)distance d on T by

$$d(s,t) = \sigma_p(X_s - X_t)\,. \tag{5.4}$$

When $p > 1$, a p-stable r.v. is integrable, and $\mathsf{E}|X|$ is proportional to $\sigma_p(X)$. Thus one can also define an equivalent distance by $d'(s,t) = \mathsf{E}|X_s - X_t|$.

A typical example of a p-stable process is given by $X_t = \sum_{i\le n} t_i Y_i$ where $t = (t_i)_{i\le n}$ and $(Y_i)_{i\le n}$ are independent p-stable r.v.s. It can in fact be shown that this example is generic in the sense that "each p-stable process (with a finite index set) can be arbitrarily well approximated by a process of this type". Assuming further that $\sigma_p(Y_i) = 1$ for each i, (5.2) implies that the distance induced by the process is then the ℓ^p distance, $d(X_s, X_t) = \|s - t\|_p$.

At the heart of this chapter is the fact that a p-stable process (X_t) can be represented as a conditionally Gaussian process. That is, we can find two probability spaces $(\Omega\,,\,\mathsf{P})$ and $(\Omega'\,,\,\mathsf{P}')$ and a family $(Y_t)_{t\in T}$ of r.v.s on $\Omega\times\Omega'$ (provided with the product probability), such that

$$\text{Given any finite subset } U \text{ of } T, \text{ the joint laws of } (Y_t)_{t\in U} \text{ and } (X_t)_{t\in U} \text{ are identical} \tag{5.5}$$

$$\text{Given } \omega\in\Omega\,, \text{ the process } \omega'\mapsto Y_t(\omega,\omega') \text{ is a centered Gaussian process.} \tag{5.6}$$

This result holds for any value of p with $1 \le p < 2$. A proof is given in Sect. C.3. Our strategy is to study the process (Y_t) as in (5.6) at given ω.[1] A remarkable fact is that we do not need to know precisely how the previous representation of the process (X_t) arises. More generally, if you are disturbed by the fact that you have no intuition about p-stable processes, do not be discouraged. Our result will not need any understanding of p-stable processes beyond what we have already explained.

[1] If you have already glanced through the rest of the book, you should be aware that a basic reason the special case of p-stable processes is simple is that these processes are conditionally Gaussian. Many more processes of interest (such as infinitely divisible processes) are not conditionally Gaussian, but are conditionally Bernoulli.

5.2 A Lower Bound for p-Stable Processes

The main goal of this chapter is to prove the following:

Theorem 5.2.1 *For $1 < p < 2$, there is a number $K(p)$ such that for any p-stable process $(X_t)_{t\in T}$, we have*

$$\gamma_q(T, d) \leq K(p)\mathsf{E}\sup_{t\in T} X_t\,, \tag{5.7}$$

where q is the conjugate exponent of p, i.e., $1/q + 1/p = 1$, and where d is as in (5.4).

Certainly this result reminds us of the inequality $\gamma_2(T, d) \leq L\mathsf{E}\sup_{t\in T} X_t$ of the majorizing measure theorem (Theorem 2.10.1). A striking difference is that the tails of p-stable r.v.s are very large (see (5.3)) and are not relevant to (5.7). The bound (5.7) cannot be reversed.

Exercise 5.2.2

(a) Consider i.i.d. p-stable r.v.s $(Y_i)_{i\leq N}$ with $\sigma_p(Y_i) = 1$. For $t \in \mathbb{R}^N$, set $X_t = \sum_{i\leq n} t_i Y_i$. Prove that the distance (5.4) is given by $d(s,t) = (\sum_{i\leq N}|s_i - t_i|^p)^{1/p}$.
(b) Let $T = \{(\pm 1, \dots, \pm 1)\}$. Prove that in this case the two sides of (5.7) are of the same order.
(c) Let now T consist of the N sequences $(0, \dots, 0, 1, 0, \dots, 0)$. Prove that the two sides of (5.7) are not of the same order.

This exercise leaves little hope to compute $\mathsf{E}\sup_{t\in T} X_t$ as a function of the geometry of (T, d) only.

The bound (5.7) cannot be reversed, but it would be a lethal mistake to think that it is "weak". It provides in fact an exact information on some aspects of the process $(X_t)_{t\in T}$, but these aspects are not apparent at first sight, and we will fully understand them only in Chap. 11.

Theorem 5.2.1 has a suitable version for $p = 1$, which we will state at the end of this chapter.

5.3 Philosophy

Let us consider the process (Y_t) as in (5.5) and (5.6). We denote by E' integration in P' only. Given ω, we consider the random distance d_ω on T given by

$$d_\omega(s,t) = \left(\mathsf{E}'(Y_s(\omega,\omega') - Y_t(\omega,\omega'))^2\right)^{1/2}\,. \tag{5.8}$$

It is the canonical distance associated with the Gaussian process $(Y_t(\omega, \cdot))$. Consider the r.v. $Z = \sup_{t \in T} Y_t$. Then Theorem 2.10.1 implies

$$\gamma_2(T, d_\omega) \le L\mathsf{E}'Z ,$$

and taking expectation gives

$$\mathsf{E}\gamma_2(T, d_\omega) \le L\mathsf{E}\mathsf{E}'Z = L\mathsf{E}\sup_{t \in T} X_t . \tag{5.9}$$

Now, from the information (5.9), how do we gain a control of the size of the metric space (T, d)? There is a very important principle at work here, which will play a major part in the book. Suppose that we have a set T and that on T we have a distance d and a random distance d_ω.

Principle A If, given $s, t \in T$, it is very rare that the distance $d_\omega(s, t)$ is very much smaller than $d(s, t)$, then some measure of size of (T, d) is controlled from above by the typical value of $\gamma_2(T, d_\omega)$.

We do not expect the reader to fully understand this principle now, but it will become clearer as we repeatedly apply it. In the present case, the property that it is very rare that $d_\omega(s, t)$ is very much smaller than $d(t, s)$ is expressed by the following:

Lemma 5.3.1 *Define α by*

$$\frac{1}{\alpha} := \frac{1}{p} - \frac{1}{2} . \tag{5.10}$$

Then for all $s, t \in T$ and $\epsilon > 0$, we have

$$\mathsf{P}(d_\omega(s, t) \le \epsilon d(s, t)) \le \exp\Big(-\frac{b_p}{\epsilon^\alpha}\Big) \tag{5.11}$$

where d is the distance (5.4) and where $b_p > 0$ depends on p only.

Thus, given a pair (s, t), it is rare that $d_\omega(s, t)$ is much smaller than $d(s, t)$. Given two pairs (s, t) and (s', t'), we however know nothing about the joint distribution of the r.v.s $d_\omega(s, t)$ and $d_\omega(s', t')$. Still the information contained in this lemma suffices to deduce Theorem 5.2.1 from the majorizing measure theorem (Theorem 2.10.1). This is an occurrence of Principle A.

Proof Since the process $Y_t(\omega, \cdot)$ is Gaussian, we have

$$\mathsf{E}' \exp i\lambda(Y_s - Y_t) = \exp\Big(-\frac{\lambda^2}{2} d_\omega^2(s, t)\Big) .$$

Taking expectation, using (5.1), and since the pair (Y_s, Y_t) has the same law as the pair (X_s, X_t), we get

$$\exp\Big(-\frac{|\lambda|^p}{2}d^p(s,t)\Big) = \mathsf{E}\exp\Big(-\frac{\lambda^2}{2}d_\omega^2(s,t)\Big) . \tag{5.12}$$

By Markov's inequality, for any r.v. Z and any u, we have

$$\mathsf{P}(Z \le u)\exp\Big(-\frac{\lambda^2 u}{2}\Big) \le \mathsf{E}\exp\Big(-\frac{\lambda^2}{2}Z\Big) .$$

Using this for $Z = d_\omega^2(s,t)$ and $u = \epsilon^2 d^2(s,t)$, we get, using (5.12),

$$\mathsf{P}(d_\omega(s,t) \le \epsilon d(s,t)) \le \exp\Big(\frac{1}{2}(\lambda^2\epsilon^2 d^2(s,t) - |\lambda|^p d^p(s,t))\Big) .$$

The conclusion follows by optimization over λ. □

5.4 Simplification Through Abstraction

Now that we have extracted the relevant features, Lemma 5.3.1 and Principle A, we can prove an abstract result.

Theorem 5.4.1 *Consider a finite metric space (T, d) and a random distance d_ω on T. Assume that for some $b > 0$, we have*

$$\forall s,t \in T\,,\ \forall \epsilon > 0\,,\ \mathsf{P}(d_\omega(s,t) \le \epsilon d(s,t)) \le \exp\Big(-\frac{b}{\epsilon^\alpha}\Big) , \tag{5.13}$$

where $\alpha > 2$. Then

$$\gamma_q(T,d) \le K\mathsf{E}\gamma_2(T,d_\omega) , \tag{5.14}$$

where

$$\frac{1}{q} = \frac{1}{2} - \frac{1}{\alpha} ,$$

and where K depends on α and b only.

Proof of Theorem 5.2.1 When T is Finite We apply Theorem 5.4.1 with the value α given by (5.10), so that the value of q of Theorem 5.4.1 satisfies $1/q = 1 - 1/p$. It follows from (5.14) that $\gamma_q(T,d) \le L\mathsf{E}\gamma_2(T,d_\omega)$, and combining with (5.9), this implies (5.7). □

Of course T need not be finite and some work would be needed to handle this case. This work is predictable and tedious, and now is not the time to be distracted by routine considerations.[2] We turn to the proof of Theorem 5.4.1. The main step of the proof is the following lemma:

Lemma 5.4.2 *Consider a probability measure μ on T. Then for each $t \in T$, with probability $\geq 15/16$, we have*

$$\int_0^\infty \Big(\log \frac{1}{\mu(B_d(t,\epsilon))}\Big)^{1/q} \mathrm{d}\epsilon \leq K \int_0^\infty \sqrt{\log \frac{1}{\mu(B_{d_\omega}(t,\epsilon))}} \mathrm{d}\epsilon + K\Delta(T,d) . \tag{5.15}$$

Proof We define $\epsilon_0 = \Delta(T,d)$, and for $n \geq 1$, we define ϵ_n by

$$\epsilon_n = \inf\{\epsilon > 0 \, ; \, \mu(B_d(t,\epsilon)) \geq 1/N_n\}$$

and we prove first that

$$\int_0^\infty \Big(\log \frac{1}{\mu(B_d(t,\epsilon))}\Big)^{1/q} \mathrm{d}\epsilon \leq K \sum_{n\geq 1} 2^{n/q} \epsilon_n + K\Delta(T,d) . \tag{5.16}$$

For this, let us set $f(\epsilon) = (\log(1/\mu(B_d(t,\epsilon))))^{1/q}$ and observe that $f(\epsilon) = 0$ for $\epsilon > \epsilon_0 = \Delta(T,d)$. Since $f(\epsilon) \leq K2^{n/q}$ for $\epsilon > \epsilon_n$, we have

$$\int_0^\infty f(\epsilon)\mathrm{d}\epsilon = \sum_{n\geq 0} \int_{\epsilon_{n+1}}^{\epsilon_n} f(\epsilon)\mathrm{d}\epsilon \leq K \sum_{n\geq 0} 2^{(n+1)/q}\epsilon_n = K \sum_{n\geq 1} 2^{n/q}\epsilon_n + K\Delta(T,d) .$$

The heart of the argument starts now. By (5.13), it holds that for any s

$$d(s,t) \geq \epsilon_n \Rightarrow \mathsf{P}\Big(d_\omega(s,t) \leq \frac{1}{K} 2^{-n/\alpha}\epsilon_n\Big) \leq \exp(-2^{n+2}) ,$$

so that by Fubini's theorem

$$\mathsf{E}\mu(\{s \, ; \, d(s,t) \geq \epsilon_n, d_\omega(s,t) \leq 2^{-n/\alpha}\epsilon_n/K\}) \leq \exp(-2^{n+2})$$

and by Markov's inequality,

$$\mathsf{P}\Big(\mu(\{s \, ; \, d(s,t) \geq \epsilon_n, d_\omega(s,t) \leq 2^{-n/\alpha}\epsilon_n/K\}) \geq 1/N_n\Big) \leq N_n \exp(-2^{n+2}) .$$

[2] It is explained in the proof of Theorem 11.7.1 how to cover the case where T is countable.

As $\sum_{n\geq 0} N_n \exp(-2^{n+2}) \leq \sum_{n\geq 0} \exp(-3\cdot 2^n) \leq 1/16$, with probability $\geq 15/16$ for each $n \geq 1$, one has

$$\mu(\{s\ ;\ d(s,t) \geq \epsilon_n, d_\omega(s,t) \leq 2^{-n/\alpha}\epsilon_n/K\}) \leq 1/N_n\ .$$

Since $\mu(\{s\ ;\ d(s,t) \leq \epsilon_n\}) \leq 1/N_n$, it follows that

$$\mu(B_{d_\omega}(t, 2^{-n/\alpha}\epsilon_n/K)) \leq 2/N_n\ .$$

Consequently for $\epsilon \leq \eta_n := 2^{-n/\alpha}\epsilon_n/K$, we have $(\log(1/\mu(B_{d_\omega}(t,\epsilon))))^{1/2} \geq 2^{n/2}/K$ so that

$$\int_0^\infty \Big(\log\frac{1}{\mu(B_{d_\omega}(t,\epsilon))}\Big)^{1/2}\mathrm{d}\epsilon \geq \sum_{n\geq 1}\int_{\eta_{n+1}}^{\eta_n}\Big(\log\frac{1}{\mu(B_{d_\omega}(t,\epsilon))}\Big)^{1/2}\mathrm{d}\epsilon$$
$$\geq \frac{1}{K}\sum_{n\geq 1}2^{n/2}(\eta_n-\eta_{n+1}) \geq \frac{1}{K}\sum_{n\geq 1}(2^{n/2}-2^{(n-1)/2})\eta_n = \frac{1}{K}\sum_{n\geq 1}2^{n/q}\epsilon_n\ ,$$

where we use in the last equality that $1/2 - 1/\alpha = 1/q$. Combining with (5.16) the proof of (5.15) is finished. □

Proof of Theorem 5.4.1 It follows from (2.7) and (5.15) that

$$\int_0^\infty \Big(\log\frac{1}{\mu(B_d(t,\epsilon))}\Big)^{1/q}\mathrm{d}\epsilon \leq K\mathsf{E}\int_0^\infty \sqrt{\log\frac{1}{\mu(B_{d_\omega}(t,\epsilon))}}\mathrm{d}\epsilon + K\Delta(T,d)\ , \tag{5.17}$$

so that, integrating with respect to μ and using linearity of expectation,

$$\int_T \mathrm{d}\mu(t)\int_0^\infty \Big(\log\frac{1}{\mu(B_d(t,\epsilon))}\Big)^{1/q}\mathrm{d}\epsilon \leq K\Delta(T,d)$$
$$+K\mathsf{E}\int_T\mathrm{d}\mu(t)\int_0^\infty\sqrt{\log\frac{1}{\mu(B_{d_\omega}(t,\epsilon))}}\mathrm{d}\epsilon\ .$$

It follows from (3.32) that the last term is $\leq K\mathsf{E}\gamma_2(T,d_\omega)$. This does not depend on μ so that we have proved that

$$\sup_\mu\int_T\mathrm{d}\mu(t)\int_0^\infty\Big(\log\frac{1}{\mu(B_d(t,\epsilon))}\Big)^{1/q}\mathrm{d}\epsilon \leq K\mathsf{E}\gamma_2(T,d_\omega) + K\Delta(T,d)\ . \tag{5.18}$$

Next, we show that the last term is of smaller order. Given $s,t \in T$, from (5.13) with probability $\geq 1/2$, one has $d_\omega(s,t) \geq d(s,t)/K$, and consequently

$\Delta(T, d_\omega) \geq \Delta(T, d)/K$ so that

$$\Delta(T, d) \leq K\Delta(T, d_\omega) \leq K\gamma_2(T, d_\omega) ,$$

and taking expectation, we obtain $\Delta(T, d) \leq K\mathsf{E}\gamma_2(T, d_\omega)$. Combining with (5.18), we finally obtain

$$\sup_\mu \int_T \mathrm{d}\mu(t) \int_0^\infty \Big(\log \frac{1}{\mu(B_d(t, \epsilon))}\Big)^{1/q} \mathrm{d}\epsilon \leq K\mathsf{E}\gamma_2(T, d_\omega) . \tag{5.19}$$

Now, just as in the case $q = 2$ of (3.31), we can use Fernique's convexity argument to prove that

$$\gamma_q(T, d) \leq K \sup_\mu \int_T \mathrm{d}\mu(t) \int_0^\infty \Big(\log \frac{1}{\mu(B_d(t, \epsilon))}\Big)^{1/q} \mathrm{d}\epsilon ,$$

and combining with (5.19), this concludes the proof. □

5.5 1-Stable Processes

In this section, we state an extension of Theorem 5.4.1 to the case $\alpha = 2$, and we explore the consequences of this result on 1-stable processes. According to (5.3), 1-stable r.v.s do not have expectation. This is the main technical difficulty: we can no longer use expectation to measure the size of 1-stable processes. It seems counterproductive to spend space and energy at this stage on such a specialized topic, so we state our results without proofs. Some proofs can be found in [132].[3] We set $M_0 = 1$, $M_n = 2^{N_n}$ for $n \geq 1$. Given a metric space (T, d), we define

$$\gamma_\infty(T, d) = \inf_{\mathcal{B}} \sup_{t \in T} \sum_{n \geq 0} 2^n \Delta(B_n(t)) , \tag{5.20}$$

where the infimum is taken over all increasing families of partitions $(\mathcal{B}_n)$ of T with card $\mathcal{B}_n \leq M_n$. This new quantity is a kind of limit of the quantities $\gamma_\alpha(T, d)$ as $\alpha \to \infty$.

Exercise 5.5.1 Consider the quantity $\gamma^*(T, d)$ defined as

$$\gamma^*(T, d) = \inf \sup_{t \in T} \sum_{n \geq 0} \Delta(A_n(t)) , \tag{5.21}$$

[3] We know now how to give much simpler proofs than those of [132].

where the infimum is computed over all admissible sequences of partitions $(\mathcal{A}_n)$. Prove that

$$\frac{1}{L}\gamma^*(T,d) \le \gamma_\infty(T,d) \le L\gamma^*(T,d) . \tag{5.22}$$

Hint: Given an increasing sequence of partitions $(\mathcal{B}_n)$ with $\operatorname{card}\mathcal{B}_n \le M_n$, consider the increasing sequence of partitions $(\mathcal{A}_m)$ given by $\mathcal{A}_m = \mathcal{B}_n$ for $2^n \le m < 2^{n+1}$.

Theorem 5.5.2 *Consider a finite metric space (T,d) and a random distance d_ω on T. Assume that*

$$\forall s,t \in T\,,\ \forall \epsilon > 0\,,\ \mathsf{P}(d_\omega(s,t) < \epsilon d(s,t)) \le \exp\Big(-\frac{1}{\epsilon^2}\Big) .$$

Then

$$\mathsf{P}\Big(\gamma_2(T,d_\omega) \ge \frac{1}{L}\gamma_\infty(T,d)\Big) \ge \frac{3}{4} .$$

Applying this result to 1-stable processes, we obtain the following:

Theorem 5.5.3 *For every 1-stable process $(X_t)_{t\in T}$ and $t_0 \in T$, we have*

$$\mathsf{P}\Big(\sup_{t\in T}(X_t - X_{t_0}) \ge \frac{1}{L}\gamma_\infty(T,d)\Big) \ge \frac{1}{L} .$$

This result looks weak, but it is hard to improve: when T consists of two points t_0 and t_1, then $\sup_{t\in T}(X_t - X_{t_0}) = \max(X_{t_1} - X_{t_0}, 0)$ is 0 when $X_{t_1} - X_{t_0} \le 0$, which happens with probability $1/2$.

Key Ideas to Remember

- We have met the powerful Principle A which lets us deduce some "smallness" information about a metric space from the existence of a random distance d_ω such that we control $\mathsf{E}\gamma_2(T,d_\omega)$ from above.
- We have seen a typical application of this principle to gain information about p-stable processes, in a way which will be vastly generalized later.

5.6 Where Do We Stand?

We have found an angle of attack on processes which are conditionally Gaussian. Unfortunately, such processes are uncommon. On the other hand, many processes are conditionally Bernoulli processes (with a meaning to be explained in the next chapter). The same line of attack will work on these processes, but this will require considerable work, as the study of Bernoulli processes is much more difficult than that of Gaussian processes.

Chapter 6
Bernoulli Processes

6.1 Bernoulli r.v.s

Throughout the book, we denote by ε_i independent Bernoulli (=coin flipping) r.v.s; that is,

$$\mathsf{P}(\varepsilon_i = \pm 1) = \frac{1}{2} .$$

(Thus ε_i is a r.v., while ϵ_i is a small positive number.)

Consider independent symmetric r.v.s ξ_i. It is fundamental that if (ε_i) denotes an independent sequence of Bernoulli r.v.s, which is independent of the sequence (ξ_i), then the sequences (ξ_i) and $(\varepsilon_i \xi_i)$ have the same distribution. This is obvious since this is already the case conditionally on the sequence (ε_i), by the very definition of the fact that the sequence (ξ_i) is symmetric. We will spend much time studying random sums (and series) of functions of the type $X(u) = \sum_i \xi_i \chi_i(u)$ where χ_i are functions on an index set U. This sum has the same distribution as the sum $\sum_i \varepsilon_i \xi_i \chi_i(u)$. Given the randomness of the ξ_i, this is a random sum of the type $\sum_i \varepsilon_i f_i(u)$ where $f_i(u)$ are functions. Then, all that matters is the set of coefficients $T = \{t = (f_i(u))_i ; u \in U\}$. This motivates the forthcoming definition of Bernoulli processes.

Given a sequence $(t_i)_{i \ge 1} \in \ell^2 = \ell^2(\mathbb{N}^*)$, we define

$$X_t = \sum_{i \ge 1} t_i \varepsilon_i . \tag{6.1}$$

Consider a subset T of ℓ^2. The *Bernoulli process* defined by T is the family $(X_t)_{t \in T}$.

M. Talagrand, *Upper and Lower Bounds for Stochastic Processes*, Ergebnisse der Mathematik und ihrer Grenzgebiete. 3. Folge / A Series of Modern Surveys in Mathematics 60, https://doi.org/10.1007/978-3-030-82595-9_6

A simple and essential fact is that the r.v. X_t "has better tails than the corresponding Gaussian r.v." as is expressed by the following, for which we refer to [53, page 90] or to Exercise 6.1.2:

Lemma 6.1.1 (The Sub-Gaussian Inequality) *Consider independent Bernoulli r.v.s ε_i and real numbers t_i. Then for each $u > 0$, we have*

$$\mathsf{P}\Big(\Big|\sum_i \varepsilon_i t_i\Big| \ge u\Big) \le 2\exp\Big(-\frac{u^2}{2\sum_i t_i^2}\Big) . \tag{6.2}$$

Exercise 6.1.2

(a) Use Taylor series to prove that for $\lambda \in \mathbb{R}$

$$\mathsf{E}\exp\lambda\varepsilon_i = \cosh\lambda \le \exp\frac{\lambda^2}{2} .$$

(b) Prove that

$$\mathsf{E}\exp\Big(\lambda\sum_i \varepsilon_i t_i\Big) \le \exp\Big(\frac{\lambda^2}{2}\sum_i t_i^2\Big) ,$$

and prove (6.2) using the formula $\mathsf{P}(X \ge u) \le \exp(-\lambda u)\mathsf{E}\exp\lambda X$ for $u > 0$ and $\lambda > 0$.

Corollary 6.1.3 (Khintchin's Inequality) *Consider complex numbers t_i, independent Bernoulli r.v.s ε_i, and $p \ge 1$. Then*

$$\Big(\mathsf{E}\Big|\sum_i \varepsilon_i t_i\Big|^p\Big)^{1/p} \le L\sqrt{p}\Big(\sum_i |t_i|^2\Big)^{1/2} . \tag{6.3}$$

Proof We reduce to the case of real numbers, and we combine the sub-Gaussian inequality with (2.24). □

In the case where $t_i = 1$ for $i \le N$ and $t_i = 0$ otherwise, (6.2) gives a strong quantitative form to the well-known statement that $\sum_{i\le N}\varepsilon_i$ is typically of order $\sqrt{N}$. The reason why the sum of N numbers of absolute value 1 can be of order $\sqrt{N}$ is because there is *cancellation* between the terms. It is completely wrong when no such cancellation occurs, e.g., in the exceptional case where all the ε_i equal 1. In contrast, a bound such as $|\sum_i \varepsilon_i t_i| \le \sum_i |t_i|$, which we will massively use below, *does not* rely on cancellation, since it holds even if all terms are of the same sign. More generally, there is typically cancellation in a sum $\sum_i \varepsilon_i t_i$ when $\sqrt{\sum_i |t_i|^2} \ll \sum_i |t_i|$.

6.2 Boundedness of Bernoulli Processes

Considering a subset T of $\ell^2 = \ell^2(\mathbb{N}^*)$ and the corresponding Bernoulli process $(X_t)_{t\in T}$, we set

$$b(T) := \mathsf{E}\sup_{t\in T} X_t = \mathsf{E}\sup_{t\in T}\sum_{i\geq 1} t_i\varepsilon_i \ . \tag{6.4}$$

We observe that $b(T) \geq 0$, that $b(T) \leq b(T')$ if $T \subset T'$, and that $b(T+t_0) = b(T)$, where $T + t_0 = \{t + t_0; t \in T\}$.

We would like to understand the value of $b(T)$ from the geometry of T, as we did in the case of Gaussian processes. Lemma 6.1.1 states that the process $(X_t)_{t\in T}$ satisfies the increment condition (2.4) so that Theorem 2.7.11 implies

$$b(T) \leq L\gamma_2(T) \ , \tag{6.5}$$

where we remind the reader that we often write $\gamma_2(T)$ instead of $\gamma_2(T, d)$ when d is the ℓ^2 distance.[1] Let us now write

$$g(T) = \mathsf{E}\sup_{t\in T}\sum_{i\geq 1} t_i g_i \ .$$

Since $\gamma_2(T) \leq Lg(T)$ by Theorem 2.10.1, Bernoulli processes "are smaller than the corresponding Gaussian processes". There is a much simpler direct proof of this fact.

Exercise 6.2.1 (Review of Jensen's Inequality) In this exercise, we review Jensen's inequality, a basic tool of probability theory. It states that if X is a r.v. valued in vector space W and Φ a convex function on W, then $\Phi(\mathsf{E}X) \leq \mathsf{E}\Phi(X)$. When using this inequality, we will use the sentence "we lower the value by taking expectation inside Φ rather than outside".

(a) If you know the Hahn-Banach theorem, convince yourself (or learn in a book) that $\Phi(x) = \sup_{f\in\mathcal{C}} f(x)$ where $\mathcal{C}$ is the class of affine functions (the sum of a constant and a linear function) which are $\leq \Phi$. Thus, for $f \in \mathcal{C}$, we have $\mathsf{E}f(X) = f(\mathsf{E}(X))$. Deduce Jensen's inequality from this fact.
(b) For a r.v. X and $p \geq 1$, prove that $|x + \mathsf{E}X|^p \leq \mathsf{E}|x + X|^p$.
(c) If X, Y and $\theta \geq 0$ are independent r.v.s, prove that $\mathsf{E}\theta|X+\mathsf{E}Y|^p \leq \mathsf{E}\theta|X+Y|^p$.

---

[1] Since (6.5) makes a massive use of the sub-Gaussian inequality (6.2) to control the increments along the chaining, it will be natural to say that this bound *relies on cancellation*, in sharp contrast with the bound (6.8) below.

Proposition 6.2.2 *It holds that*

$$b(T) \le \sqrt{\frac{\pi}{2}} g(T) . \tag{6.6}$$

Proof If $(\varepsilon_i)_{i\ge 1}$ is an i.i.d. Bernoulli sequence that is independent of the sequence $(g_i)_{i\ge 1}$, then the sequence $(\varepsilon_i |g_i|)_{i\ge 1}$ is i.i.d. standard Gaussian. Thus

$$g(T) = \mathsf{E} \sup_{t\in T} \sum_{i\ge 1} \varepsilon_i |g_i| t_i .$$

Denoting by E_g expectation in the r.v.s. g_i only (given the r.v.s ε_i) and since $\mathsf{E}_g |g_i| = \sqrt{2/\pi}$, we get

$$g(T) = \mathsf{E}\mathsf{E}_g \sup_{t\in T} \sum_{i\ge 1} t_i |g_i| \varepsilon_i \ge \sqrt{\frac{2}{\pi}} \mathsf{E} \sup_{t\in T} \sum_{i\ge 1} t_i \varepsilon_i = \sqrt{\frac{2}{\pi}} b(T) ,$$

where in the second inequality we apply Jensen's inequality to take the expectation E_g inside the supremum rather than outside. □

It is worth making a detour to state a general result in the direction of (6.6).

Lemma 6.2.3 *Consider vectors x_i in a complex Banach space and independent symmetric real-valued r.v.s ξ_i. Then, if ε_i denote independent Bernoulli r.v.s, we have*

$$\mathsf{E}\Big\| \sum_i \xi_i x_i \Big\| \ge \mathsf{E}\Big\| \sum_i (\mathsf{E}|\xi_i|) \varepsilon_i x_i \Big\| . \tag{6.7}$$

Proof Assuming without loss of generality that the r.v.s ξ_i and ε_i are independent, we use the symmetry of the r.v.s ξ_i to write

$$\mathsf{E}\Big\| \sum_i \xi_i x_i \Big\| = \mathsf{E}\Big\| \sum_i \varepsilon_i |\xi_i| x_i \Big\| .$$

Now $\mathsf{E}\| \sum_i \varepsilon_i |\xi_i| x_i \| \ge \mathsf{E}\| \sum_i (\mathsf{E}|\xi_i|) \varepsilon_i x_i \|$ as a consequence of Jensen's inequality. □

Thus to find a lower bound to $\mathsf{E}\| \sum_i \xi_i x_i \|$, we can reduce to the case where ξ_i is of the type $a_i \varepsilon_i$.[2]

We go back to Bernoulli processes. We can bound a Bernoulli process by comparing it with a Gaussian process or equivalently by using (6.6). There is

[2] But this method is not always sharp.

however a completely different method to bound Bernoulli processes. Denoting by $\|t\|_1 = \sum_{i\geq 1} |t_i|$ the ℓ^1 norm of t, the following proposition is trivial:

Proposition 6.2.4 *We have*

$$b(T) \leq \sup_{t\in T} \|t\|_1 \; . \tag{6.8}$$

We have found two very different ways to bound $b(T)$, namely, (6.6) and (6.8).

Exercise 6.2.5 Convince yourself that these two ways are really different from each other by considering the following two cases: $T = \{u, 0\}$ where $u \notin \ell^1$ and T the unit ball of ℓ^1.

We recall the Minkowski sum $T_1 + T_2 = \{t^1 + t^2 \; ; \; t^1 \in T_1, t^2 \in T_2\}$. The following definition and proposition formalize the idea that we can also bound $b(T)$ through mixtures of the previous situations.

Definition 6.2.6 For a subset T of ℓ^2, we set[3]

$$b^*(T) := \inf\left\{\gamma_2(T_1) + \sup_{t\in T_2} \|t\|_1 \; ; \; T \subset T_1 + T_2\right\} . \tag{6.9}$$

Since $X_{t^1+t^2} = X_{t^1} + X_{t^2}$, we have

$$\sup_{t\in T_1+T_2} X_t = \sup_{t\in T_1} X_t + \sup_{t\in T_2} X_t \; .$$

Taking expectation yields $b(T) \leq b(T_1+T_2) = b(T_1)+b(T_2)$. Combining with (6.5) and (6.8), we have proved the following:

Proposition 6.2.7 *We have*

$$b(T) \leq L b^*(T) \; . \tag{6.10}$$

It is natural to conjecture that the previous bound on $b(T)$ is sharp, that is, that there exist *no other way* to bound Bernoulli processes than the previous two methods and the mixtures of them. This was known as the Bernoulli conjecture. It took nearly 25 years to prove it.[4]

---

[3] You may find the notation silly, since T_1 is controlled by the ℓ^2 norm and T_2 by the ℓ^1 norm. The idea underlying my notation, here and in similar situations, is that T_1 denotes what I see as the main part of T, whereas T_2 is more like a perturbation term.

[4] Please see Footnote 2 on page 326 concerning the name I give to this result.

Theorem 6.2.8 (The Latała-Bednorz Theorem) *There exists a universal constant L such that given any subset T of ℓ^2, we have*

$$b^*(T) \le Lb(T) . \tag{6.11}$$

The proof of Theorem 6.2.8 will consist in describing a procedure to decompose each point $t \in T$ as a sum $t = t^1 + t^2$ where $\|t^2\|_1 \le Lb(T)$ and $T_1 = \{t^1 ; t \in T\}$ satisfies $\gamma_2(T_1) \le Lb(T)$. This procedure makes T naturally appear as a subset of a sum $T_1 + T_2$, even though T may be very different itself from such a sum. The intrinsic difficulty is that this decomposition is neither unique nor canonical. To illustrate the difficulty, consider a set T_1 with $\gamma_2(T_1) \le 1$, so that $b(T_1) \le L$. To each point t of T_1, let us associate a point $\varphi(t)$ with $\|\varphi(t)\|_1 \le 1$, and let $T = \{t + \varphi(t); t \in T_1\}$. Thus $b(T) \le L$. Now, we are only given the set T. How do we reconstruct the set T_1?

The proof of the Latała-Bednorz result involves a number of deep ideas. These are better presented gradually on simpler situations (following the path by which they were discovered), and the proof of the theorem is delayed to Chap. 10. In the rest of the present chapter, we build our understanding of Bernoulli processes. In the next section, we present three fundamental results for Bernoulli process, two of which have close relationships with properties of Gaussian processes. We then start developing the leading idea: to control Bernoulli processes, one has to control the index set T with respect to the supremum norm.

6.3 Concentration of Measure

The following "concentration of measure" result should be compared with Lemma 2.10.6:

Theorem 6.3.1 *Consider a subset $T \subset \ell^2$, and assume that for a certain $\sigma > 0$, we have $T \subset B(0, \sigma)$. Consider numbers $(a(t))_{t \in T}$, and let M be a median of the r.v. $\sup_{t \in T}(\sum_i \varepsilon_i t_i + a(t))$. Then*

$$\forall u > 0 \,,\; \mathsf{P}\Big(\Big|\sup_{t\in T}\Big(\sum_{i\ge 1}\varepsilon_i t_i + a(t)\Big) - M\Big| \ge u\Big) \le 4\exp\Big(-\frac{u^2}{8\sigma^2}\Big) . \tag{6.12}$$

In particular,

$$\Big|\mathsf{E}\sup_{t\in T}\Big(\sum_{i\ge 1}\varepsilon_i t_i + a(t)\Big) - M\Big| \le L\sigma \tag{6.13}$$

and

$$\forall u > 0, \mathsf{P}\Big(\Big|\sup_{t\in T}\big(\sum_{i\ge 1}\varepsilon_i t_i + a(t)\big) - \mathsf{E}\sup_{t\in T}\big(\sum_{i\ge 1}\varepsilon_i t_i + a(t)\big)\Big| \ge u\Big) \le L\exp\Big(-\frac{u^2}{L\sigma^2}\Big). \tag{6.14}$$

The proof relies on the fact that the function $\varphi(x) = \sup_{t\in T}(\sum_{i\le N} x_i t_i + a(t))$ on ℓ^2 is convex and has a Lipschitz constant $\le \sigma$. Such a function always satisfies a deviation inequality $\mathsf{P}(|\varphi((\varepsilon_i)_{i\ge 1}) - M| \ge u) \le 4\exp(-u^2/(8\sigma^2))$ when M is a median of $\varphi((\varepsilon_i)_{i\ge 1})$. This fact has a short and almost magic proof (see, for example, [53] (1.9)). We do not reproduce this proof for a good reason: the reader must face the fact that if she intends to become fluent in the area of probability theory we consider here, she must learn more about concentration of measure and that this is better done by looking, for example, at [121] and [52] rather than just at the proof of Theorem 6.3.1.

We end this section with a few simple important facts. We first recall the Paley-Zygmund inequality (sometimes called also the second moment method): for a r.v. $X \ge 0$, with $\mathsf{E}X^2 > 0$,

$$\mathsf{P}\Big(X \ge \frac{1}{2}\mathsf{E}X\Big) \ge \frac{1}{4}\frac{(\mathsf{E}X)^2}{\mathsf{E}X^2}\,. \tag{6.15}$$

Exercise 6.3.2 Prove (6.15). Hint: Let $A = \{X \ge \mathsf{E}X/2\}$. Show that $\mathsf{E}X\mathbf{1}_{A^c} \le \mathsf{E}X/2$. Show that $\mathsf{E}X/2 \le \mathsf{E}(X\mathbf{1}_A) \le (\mathsf{E}X^2\mathsf{P}(A))^{1/2}$.

Corollary 6.3.3 *If ε_i are independent Bernoulli r.v.s and t_i are numbers, it holds that*

$$\mathsf{P}\Big(\Big|\sum_{i\ge 1}\varepsilon_i t_i\Big| \ge \frac{1}{L}\big(\sum_{i\ge 1}|t_i|^2\big)^{1/2}\Big) \ge \frac{1}{L}\,. \tag{6.16}$$

Proof By the sub-Gaussian inequality and (2.24), the r.v. $X = |\sum_{i\ge 1}\varepsilon_i t_i|^2$ satisfies $\mathsf{E}X^2 \le L(\sum_{i\ge 1}|t_i|^2)^2 = L(\mathsf{E}X)^2$. We then apply the Paley-Zygmund inequality (6.15). □

Exercise 6.3.4 As a consequence of (6.16), prove that if the series $\sum_{n\ge 1}\varepsilon_i t_i$ converges a.s., then $\sum_{i\ge 1} t_i^2 < \infty$.

As a consequence of (6.16), we have

$$\mathsf{E}\Big|\sum_{i\ge 1}\varepsilon_i t_i\Big| \ge \frac{1}{L}\|t\|_2\,. \tag{6.17}$$

Exercise 6.3.5 Prove that for a r.v. $Y \ge 0$ with $\mathsf{E}Y > 0$, one has $\mathsf{E}Y\mathsf{E}Y^3 \ge (\mathsf{E}Y^2)^2$, and find another proof of (6.17).

Lemma 6.3.6 *For a subset T of ℓ^2, we have*

$$\Delta(T, d_2) \le Lb(T) \; . \tag{6.18}$$

Proof Assuming without loss of generality that $0 \in T$, we have

$$\forall t \in T \; , \; b(T) \ge \mathsf{E} \max\Big(0, \sum_{i \ge 1} \varepsilon_i t_i\Big) = \frac{1}{2} \mathsf{E} \Big| \sum_{i \ge 1} \varepsilon_i t_i \Big| \ge \frac{1}{L} \|t\|_2 \; ,$$

using that $\max(x, 0) = (|x| + x)/2$ in the equality and (6.17) in the last inequality. This proves (6.18). □

6.4 Sudakov Minoration

In this section, we prove a version of Lemma 2.10.2 (Sudakov minoration) for Bernoulli processes. This will be our first contact with the essential idea that when all the coefficients t_i are small (say, compared to $\|t\|_2$), the r.v. $\sum_{i \ge 1} t_i \varepsilon_i$ resembles a Gaussian r.v., by the central limit theorem. Therefore one expects that when, in some sense, the set T is small for the ℓ^∞ norm, $g(T)$ (or, equivalently, $\gamma_2(T)$) is not too much larger than $b(T)$. This will also be the main idea of Sect. 6.6.

Theorem 6.4.1 *Consider $t_1, \ldots, t_m$ in ℓ^2, and assume that*

$$\ell \ne \ell' \Rightarrow \|t_\ell - t_{\ell'}\|_2 \ge a \; . \tag{6.19}$$

Assume moreover that

$$\forall \ell \le m \; , \; \|t_\ell\|_\infty \le b \; . \tag{6.20}$$

Then

$$\mathsf{E} \sup_{\ell \le m} \sum_{i \ge 1} \varepsilon_i t_{\ell,i} \ge \frac{1}{L} \min\Big(a\sqrt{\log m}, \frac{a^2}{b}\Big) \; . \tag{6.21}$$

For a first understanding of this theorem, one should consider the case where t_i is the i-th element of the basis of ℓ^2. One should also compare (6.21) with Lemma 2.10.2, which in the present language asserts that

$$\mathsf{E} \sup_{\ell \le m} \sum_{i \ge 1} g_i t_{\ell,i} \ge \frac{a}{L_1} \sqrt{\log m} \; , \tag{6.22}$$

and will be the basis of the proof of (6.21). To understand the need of the minimum in (6.21), you should solve the next exercise.

Exercise 6.4.2 Convince yourself that in (6.21) the term a^2/b is of the correct order. Hint: Remember that $\sum_i \varepsilon_i t_{\ell,i} \le \sum_i |t_{\ell,i}|$. Look for examples where $t_{\ell,i} \in \{0, b\}$.

Corollary 6.4.3 *For a set $T \subset \ell^2$ such that*

$$\forall t \in T \; ; \; \|t\|_\infty \le b$$

and for any $a > 0$, we have

$$b(T) \ge \frac{1}{L} \min\left(a\sqrt{N(T, d_2, a)}, \frac{a^2}{b}\right) . \tag{6.23}$$

Proof By Lemma 2.9.3(a) for $m = N(T, d_2, a)$, we can find points $(t_\ell)_{\ell \le m}$, as in (6.20). □

We start the preparations for the proof of (6.21).

Lemma 6.4.4 (The Contraction Principle) *Consider independent and symmetric r.v.s η_i valued in a Banach space and numbers α_i with $|\alpha_i| \le 1$. Then*

$$\mathsf{E}\Big\| \sum_{i\ge1} \alpha_i \eta_i \Big\| \le \mathsf{E}\Big\| \sum_{i\ge1} \eta_i \Big\| . \tag{6.24}$$

Proof We consider the quantity $\mathsf{E}\| \sum_{i\ge1} \alpha_i \eta_i \|$ as a function of the numbers α_i. It is convex, and its domain is a convex compact set. Therefore it attains its maximum at an extreme point of its domain. For such an extreme point, $\alpha_i = \pm1$ for each i, and in that case, the left- and right-hand sides of (6.24) coincide. □

We will also need the following variation of Bernstein's inequality:

Lemma 6.4.5 *Consider centered independent r.v.s W_i and numbers a_i such that $\mathsf{E}\exp(|W_i|/a_i) \le 2$. Then for $v > 0$, we have*

$$\mathsf{P}\Big(\sum_{i\ge1} W_i \ge v\Big) \le 2\exp\Big(-\frac{1}{L}\min\Big(\frac{v^2}{\sum_{i\ge1} a_i^2}, \frac{v}{\sup_{i\ge1} a_i}\Big)\Big) . \tag{6.25}$$

Proof We write

$$|\exp x - 1 - x| \le \sum_{k\ge2} \frac{|x|^k}{k!} \le |x|^2 \exp|x| ,$$

so that, using the Cauchy–Schwarz inequality in the second inequality,

$$\mathsf{E}\exp\lambda W_i \le 1+\lambda^2\mathsf{E}W_i^2\exp|\lambda W_i| \le 1+\lambda^2(\mathsf{E}W_i^4)^{1/2}(\mathsf{E}\exp 2|\lambda W_i|)^{1/2}\,.$$

Now for $|\lambda|a_i \le 1/2$, we have $\mathsf{E}\exp 2|\lambda W_i| \le 2$. Since $\mathsf{E}\exp(|W_i|/a_i) \le 2$, we also have $\mathsf{E}W_i^4 \le La_i^4$. Thus, for $|\lambda|a_i \le 1/2$, we have

$$\mathsf{E}\exp\lambda W_i \le 1+L\lambda^2 a_i^2 \le \exp L\lambda^2 a_i^2\,,$$

from which the conclusion follows as in the proof of (4.44). □

A technical ingredient of the proof of Theorem 6.4.1 is the following consequence of Lemma 6.4.5:

Corollary 6.4.6 *Consider independent standard Gaussian r.v.s* (g_i). *Given a number* $A>0$, *we may find a number* c *large enough such that the r.v.s* $\xi_i = g_i\mathbf{1}_{\{|g_i|>c\}}$ *satisfy the following property. Consider an integer* N *and numbers* $a,b>0$ *such that*

$$\sqrt{\log N} \le \frac{a}{b}\,. \tag{6.26}$$

For $\ell \le N$, *consider* $t_\ell = (t_{\ell,i})_{i\ge 1}$ *with* $\|t_\ell\|_2 \le 2a$ *and* $\|t_\ell\|_\infty \le b$. *Then*

$$\mathsf{E}\sup_{\ell\le N}\sum_{i\ge 1}\xi_i t_{\ell,i} \le \frac{a}{A}\sqrt{\log N}\,. \tag{6.27}$$

If, instead of ξ_i we had g_i in the left-hand side, we would obtain a bound $La\sqrt{\log N}$ (see (2.15)). The content of the lemma is that we can improve on that bound by a large constant factor by taking c large.

Proof Given a number $B>0$, we have $\mathsf{E}\exp B|g_i| < \infty$, and it is obvious that for c large enough we have $\mathsf{E}\exp B|\xi_i| \le 2$. Given ℓ, we use the Bernstein-like inequality (6.25) with $W_i = \xi_i t_{\ell,i}$ and $a_i = |t_{\ell,i}|/B$, so that $\sum_{i\ge 1} a_i^2 \le 4a^2/B^2$ and $\sup_{i\ge 1}|a_i| \le b/B$, and we obtain

$$\mathsf{P}\Big(\sum_{i\ge 1}\xi_i t_{\ell,i} \ge v\Big) \le L\exp\Big(-\frac{1}{L}\min\Big(\frac{v^2B^2}{a^2},\frac{vB}{b}\Big)\Big)\,,$$

so that, using (6.26) in the first inequality, for $vB \ge a\sqrt{\log N}$, we have

$$\begin{aligned}\mathsf{P}\Big(\sup_{\ell\le N}\sum_{i\ge 1}\xi_i t_{\ell,i} \ge v\Big) &\le LN\exp\Big(-\frac{1}{L}\min\Big(\frac{v^2B^2}{a^2},\frac{vB\sqrt{\log N}}{a}\Big)\Big)\\ &\le LN\exp\Big(-\frac{vB\sqrt{\log N}}{a}\Big)\,,\end{aligned} \tag{6.28}$$

and (2.6) implies that $\mathsf{E}\sup_{\ell\le N}\sum_{i\ge 1}\xi_i t_{\ell,i} \le La\sqrt{\log N}/B$. □

We are now ready to perform the main step of the proof of Theorem 6.4.1.

Proposition 6.4.7 *Assume the hypotheses of Theorem 6.4.1 and that furthermore*

$$\sqrt{\log m} \le \frac{a}{b} \tag{6.29}$$

$$\forall \ell \le m \; ; \; \|t_\ell\|_2 \le 2a \; . \tag{6.30}$$

Then (6.21) holds.

Condition (6.29) is motivated by the important idea that the critical case of Theorem 6.4.1 is where the two terms in the minimum in the right-hand side of (6.21) are nearly equal.

Proof Consider a parameter $c > 0$ and define $\xi_i = g_i \mathbf{1}_{\{|g_i|>c\}}$ and $\xi_i' = g_i \mathbf{1}_{\{|g_i|\le c\}}$. Thus, using (6.22) in the first inequality,

$$\frac{a}{L_1}\sqrt{\log m} \le \mathsf{E} \sup_{\ell \le m} \sum_{i\ge 1} g_i t_{\ell,i} \le \mathsf{E} \sup_{\ell \le m} \sum_{i\ge 1} \xi_i' t_{\ell,i} + \mathsf{E} \sup_{\ell \le m} \sum_{i\ge 1} \xi_i t_{\ell,i} \; . \tag{6.31}$$

Now, using Corollary 6.4.6 for $A = 2L_1$ shows that if c is a large enough constant, we have

$$\mathsf{E} \sup_{\ell \le m} \sum_{i\ge 1} \xi_i t_{\ell,i} \le \frac{a}{2L_1}\sqrt{\log m} \; , \tag{6.32}$$

so that (6.31) implies

$$\frac{a}{2L_1}\sqrt{\log m} \le \mathsf{E} \sup_{\ell \le m} \sum_{i\ge 1} \xi_i' t_{\ell,i} \le c\mathsf{E} \sup_{\ell \le m} \sum_{i\ge 1} \varepsilon_i t_{\ell,i} \; , \tag{6.33}$$

where the last inequality is obtained by copying the argument of (6.24). Using (6.29) shows that $\min(a\sqrt{\log m}, a^2/b) = a\sqrt{\log m}$ so that (6.33) implies (6.21). □

Proposition 6.4.8 *The conclusion of Theorem 6.4.1 holds true if we assume moreover that* $\|t_\ell\|_2 \le 2a$ *for each* $\ell \le m$.

The improvement over Proposition 6.4.7 is that we no longer assume (6.29).

Proof It follows from Lemma 6.3.6 and (6.19) that

$$\mathsf{E} \sup_{\ell \le m} \sum_{i\ge 1} t_{\ell,i}\varepsilon_i \ge \frac{a}{L} \; , \tag{6.34}$$

Assume first that $a/b \le \sqrt{\log 2}$. Then $a \ge a^2/(Lb)$ and (6.34) implies (6.21). Thus, it suffices to prove (6.21) when $a/b \ge \sqrt{\log 2}$. For this, consider the largest integer

$N \le m$ for which

$$\sqrt{\log N} \le \frac{a}{b} \,. \tag{6.35}$$

Then $N \ge 2$. Next we prove that

$$a\sqrt{\log N} \ge \frac{1}{L} \min\left(a\sqrt{\log m}, \frac{a^2}{b}\right) . \tag{6.36}$$

Indeed this is obvious if $N = m$. When $N < m$, the definition of N shows that $\sqrt{\log(N+1)} \ge a/b$, so that $a\sqrt{\log N} \ge a^2/(Lb)$, proving (6.36). Consequently, it suffices to prove (6.21) when m is replaced by N, and then (6.29) is satisfied according to (6.35). The conclusion follows from Proposition 6.4.7. □

Proof of Theorem 6.4.1 Let $T = \{t_1, \ldots, t_m\}$, so that we want to prove (6.21), i.e.,

$$b(T) \ge \frac{1}{L} \min\left(a\sqrt{\log m}, \frac{a^2}{b}\right) . \tag{6.37}$$

We have proved in Proposition 6.4.8 that when furthermore we have $\|t_\ell\|_2 \le 2a$ for each ℓ, then

$$b(T) \ge \frac{1}{L_2} \min\left(a\sqrt{\log m}, \frac{a^2}{b}\right) . \tag{6.38}$$

To prove (6.37), we may assume that

$$b(T) \le \frac{a^2}{8L_2 b} \tag{6.39}$$

because there is nothing to prove otherwise. Consider a point $t \in T$ and an integer $k \ge 0$. The proof then relies on a simple iteration procedure. Assume that in the ball $B(t, 2^k a)$, we can find points $u_1, \ldots, u_N \in T$ with $d(u_\ell, u_{\ell'}) \ge 2^{k-1}a$ whenever $\ell \ne \ell'$. We can then use (6.38) for the points $u_1 - t, \ldots, u_N - t$, with $2^{k-1}a$ instead of a and $2b$ instead of b to obtain

$$b(T) \ge \frac{1}{L_2} \min\left(2^{k-1}a\sqrt{\log N}, \frac{2^{2k-2}a^2}{2b}\right) .$$

Using (6.39), this implies that $Lb(T) \ge 2^k a\sqrt{\log N}$. Hence, $N \le M_k := \exp(L2^{-2k}b(T)^2/a^2)$. Thus any ball $B(t, 2^k a)$ of T can be covered using at most M_k balls of T radius $2^{k-1}a$. We then iterate this result as follows: Consider any number k_0 large enough so that $T \subset B(t, 2^{k_0}a)$ for a certain $t \in T$. Then we can cover T by at most M_{k_0} balls centered in T of radius $2^{k_0-1}a$. Each of these balls can in turn be covered by at most M_{k_0-1} balls of T of radius $2^{k_0-2}a$, so that T

can be covered by at most $M_{k_0}M_{k_0-1}$ such balls. Continuing in this manner until we cover T by balls of radius $a/2$ requires at most $\prod_{k\ge 0} M_k$ balls of radius $a/2$. Since $t_\ell \notin B(t_{\ell'}, a/2)$ for $\ell \neq \ell'$, these balls of radius $a/2$ contain a single point of T, and we have shown that $m \le \prod_{k\ge 0} M_k \le \exp(Lb(T)^2/a^2)$, i.e., $b(T) \ge a\sqrt{\log m}/L$. □

Combining our results, we may now prove a version of Proposition 2.10.8 for Bernoulli processes.

Proposition 6.4.9 *There exist constants L_1 and L_2 with the following properties. Consider numbers $a, b, \sigma > 0$, vectors $t_1, \dots, t_m \in \ell^2$, that satisfy (6.19) and (6.20). For $\ell \le m$, consider sets H_ℓ with $H_\ell \subset B_2(t_\ell, \sigma)$. Then*

$$b\Big(\bigcup_{\ell\le m} H_\ell\Big) \ge \frac{1}{L_1}\min\Big(a\sqrt{\log m}, \frac{a^2}{b}\Big) - L_2\sigma\sqrt{\log m} + \min_{\ell\le m} b(H_\ell)\,. \tag{6.40}$$

The proof is identical to that of Proposition 2.10.8, if one replaces Lemmas 2.10.2 and 2.10.6, respectively, by Theorems 6.4.1 and 6.3.1.

Corollary 6.4.10 *There exists a constant L_0 with the following property. Consider a set D with $\Delta(D, d_\infty) \le 4a/\sqrt{\log m}$, and for $\ell \le m$, consider points $t_\ell \in D$ that satisfy (6.19), i.e., $\|t_\ell - t_{\ell'}\|_2 \ge a$ for $\ell \ne \ell'$. Consider moreover sets $H_\ell \subset B_2(t_\ell, a/L_0)$. Then*

$$b\Big(\bigcup_{\ell\le m} H_\ell\Big) \ge \frac{a}{L_0}\sqrt{\log m} + \min_{\ell\le m} b(H_\ell)\,. \tag{6.41}$$

Proof Since $b(T - t_1) = b(T)$, we may assume without loss of generality that $t_1 = 0$. Since $\Delta(D, d_\infty) \le b := 4a/\sqrt{\log m}$, we have $\|t_\ell\|_\infty \le b$ for all $\ell \le m$, and (6.40) used for $\sigma = a/L_0$ gives

$$b\Big(\bigcup_{\ell\le m} H_\ell\Big) \ge \frac{1}{4L_1}a\sqrt{\log m} - \frac{aL_2}{L_0}\sqrt{\log m} + \min_{\ell\le m} b(H_\ell)\,,$$

so that if $L_0 \ge 8L_1L_2$ and $L_0 \ge 8L_1$ we get (6.41). □

This corollary will be an essential tool to prove the Latała-Bednorz theorem.

6.5 Comparison Principle

Our last fundamental result is a comparison principle. Let us say that a map θ from $\mathbb{R}$ to $\mathbb{R}$ is a *contraction* if $|\theta(s) - \theta(t)| \le |s - t|$ for each $s, t \in \mathbb{R}$.

Theorem 6.5.1 *For $i \geq 1$, consider contractions θ_i with $\theta_i(0) = 0$. Then for each (finite) subset T of ℓ^2, we have*

$$\mathsf{E}\sup_{t\in T}\sum_{i\geq 1}\varepsilon_i\theta_i(t_i) \leq b(T) = \mathsf{E}\sup_{t\in T}\sum_{i\geq 1}\varepsilon_i t_i \ . \tag{6.42}$$

A more general comparison result may be found in [112, Theorem 2.1]. We give here only the simpler proof of the special case (6.42) that we need.

Proof The purpose of the condition $\theta_i(0) = 0$ is simply to ensure that $(\theta_i(t_i)) \in \ell^2$ whenever $(t_i) \in \ell^2$. A simple approximation procedure shows that it suffices to show that for each N, we have

$$\mathsf{E}\sup_{t\in T}\sum_{1\leq i\leq N}\varepsilon_i\theta_i(t_i) \leq \mathsf{E}\sup_{t\in T}\sum_{1\leq i\leq N}\varepsilon_i t_i \ .$$

By iteration, it suffices to show that $\mathsf{E}\sup_{t\in T}\sum_{1\leq i\leq N}\varepsilon_i t_i$ decreases when t_1 is replaced by $\theta_1(t_1)$. By conditioning on $\varepsilon_2, \varepsilon_3, \ldots, \varepsilon_N$, it suffices to prove that for a subset T of $\mathbb{R}^2$ and a contraction θ, we have

$$\mathsf{E}\sup_{t=(t_1,t_2)\in T}(\varepsilon_1\theta(t_1) + t_2) \leq \mathsf{E}\sup_{t=(t_1,t_2)\in T}(\varepsilon_1 t_1 + t_2) \ . \tag{6.43}$$

Now

$$2\mathsf{E}\sup_{t=(t_1,t_2)\in T}(\varepsilon_1\theta(t_1) + t_2) = \sup_{s'\in T}(\theta(s'_1) + s'_2) + \sup_{s\in T}(-\theta(s_1) + s_2) \ .$$

Thus to prove (6.43) it suffices to show that for $s, s' \in T$, we have

$$\theta(s'_1) + s'_2 - \theta(s_1) + s_2 \leq 2\mathsf{E}\sup_{t=(t_1,t_2)\in T}(\varepsilon_1 t_1 + t_2) \ . \tag{6.44}$$

To bound the right-hand side from below, we may take either $t = s'$ when $\varepsilon_1 = 1$ and $t = s$ when $\varepsilon_1 = -1$ or the opposite:

$$\begin{aligned} &2\mathsf{E}\sup_{t=(t_1,t_2)\in T}(\varepsilon_1 t_1 + t_2) \\ &\qquad \geq \max(s'_1 + s'_2 - s_1 + s_2, s_1 + s_2 - s'_1 + s'_2) = s_2 + s'_2 + |s'_1 - s_1| \ , \end{aligned}$$

so that (6.44) simply follows from the fact that $\theta(s'_1) - \theta(s_1) \leq |s'_1 - s_1|$ since θ is a contraction. □

Corollary 6.5.2 *For each subset T of ℓ^2, we have*

$$\mathsf{E}\sup_{t\in T}\Big|\sum_{i\geq 1}\varepsilon_i|t_i|\Big| \leq 2\mathsf{E}\sup_{t\in T}\Big|\sum_{i\geq 1}\varepsilon_i t_i\Big| \ .$$

Proof Writing $x^+ = \max(x, 0)$, we have $|x| = x^+ + (-x)^+$ so that by symmetry

$$\mathsf{E}\sup_{t\in T}\Big|\sum_{i\ge 1}\varepsilon_i|t_i|\Big| = 2\mathsf{E}\sup_{t\in T}\Big(\sum_{i\ge 1}\varepsilon_i|t_i|\Big)^+ = 2\mathsf{E}\sup_{t\in T'}\sum_{i\ge 1}\varepsilon_i|t_i|$$

where $T' = T \cup \{0\}$. Now, using (6.42) in the first inequality, we have

$$\mathsf{E}\sup_{t\in T'}\sum_{i\ge 1}\varepsilon_i|t_i| \le \mathsf{E}\sup_{t\in T'}\sum_{i\ge 1}\varepsilon_i t_i \le \mathsf{E}\sup_{t\in T}\Big|\sum_{i\ge 1}\varepsilon_i t_i\Big| . \qquad \square$$

6.6 Control in ℓ^∞ Norm

Bernoulli processes are much easier to understand when the index set is small in the supremum norm. The main result of this section goes in this direction. It is weaker than Theorem 6.2.8 (the Latała-Bednorz Theorem), but the proof is much easier.

Theorem 6.6.1 *There exists a universal constant L such that for any subset T of ℓ^2, we have*

$$\gamma_2(T) \le L\big(b(T) + \sqrt{b(T)\gamma_1(T, d_\infty)}\big) . \tag{6.45}$$

Corollary 6.6.2 *We have*

$$b(T) \ge \frac{1}{L}\min\Big(\gamma_2(T), \frac{\gamma_2(T)^2}{\gamma_1(T, d_\infty)}\Big) . \tag{6.46}$$

Proof Denoting by L^* the constant of (6.45), if $b(T) \le \gamma_2(T)/(2L^*)$, then (6.45) implies

$$\gamma_2(T) \le \gamma_2(T)/2 + L^*\sqrt{b(T)\gamma_1(T, d_\infty)} ,$$

hence $b(T) \ge \gamma_2(T)^2/4(L^*)^2\gamma_1(T, d_\infty)$. $\square$

Exercise 6.6.3 Find examples of situations where $\gamma_2(T) \ll \gamma_1(T, d_\infty)$ and $b(T)$ is of order $\gamma_2(T)^2/\gamma_1(T, d_\infty)$, not $\gamma_2(T)$. Hint: Try cases where $t_i \in \{0, 1\}$ for each i and each t.

We recall the constant L_0 of Corollary 6.4.10. The next lemma is our main tool.

Lemma 6.6.4 *Consider a number $r \ge 2L_0$. Consider $B \subset \ell^2$ such that $\Delta(B, d_\infty) \le 4a/\sqrt{\log m}$. Then we can find a partition $(A_\ell)_{\ell\le m}$ of B into sets which have* either *of the following properties:*

$$\Delta(A_\ell, d_2) \le 2a , \tag{6.47}$$

or else

$$t \in A_\ell \Rightarrow b(B \cap B_2(t, 2a/r)) \le b(B) - \frac{a}{L_0}\sqrt{\log m} \,. \tag{6.48}$$

Proof The proof is almost identical to that of Lemma 2.9.4, using now Corollary 6.4.10. Consider the set

$$C = \Big\{ t \in B \; ; \; b(B \cap B_2(t, 2a/r)) > b(B) - \frac{a}{L_0}\sqrt{\log m} \Big\} \,.$$

Consider points $(t_\ell)_{\ell \le m'}$ in C such that $d_2(t_\ell, t_{\ell'}) \ge a$ for $\ell \ne \ell'$. Since $2/r \le 1/L_0$, using (6.41) for the sets $H_\ell := B \cap B(t_\ell, 2a/r)$ shows that

$$\begin{aligned} b(B) \ge b\Big(\bigcup_{\ell \le m} H_\ell\Big) &\ge \frac{a}{L_0}\sqrt{\log m'} + \min_{\ell \le m'} b(B \cap B_2(t, 2a/r)) \\ &> \frac{a}{L_0}\sqrt{\log m'} + b(B) - \frac{a}{L_0}\sqrt{\log m} \end{aligned} \tag{6.49}$$

and thus $m' < m$. Consequently by Lemma 2.9.3(b), we may cover C by $m' < m$ balls B_ℓ of radius a. We then set $A_\ell = C \cap (B_\ell \setminus \cup_{\ell' < \ell} B_{\ell'})$ for $\ell \le m'$, $A_\ell = \emptyset$ for $m' < \ell < m$ and $A_m = B \setminus C$. □

Proof of Theorem 6.6.1 The reader should review the proof of Theorem 2.9.1 now. We fix r as in Lemma 6.6.4. We consider an integer $\tau \ge 2$ to be specified later and an admissible sequence of partitions $(\mathcal{D}_n)$ of T such that

$$\sup_{t \in T} \sum_{n \ge 0} 2^n \Delta(D_n(t), d_\infty) \le 2\gamma_1(T, d_\infty) \,. \tag{6.50}$$

By induction over n, we construct an admissible sequence $(\mathcal{A}_n)$ of partitions of T, and for $A \in \mathcal{A}_n$, an integer $j_n(A) \in \mathbb{Z}$ such that $\Delta(A, d_2) \le 2r^{-j_n(A)}$. For $n \le \tau$, we set $\mathcal{A}_n = \{T\}$, and for $A \in \mathcal{A}_n$, we set $j_n(A) = j_0$ where j_0 is the largest integer with $\Delta(T, d_2) \le 2r^{-j_0}$.

Assuming now that $\mathcal{A}_n$ has been constructed for some $n \ge \tau$, with card $\mathcal{A}_n \le N_n$, we consider $B \in \mathcal{A}_n$, and we proceed to partition it. First we partition B into the elements $B \cap D$ for $D \in \mathcal{D}_{n-1}$. Consider such a set $B \cap D$.

First Case Assume that

$$\Delta(D, d_\infty) \le \frac{2r^{-j_n(B)-1}}{\sqrt{\log N_{n-\tau}}} \,. \tag{6.51}$$

We may then apply Lemma 6.6.4 with $a = r^{-j_n(B)-1}$ to partition $B \cap D$ into sets $(A_{\ell,B,D})_{\ell \le N_{n-\tau}}$ such that either $\Delta(A_{\ell,B,D}, d_2) \le 2r^{-j_n(B)-1}$ (and we then set $j_{n+1}(A_{\ell,B,D}) = j_n(B) + 1$) or else

$$t \in A_{\ell,B,D} \Rightarrow b(B \cap D \cap B_2(t, 2r^{-j_n(B)-2})) \le b(B \cap D) - \frac{1}{L_0} 2^{(n-\tau)/2} r^{-j_n(B)-1} ,$$

and we then set $j_{n+1}(A_{\ell,B,D}) = j_n(B)$. In that second case, we have in particular

$$t \in A_{\ell,B,D} \Rightarrow b(A_{\ell,B,D} \cap B_2(t, 2r^{-j_n(B)-2})) \le b(B) - \frac{1}{L_0} 2^{(n-\tau)/2} r^{-j_n(B)-1} . \tag{6.52}$$

Second Case Assume that (6.51) fails. In this case, we decide that $B \cap D \in \mathcal{A}_{n+1}$, and we set $j_{n+1}(B \cap D) = j_n(B)$. Since (6.51) fails, we then have

$$2^{(n-\tau+1)/2} r^{-j_{n+1}(B\cap D)} \le Lr2^{n-\tau} \Delta(D, d_\infty) , \tag{6.53}$$

To sum up, the partition $\mathcal{A}_{n+1}$ consists of all the sets $B \cap D$ (with $B \in \mathcal{A}_n$, $D \in \mathcal{D}_{n-1}$) for which (6.51) fails, as well as of all the sets $A_{\ell,B,D} \subset B \cap D$ for pairs B, D (with $B \in \mathcal{A}_n$, $D \in \mathcal{D}_{n-1}$) which satisfy (6.51). This completes the construction. We have as desired card $\mathcal{A}_{n+1} \le N_n N_{n-1} N_{n-\tau} \le N_n N_{n-1}^2 = N_{n+1}$.

Let us fix $t \in T$ and set $j(n) = j_n(A_n(t))$. Let $a(n) = 2^{n/2} r^{-j(n)}$. We are going to prove that

$$2^{-\tau/2} \sum_{n \ge 0} a(n) \le Lr(b(T) + 2^{-\tau} \gamma_1(T, d_\infty)) . \tag{6.54}$$

It then follows that

$$\gamma_2(T, d_2) \le Lr(2^{\tau/2} b(T) + 2^{-\tau/2} \gamma_1(T, d_\infty)) ,$$

and we finish the proof by optimization over τ: if $\gamma_1(T, d_\infty) \le 4b(T)$, we take $\tau = 2$; otherwise, we take 2^τ about $\gamma_1(T, d_\infty)/b(T)$.

The key property of the construction is that if $B \in \mathcal{A}_n$, $A \in \mathcal{A}_{n+1}$, $A \subset B$, and $j_n(B) = j_{n+1}(A)$ then either (by (6.53)), there exists $D \in \mathcal{D}_{n-1}$ with $B \subset D$ and

$$2^{(n-\tau+1)/2} r^{-j_{n+1}(A)} \le Lr2^{n-\tau} \Delta(D, d_\infty) , \tag{6.55}$$

or else by (6.52)

$$t \in A \Rightarrow b(A \cap B_2(t, 2r^{-j_n(B)-2})) \le b(B) - \frac{1}{L} 2^{(n-\tau)/2} r^{-j_n(B)-1} . \tag{6.56}$$

The proof of (6.54) is nearly identical to the part of the proof of Theorem 2.9.1 following Eq. (2.88). We let the reader prove that the sequence $(a(n))$ is bounded. Consider then the set I as provided by Lemma 2.9.5 for $\alpha = \sqrt{2}$. It suffices to prove (6.54) when the sum over $n \geq 0$ is replaced by the sum over $n \in I \setminus \{0\}$. As in (2.91) for $n \in I \setminus \{0\}$, we have $j(n-1) = j(n)$ and $j(n+1) = j(n)+1$. Let us enumerate the elements[5] of $I \setminus \{0\}$ as $n_1 < n_2 < \ldots$, so that $j(n_{k+1}) \geq j(n_k)+1$.

We consider $k \geq 1$ (so that $n_k \geq 1$), and we proceed to bound $a(n_k)$. Since $j(n_{k+2}) \geq j(n_{k+1})+1 \geq j(n_k)+2$,, we have

$$\Delta(A_{n_{k+2}}(t)) \leq 2r^{-j(n_{k+2})} \leq 2r^{-j(n_k)-2} . \tag{6.57}$$

Let us define $n = n_k - 1$, so that $j(n+1) = j(n)$, and define also $B = A_n(t)$, $A = A_{n_k}(t) = A_{n+1}(t)$, so that $j_n(B) = j(n) = j(n+1) = j(n_k) = j_{n+1}(A)$. We know that either (6.55) or (6.56) hold. If (6.56) holds, we conclude by (6.57) that $A_{n_k+2}(t) \subset A \cap B_2(t, 2r^{-j_n(B)-2})$ so that

$$2^{-\tau/2} a(n_k) \leq Lr(b(A_n(t)) - b(A_{n_{k+2}}(t))) . \tag{6.58}$$

If, on the other hand, (6.55) holds, we obtain

$$a(n_k) \leq Lr2^{n-\tau/2} \Delta(D_{n-1}(t), d_\infty) . \tag{6.59}$$

As in Theorem 2.9.1, summation of these inequalities concludes the proof of (6.54). □

6.7 Peaky Parts of Functions

One basic idea underlying the Bernoulli conjecture is that a sequence (t_i) has a "spread out part" and a "peaky part". The r.v.s $\sum_{i \geq 1} \varepsilon_i t_i$ are controlled by comparison with Gaussian processes for the spread out parts and by taking absolute values for the peaky parts. The notions of "spread out" and "peaky" parts refer to the space $\ell^2(\mathbb{N})$. In this section, we study how to perform the same decomposition for functions in $L^2(\nu)$, where ν is a positive measure (which need not be a probability). The case where the measure space is $\mathbb{N}$ and ν is the counting measure $\nu(A) = \operatorname{card} A$ is the previous case of $\ell^2(\mathbb{N})$. In this section, the distances d_2 and d_∞ refer to the distances induced by the norms in $L^2(\nu)$ and $L^\infty(\nu)$.

For a single function, it is quite obvious what to do, and this is spelled out in the next lemma.

[5] We assume here that I is infinite, leaving the necessary simple modifications of the argument when I is finite to the reader.

Lemma 6.7.1 *Consider $f \in L^2(\nu)$ and $u > 0$. Then we can write $f = f_1 + f_2$ where*

$$\|f_1\|_2 \le \|f\|_2\ ,\ \|f_1\|_\infty \le u\ ;\ \|f_2\|_2 \le \|f\|_2\ ,\ \|f_2\|_1 \le \frac{\|f\|_2^2}{u}\ . \tag{6.60}$$

Proof We set $f_1 = f\mathbf{1}_{\{|f|\le u\}}$, so that the first part of (6.60) is obvious. We set $f_2 = f\mathbf{1}_{\{|f|>u\}} = f - f_1$, so that

$$u\|f_2\|_1 = \int u|f|\mathbf{1}_{\{|f|>u\}}\mathrm{d}\nu \le \int f^2\mathrm{d}\nu = \|f\|_2^2\ . \qquad \square$$

Matters are very much more difficult when one deals with a class of functions and where the goal is to simultaneously decompose all functions in the class. Our top-of-the-line result in this direction is surprisingly sophisticated, so we start here by a simpler, yet non-trivial result. This result has its own importance, as it will be used to study empirical processes. We denote by B_1 the unit ball of $L^1(\nu)$.

Theorem 6.7.2 *Consider a countable set $T \subset L^2(\nu)$ and a number $u > 0$. Assume that $S = \gamma_2(T, d_2) < \infty$. Then there is a decomposition $T \subset T_1 + T_2$ where*

$$\gamma_2(T_1, d_2) \le LS\ ;\ \gamma_1(T_1, d_\infty) \le LSu \tag{6.61}$$

$$\gamma_2(T_2, d_2) \le LS\ ;\ T_2 \subset \frac{LS}{u}B_1\ . \tag{6.62}$$

Here as usual $T_1 + T_2 = \{t_1 + t_2\ ;\ t_1 \in T_1\ ,\ t_2 \in T_2\}$. The sets T_1 and T_2 are not really larger than T with respect to γ_2. Moreover, for each of them, we have some extra information: we control $\gamma_1(T_1, d_\infty)$, and we control the L^1 norm of the elements of T_2. In some sense, Theorem 6.7.2 is an extension of Lemma 6.7.1, which deals with the case where T consists of a single function.

We will present two proofs of the theorem: The first proof is easy to discover, as it implements what is the most obvious approach. The second proof, while somewhat simpler, is far less intuitive. It is the second proof which will be useful in the long run.

Proof The idea is simply to write an element of T as the sum of the increments along a chain and to apply Lemma 6.7.1 to each of these increments. We will also take advantage of the fact that T is countable to write each element of T as the sum of the increments along a chain of finite length, but this is not an essential part of the argument.

As usual, $\Delta_2(A)$ denotes the diameter of A for the distance d_2. We consider an admissible sequence of partitions $(\mathcal{A}_n)_{n\ge0}$ with

$$\sup_{t\in T}\sum_{n\ge0} 2^{n/2}\Delta_2(A_n(t)) \le 2\gamma_2(T, d_2)\ . \tag{6.63}$$

For each $n \geq 0$ and each $A \in \mathcal{A}_n$, we are going to pick a point $t_{n,A} \in A$. It will be convenient to ensure that each point of T is of the type $t_{n,A}$ for a certain n and a certain $A \in \mathcal{A}_n$. To ensure this, we enumerate T as $(t_n)_{n \geq 0}$. For $A \in \mathcal{A}_n$, we choose for $t_{n,A} \in A$ any point we want unless $A = A_n(t_n)$, in which case we choose $t_{n,A} = t_n$. For $t \in T$ and $n \geq 0$, let us define $\pi_n(t) = t_{n,A}$ where $A = A_n(t)$. Thus, if $t = t_n$, then $A = A_n(t_n)$, and by construction, $\pi_n(t) = t_{n,A} = t_n$. For $n \geq 1$, let us set $f_{t,n} = \pi_n(t) - \pi_{n-1}(t)$. Thus

$$\|f_{t,n}\|_2 \leq \Delta_2(A_{n-1}(t)) . \tag{6.64}$$

Moreover, $f_{t,n}$ depends only on $A_n(t)$: if $A_n(s) = A_n(t)$, then $A_{n-1}(t) = A_{n-1}(s)$ and $f_{t,n} = f_{s,n}$. Thus as t varies in T, there are at most N_n different functions $f_{t,n}$. Using Lemma 6.7.1 with $2^{-n/2}u\|f_{t,n}\|_2$ instead of u, we can decompose $f_{t,n} = f_{t,n}^1 + f_{t,n}^2$ where

$$\|f_{t,n}^1\|_2 \leq \|f_{t,n}\|_2 \ , \ \ \|f_{t,n}^1\|_\infty \leq 2^{-n/2}u\|f_{t,n}\|_2 \tag{6.65}$$

$$\|f_{t,n}^2\|_2 \leq \|f_{t,n}\|_2 \ , \ \ \|f_{t,n}^2\|_1 \leq \frac{2^{n/2}}{u}\|f_{t,n}\|_2 \ . \tag{6.66}$$

To construct the sets T^1 and T^2, given $t \in T$, we set $g_{t,0}^1 = t_{0,T}$ and $g_{t,0}^2 = 0$, while if $n \geq 1$, we set

$$g_{t,n}^1 = t_{0,T} + \sum_{1 \leq k \leq n} f_{t,k}^1 \ , \ \ g_{t,n}^2 = \sum_{1 \leq k \leq n} f_{t,k}^2 \ .$$

We set

$$T_n^1 = \left\{g_{t,m}^1 \ ; \ m \leq n \ , \ t \in T\right\} \ ; \ T_n^2 = \left\{g_{t,m}^2 \ ; \ m \leq n \ , \ t \in T\right\} ,$$

so that the sequences $(T_n^1)_{n \geq 1}$ and $(T_n^2)_{n \geq 1}$ are increasing. We set

$$T_1 = \bigcup_{n \geq 0} T_n^1 \ ; \ T_2 = \bigcup_{n \geq 0} T_n^2 \ .$$

We prove now that $T \subset T_1 + T_2$. Indeed, if $t \in T$, then $t = t_n$ for some n and we have arranged that then $\pi_n(t) = t$. Since $\pi_0(t) = t_{0,T}$, we have

$$\begin{aligned} t - t_{0,T} = \pi_n(t) - \pi_0(t) &= \sum_{1 \leq k \leq n} \pi_k(t) - \pi_{k-1}(t) \\ &= \sum_{1 \leq k \leq n} f_{t,k} = \sum_{1 \leq k \leq n} f_{t,k}^1 + \sum_{1 \leq k \leq n} f_{t,k}^2 \ , \end{aligned}$$

so that $t = g_{t,n}^1 + g_{t,n}^2 \in T_1 + T_2$.

We now start the proof of (6.61). Since for $j = 1, 2$ the element $g^j_{t,n}$ depends only on $A_n(t)$, we have card $T^j_n \le N_0 + \cdots + N_n$, so that card $T^j_0 = 1$ and card $T^j_n \le N_{n+1}$. Consider $t^1 \in T_1$, so that $t^1 = g^1_{t,m}$ for some m and some $t \in T$. If $m \le n$, we have $t^1 = g^1_{t,m} \in T^1_n$ so that $d_2(t^1, T^1_n) = 0$. If $m > n$, we have $g^1_{t,n} \in T^1_n$, so that, using in succession the first part of (6.65) and (6.64) in the third inequality, we get

$$d_2(t^1, T^1_n) \le d_2(g^1_{t,m}, g^1_{t,n}) = \|g^1_{t,m} - g^1_{t,n}\|_2 \le \sum_{k>n} \|f^1_{t,k}\|_2 \le \sum_{k>n} \Delta_2(A_{k-1}(t)) . \tag{6.67}$$

Hence, using (6.63) in the last inequality,

$$\begin{aligned}\sum_{n\ge 0} 2^{n/2} d_2(t^1, T^1_n) &\le \sum_{n\ge 0, k>n} 2^{n/2} \Delta_2(A_{k-1}(t)) \\ &\le L \sum_{k\ge 1} 2^{k/2} \Delta_2(A_{k-1}(t)) \le LS .\end{aligned}$$

It then follows from Proposition 2.9.7 that $\gamma_2(T_1, d_2) \le LS$. The proof that $\gamma_2(T_2, d_2) \le LS$ is identical, using now the first part of (6.66) rather than the first part of (6.65).

To control $\gamma_1(T_1, d_\infty)$, we use the same approach. We replace (6.67) by

$$d_\infty(t^1, T^1_n) \le d_\infty(g^1_{t,m}, g^1_{t,n}) \le \sum_{k>n} \|f^1_{t,k}\|_\infty \le \sum_{k>n} 2^{-k/2} u \Delta_2(A_{k-1}(t)) .$$

Hence

$$\begin{aligned}\sum_{n\ge 0} 2^n d_\infty(t^1, T^1_n) &\le u \sum_{n\ge 0, k>n} 2^{n-k/2} \Delta_2(A_{k-1}(t)) \\ &\le Lu \sum_{k\ge 1} 2^{k/2} \Delta_2(A_{k-1}(t)) \le LuS ,\end{aligned}$$

and it follows again from (a suitable version of) Proposition 2.9.7 that $\gamma_1(T_1, d_\infty) \le LSu$. Finally, (6.66) and (6.65) yield

$$\|g^2_{t,n}\|_1 \le \sum_{k\ge 1} \|f^2_{t,k}\|_1 \le \sum_{k\ge 1} \frac{2^{k/2}}{u} \Delta_2(A_{k-1}(t)) \le \frac{LS}{u} ,$$

so that $T_2 \subset LSB_1/u$. This completes the proof. □

Later, in Theorem 9.2.4, we will prove a far-reaching generalization of Theorem 6.7.2 with sweeping consequences. Our proof of Theorem 9.2.4 will be based on a slightly different idea. To prepare for this proof, we give an alternate proof of Theorem 6.7.2 based on the same ideas.

Second proof of Theorem 6.7.2 We keep the notation of the first proof, and we construct the points $\pi_n(t)$ as we did there. For $t \in T$, we define $\delta_n(t) := \Delta(A_n(t))$, so that $\sup_{t\in T} \sum_{n\geq 0} 2^{n/2}\delta_n(t) \leq LS$. We denote by Ω the underlying measure space. Given $t \in T$ and $\omega \in \Omega$, we define

$$m(t,\omega) = \inf\left\{n \geq 0\ ;\ |\pi_{n+1}(t)(\omega) - \pi_n(t)(\omega)| \geq u2^{-n/2}\delta_n(t)\right\} \tag{6.68}$$

if the set on the right is not empty and $m(t,\omega) = \infty$ otherwise. For $n < n' \leq m(t,\omega)$, we have

$$|\pi_n(t)(\omega) - \pi_{n'}(t)(\omega)| \leq \sum_{n\leq p<n'} |\pi_{p+1}(t)(\omega) - \pi_p(t)(\omega)| \leq u \sum_{n\leq p<n'} 2^{-p/2}\delta_p(t)\ .$$

In particular when $m(t,\omega) = \infty$, the sequence $(\pi_n(t)(\omega))$ of real numbers is a Cauchy sequence and hence convergent. When $m(t,\omega) = \infty$, we define

$$t^1(\omega) = \lim_{n\to\infty} \pi_n(t)(\omega)\ .$$

When $m(t,\omega) < \infty$ we define

$$t^1(\omega) = \pi_{m(t,\omega)}(t)(\omega)\ .$$

We define

$$t^2 = t - t^1\ ;\ T_1 := \{t^1\ ;\ t \in T\}\ ;\ T_2 := \{t^2\ ;\ t \in T\}\ ,$$

and we proceed to prove (6.61) and (6.62). We define

$$t_n^1(\omega) = \pi_{m(t,\omega)\wedge n}(t)(\omega)\ .$$

To match with the previous notation for $n \geq 1$, let us define

$$\begin{aligned} f_{t,n}^1(\omega) := t_n^1(\omega) - t_{n-1}^1(\omega) &= \pi_{m(t,\omega)\wedge n}(t)(\omega) - \pi_{m(t,\omega)\wedge(n-1)}(t)(\omega) \\ &= (\pi_n(t)(\omega) - \pi_{n-1}(t)(\omega))\mathbf{1}_{\{m(t,\cdot)\geq n\}}(\omega)\ , \end{aligned}$$

so that $\|f_{t,n}^1\|_2 \leq \Delta(A_{n-1}(t)) = \delta_{n-1}(t)$ and $\|f_{t,n}^1\|_\infty \leq u2^{(-n+1)/2}\delta_{n-1}(t)$ because $|\pi_n(t)(\omega) - \pi_{n-1}(t)(\omega)| \leq u2^{(-n+1)/2}\delta_{n-1}(t)$ when $m(t,\omega) \geq n$. Also $t_n^1 = t_0^1 + \sum_{1\leq m\leq n} f_{t,n}^1$. The proof of (6.61) is then just as before. Let us now define

$$t_n^2 = t^2\mathbf{1}_{\{m(t,\cdot)=n\}} = (t - \pi_n(t))\mathbf{1}_{\{m(t,\cdot)=n\}}\ . \tag{6.69}$$

On the set $\{m(t,\omega) = \infty\}$, we have $t = t^1$ a.e. for the measure ν since $\|t - \pi_n(t)\|_2 \to 0$ as $n \to \infty$. Thus, on that set, we have $t^2 = 0$ a.e., and consequently a.e. we have

$$t^2 = t^2 \mathbf{1}_{\{m(t,.)<\infty\}} = t^2 \sum_{n\geq 0} t^2 \mathbf{1}_{\{m(t,.)=n\}} = \sum_{n\geq 0} t_n^2 .$$

Since $\|t_n^2\|_2 \leq \|t - \pi_n(t)\|_2 \leq \delta_n(t)$, the proof that $\gamma_2(T_2, d_2) \leq LS$ is as before. Furthermore, using (6.69) and the Cauchy–Schwarz inequality, we have

$$\|t_n^2\|_1 \leq \|t - \pi_n(t)\|_2 \sqrt{\nu(\{m(t,\cdot) = n\})} \leq \delta_n(t) \sqrt{\nu(\{m(t,\cdot) = n\})} .$$

Since $|\pi_{n+1}(t) - \pi_n(t)| \geq u 2^{-n/2} \delta(n)$ on the set $\{m(t,\cdot) = n\}$, Markov's inequality yields

$$\nu(\{m(t,\cdot) = n\}) \leq \frac{\|\pi_{n+1}(t) - \pi_n(t)\|_2^2}{(u 2^{-n/2} \delta_n(t))^2} \leq \frac{\delta_n(t)^2}{(u 2^{-n/2} \delta_n(t))^2} = \frac{2^n}{u^2} ,$$

and thus $\|t_n^2\|_1 \leq 2^{n/2} \delta_n(t)/u$ and hence $\|t^2\|_1 \leq LS/u$. □

6.8 Discrepancy Bounds for Empirical Processes

Throughout this section, we consider a probability space (Ω, μ) and (to avoid well-understood measurability problems) a countable bounded subset of $L^2(\mu)$, which, following the standard notation in empirical processes theory, we will denote by $\mathcal{F}$ rather than T. (Since $\mathcal{F}$ is countable, there is no need to really distinguish between actual functions on Ω and classes of functions in $L^2(\mu)$.) To lighten notation, we set

$$\mu(f) = \int f \mathrm{d}\mu .$$

Consider i.i.d. r.v.s $(X_i)_{i\geq 1}$ valued in Ω, distributed like μ and

$$S_N(\mathcal{F}) := \mathsf{E} \sup_{f\in\mathcal{F}} \Big| \sum_{i\leq N} (f(X_i) - \mu(f)) \Big| . \tag{6.70}$$

We have already seen in Chap. 4 the importance of evaluating such quantities. As in the case of Bernoulli processes, there should be two different reasons why the sum $S_N(\mathcal{F})$ should be small:

- On the one hand, there may be cancellation between the different terms.
- On the other hand, it might happen that the sum $\sum_{i\leq N} |f(X_i) - \mu(f))|$ is already small without cancellation.

More specifically, we have the inequality

$$S_N(\mathcal{F}) \le 2\mathsf{E} \sup_{f\in\mathcal{F}} \sum_{i\le N} |f(X_i)| \,. \tag{6.71}$$

To see this, we simply write

$$S_N(\mathcal{F}) \le \mathsf{E} \sup_{f\in\mathcal{F}} \sum_{i\le N} |f(X_i) - \mu(f)| \le \mathsf{E} \sup_{f\in\mathcal{F}} \sum_{i\le N} |f(X_i)| + N \sup_{f\in\mathcal{F}} |\mu(f)| \,,$$

and we observe that the first term in the previous line is $\ge$ the second term through Jensen's inequality.

We may also bound $S_N(\mathcal{F})$ using chaining as follows:

Proposition 6.8.1 *If $0 \in \mathcal{F}$, we have*

$$S_N(\mathcal{F}) \le L\big(\sqrt{N}\gamma_2(\mathcal{F}, d_2) + \gamma_1(\mathcal{F}, d_\infty)\big) \,, \tag{6.72}$$

where d_2 and d_∞ are the distances on $\mathcal{F}$ induced by the norms of L^2 and L^∞, respectively.

Proof This follows from Bernstein's inequality (4.44) and Theorem 4.5.13 just as in the case of Theorem 4.5.16. The requirement that $0 \in \mathcal{F}$ is made necessary by considering the case where $\mathcal{F}$ consists of one single function f (and because of the absolute values in (6.70)). □

The bound (6.71) does not involve cancellation, and is of a really different nature than (6.72), which involves cancellation in an essential way through Bernstein's inequality.

Having two completely different methods (6.72) and (6.71) to control $S_N(\mathcal{F})$, we can interpolate between them in the spirit of (6.9) as follows:

Proposition 6.8.2 *Consider classes $\mathcal{F}, \mathcal{F}_1$ and $\mathcal{F}_2$ of functions in $L^2(\mu)$, and assume that $\mathcal{F} \subset \mathcal{F}_1 + \mathcal{F}_2$. Assume that $0 \in \mathcal{F}_1$. Then*

$$S_N(\mathcal{F}) = \mathsf{E} \sup_{f\in\mathcal{F}} \Big|\sum_{i\le N} (f(X_i) - \mu(f))\Big| \le L\big(\sqrt{N}\gamma_2(\mathcal{F}_1, d_2) + \gamma_1(\mathcal{F}_1, d_\infty)\big) + 2\mathsf{E} \sup_{f\in\mathcal{F}_2} \sum_{i\le N} |f(X_i)| \,.$$

Proof Since $\mathcal{F} \subset \mathcal{F}_1 + \mathcal{F}_2$, it is clear that $S_N(\mathcal{F}) \le S_N(\mathcal{F}_1) + S_N(\mathcal{F}_2)$. We then use the bound (6.72) for the first term and the bound (6.71) for the second term. □

It seems worth repeating what we have just said, as this is going to be a major theme of this work. Given a sum of random functions, depending on a parameter, there are two fundamentally different methods to bound the supremum of this sum over the parameter:

- We may use chaining.
- Or we may forget about possible cancellations and bound the sum of the random functions by the sum of their absolute values.

It is a rather extraordinary fact that in a wide range of situations, there is no other way to bound the sum of random functions than interpolating between these two methods (just as we did in Proposition 6.8.2). This will be proved in Chap. 11.

Our first occurrence of this extraordinary fact is that there is no other way to control $S_N(\mathcal{F})$ than the method of Proposition 6.8.2.[6] We formalize this fundamental result as follows:

Theorem 6.8.3 (The Fundamental Theorem of Empirical Processes) *Consider a class $\mathcal{F}$ of functions in $L^2(\mu)$ with $\mu(f) = 0$ for $f \in \mathcal{F}$ and an integer N. Then we can find a decomposition $\mathcal{F} \subset \mathcal{F}_1 + \mathcal{F}_2$ with $0 \in \mathcal{F}_1$ such that the following properties hold:*

$$\gamma_2(\mathcal{F}_1, d_2) \le \frac{L}{\sqrt{N}} S_N(\mathcal{F}) ,$$

$$\gamma_1(\mathcal{F}_1, d_\infty) \le L S_N(\mathcal{F}) ,$$

$$\mathsf{E} \sup_{f \in \mathcal{F}_2} \sum_{i \le N} |f(X_i)| \le L S_N(\mathcal{F}) .$$

We are not ready yet for the proof of this result, which is delayed until Chap. 11.[7]

Exercise 6.8.4 We say that a countable class $\mathcal{F}$ of functions is a Glivenko-Cantelli class if

$$\lim_{N \to \infty} \mathsf{E} \sup_{f \in \mathcal{F}} \Big| \frac{1}{N} \sum_{i \le N} (f(X_i) - \mu(f)) \Big| = \lim_{N \to \infty} \frac{S_N(\mathcal{F})}{N} = 0 .$$

Assuming that $\mathcal{F}$ is uniformly bounded, prove that $\mathcal{F}$ is a Glivenko-Cantelli class if and only if for each $\epsilon > 0$, one can find a decomposition $\mathcal{F} \subset \mathcal{F}_1 + \mathcal{F}_2$ and an integer N_0 such that $\mathcal{F}_1$ is finite and

$$N \ge N_0 \Rightarrow \mathsf{E} \sup_{f \in \mathcal{F}_2} \frac{1}{N} \sum_{i \le N} |f(X_i)| \le \epsilon .$$

---

[6] In particular Bernstein's inequality suffices to perform the chaining.

[7] A good path to get a feeling for this theorem is to study Exercise 14.2.3 in due time.

Hint: Use Theorem 6 of [99] for the "only if" part. Warning: You need good technique to succeed.

Key Ideas to Remember

- Bernoulli random variables (i.e., independent random signs) are among the most important probability structures. Their linear combinations satisfy the fundamental sub-Gaussian inequality (6.2).
- A Bernoulli process is always smaller than the corresponding Gaussian process.
- A Bernoulli process can however be bounded in a trivial way without using cancellation.
- One may interpolate between the two previous methods to bound a Bernoulli process. That this interpolation provides the best possible method of bounding a Bernoulli process is the fundamental Latała-Bednorz theorem.
- Bernoulli processes satisfy concentration of measure properties which are even better than those of Gaussian processes.
- Bernoulli processes satisfy a suitable version of Sudakov minoration, which however requires a control in the supremum norm.
- Bernoulli processes satisfy a fundamental comparison principle: contracting the coefficients decreases the size of the process.
- Elements of a not too large set of functions on a measure space can be split in their "peaky part" and their "spread out parts", an idea which we will push very far.
- When one looks at discrepancy bounds for classes of functions in the spirit of the Latała-Bednorz theorem, one is lead to formulate amazing conjectures, which will turn out to be true as we will prove later.

6.9 Notes and Comments

A rather different proof of Proposition 6.4.8 is given in [113]. Probably the proof of [113] is more elegant and deeper than the proof we give here, but the latter has the extra advantage of showing the connection between Proposition 6.4.8 and the Marcus-Pisier theorem, Theorem 7.4.2.

Chapter 7
Random Fourier Series and Trigonometric Sums

The topic of random Fourier series illustrates well the impact of abstract methods, and it might be useful to provide an (extremely brief) history of the topic.

In a series of papers in 1930 and 1932, R. Paley and A. Zygmund [78–80] raised (among other similar problems) the question of the uniform convergence of the random series

$$\sum_{k\geq 1} a_k \varepsilon_k \exp(ikx) \tag{7.1}$$

over $x \in [0, 2\pi]$, where a_k are real numbers and ε_k are independent Bernoulli r.v.s (and where $i^2 = -1$). Considering the numbers s_p defined by $s_p^2 = \sum_{2^p \leq k < 2^{p+1}} a_k^2$, they prove in particular the necessity of the condition $\sum_p s_p < \infty$. Later, R. Salem and A. Zygmund [94] proved that if the sequence $(s_p)_{p\geq 0}$ is non-increasing, then, conversely, the condition $\sum_p s_p < \infty$ suffices for the uniform convergence of the random Fourier series. The combination of these two results is remarkably sharp, but certainly does not characterize the series (7.1) which converge uniformly.

The discovery by X. Fernique that Dudley's bound could be reversed for stationary Gaussian processes [32] was a major progress, with considerable influence. The Dudley-Fernique characterization of boundedness of stationary Gaussian processes opened the way for M. Marcus and G. Pisier to find necessary and sufficient conditions for uniform convergence of a large class of random Fourier series. The conditions of Marcus and Pisier are of the type $\gamma_2([0, 2\pi], d) < \infty$ for a certain distance d, and it is a non-trivial task (which is thoroughly performed in [61] and will not be repeated here) to show that they improve on the "classical" results of Paley, Salem, and Zygmund. The results of [61] cover not only the case of series of the type (7.1) but more general cases such as the series

$$\sum_{k\geq 1} a_k \xi_k \exp(ikx) \tag{7.2}$$

M. Talagrand, *Upper and Lower Bounds for Stochastic Processes*, Ergebnisse der Mathematik und ihrer Grenzgebiete. 3. Folge / A Series of Modern Surveys in Mathematics 60, https://doi.org/10.1007/978-3-030-82595-9_7

where the independent symmetric r.v.s ξ_k satisfy $\sup_k (\mathsf{E}\xi_k^2)^{1/2}/(\mathsf{E}|\xi_k|) < \infty$ (and many other situations).

The work of Marcus and Pisier on random Fourier series was extended by Marcus [59] to more general situations (that involve the infinitely divisible processes that we will study in Chap. 12). Marcus fails however to obtain necessary and sufficient conditions. Obtaining these requires the new idea of "families of distances" which we develop in Sect. 7.5. In the present chapter, we provide (in a far more general setting) what is in a sense the final result, necessary and sufficient conditions for the almost sure convergence of the series (7.2) assuming only that the r.v.s ξ_k are independent symmetric.

In retrospect it might be hard to understand why the topic of random Fourier series was so popular at one point. What is certain is that the interest had already waned when the author performed his work, and it is doubtful that this work has yet found even a single reader.

Why, then, should you bother to even read a single line of the present chapter? First, the title of the chapter is somewhat misleading. The main focus of it is not to decide whether certain series converge or not, but to provide upper and lower bounds on the supremum norm of certain random trigonometric sums.[1] When one has obtained upper and lower bounds which are sufficiently close to each other, it is hardly more than an exercise to obtain necessary and sufficient conditions for the convergence of random Fourier series. This exercise is carried out in Sect. 7.10. Why, then, should you be interested in random trigonometric sums? The reason is very simple. In general, the study of non-Gaussian processes (and in particular the search for lower bounds) is an order of magnitude harder than the Gaussian case, because all kinds of difficulties occur simultaneously. One of them (which already occurs in the Gaussian case) is the need to use "generic chaining ideas" because covering numbers do not suffice. A fundamental feature of random trigonometric sums is that this specific difficulty does not occur: in contrast with the case of general processes, covering numbers do suffice. Random Fourier series provide a simple setting where we can first learn to face the difficulties inherent to non-Gaussian processes, before we face these difficulties in the much harder context of the generic chaining.

The reason why covering numbers suffice to study random Fourier series is that the distances involved on the underlying group are translation-invariant. As we will learn in the next section, for translation-invariant distances, not only covering numbers suffice, but these covering numbers can basically be computed by evaluating the Haar measure of certain small balls. This is a tremendous simplification (which explains why very precise results can be proved). And how could we dream of making progress on general processes if we do not thoroughly

[1] These objects will be defined precisely in a few pages.

understand this much simpler case first? This is why the author concentrated his efforts for many years on random Fourier series. The strategy paid off: this setting turned out to be ideal to invent some of the fundamental tools on which the subsequent chapters are built and first of all the concept of a family of distances. Furthermore, the ideas used in the present chapter to obtain lower bounds on random Fourier series will be given a sweeping generalization in Chap. 11, and this will shed considerable light on the structure of several fundamental processes. So, the reader's main motivation need not be the results on random Fourier series per se, but the ideas she will learn from studying them. In fact, successfully reading the rest of Part II probably requires reading the present chapter up to Sect. 7.7 inclusive.

We start the chapter by investigating the central structure, translation-invariant distances. Our first basic results on random Fourier series are proved in Sects. 7.2–7.4. The main results are stated in Sect. 7.5, where the concept of a family of distances is also introduced.

7.1 Translation-Invariant Distances

The superiority of the generic chaining bound (2.59) over Dudley's entropy bound (2.38) is its ability to take advantage of the fact that the metric space (T, d) (where the distance d controls the increments of the process as in (2.4)) need not be "homogeneous" in the sense that at a given scale different regions of the space may look very different from each other. When, however, this is not the case, the situation should be simpler and Dudley's bound should be optimal. A typical such case is when T is a compact metrizable Abelian group[2] and d is a translation-invariant distance

$$\forall s, t, v \in T \ , \ d(s + v, t + v) = d(s, t) \ .$$

We denote by μ the normalized Haar measure of T, that is, $\mu(T) = 1$ and μ is translation-invariant. Thus, *all balls for d with a given radius have the same Haar measure.*[3]

To study the size of the space (T, d), it is very convenient in this setting to use as a "main parameter" the function $\epsilon \mapsto \mu(B_d(0, \epsilon))$. We recall that we defined $N_0 = 1$ and $N_n = 2^{2^n}$ for $n \geq 1$.

[2] At the expense of minor complications, the same methods cover the case where T is subset with non-empty interior of a locally compact group.

[3] If you find this setting too abstract, you may assume that T is $\mathbb{R}/(2\pi\mathbb{Z})$. The proofs are identical, but you will soon realize that cluttering your mind with irrelevant information makes things harder, illustrating the wisdom of the advice of Gustave Choquet given on page 1.

Theorem 7.1.1 *Consider a continuous*[4] *translation-invariant distance d on T. For $n \geq 0$ define*

$$\epsilon_n = \inf\left\{\epsilon > 0 \; ; \; \mu(B_d(0,\epsilon)) \geq 2^{-2^n} = N_n^{-1}\right\} . \tag{7.3}$$

Then

$$\frac{1}{L}\sum_{n\geq 0} \epsilon_n 2^{n/2} \leq \gamma_2(T,d) \leq L \sum_{n \geq 0} \epsilon_n 2^{n/2} . \tag{7.4}$$

Our first lemma shows that the numbers ϵ_n are basically the entropy numbers of Sect. 2.5, so that (7.4) simply states (as expected in this homogeneous case) that $\gamma_2(T,d)$ is equivalent to Dudley's integral.

Lemma 7.1.2 *The entropy numbers $e_n(T) = e_n(T,d)$ satisfy*

$$\epsilon_n \leq e_n(T) \leq 2\epsilon_n . \tag{7.5}$$

This near trivial lemma has staggering consequences: the only characteristic of the balls $B_d(0,\epsilon)$ which influences the entropy numbers is their measure, *entirely irrespective of their shape.* As we will explain soon, it is really child's play to control this measure.

Lemma 7.1.2 is in turn based on an even more trivial "volume argument" which we state separately for further use.

Lemma 7.1.3 *Consider a subset B of T. Then there exists a subset U of T with* card $U \leq 1/\mu(B)$ *such that whenever $t \in T$ we can find $s \in U$ with $t \in s + B - B$, where $B - B = \{t_1 - t_2; t_1, t_2 \in B\}$.*

Proof Any set U such that the sets $s + B$ are disjoint for $s \in U$ satisfies card $U \cdot \mu(B) \leq 1$ because $\mu(s+B) = \mu(B)$. Thus exists such a set U whose cardinality is as large as possible. Then for each $t \in T$, there exists $s \in U$ for which $(t+B) \cap (s+B) \neq \emptyset$ (for otherwise we could add the point t to U). Then $t \in s + B - B$. □

Corollary 7.1.4 *If $\mu(B) > 1/2$ then $T = B - B$.*

Proof In that case card $U = 1$ so that U consists of a single point u, and $t - u \in B - B$ for each t in T, so that $B - B = T$. □

Exercise 7.1.5 Find a direct argument. Convince yourself that the conclusion utterly fails when the distance is not required to be translation-invariant.

Lemma 7.1.6 *For any $\epsilon > 0$, we have $B(0,\epsilon) - B(0,\epsilon) \subset B(0,2\epsilon)$.*

---

[4] That is, the function $(s,t) \mapsto d(s,t)$ is continuous on T^2. This is the case of interest, but much weaker regularity properties suffice. All we really need is that the balls are measurable.

Proof If $t, t' \in B(0, \epsilon)$, then $d(t - t', 0) \le d(t - t', -t') + d(-t', 0) = d(t, 0) + d(0, t') \le 2\epsilon$ using twice that the distance is translation-invariant. □

The previous lemmas play a fundamental role throughout the chapter, where translation-invariance is constant feature.

Proof of Lemma 7.1.2 Since μ is translation-invariant, all the balls of T with the same radius have the same measure. Consequently if one can cover T by N_n balls with radius ϵ, then $\epsilon_n \le \epsilon$, and this proves the left-hand side inequality.

To prove the right-hand side, we use Lemma 7.1.3 for $B = B(0, \epsilon_n)$. Thus we can cover T by at most $\operatorname{card} U \le \mu(B)^{-1} \le N_n$ translates $s + B - B$ of the set $B - B$. Furthermore $s + B - B \subset s + B(0, 2\epsilon_n) = B(s, 2\epsilon_n)$ by Lemma 7.1.6. □

Proof of Theorem 7.1.1 The right-hand side inequality follows from (7.5) and (2.56). To prove the left-hand side inequality, we consider an admissible sequence $(\mathcal{A}_n)$ of partitions of T with $\sup_{t\in T} \sum_{n\ge 0} 2^{n/2} \Delta(A_n(t), d) \le 2\gamma_2(T, d)$. We construct by induction a decreasing sequence $C_n \in \mathcal{A}_n$ with $\mu(C_n) \ge N_{n+1}^{-1}$ as follows. First we choose $C_0 = T$. Having constructed $C_n \in \mathcal{A}_n$, we note that

$$N_{n+1}^{-1} \le \mu(C_n) = \sum \{\mu(A), A \in \mathcal{A}_{n+1}, A \subset C_n\} ,$$

and since the sum has at most N_{n+1} terms, one of these is $\ge N_{n+1}^{-2} = N_{n+2}^{-1}$. Thus there exists $A \in \mathcal{A}_{n+1}$ with $A \subset C_n$ and $\mu(A) \ge N_{n+2}^{-1}$. We choose for C_{n+1} such a set A, completing the induction.

Since d is translation-invariant, it follows from (7.3) that C_n cannot be contained in a ball with radius $< \epsilon_{n+1}$ and thus that $\Delta(C_n, d) \ge \epsilon_{n+1}$.

Consider now $t \in C_k$. For $0 \le n \le k$, we have $t \in C_n \in \mathcal{A}_n$ so that $A_n(t) = C_n$ and thus

$$\sum_{0\le n\le k} \epsilon_{n+1} 2^{n/2} \le \sum_{0\le n\le k} 2^{n/2} \Delta(C_n, d) = \sum_{0\le n\le k} 2^{n/2} \Delta(A_n(t), d) \le 2\gamma_2(T, d) .$$

Since $\epsilon_0 \le \Delta(A_0, d)$ this completes the proof of the left-hand side inequality of (7.4).[5] □

Recalling the numbers ϵ_n of (7.3), it is very useful for the sequel to form the following mental picture:

For our purposes, it is the number $\sum_{n\ge 0} \epsilon_n 2^{n/2}$

which determines the size of the space (T, d) .

[5] There is no reason for which the sets of $\mathcal{A}_n$ should be measurable for μ, but our argument works anyway replacing "measure" by "outer measure".

All we will have to do to understand very general random Fourier series is to discover the proper numerical series whose convergence is equivalent to the almost sure convergence of the random Fourier series!

Exercise 7.1.7 With the notation of Theorem 7.1.1 prove that for a constant K depending only on α, for $\alpha \geq 1$ we have

$$\frac{1}{K}\sum_{n\geq 0}\epsilon_n 2^{n/\alpha} \leq \gamma_\alpha(T,d) \leq K\sum_{n\geq 0}\epsilon_n 2^{n/\alpha} \,. \tag{7.6}$$

The following will be used many times:

Exercise 7.1.8 Assume that for $n \geq 0$, we have a set $D_n \subset T$ with $\mu(D_n) \geq N_n^{-1}$ and

$$s \in D_n \Rightarrow d(s,0) \leq \epsilon_n \,.$$

Then $\gamma_2(T,d) \leq L\sum_{n\geq 0} 2^{n/2}\epsilon_n$.

7.2 Basics

7.2.1 Simplification Through Abstraction

How should one approach the study of the random series (7.2)? To study the uniform convergence of such a series for $x \in [0, 2\pi]$, we will have to control quantities such as

$$\sup_{0\leq x\leq 2\pi}\Big|\sum_{k\leq n}\xi_k \exp(ikx)\Big| \,.$$

Let us observe that $t := \exp(ix)$ is a complex number of modulus 1 and that $\exp(ikx) = t^k$ so the above quantity is

$$\sup_{t\in\mathbb{U}}\Big|\sum_{k\leq n}\xi_k t^k\Big| \,,$$

where $\mathbb{U}$ is the set of complex numbers of modulus 1. Provided with the multiplication, $\mathbb{U}$ is a compact metrizable group. The functions $\chi_k(t) = t^k$ have a very special property with respect to the group operation: $\chi_k(st) = \chi_k(s)\chi_k(t)$.[6]

These remarks suggest to think of the series (7.2) as a series $\sum_{k\geq 1}\xi_k\chi_k(t)$ of random functions on $\mathbb{U}$. This abstract point of view is extremely fruitful, as

[6] If you like big words, they are group homomorphisms from $\mathbb{U}$ to $\mathbb{U}$.

otherwise it would be impossible to resist the temptation to use the special structure of the set $[0, 2\pi]$ and its natural distance.[7] It took indeed a very long time to understand that this natural distance is *not* what is relevant here.

7.2.2 Setting

We consider a compact metrizable Abelian group T.[8] Since T is Abelian, we follow the tradition to denote the group operation additively.[9] A *character* χ on T is a *continuous* map from T to $\mathbb{C}$ such that $|\chi(t)| = 1$ for each t and $\chi(s+t) = \chi(s)\chi(t)$ for any $s, t \in T$.[10] In particular $\chi(0) = 1$. Under pointwise multiplication, the set of characters on T form a group G called the *dual group* of T. The unit element of this group is the character $\mathbf{1}$ which takes the value 1 everywhere. If this abstract setting bothers you, you will lose nothing by assuming everywhere $T = \mathbb{U}$ and $G = \mathbb{Z}$,[11] except that the extra information will needlessly clutter your mind.

For our purpose, the fundamental property of characters is that for $s, t, u \in T$, we have $\chi(s+u) - \chi(t+u) = \chi_i(u)(\chi(s) - \chi(t))$ so that

$$|\chi(s+u) - \chi(t+u)| = |\chi(s) - \chi(t)| \,. \tag{7.7}$$

Taking $u = -t$ in (7.7) we get

$$|\chi(s) - \chi(t)| = |\chi(s-t) - 1| \,. \tag{7.8}$$

A random Fourier series is a series

$$\sum_{i \geq 1} \xi_i \chi_i$$

where ξ_i is a complex-valued r.v. and χ_i is a (nonrandom) character. We assume that $(\xi_i)_{i \geq 1}$ are symmetric independent r.v.s. It is not required that $\chi_i \neq \chi_j$ for $i \neq j$, although one may assume this condition without loss of generality. We will study the convergence of such series in Sect. 7.10, and for now we concentrate on the central part of this study, i.e., the study of *finite* sums $\sum_i \xi_i \chi_i$, which we call *random trigonometric sums*. Thus finite sums are denoted by $\sum_i$, as a short hand

---

[7] Let us remember that going to an abstract setting was also a very important step in the understanding of the structure of Gaussian processes.

[8] It requires only minor complications (but no new idea) to develop the theory in locally compact groups.

[9] In contrast with the case of the group $\mathbb{U}$ where multiplicative notation is more natural.

[10] So that χ is simply a group homomorphism form T to $\mathbb{U}$.

[11] To $k \in Z$ corresponds the character $\chi_k(t) = t^k$.

for $\sum_{i\in I}$ where I is a set of indices, whereas infinite series are denoted by $\sum_{i\geq 1}$. We denote by $\|\cdot\|$ the supremum norm of such a sum, so that

$$\Big\|\sum_i \xi_i \chi_i\Big\| = \sup_{t\in T}\Big|\sum_i \xi_i \chi_i(t)\Big| \,.$$

This notation is used throughout this chapter and must be learned now. Our ultimate goal is to find upper and lower bounds for the quantity $\|\sum_i \xi_i \chi_i\|$ that are of the same order in full generality, but we first state some simple facts.

7.2.3 *Upper Bounds in the Bernoulli Case*

Particularly important random trigonometric sums are sums of the type $\sum_i a_i \varepsilon_i \chi_i$, where a_i are numbers, χ_i are characters, and ε_i are independent Bernoulli r.v.s.[12] For such a sum let us consider the distance d on T defined by

$$d(s,t)^2 = \sum_i |a_i|^2 |\chi_i(s) - \chi_i(t)|^2 \,, \tag{7.9}$$

which is translation-invariant according to (7.7).

Exercise 7.2.1 Convince yourself that in the case where $T = \mathbb{U}$ is the group of complex numbers of modulus 1, there need not by a simple relation between the distance of (7.9) and the natural distance of $\mathbb{U}$ induced by the distance on $\mathbb{C}$. Hint: If $\chi_i(t) = t^i$ consider the case where only one of the coefficients a_i is not zero.

Proposition 7.2.2 *Consider the numbers ϵ_n defined in (7.3) with respect to the distance $d(s,t)$ of (7.9). Then*[13]

$$\Big(\mathsf{E}\Big\|\sum_i a_i \varepsilon_i \chi_i\Big\|^2\Big)^{1/2} \leq L\sum_{n\geq 0} 2^{n/2}\epsilon_n + \Big(\sum_i |a_i|^2\Big)^{1/2} \,. \tag{7.10}$$

Proof Consider real-valued functions η_i on T and the process

$$X_t = \sum_i \eta_i(t)\varepsilon_i \,. \tag{7.11}$$

[12] So that, in the previous notation, $\xi_i = a_i\varepsilon_i$.

[13] There is nothing magical in the fact that we use the L^2 norm rather than the L^1 norm in the left-hand side of (7.10). We will need this once later. It is known from general principles that these two norms are equivalent for random trigonometric sums of type $\sum_i a_i \varepsilon_i \chi_i$.

Using the subgaussian inequality (6.2), this process satisfies the increment condition (2.4) with respect to the distance d^* given by

$$d^*(s,t)^2 := \sum_i |\eta_i(s) - \eta_i(t)|^2 ,$$

and therefore from (2.66),

$$\Big(\mathsf{E} \sup_{s,t\in T} |X_s - X_t|^2\Big)^{1/2} \le L\gamma_2(T,d^*) . \tag{7.12}$$

Furthermore (7.12) holds also for complex-valued functions η_i. This is seen simply by considering separately the real and imaginary parts. When $\eta_i = a_i\chi_i$ we have $d^* = d$, and the result follows from the right-hand side of (7.4), using also that $(\mathsf{E}\sup_{t\in T}|X_t|^2)^{1/2} \le (\mathsf{E}\sup_{s,t\in T}|X_s - X_t|^2)^{1/2} + (\mathsf{E}|X_0|^2)^{1/2}$ and $\mathsf{E}|X_0|^2 = \sum_i |a_i|^2$.[14] □

The previous result may sound simple, but it is essential to fully understand what was the central step in the argument, because it is the very same phenomenon which is at the root of the upper bounds we will prove later. This central step is Lemma 7.1.3: when the Haar measure of a set B is not too small, one may cover T by not too many translates of $B - B$. This uses translation invariance in a crucial way.

7.2.4 Lower Bounds in the Gaussian Case

We turn to a lower bound in the case where the r.v.s ξ_i are i.i.d. Gaussian. It is a simple consequence of the majorizing measure theorem, Theorem 2.10.1.

Lemma 7.2.3 *Consider a finite number of independent standard normal r.v.s g_i and complex numbers a_i. Then*

$$\frac{1}{L}\gamma_2(T,d) \le \mathsf{E}\Big\|\sum_i a_i g_i \chi_i\Big\| , \tag{7.13}$$

[14] Exactly the same result holds when we replace the independent Bernoulli r.v.s ε_i *by independent Gaussian r.v.s.* Moreover, as we will see in (7.40), it is a general fact that "Gaussian sums are larger than Bernoulli sums", so that considering Gaussian r.v.s yields a *stronger* result than considering Bernoulli r.v.s. We are here using Bernoulli r.v.s as we will apply the exact form (7.10) .

where d is the distance on T given by

$$d(s,t)^2 = \sum_i |a_i|^2 |\chi_i(s) - \chi_i(t)|^2 . \tag{7.14}$$

We cannot immediately apply the majorizing measure theorem here, because it deals with real-valued processes, while here we deal with complex-valued ones. To fix this, we denote by $\Re z$ and $\Im z$ the real part and the imaginary part of a complex number z.

Lemma 7.2.4 *Consider a complex-valued process $(X_t)_{t\in T}$, and assume that both $(\Re X_t)_{t\in T}$ and $(\Im X_t)_{t\in T}$ are Gaussian processes. Consider the distance $d(s,t) = (\mathsf{E}|X_s - X_t|^2)^{1/2}$ on T. Then*

$$\frac{1}{L}\gamma_2(T,d) \le \mathsf{E} \sup_{s,t\in T} |X_s - X_t| \le L\gamma_2(T,d) . \tag{7.15}$$

Proof Consider the distances d_1 and d_2 on T given respectively by

$$d_1(s,t)^2 = \mathsf{E}\big(\Re(X_s - X_t)\big)^2$$

and

$$d_2(s,t)^2 = \mathsf{E}\big(\Im(X_s - X_t)\big)^2 .$$

Combining the left-hand side of (2.114) with Lemma 2.2.1 implies

$$\gamma_2(T,d_1) \le L\mathsf{E} \sup_{s,t\in T} |\Re X_s - \Re X_t| \le L\mathsf{E} \sup_{s,t\in T} |X_s - X_t|$$

and similarly $\gamma_2(T,d_2) \le L\mathsf{E}\sup_{s,t\in T} |X_s - X_t|$. Since $d \le d_1 + d_2$, (4.55) implies that $\gamma_2(T,d) \le L\mathsf{E}\sup_{s,t\in T} |X_s - X_t|$. To prove the right-hand side inequality of (7.15), we simply use (2.59) separately for the real and the imaginary part (keeping Lemma 2.2.1 in mind). □

Proof of Lemma 7.2.3 It follows from (7.15), since $\mathsf{E}\sup_{s,t\in T} |X_s - X_t| \le 2\mathsf{E}\sup_{t\in T} |X_t|$. □

7.3 Random Distances

7.3.1 Basic Principles

Without giving details yet, let us sketch some of the main features of our approach to a trigonometric sum $\sum_i \xi_i \chi_i$. It has the same distribution as the sum $\sum_i \varepsilon_i \xi_i \chi_i$

where (ε_i) are independent Bernoulli r.v.s which are also independent of the ξ_i. Such a random series will be studied *conditionally on the values of the r.v.s* ξ_i. This will bring in the distance corresponding to (6.2), namely, the distance d_ω (where the subscript ω symbolizes the randomness of the ξ_i) given by

$$d_\omega(s,t)^2 = \sum_i |\xi_i|^2 |\chi_i(s) - \chi_i(t)|^2 \,. \tag{7.16}$$

We will then try to relate the typical properties of the metric space (T, d_ω) with the properties of the metric space (T, d) where

$$d(s,t)^2 = \mathsf{E} d_\omega(s,t)^2 = \sum_i \mathsf{E}|\xi_i|^2 |\chi_i(s) - \chi_i(t)|^2 \,. \tag{7.17}$$

We now try to formulate in a rather imprecise way two key ideas of our approach.[15] You should not expect at this stage to fully understand them. Real understanding will come only after a detailed analysis of the forthcoming proofs. Yet, keeping these (imprecise) ideas in your mind may help to grasp the overall directions of these proofs. We have already met the first principle on page 168, but it bears repetition.

Principle A If, given $s, t \in T$, it is very rare that the distance $d_\omega(s,t)$ is very much smaller than $d(s,t)$, some measure of size of (T, d) is controlled from above by the typical value of $\gamma_2(T, d_\omega)$.

In other words, the balls $B_d(0, \epsilon)$ cannot be too small. The reason for this is simple. If this were the case, since $d_\omega(s,t)$ is not much smaller than $d(s,t)$, the balls $B_{d_\omega}(0, \epsilon)$ would also typically be very small, and then $\gamma_2(T, d_\omega)$ would typically be very large. This principle is adapted to finding lower bounds.

We will be able to use a suitable version of Principle A in cases where there is no translation invariance, and it is at the root of the results of Chap. 11.

An equally simple principle works the other way around and is adapted to find upper bounds on trigonometric sums.

Principle B If, given $s, t \in T$, the distance $d_\omega(s,t)$ is typically not too much larger than $d(s,t)$, then the typical value of $\gamma_2(T, d_\omega)$ is controlled from above by $\gamma_2(T, d)$.

The reason is that $\gamma_2(T, d)$ controls from below the size of the balls $B_d(0, \epsilon)$, in the sense that at a given value of $\gamma_2(T, d)$ the measure of these balls cannot be very small. This in turn implies that the balls B_{d_ω} cannot be too small because d_ω is typically not much larger than d. But this controls from above the size of $\gamma_2(T, d_\omega)$.[16]

---

[15] The setting for these ideas is somewhat general that the specific situation considered above.

[16] This last step is unfortunately specific to the case of translation-invariant distances.

The following theorem is an implementation of Principle B and is at the root of a main result of this chapter, the Marcus-Pisier theorem, Theorem 7.4.2:[17]

Theorem 7.3.1 *Consider a translation-invariant distance d_ω on T that depends on a random parameter ω. Assuming enough measurability and integrability, consider the distance $\bar{d}$ given by*

$$\bar{d}(s,t) = \mathsf{E}d_\omega(s,t) . \tag{7.18}$$

Then

$$\mathsf{E}\gamma_2(T,d_\omega) \le L\gamma_2(T,\bar{d}) + L\mathsf{E}\Delta(T,d_\omega) . \tag{7.19}$$

Proof It is obvious that $\bar{d}$ is a distance. Consider the corresponding numbers $\bar{\epsilon}_n$ as in (7.3). For each $n \ge 1$ let us set $B_n := B_{\bar{d}}(0,\bar{\epsilon}_n)$, so that $\mu(B_n) \ge N_n^{-1}$ by definition of $\bar{\epsilon}_n$. Let us define

$$b_n(\omega) = \frac{1}{\mu(B_n)}\int_{B_n} d_\omega(0,t)\mathrm{d}\mu(t) . \tag{7.20}$$

Markov's inequality used for the measure μ at the given ω implies $\mu(\{t \in B_n ; d_\omega(0,t) \ge 2b_n(\omega)\}) \le \mu(B_n)/2$. Consequently

$$\mu(\{t \in B_n \; ; \; d_\omega(0,t) \le 2b_n(\omega)\}) \ge \frac{1}{2}\mu(B_n) \ge \frac{1}{2}N_n^{-1} \ge N_{n+1}^{-1} ,$$

so that $\epsilon_{n+1}(\omega) \le 2b_n(\omega)$, where of course $\epsilon_{n+1}(\omega)$ is defined as in (7.3) for the distance d_ω. Also, $\epsilon_0(\omega) \le \Delta(T,d_\omega)$, so that

$$\begin{aligned}\sum_{n\ge0}\epsilon_n(\omega)2^{n/2} &\le L\Delta(T,d_\omega) + L\sum_{n\ge1}\epsilon_n(\omega)2^{n/2} \\ &= L\Delta(T,d_\omega) + L\sum_{n\ge0}\epsilon_{n+1}(\omega)2^{(n+1)/2} \\ &\le L\Delta(T,d_\omega) + L\sum_{n\ge0}b_n(\omega)2^{n/2} .\end{aligned}$$

Thus (7.4) implies

$$\gamma_2(T,d_\omega) \le L\Delta(T,d_\omega) + L\sum_{n\ge0}b_n(\omega)2^{n/2} .$$

[17] We will not need Principle A for this result, as in the special situation considered there one may find specific arguments for lower bounds.

Taking expectations yields

$$\mathsf{E}\gamma_2(T,d_\omega) \le L\mathsf{E}\Delta(T,d_\omega) + L\sum_{n\ge 0}\mathsf{E}b_n(\omega)2^{n/2}\,. \tag{7.21}$$

For $t \in B_n$ we have $\mathsf{E}d_\omega(0,t) = \bar{d}(0,t) \le \bar{\epsilon}_n$ so that taking expectation in (7.20), we obtain $\mathsf{E}b_n(\omega) \le \bar{\epsilon}_n$. Thus (7.19) follows from (7.21) and (7.4). □

Exercise 7.3.2

(a) Use Lemma 7.1.3 to prove that $\mu(A_\omega) \le 1/2$, where $A_\omega = \{t \in T;\, d_\omega(0,t) \le \Delta(T,d_\omega)/4\}$.
(b) Prove that the last term is not necessary in (7.19). Hint: This is harder.

Exercise 7.3.3 Show that if T is an arbitrary metric space and d_ω an arbitrary random metric, then (7.19) need not hold. Caution: This is not trivial.

7.3.2 A General Upper Bound

We turn to upper bounds, which are a rather simple consequence of the work of the previous subsection. Not only these bounds are interesting in their own right, they are a basic ingredient of the Marcus-Pisier theorem, the central result of next section.

Theorem 7.3.4 *Assume that the r.v.s ξ_i are symmetric and independent and have a second moment, and consider on T the distance d given by*

$$d(s,t)^2 = \sum_i \mathsf{E}|\xi_i|^2|\chi_i(s) - \chi_i(t)|^2\,. \tag{7.22}$$

Then

$$\mathsf{E}\Big\|\sum_i \xi_i\chi_i\Big\| \le L\gamma_2(T,d) + L\Big(\sum_i \mathsf{E}|\xi_i|^2\Big)^{1/2}\,. \tag{7.23}$$

If $X_t = \sum_i \xi_i\chi_i(t)$, (7.22) implies $\mathsf{E}|X_s - X_t|^2 \le d(s,t)^2$, but it does not seem possible to say much more (such as controlling higher moments of the r.v.s $|X_s - X_t|$) unless one assumes more on the r.v.s ξ_i, e.g., that they are Gaussian. Also, as we will learn in Sect. 16.8, the condition $\mathsf{E}|X_s - X_t|^2 \le d(s,t)^2$ is way too weak by itself to ensure the regularity of the process. Therefore it is at first surprising to obtain a conclusion as strong as (7.23). Theorem 7.3.4 is another deceptively simple-looking result on which the reader should meditate.

Proof of Theorem 7.3.4 We write

$$\mathsf{E}\sup_t \Big|\sum_i \xi_i \chi_i(t)\Big| \;\le\; \mathsf{E}\sup_t \Big|\sum_i (\xi_i \chi_i(t) \;-\; \xi_i \chi_i(0))\Big| \;+\; \mathsf{E}\Big|\sum_i \xi_i \chi_i(0)\Big| \;\;. \tag{7.24}$$

Since for each character χ we have $\chi(0) = 1$, using the Cauchy-Schwarz inequality we have $\mathsf{E}|\sum_i \xi_i \chi_i(0)| \le (\mathsf{E}|\sum_i \xi_i|^2)^{1/2} = (\sum_i \mathsf{E}|\xi_i|^2)^{1/2}$ so that it suffices to prove that

$$\mathsf{E}\sup_{t,s} \Big|\sum_i (\xi_i \chi_i(t) - \xi_i \chi_i(s))\Big| \le L\gamma_2(T,d) + L\Big(\sum_i \mathsf{E}|\xi_i|^2\Big)^{1/2} . \tag{7.25}$$

Since the r.v.s ξ_i are symmetric, the sum $\sum_i \xi_i \chi_i$ has the same distribution as the sum $\sum_i \varepsilon_i \xi_i \chi_i$, where the Bernoulli r.v.s ε_i are independent and independent of the r.v.s ξ_i, and in particular[18]

$$\mathsf{E}\sup_{t,s} \Big|\sum_i (\xi_i \chi_i(t) - \xi_i \chi_i(s))\Big| = \mathsf{E}\sup_{t,s} \Big|\sum_i (\varepsilon_i \xi_i \chi_i(t) - \varepsilon_i \xi_i \chi_i(s))\Big| . \tag{7.26}$$

For clarity let us assume that the underlying probability space is a product $\Omega \times \Omega'$, with a product probability $\mathsf{P} = \mathsf{P}_\xi \otimes \mathsf{P}_\varepsilon$, and that if (ω, ω') is the generic point of this product, then $\xi_i = \xi_{i,\omega}$ depends on ω only and $\varepsilon_i = \varepsilon_{i,\omega'}$ depends on ω' only. For each ω define the distance d_ω on T by

$$d_\omega(s,t)^2 = \sum_i |\xi_{i,\omega}|^2 |\chi_i(s) - \chi_i(t)|^2 ,$$

so that

$$\Delta(T, d_\omega)^2 \le 4 \sum_i |\xi_{i,\omega}|^2 \tag{7.27}$$

and

$$\mathsf{E}\, d_\omega(s,t)^2 = \sum_i \mathsf{E}|\xi_i|^2 |\chi_i(s) - \chi_i(t)|^2 = d(s,t)^2 . \tag{7.28}$$

---

[18] Maybe here is a place to stress the obvious. When there are several sources of randomness, such as in the second term below, the operator E takes expectation over all of them.

The Cauchy-Schwarz inequality shows that the distance $\bar{d}$ given by $\bar{d}(s,t) = \mathsf{E}d_\omega(s,t)$ satisfies $\bar{d} \le d$ and also from (7.27) that

$$\mathsf{E}\Delta(T, d_\omega) \le 2\Big(\sum_i \mathsf{E}|\xi_i|^2\Big)^{1/2} . \tag{7.29}$$

Next, denoting by E_ε expectation in ω' only,[19] we use (7.12) with $\eta_i(s) = \xi_{i,\omega}\chi_i(s)$ (so that then $d^* = d_\omega$) to obtain that for each ω we have

$$\mathsf{E}_\varepsilon \sup_{t,s} \Big|\sum_i \big(\varepsilon_{i,\omega'}\xi_{i,\omega}\chi_i(t) - \varepsilon_{i,\omega'}\xi_{i,\omega}\chi_i(s)\big)\Big| \le L\gamma_2(T, d_\omega) .$$

Taking expectation and using (7.26) we obtain

$$\mathsf{E} \sup_{t,s} \Big|\sum_i (\xi_i\chi_i(t) - \xi_i\chi_i(s))\Big| \le L\mathsf{E}\gamma_2(T, d_\omega) . \tag{7.30}$$

The distances d_ω are translation-invariant, as follows from the facts that $\chi_i(s+u) = \chi_i(s)\chi_i(u)$ and $|\chi_i(u)| = 1$ for each i, so that (7.19) implies

$$\mathsf{E}\gamma_2(T, d_\omega) \le L\gamma_2(T, \bar{d}) + L\mathsf{E}\Delta(T, d_\omega) \le L\gamma_2(T, d) + L\mathsf{E}\Delta(T, d_\omega) .$$

The desired inequality (7.25) then follows combining this with (7.29). □

7.3.3 A Side Story

Let us now start a side story and discuss the second term in the right-hand side of (7.23). If we consider the case where for all i, $\chi_i = \mathbf{1}$, the character taking value 1 everywhere, then $d \equiv 0$ and this second term is really needed. This is however basically the only case where this term is needed, as the following shows:

Lemma 7.3.5 *Assume*

$$\forall i \ ; \ \chi_i \ne \mathbf{1} . \tag{7.31}$$

Then, recalling the distance d of (7.22),

$$2\sum_i \mathsf{E}|\xi_i|^2 \le \Delta(T, d)^2 \tag{7.32}$$

[19] The idea behind the notation E_ε is that "we take expectation in the randomness of the ε_i only".

and in particular we may replace (7.23) by

$$\mathsf{E}\Big\|\sum_i \xi_i \chi_i\Big\| \le L\gamma_2(T,d) . \tag{7.33}$$

Lemma 7.3.6 *Two different characters are orthogonal in $L^2(T, \mathrm{d}\mu)$.*

Proof Consider two such characters χ and γ. Then for any $s \in T$,

$$\int \bar{\chi}(t)\gamma(t)\mathrm{d}\mu(t) = \int \bar{\chi}(s+t)\gamma(s+t)\mathrm{d}\mu(t) = \bar{\chi}(s)\gamma(s)\int \bar{\chi}(t)\gamma(t)\mathrm{d}\mu(t) .$$

Thus, either χ and γ are orthogonal in $L^2(T, \mathrm{d}\mu)$ or else $\bar{\chi}(s)\gamma(s) = 1$ for all s, i.e., $\chi = \gamma$. □

Corollary 7.3.7 *For each character $\chi \neq 1$ it holds that*

$$\int |\chi(s) - 1|^2 \mathrm{d}\mu(s) = 2 . \tag{7.34}$$

***Proof of Lemma** 7.3.5* We integrate in s the equality $\sum_i \mathsf{E}|\xi_i|^2 |\chi_i(s) - \chi_i(0)|^2 = d(s,0)^2$ to obtain

$$2\sum_i \mathsf{E}|\xi_i|^2 = \int_T d(s,0)^2 \mathrm{d}\mu(s) \le \Delta(T,d)^2 .$$ □

We lighten the exposition by always assuming (7.31) .

This holds even when we forget to repeat it. The only difference this assumption makes is that we no longer have to bother writing the term $(\sum_i \mathsf{E}|\xi_i|^2)^{1/2}$. For example, (7.23) becomes

$$\mathsf{E}\Big\|\sum_i \xi_i \chi_i\Big\| \le L\gamma_2(T,d) . \tag{7.35}$$

The following exercise insists on the fact that assuming (7.31) looses no real information. It simply avoids writing an extra term $\mathsf{E}|\xi_{i_0}|$ both in the upper and lower bounds.

Exercise 7.3.8 Assume that $\chi_{i_0} = 1$ and that $\chi_i \neq 1$ when $i \neq i_0$. Prove that

$$\mathsf{E}|\xi_{i_0}| \le \mathsf{E}\Big\|\sum_i \xi_i \chi_i\Big\| \le \mathsf{E}|\xi_{i_0}| + L\gamma_2(T,d) . \tag{7.36}$$

Exercise 7.3.9 The present exercise deduces classical bounds for trigonometric sums from (7.33). (If it is not obvious that it deals with these, please read Sect. 7.2.1

again.) We consider the case where $T = \mathbb{U}$ is the set of complex numbers of modulus 1 and where $\chi_i(t) = t^i$, the i-th power of t. We observe the bound

$$|s^i - t^i| \le \min(2, |i||s-t|) . \tag{7.37}$$

Let $c_i = \mathsf{E}|\xi_i|^2$, and consider the distance d of (7.22), $d(s,t)^2 = \sum_i c_i |s^i - t^i|^2$. Let $b_0 = \sum_{|i|\le 3} c_i$ and for $n \ge 1$, let $b_n = \sum_{N_n \le |i| \le N_{n+1}} c_i$. Prove that

$$\gamma_2(T,d) \le L \sum_{n\ge 0} 2^{n/2}\sqrt{b_n} , \tag{7.38}$$

and consequently from (7.23)

$$\mathsf{E}\Big\| \sum_i \xi_i \chi_i \Big\| \le L \sum_{n\ge 0} 2^{n/2}\sqrt{b_n} . \tag{7.39}$$

Hint: Here since the group is in multiplicative form, the unit is 1 rather than 0. Observe that $d(t,1)^2 \le \sum_i c_i \min(4, |i|^2|t-1|^2)$. Use this bound to prove that the quantity ϵ_n of Theorem 7.1.1 satisfies $\epsilon_n^2 \le L \sum_i c_i \min(1, |i|^2 2^{-2^{n+1}})$ and conclude using (7.4). If you find this exercise too hard, you will find its solution in Sect. 7.12.

Exercise 7.3.10 For a trigonometric polynomial $A = \sum_i a_i \chi_i$ (where the χ_i are all distinct), let us set $\mathcal{N}(A) = \mathsf{E}\| \sum_i a_i g_i \chi_i \|$ where the g_i are independent standard r.v.s and $\|A\|_P = \mathcal{N}(A) + \|A\|$. This exercise is devoted to the proof that $\|AB\|_P \le L\|A\|_P\|B\|_P$, a result of G. Pisier proved in [83].

(a) Prove that the distance (7.14) satisfies $d(s,t) = \|A^s - A^t\|_2$ where $A^s(u) = A(s+u)$ for $s, x \in T$.
(b) If $\chi_{i_0} = \mathbf{1}$ prove that $|a_{i_0}| \le \|A\|$.
(c) Prove that $\|A\| + \gamma_2(T,d) \le L\|A\|_P \le L(\|A\| + \gamma_2(T,d))$. Hint: Use (7.36).
(d) Prove the desired result. Hint $A^sB^s - A^tB^t = (A^s - A^t)B^s + A^t(B^s - B^t)$. Use also (4.55) and Exercise 2.7.4.

Exercise 7.3.11 This exercise continues the previous one. It contains part of the work needed to compute the dual norm of $\mathcal{N}$, as also achieved in [83].

(a) Prove that if a linear operator U on the space of trigonometric polynomials into itself commutes with translations (in the sense that $(U(A))^s = U(A^s)$ with the notation of the previous exercise, then for each character χ one has $U(\chi) = u_\chi \chi$ for a certain u_χ. (In words: U is a "multiplier".)
(b) For a function f on T consider the norm

$$\|f\|_{\psi_2} = \inf\left\{c > 0; \int \exp(|f|^2/c^2)\mathrm{d}\mu \le 2\right\} .$$

and for a linear operator U as in (a) let

$$\|U\|_{2,\psi_2} = \inf\left\{C > 0\;;\;\forall A\;,\;\|U(A)\|_{\psi_2} \le C\|A\|_2\right\}.$$

Given a trigonometric polynomial A, think of $X_t := U(A)^t$ as a r.v. on the space (T, μ). Use (a) of the previous exercise to prove that it satisfies the increment condition $\|X_s - X_t\|_{\psi_2} \le L\|U\|_{2,\psi_2} d(s,t)$. Relate this to (2.4) and use (2.60) to prove that $\sup_{s,t\in T} |U(A)(s) - U(A)(t)| \le L\mathcal{N}(A)\|U\|_{2,\psi_2}$.

(c) Prove that $|U(A)(0)| \le L\mathcal{N}(A)\|U\|_{2,\psi_2}$.

7.4 The Marcus-Pisier Theorem

7.4.1 The Marcus-Pisier Theorem

As a special case of (6.7), since $\mathsf{E}|g| = \sqrt{2/\pi}$ when g is a standard Gaussian r.v., we get the following version of (6.6):

$$\mathsf{E}\Big\|\sum_i \varepsilon_i x_i\Big\| \le \sqrt{\frac{\pi}{2}}\,\mathsf{E}\Big\|\sum_i g_i x_i\Big\|. \tag{7.40}$$

Exercise 7.4.1 Prove that the inequality (7.40) cannot be reversed in general. More precisely find a situation where the sum is of length n and the right-hand side is about $\sqrt{\log n}$ times larger than the left-hand side.

In this subsection we prove the fundamental fact that in the setting of random trigonometric sums, where the Banach space is the space of continuous functions of T provided with the supremum norm, and when x_i is the function $a_i\chi_i$, we can reverse the general inequality (7.40). As a consequence we obtain in Corollary 7.4.5 below the estimate for $\mathsf{E}\|\sum_i a_i\varepsilon_i\chi_i\|$ on which all our further work relies.

Once this difficult result has been obtained, it is a very easy matter to achieve our goal of finding upper and lower bounds on the quantities $\mathsf{E}\|\sum_i \xi_i\chi_i\|$ under the extra condition that the r.v.s "ξ_i have L^1 and L^2 norms of the same order". This assumption will later be removed, but this requires considerable work.

Theorem 7.4.2 (The Marcus-Pisier Theorem [61]) *Consider complex numbers a_i, independent Bernoulli r.v.s ε_i, and independent standard Gaussian r.v.s g_i. Then*

$$\mathsf{E}\Big\|\sum_i a_i g_i\chi_i\Big\| \le L\mathsf{E}\Big\|\sum_i a_i\varepsilon_i\chi_i\Big\|. \tag{7.41}$$

Proof The argument resembles that of Theorem 6.4.1.[20] Consider a number $c > 0$. Then

$$\mathsf{E}\Big\| \sum_i a_i g_i \chi_i \Big\| \le \mathrm{I} + \mathrm{II}\,, \tag{7.42}$$

where

$$\mathrm{I} = \mathsf{E}\Big\| \sum_i a_i g_i \mathbf{1}_{\{|g_i| \le c\}} \chi_i \Big\|$$

and

$$\mathrm{II} = \mathsf{E}\Big\| \sum_i a_i g_i \mathbf{1}_{\{|g_i| > c\}} \chi_i \Big\|\,.$$

Let us define $u(c) = (\mathsf{E} g^2 \mathbf{1}_{\{|g| \ge c\}})^{1/2}$. Consider the distance d given by (7.14). When $\xi_i = a_i g_i \mathbf{1}_{\{|g_i| \ge c\}}$, we have $\mathsf{E}|\xi_i|^2 = |a_i|^2 u(c)^2$ so that the distance d' given by (7.22) satisfies $d' = u(c) d$. Thus $\gamma_2(T, d') = u(c)\gamma_2(T, d)$ and (7.35) implies

$$\mathrm{II} \le L u(c) \gamma_2(T, d)\,. \tag{7.43}$$

Recalling the lower bound (7.13), it follows that we can choose c a universal constant large enough that $\mathrm{II} \le (1/2)\mathsf{E}\| \sum_i a_i g_i \chi_i \|$. We fix such a value of c. Then (7.42) entails

$$\mathsf{E}\Big\| \sum_i a_i g_i \chi_i \Big\| \le 2\,\mathrm{I}\,.$$

Consider independent Bernoulli r.v.s ε_i that are independent of the r.v.s g_i, so that by symmetry

$$\mathrm{I} = \mathsf{E}\Big\| \sum_i a_i \varepsilon_i g_i \mathbf{1}_{\{|g_i| < c\}} \chi_i \Big\|\,.$$

The contraction principle (Lemma 6.4.4) used given the randomness of the variables g_i yields $\mathrm{I} \le c\mathsf{E}\| \sum_i a_i \varepsilon_i \chi_i \|$, which completes the proof. □

Exercise 7.4.3 Show that (7.41) does not hold when χ_i are general maps from T to $\mathbb{C}$ with $|\chi_i(t)| = 1$.

Exercise 7.4.4 In this exercise we deduce the Marcus-Pisier theorem from Theorem 6.2.8 (the Latała-Bednorz theorem). This is not an economical way to proceed,

[20] This is not a coincidence. I studied the Marcus-Pisier theorem before I invented Theorem 6.4.1.

since the proof of the Latała-Bednorz theorem is very much harder than the proof of the Marcus-Pisier theorem. Nonetheless the argument is very instructive as to what role translation invariance plays. We set $S = \mathsf{E}\|\sum_i a_i \varepsilon_i \chi_i\|$.

(a) Use Theorem 6.2.8 to show that we have a decomposition $a_i \chi_i(t) = u_i(t) + v_i(t)$ where $\sum_i |v_i(t)| \le LS$ and $\mathsf{E}\sup_{t\in T} |\sum_i g_i u_i(t)| \le LS$.
(b) Apply the previous decomposition to $t + s$ instead of t and average over s to prove that we have $a_i = u_i + v_i$ where $\sum_i |v_i| \le LS$ and $\mathsf{E}\sup_{t\in T} |\sum_i g_i u_i \chi_i(t)| \le LS$. Conclude.

Combining (7.41) with (7.13), recalling the distance d of (7.14), and using (7.33) for the upper bound, we obtain the following fundamental result:

Corollary 7.4.5 *We have*

$$\frac{1}{L}\gamma_2(T,d) \le \mathsf{E}\Big\|\sum_i a_i \varepsilon_i \chi_i\Big\| \le L\gamma_2(T,d)\,. \tag{7.44}$$

The next technical lemma makes the left-hand side of (7.44) more precise. Its proof reveals why we controlled the squares in (7.10). The relevance of this lemma will become clear only later.

Lemma 7.4.6 *We have*

$$\mathsf{P}\Big(\Big\|\sum_i a_i \varepsilon_i \chi_i\Big\| \ge \frac{1}{L}\gamma_2(T,d)\Big) \ge \frac{1}{L}\,. \tag{7.45}$$

The r.v. $X = \|\sum_i \varepsilon_i a_i \chi_i\|$ satisfies $\mathsf{E}X \ge \gamma_2(T, d_\omega)/L$ by (7.44). Combining (7.6), (7.10), and (7.32), it also satisfies $\mathsf{E}X^2 \le L\gamma_2(T,d)^2$. The conclusion follows from the Paley-Zygmund inequality (6.15).

7.4.2 *Applications of the Marcus-Pisier Theorem*

It is now easy to complete the goal of providing upper and lower bounds of $\|\sum_i \xi_i \chi_i\|$ which are of the same order when the L^1 and L^2 norms of the ξ_i are of the same order.

Proposition 7.4.7 *Consider independent symmetric real valued random variables ξ_i and characters χ_i. Consider on T the two distances given by*

$$d_1(s,t)^2 = \sum_i (\mathsf{E}|\xi_i|)^2 |\chi_i(s) - \chi_i(t)|^2$$

and

$$d_2(s,t)^2 = \sum_i (\mathsf{E}\,\xi_i^2)|\chi_i(s) - \chi_i(t)|^2 \,.$$

Then, assuming (7.31), we have

$$\frac{1}{L}\gamma_2(T,d_1) \le \mathsf{E}\Big\|\sum_i \xi_i\chi_i\Big\| \le L\gamma_2(T,d_2)\,. \tag{7.46}$$

Proof The right-hand side of (7.46) simply reproduces (7.33). The left-hand side follows by combining (6.7) and (7.44) for $a_i = \mathsf{E}|\xi_i|$. □

Let us set

$$A = \sup_i \frac{(\mathsf{E}\xi_i^2)^{1/2}}{\mathsf{E}|\xi_i|}\,. \tag{7.47}$$

Then $\gamma_2(T,d_2) \le A\gamma_2(T,d_1)$ and we have obtained upper and lower estimates of the quantity $\mathsf{E}\|\sum_i a_i\xi_i\chi_i\|$ whose ratio is $\le LA$. In the case where the r.v.s ξ_i are not square-integrable, we will obviously need other methods. We shall return to this topic later, where we shall be able to estimate the quantity $\mathsf{E}\|\sum_i a_i\xi_i\chi_i\|$ under the only assumption that the r.v.s ξ_i are independent and symmetric. We shall also investigate the convergence of random Fourier series. We simply mention here that for such a series where the quantity A of (7.47) is finite, Proposition 7.4.7 allows us to show that the necessary and sufficient condition for convergence is $\gamma_2(T,d_2) < \infty$. For further use, let us draw a simple consequence of Proposition 7.4.7, whose proof should now be obvious.

Corollary 7.4.8 *With the notations of Proposition 7.4.7 consider also independent symmetric r.v.s (θ_i). Then we have*

$$\mathsf{E}\Big\|\sum_i \xi_i\chi_i\Big\| \le L\sup_i \frac{(\mathsf{E}\xi_i^2)^{1/2}}{\mathsf{E}|\theta_i|}\mathsf{E}\Big\|\sum_i \theta_i\chi_i\Big\|\,. \tag{7.48}$$

It is not always easy to estimate the quantity $\gamma_2(T,d)$ in concrete situations. The book of Marcus and Pisier [61] contains a thorough account (which we will not reproduce) of the link between the present results and the "classical ones". To illustrate the problems that arise, consider, for example, the case where $T = \{-1,1\}^N$ and for $i \le N$ and $t = (t_i)_{i\le N} \in T$, let $\chi_i(t) = t_i$. Since $|t_i| = 1$, for real numbers, a_i it holds that $\|\sum_{i\le N} a_i\varepsilon_i t_i\| = \sum_{i\le N}|a_i|$. Combining with (7.23) and (7.44), we get

$$\frac{1}{L}\sum_{i\le N}|a_i| \le \gamma_2(T,d) \le L\sum_{i\le N}|a_i|\,, \tag{7.49}$$

where $d(s,t)^2 = \sum_{i\le N} a_i^2|\chi_i(s) - \chi_i(t)|^2 = 4\sum_{i\le N} a_i^2 \mathbf{1}_{\{t_i\ne s_i\}}$. The following exercise is in fact quite challenging:

Exercise 7.4.9 Find a direct proof of (7.49).

Let us end this section by a comparison theorem, which is a rather direct consequence of Proposition 7.4.7.

Proposition 7.4.10 *Consider independent symmetric r.v.s ξ_i, θ_i, and characters χ_i. Let us assume (7.31) and that the following holds for a certain constant C*

$$\forall i\ ,\ \forall u > C\ ,\ \mathsf{P}(|\theta_i| \ge u) \ge \mathsf{P}(|\xi_i| \ge Cu)\ , \tag{7.50}$$

$$\mathsf{E}|\theta_i| \ge 1/C\ . \tag{7.51}$$

Then for numbers (a_i) we have

$$\mathsf{E}\Big\|\sum_i a_i\xi_i\chi_i\Big\| \le K\mathsf{E}\Big\|\sum_i a_i\theta_i\chi_i\Big\| \tag{7.52}$$

where K depends on C only.

Proof A main ingredient of the proof is that (7.50) implies that there is a joint realization[21] of the pairs $(|\xi_i|, |\theta_i|)$ such that $|\xi_i| \le K(|\theta_i| + 1)$. As I do not want to struggle on irrelevant technicalities, I will prove this only when the distributions of $|\xi_i|$ and $|\theta_i|$ have no atom. Then one simply takes $|\xi_i| = f_i(|\theta_i|)$ where $f_i(t)$ is the smallest number such $\mathsf{P}(|\xi_i| \ge f_i(t)) = \mathsf{P}(|\theta_i| \ge t)$. Thus it follows from (7.50) that $f_i(t) \le Ct$ for $t \ge C$, and since f_i is increasing, we have $f_i(t) \le C^2$ for $t \le C$ so that $f_i(t) \le K(t+1)$.

A second main ingredient is that if $|b_i| \le |c_i|$ then

$$\mathsf{E}\Big\|\sum_i b_i\varepsilon_i\chi_i\Big\| \le K\mathsf{E}\Big\|\sum_i c_i\varepsilon_i\chi_i\Big\|\ . \tag{7.53}$$

This follows from (7.48). Let us denote by E_ε expectation in the Bernoulli r.v.s (ε_i) only. We then write, using (7.53) in the first inequality and the triangle inequality in the second one

$$\begin{aligned}\mathsf{E}_\varepsilon\Big\|\sum_i a_i\varepsilon_i|\xi_i|\chi_i\Big\| &= \mathsf{E}_\varepsilon\Big\|\sum_i a_i\varepsilon_i f_i(|\theta_i|)\chi_i\Big\| \le K\mathsf{E}_\varepsilon\Big\|\sum_i a_i\varepsilon_i(|\theta_i|+1)\chi_i\Big\| \\ &\le K\mathsf{E}_\varepsilon\Big\|\sum_i a_i\varepsilon_i|\theta_i|\chi_i\Big\| + K\mathsf{E}_\varepsilon\Big\|\sum_i a_i\varepsilon_i\chi_i\Big\|\ . \end{aligned} \tag{7.54}$$

[21] Also called a *coupling*.

Now, we assume in (7.51) that $\mathsf{E}|\theta_i| \geq 1/K$ and it then follows from (7.48) that $\mathsf{E}\big\|\sum_i a_i\varepsilon_i\chi_i\big\| \leq K\mathsf{E}\big\|\sum_i a_i\theta_i\chi_i\big\|$. Finally since (ξ_i) are independent symmetric, the sequences (ξ_i) and $(\varepsilon_i|\xi_i|)$ have the same distribution (and similarly for θ_i) so that taking expectation in (7.54) finishes the proof. □

7.5 Statement of Main Results

7.5.1 General Setting

In the rest of this chapter, we complete the program outlined in Sect. 7.2 of finding upper and lower bounds of the same order for the quantities $\|\sum_i \xi_i\chi_i\|$ where χ_i are characters and ξ_i are independent symmetric r.v.s. (Let us stress that *no* moment conditions whatsoever are now required on the variables ξ_i.) As a consequence we obtain necessary and sufficient conditions for the convergence of random Fourier series in a very general setting (and in particular the series (7.2)). These characterizations are in essence of the same nature as the results of Marcus and Pisier. Unfortunately this means that it is not always immediate to apply them in concrete situations, but we will illustrate at length how this can be done. Fulfilling this program requires a key conceptual advance compared to the work of Sect. 7.3.2: the idea of "families of distances", which is one of the central themes of this work.

We will consider random sums of functions on T which are more general than the sums $\sum_i \xi_i\chi_i$ (where the ξ_i are independent symmetric r.v.s) which we have been considering up to this point. This extra generality offers no difficulty whatsoever while covering other interesting situations, such as the case of "harmonizable infinitely divisible processes" to be considered in Chap. 12.

We describe our setting now. We assume as in Sect. 7.2 that T is a compact metrizable Abelian group with Haar measure μ.[22] We denote by G is dual group, i.e., the set of continuous characters on T, and by $\mathbb{C}G$ the set of functions of the type $a\chi$ where $a \in \mathbb{C}$ and $\chi \in G$.[23]

Consider independent r.v.s Z_i valued in $\mathbb{C}G$, so that Z_i is a random function on T. The crucial property is

$$\forall s, t \in T\ ,\ |Z_i(s) - Z_i(t)| = |Z_i(s - t) - Z_i(0)|\ , \tag{7.55}$$

which holds since it holds for characters by (7.8).

Our purpose is to study random trigonometric sums of the type $\sum_i \varepsilon_i Z_i$ where ε_i are independent Bernoulli r.v.s, independent of the Z_i. This amounts to considering

[22] It requires only simple changes to treat the case where T is only locally compact and not necessarily metrizable.

[23] Please note that $\mathbb{C}G$ is not a vector space!

sums of the type $\sum_i Z_i$ where the r.v.s Z_i are independent *symmetric*.[24] We set

$$X_t = \sum_i \varepsilon_i Z_i(t) . \tag{7.56}$$

In particular we aim to study the r.v.

$$\sup_{t \in T} |X_t| = \| \sum_i \varepsilon_i Z_i \| , \tag{7.57}$$

where $\| \cdot \|$ denotes the supremum norm in the space of continuous functions on T and specifically to find "upper and lower bounds" on this r.v.

Recalling that $\mathbf{1}$ denotes the character everywhere equal to 1, in order to avoid trivial situations, we *always* assume the following:

$$\forall i \ , \ Z_i \notin \mathbb{C}\mathbf{1} \ a.s. \tag{7.58}$$

This exactly corresponds to our condition (7.31) of assuming $\chi_i \neq \mathbf{1}$ in the preceding section.

To give a concrete example, let us consider characters χ_i with $\chi_i \neq \mathbf{1}$ and real-valued symmetric r.v.s ξ_i, only finitely many of which are not 0. The r.v.s $Z_i = \xi_i \chi_i$ are valued in $\mathbb{C}G$, independent symmetric (since it is the case for the r.v.s ξ_i), so that the quantity (7.57) reduces to

$$\| \sum_i \varepsilon_i \xi_i \chi_i \| . \tag{7.59}$$

We provided a partial answer to the question of bounding from above the quantity (7.59) in Sect. 7.3.2 under the condition that $\mathsf{E}\xi_i^2 < \infty$ for each i. However the r.v.s ξ_i might have "fat tails", and it will turn out that the size of these tails governs the size of the quantity (7.59). The results we will prove allow to control the quantity (7.59) *without* assuming that $\mathsf{E}\xi_i^2 < \infty$.

The leading idea of our approach is to work *conditionally on the r.v.s* Z_i. We will detail later how this is done (following the procedure described on page 214), but the point is simple: when we fix the points $Z_i \in \mathbb{C}G$, then $Z_i = a_i \chi_i$ where a_i is a given (=nonrandom) number and χ_i is a given (=nonrandom) character. This is the *essential property* of the Z_i. In this manner we basically reduce the study of the general sums $\sum_i \varepsilon_i Z_i$ to the study of the much simpler sums $\sum_i \varepsilon_i a_i \chi_i$ (but this reduction is by no means routine).

[24] We refer the reader to, e.g., Proposition 8.1.5 of [33] for a detailed study of a "symmetrization procedure" in the setting of random Fourier series, showing that the symmetric case is the important one.

This is how we will obtain lower bounds, using (7.44). Upper bounds are more delicate. We originally discovered our upper bounds using chaining, but we will present a simpler argument (which is somewhat specific to random Fourier series).

We end this section by commenting a bit on what the random function Z_i valued in $\mathbb{C}G$ looks like. By definition of $\mathbb{C}G$, we have $Z_i = \xi_i \chi_i$ where ξ_i is a complex-valued r.v. and χ_i is a random character, but we do *not* assume that ξ_i and χ_i are independent r.v.s. Since for a character χ we have $\chi(0) = 1$, we simply have $\xi_i = Z_i(0)$. Let us describe the situation more precisely.[25] Since we assume that T is metrizable, its dual group G is countable (as follows from Lemma 7.3.6 and the fact that $L^2(T,\mu)$ is separable). Recalling that $\mathbf{1}$ denotes the character equal to 1 everywhere[26] we can enumerate $G \setminus \{\mathbf{1}\}$ as a sequence $(\chi_\ell)_{\ell \ge 1}$, and (7.58) implies that a.s. we have $Z_i \in \cup_\ell \mathbb{C}\chi_\ell$. Unless $Z_i = 0$, there is a unique $\ell \ge 1$ such that $Z_i \in \mathbb{C}\chi_\ell$. Therefore $Z_i = \sum_{\ell \ge 1} Z_i \mathbf{1}_{\{Z_i \in \mathbb{C}\chi_\ell\}}$. When $Z_i \in \mathbb{C}\chi_\ell$ we have $Z_i = Z_i(0)\chi_\ell$, so that we have the expression

$$Z_i = \sum_{\ell \ge 1} \xi_{i,\ell}\chi_\ell \,, \tag{7.60}$$

where

$$\xi_{i,\ell} = Z_i(0)\mathbf{1}_{\{Z_i \in \mathbb{C}\chi_\ell\}} \,.$$

The important point in (7.60) is that the r.v.s $(\xi_{i,\ell})_{\ell \ge 1}$ "have disjoint support": if $\ell \ne \ell'$ then $\xi_{i,\ell}\xi_{i,\ell'} = 0$ a.s, so for any realization of the randomness, in the sum $Z_i = \sum_{\ell \ge 1} \xi_{i,\ell}\chi_\ell$, there is at most one non-zero term.

7.5.2 Families of Distances

We expect that a control from above of the quantity (7.57) implies a kind of smallness of T. But smallness with respect to what? An obvious idea would be to consider the distance defined by

$$d(s,t)^2 = \sum_i \mathsf{E}|Z_i(s) - Z_i(t)|^2 \,. \tag{7.61}$$

Unfortunately such a formula gives too much weight to the large values of Z_i. To remove the influences of these large values, we need to truncate. At which level should we truncate? As you might guess, there is no free lunch, and we need to consider truncations at all possible levels. That is, for $s,t \in T$ and $u \ge 0$, we

[25] This description will not be used before Sect. 7.9.

[26] Which happens to be the unit of G.

consider the quantities

$$\varphi(s,t,u) = \sum_i \mathsf{E}(|u(Z_i(s) - Z_i(t))|^2 \wedge 1) , \tag{7.62}$$

where $x \wedge 1 = \min(x, 1)$.[27] Given a number $r \geq 2$, for $j \in \mathbb{Z}$, we define

$$\varphi_j(s,t) = \varphi(s,t,r^j) = \sum_i \mathsf{E}(|r^j(Z_i(s) - Z_i(t))|^2 \wedge 1) . \tag{7.63}$$

Thus, φ_j is the square of a translation-invariant distance on T. The "family of distances"[28] (φ_j) is appropriate to estimate the quantity (7.57). For the purposes of this section, it suffices to consider the case $r = 2$. Other values of r are useful for related purposes, so for consistency we allow the case $r > 2$ (which changes nothing to the proofs). We observe that $\varphi_{j+1} \geq \varphi_j$.

The concept of family of distances may be disturbing at first, but once one gets used to it, it is not any harder than working with one single distance. There is a foolproof rule:

To make sense of a statement involving a family of distances,

pretend that $\varphi_j(s,t) = r^{2j} d(s,t)^2$ for a given distance d . (7.64)

To motivate this rule, observe that if we should disregard the truncation in (7.62) by (7.61), we would have indeed $\varphi_j(s,t) = r^{2j} d(s,t)^2$.

It is clear at least since we stated Theorem 2.7.14 that families of distances will be relevant in bounding stochastic processes. One may ask then if there is a simple way to understand why the family of distances (7.63) is relevant here. This will become apparent while going through the mechanism of the proofs, but we can already stress some features. The right-hand side of (7.63) is the expected value of a sum of an independent family of positive r.v.s, all bounded by 1. Elementary considerations show that such sums are strongly concentrated around their expectations (as expressed in Lemma 7.7.2). Taking advantage of that fact alone, we have a control (either from above or from below) of the size of T as measured in an appropriate way by the family of distances (7.63); we will show that for the typical choice of the randomness of the Z_i, we also have a similar control of the size of T for the family of random distances ψ_j given by $\psi_j(s,t) = \sum_i |r^j(Z_i(s) - Z_i(t))|^2 \wedge 1$. In this manner we really reduce the entire problem of studying random trigonometric sums to the case of the sums $\sum_i \varepsilon_i a_i \chi_i$.

[27] As you notice, writing (7.62) is not the most immediate way to truncate, but you will understand soon the advantages of using this formulation.

[28] This is a very dumb name since they are not distances but squares of distances and since it is more a sequence than a family. I am not going to change it.

It is also apparent how to produce lower bounds. Recalling that when we work given the Z_i these are of the type $a_i\chi_i$, this is based on the obvious fact

$$\psi_j(s,t) = \sum_i |r^j(Z_i(s) - Z_i(t))|^2 \wedge 1 = \sum_i |r^j a_i(\chi_i(s) - \chi_i(t)|^2 \le r^{2j} d(s,t)^2 ,$$

where $d(s,t)^2 = \sum_i |a_i(\chi_i(s) - \chi_i(t)|^2$ is the distance (7.14). We can then expect that lower bounds on the family $\psi_j(s,t)$ will produce lower bounds on the distance d and in turn through (7.44) lower bounds on the random trigonometric sum $\sum_i \varepsilon_i a_i \chi_i$.

It is far less apparent why the distances $\psi_j(s,t)$ also suffice to obtain upper bounds, and we have no magic explanation to offer here. We know two very different proofs of this fact, and neither of them is very intuitive. Let us only observe that in writing (7.62) we disregard some information about the large values of Z_i, but we will control these large values simply because there are very few of them (a finite number in the setting of infinite series).

7.5.3 Lower Bounds

Our first result is a lower bound for the sum $\|\sum_i \varepsilon_i Z_i\|$ (although it will not be immediately obvious that this is a lower bound).

Theorem 7.5.1 *There exists a universal constant α_0 with the following property. Assume that for some $M > 0$ we have*

$$\mathsf{P}\Big(\Big\|\sum_i \varepsilon_i Z_i\Big\| \ge M\Big) \le \alpha_0 . \tag{7.65}$$

Then for $n \ge 0$ we can find numbers $j_n \in \mathbb{Z}$ such that[29]

$$\forall s,t \in T ,\ \varphi_{j_0}(s,t) \le 1 \tag{7.66}$$

and that for each $n \ge 1$

$$\mu(\{s\ ;\ \varphi_{j_n}(s,0) \le 2^n\}) \ge 2^{-2^n} = N_n^{-1} . \tag{7.67}$$

[29] The number 1 in the right-hand side of (7.66) does not play any special role and can be replaced by any other constant > 0.

for which

$$\sum_{n\geq 0} 2^n r^{-j_n} \leq KM \,. \tag{7.68}$$

For a first understanding of this result, recall that $\varphi_j \leq \varphi_{j+1}$ so that the conditions (7.66) and (7.67) are easier to satisfy for small values of the j_n. On the other hand, (7.68) states that we can satisfy these conditions by values of the j_n which are not too small.

Conditions (7.66) and (7.67) are very important, and it is extremely useful to consider the largest integers which satisfy them. Throughout this chapter, when dealing with trigonometric sums, we will use the notation

$$\bar{j}_0 = \sup\big\{j \in \mathbb{Z}\;;\; \forall s,t \in T\;;\; \varphi_j(s,t) \leq 1\big\} \in \mathbb{Z} \cup \{\infty\}\,. \tag{7.69}$$

This definition makes sense because the set on the right is never empty. Indeed, since $|Z_i(t)| = |Z_i(0)|$,

$$\varphi_j(s,t) \leq \mathsf{E}\sum_i |2r^j Z_i(0)|^2 \wedge 1\,,$$

and since the sum is finite, it follows from dominated convergence that the limit of the right-hand side as $j \to -\infty$ is zero, and thus there exists j for which $\sup_{s,t\in T} \varphi_j(s,t) \leq 1$. Similarly for $n \geq 1$, we define

$$\bar{j}_n = \sup\big\{j \in \mathbb{Z}\;;\; \mu(\{s\;;\;\varphi_j(s,0) \leq 2^n\}) \geq 2^{-2^n} = N_n^{-1}\big\} \in \mathbb{Z}\cup\{\infty\}\,. \tag{7.70}$$

One may like to think of $r^{-\bar{j}_0}$ as a substitute for the diameter of T and as $2^{n/2}r^{-\bar{j}_n}$ as a substitute for the entropy number e_n (as will become gradually apparent). We can now make clear why Theorem 7.5.1 is a lower bound.

Corollary 7.5.2 *There exists a universal constant $\alpha_0 > 0$ such that*

$$\mathsf{P}\Big(\Big\|\sum_i \varepsilon_i Z_i\Big\| > \frac{1}{K}\sum_{n\geq 0} 2^n r^{-\bar{j}_n}\Big) \geq \alpha_0\,, \tag{7.71}$$

where K depends on r only.

Proof Set $M = (2K_0)^{-1}\sum_{n\geq 0} 2^n r^{-\bar{j}_n}$ where K_0 the constant of (7.68). Assume for contradiction that (7.65) holds, and consider the numbers j_n provided by Theorem 7.5.1 so that $j_n \leq \bar{j}_n$ and

$$K_0 M = \frac{1}{2}\sum_{n\geq 0} 2^n r^{-\bar{j}_n} < \sum_{n\geq 0} 2^n r^{-j_n}\,.$$

This contradicts (7.68), proving that (7.65) does not hold, so that (7.71) holds. □

Let us note the following consequence of (7.71) and (2.7):

$$\frac{1}{K}\sum_{n\geq 0} 2^n r^{-\bar{j}_n} \leq \mathsf{E}\Big\|\sum_i \varepsilon_i Z_i\Big\| . \tag{7.72}$$

To understand the nature of the sum $\sum_{n\geq 0} 2^n r^{-\bar{j}_n}$, let us apply the strategy (7.64), assuming that $\varphi_j(s,t) = r^{2j} d(s,t)^2$. Then

$$\bar{j}_n = \sup\left\{ j \in \mathbb{Z}\ ;\ \mu(\{s\ ;\ d(s,0) \leq 2^{n/2} r^{-j}\}) \geq N_n^{-1} \right\} ,$$

and thus if ϵ_n is defined as in (7.3), we see that $2^{n/2} r^{-\bar{j}_n} \simeq \epsilon_n$, so that $2^n r^{-\bar{j}_n} \simeq \epsilon_n 2^{n/2}$.

$$\textit{The quantity } \sum_{n\geq 0} 2^n r^{-\bar{j}_n} \textit{ appears as the natural substitute}$$

$$\textit{for the familiar entropy integral } \sum_{n\geq 0} \epsilon_n 2^{n/2} . \tag{7.73}$$

In (7.69), in the condition $\varphi_j(s,t) \leq 1$ the number 1 can be replaced by any other provided of course that one changes the constant in (7.71). On page 232 the reader will find a computation of the quantity $\sum_{n\geq 0} 2^n r^{-\bar{j}_n}$ in some simple cases.

It is important to understand the next example and its consequences.

Exercise 7.5.3 Assume that $\sum_i \mathsf{P}(|Z_i| \neq 0) \leq 1$. Prove that $\bar{j}_n = \infty$ for each n and that (7.71) brings no information.

It would be a devastating misunderstanding to conclude from this example that (7.71) is a "weak result". The real meaning is more subtle: (7.71) does not bring much information on certain sums, but these are of a very special type. The decomposition theorem (Theorem 7.5.14) states that a general random trigonometric sum is the sum of two pieces. For one of these (7.71) captures exact information, and the other piece is of a very special type. This will be a feature of our main results in Chap. 11.

One of the main results of Chap. 11 will be a considerable generalization of (7.72) to a setting where there is no translation invariance. Then we will not be able to use covering numbers (as is being done in a sense in (7.72)). The next exercise will help you understand the formulation we will adopt there.

Exercise 7.5.4

(a) Prove that for $s, t, u \in T$ and any j, we have

$$\varphi_j(s,t) \leq 2(\varphi_j(s,u) + \varphi_j(u,t)) .$$

(b) Consider a subset D of T as assume that for any $s \in D$ we have $\varphi_j(s,0) \le d$ for a certain number d. Prove that if s and t belong to the same translate of $D - D$, we have $\varphi_j(s,t) \le 4d$.

(c) Prove that under the conditions (7.66)–(7.68), we can find an admissible sequence $(\mathcal{A}_n)$ of partitions of T with the following property: If $n \ge 1$,

$$s, t \in A \in \mathcal{A}_n \Rightarrow \varphi_{j_{n-1}}(s,t) \le 2^{n+1} . \tag{7.74}$$

Hint: Use Exercise 2.7.6 and Lemma 7.1.3.

7.5.4 Upper Bounds

Let us turn to upper bounds. In order to avoid technical statements at this stage, we will assume that there is enough integrability that the size of $\|\sum_i \varepsilon_i Z_i\|$ can be measured by its expectation $\mathsf{E}\|\sum_i \varepsilon_i Z_i\|$.[30] Corollary 7.5.2 states that the typical value of $\|\sum_i \varepsilon_i Z_i\|$ controls from above the "entropy integral" $\sum_{n\ge 0} 2^n r^{-\bar{j}_n}$. In the reverse direction, since the quantities $\bar{j}_n$ say nothing about the large values of Z_i, we cannot expect that the "entropy integral" $\sum_{n\ge 0} 2^n r^{-\bar{j}_n}$ will control the tails of the r.v. $\|\sum_i \varepsilon_i Z_i\|$. However, as the following expresses, we control the size of $\|\sum_i \varepsilon_i Z_i\|$ as soon as we control the "entropy integral" $\sum_{n\ge 0} 2^n r^{-\bar{j}_n}$ *and* the size of the single r.v. $\sum_i \varepsilon_i Z_i(0)$.

Theorem 7.5.5 *For $n \ge 0$ consider numbers $j_n \in \mathbb{Z}$, which satisfy (7.66) and (7.67). Then, for any $p \ge 1$, we have*

$$\Big(\mathsf{E}\Big\|\sum_i \varepsilon_i Z_i\Big\|^p\Big)^{1/p} \le K\Big(\sum_{n\ge 0} 2^n r^{-j_n} + \Big(\mathsf{E}\Big|\sum_i \varepsilon_i Z_i(0)\Big|^p\Big)^{1/p}\Big) , \tag{7.75}$$

where K depends only on r and p.

Of course in (7.75), the larger j_n, the better bound we get. The best bound is obtained for $j_n = \bar{j}_n$.

7.5.5 Highlighting the Magic of Theorem 7.5.5

The key to Theorem 7.5.5 is the case where $Z_i = a_i \chi_i$ is a constant multiple of a nonrandom character. We investigated this situation in detail in Sect. 7.4, but Theorem 7.5.5 provides crucial new information, and we explain this now. In that

[30] More general statements which assume no integrability will be given in Sect. 7.8.

case, the function φ_j of (7.63) are given by

$$\varphi_j(s,t) = \sum_i |r^j a_i(\chi_i(s) - \chi_i(t))|^2 \wedge 1 \,. \tag{7.76}$$

Thus, assuming the conditions (7.66) and (7.67) and using (7.75) for $p = 1$ yields the bound

$$\mathsf{E}\Big\| \sum_i a_i \varepsilon_i \chi_i \Big\| \le K \sum_{n \ge 0} 2^n r^{-j_n} + K\Big(\sum_i |a_i|^2\Big)^{1/2} \,. \tag{7.77}$$

Consider the distance d of (7.22), given by $d(s,t)^2 = \sum_{i \ge 1} |a_i|^2 |\chi_i(s) - \chi_i(t)|^2$, and define

$$\epsilon_n = \inf\big\{\epsilon > 0 \,;\ \mu(B_d(0,\epsilon)) \ge N_n^{-1}\big\} \,. \tag{7.78}$$

It then follows from (7.44) and (7.4) that

$$\frac{1}{K} \sum_{n \ge 0} 2^{n/2} \epsilon_n \le \mathsf{E}\Big\| \sum_i a_i \varepsilon_i \chi_i \Big\| \le K \sum_{n \ge 0} 2^{n/2} \epsilon_n \,. \tag{7.79}$$

We have reached here a very unusual situation: it is by no means obvious to compare the upper bounds in (7.77) and (7.79). In particular, combining these inequalities we reach the following:

$$\sum_{n \ge 0} 2^{n/2} \epsilon_n \le K \sum_{n \ge 0} 2^n r^{-j_n} + K\Big(\sum_i |a_i|^2\Big)^{1/2} \,, \tag{7.80}$$

where K depends on r only, but we do not know how to give a direct proof of this inequality. The quantities appearing in this inequality involve only characters and complex numbers. There are no random variables involved, so why should one use random trigonometric sums to prove it?

Research Problem 7.5.6 Find a proof of (7.80) which does not use random trigonometric sums.

While the inequality (7.80) is mysterious, the reverse inequality is very clear. First, we proved in Sect. 7.3.3 (assuming $\chi_i \ne 1$ for each i) that $\sum_i |a_i|^2 \le L\Delta(T,d)^2 = L\epsilon_0^2$. Next, keeping (7.32) in mind, we prove that

$$\sum_{n \ge 0} 2^n r^{-\bar{j}_n} \le K \sum_{n \ge 0} 2^{n/2} \epsilon_n \,. \tag{7.81}$$

When $\bar{j}_n$ is finite, we have

$$\mu(\{s \in T \; ; \; \varphi_{\bar{j}_n+1}(s,0) \le 2^n\}) < N_n^{-1} \,. \tag{7.82}$$

Since obviously $\varphi_j(s,t) \le r^{2j} d(s,t)^2$ we have

$$B_d(0, 2^{n/2} r^{-\bar{j}_n - 1}) \subset \{s \in T \; ; \; \varphi_{\bar{j}_n+1}(s,0) \le 2^n\} \,.$$

Combining with (7.82) and (7.78), this proves that $2^{n/2} r^{-\bar{j}_n - 1} \le \epsilon_n$ and (7.81).

In summary of this discussion, we could argue that even though in retrospect (7.77) does not improve on (7.79), it is a better inequality, because it is quite obvious that its right-hand side is of smaller order than the right-hand side of (7.79), whereas the opposite is absolutely not obvious.

Exercise 7.5.7 Assume that $\varphi_j(s,t) < 1$ for each $s, t \in T$ and that $\chi_i \neq 1$ for all i. Prove that $\sum_i a_i^2 \le r^{-2j}/2$.

7.5.6 Combining Upper and Lower Bounds

We can combine Corollary 7.5.2 and Theorem 7.5.5 to provide upper and lower bounds for $(\mathsf{E}\|\sum_i \varepsilon_i Z_i\|^p)^{1/p}$ that are of the same order. Let us state the result in the case of (7.59). From now on, K denotes a number that depends only on r and p and that need not be the same on each occurrence.

Theorem 7.5.8 *Assume that the r.v.s ξ_i are independent symmetric. If the numbers $\bar{j}_n$ are as in (7.69) and (7.70), then, for each $p \ge 1$,*

$$\frac{1}{K}\Big(\sum_{n\ge 0} 2^n r^{-\bar{j}_n} + \Big(\mathsf{E}\Big|\sum_i \xi_i\Big|^p\Big)^{1/p}\Big) \le \Big(\mathsf{E}\Big\|\sum_i \xi_i \chi_i\Big\|^p\Big)^{1/p}$$
$$\le K\Big(\sum_{n\ge 0} 2^n r^{-\bar{j}_n} + \Big(\mathsf{E}\Big|\sum_i \xi_i\Big|^p\Big)^{1/p}\Big) \,. \tag{7.83}$$

Not the least remarkable feature of this result is that it assumes nothing (beyond independence and symmetry) on the r.v.s ξ_i.

7.5.7 An Example: Tails in u^{-p}

Explicit examples of application of these abstract theorems will be given in Sect. 7.12, but right away we illustrate Theorem 7.5.8 in some cases (investigated first by M. Marcus and G. Pisier [62]): having upper and lower bounds which hold in great generality unfortunately does not mean that it is always obvious to relate them to other known bounds. We consider complex numbers a_i and the distance d_p on T defined by

$$d_p(s,t)^p = \sum_i |a_i(\chi_i(s) - \chi_i(t))|^p \;. \tag{7.84}$$

Proposition 7.5.9 *Consider symmetric r.v.s θ_i which satisfy for certain numbers $1 < p < 2$ and $C > 0$ and for all $u \ge 0$*

$$\mathsf{P}(|\theta_i| \ge u) \le Cu^{-p} \;. \tag{7.85}$$

Assume that $\xi_i = a_i\theta_i$. Then we have

$$\sum_{n\ge 0} 2^n r^{-\bar{j}_n} \le K\gamma_q(T, d_p) \;. \tag{7.86}$$

Here we use the notation of Corollary 7.5.2, K denotes a constant depending only on C, r and p, and $1/p + 1/q = 1$. The point of (7.86) is that it relates the quantity $\sum_{n\ge 0} 2^n r^{-\bar{j}_n}$ with the more usual quantity $\gamma_q(T, d_p)$. The proof depends on two simple lemmas.

Lemma 7.5.10 *Under (7.85) for any $j \in \mathbb{Z}$ we have*

$$\varphi_j(s,t) \le Kr^{jp} d_p(s,t)^p \;. \tag{7.87}$$

Proof Using (7.85) in the second line we obtain that for $v \neq 0$,

$$\begin{aligned} \mathsf{E}(|v\theta_i|^2 \wedge 1) &= \int_0^1 \mathsf{P}(|v\theta_i|^2 \ge t)\mathrm{d}t = \int_0^1 \mathsf{P}\Big(|\theta_i| \ge \frac{t^{1/2}}{|v|}\Big)\mathrm{d}t \\ &\le \int_0^1 C\frac{|v|^p}{t^{p/2}}\mathrm{d}t = K|v|^p \;. \end{aligned} \tag{7.88}$$

Since $\xi_i = a_i\theta_i$ this implies $\mathsf{E}(|r^j\xi_i(\chi_i(s)-\chi_i(t))|^2 \wedge 1) \le Kr^{jp}|a_i(\chi_i(s)-\chi_i(t))|^p$ and summation over i yields the result. □

Lemma 7.5.11 *Consider for $n \ge 0$ the numbers ϵ_n as in Theorem 7.1.1, for the distance d_p, i.e., $\epsilon_n = \inf\{\epsilon > 0; \mu(\{s; d_p(s,0) \le \epsilon\}) \ge N_n^{-1}$. Then*

$$2^{n/p} r^{-\bar{j}_n} \le K\epsilon_n \;. \tag{7.89}$$

Proof We may assume that $\bar{j}_n < \infty$. Since $\{s; d_p(s,0) \le 2^{n/p}r^{-\bar{j}_n}/K\} \subset \{s; \varphi_{\bar{j}_n+1}(s,0) \le 2^n\}$ and since $\mu(\{s; \varphi_{\bar{j}_n+1}(s,0) \le 2^n\}) < N_n^{-1}$, we have $\mu(\{s; d_p(s,0) \le 2^{n/p}r^{-\bar{j}_n}/K\}) < N_n^{-1}$ so that $2^{n/p}r^{-\bar{j}_n}/K \le \epsilon_n$. □

Proof of Proposition 7.5.9 Consider for $n \geq 0$ the numbers ϵ_n as above, so that

$$\sum_{n \geq 0} \epsilon_n 2^{n/q} \leq K \gamma_q(T, d_p)$$

by (7.6). The result follows by (7.89). □

Exercise 7.5.12 Use (7.83) (taking $p = 1$ there) to prove that when $\chi_i \neq \mathbf{1}$ for each i then

$$\mathsf{E}\Big\| \sum_i a_i \theta_i \chi_i \Big\| \leq K \gamma_q(T, d_p) \,. \tag{7.90}$$

Hint: You have to prove that $\mathsf{E}|\sum_i a_i \theta_i| \leq K(\sum_i |a_i|^p)^{1/p}$ and $(\sum_i |a_i|^p)^{1/p} \leq K\Delta(T, d_p)$. The first inequality is elementary but rather tricky, and the second uses the methods of Sect. 7.3.3.

The following is a kind of converse to (7.90):

Proposition 7.5.13 *Consider* $1 < p < 2$ *and its conjugate number* q*. Consider independent symmetric r.v.s* (θ_i)*, and assume that for some constant* C *and all* i*, we have*

$$u \geq C \Rightarrow \mathsf{P}(|\theta_i| \geq u) \geq \frac{1}{Cu^p} \,. \tag{7.91}$$

Assume also that $\chi_i \neq 1$ *for each* i*. Then*

$$\gamma_q(T, d_p) \leq K \mathsf{E}\Big\| \sum_i a_i \theta_i \chi_i \Big\| \,, \tag{7.92}$$

where K *depends only on* C*.*

Magic proof of Proposition[31] 7.5.13 This proof uses the concept of p-stable r.v. which was described in Sect. 5.1. Consider an i.i.d. sequence (ξ_i) of p-stable r.v.s ξ_i with $\mathsf{E}|\xi_i| = 1$. It is then known that $\mathsf{P}(|\xi_i| \geq u) \leq Ku^{-p}$ for $u > 0$ so that (7.50) holds, and therefore (7.52) holds. Now, Theorem 5.2.1 asserts that $\gamma_q(T, d_p) \leq K\mathsf{E}\|\sum_i \xi_i \chi_i\|$, and combining with (7.52) this finishes the proof. □

In Sect. 7.11 we give an arguably more natural proof which does not use p-stable r.v.s.

7.5.8 The Decomposition Theorem

In this section we go back to the theme started in Sect. 6.8. We will prove later, in Theorem 11.10.3, that under rather general circumstances, a random series can be decomposed into two pieces, one of which can be controlled by

[31] Please see the comments about this proof in Sect. 7.14.

chaining and one which can be controlled by ignoring cancellations between the terms of the series. This theorem applies in particular in the case of the random trigonometric sums we are considering here.[32] Unfortunately, the two terms of the decomposition of a random trigonometric sum constructed in Theorem 11.10.3 are *not themselves random trigonometric sums*. The following asserts that we can obtain a similar decomposition where the two terms of the decomposition *are themselves random trigonometric sums*. To understand this result, the reader should review Theorem 6.2.8, which is of a strikingly similar nature (but much more difficult).

Theorem 7.5.14 *Consider independent r.v.s Z_i valued in $\mathbb{C}G$, and assume (7.58). Set $S = \mathsf{E}\|\sum_i \varepsilon_i Z_i\|$. Then there is a decomposition $Z_i = Z_i' + Z_i''$, where both Z_i' and Z_i'' are in $\mathbb{C}G$, where each of the sequences (Z_i') and (Z_i'') is independent,*[33] *and satisfy*

$$\mathsf{E}\sum_i |Z_i'(0)| \le LS \tag{7.93}$$

and

$$\gamma_2(T, d) \le LS \ , \tag{7.94}$$

where the distance d is given by $d(s,t)^2 = \sum_i \mathsf{E}|Z_i''(s) - Z_i''(t)|^2$. Furthermore in the case of usual random trigonometric sums, when $Z_i = \xi_i \chi_i$, the decomposition takes the form $Z_i' = \xi_i' \chi_i$ and $Z_i'' = \xi_i'' \chi_i$.

In the case of usual random series, this decomposition witnesses in a transparent way the size of $S = \mathsf{E}\|\sum_i \xi_i \chi_i\|$. Indeed, (7.93) makes it obvious that $\mathsf{E}\|\sum_i \xi_i' \chi_i\| \le LS$, whereas (7.33) and (7.94) imply that $\mathsf{E}\|\sum_i \xi_i'' \chi_i\| \le L\gamma_2(T, d) \le LS$. The argument works just the same in general, as the following shows:

Exercise 7.5.15 Generalize (7.23) to the case of sums $\sum_i \varepsilon_i Z_i$. That is, prove that

$$\mathsf{E}\Big\|\sum_i \varepsilon_i Z_i\Big\| \le L\gamma_2(T, d) + L\Big(\sum_i |Z_i(0)|^2\Big)^{1/2}$$

where

$$d(s,t)^2 = \sum_i |Z_i(s) - Z_i(t)|^2 \ .$$

---

[32] The main ingredient of the proof in the case of random trigonometric series is Theorem 7.5.1 which is far easier than the corresponding result Theorem 11.7.1 in the general case.

[33] But the sequences are not independent of each other.

7.5.9 Convergence

We state now our convergence theorems. We consider independent r.v.s $(Z_i)_{i\geq 1}$ with $Z_i \in \mathbb{C}G$ and independent Bernoulli r.v.s ε_i independent of the randomness of the Z_i. Throughout this section we say that the series $\sum_{i\geq 1} \varepsilon_i Z_i$ converges a.s. if it converges a.s. in the Banach space of continuous functions on T provided with the uniform norm.[34] We also recall that we assume (7.58).

Theorem 7.5.16 *The series $\sum_{i\geq 1} \varepsilon_i Z_i$ converges a.s. if and only if the following occurs: There exists j_0 such that*

$$\forall\, s, t \in T\,,\ \sum_{i\geq 1} \mathsf{E}(|r^{j_0}(Z_i(s) - Z_i(t))|^2 \wedge 1) \leq 1\,, \tag{7.95}$$

and for $n \geq 1$ there exists $j_n \in \mathbb{Z}$ for which

$$\mu\Big(\Big\{s \in T\,;\ \sum_{i\geq 1} \mathsf{E}(|r^{j_n}(Z_i(s) - Z_i(0))|^2 \wedge 1) \leq 2^n\Big\}\Big) \geq \frac{1}{N_n}\,, \tag{7.96}$$

such that

$$\sum_{n\geq 0} 2^n r^{-j_n} < \infty\,. \tag{7.97}$$

Moreover, when these conditions are satisfied, for each $p \geq 1$ we have

$$\mathsf{E}\Big\|\sum_{i\geq 1} \varepsilon_i Z_i\Big\|^p < \infty \Leftrightarrow \mathsf{E}\Big|\sum_{i\geq 1} \varepsilon_i Z_i(0)\Big|^p < \infty\,.$$

We have also the following, less concrete but more spectacular:

Theorem 7.5.17 *The series $\sum_{i\geq 1} \varepsilon_i Z_i$ converges almost surely if and only if one may find a decomposition $Z_i = Z_i^1 + Z_i^2 + Z_i^3$ with the following properties. First, each of the sequences (Z_i^ℓ) for $\ell = 1, 2, 3$ is independent and valued in $\mathbb{C}G$. Next*

$$\sum_{i\geq 1} \mathsf{P}(Z_i^1 \neq 0) < \infty\,, \tag{7.98}$$

$$\sum_{i\geq 1} \mathsf{E}|Z_i^2(0)| < \infty\,, \tag{7.99}$$

$$\gamma_2(T, d) < \infty\,, \tag{7.100}$$

[34] This is the most natural notion of convergence. A classical theorem of Billard (see [40, Theorem 3, p. 58]) relates different notions of convergence of a random Fourier series.

where the distance d is given by $d(s,t)^2 = \sum_{i\ge 1} \mathsf{E}|Z_i^3(s) - Z_i^3(t)|^2$. Furthermore, when $Z_i = \xi_i \chi_i$ there exists a decomposition $\xi_i = \xi_i^1 + \xi_i^2 + \xi_i^3$ such that for $\ell \le 3$ we have $Z_i^\ell = \xi_i^\ell \chi_i$.

The necessary conditions and the sufficient conditions stated in the next theorem are due to Marcus and Pisier [62] and were known much before the more general Theorem 7.5.16. We will show how to deduce them from that result.[35]

Theorem 7.5.18

(a) *For $i \ge 1$ consider characters χ_i and numbers a_i. Then the series $\sum_{i\ge 1} a_i \varepsilon_i \chi_i$ converges almost surely if and only if $\gamma_2(T, d_2) < \infty$ where d_2 the distance given by*

$$d_2(s,t)^2 = \sum_{i\ge 1} |a_i|^2 |\chi_i(s) - \chi_i(t)|^2 \,. \tag{7.101}$$

(b) *Consider independent symmetric r.v.s $(\theta_i)_{i\ge 1}$ and numbers $(a_i)_{i\ge 1}$. Consider a number $1 < p < 2$ and the distance d_p on T given by*

$$d_p(s,t)^p = \sum_{i\ge 1} |a_i|^p |\chi_i(s) - \chi_i(t)|^p \,. \tag{7.102}$$

Assuming that the r.v.s $(\theta_i)_{i\ge 1}$ satisfy (7.85) then if $\gamma_q(T, d_p) < \infty$ (where q is the conjugate exponent of p), the series $\sum_{i\ge 1} a_i \theta_i \chi_i$ converges a.s.

(c) *With the same notation as in (b) if the r.v.s $(\theta_i)_{i\ge 1}$ satisfy (7.92) and if the series $\sum_{i\ge 1} a_i \theta_i \chi_i$ converges a.s., then $\gamma_q(T, d_p) < \infty$.*

7.6 A Primer on Random Sets

The purpose of the present section is to bring forward several very elementary facts which will be used on several occasions in the rest of the present chapter and Chap. 11. As these facts will be part of non-trivial proofs, it may help the reader to meet them first in their simplest setting. We consider a probability space (T, μ) and a random subset T_ω of T where we symbolize the randomness by a point ω in a certain probability space Ω. We are mostly interested in the case where these random subsets are typically very small. (Going to complements this also covers the case where the complements of these sets are very small.) We assume enough measurability, and for $s \in T$ we define

$$p(s) = \mathsf{P}(s \in T_\omega) \,.$$

[35] Unfortunately, this is not entirely obvious.

Lemma 7.6.1 *We have*

$$\mathsf{E}\mu(T_\omega) = \int_T p(s)\mathrm{d}\mu(s) \ .$$

Proof This is simply Fubini's theorem. Consider the set $\Theta = \{(s,\omega); s \in T_\omega\} \subset T \times \Omega$. Then

$$\mu \otimes \mathsf{P}(\Theta) = \int_T \mathsf{P}(s \in T_\omega)\mathrm{d}\mu(s) = \int_T p(s)\mathrm{d}\mu(s)$$

and also

$$\mu \otimes \mathsf{P}(\Theta) = \int_\Omega \mu(T_\omega)\mathrm{d}\mathsf{P}(\omega) = \mathsf{E}\mu(T_\omega) \ . \qquad \square$$

The following result quantifies that $\mathsf{E}\mu(T_\omega)$ is small if $p(s)$ is typically small:

Lemma 7.6.2 *Consider a subset A of T and assume that for a certain $\epsilon > 0$ we have $s \notin A \Rightarrow p(s) \le \epsilon$. Then*

$$\mathsf{E}\mu(T_\omega) \le \mu(A) + \epsilon \ . \tag{7.103}$$

Proof Since $p(s) \le 1$ for any $s \in T$ and $p(s) \le \epsilon$ for $s \in A^c$, we have

$$\int_T p(s)\mathrm{d}\mu(s) = \int_A p(s)\mathrm{d}\mu(s) + \int_{A^c} p(s)\mathrm{d}\mu(s) \le \mu(A) + \epsilon \ . \qquad \square$$

We will use this result when ϵ is overwhelmingly small, say $\epsilon = 1/N_n$. In that case we will be able to show that $\mu(T_\omega)$ is small with overwhelming probability simply by using Markov's inequality as in the following:

Lemma 7.6.3 *Assume that for some $\epsilon > 0$, there is a subset A of T with $\mu(A) \le \epsilon$ and $p(s) \le \epsilon$ for $s \notin A$. Then*

$$\mathsf{P}(\mu(T_\omega) \ge 2\sqrt{\epsilon}) \le \sqrt{\epsilon} \ .$$

Proof Indeed $\mathsf{E}\mu(T_\omega) \le 2\epsilon$ by Lemma 7.6.2, and the conclusion by Markov's inequality.

Generally speaking, the use of Fubini's theorem in the present situation is often precious, as in the following result:

Lemma 7.6.4 *Let $c = \mathsf{E}\mu(T_\omega)$. Then for $b < c$ we have*

$$\mathsf{P}(\mu(T_\omega) \ge b) \ge \frac{c-b}{1-b} \ .$$

Proof Denoting by Ω_0 the event $\mu(T_\omega) \ge b$ we write (using also that $\mu(T_\omega) \le 1$)

$$c = \mathsf{E}\mu(T_\omega) = \mathsf{E}\mathbf{1}_{\Omega_0}\mu(T_\omega) + \mathsf{E}\mathbf{1}_{\Omega_0^c}\mu(T_\omega) \le \mathsf{P}(\Omega_0) + b(1 - \mathsf{P}(\Omega_0)) . \qquad \square$$

Exercise 7.6.5 Consider numbers $0 < b, c \le 1$, a set T and a probability measure μ on T. For each $t \in T$ consider an event Ξ_t with $\mathsf{P}(\Xi_t) \ge c$. Then the event Ξ defined by $\mu(\{t; \omega \in \Xi_t\}) \ge b$ satisfies $\mathsf{P}(\Xi) \ge (c - b)/(1 - b)$.

7.7 Proofs, Lower Bounds

As we already explained, the main idea is to work given the r.v.s Z_i. The way to explain the proof strongly depends on whether one assumes that the reader has internalized the basic mechanisms of dealing with two independent sources of randomness or whether one does not make this assumption. Trying to address both classes of readers, we will give full technical details in the first result where the two sources of randomness really occur, Lemma 7.7.5.[36]

Let us define the distance $d_\omega(s, t)$ on T by

$$d_\omega(s, t)^2 = \sum_i |Z_i(s) - Z_i(t)|^2 . \tag{7.104}$$

This distance depends on the random quantities Z_i, so it is a random distance. Here and in the rest of the chapter, the letter ω symbolizes the randomness of the Z_i, so an implicit feature of the notation $d_\omega(s, t)$ is that this distance depends only on the randomness of the Z_i but not on the randomness of the ε_i. One should form the following mental picture: working given the r.v.s Z_i means working at a fixed value of ω.

Our goal now is to control $\gamma_2(T, d_\omega)$ from below.[37] The plan is to control d_ω from below and to show that consequently the balls with respect to d_ω are small. The basic estimate is as follows:

Lemma 7.7.1 *Assume that for a certain $j \in \mathbb{Z}$, the point $s \in T$ satisfies*

$$\varphi_{j+1}(s, 0) \ge 2^n . \tag{7.105}$$

[36] The mathematics involved are no more complicated than Fubini's theorem.

[37] This quantity involves only the randomness of the Z_i.

Then

$$\mathsf{P}(d_\omega(s,0) \le 2^{n/2-1} r^{-j-1}) \le \mathsf{P}\Big(\sum_i |r^{j+1}(Z_i(s) - Z_i(0))|^2 \wedge 1 \le 2^{n-2}\Big)$$
$$\le N_{n-2}^{-1} . \qquad (7.106)$$

The first inequality is obvious from the definition (7.104) of d_ω since when $d_\omega(s,0) \le r^{-j-1}2^{n/2-1}$ we have $\sum_i |r^{j+1}(Z_i(s) - Z_i(0))|^2 \le 2^{n-2}$. The proof of the second inequality requires elementary probabilistic inequalities to which we turn now.[38] These estimates will be used many times.

Lemma 7.7.2 *Consider independent r.v.s* $(W_i)_{i\ge1}$, *with* $0 \le W_i \le 1$.

(a) If $4A \le \sum_{i\ge1} \mathsf{E}W_i$, *then*

$$\mathsf{P}\Big(\sum_{i\ge1} W_i \le A\Big) \le \exp(-A) .$$

(b) If $A \ge 4\sum_{i\ge1} \mathsf{E}W_i$, *then*

$$\mathsf{P}\Big(\sum_{i\ge1} W_i \ge A\Big) \le \exp\Big(-\frac{A}{2}\Big) .$$

Proof

(a) Since $1 - x \le e^{-x} \le 1 - x/2$ for $0 \le x \le 1$, we have

$$\mathsf{E}\exp(-W_i) \le 1 - \frac{\mathsf{E}W_i}{2} \le \exp\Big(-\frac{\mathsf{E}W_i}{2}\Big)$$

and thus

$$\mathsf{E}\exp\Big(-\sum_{i\ge1} W_i\Big) \le \exp\Big(-\frac{1}{2}\sum_{i\ge1}\mathsf{E}W_i\Big) \le \exp(-2A) .$$

We conclude with the inequality $\mathsf{P}(Z \le A) \le \exp A\, \mathsf{E}\exp(-Z)$.

(b) Observe that $1 + x \le e^x \le 1 + 2x$ for $0 \le x \le 1$, so, as before,

$$\mathsf{E}\exp\sum_{i\ge1} W_i \le \exp 2\sum_{i\ge1}\mathsf{E}W_i \le \exp\frac{A}{2}$$

and we use now that $\mathsf{P}(Z \ge A) \le \exp(-A)\mathsf{E}\exp Z$. □

[38] Much more general and sharper results exist in the same direction, but the simple form we provide suffices for our needs.

Proof of Lemma 7.7.1 Let $W_i = |r^{j+1}(Z_i(s) - Z_i(0))|^2 \wedge 1$ and $A = 2^{n-2}$. Then $\sum_i \mathsf{E}W_i = \varphi_{j+1}(s,0) \geq 2^n = 4A$. The result then follows from Lemma 7.7.2(a). □

We can now give the central argument.

Lemma 7.7.3 *With probability* $\geq 1/2$ *it occurs that*

$$\sum_{n\geq 5} 2^n r^{-\bar{j}_n} \leq L r \gamma_2(T, d_\omega) . \tag{7.107}$$

The idea is very simple. Assuming $\bar{j}_n < \infty$, if a point $s \in T$ satisfies

$$\varphi_{\bar{j}_n+1}(s,0) \geq 2^n , \tag{7.108}$$

(7.106) shows that for most of the choices of the randomness of ω it holds that $d_\omega(s,0) \geq 2^{n/2-1} r^{-\bar{j}_n-1}$. The definition of $\bar{j}_n$ shows that all but very few of the points s satisfy (7.108). Thus for most of the choices of the randomness of the Z_i, there will be only few points in T which satisfy $d_\omega(s,0) \leq 2^{n/2-1} r^{-\bar{j}_n-1}$, and this certainly contributes to make $\gamma_2(T, d_\omega)$ large. Using this information for many values of n at the same time carries the day. The magic is that all the estimates fall very nicely into place.

We now start the proof of Lemma 7.7.3. The sequence $(\bar{j}_n)_{n\geq 0}$ is obviously non-decreasing and $\bar{j}_n = \infty$ for n large enough, because since we consider a finite sum, say of N terms, for any value of j we have $\varphi_j(s,t) \leq N$. There is nothing to prove if $\bar{j}_n = \infty$ for $n \geq 5$, so we may consider the largest integer $n_0 \geq 5$ such that $\bar{j}_n < \infty$ for $n \leq n_0$. Then $\sum_{n\geq 5} 2^n r^{-\bar{j}_n} = \sum_{5\leq n\leq n_0} 2^n r^{-\bar{j}_n}$.

Lemma 7.7.4 *For* $n \leq n_0$ *the event* Ξ_n *defined by*

$$\mu\Big(\Big\{s \in T \,;\, \sum_i |r^{\bar{j}_n+1}(Z_i(s) - Z_i(0))|^2 \wedge 1 \leq 2^{n-2}\Big\}\Big) < \frac{1}{N_{n-3}} \tag{7.109}$$

satisfies $\mathsf{P}(\Xi_n) \geq 1 - 2/N_{n-3}$.

Proof We think of n as fixed and we follow the strategy of Lemma 7.6.3 for the set $T_\omega = \{s \in T \,;\, \sum_i |r^{\bar{j}_n+1}(Z_i(s) - Z_i(0))|^2 \wedge 1 \leq 2^{n-2}\}$, where ω symbolizes the randomness of the Z_i. That is, we bound $\mathsf{E}\mu(T_\omega)$ and we use Markov's inequality.

The definition of $\bar{j}_n$ implies that the set $A = \{s \,;\, \varphi_{\bar{j}_n+1}(s,0) \leq 2^n\}$ satisfies

$$\mu(A) < N_n^{-1} . \tag{7.110}$$

Next, if we assume that $s \notin A$ then

$$\varphi_{\bar{j}_n+1}(s,0) = \sum_i \mathsf{E}(|r^{\bar{j}_n+1}(Z_i(s) - Z_i(0))|^2 \wedge 1) \geq 2^n ,$$

and by (7.106) it holds that $p(s) := \mathsf{P}(s \in T_\omega) \le 1/N_{n-2}$. Using (7.103) for $\epsilon = 1/N_{n-2}$, we then have

$$\mathsf{E}\mu(T_\omega) \le \mu(A) + \epsilon \le \frac{1}{N_n} + \frac{1}{N_{n-2}} < \frac{2}{N_{n-2}} , \tag{7.111}$$

where we have also used (7.110). Thus, by Markov's inequality, we have $\mathsf{P}(\mu(T_\omega) \ge 1/N_{n-3}) \le 2N_{n-3}/N_{n-2}$ and then

$$\mathsf{P}(\mu(T_\omega) < 1/N_{n-3}) \ge 1 - 2N_{n-3}/N_{n-2} = 1 - 2/N_{n-3} . \qquad \square$$

***Proof of Lemma** 7.7.3* As a consequence of Lemma 7.7.4 the event

$$\Xi := \bigcap_{5 \le n \le n_0} \Xi_n \tag{7.112}$$

satisfies $\mathsf{P}(\Xi^c) \le \sum_{n\ge 5} \mathsf{P}(\Xi_n^c) \le \sum_{n\ge 5} N_{n-3}^{-1} \le 1/2$ so that $\mathsf{P}(\Xi) \ge 1/2$. Moreover, since

$$\sum_i |r^{\bar{j}_n+1}(Z_i(s) - Z_i(t))|^2 \wedge 1 \le \sum_i |r^{\bar{j}_n+1}(Z_i(s) - Z_i(t))|^2 = r^{2\bar{j}_n+2} d_\omega(s,t)^2 ,$$

(7.109) yields

$$\mu\big(\{s \in T \,;\, d_\omega(s,0) \le r^{-\bar{j}_n-1}2^{n/2-1}\}\big) < \frac{1}{N_{n-3}} .$$

It follows that when Ξ occurs the number $\epsilon_n = \epsilon_n(\omega)$ as in (7.3) satisfies $\epsilon_{n-3}(\omega) \ge r^{-\bar{j}_n-1}2^{n/2-1}$. Consequently

$$\sum_{5\le n\le n_0} 2^n r^{-\bar{j}_n} \le Lr \sum_{n\ge 0} 2^{n/2}\epsilon_n(\omega) \le Lr\gamma_2(T, d_\omega) ,$$

where we use (7.4) in the last equality. We have proved that (7.107) holds for $\omega \in \Xi$ and hence with probability $\ge 1/2$. $\square$

Lemma 7.7.5 *There exists a constant $\alpha_1 > 0$ such that if*

$$\mathsf{P}\Big(\Big\| \sum_i \varepsilon_i Z_i \Big\| \ge M\Big) < \alpha_1 ,$$

then

$$\sum_{n\ge 5} 2^n r^{-\bar{j}_n} \le LrM . \tag{7.113}$$

This is the first result involving both the randomness of the Z_i and the randomness of the ε_i. The proof consists in proving the existence of a constant $\alpha_1 > 0$ such that

$$\mathsf{P}\Big(\Big\|\sum_i \varepsilon_i Z_i\Big\| \geq \frac{1}{Lr}\sum_{n\geq 5} 2^n r^{-\bar{j}_n}\Big) \geq \alpha_1 . \tag{7.114}$$

Proof for the Probabilist The inequality (7.114) follows by using (7.7.3) first and then Lemma 7.4.6 given the r.v.s Z_i. □

Proof for the Novice in Probability This proof consists in detailing the mechanism at hand in the previous argument. We assume as on page 214 that the underlying probability space is a product $\Omega \times \Omega'$, with a product probability $\mathsf{P} = \mathsf{P}_Z \otimes \mathsf{P}_\varepsilon$, and that if (ω, ω') is the generic point of this product, then $Z_i = Z_{i,\omega}$ depends on ω only and $\varepsilon_i = \varepsilon_{i,\omega'}$ depends on ω' only. By Fubini's theorem, for a set $A \subset \Omega \times \Omega'$,

$$\mathsf{P}(A) = \int \mathsf{P}_\varepsilon(\{\omega' \in \Omega' \,;\, (\omega, \omega') \in A\}) d\mathsf{P}_Z(\omega) .$$

In particular, for any set $B \subset \Omega$ we have

$$\mathsf{P}(A) \geq \mathsf{P}_Z(B) \inf_{\omega \in B} \mathsf{P}_\varepsilon(\{\omega' \in \Omega' \,;\, (\omega, \omega') \in A\}) . \tag{7.115}$$

By Lemma 7.7.3 the set $B = \{\omega \in \Omega; \sum_{n\geq 5} 2^n r^{-\bar{j}_n} \leq Lr\gamma_2(T, d_\omega)\}$ satisfies $\mathsf{P}_Z(B) \geq 1/2$. Using (7.45) at a given value of ω we obtain

$$\mathsf{P}_\varepsilon\Big(\Big\{\omega' \in \Omega' \,;\, \Big\|\sum_i \varepsilon_i Z_i\Big\| \geq \frac{1}{L}\gamma_2(T, d_\omega)\Big\}\Big) \geq \frac{1}{L} ,$$

so that for $\omega \in B$ we have

$$\mathsf{P}_\varepsilon\big(\big\{\omega' \in \Omega' \,;\, (\omega, \omega') \in A\big\}\big) \geq \frac{1}{L} ,$$

where $A = \{(\omega, \omega'); \|\sum_i \varepsilon_i Z_i\| \geq (1/Lr)\sum_{n\geq 5} 2^n r^{-\bar{j}_n}\}$ and (7.115) proves (7.114). □

It is unfortunate that in (7.113) the summation starts at $n = 5$ for otherwise we would be done. The next result addresses this problem.

Proposition 7.7.6 *There exists a constant $\alpha_2 > 0$ with the following property. Assume that for a certain number M we have*

$$\mathsf{P}\Big(\Big\|\sum_i \varepsilon_i Z_i\Big\| > M\Big) \leq \alpha_2 . \tag{7.116}$$

Then the number $\bar{j}_0$ of (7.69) satisfies $r^{-\bar{j}_0} \leq LrM$.

Lemma 7.7.7 *Consider independent complex-valued r.v.s U_i and independent Bernoulli r.v.s ε_i that are independent of the r.v.s U_i. Assume that*

$$\sum_{i\geq 1} \mathsf{E}(|U_i|^2 \wedge 1) \geq 1 . \tag{7.117}$$

Then

$$\mathsf{P}\Big(\Big|\sum_{i\geq 1} \varepsilon_i U_i\Big| \geq \frac{1}{L}\Big) \geq \frac{1}{L} . \tag{7.118}$$

Proof We use Lemma 7.7.2(a) with $W_i = |U_i|^2 \wedge 1$ and $A = 1/4$ to obtain

$$\mathsf{P}\Big(\sum_{i\geq 1} |U_i|^2 \wedge 1 \geq \frac{1}{4}\Big) \geq \frac{1}{L} . \tag{7.119}$$

Now, (6.16) implies that $\mathsf{P}_\varepsilon(|\sum_i \varepsilon_i U_i| \geq (\sum_i |U_i|^2)^{1/2}/L) \geq 1/L$, where P_ε denotes the conditional probability given the randomness of the $(U_i)_{i\geq 1}$. Combining with (7.119) concludes the proof. □

Exercise 7.7.8 Write all details of the previous argument in the spirit of the second proof of Lemma 7.7.5.

Exercise 7.7.9 Assuming $\sum_{i\geq 1} \mathsf{E}(|U_i|^2 \wedge 1) \geq \beta > 0$ prove that then $\mathsf{P}(|\sum_{i\geq 1} \varepsilon_i U_i| \geq 1/K) \geq 1/K$ where K depends on β only.

Proof of Proposition 7.7.6 Let us denote by L_0 the constant of (7.118). We will show that $\alpha_2 = 1/L_0$ works. Since

$$\Big|\sum_i \varepsilon_i (Z_i(s) - Z_i(t))\Big| \leq 2\Big\|\sum_i \varepsilon_i Z_i\Big\| ,$$

(7.116) implies

$$\forall\, s, t \in T\,,\ \mathsf{P}\Big(\Big|\sum_i \varepsilon_i (Z_i(s) - Z_i(t))\Big| \geq 2M\Big) \leq \mathsf{P}\Big(\Big\|\sum_i \varepsilon_i Z_i\Big\| \geq M\Big) < \frac{1}{L_0}. \tag{7.120}$$

The condition $\mathsf{P}(|\sum_i \varepsilon_i (Z_i(s) - Z_i(t))| \geq 2M) < 1/L_0$ means that if we set $U_i = w(Z_i(s) - Z_i(t))$ where $w = 1/(2L_0 M)$, we have $\mathsf{P}(\sum_i \varepsilon_i U_i \geq 1/L_0) < 1/L_0$. Consequently (7.118) fails when $U_i = w(Z_i(s) - Z_i(t))$ and therefore Lemma 7.7.7 implies

$$\forall\, s, t \in T\,,\ \sum_i \mathsf{E}(|w(Z_i(s) - Z_i(t))|^2 \wedge 1) < 1 . \tag{7.121}$$

Let j^* be the largest integer with $r^{j^*} \le w$, so that $r^{j^*+1} > w$ and thus $r^{-j^*} \le r/w \le 2L_0 rM = LrM$. Since $r^{j^*} \le w$, (7.121) implies

$$\forall s, t \in T\ ,\ \varphi_{j^*}(s,t) < 1\ .$$

Consequently, $\bar{j}_0 \ge j^*$ and therefore $r^{-\bar{j}_0} \le r^{-j^*} \le LrM$. □

Exercise 7.7.10 Given a number $\beta > 0$ prove that there is a number $\alpha(\beta)$ depending on β only such that if $\mathsf{P}(\|\sum_{i\ge i} \varepsilon_i Z_i\| \ge M) \ge \alpha(\beta)$, there exists j_0 with $\varphi_{j_0}(s,t) \le \beta$ for $s, t \in T$ and $r^{-j_0} \le KM$ where K depends on β only.

Proof of Theorem 7.5.1 We show that any constant $\alpha_0 < \min(\alpha_1, \alpha_2)$ works. Let us assume that for a certain number M we have

$$\mathsf{P}\Big(\Big\|\sum_i \varepsilon_i Z_i\Big\| \ge M\Big) \le \alpha_0\ .$$

It then follows from (7.113) that $\sum_{n\ge 5} 2^n r^{-\bar{j}_n} \le LrM$ and from Proposition 7.7.6 that $r^{-\bar{j}_0} \le LrM$. Since $\sum_{n\le 4} 2^n r^{-\bar{j}_n} \le Lr^{-\bar{j}_0}$, we have proved that $\sum_{n\ge 0} 2^n r^{-\bar{j}_n} \le LrM$. □

7.8 Proofs, Upper Bounds

7.8.1 *Road Map*

We are done with the lower bounds. It is not obvious yet, but the arguments for these lower bounds are potentially very general, and we will meet them later in Chap. 11. Our goal now is to work toward the upper bounds, proving Theorem 7.5.5. The crucial case of the result is when the Z_i are not random, $Z_i = a_i \chi_i$ for a complex number a_i and a character χ_i.

Starting now, the arguments are somewhat specific to random trigonometric sums and use translation invariance in a fundamental way. Thus, the remainder of this chapter may be skipped by the reader who is not interested in random Fourier series per se, although it should certainly be very helpful for the sequel to understand Theorem 7.5.14 (the decomposition theorem) since similar but far more difficult results are the object of much later work.

The crucial inequality (7.145) below pertains to the case $Z_i = a_i \chi_i$ for a complex number a_i and a character χ_i. The basic mechanism at work in the proof of (7.145) is a "miniature version" of Theorem 7.5.14. Once we have this inequality, more general upper bounds we need will be obtained by using it at given values of the Z_i. This is the object of Sect. 7.8.6. We will still need to complement these bounds by very classical considerations in Sect. 7.8.7 before we can finally prove Theorem 7.5.8.

7.8.2 A Key Step

The importance of the following result will only become clear later.

Theorem 7.8.1 *Consider characters χ_i and assume that for a certain subset A of T and each i we have $|\int_A \chi_i(s)d\mu(s)| \ge \mu(A)/2$. Then given numbers a_i and independent standard Gaussian r.v.s g_i, we have*

$$\mathsf{E}\Big\|\sum_i a_i g_i \chi_i\Big\| \le L\Big(\sum_i |a_i|^2\Big)^{1/2}\sqrt{\log(2/\mu(A))}\,. \tag{7.122}$$

Having analyzed that this property was central for the decomposition theorem but being unable to prove it, I submitted this question to Gilles Pisier. He immediately pointed out that Theorem 7.1 of his paper [86] provides a positive answer in the case where the a_i are all equal. Analysis of the arguments of [86] then easily led to the proof which we present here. To prove (7.122) we can assume that $\chi_i \neq \mathbf{1}$ for each i by bounding separately the term where $\chi_i = \mathbf{1}$. Setting $v_i = \int_A \chi_i(s)d\mu(s)$, so that $|v_i| \ge \mu(A)/2$, we will compare suitable upper and lower bounds for the quantity $\mathsf{E}\|\sum_i a_i v_i g_i \chi_i\|$.

The lower bound is easy. Consider the distances d and $\bar{d}$ given by $d(s,t)^2 = \sum_i |a_i|^2|\chi_i(s) - \chi_i(t)|^2$ and $\bar{d}(s,t)^2 = \sum_i |a_i v_i|^2|\chi_i(s) - \chi_i(t)|^2$. Then since $|v_i| \ge \mu(A)/2$ we have $\bar{d}(s,t) \ge \mu(A)d(s,t)/2$ and thus $\mu(A)\gamma_2(T,d) \le L\gamma_2(T,\bar{d})$.[39] Furthermore $\mathsf{E}\|\sum_i a_i g_i \chi_i\| \le L\gamma_2(T,d)$ by (7.33) and $\gamma_2(T,\bar{d}) \le L\mathsf{E}\|\sum_i a_i v_i g_i \chi_i\|$ by (7.13). Consequently,

$$\frac{\mu(A)}{L}\mathsf{E}\Big\|\sum_i a_i g_i \chi_i\Big\| \le \mathsf{E}\Big\|\sum_i a_i v_i g_i \chi_i\Big\|\,. \tag{7.123}$$

The upper bound on the quantity $\mathsf{E}\|\sum_i a_i v_i g_i \chi_i\|$ is provided by the following:

Lemma 7.8.2 *We have*

$$\mathsf{E}\Big\|\sum_i a_i v_i g_i \chi_i\Big\| \le L\Big(\sum_i |a_i|^2\Big)^{1/2}\mu(A)\sqrt{\log(2/\mu(A))}\,. \tag{7.124}$$

Proof of Theorem 7.8.1 Combine (7.123) and (7.124). □

Before we can prove (7.124) we need two simple lemmas.

Lemma 7.8.3 *Consider complex numbers b_i with $\sum_i |b_i|^2 \le 1/8$. Then $\mathsf{E}\exp|\sum_i b_i g_i|^2 \le L$.*

Proof Separating the real and imaginary parts, $|\sum_i b_i g_i|^2 = h_1^2 + h_2^2$ where h_1 h_2 are Gaussian (not necessarily independent) with $\mathsf{E}h_1^2 \le 1/8$ and

[39] See Exercise 2.7.4.

$\mathsf{E}h_2^2 \le 1/8$ and we simply use the Cauchy-Schwarz inequality $\mathsf{E}\exp(h_1^2 + h_2^2) \le (\mathsf{E}\exp 2h_1^2)^{1/2}(\mathsf{E}\exp 2h_2^2)^{1/2}$. □

Lemma 7.8.4 *For any function $f \ge 0$ on G,*[40] *any set A with $\mu(A) > 0$ and any number $C > 0$ we have*

$$\int_A f\mathrm{d}\mu \le 4C\mu(A)\sqrt{\log(2/\mu(A))}\int \exp(f^2/C^2)\mathrm{d}\mu\ . \tag{7.125}$$

Proof The function $x \mapsto x^{-1}\exp x^2$ increases for $x \ge 1$, so that given a number $0 < a \le 1$, for $x \ge 2\sqrt{\log(2/a)}$, we have $x \le a\exp x^2$ because this is true for $x = 2\sqrt{\log(2/a)}$. Consequently for $x \ge 0$ we have $x \le 2\sqrt{\log(2/a)} + a\exp x^2$. Therefore we have

$$f \le 2\sqrt{\log(2/\mu(A))} + \mu(A)\exp f^2\ .$$

Integration over A gives

$$\begin{aligned}\int_A f\mathrm{d}\mu &\le 2\mu(A)\sqrt{\log(2/\mu(A))} + \mu(A)\int \exp f^2\mathrm{d}\mu\\ &\le 4\mu(A)\sqrt{\log(2/\mu(A))}\int \exp f^2\mathrm{d}\mu\ ,\end{aligned}$$

where we have used that $1 \le \int \exp f^2\mathrm{d}\mu$ for the first term and $1 \le 2\sqrt{\log(2/\mu(A))}$ for the second term. This proves (7.125) by replacing f by f/C. □

Proof of Lemma 7.8.2 Recalling the value of v_i and that $\chi_i(s)\chi_i(t) = \chi_i(s+t)$ we have

$$\sum_i a_i v_i g_i \chi_i(t) = \int_A \sum_i a_i g_i \chi_i(t+s)\mathrm{d}\mu(s) = \int_{A+t}\sum_i a_i g_i \chi_i(s)\mathrm{d}\mu(s)\ .$$

Since $\mu(A) = \mu(A+t)$, using (7.125) in the inequality below we obtain

$$\begin{aligned}\Big\|\sum_i a_i v_i g_i \chi_i\Big\| &= \sup_t\Big|\sum_i a_i v_i g_i \chi_i(t)\Big| = \sup_t\Big|\int_{A+t}\sum_i a_i g_i\chi_i(s)\mathrm{d}\mu(s)\Big|\\ &\le LC\mu(A)\sqrt{\log(2/\mu(A))}\int \exp\Big(\Big|\sum_i a_i g_i \chi_i(s)\Big|^2/C^2\Big)\mathrm{d}\mu(s)\ ,\end{aligned} \tag{7.126}$$

[40] The group structure is not being used here, and this lemma is a general fact of measure theory.

where $C = 4(\sum_i |a_i|^2)^{1/2}$. Lemma 7.8.3 used for $b_i = a_i/C$ shows that for each s we have $\mathsf{E}\exp(|\sum_i a_i g_i \chi_i(s)|^2/C^2) \le L$ so that taking expectation in (7.126) concludes the proof of (7.124). □

The result proved in the following exercise was discovered while trying to prove (7.122) and is in the same general direction:

Exercise 7.8.5

(a) Prove that there is a number $\alpha_0 > 0$ such that if z, z' are complex numbers of modulus 1 with $|z-1| \le \alpha_0$ and $|z'-1| \le \alpha_0$ then $|z^5 - z'^5| \ge 4|z-z'|$.
(b) Consider $0 < \alpha \le \alpha_0$, characters $(\chi_i)_{i \le N}$ on a compact group T and the set $A = \{t \in T, \forall i \le N, |\chi_i(t) - 1| \le \alpha\}$. Consider the set $B = \{t \in T, \forall i \le N, |\chi_i(t) - 1| \le \alpha/2\}$. Consider a subset U of A such that $s, t \in U, s \ne t \Rightarrow s - t \notin B$. Prove that the sets $5t + A$ for $t \in U$ are disjoint.
(c) Prove that $\mu(B) \ge \mu(A)^2$.

7.8.3 *Road Map: An Overview of Decomposition Theorems*

Given a stochastic process $(X_t)_{t\in T}$, we would like to decompose it as a sum of two (or more) simpler pieces, $X_t = X_t^1 + X_t^2$, where, say, it will be far easier to control the size of each of the processes $(X_t^1)_{t\in T}$ and $(X_t^2)_{t\in T}$ than that of $(X_t)_{t\in T}$ because each of the two pieces can be controlled by a specific method. Furthermore, when the process $(X_t)_{t\in T}$ has a certain property, say $X_t = \sum_i Z_i(t)$ is the sum of independent terms, we would like each of the pieces X_t^1 and X_t^2 to have the same property. This will be achieved in Chap. 11 in a very general setting and in the present chapter in the case where $Z_i \in \mathbb{C}G$. In the next section, the decomposition will take a very special form, of the type $X_t = \sum_{i\in I} a_i \chi_i(t) = X_t^1 + X_t^2$ where $X_t^j = \sum_{i\in I_j} a_i \chi_i(t)$ for two disjoint sets I_1, I_2 with $I = I_1 \cup I_2$. This is accidental. In general, the decomposition is more complicated, each piece Z_i has to be split in a non-trivial manner, as will be done in Sect. 7.9.

Although this will not be used before Chap. 11, to help the reader form a correct picture, let us explain that the decomposition of stochastic processes in a sum of simpler pieces is also closely related to the decomposition of sets of functions as in Sect. 6.7 or as in the more complex results of Chap. 9. This is because many of the processes we consider in this book (but *not* in this chapter) are naturally indexed by sets of functions (in this chapter the process is indexed by the group T). A prime example of this is the process $X_t = \sum_{i\in I} g_i t_i$ (where (g_i) are independent standard Gaussian r.v.s) which is indexed by $\ell^2(I)$ (where I is a finite set). For such processes the natural way to decompose X_t into a sum of two pieces is to decompose t itself into such a sum. The main theorems of Chap. 11 are precisely obtained by an abstract version of this method.

7.8.4 Decomposition Theorem in the Bernoulli Case

Our main result in this section is closely related to Theorem 7.5.14 in the case where Z_i is nonrandom, $Z_i = a_i \chi_i$. It will be the basis of our upper bounds. We consider a finite set I, numbers $(a_i)_{i\in I}$, characters $(\chi_i)_{i\in I}$ with $\chi_i \neq \mathbf{1}$, and for $j \in \mathbb{Z}$ we define

$$\psi_j(s,t) = \sum_i |r^j a_i(\chi_i(s) - \chi_i(t))|^2 \wedge 1 \;, \tag{7.127}$$

where as usual $\sum_i$ is a shorthand for $\sum_{i\in I}$. Consider a parameter $w \geq 1$ and for $n \geq 0$ integers $j_n \in \mathbb{Z}$. Consider the set

$$D_n = \{t \in T \;;\; \psi_{j_n}(t,0) \leq w2^n\} \;. \tag{7.128}$$

Theorem 7.8.6 *Assume the following conditions:*

$$\mu(D_0) \geq \frac{3}{4} \;, \tag{7.129}$$

$$\forall n \geq 1 \;,\; \mu(D_n) \geq N_n^{-1} \;. \tag{7.130}$$

Then we can decompose I as a disjoint union of three subsets I_1, I_2, I_3 with the following properties:

$$I_1 = \{i \in I \;;\; |a_i| \geq r^{-j_0}\} \;, \tag{7.131}$$

$$\sum_{i\in I_2} |a_i| \leq Lw \sum_{n\geq 0} 2^n r^{-j_n} \;, \tag{7.132}$$

$$\mathsf{E}\Big\| \sum_{i\in I_3} a_i g_i \chi_i \Big\| \leq L\sqrt{w} \sum_{n\geq 0} 2^n r^{-j_n} \;, \tag{7.133}$$

where $(g_i)_{i\in I_3}$ are independent standard Gaussian r.v.s.

To prepare for the proof, a basic observation is that as a consequence of (7.128), we have

$$\int_{D_n} \psi_{j_n}(s,0)\mathrm{d}\mu(s) \leq w2^n \mu(D_n) \;,$$

and using the definition of ψ_j this means that

$$\int_{D_n} \sum_i |r^{j_n} a_i(\chi_i(s) - 1)|^2 \wedge 1 \mathrm{d}\mu(s) \leq w2^n \mu(D_n) \;. \tag{7.134}$$

For each $n \ge 0$ we define

$$U_n = \left\{ i \in I ; \int_{D_n} |\chi_i(s) - 1|^2 \mathrm{d}\mu(s) \ge \mu(D_n) \right\} . \tag{7.135}$$

The idea of this definition is that if $i \in U_n$, then $\chi_i(s)$ is not too close to 1 on D_n so that $\int_{D_n} |r^{j_n} a_i(\chi_i(s) - 1)|^2 \wedge 1 \mathrm{d}\mu(s)$ should be about $\mu(D_n)|(r^{j_n} a_i|^2 \wedge 1)$. We will make this explicit in Lemma 7.8.9, but before that we stress the miracle of this definition: if $i \notin U_n$, we have exactly the information we need to appeal to Theorem 7.8.1.

Lemma 7.8.7 *If $i \notin U_n$ then $|\int_{D_n} \chi_i(s) \mathrm{d}\mu(s)| \ge \mu(D_n)/2$.*

Proof Indeed, since $i \notin U_n$,

$$\mu(D_n) > \int_{D_n} |\chi_i(s) - 1|^2 \mathrm{d}\mu(s) = 2\mu(D_n) - 2\mathrm{Re} \int_{D_n} \chi_i(s) \mathrm{d}\mu(s) . \qquad \square$$

Lemma 7.8.8 *For complex numbers x, y with $|y| \le 4$ we have*

$$|xy| \wedge 1 \ge |y|(|x| \wedge 1)/4 . \tag{7.136}$$

Proof We have $|xy| \wedge 1 \ge ((|x| \wedge 1)|y|) \wedge 1$ and $|y|(|x| \wedge 1) \le 4$. We then use that for $0 \le a \le 4$ we have $a \wedge 1 \ge a/4$. $\square$

Lemma 7.8.9 *We have*

$$\sum_{i \in U_n} |r^{j_n} a_i|^2 \wedge 1 \le 4w2^n . \tag{7.137}$$

Proof According to (7.136) we have

$$|r^{j_n} a_i(\chi_i(s) - 1)|^2 \wedge 1 \ge \frac{1}{4} |\chi_i(s) - 1|^2 (|r^{j_n} a_i|^2 \wedge 1) ,$$

so that (7.134) implies

$$\sum_i (|r^{j_n} a_i|^2 \wedge 1) \int_{D_n} |\chi_i(s) - 1|^2 \mathrm{d}\mu(s) \le 4w2^n \mu(D_n) ,$$

from which the result follows since $\int_{D_n} |\chi_i(s) - 1|^2 \mathrm{d}\mu(s) \ge \mu(D_n)$ for $i \in U_n$. $\square$

The next task is to squeeze out all the information we can from (7.137).

Lemma 7.8.10 *We have $U_0 = I$.*

Proof We have $\int_{T\setminus D_0}|\chi_i(t)-1|^2\mathrm{d}\mu(s)\le 4\mu(T\setminus D_0)\le 1$ since $|\chi_i(s)-1|^2\le 4$ and $\mu(T\setminus D_0)\le 1/4$ by (7.129). Thus $\int_{D_0}|\chi_i(s)-1|^2\mathrm{d}\mu(s)\ge\int_T|\chi_i(s)-1|^2\mathrm{d}\mu(s)-1\ge 1\ge\mu(D_0)$ because $\int_T|\chi_i(s)-1|^2\mathrm{d}\mu(s)=2$. □

Let us define

$$V_n=\bigcap_{0\le k\le n}U_k$$

so that $V_0=U_0=I$ and for $i\in I$ it makes sense to define

$$u_i=\inf\{r^{-j_n}\ ;\ i\in\mathbb{N}\ ,\ i\in V_n\}\ . \tag{7.138}$$

We then define

$$I_2=\{i\in I\ ;\ u_i\le|a_i|<r^{-j_0}\}\ .$$

Lemma 7.8.11 *We have*

$$\sum_{i\in I_2}|a_i|\le Lw\sum_{n\ge 0}2^n r^{-j_n}\ . \tag{7.139}$$

Proof It follows from (7.137) that

$$\operatorname{card}\{i\in U_n\ ;\ |a_i|\ge r^{-j_n}\}\le 4w2^n \tag{7.140}$$

and consequently

$$\sum_{n\ge 1}r^{-j_{n-1}}\operatorname{card}\{i\in U_n\ ;\ |a_i|\ge r^{-j_n}\}\le Lw\sum_{n\ge 1}2^n r^{-j_{n-1}}\le Lw\sum_{n\ge 0}2^n r^{-j_n}\ . \tag{7.141}$$

Now, we have

$$\begin{aligned}\sum_{n\ge 1}r^{-j_{n-1}}\operatorname{card}\{i\in U_n\ ;\ |a_i|\ge r^{-j_n}\} &= \sum_{i\in I,n\ge 1}r^{-j_{n-1}}\mathbf{1}_{\{i\in U_n,|a_i|\ge r^{-j_n}\}} \qquad (7.142)\\ &\ge\sum_{i\in I_2}\sum_{n\ge 1}r^{-j_{n-1}}\mathbf{1}_{\{i\in U_n,|a_i|\ge r^{-j_n}\}}\ .\end{aligned}$$

For $i\in I_2$ we have $|a_i|\ge u_i$, and by definition of u_i, there exists n with $r^{-j_n}\le|a_i|$ and $i\in V_n$. Consider the smallest integer $k\ge 0$ such that $r^{-j_k}\le|a_i|$. Since $|a_i|<r^{-j_0}$ we obtain $k\ge 1$ so that $|a_i|\le r^{-j_{k-1}}$. Since $i\in V_n\subset U_k$ this shows that $\sum_{n\ge 1}r^{-j_{n-1}}\mathbf{1}_{\{i\in U_n,|a_i|\ge r^{-j_n}\}}\ge r^{-j_{k-1}}\ge|a_i|$. Thus the right-hand side of (7.142) is $\ge\sum_{i\in I_2}|a_i|$, completing the proof. □

Proof of Theorem 7.8.6 Let $I_3 = \{i \in I; |a_i| < u_i\}$ so that I is the disjoint union of I_1, I_2, I_3. It remains only to prove the inequality (7.133) involving the set I_3. For $n \geq 0$ let us set $W_n = I_3 \cap (V_n \setminus V_{n+1})$ so that for $i \in W_n$ we have $i \notin U_{n+1}$, and thus $|\int_{D_{n+1}} \chi_i(s) d\mu(s)| \geq \mu(D_{n+1})/2$ by Lemma 7.8.7. Since $\sqrt{\log(2/\mu(D_{n+1}))} \leq L2^{n/2}$ by (7.130), it then follows from (7.122) used with $A = D_{n+1}$ that

$$\mathsf{E}\Big\| \sum_{i \in W_n} a_i g_i \chi_i \Big\| \leq L2^{n/2} \Big(\sum_{i \in W_n} |a_i|^2 \Big)^{1/2} . \tag{7.143}$$

Let us bound the right-hand side. For $i \in W_n$ we have $i \in I_3$ so that $|a_i| < u_i$. Since $i \in V_n$ we have also $u_i \leq r^{-j_n}$. Thus $|a_i r^{j_n}| \leq 1$ and (7.137) implies $\sum_{i \in W_n} |a_i|^2 \leq Lw2^n r^{-2j_n}$. Finally (7.143) implies

$$\mathsf{E}\Big\| \sum_{i \in W_n} a_i g_i \chi_i \Big\| \leq L\sqrt{w} 2^n r^{-j_n} .$$

For $i \in I_3$ we have $u_i > 0$ since $|a_i| < u_i$ so that there is a largest n for which $i \in V_n$. Then $i \notin V_{n+1}$ so that $i \in V_n \setminus V_{n+1}$ and thus $i \in W_n$. We have proved that $I_3 = \cup_n W_n$. Use of the triangle inequality then implies

$$\mathsf{E}\Big\| \sum_{i \in I_3} a_i g_i \chi_i \Big\| \leq L\sqrt{w} \sum_{n \geq 0} 2^n r^{-j_n} .$$ □

Corollary 7.8.12 *Under the conditions of Theorem 7.8.6, we have*

$$\gamma_2(T, d_2) \leq Lw \sum_{n \geq 0} 2^n r^{-j_n} + L \sum_i |a_i| \mathbf{1}_{\{|a_i| \geq r^{-j_0}\}} , \tag{7.144}$$

where the distance d_2 is given by $d_2(s,t)^2 = \sum_i |a_i|^2 |\chi_i(s) - \chi_i(t)|^2$.

Proof For any set J we have $\mathsf{E}\| \sum_{i \in J} a_i g_i \chi_i \| \leq L \sum_{i \in J} |a_i|$. Using this for $J = I_1$ gives $\mathsf{E}\| \sum_{i \in I_1} a_i g_i \chi_i \| \leq L \sum_i |a_i| \mathbf{1}_{\{|a_i| \geq r^{-j_0}\}}$. Using it again for $J = I_2$ and combining with (7.132) gives $\mathsf{E}\| \sum_{i \in I_2} a_i g_i \chi_i \| \leq \sum_{i \in I_2} |a_i| \leq Lw \sum_{n \geq 0} 2^n r^{-j_n}$. Combining these two inequalities with (7.133) yields

$$\mathsf{E}\Big\| \sum_i a_i g_i \chi_i \Big\| \leq Lw \sum_{n \geq 0} 2^n r^{-j_n} + \sum_i |a_i| \mathbf{1}_{\{|a_i| \geq r^{-j_0}\}} .$$

The result then follows from (7.13). □

7.8.5 Upper Bounds in the Bernoulli Case

Theorem 7.8.13 *Under the conditions of Theorem 7.8.6 for each $p \geq 1$, it holds that*

$$\Big(\mathsf{E}\sup_{s\in T}\Big|\sum_i \varepsilon_i a_i(\chi_i(s)-\chi_i(0))\Big|^p\Big)^{1/p}$$
$$\leq K(r,p)w\sum_{n\geq 0}2^n r^{-j_n} + K(r)\sum_i |a_i|\mathbf{1}_{\{|a_i|\geq r^{-j_0}\}}\,. \tag{7.145}$$

Proof For a subset J of I define $\varphi(J) = (\mathsf{E}\sup_{s\in T}|\sum_{i\in J}\varepsilon_i a_i(\chi_i(s)-\chi_i(0))|^p)^{1/p}$. Observe first that as a consequence of the trivial fact that $|\sum_{i\in J}\varepsilon_i a_i(\chi_i(s)-\chi_i(0))| \leq 2\sum_{i\in J}|a_i|$, we have $\varphi(J) \leq 2\sum_{i\in J}|a_i|$. Thus $\varphi(I_1) \leq 2\sum_i |a_i|\mathbf{1}_{\{|a_i|\geq r^{-j_0}\}}$ and $\varphi(I_2) \leq Lw\sum_{n\geq 0}2^n r^{-j_n}$. Furthermore, by the triangle inequality if J_1 and J_2 are two disjoint subsets of I, we have $\varphi(J_1\cup J_2) \leq \varphi(J_1)+\varphi(J_2)$. Thus it suffices to prove (7.145) when the summation is restricted to I_3. It is then is a consequence of (2.66) applied to the process $X_t = \sum_{i\in I_3}\varepsilon_i a_i\chi_i(t)$. Indeed, according to the subgaussian inequality (6.2), this process satisfies the increment condition (2.4) with respect to the distance d_3 given by $d_3(s,t)^2 = \sum_{i\in I_3}|a_i(\chi_i(s)-\chi_i(t))|^2$. Furthermore $\gamma_2(T,d_3) \leq L\sqrt{w}\sum_{n\geq 0}2^n r^{-j_n}$ as follows from (7.133) and (7.13). □

7.8.6 The Main Upper Bound

In this section we state and prove our main upper bound, Theorem 7.8.14. It will follow from (7.145) given the randomness of the Z_i. We recall that E_ε denotes expectation in the r.v.s ε_i only. We recall that

$$\varphi_j(s,t) = \sum_{i\geq 1}\mathsf{E}(|r^j(Z_i(s)-Z_i(t))|^2\wedge 1)\,.$$

Theorem 7.8.14 *For $n \geq 0$ consider numbers $j_n \in \mathbb{Z}$. Assume that*

$$\forall s,t\in T\;,\;\varphi_{j_0}(s,t)\leq 1 \tag{7.146}$$

$$\forall n\geq 1\;,\;\mu(\{s\in T\;;\;\varphi_{j_n}(s,0)\leq 2^n\})\geq 2^{-2^n} = N_n^{-1}\,. \tag{7.147}$$

Then for each $p \geq 1$ we can write

$$\Big(\mathsf{E}_\varepsilon\sup_{s\in T}\Big|\sum_{i\geq 1}\varepsilon_i(Z_i(s)-Z_i(0))\Big|^p\Big)^{1/p} \leq Y_1+Y_2\,, \tag{7.148}$$

where

$$(\mathsf{E}Y_1^p)^{1/p} \le K(r,p)\sum_{n\ge 0} 2^n r^{-j_n} \;, \tag{7.149}$$

and

$$Y_2 \le K(r)\sum_{i\ge 1} |Z_i(0)|\mathbf{1}_{\{|Z_i(0)|\ge r^{-j_0}\}} \;. \tag{7.150}$$

In words, Y_2 collects the contributions of the "large" terms, and we control well all moments of Y_1. It is by no means obvious how to control the term Y_2, and it will be a separate task to learn how to do this. The difficulty of that task will depend on our goal. The simplest situation will be when we will apply this result to study the convergence of series: then $\mathsf{P}(Y_2 \ne 0)$ will be small.

The reader is invited to meditate on the strength of this result. In particular, we do not know how to deduce it from the decomposition theorem, even in the precise form of Corollary 7.9.2.[41]

The main step of the proof is as follows. It will allow us to use Theorem 7.8.13 given the randomness of the Z_i, which is symbolized by ω.

Proposition 7.8.15 *Under the conditions of Theorem 7.8.14 denote by $w(\omega) \in \mathbb{R}^+ \cup \{\infty\}$ the smallest number ≥ 1 for which*

$$\mu\Big(\Big\{s \in T \;;\; \sum_{i\ge 1} |r^{j_0}(Z_i(s) - Z_i(0))|^2 \wedge 1 \le w(\omega)\Big\}\Big) \ge 3/4 \;, \tag{7.151}$$

$$\forall n \ge 1 \;,\; \mu\Big(\Big\{s \in T \;;\; \sum_{i\ge 1} |r^{j_n}(Z_i(s) - Z_i(0))|^2 \wedge 1 \le w(\omega)2^n\Big\}\Big) \ge N_{n+1}^{-1} \;. \tag{7.152}$$

Then the r.v. $w(\omega)$ satisfies $\mathsf{P}(w(\omega) > u) < L\exp(-u/L)$. In particular $\mathsf{E}w(\omega)^p \le K(r,p)$ for each p.

Proof of Theorem 7.8.14 Given the r.v.s Z_i, consider the quantities $\psi_j(s,t) = \sum_i |r^j(Z_i(s) - Z_i(t))|^2 \wedge 1$. As we forcefully pointed out, given the randomness of the Z_i, then Z_i is of the type $a_i\chi_i$ for a number a_i and a character χ_i. Thus (7.151) implies (7.129) and (7.152) implies (7.130) with $w = w(\omega)$. We are then within the hypotheses of Theorem 7.8.13 and (7.145) implies (7.148) where Y_2 is as in (7.150) and where $Y_1 \le K(r,p)w(\omega)\sum_{n\ge 0} 2^n r^{-j_n}$, from which (7.149) follows using the previous proposition. □

[41] The specific problem is to show that the terms Z_i^3 satisfy (7.148) for $p > 2$. For $p \le 2$ this can be shown using the same arguments as in Theorem 7.3.4.

To prepare the proof of Proposition 7.8.15 we set $D_0 = T$ and for $n \geq 1$ we set

$$D_n = \{s \in T \; ; \; \varphi_{j_n}(s, 0) \leq 2^n\} ,$$

so that by (7.147) for $n \geq 0$ we have $\mu(D_n) \geq 1/N_n$. The strategy to follow is then obvious: given n, it follows from Lemma 7.7.2 that for any point $s \in D_n$ it is very rare that $\sum_{i \geq 1} |r^{j_n}(Z_i(s) - Z_i(0))|^2 \wedge 1$ is much larger than 2^n. Thus by Fubini's theorem, it is very rare that the set of points $s \in D_n$ with this property comprises more than $1/4$ of the points of D_n (this event is the complement of the event $\Omega_{n,u}$ below). Thus with probability close to one, this should occur for all n.

Lemma 7.8.16 *Consider a parameter $u \geq 1$. For each $n \geq 0$ consider the* random *subset $B_{n,u}$ of D_n defined as follows:*

$$B_{n,u} := \Big\{ s \in D_n \; ; \; \sum_{i \geq 1} |r^{j_n}(Z_i(s) - Z_i(0))|^2 \wedge 1 \leq u 2^{n+2} \Big\} . \tag{7.153}$$

Then the event $\Omega_{n,u}$ defined by

$$\Omega_{n,u} = \{\omega \in \Omega \; ; \; \mu(B_{n,u}) \geq 3\mu(D_n)/4\}$$

(where as usual ω symbolizes the randomness of the Z_i) satisfies

$$\mathsf{P}(\Omega_{n,u}) \geq 1 - 4\exp(-u2^{n+1}) . \tag{7.154}$$

Proof Consider $s \in D_n$ and $W_i = |r^{j_n}(Z_i(s) - Z_i(0))|^2 \wedge 1$, so that $\sum_{i \geq 1} \mathsf{E} W_i = \varphi_{j_n}(s, 0) \leq 2^n$. It follows from Lemma 7.7.2 (b), used with $A = u2^{n+2}$, that

$$\begin{aligned} \mathsf{P}(s \notin B_{n,u}) &= \mathsf{P}\Big(\sum_{i \geq 1} |r^{j_n}(Z_i(s) - Z_i(0))|^2 \wedge 1 > u2^{n+2}\Big) \\ &\leq \delta_n := \exp(-u2^{n+1}) . \end{aligned}$$

Then we have

$$\mathsf{E}\mu(D_n \backslash B_{n,u}) = \int_{D_n} \mathsf{P}(s \notin B_{n,u}) \mathrm{d}\mu(s) \leq \delta_n \mu(D_n) . \tag{7.155}$$

Consequently $\mathsf{P}(\mu(D_n \backslash B_{n,u}) \geq \mu(D_n)/4) \leq 4\delta_n$ by Markov's inequality. Therefore the event $\Omega_{n,u}$ defined by $\mu(B_{n,u}) \geq 3\mu(D_n)/4$ satisfies

$$\mathsf{P}(\Omega_{n,u}) = \mathsf{P}\Big(\mu(B_{n,u}) \geq \frac{3}{4}\mu(D_n)\Big) = \mathsf{P}\Big(\mu(D_n \setminus B_{n,u}) \leq \frac{1}{4}\mu(D_n)\Big) \geq 1 - 4\delta_n$$

and we have proved (7.154). □

Exercise 7.8.17 Write a detailed proof of (7.155).

Proof of Proposition 7.8.15 We recall the sets $B_{n,u}$ and the event $\Omega_{n,u}$ of Lemma 7.8.16, and we define

$$\Omega_u = \bigcap_{n \geq 0} \Omega_{n,u} , \tag{7.156}$$

so that from (7.154)

$$\mathsf{P}(\Omega_u) \geq 1 - L \exp(-u) . \tag{7.157}$$

For $\omega \in \Omega_u$ we have $\mu(B_{0,u}) \geq 3/4$ (since $D_0 = T$) and $\mu(B_{n,u}) \geq 3\mu(D_n)/4 \geq N_{n+1}^{-1}$ for $n \geq 1$. Thus by the very definition of $w(\omega)$, we then have $w(\omega) \leq 4u$. Thus $\mathsf{P}(w(\omega) > 4u) \leq \mathsf{P}(\Omega_u^c) \leq \exp(-u)$. □

7.8.7 *Sums with Few Non-zero Terms*

To complete the proof of Theorem 7.5.5 using Theorem 7.8.14, we need to learn how to control $\mathsf{E}Y_2^p$. This is the content of the next result.

Theorem 7.8.18 *Assume (7.66), i.e.*

$$\forall s, t \in T \;,\; \varphi_{j_0}(s,t) \leq 1 . \tag{7.66}$$

Then for each $p \geq 1$ the variables $\zeta_i := |Z_i(0)|\mathbf{1}_{\{|Z_i(0)| \geq r^{-j_0}\}}$ satisfy

$$\Big(\mathsf{E}\big(\sum_{i \geq 1} \zeta_i\big)^p\Big)^{1/p} \leq K\Big(r^{-j_0} + \Big(\mathsf{E}\Big|\sum_{i \geq 1} \varepsilon_i Z_i(0)\Big|^p\Big)^{1/p}\Big) . \tag{7.158}$$

Proof of Theorem 7.5.5 We use Theorem 7.8.14. We raise (7.148) to the power p, we use that $(Y_1 + Y_2)^p \leq K(p)(Y_1^p + Y_2^p)$, and we take expectation. The control of $\mathsf{E}Y_1^p$ is provided by (7.149), and the control of $\mathsf{E}Y_2^p$ is provided by (7.158). □

The basic reason we shall succeed in proving Theorem 7.8.18 is that typically only a few of the r.v.s ζ_i will be non-zero, a fact which motivates the title of this section. Our first goal is to prove this. We start with a simple fact.

Lemma 7.8.19 *For any $j \in \mathbb{Z}$ we have*

$$\sum_i \mathsf{E}(|2r^j Z_i(0)|^2 \wedge 1) \leq 2 \sup_{s,t \in T} \sum_i \mathsf{E}(|r^j (Z_i(s) - Z_i(t))|^2 \wedge 1) . \tag{7.159}$$

Thus if j_0 satisfies (7.66) then

$$\sum_i \mathsf{E}(|2r^{j_0} Z_i(0)|^2 \wedge 1) \le 2 \,, \tag{7.160}$$

and in particular

$$\sum_i \mathsf{P}(|Z_i(0)| \ge r^{-j_0}) \le 2 \,. \tag{7.161}$$

One should stress the interesting nature of this statement: a control on the size of the *differences* $Z_i(s) - Z_i(t)$ implies a control of the size of $Z_i(0)$. The hypothesis (7.58) that $Z_i \notin \mathbb{C}\mathbf{1}$ a.e. is essential here.

Proof Since $Z_i \in \mathbb{C}G$, we have $Z_i(s) = \chi(s) Z_i(0)$ for a certain character χ, and since by (7.58) $\chi \neq \mathbf{1}$ a.e., (7.34) implies that a.e.

$$\int |Z_i(s) - Z_i(0)|^2 \mathrm{d}\mu(s) = 2|Z_i(0)|^2 \,, \tag{7.162}$$

whereas $|Z_i(s) - Z_i(0)|^2 \le 2|Z_i(s)|^2 + 2|Z_i(0)|^2 = 4|Z_i(0)|^2$. Now for $x \ge 0$ the function $\psi(x) = (r^{2j} x) \wedge 1$ is concave with $\psi(0) = 0$, so it satisfies $x\psi(y) \le y\psi(x)$ for $x \le y$. Using this for $x = |Z_i(s) - Z_i(0)|^2$, $y = 4|Z_i(0)|^2 = |2Z_i(0)|^2$, and integrating in s with respect to μ, we obtain $\psi(|2Z_i(0)|^2) \le 2\int \psi(|Z_i(s) - Z_i(0)|^2)\mathrm{d}\mu(s)$, and taking expectation, we obtain

$$\mathsf{E}(|2r^j Z_i(0)|^2 \wedge 1) \le 2 \int \mathsf{E}\big((r^{2j}|Z_i(s) - Z_i(0)|^2) \wedge 1\big)\mathrm{d}\mu(s) \,.$$

Summation over i then makes (7.159) obvious, and (7.161) follows since $\mathsf{P}(|Z_i(0)| \ge r^{-j_0}) \le \mathsf{E}(|2r^{j_0} Z_i(0)|^2 \wedge 1)$. □

Exercise 7.8.20 Instead of (7.66) assume that $\mu(\{s \in T; \varphi_{j_0}(s, 0) \le 1\}) \ge 3/4$. Prove that $\sum_i \mathsf{E}(|2r^{j_0} Z_i(0)|^2 \wedge 1) \le 4$.

So, as promised earlier, (7.161) means that typically only a few of the r.v.s $\zeta_i = |Z_i(0)|\mathbf{1}_{\{|Z_i(0)| \ge r^{-j_0}\}}$ can be non-zero at the same time. To lighten notation, in the next few pages, $K = K(p)$ denotes a number depending on p only. Also, even though the sums we consider are finite sums $\sum_i$, it is convenient for notation to write them as infinite sums $\sum_{i \ge 1}$, with terms which are eventually zero.

We start the study of sums of independent r.v.s with only a few non-zero terms.

Lemma 7.8.21 *Consider independent centered complex-valued r.v.s θ_i. Assume that*[42]

$$\sum_{i\geq 1} \mathsf{P}(\theta_i \neq 0) \leq 8 \,. \tag{7.163}$$

Then, for each $p \geq 1$, we have

$$\sum_{i\geq 1} \mathsf{E}|\theta_i|^p \leq K\mathsf{E}\Big|\sum_{i\geq 1}\theta_i\Big|^p \,. \tag{7.164}$$

The intuition here is that because the typical number of non-zero values of θ_i is about 1, it is not surprising that $|\sum_{i\geq 1}\theta_i|^p$ should be comparable to $\sum_{i\geq 1}|\theta_i|^p$.

Lemma 7.8.22 *Consider independent events Ξ_i with $\mathsf{P}(\Xi_i) \leq 1/2$ and $\sum_i \mathsf{P}(\Xi_i) \leq 8$. Then $\mathsf{P}(\bigcup_i \Xi_i) \leq 1 - e^{-16}$.*

Proof We use that $1 - x \geq \exp(-2x)$ for $0 \leq x \leq 1/2$, so that

$$1 - \mathsf{P}\Big(\bigcup_i \Xi_i\Big) = \prod_i (1 - \mathsf{P}(\Xi_i)) \geq \prod_i \exp(-2\mathsf{P}(\Xi_i)) \geq \exp(-16) \,. \qquad \square$$

Proof of Lemma 7.8.21 From (7.163) there are at most 16 indices i with $\mathsf{P}(\theta_i \neq 0) \geq 1/2$. We can assume without loss of generality that for $i \geq 17$ we have $\mathsf{P}(\theta_i \neq 0) \leq 1/2$. As a consequence of Jensen's inequality[43] for any set J of indices, we have $\mathsf{E}|\sum_{i\in J}\theta_i|^p \leq \mathsf{E}|\sum_{i\geq 1}\theta_i|^p$. In particular for any index i_0, we have $\mathsf{E}|\theta_{i_0}|^p \leq \mathsf{E}|\sum_{i\geq 1}\theta_i|^p$. Thus $\sum_{i\leq 16}\mathsf{E}|\theta_i|^p \leq 16\mathsf{E}|\sum_{i\geq 1}\theta_i|^p$ and $\mathsf{E}|\sum_{i\geq 17}\theta_i|^p \leq \mathsf{E}|\sum_{i\geq 1}\theta_i|^p$. Therefore it suffices to prove that $\mathsf{E}\sum_{i\geq 17}|\theta_i|^p \leq L\mathsf{E}|\sum_{i\geq 17}\theta_i|^p$. Consequently, we may assume $\mathsf{P}(\theta_i \neq 0) \leq 1/2$ for each i.

For $n \geq 1$ consider $\Omega_n = \{\exists i \leq n, \theta_i \neq 0\}$. Then $\mathsf{P}(\Omega_n) \leq 1 - e^{-16}$ by Lemma 7.8.22, so that $\mathsf{P}(\Omega_n^c) \geq e^{-16}$.

We prove by induction on n that

$$\sum_{i\leq n} e^{-16}\mathsf{E}|\theta_i|^p \leq \mathsf{E}\Big|\sum_{i\leq n}\theta_i\Big|^p \,. \tag{7.165}$$

It is obvious that (7.165) holds for $n = 1$. Assuming it holds for n, we have

$$\mathsf{E}\Big|\sum_{i\leq n+1}\theta_i\Big|^p = \mathsf{E}\mathbf{1}_{\Omega_n}\Big|\sum_{i\leq n+1}\theta_i\Big|^p + \mathsf{E}\mathbf{1}_{\Omega_n^c}\Big|\sum_{i\leq n+1}\theta_i\Big|^p \,. \tag{7.166}$$

[42] The number 8 does not play a special role and can be replaced by any other number.

[43] If this is not obvious to you, please review Exercise 6.2.1 (c).

Now, since θ_{n+1} is centered and independent of both Ω_n and $\sum_{i\le n}\theta_i$, Jensen's inequality implies

$$\mathsf{E}\mathbf{1}_{\Omega_n}\Big|\sum_{i\le n+1}\theta_i\Big|^p \ge \mathsf{E}\mathbf{1}_{\Omega_n}\Big|\sum_{i\le n}\theta_i\Big|^p = \mathsf{E}\Big|\sum_{i\le n}\theta_i\Big|^p \,. \tag{7.167}$$

Since for $i \le n$ we have $\theta_i = 0$ on Ω_n^c,

$$\mathsf{E}\mathbf{1}_{\Omega_n^c}\Big|\sum_{i\le n+1}\theta_i\Big|^p = \mathsf{E}\mathbf{1}_{\Omega_n^c}|\theta_{n+1}|^p = \mathsf{P}(\Omega_n^c)\mathsf{E}|\theta_{n+1}|^p \ge e^{-16}\mathsf{E}|\theta_{n+1}|^p \,,$$

using independence in the second equality and that $\mathsf{P}(\Omega_n^c) \ge e^{-16}$. Combining with (7.166) and (7.167) and using the induction hypothesis completes the induction. □

Lemma 7.8.23 *Consider independent r.v.s* $\eta_i \ge 0$ *with* $\sum_{i\ge 1}\mathsf{P}(\eta_i > 0) \le 8$. *Then for each* $p \ge 1$,

$$\mathsf{E}\Big(\sum_{i\ge 1}\eta_i\Big)^p \le K\sum_{i\ge 1}\mathsf{E}\eta_i^p \,. \tag{7.168}$$

Again, the intuition here is that the typical number of non-zero values of η_i is about 1, so that $(\sum_{i\ge 1}\eta_i)^p$ is not much larger than $\sum_{i\ge 1}\eta_i^p$.

Proof The starting point of the proof is the inequality

$$(a+b)^p \le a^p + K(a^{p-1}b + b^p)\,, \tag{7.169}$$

where $a, b \ge 0$. This is elementary, by distinguishing the cases $b \le a$ and $b \ge a$. Let $S_n = \sum_{i\le n}\eta_i$, so that using (7.169) for $a = S_n$ and $b = \eta_{n+1}$ and taking expectation, we obtain

$$\mathsf{E}S_{n+1}^p \le \mathsf{E}S_n^p + K(\mathsf{E}S_n^{p-1}\eta_{n+1} + \mathsf{E}\eta_{n+1}^p)\,. \tag{7.170}$$

Let $a_n = \mathsf{P}(\eta_n > 0)$. From Hölder's inequality, we get

$$\mathsf{E}S_n^{p-1} \le (\mathsf{E}S_n^p)^{(p-1)/p} \,;\; \mathsf{E}\eta_{n+1} \le a_{n+1}^{(p-1)/p}(\mathsf{E}\eta_{n+1}^p)^{1/p} \,.$$

Using independence then implies

$$\mathsf{E}S_n^{p-1}\eta_{n+1} = \mathsf{E}S_n^{p-1}\mathsf{E}\eta_{n+1} \le (\mathsf{E}S_n^p)^{(p-1)/p}a_{n+1}^{(p-1)/p}(\mathsf{E}\eta_{n+1}^p)^{1/p} \,.$$

Now, for numbers $a, b > 0$, Young's inequality implies that $a^{(p-1)/p}b^{1/p} \le a + b$ and consequently

$$\mathsf{E}S_n^{p-1}\eta_{n+1} \le a_{n+1}\mathsf{E}S_n^p + \mathsf{E}\eta_{n+1}^p .$$

Combining with (7.170) yields

$$\mathsf{E}S_{n+1}^p \le \mathsf{E}S_n^p(1 + Ka_{n+1}) + K\mathsf{E}\eta_{n+1}^p \le (\mathsf{E}S_n^p + K\mathsf{E}\eta_{n+1}^p)(1 + Ka_{n+1}) .$$

In particular we obtain by induction on n that

$$\mathsf{E}S_n^p \le K\Big(\sum_{i\le n}\mathsf{E}\eta_i^p\Big)\prod_{i\le n}(1 + Ka_i) ,$$

which concludes the proof since $\sum_{i\ge 1} a_i \le 8$ by hypothesis. □

We now have all the tools to prove our main result.

Proof of Theorem 7.8.18 Let us define $\theta_i := Z_i(0)\mathbf{1}_{\{|Z_i(0)|\ge r^{-j_0}\}}$ so that $\zeta_i = |\theta_i| = |\varepsilon_i\theta_i|$, and (7.161) implies

$$\sum_{i\ge 1}\mathsf{P}(\zeta_i \ne 0) = \sum_{i\ge 1}\mathsf{P}(\theta_i \ne 0) \le 2 . \tag{7.171}$$

Using (7.168) in the first inequality and (7.164) (for $\varepsilon_i\theta_i$ rather than θ_i) in the second one, we obtain

$$\mathsf{E}\Big(\sum_{i\ge 1}\zeta_i\Big)^p \le K\sum_{i\ge 1}\mathsf{E}\zeta_i^p \le K\mathsf{E}\Big|\sum_{i\ge 1}\varepsilon_i\theta_i\Big|^p .$$

Let $\theta_i' := Z_i(0) - \theta_i = Z_i(0)\mathbf{1}_{\{|Z_i(0)|<r^{-j_0}\}}$, so that $\theta_i = Z_i(0) - \theta_i'$ and thus

$$\mathsf{E}\Big|\sum_{i\ge 1}\varepsilon_i\theta_i\Big|^p \le K\mathsf{E}\Big|\sum_{i\ge 1}\varepsilon_i Z_i(0)\Big|^p + K\mathsf{E}\Big|\sum_{i\ge 1}\varepsilon_i\theta_i'\Big|^p ,$$

and in order to prove (7.158), it suffices to prove that

$$\mathsf{E}\Big|\sum_{i\ge 1}\varepsilon_i\theta_i'\Big|^p \le Kr^{-j_0 p} . \tag{7.172}$$

First, Khinchin's inequality (6.3) implies

$$\mathsf{E}_\varepsilon\Big|\sum_{i\ge 1}\varepsilon_i\theta_i'\Big|^p \le K\Big(\sum_{i\ge 1}|\theta_i'|^2\Big)^{p/2} . \tag{7.173}$$

The r.v.s $W_i = r^{2j_0}|\theta_i'|^2$ satisfy $0 \le W_i \le 1$ and $\sum_{i\ge1} \mathsf{E}W_i \le 2$ by (7.160). Lemma 7.7.2 (b) provides the estimate $\mathsf{P}(\sum_{i\ge1} W_i \ge t) \le \exp(-t/2)$ for $t \ge 8$, and this implies $\mathsf{E}\big(\sum_{i\ge1} W_i\big)^{p/2} \le K$. Consequently taking expectation in (7.173) yields (7.172) and completes the proof. □

7.9 Proof of the Decomposition Theorem

7.9.1 Constructing Decompositions

Consider independent r.v.s $Z_i \in \mathbb{C}G$ and assume (7.58) as always. Our approach will be based on the decomposition (7.60): $Z_i = \sum_{\ell\ge1} \xi_{i,\ell}\chi_\ell$ where the r.v.s $(\xi_{i,\ell})_{\ell\ge1}$ are valued in $\mathbb{C}$ and have "disjoint supports".[44] We will use many times that for a function h with $h(0) = 0$ and r.v.s f_ℓ with disjoint support, we have

$$h\Big(\sum_\ell f_\ell\Big) = \sum_\ell h(f_\ell) .$$

In particular, using that for each i the functions $(\xi_{i,\ell})_{\ell\ge1}$ have disjoint support, (7.63) becomes

$$\varphi_j(s,t) = \sum_i \mathsf{E}(|r^j(Z_i(s) - Z_i(t))|^2 \wedge 1) = \sum_{i,\ell} \mathsf{E}(|r^j\xi_{i,\ell}(\chi_\ell(s) - \chi_\ell(t))|^2 \wedge 1) . \tag{7.174}$$

Consider a parameter w (which will be useful for later purposes) and integers $(j_n)_{n\ge0}$. For $n \ge 0$, we set

$$D_n = \{s \in T \ ; \ \varphi_{j_n}(s,0) \le w2^n\} . \tag{7.175}$$

Let us then assume that

$$\mu(D_0) \ge \frac{3}{4} \ ; \ \forall n \ge 1 \ , \ \mu(D_n) \ge \frac{1}{N_n} . \tag{7.176}$$

Proposition 7.9.1 *Under the preceding condition (7.176), there exist truncation levels $u_\ell \ge 0$ with the following properties. First,*

$$\sum_i \sum_{\ell\ge1} \mathsf{E}|\xi_{i,\ell}|\mathbf{1}_{\{u_\ell \le |\xi_{i,\ell}| \le r^{-j_0}\}} \le Lw \sum_{n\ge0} 2^n r^{-j_n} . \tag{7.177}$$

[44] We use this expression as a shorthand for the following property: For each i we have $\xi_{i,\ell}\xi_{i,\ell'} = 0$ a.s. if $\ell \ne \ell'$. In any given realization of the sequence, $\xi_{i,\ell}$ at most one term is not zero.

Moreover, setting

$$\tilde{a}_\ell := \Big(\sum_i \mathsf{E}|\xi_{i,\ell}|^2 \mathbf{1}_{\{|\xi_{i,\ell}|<u_\ell\}}\Big)^{1/2} \tag{7.178}$$

then

$$\mathsf{E}\Big\|\sum_{\ell\ge 1} \tilde{a}_\ell g_\ell \chi_\ell\Big\| \le L\sqrt{w}\sum_{n\ge 0} 2^n r^{-j_n}\ , \tag{7.179}$$

where $(g_\ell)_{\ell\ge 1}$ *are standard independent Gaussian r.v.s. Finally*

$$\sum_i \mathsf{P}(|Z_i(0)| \ge r^{-j_0}) \le 4w\ . \tag{7.180}$$

The proof of Proposition 7.9.1 occupies Sect. 7.9.2.

Corollary 7.9.2 *Under Condition (7.176) we can find a decomposition* $Z_i = Z_i^1 + Z_i^2 + Z_i^3$ *such that for each* $s = 1, 2, 3$ *the sequence* $(Z_i^s)_i$ *is independent, valued in* $\mathbb{C}G$ *and moreover*

$$\sum_i \mathsf{P}(Z_i^1 \ne 0) \le 4w\ , \tag{7.181}$$

$$\sum_i \mathsf{E}|Z_i^2(0)| \le Lw\sum_{n\ge 0} 2^n r^{-j_n}\ , \tag{7.182}$$

$$\gamma_2(T,d) \le L\sqrt{w}\sum_{n\ge 0} 2^n r^{-j_n}\ , \tag{7.183}$$

where the distance d *is given by* $d(s,t)^2 = \sum_i \mathsf{E}|Z_i^3(s) - Z_i^3(t)|^2$.

Proof We write

$$Z_i^1 = Z_i \mathbf{1}_{\{|Z_i(0)|>r^{-j_0}\}}\ , \tag{7.184}$$

$Z_i^2 = \sum_{\ell\ge 1} \xi_{i,\ell}\mathbf{1}_{\{u_\ell\le|\xi_{i,\ell}|\le r^{-j_0}\}}\chi_\ell$ and $Z_i^3 = \sum_{\ell\ge 1}\xi_{i,\ell}\mathbf{1}_{\{|\xi_{i,\ell}|<u_\ell\}}\chi_\ell$,[45] so that using (7.178) and the fact that the r.v.s $(\xi_{i,\ell})_{\ell\ge 1}$ have a disjoint support, we have

$$\begin{aligned} d(s,t)^2 &= \sum_i \mathsf{E}|Z_i^3(s) - Z_i^3(t)|^2 = \sum_i\sum_{\ell\ge 1} \mathsf{E}\big|\xi_{i,\ell}\mathbf{1}_{\{|\xi_{i,\ell}|<u_\ell\}}(\chi_\ell(s)-\chi_\ell(t))\big|^2 \\ &= \sum_{\ell\ge 1}\sum_i \mathsf{E}\big|\xi_{i,\ell}\mathbf{1}_{\{|\xi_{i,\ell}|<u_\ell\}}\big|^2 |\chi_\ell(s)-\chi_\ell(t)|^2 = \sum_{\ell\ge 1}\tilde{a}_\ell^2|\chi_\ell(s)-\chi_\ell(t)|^2\ . \end{aligned}$$

[45] Thus, if $Z_i = \xi_i\chi_i$, then each Z_i^1, Z_i^2, Z_i^3 is of the type $\eta_i\chi_i$.

Thus d is the canonical distance associated with the process $\sum_{\ell\geq 1}\tilde{a}_\ell g_\ell \chi_\ell$, and (7.183) follows from (7.179) and (7.13). The rest is obvious: (7.181) follows from (7.180) and (7.182) follows from (7.177). □

Proof of Theorem 7.5.14 Let us set $S = \mathsf{E}\|\sum_i \varepsilon_i Z_i\|$ so that by Markov's inequality (7.65) holds for $M = S/\alpha_0$. We then deduce from Theorem 7.5.1 that there exist integers $(j_n)_{n\geq 0}$ for which $\sum_{n\geq 0} 2^n r^{-j_n} \leq LS$ and which satisfy (7.66) and (7.67). In particular, (7.176) holds for $w = 1$. We then apply Corollary 7.9.2 (still with $w = 1$) to obtain a decomposition $Z_i = Z_i^1 + Z_i^2 + Z_i^3$. We set $Z_i' = Z_i^1 + Z_i^2$, $Z_i'' = Z_i^3$. We note that (7.94) follows from (7.184). To prove Theorem 7.5.14, it therefore suffices to prove that $\mathsf{E}\sum_i |Z_i'(0)| \leq LS$. Since $\mathsf{E}\sum_i |Z_i^2(0)| \leq LS$ by (7.182), it suffices to prove that $\mathsf{E}\sum_i |Z_i^1(0)| \leq LS$. Recalling the expression (7.184) of Z_i^1 and setting $\theta_i = Z_i(0)\mathbf{1}_{\{|Z_i(0)|>r^{-j_0}\}}$ it suffices to prove that

$$\mathsf{E}\sum_i |\theta_i| \leq LS\,. \tag{7.185}$$

Now $\sum_i \mathsf{P}(\theta_i \neq 0) \leq 4$ by (7.181). Then using (7.164) for $p = 1$ proves that $\mathsf{E}\sum_i |\theta_i| \leq L\mathsf{E}|\sum_i \varepsilon_i\theta_i|$. But, using (6.17) in the third inequality,

$$\begin{aligned}\mathsf{E}\Big|\sum_i \varepsilon_i\theta_i\Big| = \mathsf{E}\mathsf{E}_\varepsilon\Big|\sum_i \varepsilon_i\theta_i\Big| &\leq \mathsf{E}\Big(\sum_i |\theta_i|^2\Big)^{1/2} \leq \mathsf{E}\Big(\sum_i |Z_i(0)|^2\Big)^{1/2} \\ &\leq L\mathsf{E}\mathsf{E}_\varepsilon\Big|\sum_i \varepsilon_i Z_i(0)\Big| = L\mathsf{E}\Big|\sum_i \varepsilon_i Z_i(0)\Big| \leq LS\,. \end{aligned}\tag{7.186}$$

We have proved (7.185) and completed the proof. □

7.9.2 *Proof of Proposition 7.9.1*

The reader should master the proof of the simpler Theorem 7.8.6 before attempting to read this more complicated argument. It uses essentially the same idea, which we spell out in the simpler case where $Z_i = \xi_i\chi_i$ for a nonrandom character χ_i. The essential step is to construct the truncation level u_i at which we truncate ξ_i. It is given by formula $u_i = \inf\{r^{-j_n}\}$ where the infimum is taken over the values of n such that for each $k \leq n$ we have $\int_{D_k} |\chi_i(s) - 1|^2 \mathrm{d}\mu(s) \geq \mu(D_k)$.

To start the proof of Proposition 7.9.1, for $n \geq 0$ we define

$$U_n = \Big\{\ell \geq 1\,;\ \int_{D_n} |\chi_\ell(s) - 1|^2 \mathrm{d}\mu(s) \geq \mu(D_n)\Big\}\,. \tag{7.187}$$

As in Lemma 7.8.10 we have

$$U_0 = \mathbb{N}^* . \tag{7.188}$$

Lemma 7.9.3 *If $n \geq 0$ and $\ell \in U_n$ for each i we have*

$$\mu(D_n)\mathsf{E}(|r^{j_n}\xi_{i,\ell}|^2 \wedge 1) \leq 4 \int_{D_n} \mathsf{E}(|r^{j_n}\xi_{i,\ell}(\chi_\ell(s) - 1)|^2 \wedge 1) d\mu(s) . \tag{7.189}$$

Proof We deduce from (7.136) that

$$|r^{j_n}\xi_{i,\ell}(\chi_\ell(s) - 1)|^2 \wedge 1 \geq |\chi_\ell(s) - 1|^2(|r^{j_n}\xi_{i,\ell}|^2 \wedge 1)/4 ,$$

we take expectation, and we integrate over D_n, using the $\int_{D_n} |\chi_\ell(s) - 1|^2 d\mu(s) \geq \mu(D_n)$ since $\ell \in U_n$. □

Corollary 7.9.4 *If $n \geq 0$ we have*

$$\sum_i \sum_{\ell \in U_n} \mathsf{E}(|r^{j_n}\xi_{i,\ell}|^2 \wedge 1) \leq 2^{n+2} w . \tag{7.190}$$

Proof Summing the inequalities (7.189) over i and $\ell \in U_n$, we obtain

$$\mu(D_n)\sum_i \sum_{\ell \in U_n} \mathsf{E}(|r^{j_n}\xi_{i,\ell}|^2 \wedge 1)$$

$$\leq 4 \sum_i \int_{D_n} \sum_{\ell \geq 1} \mathsf{E}(|r^{j_n}\xi_{i,\ell}(\chi_\ell(s) - 1)|^2 \wedge 1) d\mu(s) . \tag{7.191}$$

Since the r.v.s $(\xi_{i,\ell})_{\ell \geq 1}$ have disjoint supports, as in (7.174), we have

$$\sum_{\ell \geq 1} \mathsf{E}(|r^{j_n}\xi_{i,\ell}(\chi_\ell(s) - 1)|^2 \wedge 1) = \mathsf{E}(|r^{j_n}(Z_i(s) - Z_i(0))|^2 \wedge 1) .$$

Recalling the definition (7.174) of φ_j and that $\varphi_{j_n}(s, 0) \leq w2^n$ for $s \in D_n$ we conclude that

$$\sum_i \int_{D_n} \mathsf{E}(|r^{j_n}(Z_i(s) - Z_i(0))|^2 \wedge 1) d\mu(s) = \int_{D_n} \varphi_{j_n}(s, 0) d\mu(s) \leq 2^n w\mu(D_n) .$$

Therefore we deduce from (7.191) that

$$\mu(D_n)\sum_i \sum_{\ell \in U_n} \mathsf{E}(|r^{j_n}\xi_{i,\ell}|^2 \wedge 1) \leq 2^{n+2} w\mu(D_n) ,$$

which concludes the proof. □

Proof of (7.180) We have

$$\mathsf{P}(|Z_i(0)| \geq r^{-j_0}) \leq \mathsf{E}(|r^{j_0} Z_i(0)|^2 \wedge 1) = \sum_{\ell \geq 1} \mathsf{E}(|r^{j_0} \xi_{i,\ell}|^2 \wedge 1) ,$$

where we use in the equality that the r.v.s $(\xi_{i,\ell})_{\ell \geq 1}$ have disjoint supports. Since $U_0 = \mathbb{N}^*$ by (7.188) the case $n = 0$ of (7.190) proves (7.180). □

We set $V_n = \bigcap_{0 \leq k \leq n} U_k$, so that $V_0 = \mathbb{N}^*$ by (7.188). We define

$$u_\ell = \inf\{r^{-j_n} \ , \ \ell \in V_n\} , \tag{7.192}$$

and we keep in mind that, by definition for $\ell \in V_n$, we have $u_\ell \leq r^{-j_n}$.

Lemma 7.9.5 *For each i and each ℓ, we have*

$$|\xi_{i,\ell}| \mathbf{1}_{\{u_\ell < |\xi_{i,\ell}| \leq r^{-j_0}\}} \leq \sum_{\{n \geq 1; \ell \in U_n\}} r^{-j_{n-1}} \mathbf{1}_{\{|\xi_{i,\ell}| \geq r^{-j_n}\}} . \tag{7.193}$$

The sum on the right is over the values of $n \geq 1$ such that U_n contains ℓ.

Proof Consider ω with $u_\ell < |\xi_{i,\ell}(\omega)| \leq r^{-j_0}$. By the definition (7.192) of u_ℓ, there exist n such that $\ell \in V_n$ and $r^{-j_n} < |\xi_{i,\ell}(\omega)|$. Consider the smallest integer $k \leq n$ such that $r^{-j_k} < |\xi_{i,\ell}(\omega)|$. Then since $r^{-j_k} < |\xi_{i,\ell}(\omega)| \leq r^{-j_0}$ we have $k \geq 1$. Thus $|\xi_{i,\ell}(\omega)| \leq r^{-j_{k-1}}$, for otherwise k would not be the smallest possible. Thus $|\xi_{i,\ell}(\omega)| \leq r^{-j_{k-1}} \mathbf{1}_{\{|\xi_{i,\ell}| \geq r^{-j_k}\}}(\omega)$. Since $\ell \in V_n$ and $k \leq n$ by definition of V_n, we have $\ell \in U_k$. The result follows by considering the term for $n = k$ in the sum in the right-hand side of (7.193). □

Proof of (7.177) Taking expectation in (7.193) and summing over i and $\ell \geq 1$ shows that the left-hand side of (7.177) is bounded by

$$\sum_{n \geq 1} \sum_i \sum_{\ell \in U_n} r^{-j_{n-1}} \mathsf{P}(|\xi_{i,\ell}| \geq r^{-j_n}) \leq Lw \sum_{n \geq 1} 2^n r^{-j_{n-1}} \leq Lw \sum_{n \geq 0} 2^n r^{-j_n} ,$$

where we have used that $\mathsf{P}(|\xi_{i,\ell}| \geq r^{-j_n}) \leq \mathsf{E}(|r^{j_n} \xi_{i,\ell}|^2 \wedge 1)$ and (7.190) in the first inequality. □

Lemma 7.9.6 *Recalling the quantities $\tilde{a}_\ell$ of (7.178), for each $n \geq 0$ we have*

$$\sum_{\ell \in V_n} \tilde{a}_\ell^2 \leq Lw 2^n r^{-2j_n} . \tag{7.194}$$

Proof Let us write $\eta_{i,\ell} := \xi_{i,\ell} \mathbf{1}_{\{|\xi_{i,\ell}| < u_\ell\}}$, so that by definition (7.178), we have $\tilde{a}_\ell^2 = \sum_i \mathsf{E}|\eta_{i,\ell}|^2$. When $\ell \in V_n$, as we noted we have $u_\ell \leq r^{-j_n}$ so that since $|\eta_{i,\ell}| \leq u_\ell$ we have $|r^{j_n} \eta_{i,\ell}|^2 \leq 1$. Thus $\mathsf{E}|r^{j_n} \eta_{i,\ell}|^2 = \mathsf{E}(|r^{j_n} \eta_{i,\ell}|^2 \wedge 1) \leq \mathsf{E}(|r^{j_n} \xi_{i,\ell}|^2 \wedge 1)$, and (7.190) implies

$$\sum_i \sum_{\ell \in V_n} \mathsf{E}|r^{j_n} \eta_{i,\ell}|^2 \le 4w2^n .$$

The left-hand side above is $r^{2j_n} \sum_{\ell \in V_n} \sum_i \mathsf{E}|\eta_{i,\ell}|^2$, and recalling (7.178) this is $r^{2j_n} \sum_{\ell \in V_n} \tilde{a}_\ell^2$ so that (7.194) is proved. □

Lemma 7.9.7 *For $n \ge 0$ let $W_n = V_n \setminus V_{n+1}$. Then*

$$\mathsf{E}\| \sum_{\ell \in W_n} \tilde{a}_\ell g_\ell \chi_\ell \| \le L\sqrt{w} 2^n r^{-j_n} . \tag{7.195}$$

Proof Since $V_{n+1} = V_n \cap U_{n+1}$ for $\ell \in W_n$, we have $\ell \notin U_{n+1}$ so that by definition of that set,

$$\mu(D_{n+1}) > \int_{D_{n+1}} |\chi_\ell(s) - 1|^2 \mathrm{d}\mu(s) = 2\mu(D_{n+1}) - 2\mathrm{Re} \int_{D_{n+1}} \chi_\ell(s) \mathrm{d}\mu(s) ,$$

and in particular $|\int_{D_{n+1}} \chi_\ell(s) \mathrm{d}\mu(s)| \ge \mu(D_{n+1})/2$. Thus (7.195) follows from (7.122) used with $A = D_{n+1}$ and (7.194). □

Proof of Proposition 7.9.1 It remains only to prove (7.179). For this we write

$$\mathsf{E}\| \sum_{\ell \in \cup_{n \ge 0} W_n} \tilde{a}_\ell g_\ell \chi_\ell \| \le \sum_{n \ge 0} \mathsf{E}\| \sum_{\ell \in W_n} \tilde{a}_\ell g_\ell \chi_\ell \| \le L\sqrt{w} \sum_{n \ge 0} 2^n r^{-j_n} ,$$

where we have used (7.195) in the last inequality, and we observe that by (7.192) for $\ell \notin \cup_{n \ge 0} W_n$ i.e., $\ell \in \cap_{n \ge 0} U_n$ we have $u_\ell = 0$, so that $\tilde{a}_\ell = 0$. □

Exercise 7.9.8 Consider a random trigonometric sum $\sum_{i \in I} a_i g_i \chi_i$ and the associated distance d. Consider numbers ϵ_n such that $\mu(\{s \in T; d(s,0) \le \epsilon_n\}) \ge N_n^{-1}$. Find a partition $I = \cup_n I_n$ such that $\mathsf{E}\| \sum_{i \in I_n} a_i g_i \chi_i \| \le L2^{-n/2} \epsilon_n$. The point of this result is that if ϵ_n is as small as possible, then by (7.4) and (7.13), we have $\sum_{n \ge 0} 2^{n/2} \epsilon_n \le L\mathsf{E}\| \sum_i a_i g_i \chi_i \|$. Then in the bound $\mathsf{E}\| \sum_i a_i g_i \chi_i \| \le \sum_{n \ge 0} \mathsf{E}\| \sum_{i \in I_n} a_i g_i \chi_i \|$, the right- and the left-hand sides are of the same order. Hint: Copy the previous arguments. If $D_n = \{s \in T; d(s,0) \le \epsilon_n\}$ and U_n is given by (7.187), then define $I_n = \{i \in I, \forall k \le n, i \in U_k, i \notin U_{n+1}\}$.

Exercise 7.9.9 ([86] Proposition 4.5) Consider characters $(\chi_i)_{i \le N}$. Assume that $\int \exp(|\sum_{i \le N} \chi_i|^2/CN) \mathrm{d}\mu \le 2$. Prove that $\mathsf{E}\| \sum_{i \le N} g_i \chi_i \| \ge N/K(C)$. Hint: Prove that if a set D satisfies $\sqrt{\log(2/\mu(D))} \le N/K(C)$ then $\sup_{s \in D} \sum_{i \le N} |\chi_i(s) - 1|^2 \ge N$ by proving that $\int_D \sum_i |\chi_i(s) - 1|^2 \mathrm{d}\mu(s) \ge N\mu(D)$.

7.10 Proofs, Convergence

After the hard work of proving inequalities such as (7.68) and (7.148) has been completed, the proof of Theorem 7.5.16, which is the goal of this section, involves only "soft arguments". To prove convergence of a series of independent symmetric r.v.s, we shall use the following general principle, which relates the convergence a.s. of a random series with its convergence in probability.

Lemma 7.10.1 *Consider independent symmetric Banach space valued r.v.s W_i. Then the series $\sum_{i\geq 1} W_i$ converges a.s. if and only if it is a Cauchy sequence in probability, i.e.*

$$\forall\, \delta > 0\,,\ \exists k_0\,,\ k_0 \leq k \leq n \Rightarrow \mathsf{P}\Big(\| \sum_{k\leq i\leq n} W_i \| \geq \delta\Big) \leq \delta\,. \tag{7.196}$$

Proof It suffice to prove that (7.196) implies convergence. Let $S_k = \sum_{i\leq k} W_i$. Then the Lévy inequality

$$\mathsf{P}\Big(\sup_{k\leq n} \|S_k\| \geq a\Big) \leq 2\mathsf{P}(\|S_n\| \geq a)$$

(see [53], page 47, equation (2.6)) implies

$$\mathsf{P}\Big(\sup_{k} \|S_k\| \geq a\Big) \leq 2 \sup_n \mathsf{P}(\|S_n\| \geq a)\ ,$$

and starting the sum at an integer k_0 as in (7.196) rather than at 1, we obtain

$$\mathsf{P}\Big(\sup_{k} \|S_k - S_{k_0}\| \geq a\Big) \leq 2 \sup_n \mathsf{P}(\|S_n - S_{k_0}\| \geq a)\ .$$

For $a = \delta$ the right-hand side above is $\leq 2\delta$. Since $\|S_n - S_k\| \leq \|S_n - S_{k_0}\| + \|S_k - S_{k_0}\|$, this proves that

$$\mathsf{P}\big(\sup_{k_0\leq k\leq n} \|S_n - S_k\| \geq 2\delta\big) \leq 4\delta\ ,$$

and in turn that a.s. the sequence $(S_k(\omega))_{k\geq 1}$ is a Cauchy sequence in probability. □

Exercise 7.10.2

(a) Let $(W_i)_{i\geq 1}$ be independent symmetric real-valued r.v.s. Assume for some $a > 0$ (or, equivalently, all $a > 0$) we have

$$\sum_{i\geq 1} \mathsf{E}(W_i^2 \wedge a^2) < \infty\ . \tag{7.197}$$

Prove that the series $\sum_{i\geq 1} W_i$ converges a.s.

(b) Prove the converse.

Exercise 7.10.3 The neighborhoods of zero for the convergence in probability are the sets of functions such that $\mathsf{P}(|f| \geq \delta) \leq 1 - \delta$ for some $\delta > 0$. Prove that a function is "small in probability" if and only if there is a set a probability almost 1 on which the integral of the function is small. Prove that the convergence in L^p $(p \geq 1)$ is stronger than convergence in probability.

We will prove Theorems 7.5.16 and 7.5.17 at the same time by proving the following statements. In each of them, $(Z_i)_{i\geq 1}$ is an independent sequence with $Z_i \in \mathbb{C}G$ and $(\varepsilon_i)_{i\geq 1}$ are independent Bernoulli r.v.s. independent of the sequence (Z_i).

Lemma 7.10.4 *If the series $\sum_{i\geq 1} \varepsilon_i Z_i$ converges a.s., then for each $\alpha > 0$ there exists M such that for each k we have $\mathsf{P}(\|S_k\| \geq M) \leq \alpha$, where S_k is the partial sum, $S_k = \sum_{1\leq i\leq k} \varepsilon_i Z_i$.*

Proof Denoting by S the sum of the series, given $\alpha > 0$ there exists k_0 such that $\mathsf{P}(\|S - S_k\| \geq 1) \leq \alpha/2$ for $k \geq k_0$. Consider then M_0 such that $\mathsf{P}(\|S\| \geq M_0) < \alpha/2$, so that for $k \geq k_0$ we have $\mathsf{P}(\|S_k\| \geq M_0 + 1) < \alpha$, from which the result follows. □

The next result is a version of Theorem 7.5.1 adapted to infinite sums. We recall the number α_0 of this theorem.

Lemma 7.10.5 *Consider an independent sequence $(Z_i)_{i\geq 1}$ with $Z_i \in \mathbb{C}G$, and let $S_k = \sum_{1\leq i\leq k} \varepsilon_i Z_i$, where the Bernoulli r.v.s ε_i are independent of the Z_i. Assume that for each k we have*

$$\mathsf{P}(\|S_k\| \geq M) \leq \alpha_0 \,. \tag{7.198}$$

For $j \in \mathbb{Z}$ we define

$$\varphi_j(s,t) = \sum_{i\geq 1} \mathsf{E}(|r^j(Z_i(s) - Z_i(t))|^2 \wedge 1) \,. \tag{7.199}$$

Then we can find integers $(j_n)_{n\geq 0}$ such that

$$\forall s,t \in T \,,\ \varphi_{j_0}(s,t) \leq 1 \tag{7.200}$$

$$\mu(\{s\,;\ \varphi_{j_n}(s,0) \leq 2^n\}) \geq N_n^{-1} \,, \tag{7.201}$$

and

$$\sum_{n\geq 0} 2^n r^{-j_n} \leq KM \,. \tag{7.202}$$

Lemma 7.10.6 *Assume that there exists integers $(j_n)_{n\geq 0}$ as in (7.200)–(7.202). Then there exists a decomposition of Z_i as in Corollary 7.9.2. except that all finite sums $\sum_i$ are replaced by infinite sums $\sum_{i\geq 1}$.*

Lemma 7.10.7 *When there is a decomposition of Z_i as in Lemma 7.10.6 the series $\sum_{i\geq 1} \varepsilon_i Z_i$ converges a.s.*

Proof of Theorems 7.5.16 and 7.5.17 The theorems follow from these lemmas using also that the last statement of Theorem 7.5.16 follows from Theorem 7.5.5. □

Proof of Lemma 7.10.5 The reader should review Theorem 7.5.1 at this stage, as our proof consists of using this result for each k and a straightforward limiting argument. Let us define

$$\varphi_{k,j}(s,t) = \sum_{i\leq k} \mathsf{E}(|r^j(Z_i(s) - Z_i(t))|^2 \wedge 1) \;,$$

so that

$$\varphi_j(s,t) = \lim_{k\to\infty} \varphi_{k,j}(s,t) \;. \tag{7.203}$$

Theorem 7.5.1 implies that for each k we can find numbers $(j_{k,n})_{n\geq 0}$ for which

$$\forall s,t \in T \;;\; \varphi_{k,j_{k,0}}(s,t) \leq 1 \;, \tag{7.204}$$

and, for $n \geq 0$,

$$\mu(\{s \;;\; \varphi_{k,j_{k,n}}(s,0) \leq 2^n\}) \geq N_n^{-1} \tag{7.205}$$

such that the following holds:

$$\sum_{n\geq 0} 2^n r^{-j_{k,n}} \leq LM \;. \tag{7.206}$$

The conclusion will then follow by a limiting argument that we detail now. The plan is to take a limit $k \to \infty$ in (7.204) and (7.205). As a first step, for each n we would like to take the limit $\lim_{k\to\infty} j_{k,n}$. We will ensure that the limit exists by taking a subsequence. It follows from (7.206) that for each n the sequence $(j_{k,n})_k$ is bounded from below. To ensure that it is also bounded from above, we consider any sequence (j_n^*) such that $\sum_{n\geq 0} 2^n r^{-j_n^*} \leq LM$, and we replace $j_{k,n}$ by $\min(j_{k,n}, j_n^*)$. Thus $j_{n,k}$ is now bounded from above by j_n^* and (7.205) and (7.206) still hold. Thus we can find a sequence $(k(q))$ with $k(q) \to \infty$ such that for each n, $j_n = \lim_{q\to\infty} j_{k(q),n}$ exists, i.e., for each n, $j_{k(q),n} = j_n$ for q large enough. By taking a further subsequence if necessary, we may assume that for each $n \geq 0$ we have

$$q \geq n \Rightarrow j_{k(q),n} = j_n ,$$

so that then $\varphi_{k(q),j_n} = \varphi_{k(q),j_{k(q),n}}$. Consequently (7.204) implies

$$\forall s,t \in T , \ \varphi_{k(q),j_0}(s,t) \leq 1 , \tag{7.207}$$

and (7.205) implies that for $n \geq 1$ and $q \geq n$

$$\mu(\{s \ ; \ \varphi_{k(q),j_n}(s,0) \leq 2^n\}) \geq N_n^{-1} , \tag{7.208}$$

while, from (7.206),

$$\sum_{0 \leq n \leq q} 2^n r^{-j_n} = \sum_{0 \leq n \leq q} 2^n r^{-j_{k(q),n}} \leq KM .$$

Letting $q \to \infty$ proves that $\sum_{n \geq 0} 2^n r^{-j_n} \leq LM$. On the other hand, (7.203) implies $\varphi_j(s,t) = \lim_{q \to \infty} \varphi_{k(q),j}(s,t)$. Together with (7.207) and (7.208), this proves that

$$\forall s,t \in T \ , \ \varphi_{j_0}(s,t) \leq 1 ,$$

and for each n,

$$\mu(\{s \ ; \ \varphi_{j_n}(s,0) \leq 2^n\}) \geq N_n^{-1} .$$

□

Proof of Lemma 7.10.6 Copy the proof of Corollary 7.9.2 verbatim. □

Before we prove Lemma 7.10.7 we need another simple result.

Lemma 7.10.8 *Consider a decreasing sequence of translation-invariant distances $(d_k)_{k \geq 1}$ on T. Assume $\gamma_2(T,d_1) < \infty$ and that for each $s \in T$ we have $\lim_{k \to \infty} d_k(s,0) = 0$. Then $\lim_{k \to \infty} \gamma_2(T,d_k) = 0$.*

Proof Given $\epsilon > 0$, since $\lim_{k \to \infty} d_k(s,0) = 0$ for each $s \in T$, we have $T = \cup_k B_k$ where $B_k = \{s \in T; \forall n \geq k, d_n(s,0) \leq \epsilon\}$. Thus for k large enough we have $\mu(B_k) > 1/2$. Corollary 7.1.4 and Lemma 7.1.6 prove then that $\Delta(T,d_k) \leq 4\epsilon$. We have shown that $\lim_{k \to \infty} \Delta(T,d_k) = 0$.

Next, according to (7.4) we can find numbers ϵ_n with $\mu(\{s; d_1(s,0) \leq \epsilon_n\}) \geq N_n^{-1}$ and $\sum_{n \geq 0} 2^{n/2} \epsilon_n < \infty$. Let $\epsilon_{n,k} = \min(\epsilon_n, \Delta(T,d_k))$. Then since $d_k \leq d_1$ we have $\{s; d_1(s,0) \leq \epsilon_n\} \subset \{s; d_k(s,0) \leq \epsilon_{n,k}\}$ so that this latter set has measure $\geq N_n^{-1}$, and by (7.4) again we have $\gamma_2(T,d_k) \leq L \sum_{n \geq 0} 2^{n/2} \epsilon_{n,k}$. The right-hand side goes to 0 as $k \to \infty$ by dominated convergence. □

Proof of Lemma 7.10.7 For each $\ell = 1, 2, 3$, we will prove that the series $\sum_{i\geq 1} Z_i^\ell$ converges a.s. For $\ell = 1$ this is obvious since by (7.181) a.s. only finitely many of the terms are $\neq 0$. For $\ell = 2$ and $\ell = 3$, we will deduce this from Lemma 7.10.1. For $\ell = 2$ this follows from the fact that $\| \sum_{k\leq i\leq n} \varepsilon_i Z_i^2 \| \leq \sum_{i\geq k} |Z_i^2(0)|$ since $|Z_i(t)| = |Z_i(0)|$ because $Z_i \in \mathbb{C}G$. So let us turn to the hard case $\ell = 3$. For $k \geq 1$ consider the distance d_k on T defined by $d_k(s,t)^2 = \sum_{i\geq k} \mathsf{E}|Z_i^3(s) - Z_i^3(t)|^2$. By the version of (7.33) proposed in Exercise 7.5.15 we have $\mathsf{E}\| \sum_{k\leq i\leq n} \varepsilon_i Z_i \| \leq L\gamma_2(T, d_k)$, so that using Markov's inequality, to obtain (7.196) it suffices to prove that $\lim_{k\to\infty} \gamma_2(T, d_k) = 0$. But this follows from Lemma 7.10.8 since by (7.183) we have $\gamma_2(T, d_1) < \infty$. □

7.11 Further Proofs

The proof we gave of Proposition 7.5.13 uses p-stable r.v.s. It is a matter of taste, but for the author this feels like an unnatural trick. Our first goal in the present section is to provide a proof of Proposition 7.11.3, an extension of Proposition 7.5.13 which does not use this trick and which brings forward the combinatorics of the situation. Finally we will prove Theorem 7.5.18.

7.11.1 Alternate Proof of Proposition 7.5.13

Consider $1 < p < 2$. Consider numbers a_i and characters χ_i, and we recall the distance d_p of (7.84). We assume that the independent symmetric r.v.s θ_i satisfy (7.91), i.e., $\mathsf{P}(|\theta_i| \geq u) \geq 1/(Cu^p)$ for $u \geq C$, and we set

$$\varphi_j(s,t) = \sum_i \mathsf{E}|r^j a_i \theta_i (\chi_i(s) - \chi_i(t))|^2 \wedge 1 \,. \tag{7.209}$$

Proposition 7.11.1 *Consider a sequence $(j_n)_{n\geq 0}$ and assume that*

$$\forall s,t \in T \,,\; \varphi_{j_0}(s,t) \leq 1 \,, \tag{7.210}$$

$$\forall n \geq 1 \,,\; \mu(\{s \in T \,;\, \varphi_{j_n}(s,0) \leq 2^n\}) \geq N_n^{-1} \,. \tag{7.211}$$

Then

$$\gamma_q(T, d_p) \leq K\gamma_2(T, d_2) + K \sum_{n\geq 0} 2^n r^{-j_n} \,, \tag{7.212}$$

where q is the conjugate exponent of p and where d_2 is the distance given by (7.84) for $p = 2$. Here K depends on p, r, and C.

Proof Combining (7.5) and (7.6), we obtain

$$\sum_{n\geq 0} 2^{n/2} e_n(T, d_2) \leq K\gamma_2(T, d_2) .$$

Let us set $\alpha_n = 2^{n/2} e_n(T, d_2)$. The first step is to reduce to the case where

$$\forall n \geq 0 \; , \; \alpha_n \leq 2^n r^{-j_n} . \tag{7.213}$$

The purpose of this condition is certainly not obvious now and will become apparent only later in the proof. To obtain this condition, we construct a sequence (j'_n) as follows. For each $n \geq 0$, if $2^n r^{-j_n} \geq \alpha_n$ we set $j'_n = j_n$. Otherwise, we define j'_n as the largest integer such that $2^n r^{-j'_n} \geq \alpha_n$. Thus $j'_n \leq j_n$ and $2^n r^{-j'_n} \leq r\alpha_n$. We then have

$$\forall n \geq 0 \; , \; \alpha_n \leq 2^n r^{-j'_n} , \tag{7.214}$$

and since $\sum_{n\geq 0} \alpha_n \leq K\gamma_2(T, d_2)$ this yields

$$\sum_{n\geq 0} 2^n r^{-j'_n} \leq \sum_{n\geq 0} 2^n r^{-j_n} + Kr\gamma_2(T, d_2) . \tag{7.215}$$

Since $\varphi_j(s, t)$ is increasing with j, (7.210) and (7.211) hold for j'_n instead of j_n. That is, replacing the sequence (j_n) by the sequence (j'_n), we may assume that (7.213) holds.

The main argument starts now. For $n \geq 0$ we construct sets B_n. This idea is that these sets are of rather large measure while being small for both d_2 and φ_{j_n} (following the philosophy of Theorem 4.5.13). We will then show that these sets are also small for d_p and this will yield (7.212). We choose $B_0 = T$. By (7.211) for $n \geq 1$, the set $A_n := \{s \in T \; ; \; \varphi_{j_n}(s, 0) \leq 2^n\}$ satisfies $\mu(A_n) \geq 1/N_n$. Furthermore we can cover T by N_n balls $(C_j)_{j\leq N_n}$ of radius $\leq 2e_n(T, d_2)$. The sets $A_n \cap C_j$ for $j \leq N_n$ cover A_n, so that $\mu(A_n) \leq \sum_{i\leq N_n} \mu(A_n \cap C_j)$. Thus one of the sets $A_n \cap C_j$ (call it B_n) is such that $\mu(B_n) \geq \mu(A_n)/N_n \geq 1/N_n^2 = 1/N_{n+1}$. Since $B_n \subset C_j$ we have

$$\Delta(B_n, d_2) \leq 4e_n(T, d_2) . \tag{7.216}$$

Our goal next is to prove that for $n \geq 0$ we have

$$s, t \in B_n \Rightarrow d_p(s, t) \leq K2^{n/p} r^{-j_n} . \tag{7.217}$$

Since φ_j is the square of a distance, and since $\varphi_{j_n}(s, 0) \leq 2^n$ for $s \in A_n$, we have

$$s, t \in B_n \Rightarrow \varphi_{j_n}(s, t) \leq 2(\varphi_{j_n}(s, 0) + \varphi_{j_n}(0, t)) \leq 2^{n+2} . \tag{7.218}$$

Next, for any number b with $|b| \le 1/C$, using (7.91) we have

$$\mathsf{E}|b\theta_i|^2 \wedge 1 \ge \mathsf{P}(|\theta_i| \ge 1/|b|) \ge |b|^p/K \ . \tag{7.219}$$

Thus for $|b| \le 1/C$ we have $|b|^p \le K\mathsf{E}|b\theta_i|^2 \wedge 1$. Consequently, since $|b|^p \le K|b|^2$ for $|b| \ge 1/C$ we have

$$|b|^p \le K\mathsf{E}|b\theta_i|^2 \wedge 1 + K|b|^2 \ .$$

Using this for $b = r^{j_n} a_i(\chi_i(s) - \chi_i(t))$ and summing over i we get

$$r^{pj_n} d_p(s,t)^p \le K\varphi_{j_n}(s,t) + Kr^{2j_n} d_2(s,t)^2 \ . \tag{7.220}$$

Using (7.218), and recalling that by (7.216) we have $d_2(s,t) \le 4e_n(T,d_2)$, we have proved that

$$s,t \in B_n \Rightarrow r^{pj_n} d_p(s,t)^p \le K2^n + Kr^{2j_n} e_n(T,d_2)^2 \le K2^n \ ,$$

where we have used in the last inequality that $2^{n/2} e_n(T,d_2) \le 2^n r^{-j_n}$ by (7.213), i.e., $r^{j_n} e_n(T,d_2) \le 2^{n/2}$. We have proved (7.217).

To finish the proof, using the translation invariance of d_p and μ, it is then true from (7.217) that for $n \ge 1$

$$\mu(\{s \in T \ ; \ d_p(s,0) \le K2^{n/p} r^{-j_n}\}) \ge \mu(B_n) \ge N_{n+1}^{-1} \ .$$

Then (7.5) and the definition (7.3) of ϵ_n imply that $e_{n+1}(T,d_p) \le K2^{n/p} r^{-j_n}$. Since $e_0(T,d_p) \le \Delta(T,d_p) \le Kr^{-j_0}$ by (7.217) used for $n = 0$, we then obtain (7.212) (using (7.6) in the first inequality):

$$\gamma_q(T,d_p) \le K \sum_{n \ge 0} 2^{n/q} e_n(T,d_p) \le K \sum_{n \ge 0} 2^n r^{-j_n} \ . \qquad \square$$

Corollary 7.11.2 *Under the conditions of Proposition 7.11.1 we have*

$$\gamma_q(T,d_p) \le K \sum_{n \ge 0} 2^n r^{-j_n} + K \sum_i |a_i| \mathbf{1}_{\{|a_i| > r^{-j_0}|\}} \ . \tag{7.221}$$

Proof The idea is to use Corollary 7.8.12 to control the term $\gamma_2(T,d_2)$ of (7.212). First

$$\mathsf{E}|r^j a_i \theta_i(\chi_i(s) - \chi_i(t))|^2 \wedge 1 \ge \mathsf{P}(|\theta_i| \ge 1)|r^j a_i(\chi_i(s) - \chi_i(t))|^2 \wedge 1 \ . \tag{7.222}$$

Let us set $W_0 = \max_i \mathsf{P}(|\theta_i| \ge 1)^{-1}$, so that by (7.91) we have $W_0 \le K$ where K depends on C only. Let us set $\psi_j(s,t) = \sum_i |r^j a_i(\chi_i(s) - \chi_i(t))|^2 \wedge 1$. Thus, recalling (7.209), it follows from (7.222) that $\psi_j(s,t) \le W_0 \varphi_j(s,t)$. Setting

$D_n = \{s \in T; \psi_j(s,0) \le W_0 2^n\}$ it follows from (7.210) and (7.211) that $D_0 = T$ and $\mu(D_n) \ge N_n^{-1}$ for $n \ge 1$. We appeal to Corollary 7.8.12 to obtain that $\gamma_2(T, d_2) \le K \sum_{n\ge 0} 2^n r^{-j_n} + K \sum_i |a_i| \mathbf{1}_{\{|a_i| > r^{-j_0}|\}}$. Combining this with (7.212) implies (7.221). □

We are now ready to prove the following generalization of Proposition 7.5.13, of independent interest:

Proposition 7.11.3 *Consider* $1 < p < 2$.[46] *Consider independent symmetric r.v.s* θ_i *which satisfy (7.91). Then there is a constant* α *depending only on* C *such that for any numbers* a_i *and characters* χ_i*, we have*

$$\mathsf{P}\Big(\Big\| \sum_i a_i \theta_i \chi_i \Big\| \ge M\Big) < \alpha \Rightarrow \gamma_q(T, d_p) \le KM \ , \tag{7.223}$$

where d_p *is the distance (7.84) and* q *is the conjugate of* p.

Proof Define as usual $\varphi_j(s,t) = \sum_i \mathsf{E}|r^j a_i \theta_i (\chi_i(s) - \chi_i(t))|^2 \wedge 1$. According to Theorem 7.5.1 if α is small enough we can find numbers $j_n \in \mathbb{Z}$ such that $D_0 = T$, $\mu(D_n) \ge N_n^{-1}$ for $n \ge 1$ and $\sum_{n\ge 0} 2^n r^{-j_n} \le KM$, where $D_n = \{s \in T; \varphi_{j_n}(s,0) \le 2^n\}$. The conditions (7.210) and (7.211) of Proposition 7.11.1 are then satisfied, so that (7.221) of Corollary 7.11.2 holds, and this inequality implies

$$\gamma_q(T, d_p) \le KM + K \sum_i |a_i| \mathbf{1}_{\{|a_i| > r^{-j_0}\}} \ . \tag{7.224}$$

To control the last term, we will prove that

$$\operatorname{card}\{i; |a_i| \ge r^{-j_0}\} \le K \ ; \ \max_i |a_i| \le KM \ , \tag{7.225}$$

which will end the proof. We appeal to Lemma 7.8.19: (7.160) implies $\sum_i \mathsf{E}(|r^{j_0} a_i \theta_i|^2 \wedge 1) \le 2$. Since $\mathsf{E}(|r^{j_0} a_i \theta_i|^2 \wedge 1) \ge 1/K$ for $|a_i| \ge r^{-j_0}$ this proves the first part of (7.225). Consider now a certain index i_0. We are going to prove that if α is small enough then $|a_{i_0}| \le M/C$, concluding the proof. Let $\eta_i = -1$ if $i \ne i_0$ and $\eta_{i_0} = 1$ so that the sequence $(\eta_i \theta_i)$ has the same distribution as the sequence (θ_i) and thus when $\mathsf{P}(|\sum_i a_i \theta_i| \le M) \ge 1 - \alpha$ we also have $\mathsf{P}(|\sum_i \eta_i a_i \theta_i| \le M) \ge 1 - \alpha$. Since $2|a_{i_0}\theta_{i_0}| \le |\sum_i a_i \theta_i| + |\sum_i \eta_i a_i \theta_i|$ we then have $\mathsf{P}(|a_{i_0}\theta_{i_0}| \ge M) \le \mathsf{P}(|\sum_i a_i \theta_i| \ge M) + \mathsf{P}(|\sum_i \eta_i a_i \theta_i| \ge M) \le 2\alpha$. On the other hand from (7.91) we have $\mathsf{P}(|\theta_{i_0}| \ge C) \ge 1/C^{p+1}$. Assuming that we have chosen α small enough that $2\alpha < 1/C^{p+1}$, we then conclude that $C < M/|a_{i_0}|$. □

[46] We leave as a challenge to the reader to consider the case $p = 2$. In that case it suffices to assume that for a certain $\beta > 0$ we have $\mathsf{P}(|\theta_i| \ge \beta) \ge \beta$.

7.11.2 Proof of Theorem 7.5.18

Proof of Theorem 7.5.18 (b) The condition $\gamma_q(T, d_p)$ is stronger than the sufficient condition of Theorem 7.5.16. This is shown by Proposition 7.5.9.[47] □

Proof of Theorem 7.5.18 (c) Combine Proposition 7.11.3, Lemma 7.10.4, and the next lemma. □

Lemma 7.11.4 *Consider an increasing sequence (d_k) of translation-invariant (quasi) distances on T. Assume that the limiting distance $d(s,t) = \lim_{k\to\infty} d_k(s,t)$ is finite. Then $\gamma_q(T,d) \le K \sup_k \gamma_q(T, d_k)$.*

Proof Combining (7.5) and (7.6), we obtain that for any translation-invariant distance δ, we have $\sum_{n\ge 0} 2^{n/q} e_n(T,\delta) \le \gamma_q(T,\delta) \le K \sum_{n\ge 0} 2^{n/q} e_n(T,\delta)$, so that it suffices to prove that $e_n(T,d) \le 2\lim_{k\to\infty} e_n(T, d_k)$. According to Lemma 2.9.3 (a) given $a < e_n(T,d)$ we can find points $(t_\ell)_{\ell \le N_n}$ such that $d(t_\ell, t_{\ell'}) > a$ for $\ell \ne \ell'$. Then for k large enough we also have $d_k(t_\ell, t_{\ell'}) > a$ and thus $e_n(T, d_k) > a/2$. □

Exercise 7.11.5 Complete the proof of Theorem 7.5.18 (a) using similar but much easier arguments.

7.12 Explicit Computations

In this section we give some examples of concrete results that follow from the abstract theorems that we stated. The link between the abstract theorems and the classical results of Paley and Zygmund and Salem and Zygmund has been thoroughly investigated by Marcus and Pisier [61], and there is no point reproducing it here. Rather, we develop a specific direction that definitely goes beyond these results. It was initiated in [118] and generalized in [33]. There is a seemingly infinite number of variations on the same theme. The one variation we present has no specific importance but illustrates how precisely these matters are now understood; see Theorem 7.12.5 as a vivid example.

We shall consider only questions of convergence. We use the notation of Exercise 7.3.9, so that T is the group of complex numbers of modulus 1, and for $t \in T$, $\chi_i(t) = t^i$ is the i-th power of t. We consider independent r.v.s $(X_i)_{i\ge 1}$ and complex numbers $(a_i)_{i\ge 1}$, and we are interested in the case where[48]

$$Z_i(t) = a_i X_i \chi_i(t) = a_i X_i t^i\ . \tag{7.226}$$

[47] Or, more accurately, by the version of the proposition when finite sums are replaced by series.

[48] The reason behind our formulation is that soon the r.v.s (X_i) will be assumed to be i.i.d.

We make the following assumption:

$$\sum_{i\geq 1} \mathsf{E}(|a_i X_i|^2 \wedge 1) < \infty \,. \tag{7.227}$$

To study the convergence of the series, without loss of generality, we assume that $a_i \neq 0$ for each i.

Theorem 7.12.1 *Under the previous conditions, for each $n \geq 0$ there exists a number λ_n such that*

$$\sum_{i\geq N_n} \mathsf{E}\Big(\frac{|a_i X_i|^2}{\lambda_n^2} \wedge 1\Big) = 2^n \,, \tag{7.228}$$

and the series $\sum_{i\geq 1} a_i \varepsilon_i X_i \chi_i$ converges uniformly a.s. whenever

$$\sum_{n\geq 0} 2^n \lambda_n < \infty \,. \tag{7.229}$$

As a consequence we obtain the following:

Corollary 7.12.2 *If*

$$\sum_{n\geq 0} 2^{n/2}\Big(\sum_{i\geq N_n} |a_i|^2\Big)^{1/2} < \infty \,, \tag{7.230}$$

then the series $\sum_{i\geq 1} a_i \varepsilon_i \chi_i$ converges uniformly a.s.

Proof Since $|\varepsilon_i| = 1$, (7.228) holds for $\lambda_n^2 \leq 2^{-n}\sum_{i\geq N_n} |a_i|^2$, and under (7.230) such a sequence satisfies (7.229). □

Exercise 7.12.3 Compare this result with (7.38).

Proof of Theorem 7.12.1 First we observe from (7.227) that for any N the function $\Psi(y) := \sum_{i\geq N} \mathsf{E}(|y a_i X_i|^2 \wedge 1)$ is continuous and satisfies $\lim_{y\to 0} \Psi(y) = 0$ and $\lim_{y\to\infty} \Psi(y) = \infty$, and this proves the existence of λ_n. The proof will then rely on Theorem 7.5.16. *For a change, throughout this section, we use the value $r = 2$.* Let us consider $s \in T$, and let us assume that for some integer $n \geq 1$, we have

$$|s - 1| \leq \frac{1}{N_{n+1}} \,. \tag{7.231}$$

Let us observe the following inequality: For $i \geq 1$,

$$|s^i - 1| \leq i|s - 1| \,. \tag{7.232}$$

We then write, for any integer $j \in \mathbb{Z}$, using also that $|s^i - 1| \le |s|^i + 1 \le 2$ in the last line,

$$\sum_{i\ge1} \mathsf{E}(|2^j(Z_i(s) - Z_i(0))|^2 \wedge 1) = \sum_{i\ge1} \mathsf{E}(|2^j a_i X_i(s^i - 1)|^2 \wedge 1)$$

$$\le \sum_{0\le m<n} \sum_{N_m\le i<N_{m+1}} \mathsf{E}(|2^j i a_i X_i(s - 1)|^2 \wedge 1)$$

$$+ \sum_{i\ge N_n} \mathsf{E}(|2^{j+1} a_i X_i|^2 \wedge 1) . \tag{7.233}$$

From (7.228) we observe that

$$\lambda_n 2^{j+1} \le 1 \Rightarrow \sum_{i\ge N_n} \mathsf{E}(|2^{j+1} a_i X_i|^2 \wedge 1) \le \sum_{i\ge N_n} \mathsf{E}\Big(\frac{|a_i X_i|^2}{\lambda_n^2} \wedge 1\Big) = 2^n . \tag{7.234}$$

Also, for $i \le N_{m+1}$ and $m < n$, (7.231) implies $i|s - 1| \le N_{m+1}/N_{n+1} \le N_n/N_{n+1} = 1/N_n$. Consequently, it follows from (7.228) again that

$$\lambda_m 2^j \le N_n \Rightarrow \sum_{N_m\le i<N_{m+1}} \mathsf{E}(|2^j i a_i X_i(s - 1)|^2 \wedge 1) \le \sum_{i\ge N_m} \mathsf{E}\Big(\frac{|a_i X_i|^2}{\lambda_m^2} \wedge 1\Big) = 2^m . \tag{7.235}$$

Consider the largest integer j_n which satisfies both $\lambda_n 2^{j_n+1} \le 1$ and $\lambda_m 2^{j_n} \le N_n$ for each $m < n$. Using (7.233), (7.234), and (7.235), we then get

$$\sum_{i\ge1} \mathsf{E}(|2^{j_n}(Z_i(s) - Z_i(0))|^2 \wedge 1) \le \sum_{0\le m<n} 2^m + 2^n \le 2^{n+1} . \tag{7.236}$$

Moreover the definition of j_n shows that either $\lambda_n 2^{j_n+2} \ge 1$ (in which case $2^{-j_n} \le 4\lambda_n$) or else $\lambda_m 2^{j_n+1} \ge N_n$ for some $m < n$ (in which case $2^{-j_n} \le 2\lambda_m/N_n$), so that

$$2^{-j_n} \le 4\lambda_n + 2 \sum_{0\le m<n} \frac{\lambda_m}{N_n} . \tag{7.237}$$

Let us denote by U_n the set of points s that satisfy (7.231). Thus U_n is an "interval" on the unit circle, the set of points of the type $\exp(ix)$ where $|x| \le \tau_n$, where $0 < \tau_n < \pi/2$ satisfies $2\sin(\tau_n/2) = 1/N_{n+1}$. Denoting by μ the Haar measure of T, for $n \ge 1$ we have $\mu(U_n) = \tau_n/\pi$, so that we certainly have

$\mu(U_n) \ge 1/N_{n+2}$. Recalling (7.236) we have proved that

$$\mu\Big(\Big\{s \in T \ ; \ \sum_{i\ge 1} \mathsf{E}(|2^{j_n}(Z_i(s) - Z_i(0))|^2 \wedge 1) \le 2^{n+1}\Big\}\Big) \ge \frac{1}{N_{n+2}} ,$$

while (7.229) and (7.237) imply that $\sum_{n\ge 0} 2^{n-j_n} < \infty$. Using Theorem 7.5.16 this completes the proof. □

The following provides a converse of Theorem 7.12.1 under a mild regularity condition:

Theorem 7.12.4 *Under the conditions of Theorem 7.12.1, assume moreover that the sequence (X_i) is i.i.d. and that for a certain number $C > 0$, one has*

$$k \le m \le 2k \Rightarrow |a_k| \le C|a_m| . \tag{7.238}$$

Then (7.229) holds whenever the series $\sum_{i\ge 1} a_i \varepsilon_i X_i \chi_i$ converges uniformly a.s.

Proof We use Theorem 7.5.16 to obtain a sequence (j_n) with $\sum_{n\ge 0} 2^{n-j_n} < \infty$ and

$$\forall n \ge 1 \ , \ \mu\Big(\Big\{s \in T \ ; \ \sum_{i\ge 1} \mathsf{E}(|2^{j_n}(Z_i(s) - Z_i(0))|^2 \wedge 1) \le 2^{n}\Big\}\Big) \ge \frac{1}{N_n} . \tag{7.239}$$

We will prove that (7.239) implies that

$$\lambda_{n+3} \le LC^2 2^{-j_n} , \tag{7.240}$$

completing the proof. The set of $s \in T$ such that $|s - 1| \le 1/(2N_n)$ is of measure $1/\pi N_n$, so it cannot contain the set considered in (7.239). Thus we can find $s \in T$ with

$$|s - 1| \ge \frac{1}{2N_n} \tag{7.241}$$

and

$$\sum_{i\ge 1} \mathsf{E}(|2^{j_n} a_i X_i (s^i - 1)|^2 \wedge 1) \le 2^n , \tag{7.242}$$

where we have been also using that $Z_i(s) = a_i X_i s^i$. Now let $J = \{i \ge 1; |s^i - 1| \ge 1/4\}$, so that (7.242) yields

$$\sum_{i\in J} \mathsf{E}(|2^{j_n-2} a_i X_i|^2 \wedge 1) \le 2^n , \tag{7.243}$$

and the idea is to compare with (7.228) to bound λ_n from above. To implement the idea, we will show that there are many values of $i \geq 2^{2^n+3}$ in J. Indeed we have

$$\sum_{2^p \le i < 2^{p+1}} s^i = s^{2^p} \frac{s^{2^p} - 1}{s - 1} ,$$

so that using (7.241)

$$\Big| \sum_{2^p \le i < 2^{p+1}} s^i \Big| \le 4N_n .$$

Now we have $\sum_{2^p \le i < 2^{p+1}} 1 = 2^p$, so that

$$\Big| \sum_{2^p \le i < 2^{p+1}} (s^i - 1) \Big| \ge 2^p - 4N_n .$$

If $p \geq 2^n + 3$ we have $2^p \geq 8N_n$ and thus

$$2^{p-1} \le 2^p - 4N_n \le \sum_{2^p \le i < 2^{p+1}} |s^i - 1| . \tag{7.244}$$

Let now

$$I_p = \{i \ ; \ 2^p \le i < 2^{p+1} \ , \ |s^i - 1| \ge 1/4\} . \tag{7.245}$$

Since there are 2^p terms on the right-hand side of (7.244), each of which is ≤ 2, it follows that

$$2^{p-1} \le \sum_{2^p \le i < 2^{p+1}} |s^i - 1| \le 2 \operatorname{card} I_p + 2^p \frac{1}{4} .$$

so that

$$\operatorname{card} I_p \ge 2^{p-3} . \tag{7.246}$$

From (7.238) for $2^p \leq i < 2^{p+1}$, we have $|a_i| \geq |a_{2^p}|/C$ and combining with (7.246),

$$2^{p-3} \mathsf{E}\Big(\Big| \frac{2^{j_n - 2}}{C} a_{2^p} X_{2^p} \Big|^2 \wedge 1 \Big) \le \sum_{2^p \le i < 2^{p+1}} \mathsf{E}(|2^{j_n} a_i X_i (s^i - 1)|^2 \wedge 1) . \tag{7.247}$$

Using (7.238) again, for $2^{p-1} \le i < 2^p$, we have $|a_{2^p}| \ge |a_i|/C$ and thus

$$2^{-3} \sum_{2^{p-1} \le i < 2^p} \mathsf{E}\Big(\Big|\frac{2^{j_n-2}}{C^2} a_i X_i\Big|^2 \wedge 1\Big) \le 2^{p-3} \mathsf{E}\Big(\Big|\frac{2^{j_n-2}}{C} a_{2^p} X_{2^p}\Big|^2 \wedge 1\Big) . \tag{7.248}$$

Combining with (7.247), summing over $p \ge 2^n + 3$ and combining with (7.242) yields

$$2^{-3} \sum_{i \ge 2^{2^n}+2} \mathsf{E}\Big(\Big|\frac{2^{j_n-2}}{C^2} a_i X_i\Big|^2 \wedge 1\Big) \le 2^n , \tag{7.249}$$

and, in particular,

$$\sum_{i \ge N_{n+3}} \mathsf{E}\Big(\Big|\frac{2^{j_n-2}}{C^2} a_i X_i\Big|^2 \wedge 1\Big) \le 2^{n+3} .$$

By definition of λ_n this implies

$$\frac{2^{j_n-2}}{C^2} \le \frac{1}{\lambda_{n+3}} .$$

This proves (7.240). □

To give a still more explicit example, we mention the following:

Theorem 7.12.5 *If (X_i) denotes an i.i.d. sequence distributed like X, the series $\sum_{i \ge 1} \frac{1}{i} \varepsilon_i X_i \chi_i$ converges uniformly a.s. if and only if*

$$\mathsf{E}|X| \log\log(|X| + 3) < \infty . \tag{7.250}$$

Proof Since the sequence $a_k = 1/k$ satisfies (7.238), it suffices from Theorems 7.12.1 and 7.12.4 to prove that (7.250) is equivalent to (7.229). The proof uses standard methods that are not related to the ideas of this work. It can be found in Lemma 2.1 of [118]. □

7.13 Vector-Valued Series: A Theorem of Fernique

This section illustrates X. Fernique's decisive contributions to the ideas presented in this volume. It is a side story, which can be skipped at first reading. We assume that the reader has some very basic knowledge about Banach spaces, such as formula (19.1).

We consider a compact Abelian group T and a complex Banach space E (nothing is lost by assuming that E is finite-dimensional). We denote by $\|\cdot\|$ the norm of E. Consider (finitely many) vectors a_i of E and characters χ_i on T. Consider independent standard Gaussian r.v.s g_i. We are interested in the sum $\sum_i a_i g_i \chi_i(t)$ and more specifically in estimating the quantity

$$\mathsf{E} \sup_{t\in T} \Big\| \sum_i a_i g_i \chi_i(t) \Big\| . \tag{7.251}$$

We denote by x^* the generic element of the dual E^* of E.

Theorem 7.13.1 ([33]) *We have*

$$\mathsf{E} \sup_{t\in T} \Big\| \sum_i a_i g_i \chi_i(t) \Big\| \le L\Big(\mathsf{E}\Big\| \sum_i a_i g_i \Big\| + \sup_{\|x^*\|\le 1} \mathsf{E} \sup_{t\in T} \Big| \sum_i x^*(a_i) g_i \chi_i(t) \Big|\Big) . \tag{7.252}$$

Here $\|x^*\|$ denotes the (dual) norm of x^*. It is obvious that

$$\mathsf{E}\Big\| \sum_i a_i g_i \Big\| \le \mathsf{E} \sup_{t\in T} \Big\| \sum_i a_i g_i \chi_i(t) \Big\| ,$$

$$\sup_{\|x^*\|\le 1} \mathsf{E} \sup_{t\in T} \Big| \sum_i x^*(a_i) g_i \chi_i(t) \Big| \le \mathsf{E} \sup_{t\in T} \Big\| \sum_i a_i g_i \chi_i(t) \Big\| .$$

Thus the bound (7.252) is of the correct order. Furthermore the quantities in the left-hand side are simpler than the right-hand side.

Proof The overall strategy of proof is the obvious one. We know how to estimate the supremum of a Gaussian process from the value of the functional $\gamma_2(T, d)$, and we have to relate the distances corresponding to the different Gaussian processes occurring in (7.252).

Let us denote by E_1^* the unit ball of E^*. For $(x^*, t) \in E_1^* \times T$ we set $X_{x^*,t} = \sum_i x^*(a_i) g_i \chi_i(t)$, so that

$$\mathsf{E} \sup_{t\in T} \Big\| \sum_i a_i g_i \chi_i(t) \Big\| = \mathsf{E} \sup_{(x^*,t)\in E_1^* \times T} |X_{x^*,t}| . \tag{7.253}$$

The canonical distance on $E_1^* \times T$ associated with the Gaussian process $(X_{x^*,t})$ is given by

$$d((x^*, s), (y^*, t))^2 = \sum_i |x^*(a_i)\chi_i(s) - y^*(a_i)\chi_i(t)|^2 . \tag{7.254}$$

Denoting by LS the right-hand side of (7.252), the goal is to prove that

$$\gamma_2(E_1^* \times T, d) \le LS \,. \tag{7.255}$$

On E_1^* we consider the distance δ given by

$$\delta(x^*, y^*)^2 = \sum_i |x^*(a_i) - y^*(a_i)|^2 \,. \tag{7.256}$$

Using Lemma 7.2.4 we obtain

$$\gamma_2(E_1^*, \delta) \le L\mathsf{E} \sup_{x^* \in E_1^*} |X_{x^*,0}| = L\mathsf{E}\Big\| \sum_i a_i g_i \Big\| \le LS \,. \tag{7.257}$$

Given $z^* \in E_1^*$ we consider the following distance on T:

$$d_{z^*}(s,t)^2 = \sum_i |z^*(a_i)\chi_i(s) - z^*(a_i)\chi_i(t)|^2 \,, \tag{7.258}$$

so that by Lemma 7.2.4 again we obtain

$$\forall z^* \in E_1^* \,, \gamma_2(T, d_{z^*}) \le \mathsf{E} \sup_{t \in T} \Big| \sum_i z^*(a_i) g_i \chi_i(t) \Big| \le LS \,. \tag{7.259}$$

Since the distance d_{z^*} is translation-invariant, combining (7.4) and (7.5) yields

$$\sum_{n \ge 0} 2^{n/2} e_n(T, d_{z^*}) \le LS \,. \tag{7.260}$$

The next task is to relate the distance d with the distances δ and d_{z^*}. First, since $|\chi_i(t)| = 1$, we have

$$d((x^*, t), (y^*, t)) = \delta(x^*, y^*) \,, \tag{7.261}$$

and also

$$d((x^*, s), (x^*, t)) = d_{x^*}(s, t) \,. \tag{7.262}$$

Given $x^*, y^*, z^* \in E_1^*$, and $s, t \in T$, we then have

$$\begin{aligned} d((x^*, s), (y^*, t)) &\le d((x^*, s), (z^*, s)) + d((z^*, s), (z^*, t)) + d((z^*, t), (y^*, t)) \\ &= \delta(x^*, z^*) + d_{z^*}(s, t) + \delta(y^*, z^*) \,. \end{aligned} \tag{7.263}$$

We first note that this implies

$$\Delta(E_1^* \times T, d) \le 2\Delta(E_1^*, \delta) + \sup_{z^* \in E_1^*} \Delta(T, d_{z^*}) \le LS\,, \tag{7.264}$$

using (7.257) and (7.259) in the last inequality. In the remainder of the proof, we deduce (7.255) from (7.257), (7.263), and (7.260), which finishes the proof using Theorem 2.7.11. Let us consider an admissible sequence $(\mathcal{A}_n)$ of partitions of E_1^* such that

$$\sup_{x^* \in E_1^*} \sum_{n \ge 0} 2^{n/2} \Delta(A_n(x^*), \delta) \le LS\,. \tag{7.265}$$

Given $A \in \mathcal{A}_n$ let us select a point $z^*(n, A) \in A$ for which

$$e_n(T, d_{z^*(n,A)}) \le 2 \inf\{e_n(T, d_{z^*})\,;\ z^* \in A\}\,. \tag{7.266}$$

We then construct a partition $\mathcal{C}_{A,n}$ of T into N_n sets, each of which are of diameter $\le 4e_n(T, d_{z^*(n,A)})$ for the distance $d_{z^*(n,A)}$. We consider the partition $\mathcal{B}'_n$ of $E_1^* \times T$ in sets of the type $A \times C$ where $A \in \mathcal{A}_n$ and $C \in \mathcal{C}_{A,n}$. Its cardinality is $\le N_n^2 = N_{n+1}$. Let us define $\mathcal{B}_n$ as the partition of $E_1^* \times T$ generated by $\mathcal{B}'_1, \ldots, \mathcal{B}'_n$ so that as usual the sequence $(\mathcal{B}_n)$ increases and $\operatorname{card} \mathcal{B}_n \le N_{n+2}$. Consider a point $(x^*, t) \in E_1^* \times T$. Then, denoting by $B_n(x^*, t)$ the set of $\mathcal{B}_n$ which contains this point, we have

$$B_n(x^*, t) \subset A \times C\,,$$

where $A = A_n(x^*)$ and C is the element of the partition $\mathcal{C}_{A,n}$ that contains t. For any z^*, (7.263) implies

$$\Delta(B_n(x^*, t), d) \le L(\Delta(A_n(x^*), \delta) + \Delta(C, d_{z^*}))\,. \tag{7.267}$$

Using the definition of the partition $\mathcal{C}_{A,n}$ in the first inequality and the choice of $z^*(n, A)$ in the second one, we obtain

$$\Delta(C, d_{z^*(n,A)}) \le 4e_n(T, d_{z^*(n,A)}) \le 8e_n(T, d_{x^*})\,,$$

and therefore using (7.267) for $z^* = z^*(n, A)$ we get

$$\Delta(B_n(x^*, t), d) \le L(\Delta(A_n(x^*), \delta) + e_n(T, d_{x^*}))\,.$$

It then follows from (7.260) and (7.265) that

$$\sum_{n \ge 0} 2^{n/2} \Delta(B_n(x^*, t), d) \le LS\,,$$

so that combining with Lemma 2.9.10 for $\tau = 2$ and using also (7.264) yields (7.255) and finishes the proof. □

Does (7.252) remain true when the Gaussian r.v.s are replaced by Bernoulli r.v.s? That is, is it true that

$$\mathsf{E}\sup_t \Big\| \sum_i \varepsilon_i a_i \chi_i(t) \Big\| \le L\mathsf{E}\Big\| \sum_i \varepsilon_i a_i \Big\| + L \sup_{x^* \in E^*} \mathsf{E}\sup_{t\in T} \Big| \sum_i \varepsilon_i x^*(a_i)\chi_i(t) \Big| \ ? \tag{7.268}$$

It is while pondering this question that the author formulated the Bernoulli conjecture.

Exercise 7.13.2 Use Theorem 6.2.8 to prove (7.268).

If you find this exercise too difficult, its solution can be found in [53].

Key Ideas to Remember

- For a translation-invariant distance on a compact group T, the entropy numbers are basically determined by the Haar measure of the balls of a given radius, irrelevant of their shape. This is a tremendous simplification. The generic chaining is not needed and entropy number suffices.
- Consider characters χ_i (none of them constant), and consider (finitely many) complex numbers a_i. Consider the distance on T given by $d(s,t)^2 = \sum_i |a_i|^2 |\chi_i(s) - \chi_i(t)|^2$. Denote by ε_i independent signs and by g_i independent standard Gaussian r.v.s. Then the contraction principle ensures that

$$\mathsf{E}\sup_{t\in T} \Big| \sum_i a_i \varepsilon_i \chi_i(t) \Big| \le L\mathsf{E}\sup_{t\in T} \Big| \sum_i a_i g_i \chi_i(t) \Big| \ .$$

  The Marcus-Pisier theorem asserts that

$$\mathsf{E}\sup_{t\in T} \Big| \sum_i a_i g_i \chi_i(t) \Big| \le L\gamma_2(T,d) \le L\mathsf{E}\sup_{t\in T} \Big| \sum_i a_i \varepsilon_i \chi_i(t) \Big| \ .$$

- From now on ξ_i denote independent symmetric r.v.s. One has the general bound

$$\mathsf{E}\sup_{t\in T} \Big| \sum_i \xi_i \chi_i(t) \Big| \le L\gamma_2(T,d) \ , \tag{7.269}$$

  where now the distance d is given by $d(s,t)^2 = \sum_i \mathsf{E}|\xi_i|^2 |\chi_i(s) - \chi_i(t)|^2$.
- When the variables ξ_i are not square-integrable, the main problem in controlling the quantity $\mathsf{E}\sup_{t\in T} |\sum_i \xi_i \chi_i(t)|$ is to control the typical value of $\gamma_2(T, d_\omega)$ where d_ω is the random distance given by $d_\omega(s,t)^2 = \sum_i |\xi_i|^2 |\chi_i(s) - \chi_i(t)|^2$. The characteristics of T which make such a control possible cannot apparently be described using a single distance, but can be described using a "family of distances". This feature will occur in many problems we will study later.

- Through rather general principles (which will receive later a considerable generalization), one proves that the control of the typical value of $\gamma_2(T, d_\omega)$ implies a suitable smallness of T (as appropriately measured through a certain family of distances). Thus a control from above of $\mathsf{E}\sup_{t\in T}|\sum_i \xi_i \chi_i(t)|$ implies a suitable "smallness" of T.
- Besides the bound (7.269), one has the trivial bound

$$\mathsf{E}\sup_{t\in T}\Big|\sum_i \xi_i \chi_i(t)\Big| \le \sum_i \mathsf{E}|\xi_i| . \tag{7.270}$$

In a precise way, as stated in Theorem 7.5.14, every situation is a mixture of (7.269) and (7.270): we can find a decomposition $\xi_i = \xi_i' + \xi_i''$ such that $\sum_i \mathsf{E}|\xi_i'| \le L\mathsf{E}\sup_{t\in T}|\sum_i \xi_i\chi_i(t)|$ and $\gamma_2(T, d) \le L\mathsf{E}\sup_{t\in T}|\sum_i \xi_i\chi_i(t)|$ where the distance d is given by $d(s,t)^2 = \sum_i \mathsf{E}|\xi_i''|^2|\chi_i(s) - \chi_i(t)|^2$.
- Consider independent symmetric r.v.s. $(\xi_i)_{i\ge 1}$. The historically important problem of the uniform a.s. convergence of random Fourier series of the type $\sum_{i\ge 1}\xi_i\chi_i(t)$ is now completely understood, and the solution is unexpectedly simple. There are three rather different cases where this convergence holds. There is the case $\sum_{i\ge 1}\mathsf{P}(|\xi_i| \ne 0) < \infty$, the case $\sum_{i\ge 1}\mathsf{E}|\xi_i| < \infty$ and the case $\gamma_2(T, d) < \infty$ where the distance d is defined by $d(s,t)^2 = \sum_{i\ge 1}\mathsf{E}|\xi_i|^2|\chi_i(s) - \chi_i(t)|^2$. Conversely, every case where this convergence holds is in a precise sense a mixture of the previous three cases; see Theorem 7.5.17.

7.14 Notes and Comments

The characterization of convergence of random Fourier series is almost achieved in the paper [108]. This paper uses the most natural approach to upper bounds: chaining arguments for Bernoulli processes. The paper [108] still required some weak but unnecessary tail conditions, because the chaining was not organized in an optimal way. It is only while writing [132] that the author finally succeeded in removing all extraneous conditions, by organizing the chaining as is now done in Theorem 9.2.1. Random Fourier series have the particularity that the proof of upper bounds is much more difficult than the proof of lower bounds, while often it is the opposite which happens. The simpler arguments we present now were discovered much later.

I have given the "magic proof" of Proposition 7.5.13 as an homage to the paper [62] of Marcus and Pisier, which had a considerable influence on my own research. However, now that I understand things better, I feel that p-stable r.v.s are not intrinsically related to this problem and that it is simply a coincidence that they happen to have a tail in u^{-p}, so that one could argue that bringing them to bear on this question is a "trick" rather than a method and is somewhat misleading.

Consider independent symmetric r.v.s η_i. We have (basically) controlled $\mathsf{E}\sup_{t\in T}|\sum_i a_i\eta_i\chi_i(t)|$ using characteristics of T which involve a family of distances. One could ask for which r.v.s η_i these characteristics can be expressed in function of a single distance. While we have not tried to prove this, it seems that, besides the case $\mathsf{E}|\eta_i|^2 < \infty$, the only case is the case of the tails in u^{-p} (the characteristics are then expressed using the distance d_p; see (7.90) and (7.92)). This is one reason why this case has some importance.

I will end by a personal touch. I have been thinking about random Fourier series for over 35 years, and it is quite amazing that I could still make progress after all these efforts.

Chapter 8
Partitioning Scheme and Families of Distances

In the previous chapter, in the setting of random Fourier series, we introduced the idea that it does not suffice to use one single distance to control certain stochastic processes, but that a "whole family of distances is required"; see (7.63). The situation was however made easier by translation invariance, in the sense that covering numbers provide an accurate description of the "size" of the space with respect to these distances. This will no longer be the case in general. For an accurate description, we need to generalize to tools of Chap. 2 to "families of distances". In Sect. 8.1 we generalize to the setting of "families of distances" the first partitioning scheme of Sect. 2.9, and the reader needs first to master that result. In Sect. 8.3 we will apply this tool to the study of "canonical processes". In order to study canonical processes, we first need precise estimates on the tails of certain r.v.s, and these are the goal of Sect. 8.2. The present section can be seen as a far-reaching generalization of the majorizing measure theorem 2.10.1, but none of the further material depends on it.

8.1 The Partitioning Scheme

We consider a family of maps $(\varphi_j)_{j\in\mathbb{Z}}$, with the following properties:

$$\varphi_j : T \times T \to \mathbb{R}^+ \cup \{\infty\}\,,\ \varphi_{j+1} \geq \varphi_j \geq 0\,,\ \varphi_j(s,t) = \varphi_j(t,s)\,. \tag{8.1}$$

Such maps were of fundamental use in the previous chapter; see (7.63). In many of our applications, the maps φ_j will be squares of distances and will satisfy a version of the triangle inequality. We however *do not* assume that this is the case: in the setting of Sect. 8.3 such an inequality is not satisfied.

M. Talagrand, *Upper and Lower Bounds for Stochastic Processes*, Ergebnisse der Mathematik und ihrer Grenzgebiete. 3. Folge / A Series of Modern Surveys in Mathematics 60, https://doi.org/10.1007/978-3-030-82595-9_8

We define

$$B_j(t,c) = \{s \in T\ ;\ \varphi_j(t,s) \le c\}\ .$$

We recall that a functional F on a set T is a non-decreasing map from the subsets of T to $\mathbb{R}^+$. We consider functionals $F_{n,j}$ on T for $n \ge 0$, $j \in \mathbb{Z}$. We assume

$$F_{n+1,j} \le F_{n,j}\ ;\ F_{n,j+1} \le F_{n,j}\ . \tag{8.2}$$

We will assume that the functionals $F_{n,j}$ satisfy a "growth condition" very similar in spirit to Definition 2.8.3. This condition involves as main parameter an integer $\kappa \ge 5$. We set $r = 2^{\kappa-3}$, so that $r \ge 4$. The role of r is as in (2.76), the larger r, the weaker the growth condition.[1]

Definition 8.1.1 We say that the functionals $F_{n,j}$ satisfy the *growth condition* (for r) if the following occurs. Consider any $j \in \mathbb{Z}$, any $n \ge 1$ and $m = N_n$. Consider any sets $(H_\ell)_{1\le\ell\le m}$ that are separated in the following sense: There exist points $u, t_1, \dots, t_m$ in T for which $H_\ell \subset B_{j+2}(t_\ell, 2^{n+\kappa})$ and

$$\forall \ell, \ell' \le m\ ,\ \ell \ne \ell'\ ,\ \varphi_{j+1}(t_\ell, t_{\ell'}) \ge 2^{n+1}\ , \tag{8.3}$$

$$\forall \ell \le m\ ,\ t_\ell \in B_j(u, 2^n)\ . \tag{8.4}$$

Then

$$F_{n,j}\Big(\bigcup_{\ell\le m} H_\ell\Big) \ge 2^n r^{-j-1} + \min_{\ell\le m} F_{n+1,j+1}(H_\ell)\ . \tag{8.5}$$

We have not made assumptions on how φ_j relates to φ_{j+1}; but we have little chance to prove (8.5) unless $B_{j+2}(t_\ell, 2^{n+\kappa})$ is quite smaller than $B_{j+1}(t_\ell, 2^{n+1})$.

As we already stressed, the best way to illustrate a statement about families of distances is to carry out the case where

$$\varphi_j(s,t) = r^{2j} d(s,t)^2 \tag{8.6}$$

for a distance d on T. Denoting by $B(t,b)$ the ball for d of center t and radius b, we then have

$$B_j(t,c) = B(t, r^{-j}\sqrt{c})\ .$$

[1] The reason why we take r of the type $r = 2^{\kappa-3}$ for an integer κ is purely for technical convenience.

Thus in (8.3) we require that

$$\forall \ell, \ell' \le m \,,\; \ell \ne \ell' \,,\; d(t_\ell, t_{\ell'}) \ge 2^{(n+1)/2} r^{-j-1} := a \,. \tag{8.7}$$

On the other hand,

$$B_{j+2}(t_\ell, 2^{n+\kappa}) = B(t_\ell, 2^{(n+\kappa)/2} r^{-j-2}) = B(t_\ell, \eta a) \,,$$

where $\eta := 2^{(\kappa-1)/2}/r = 2/\sqrt{r}$. Thus the condition $H_\ell \subset B_{j+2}(t_\ell, 2^{n+\kappa})$ means that $H_\ell \subset B(t_\ell, \eta a)$. As r gets larger, η gets smaller, and recalling (8.7), this means that the sets H_ℓ become better separated, in the sense that they become smaller compared to their mutual distances. Also, (8.5) reads as

$$F_{n,j}\Big(\bigcup_{\ell \le m} H_\ell\Big) \ge 2^{(n-1)/2} a + \min_{\ell \le m} F_{n+1,j+1}(H_\ell) \,,$$

which strongly resembles (2.77). Thus, we should think of the term r^{-j-1} in the right-hand side of (8.5) as a normalization factor and the condition (8.5) as being uniform over j.

Theorem 8.1.2 *Assume that the functionals $F_{n,j}$ are as above and in particular satisfy the growth condition of Definition 8.1.1 and that, for some $j_0 \in \mathbb{Z}$, we have*

$$\forall s, t \in T \,,\; \varphi_{j_0}(s,t) \le 1 \,. \tag{8.8}$$

Assume also that[2]

$$\forall s, t \in T \,,\; \forall j \in \mathbb{Z} \,,\; \varphi_{j+1}(s,t) \ge r\varphi_j(s,t) \,. \tag{8.9}$$

Then there exists an admissible sequence $(\mathcal{A}_n)$ and for each $A \in \mathcal{A}_n$ an integer $j_n(A) \in \mathbb{Z}$ and a point $t_{n,A} \in T$ such that

$$A \in \mathcal{A}_n \,,\; C \in \mathcal{A}_{n-1} \,,\; A \subset C \Rightarrow j_{n-1}(C) \le j_n(A) \le j_{n-1}(C) + 1 \tag{8.10}$$

$$\forall t \in T \,,\; \sum_{n \ge 0} 2^n r^{-j_n(A_n(t))} \le L(rF_{0,j_0}(T) + r^{-j_0}) \tag{8.11}$$

$$\forall n \ge 0 \,,\; \forall A \in \mathcal{A}_n \,,\; A \subset B_{j_n(A)}(t_{n,A}, 2^n) \,. \tag{8.12}$$

[2] In [132] the present theorem is stated without assuming this condition but the proof given there is in error. The condition (8.9) is a very mild extra hypothesis, since in the separation condition, we have already implicitly assumed that $B_{j+2}(t_\ell, 2^{n+\kappa}) = B_{j+2}(t_\ell, (4r)2^{n+1})$ is quite smaller than $B_{j+1}(t_\ell, 2^{n+1})$.

Let us stress that we do not require that $t_{n,A} \in A$. Let us also note the new feature of (8.12) compared to our previous constructions. We do not control the size of the elements A of $\mathcal{A}_n$ by requiring that they are of "small diameter" (in the sense of controlling $\varphi_{j_n(A)}(s,t)$ from above for all $s, t \in A$), but by the condition (8.12), requiring that they are contained "in a small ball". This twist is required due to the possible failure of (any form of) the triangle inequality for the "distance" $\varphi_{j_n(A)}$.

To illustrate this result, we again carry out the case (8.6) (although in that case we do not have problems with the triangle inequality). Then (8.8) means that $\Delta(T, d) \le r^{-j_0}$, while (8.12) implies $\Delta(A, d) \le 2r^{-j_n(A)} 2^{n/2}$. Moreover (8.11) implies

$$\forall t \in T\,,\ \sum_{n \ge 0} 2^{n/2} \Delta(A_n(t), d) \le L(r F_{0,j_0}(T) + r^{-j_0})\,.$$

Taking for j_0 the largest integer such that $\Delta(T, d) \le r^{-j_0}$, we get

$$\gamma_2(T, d) \le Lr\big(F_{0,j_0}(T) + \Delta(T, d)\big)\,,$$

which is very similar to Theorem 2.9.1.

The proof of Theorem 8.1.2 relies on the following, where again the functionals are as above:

Lemma 8.1.3 *Consider a set $C \subset T$, and assume that for some integers $j \in \mathbb{Z}$ and $n \ge 1$ and for some $u \in T$, we have $C \subset B_j(u, 2^n)$. Then we can find a partition $(A_\ell)_{\ell \le m}$ of C, where $m = N_n$, such that for each $\ell \le m$ we have* either

$$\exists t_\ell \in C,\ A_\ell \subset B_{j+1}(t_\ell, 2^{n+1}) \tag{8.13}$$

or else

$$2^{n-1} r^{-j-1} + \sup_{s \in A_\ell} F_{n+1,j+1}(A_\ell \cap B_{j+2}(s, 2^{n+\kappa})) \le F_{n,j}(C)\,. \tag{8.14}$$

Proof Consider the set

$$D := \{s \in C\ ;\ 2^{n-1} r^{-j-1} + F_{n+1,j+1}(C \cap B_{j+2}(s, 2^{n+\kappa})) > F_{n,j}(C)\}\,.$$

As in Lemma 2.9.4, it follows from (8.5) that D can be covered by $< m$ balls of the type $B_{j+1}(t_\ell, 2^{n+1})$. Thus we can partition D in $< m$ sets A_ℓ which satisfy (8.13): $A_\ell \subset B_{j+1}(t_\ell, 2^{n+1})$, $t_\ell \in C$. The required partition consists of these sets together with the set $C \setminus D$, which automatically satisfies (8.14). □

Proof of Theorem 8.1.2 Let us repeat that the reader should be comfortable with the proof of Theorem 2.9.1 as many features here are nearly identical. To start the construction, we define $\mathcal{A}_0 = \{T\}$, $j_0(T) = j_0$, and take any point of T for t_{0,A_0}.

To construct $\mathcal{A}_{n+1}$ once $\mathcal{A}_n$ has been constructed, to each element C of $\mathcal{A}_n$, we apply Lemma 8.1.3 with $j = j_n(C)$ and $u = t_{n,C}$ to split C into $m = N_n$ pieces

$A_1, \ldots, A_m$. (Thus, the sequence $(\mathcal{A}_n)$ is admissible since $N_n^2 \le N_{n+1}$.) Let A be one of these sets.

When A satisfies (8.14), we set $j_{n+1}(A) = j = j_n(C)$ and $t_{n+1,A} = t_{n,C}$ so that (8.4) for A follows from the same relation for C.

When $A = A_\ell$ satisfies (8.13), we define instead $j_{n+1}(A) = j + 1$ and $t_{n+1,A} = t_\ell$. Thus (8.12) holds for A and $n+1$. Our construction satisfies the further important property that

$$A \in \mathcal{A}_{n+1}, C \in \mathcal{A}_n, A \subset C, j_{n+1}(A) = j_n(C) \Rightarrow t_{n+1,A} = t_{n,C} \tag{8.15}$$

$$A \in \mathcal{A}_{n+1}, C \in \mathcal{A}_n, A \subset C, j_{n+1}(A) = j_n(C) + 1 \Rightarrow t_{n+1,A} \in C\ . \tag{8.16}$$

Let us now prove that

$$A \in \mathcal{A}_{n'}, C \in \mathcal{A}_n, n' > n, A \subset C, j_{n'}(A) > j_n(C) \Rightarrow t_{n',A} \in C\ . \tag{8.17}$$

For $n \le s \le n'$ let us denote by A_s the unique element of $\mathcal{A}_s$ with $A \subset A_s$, so that $A_{n'} = A$ and $A_n = C$. Let n'' be the largest integer with $j_{n''}(A_{n''}) < j_{n'}(A)$, so that $n \le n'' < n'$. Thus for $n'' + 1 \le k \le n'$, we have $j_k(A_k) = j_{n'}(A_{n'})$. The value of $j_k(A_k)$ does not increase over this interval, and as a consequence (8.15), the value of t_{k,A_k} does not change over this interval, i.e., it holds that $t_{n',A} = t_{n''+1,A_{n''+1}}$. Furthermore from (8.16), we have $t_{n''+1,A_{n''+1}} \in A_{n''} \subset A_n = C$, proving (8.17).

Since (8.10) holds by construction, it remains only to prove (8.11). Let us fix once and for all a point $t \in T$, and to lighten notation, let $j(n) = j_n(A_n(t))$ and $a(n) = 2^n r^{-j(n)}$, so that we have to bound $\sum_{n \ge 0} a(n)$. Consider the set

$$J = \{0\} \cup \big\{n > 0\ ;\ j(n-1) = j(n)\ ,\ j(n+1) = j(n) + 1\big\}\ ,$$

and let us enumerate J as $0 = n_0 < n_1 < n_2 \ldots$, so that $j(n_{k+1}) \ge j(n_k + 1) = j(n_k) + 1$. Since $a(n+1) = 2r^{j(n)-j(n+1)}a(n)$, Lemma 2.9.5 used for $\alpha = 2$ implies that $\sum_{n \ge 0} a(n) \le L \sum_{n \in J} a(n)$ (as in (2.91)). We apply a second time Lemma 2.9.5 with $\alpha = 2$ to the sequence $(a(n))_{n \in J}$. Defining

$$I = \{0\} \cup \{n_k\ ,\ k \ge 1\ ;\ \forall \ell > 1\ ,\ \ell \ne k\ ,\ a(n_\ell) \le a(n_k) 2^{|k-\ell|}\}\ ,$$

Lemma 2.9.5 implies $\sum_{n \in J} a(n) \le 4 \sum_{n \in I} a(n)$, and it suffices to bound this latter sum.

Consider then $n_k \in I$, so that $a(n_{k+1}) \le 2a(n_k)$ i.e.

$$2^{n_{k+1}} r^{-j(n_{k+1})} \le 2^{n_k+1} r^{-j(n_k)}\ , \tag{8.18}$$

Thus $n^* := n_{k+1} + 1$ satisfies

$$j(n^*) = j(n_{k+1} + 1) = j(n_{k+1}) + 1 \ge j(n_k) + 2 \tag{8.19}$$

and

$$A_{n^*}(t) \subset B_{j(n^*)}(s, 2^{n^*}) ,$$

where $s = t_{n^*, A_{n^*}(t)} \in A_{n_k}(t)$ by (8.17). We can also rewrite (8.18) as

$$2^{n^*} r^{-j(n^*)+j(n_k)+2} \le r 2^{n_k+2} = 2^{n_k+\kappa-1} . \tag{8.20}$$

Now, as a consequence of (8.9), we have $B_{j+1}(s, u) \subset B_j(s, u/r)$ and thus, using also (8.20) in the last inequality,

$$B_{j(n^*)}(s, 2^{n^*}) \subset B_{j(n_k)+2}(s, 2^{n^*} r^{-j(n^*)+j(n_k)+2}) \subset B_{j(n_k)+2}(s, 2^{n_k+\kappa-1}) .$$

Since $A_{n^*}(t) \subset A_{n_k}(t)$, we then have

$$F_{n_k, j(n_k)+1}(A_{n^*}(t)) \le F_{n_k, j(n_k)+1}(A_{n_k}(t) \cap B_{j(n_k)+2}(s, 2^{n_k+\kappa-1})) . \tag{8.21}$$

Assuming now $k \ge 1$, we have $j(n_k - 1) = j(n_k)$ so that setting $n = n_k - 1$, we have $j(n+1) = j(n)$. It follows by construction that when we split $C = A_n(t)$ according to Lemma 8.1.3, $A_{n_k}(t) = A_{n+1}(t)$ is a piece A_ℓ that satisfies (8.14), so that in particular

$$\frac{1}{4r} a(n_k) + \sup_{s \in A_{n_k}(t)} F_{n_k, j(n_k)+1}(A_{n_k}(t) \cap B_{j(n_k)+2}(s, 2^{n_k+\kappa-1}))$$
$$\le F_{n_k-1, j(n_k)}(A_{n_k-1}(t)) . \tag{8.22}$$

It then follows from (8.21) that

$$\frac{1}{4r} a(n_k) \le F_{n_k-1, j(n_k)}(A_{n_k-1}(t)) - F_{n_k, j(n_k)+1}(A_{n^*}(t)) . \tag{8.23}$$

Let now $f(k) := F_{n_k, j(n_k)}(A_{n_k}(t))$. Since $n_{k-1} \le n_k - 1$ we have $A_{n_k-1}(t) \subset A_{n_{k-1}}(t)$. Since $j(n_k) \ge j(n_{k-1})$, we have, using (8.2),

$$F_{n_k-1, j(n_k)}(A_{n_k-1}(t)) \le f(k-1) .$$

Since $n^* \le n_{k+2}$ we have $A_{n_{k+2}}(t) \subset A_{n^*}(t)$. Since $n_k \le n_{k+2}$ and $j(n_k) + 1 = j(n_{k+1}) \le j(n_{k+2})$ it holds that $F_{n_k, j(n_k)+1}(A_{n^*}(t)) \ge f(k+2)$, so that (8.23) implies

$$\frac{1}{4r} a(n_k) \le f(k-1) - f(k+2) ,$$

and the proof follows as usual, by summing this inequality for $n_k \in I$ to bound $\sum_{n\in I} a(n)$, using also that $j(0) = j_0$ and using also $a(0) = r^{-j_0}$. □

8.2 Tail Inequalities

Consider independent symmetric r.v.s $(Y_i)_{i\geq 1}$. Assume that we control the tails of each of them. How do we control the tails of a sum $\sum_{i\geq 1} a_i Y_i$? Let us start with a particularly instructive case. The following is a simple consequence of Lemma 6.4.5:

Lemma 8.2.1 *Consider i.i.d. copies $(Y_i)_{i\geq 1}$ of a symmetric r.v. Y which satisfies the following condition:*

$$\forall u \geq 0\,,\ \mathsf{P}(|Y| \geq u) \leq 2\exp(-u)\,. \tag{8.24}$$

Then for numbers $(t_i)_{i\geq 1}$ and any $u > 0$ we have

$$\mathsf{P}\Big(\Big|\sum_{i\geq 1} t_i Y_i\Big| \geq u\Big) \leq 2\exp\Big(-\frac{1}{L}\min\Big(\frac{u^2}{\sum_{i\geq 1} t_i^2}, \frac{u}{\max_{i\leq k}|t_i|}\Big)\Big)\,. \tag{8.25}$$

Exercise 8.2.2 Assuming now that rather than (8.24), we have

$$\mathsf{P}(|Y| \geq u) \leq 2\exp(-u^p) \tag{8.26}$$

for some $p \geq 1$. Denote by q the conjugate exponent of p. Prove that for $p \leq 2$ we have

$$\mathsf{P}\Big(\Big|\sum_{i\geq 1} t_i Y_i\Big| \geq u\Big) \leq 2\exp\Big(-\frac{1}{K}\min\Big(\frac{u^2}{\sum_{i\geq 1} t_i^2}, \frac{u^p}{(\sum_{i\geq 1}|t_i|^q)^{p/q}}\Big)\Big) \tag{8.27}$$

whereas if $p > 2$

$$\mathsf{P}\Big(\Big|\sum_{i\geq 1} t_i Y_i\Big| \geq u\Big) \leq 2\exp\Big(-\frac{1}{K}\max\Big(\frac{u^2}{\sum_{i\geq 1} t_i^2}, \frac{u^p}{(\sum_{i\geq 1}|t_i|^q)^{p/q}}\Big)\Big)\,. \tag{8.28}$$

In parallel with the way we defined Bernoulli processes, one may now define a *canonical process based on the r.v.s Y_i* by

$$X_t = \sum_{i\geq 1} t_i Y_i \tag{8.29}$$

for $t \in \ell^2$. Assuming (8.24), combining (8.25) and Theorem 4.5.13 we obtain

$$\mathsf{E} \sup_{t \in T} X_t \le L(\gamma_2(T, d_2) + \gamma_1(T, d_\infty)) . \tag{8.30}$$

When $1 \le p \le 2$, under (8.26) one obtains similarly

$$\mathsf{E} \sup_{t \in T} X_t \le L(\gamma_2(T, d_2) + \gamma_p(T, d_q)) . \tag{8.31}$$

When $p \ge 2$, under (8.26), we obtain the bounds $\mathsf{E} \sup_{t \in T} X_t \le K\gamma_2(T, d_2)$ and $\mathsf{E} \sup_{t \in T} X_t \le K\gamma_p(T, d_q)$. One may then interpolate between these bounds to obtain

$$\mathsf{E} \sup_{t \in T} X_t \le K \inf\left\{\gamma_2(T_1, d_2) + \gamma_p(T_2, d_q) \; ; \; T \subset T_1 + T_2\right\} . \tag{8.32}$$

This is very similar to the bound (6.10) (see (6.9)) on Bernoulli processes.[3] The obvious question is whether the previous bounds can be reversed when (8.26) is optimal, say

$$\mathsf{P}(|Y| \ge u) = \exp(-u^p) . \tag{8.33}$$

The author proved this in [113]. These results were then generalized by R. Latała [48], who considers r.v.s with far more general tail conditions than (8.26). Latała's results are the object of the rest of this chapter. Latała's work often displays a very high level of sophistication, and this is certainly the case here.

Throughout this section and the next, we consider independent symmetric r.v.s $(Y_i)_{i \ge 1}$. We assume that the functions

$$U_i(x) = -\log \mathsf{P}(|Y_i| \ge x) \tag{8.34}$$

are convex. In the important special case (8.33), we have $U_i(x) = x^p$. Since it is only a matter of normalization, we assume that $U_i(1) = 1$. Since $U_i(0) = 0$ we then have $U_i'(1) \ge 1$ by convexity.

In the remaining of this section, we provide the proper generalization of (8.27) and (8.28). A first idea "is to redefine the function U_i as x^2 for $-1 \le x \le 1$". In order to preserve convexity, we consider the function $\hat{U}_i(x)$ (defined on all $\mathbb{R}$) given by

$$\hat{U}_i(x) = \begin{cases} x^2 \text{ if } 0 \le |x| \le 1 \\ 2U_i(|x|) - 1 \text{ if } |x| \ge 1 , \end{cases} \tag{8.35}$$

[3] Bernoulli process, which can be thought of as the "limiting case $p = \infty$", motivated the present investigation.

so that this function is convex. Given $u > 0$, we define

$$\mathcal{N}_u(t) = \sup\Big\{\sum_{i\geq 1} t_i a_i \,;\, \sum_{i\geq 1} \hat{U}_i(a_i) \leq u\Big\}\,. \tag{8.36}$$

Proposition 8.2.3 *If $u > 0$, $v \geq 1$, the r.v. $X_t = \sum_{i\geq 1} t_i Y_i$ satisfies*

$$\mathsf{P}(X_t \geq Lv\mathcal{N}_u(t)) \leq \exp(-uv)\,. \tag{8.37}$$

To get a feeling of what happens, let us first carry out the meaning of $\mathcal{N}_u(t)$ in simple cases. The simplest case is when $U_i(x) = x^2$ for all i. It is rather immediate then that $x^2 \leq \hat{U}_i(x) \leq 2x^2$ and

$$\sqrt{u/2}\|t\|_2 \leq \mathcal{N}_u(t) \leq \sqrt{u}\|t\|_2, \tag{8.38}$$

and (8.37) takes the less mysterious form $\mathsf{P}(X_t \geq Lv\sqrt{u}\|t\|_2) \leq \exp(-uv)$.

The second simplest example is the case where for all i we have $U_i(x) = x$ for $x \geq 0$. In that case we have $|x| \leq \hat{U}_i(x) = 2|x| - 1 \leq x^2$ for $|x| \geq 1$. Thus $\hat{U}_i(x) \leq x^2$ and $\hat{U}_i(x) \leq 2|x|$ for all $x \geq 0$, and hence

$$\sum_{i\geq 1} a_i^2 \leq u \Rightarrow \sum_{i\geq 1} \hat{U}_i(a_i) \leq u$$

and

$$\sum_{i\geq 1} 2|a_i| \leq u \Rightarrow \sum_{i\geq 1} \hat{U}_i(a_i) \leq u\,.$$

Consequently, we have $\mathcal{N}_u(t) \geq \sqrt{u}\|t\|_2$ and $\mathcal{N}_u(t) \geq u\|t\|_\infty/2$. Moreover, if $\sum_{i\geq 1} \hat{U}_i(a_i) \leq u$, writing $b_i = a_i \mathbf{1}_{\{|a_i|\geq 1\}}$ and $c_i = a_i \mathbf{1}_{\{|a_i|<1\}}$ we have $\sum_{i\geq 1} |b_i| \leq u$ (since $\hat{U}_i(x) \geq |x|$ for $|x| \geq 1$) and $\sum_{i\geq 1} c_i^2 \leq u$ (since $\hat{U}_i(x) \geq x^2$ for $|x| \leq 1$). Consequently

$$\sum_{i\geq 1} t_i a_i = \sum_{i\geq 1} t_i b_i + \sum_{i\geq 1} t_i c_i \leq u\|t\|_\infty + \sqrt{u}\|t\|_2\,,$$

and we have shown that

$$\frac{1}{2}\max(u\|t\|_\infty, \sqrt{u}\|t\|_2) \leq \mathcal{N}_u(t) \leq (u\|t\|_\infty + \sqrt{u}\|t\|_2)\,, \tag{8.39}$$

and (8.37) means that

$$\mathsf{P}(X_t \geq Lv(\sqrt{u}\|t\|_2 + u\|t\|_\infty)) \leq \exp(-uv)\,,$$

which is just another way to write (8.25).

We start the proof of the tail estimate (8.37) along the standard Cramer-Chernoff method:

$$\mathsf{P}\Big(\sum_{i\geq 1} t_i Y_i \geq u\Big) \leq \inf_{\lambda>0} \exp\Big(-u\lambda + \sum_{i\geq 1} \log \mathsf{E}\exp \lambda t_i Y_i\Big)\,, \tag{8.40}$$

but the rest of the argument is not standard. To use (8.40) we first need to estimate $\mathsf{E}\exp \lambda Y_i$. Since we control the tails of Y_i, this is not going to be very difficult. For $\lambda \geq 0$ we define

$$V_i(\lambda) = \sup_x(\lambda x - \hat{U}_i(x))\,. \tag{8.41}$$

Since the function $\hat{U}_i$ is convex, the limit $\lambda_i := \lim_{x\to\infty} \hat{U}_i(x)/x \in [1,\infty]$ exists and $V_i(\lambda) < \infty$ for $\lambda < \lambda_i$. Note also that obviously V_i is an increasing function of λ.

Lemma 8.2.4 *For $\lambda \geq 0$ we have*

$$\mathsf{E}\exp \lambda Y_i \leq \exp V_i(L\lambda)\,. \tag{8.42}$$

Proof Let us first observe (taking $x = 0$ in (8.41)) that $V_i \geq 0$, and V_i is convex with $V_i(0) = 0$. Taking $x = \lambda/2$, and since $\hat{U}_i(x) = x^2$ for $|x| < 1$, we get

$$\lambda \leq 2 \Rightarrow V_i(\lambda) \geq \frac{\lambda^2}{4} \tag{8.43}$$

and taking $x = 1$ that

$$V_i(\lambda) \geq \lambda - 1\,. \tag{8.44}$$

Since $U_i'(1) \geq 1$, for $x \geq 1$ we have $U_i(x) \geq x$, so that by (8.34) we have $\mathsf{P}(|Y_i| \geq x) \leq e^{-x}$ and hence (using, e.g., that $x^2 \leq L\exp|x|/6$),

$$\mathsf{E}Y_i^2 \exp\frac{|Y_i|}{2} \leq L\,.$$

The elementary inequality $e^x \leq 1 + x + x^2 e^{|x|}$ yields that, if $\lambda \leq 1/2$,

$$\mathsf{E}\exp \lambda Y_i \leq 1 + \lambda^2 \mathsf{E}Y_i^2 \exp\lambda|Y_i| \leq 1 + L\lambda^2 \leq \exp L\lambda^2\,. \tag{8.45}$$

Now since $\lambda \leq 1/2$, we have $\lambda^2 \leq 4V_i(\lambda)$, and since V_i is convex, $V_i \geq 0$, and $V_i(0) = 0$, we have $4LV_i(\lambda) \leq V_i(4L\lambda)$, so that $L\lambda^2 \leq V_i(4L\lambda)$. This completes the proof of (8.42) in the case $\lambda \leq 1/2$.

Assume now that $\lambda \geq 1/2$, and observe that

$$\mathsf{E}\exp\lambda|Y_i| = 1 + \lambda\int_0^\infty \exp\lambda x\,\mathsf{P}(|Y_i| \geq x)\mathrm{d}x$$
$$= 1 + \lambda\int_0^\infty \exp(\lambda x - U_i(x))\mathrm{d}x\,. \qquad (8.46)$$

We will prove that, for $x \geq 0$,

$$\lambda x - U_i(x) \leq \frac{V_i(6\lambda)}{2} - \lambda x\,. \qquad (8.47)$$

Combining with (8.46), this yields

$$\mathsf{E}\exp\lambda|Y_i| \leq 1 + \lambda\int_0^\infty \exp\Big(\frac{V_i(6\lambda)}{2} - \lambda x\Big)\mathrm{d}x = 1 + \exp\frac{V_i(6\lambda)}{2}\,.$$

Now $V_i(6\lambda) \geq V_i(3) \geq 2$ (using (8.44) in the last inequality), so that $1 + \exp(V_i(6\lambda))/2 \leq \exp(V_i(6\lambda))$, completing the proof of (8.42).

To prove (8.47) we first consider the case where $x \leq 1$. Then $4\lambda x \leq 4\lambda$, $4\lambda \leq 6\lambda - 1$ (since $\lambda \geq 1/2$), and $6\lambda - 1 \leq V_i(6\lambda)$ by (8.44), so that $4\lambda x \leq V_i(6\lambda)$. Thus $\lambda x \leq V_i(6\lambda)/2 - \lambda x$, and we have

$$\lambda x - U_i(x) \leq \lambda x \leq \frac{V_i(6\lambda)}{2} - \lambda x\,.$$

When $x \geq 1$ we have $U_i(x) \geq \hat{U}_i(x)/2$ and then

$$\lambda x - U_i(x) \leq \lambda x - \frac{\hat{U}_i(x)}{2} \leq \frac{V_i(4\lambda)}{2} - \lambda x$$

because $4\lambda x - \hat{U}_i(x) \leq V_i(4\lambda)$ by definition of V_i. Since $V_i(4\lambda) \leq V_i(6\lambda)$ the proof is complete. □

Lemma 8.2.5 *For any $u > 0$ we have*

$$\sum_{i\geq 1} V_i\Big(\frac{u|t_i|}{\mathcal{N}_u(t)}\Big) \leq u\,.$$

Proof Recalling the definition (8.41) of V_i, it suffices to show that given numbers $x_i \geq 0$, we have

$$\sum_{i\geq 1}\frac{u|t_i|x_i}{\mathcal{N}_u(t)} - \sum_{i\geq 1}\hat{U}_i(x_i) \leq u\,. \qquad (8.48)$$

If $\sum_{i\geq 1} \hat{U}_i(x_i) \leq u$, then by definition (8.36) of $\mathcal{N}_u(t)$, we have $\sum_{i\geq 1} |t_i|x_i \leq \mathcal{N}_u(t)$ so we are done since $\sum_{i\geq 1} \hat{U}_i(x_i) \geq 0$. If $\sum_{i\geq 1} \hat{U}_i(x_i) = \theta u$ with $\theta > 1$, then (since $\hat{U}_i(0) = 0$ and $\hat{U}_i$ is convex) we have $\sum_{i\geq 1} \hat{U}_i(x_i/\theta) \leq u$, so that by definition of $\mathcal{N}_u(t)$, $\sum_{i\geq 1} |t_i|x_i \leq \theta\mathcal{N}_u(t)$ and the left-hand side of (8.48) is in fact ≤ 0. □

Lemma 8.2.6 *If $v \geq 1$ we have*

$$\mathcal{N}_{uv}(t) \leq v\mathcal{N}_u(t) . \tag{8.49}$$

Proof Consider numbers a_i with $\sum_{i\geq 1} \hat{U}_i(a_i) \leq uv$. By convexity of $\hat{U}$ for $v \geq 1$, we have $\hat{U}_i(a_i/v) \leq \hat{U}_i(a_i)/v$, so that $\sum_{i\geq 1} \hat{U}_i(a_i/v) \leq u$. By definition of $\mathcal{N}_u(t)$, we then have $\sum_{i\geq 1} t_i a_i/v \leq \mathcal{N}_u(t)$, i.e., $\sum_{i\geq 1} t_i a_i \leq v\mathcal{N}_u(t)$. The definition of $\mathcal{N}_{uv}(t)$ then implies (8.49). □

Proof of Proposition 8.2.3 Since by Lemma 8.2.6 we have $v\mathcal{N}_u(t) \geq \mathcal{N}_{vu}(t)$, we can assume $v = 1$. Lemma 8.2.4 implies

$$\begin{aligned}\mathsf{P}(X_t \geq y) &\leq \exp(-\lambda y)\mathsf{E}\exp\lambda X_t \\ &\leq \exp\Big(-\lambda y + \sum_{i\geq 1} V_i(L_0\lambda|t_i|)\Big) .\end{aligned}$$

We choose $y = 2L_0\mathcal{N}_u(t)$, $\lambda = 2u/y$, and we apply Lemma 8.2.5 to get

$$-\lambda y + \sum_{i\geq 1} V_i(L_0\lambda|t_i|) \leq -2u + u = -u .$$ □

We now define

$$B(u) = \{t \; ; \; \mathcal{N}_u(t) \leq u\} . \tag{8.50}$$

These sets will play an essential part in the rest of the chapter.

Corollary 8.2.7 *If $u \geq 1$ and $t \in B(u)$, we have $\|X_t\|_u \leq Lu$.*[4]

Proof By definition of $B(u)$ we have $\mathcal{N}_u(t) \leq u$. From (8.37) for $v \geq 1$ we have $\mathsf{P}(|X_t| \geq L_1 vu) \leq 2\exp(-uv)$. The r.v. $Y = |X_t|/L_1$ then satisfies $P(Y \geq w) \leq 2\exp(-w)$ for $w \geq u$. We write $Y = Y_1 + Y_2$ where $Y_1 = Y\mathbf{1}_{\{Y\leq u\}}$ and $Y_2 = Y\mathbf{1}_{\{Y>u\}}$. Thus $\mathsf{P}(Y_2 \geq w) \leq 2\exp(-w)$ for $w > 0$, so that $\|Y_2\|_u \leq Lu$ by (2.22). And we have $|Y_1| \leq u$ so that $\|Y_1\|_u \leq u$. □

[4] Here $\|X_t\|_u$ is the L^p norm for $p = u$.

Since the sets $B(u)$ are so important in the sequel, let us describe them in some simple cases. We denote by $B_p(0, v)$ the balls in ℓ^p for $1 \le p \le \infty$. When $U_i(x) = x^2$ for all i and x, (8.38) implies

$$B_2(0, \sqrt{u}) \subset B(u) \subset B_2(0, \sqrt{2u}) . \tag{8.51}$$

When $U_i(x) = x$ for all i and x, (8.39) implies

$$\frac{1}{2}(B_\infty(0, 1) \cap B_2(0, \sqrt{u})) \subset B(u) \subset 2(B_\infty(0, 1) \cap B_2(0, \sqrt{u})) . \tag{8.52}$$

The third simplest example is the case where for some $p \ge 1$ and for all i, we have $U_i(x) = U(x) = x^p$ for $x \ge 0$. The case $1 \le p \le 2$ is very similar to the case $p = 1$, so we consider only the case $p > 2$, which is a little bit trickier. Then $\hat{U}(x) = 2|x|^p - 1$ for $|x| \ge 1$. In particular $\hat{U}(x) \ge x^2$ and $\hat{U}(x) \ge |x|^p$. Therefore

$$C_u := \Big\{(a_i)_{i\ge 1} \; ; \; \sum_{i\ge 1} \hat{U}(a_i) \le u\Big\} \subset B_p(0, u^{1/p}) \cap B_2(0, u^{1/2}) . \tag{8.53}$$

Using now that $\hat{U}(x) \le 2|x|^p + x^2$ we obtain

$$\frac{1}{2}B_2(0, \sqrt{u}) \cap \frac{1}{2}B_p(0, u^{1/p}) \subset C_u . \tag{8.54}$$

Thus, from (8.53) and (8.54) we have obtained a pretty accurate description of C_u, and the task is now to translate it in a description of $B(u)$. For this we will appeal to duality and the Hahn-Banach theorem. In order to minimize technicalities we will now pretend that we work only with finite sums $\sum_{i\le n} Y_i$ (which is the important case). Let us denote by (x, y) the canonical duality of $\mathbb{R}^n$ with itself, and for a set $A \subset \mathbb{R}^n$, let us define its *polar set* A°[5] by

$$A^\circ = \{y \in \mathbb{R}^n \; ; \; \forall x \in A, (x, y) \le 1\} . \tag{8.55}$$

Lemma 8.2.8 *We have* $B(u) = uC_u^\circ$.

Proof Combine the definitions (8.36) and (8.50). □

Lemma 8.2.9 *If* A *and* B *are closed balls*[6] *in* $\mathbb{R}^n$ *and if* $A^\circ + B^\circ$ *is closed then*

$$\frac{1}{2}(A^\circ + B^\circ) \subset (A \cap B)^\circ \subset A^\circ + B^\circ . \tag{8.56}$$

---

[5] Not to be confused with the interior $\overset{\circ}{A}$ of A!

[6] A ball A is convex set with non-empty interior and $A = -A$.

Proof The proof relies on the so-called bipolar theorem (a consequence of the Hahn-Banach theorem). This theorem states that for any set $A \subset \mathbb{R}^n$, $(A^\circ)^\circ$ is the closed convex hull of A. It is obvious that $(C \cup D)^\circ = C^\circ \cap D^\circ$. Using this formula for $C = A^\circ$ and $D = B^\circ$ when A and B are closed convex sets yields $A \cap B = (A^\circ \cup B^\circ)^\circ$, so that $(A \cap B)^\circ$ is the closed convex hull of A° and B°. Let us note that for a ball A, A° is a ball so that $\lambda A^\circ \subset A^\circ$ for $0 \le \lambda \le 1$. Then (8.56) follows. □

Since $B(u) = uC_u^\circ$, denoting by q the conjugate exponent of p, it then follows from (8.53), (8.54), and (8.56) that we have

$$\frac{1}{2}(B_2(0, \sqrt{u}) + B_q(0, u^{1/q})) \subset B(u) \subset 2(B_2(0, \sqrt{u}) + B_q(0, u^{1/q})) \,. \tag{8.57}$$

Exercise 8.2.10 Find a complete proof of (8.57), when we no longer deal with finite sums and which does not use the Hahn-Banach theorem.

8.3 The Structure of Certain Canonical Processes

In this section we prove a far-reaching generalization of Theorem 2.10.1. Recalling the r.v.s Y_i of the previous section, and the definition $X_t = \sum_{i \ge 1} t_i Y_i$ of the canonical process, we "compute $\mathsf{E} \sup_{t \in T} X_t$ as a function of the geometry of T".

Recalling (8.50), given a number $r \ge 4$, we define

$$\varphi_j(s,t) = \inf\{u > 0 \,;\; s - t \in r^{-j} B(u)\} \tag{8.58}$$

when the set in the right-hand side is not empty and $\varphi_j(s,t) = \infty$ otherwise. This "family of distances" is the right tool to describe the geometry of T.

We first provide upper bounds for $\mathsf{E} \sup_{t \in T} X_t$.

Theorem 8.3.1 *Assume that there exists an admissible sequence $(\mathcal{A}_n)$ of $T \subset \ell^2$, and for $A \in \mathcal{A}_n$ an integer $j_n(A) \in \mathbb{Z}$ such that*

$$\forall A \in \mathcal{A}_n \,,\, \forall s, s' \in A \,,\, \varphi_{j_n(A)}(s, s') \le 2^n \,. \tag{8.59}$$

Then

$$\mathsf{E} \sup_{t \in T} X_t \le L \sup_{t \in T} \sum_{n \ge 0} 2^n r^{-j_n(A_n(t))} \,. \tag{8.60}$$

Proof For $s, t \in A \in \mathcal{A}_n$ by (8.59) we have $s - t \in r^{-j_n(A)} B(2^n)$, so that by Corollary 8.2.7 we have $\|X_s - X_t\|_{2^n} = \|X_{s-t}\|_{2^n} \le L 2^n r^{-j_n(A)}$. This means that then diameter of $A_n(t)$ for the distance of L^p with $p = 2^n$ is $\le L 2^n r^{-j_n(A)}$. The result then follows from Theorem 2.7.14. □

To illustrate this statement assume first $U_i(x) = x^2$ for each i. Then (and more generally when $U_i(x) \ge x^2/L$ for $x \ge 1$) by (8.51) we have $\varphi_j(s,t) \le Lr^{2j}\|s-t\|_2^2$, so that (8.59) holds as soon as $r^{2j_n(A)}\Delta(A,d_2)^2 \le 2^n/L$, where d_2 denotes the distance induced by the norm of ℓ^2. Taking for $j_n(A)$ the largest integer that satisfies this inequality implies that the right-hand side of (8.60) is bounded by $Lr\sup_{t\in T}\sum_{n\ge 0}2^{n/2}\Delta(A_n(t),d_2)$. Taking the infimum over the admissible sequences $(\mathcal{A}_n)$, this yields

$$\mathsf{E}\sup_{t\in T} X_t \le Lr\gamma_2(T,d_2)\,.$$

Assume next that $U_i(x) = x$ for each i. When $\|s-t\|_\infty \le r^{-j}/L$, (8.52) implies $\varphi_j(s,t) \le Lr^{2j}\|s-t\|_2^2$, so that (8.59) holds whenever $r^{j_n(A)}\Delta(A,d_\infty) \le 1/L$ and $r^{2j_n(A)}\Delta(A,d_2)^2 \le 2^n/L$, where d_∞ denotes the distance induced by the norm of ℓ^∞. Taking for $j_n(A)$ the largest integer that satisfies both conditions yields

$$r^{-j_n(A)} \le Lr\big(\Delta(A,d_\infty) + 2^{-n/2}\Delta(A,d_2)\big)\,,$$

so that (8.60) implies

$$\mathsf{E}\sup_{t\in T} X_t \le Lr\sup_{t\in T}\sum_{n\ge 0}\big(2^n\Delta(A_n(t),d_\infty) + 2^{n/2}\Delta(A_n(t),d_2)\big)\,, \tag{8.61}$$

and copying the beginning of the proof of Theorem 4.5.13 this implies

$$\mathsf{E}\sup_{t\in T} X_t \le Lr\big(\gamma_2(T,d_2) + \gamma_1(T,d_\infty)\big)\,. \tag{8.62}$$

Let us now turn to the converse of Theorem 8.3.1. Since U_i is convex, $U_i(x)$ grows at least as fast as linear function. We will assume the following regularity conditions, which ensures that U_i does not grow too fast: For some constant C_0, we have

$$\forall i \ge 1\,,\ \forall s \ge 1\,,\ U_i(2s) \le C_0 U_i(s)\,. \tag{8.63}$$

This condition is often called "the Δ_2-condition". We will also assume that

$$\forall i \ge 1\,,\ U_i'(0) \ge 1/C_0\,. \tag{8.64}$$

Here, $U_i'(0)$ is the right derivative at 0 of the function $U_i(x)$.

Theorem 8.3.2 *Under conditions (8.63) and (8.64), we can find r_0 (depending on C_0 only) and a number $K = K(C_0)$ such that when $r \ge r_0$, for each subset T of ℓ^2 there exists an admissible sequence $(\mathcal{A}_n)$ of T and for $A \in \mathcal{A}_n$ an integer $j_n(A) \in \mathbb{Z}$*

such that (8.59) holds together with

$$\sup_{t\in T}\sum_{n\geq 0} 2^n r^{-j_n(A_n(t))} \leq K(C_0) r \mathsf{E} \sup_{t\in T} X_t \,. \tag{8.65}$$

Together with Theorem 8.3.1, this essentially allows the computation of $\mathsf{E}\sup_{t\in T} X_t$ as a function of the geometry of T. It is not very difficult to prove that Theorem 8.3.2 still holds true without condition (8.64), and this is done in [47]. But it is an entirely different matter to remove condition (8.63). The difficulty is of the same nature as in the study of Bernoulli processes. Now that this difficulty has been solved for Bernoulli processes, by solving the Bernoulli conjecture, one may hope that eventually condition (8.63) will be removed.

Let us interpret Theorem 8.3.2 in the case where $U_i(x) = x^2$ for $x \geq 1$. In that case (and more generally when $U_i(x) \leq x^2/L$ for $x \geq 1$), we have

$$\varphi_j(s,t) \geq r^{2j}\|s-t\|_2^2/L \,, \tag{8.66}$$

so that (8.59) implies that $\Delta(A, d_2) \leq L2^{n/2} r^{-j_n(A)}$ and (8.65) implies

$$\sup_{t\in T}\sum_{n\geq 0} 2^{n/2}\Delta(A_n(t), d_2) \leq Lr\mathsf{E}\sup_{t\in T} X_t \,,$$

and hence

$$\gamma_2(T, d_2) \leq Lr\mathsf{E}\sup_{t\in T} X_t \,. \tag{8.67}$$

Thus Theorem 8.3.2 extends Theorem 2.10.1.

Next consider the case where $U_i(x) = x$ for all x. Then (8.52) implies (8.66) and thus (8.67). It also implies that $\varphi_j(s,t) = \infty$ whenever $\|s-t\|_\infty > 2r^{-j}$, because then $r^j(s-t) \notin B(u)$ whatever the value of u. Consequently, (8.59) implies that $\Delta(A, d_\infty) \leq Lr^{-j_n(A)}$, and (8.65) yields

$$\gamma_1(T, d_\infty) \leq Lr\mathsf{E}\sup_{t\in T} X_t \,.$$

Recalling (8.62) (and since here r is a universal constant), we thus have proved the following very pretty fact:

Theorem 8.3.3 *Assume that the r.v.s Y_i are independent and symmetric and satisfy $\mathsf{P}(|Y_i| \geq x) = \exp(-x)$. Then*

$$\frac{1}{L}(\gamma_2(T, d_2) + \gamma_1(T, d_\infty)) \leq \mathsf{E}\sup_{t\in T} X_t \leq L(\gamma_2(T, d_2) + \gamma_1(T, d_\infty)) \,. \tag{8.68}$$

Corollary 8.3.4 *If* $T \subset \ell^2$ *then*

$$\gamma_2(\operatorname{conv} T, d_2) + \gamma_1(\operatorname{conv} T, d_\infty) \le L(\gamma_2(T, d_2) + \gamma_1(T, d_\infty)) \,. \tag{8.69}$$

Proof Combine the trivial relation $\mathsf{E}\sup_{t \in \operatorname{conv} T} X_t = \mathsf{E}\sup_{t \in T} X_t$ with (8.68). □

Research Problem 8.3.5 Give a geometrical proof of (8.69).

A far more general question occurs in Problem 8.3.12. The next two exercises explore the subtlety of the behavior of the operation "taking the convex hull" with respect to the functional $\gamma_1(\cdot, d_\infty)$, but the third exercise, Exercise 8.3.8, is really important. It makes the Bernoulli conjecture plausible by exhibiting a similar phenomenon in an easier setting.

Exercise 8.3.6 Consider the canonical basis (t_n) of ℓ^2, $t_n = (t_{n,i})_{i \ge 1}$ with $t_{n,i} = 0$ for $i \ne n$ and $t_{n,n} = 1$. Give a geometrical proof that if $T = \{t_1, \dots, t_N\}$ then $\gamma_1(\operatorname{conv} T, d_\infty)$ is of the same order as $\gamma_1(T, d_\infty)$ (i.e., $\log N$). Caution: This is not very easy.

Exercise 8.3.7 Prove that it is not true that for a set T of sequences one has

$$\gamma_1(\operatorname{conv} T, d_\infty) \le L\gamma_1(T, d_\infty) \,.$$

Hint: Consider the set T of coordinate functions on $\{-1, 1\}^k$.

Exercise 8.3.8 Use (8.57) to prove that in the case $U_i(x) = x^p$, $p > 2$, the inequality (8.32) can be reversed. Hint: You need to master the proof of Theorem 6.7.2.

We now prepare for the proof of Theorem 8.3.2.

Lemma 8.3.9 *Under (8.63), given* $\rho > 0$ *we can find* r_0*, depending on* C_0 *and* ρ *only, such that if* $r \ge r_0$*, for* $u \in \mathbb{R}^+$ *we have*

$$B(8ru) \subset \rho r B(u) \,. \tag{8.70}$$

Proof We claim that for a constant C_1 depending only on C_0 we have

$$\forall u > 0 \,,\; \hat{U}_i(2u) \le C_1 \hat{U}_i(u) \,. \tag{8.71}$$

Indeed, it suffices to prove this for u large, where this follows from the Δ_2 condition (8.63). Consider an integer k large enough that $2^{-k+3} \le \rho$ and let $r_0 = C_1^k$. Assuming that $r \ge r_0$, we prove (8.70).

Consider $t \in B(8ru)$. Then $\mathcal{N}_{8ru}(t) \le 8ru$ by definition of $B(8ru)$, so that according to (8.36) for any numbers $(a_i)_{i \ge 1}$ we have

$$\sum_{i \ge 1} \hat{U}_i(a_i) \le 8ru \Rightarrow \sum_{i \ge 1} a_i t_i \le 8ru \,. \tag{8.72}$$

Consider numbers b_i with $\sum_{i\ge 1} \hat{U}_i(b_i) \le u$. Then by (8.71) we have $\hat{U}_i(2^k b_i) \le C_1^k \hat{U}_i(b_i) \le r\hat{U}_i(b_i)$, so that $\sum_{i\ge 1} \hat{U}_i(2^k b_i) \le ru \le 8ru$, and (8.72) implies $\sum_{i\ge 1} 2^k b_i t_i \le 8ru$. Since $2^k \ge 8/\rho$ we have shown that

$$\sum_{i\ge 1} \hat{U}_i(b_i) \le u \Rightarrow \sum_{i\ge 1} b_i \frac{t_i}{\rho r} \le u ,$$

so that $\mathcal{N}_u(t/\rho r) \le u$ and thus $t/\rho r \in B(u)$, i.e., $t \in r\rho B(u)$. □

Lemma 8.3.10 *If (8.70) holds for $\rho \le 1$, then for all $s, t \in T$ and all $j \in \mathbb{Z}$, we have $\varphi_{j+1}(s,t) \ge 8r\varphi_j(s,t)$.*

Proof If $\varphi_{j+1}(s,t) < u$ then $s - t \in r^{-j-1}B(u) \subset r^{-j}B(u/(8r))$ and thus $\varphi_j(s,t) \le u/(8r)$. Thus $\varphi_j(s,t) \le \varphi_{j+1}(s,t)/(8r)$. □

Theorem 8.3.11 *Under Condition (8.64) we can find a number $0 < \rho \le 1$ with the following property. Consider an integer $m \ge 2$. Given any points $t_1, \dots, t_m$ in ℓ^2 and $a > 0$ such that*

$$\ell \ne \ell' \Rightarrow t_\ell - t_{\ell'} \notin aB(u) \tag{8.73}$$

and given any sets $H_\ell \subset t_\ell + \rho a B(u)$, we have

$$\mathsf{E} \sup_{t\in\bigcup H_\ell} X_t \ge \frac{a}{L} \min(u, \log m) + \min_{\ell\le m} \mathsf{E} \sup_{t\in H_\ell} X_t . \tag{8.74}$$

The proof of this statement parallels that of (2.120). The first ingredient is a suitable version of Sudakov minoration, proved by R. Latała, [47] asserting that, under (8.73)

$$\mathsf{E} \sup_{\ell\le m} X_{t_\ell} \ge \frac{a}{L} \min(u, \log m) . \tag{8.75}$$

The second is a "concentration of measure" result quantifying the deviation of $\sup_{t\in H_\ell} X_t$ from its mean, in the spirit of (2.118) and (6.12). This result relies on a concentration of measure property for the probability ν of density $e^{-2|x|}$ with respect to Lebesgue measure and its powers, which was discovered in [106]. Condition (8.64) is used here, to assert that the law of Y_i is the image of ν by a Lipschitz map.

Both of the above results are fairly deep, and none of the arguments required is closely related to our main topic, so we refer the reader to [113] and [47].

Proof of Theorem 8.3.2 Consider ρ as in Theorem 8.3.11. If $r = 2^{\kappa-3}$, where κ is large enough (depending on C_0 only), Lemma 8.3.9 shows that (8.70) holds for each $u > 0$. We fix such a value of r, and we prove that the functionals $F_{n,j}(A) = 2L_0 \mathsf{E} \sup_{t\in A} X_t$, where L_0 is the constant of (8.74), satisfy the growth condition of

Definition 8.1.1. Consider $n \geq 1$ and points (t_ℓ) for $\ell \leq m = N_n$ as in (8.3). By definition of φ_{j+1} we have

$$\ell \neq \ell' \Rightarrow t_\ell - t_{\ell'} \notin r^{-j-1} B(2^{n+1}) . \tag{8.76}$$

Consider then sets $H_\ell \subset B_{j+2}(t_\ell, 2^{\kappa+n})$. By definition of φ_{j+2}, we have $B_{j+2}(t_\ell, 2^{\kappa+n}) = t_\ell + r^{-j-2} B(2^{\kappa+n})$. Using (8.70) for $u = 2^n$ (and since $2^\kappa = 8r$), we obtain that $B(2^{\kappa+n}) \subset \rho r B(2^n)$ and therefore $H_\ell \subset t_\ell + \rho r^{-j-1} B(2^n)$. Since $\log m = 2^n \log 2 \geq 2^{n-1}$, we can then appeal to (8.74) with $a = r^{-j-1}$ to obtain the desired relation

$$F_{n,j}\Big(\bigcup_{\ell \leq m} H_\ell\Big) \geq 2^n r^{-j-1} + \min_{\ell \leq m} F_{n+1,j+1}(H_\ell)$$

that completes the proof of the growth condition.

Let us denote by j_0 the largest integer such that

$$r^{-j_0} > L_0 \mathsf{E} \sup_{t \in T} X_t , \tag{8.77}$$

so that

$$r^{-j_0} \leq L_0 r \mathsf{E} \sup_{t \in T} X_t . \tag{8.78}$$

We prove that (8.8) holds for this value of j_0. Indeed, supposing that this is not the case, and recalling (8.58), we can find $t_1, t_2 \in T$ with $t_1 - t_2 \notin aB(1)$ for $a = r^{-j_0}$. Then using (8.74) for $m = 2$ and $H_1 = \{t_1/a\}$, $H_2 = \{t_2/a\}$ together with the fact that $X_{at} = aX_t$ yields

$$\frac{a}{L_0} \leq \mathsf{E} \max(X_{t_1}, X_{t_2}) \leq \mathsf{E} \sup_{t \in T} X_t , \tag{8.79}$$

which contradicts (8.77) and proves the claim.

Taking into account Lemma 8.3.10 (which ensures that (8.9) is satisfied), we are thus in a position to apply Theorem 8.1.2 to construct an admissible sequence $(\mathcal{A}_n)$. Using (8.78), (8.11) implies

$$\forall t \in T , \sum_{n \geq 0} 2^n r^{-j_n(A_n(t))} \leq L r \mathsf{E} \sup_{t \in T} X_t .$$

To finish the proof, it remains to prove (8.59). By definition of $B_j(t, u)$ and of φ_j, we have

$$s \in B_j(t, u) \Rightarrow \varphi_j(s, t) \leq u \Rightarrow s - t \in r^{-j} B(u) .$$

Thus (8.12) implies

$$\forall n \geq 1 \,,\; \forall A \in \mathcal{A}_n \,,\; \forall s \in A \,,\; s - t_{n,A} \in r^{-j_n(A)} B(2^n) \,.$$

Since $B(u)$ is a convex symmetric set, we have

$$s - t_{n,A} \in r^{-j_n(A)} B(2^n),\; s' - t_{n,A} \in r^{-j_n(A)} B(2^n) \Rightarrow \frac{s - s'}{2} \in r^{-j_n(A)} B(2^n)$$

$$\Rightarrow \varphi_{j_n(A)}\Big(\frac{s}{2}, \frac{s'}{2}\Big) \leq 2^n \,.$$

Thus we have shown that

$$\forall n \geq 1 \,,\; \forall A \in \mathcal{A}_n \,,\; \forall s, s' \in A \,,\; \varphi_{j_n(A)}\Big(\frac{s}{2}, \frac{s'}{2}\Big) \leq 2^n \,.$$

This is not exactly (8.59), but to get rid of the factor $1/2$, it would have sufficed to apply the above proof to $2T = \{2t; t \in T\}$ instead of T. □

As a consequence of Theorems 8.3.1 and 8.3.2, we have the following geometrical result. Consider a set $T \subset \ell^2$, an admissible sequence $(\mathcal{A}_n)$ of T and for $A \in \mathcal{A}_n$ an integer $j_n(A)$ such that (8.59) holds true. Then there is an admissible sequence $(\mathcal{B}_n)$ of $\operatorname{conv} T$ and for $B \in \mathcal{B}_n$ an integer $j_n(B)$ that satisfies (8.59) and

$$\sup_{t \in \operatorname{conv} T} \sum_{n \geq 0} 2^n r^{-j_n(B_n(t))} \leq K(C_0) r \sup_{t \in T} \sum_{n \geq 0} 2^n r^{-j_n(A_n(t))} \,. \tag{8.80}$$

Research Problem 8.3.12 Give a geometrical proof of this fact.

This is a far-reaching generalization of Research Problem 2.11.2.

The following generalizes Theorem 2.11.9:

Theorem 8.3.13 *Assume (8.63) and (8.64). Consider a countable subset T of ℓ^2, with $0 \in T$. Then we can find a sequence (x_n) of vectors of ℓ^2 such that*

$$T \subset (K(C_0) \mathsf{E} \sup_{t \in T} X_t) \operatorname{conv}(\{x_n \,;\, n \geq 2\} \cup \{0\})$$

and, for each n,

$$\mathcal{N}_{\log n}(x_n) \leq 1 \,.$$

The point of this result is that, whenever the sequence $(x_n)_{n\ge 2}$ satisfies $\mathcal{N}_{\log n}(x_n) \le 1$, then $\mathsf{E}\sup_{n\ge 2} X_{x_n} \le L$. To see this, we simply write, using (8.37) with $u = \log n$ in the second inequality, that for $v \ge 2$, we have

$$\mathsf{P}\Big(\sup_{n\ge 2} |X_{x_n}| \ge Lv\Big) \le \sum_{n\ge 2} \mathsf{P}(|X_{x_n}| \ge Lv\mathcal{N}_{\log n}(x_n))$$
$$\le \sum_{n\ge 2} \exp(-v\log n) \le L\exp(-v/2) . \tag{8.81}$$

Proof We choose $r = r_0$ depending on C_0 only and we consider a sequence of partitions of T as provided by Theorem 8.3.2. We choose $t_{0,T} = 0$, and for $A \in \mathcal{A}_n$, $n \ge 1$ we select $t_{n,A} \in \mathcal{A}_n$, making sure (as in the proof of Theorem 2.11.9) that each point of T is of the form $t_{n,A}$ for a certain A and a certain n. For $A \in \mathcal{A}_n$, $n \ge 1$, we denote by A' the unique element of $\mathcal{A}_{n-1}$ that contains A. We define

$$u_A = \frac{t_{n,A} - t_{n-1,A'}}{2^{n+1} r^{-j_{n-1}(A')}}$$

and $U = \{u_A;\ A \in \mathcal{A}_n, n \ge 1\}$. Consider $t \in T$, so that $t = t_{n,A}$ for some n and some $A \in \mathcal{A}_n$, and, since $A_0(t) = T$ and $t_{0,T} = 0$,

$$t = t_{n,A} = \sum_{1\le k\le n} t_{k,A_k(t)} - t_{k-1,A_{k-1}(t)} = \sum_{1\le k\le n} 2^{k+1} r^{-j_{k-1}(A_{k-1}(t))} u_{A_k(t)} .$$

Since $\sum_{k\ge 0} 2^k r^{-j_k(A_k(t))} \le K(C_0)\mathsf{E}\sup_{t\in T} X_t$ by (8.65), this shows that

$$T \subset (K(C_0)\mathsf{E}\sup_{t\in T} X_t) \operatorname{conv} U .$$

Next, we prove that $\mathcal{N}_{2^{n+1}}(u_A) \le 1$ whenever $A \in \mathcal{A}_n$. The definition of φ_j and (8.59) imply

$$\forall s\,,\ s' \in A\,,\ s - s' \in r^{-j_n(A)} B(2^{n+1})\,,$$

and the homogeneity of $\mathcal{N}_u$ yields

$$\forall s\,,\ s' \in A\,,\ \mathcal{N}_{2^{n+1}}(s - s') \le r^{-j_n(A)} 2^{n+1} .$$

Since $t_{n,A}, t_{n-1,A'} \in A'$, using the preceding inequality for $n-1$ rather than n and A' instead of A, we get

$$\mathcal{N}_{2^n}(t_{n,A} - t_{n-1,A'}) \le 2^n r^{-j_{n-1}(A')}\,,$$

and thus $\mathcal{N}_{2^n}(u_A) \le 1/2$, so that $\mathcal{N}_{2^{n+1}}(u_A) \le 1$ using (8.49).

Let us enumerate $U = (x_n)_{n \geq 2}$ in such a manner that the points of the type u_A for $A \in \mathcal{A}_1$ are enumerated before the points of the type u_A for $A \in \mathcal{A}_2$, etc. Then if $x_n = u_A$ for $A \in \mathcal{A}_k$, we have $n \leq N_0 + N_1 + \cdots + N_k \leq N_k^2$ and therefore $\log n \leq 2^{k+1}$. Thus $\mathcal{N}_{\log n}(x_n) \leq \mathcal{N}_{2^{k+1}}(x_n) = \mathcal{N}_{2^{k+1}}(u_A) \leq 1$. □

Key Ideas to Remember

- The ideas of Chap. 2 on how to measure the size of metric space smoothly extend to the setting of sets provided with families of distances, provided this families of distances satisfy suitable regularity conditions. These regularity conditions unfortunately are not satisfied in the most interesting case, that of Bernoulli processes.
- Nonetheless we can obtain far-reaching generalizations of the majorizing measure theorem and reach a complete understanding of the size of certain "canonical processes" which are linear combination of well-behaved (e.g., symmetric exponential) r.v.s.

Chapter 9
Peaky Part of Functions

9.1 Road Map

The results of this chapter will look technical at first, but they are of central importance. We introduce a way to measure the size of a set of functions, which is in a sense a weakening of the quantity $\gamma_2(T, d_2)$. The main idea is to replace the L^2 distance by a family of distances obtained by suitable truncation, very much in the spirit of (7.63). This new measure of size will look mysterious at first, but we will eventually prove a structure theorem which gives (essentially) equivalent more geometrical ways to understand it. The first part of this structure theorem is Theorem 9.2.1 which asserts that controlling the size of the set of functions implies that the set can be decomposed into a sum of simpler pieces. The converse is stated in Proposition 9.4.4.

For certain processes which are indexed by a class of functions (such as the empirical processes of Sect. 6.8), we will later prove that a control of the size of the process implies a control of the size of the class of functions in precisely the manner we are going to introduce now. The structure theorems of the present chapter will then be instrumental in making a more complete description. Furthermore, such a structure theorem is a key step of the proof of the Latała-Bednorz theorem, the towering result of this work which we prove in the next chapter.

We advise the reader to review Theorem 6.7.2 at this point.

9.2 Peaky Part of Functions, II

Let us consider a measurable space Ω provided with a positive measure ν.

Theorem 9.2.1 *Consider a countable set T of measurable functions on Ω, a number $r \geq 2$, and assume that $0 \in T$. Consider an admissible sequence of*

M. Talagrand, *Upper and Lower Bounds for Stochastic Processes*, Ergebnisse der Mathematik und ihrer Grenzgebiete. 3. Folge / A Series of Modern Surveys in Mathematics 60, https://doi.org/10.1007/978-3-030-82595-9_9

partitions $(\mathcal{A}_n)$ *of* T, *and for* $A \in \mathcal{A}_n$ *consider* $j_n(A) \in \mathbb{Z}$, *with the following property, where* $u > 0$ *is a parameter:*

$$\forall n \geq 0 \;,\; \forall\, s,t \in A \in \mathcal{A}_n \;,\; \int |r^{j_n(A)}(s(\omega) - t(\omega))|^2 \wedge 1 d\nu(\omega) \leq u2^n \;. \tag{9.1}$$

Let $S := \sup_{t \in T} \sum_{n \geq 0} 2^n r^{-j_n(A_n(t))}$.[1] *Then we can write* $T \subset T_1 + T_2 + T_3$, *where* $0 \in T_1$, *where*

$$\gamma_2(T_1, d_2) \leq L\sqrt{u}S \;, \tag{9.2}$$

$$\gamma_1(T_1, d_\infty) \leq LS \;, \tag{9.3}$$

$$\forall t \in T_2 \;,\; \|t\|_1 \leq LuS \;, \tag{9.4}$$

and where moreover

$$\forall t \in T_3 \;,\; \exists s \in T \;,\; |t| \leq 5|s|\mathbf{1}_{\{2|s| \geq r^{-j_0(t)}\}} \;. \tag{9.5}$$

To illustrate the meaning of this theorem, we replace (9.1) by the stronger condition

$$\forall\, s,t \in A \;,\; \int |r^{j_n(A)}(s(\omega) - t(\omega))|^2 d\nu(\omega) \leq u2^n \;, \tag{9.6}$$

which simply means that $\Delta(A, d_2) \leq \sqrt{u}2^{n/2}r^{-j_n(A)}$, so that $\gamma_2(T, d_2) \leq \sqrt{u}S$.[2] Then the previous decomposition is provided by Theorem 6.7.2, and we may even take $T_3 = \{0\}$. The point of Theorem 9.2.1 is that (9.1) requires a much weaker control of the large values of $s - t$ than (9.6). Equation (9.1) says little about the functions $|s|\mathbf{1}_{\{|s| \geq r^{-j_0(T)}\}}$. This is why the term T_3 of the decomposition is required. This term is of secondary importance, and in all our applications it will be easy to control. It is however *very important* that the condition (9.5) does not depend on u. It is also instructive to convince yourself that the case $u = 1$ implies the full statement of Theorem 9.2.1, by applying this case to the measure $\nu' = \nu/u$. Let us also observe that the important case of Theorem 9.2.1 is for ν an infinite measure, the prime example of which is the counting measure on $\mathbb{N}$. (When, say, ν is a probability, it is too easy to satisfy (9.1), and it is too easy for a function to be integrable.)

It is not apparent yet that Theorem 9.2.1 is a sweepingly powerful method to perform chaining for a Bernoulli process.

---

[1] The idea is of course that the smallest value of S over the preceding choices is the appropriate measure of the size of T.

[2] It is fruitful to think of the quantity S as a generalization of $\sup_t \sum 2^{n/2} \Delta(A_n(t), d_2))$.

The bad news is that the proof of Theorem 9.2.1 is definitely not appealing. The principle of the proof is clear, and it is not difficult to follow line by line, but the overall picture is far from being transparent.

Research Problem 9.2.2 Find a proof of Theorem 9.2.1 that you can explain to your grandmother.[3]

Exercise 9.2.3 Isn't it surprising that there is no dependence in r in (9.2) to (9.5)? Show that in fact the result for general r can be deduced from the result for $r = 2$. Hint: Define $j'_n(A)$ as the largest integer with $2^{j'_n(A)} \le r^{j_n(A)}$.

When we will prove the Bernoulli conjecture, we will need Theorem 9.2.4, a more general version of Theorem 9.2.1. The statement of Theorem 9.2.4 involves some technical conditions whose purpose will only become apparent later, but its proof is exactly the same as that of Theorem 9.2.1. In order to avoid repetition, we will deduce Theorem 9.2.1 from Theorem 9.2.4 and then prove Theorem 9.2.4.

Theorem 9.2.4 *Consider a countable set T of measurable functions on a measured space (Ω, ν), a number $r \ge 2$, and assume that $0 \in T$. Consider an admissible sequence of partitions $(\mathcal{A}_n)$ of T. For $t \in T$ and $n \ge 0$ consider $j_n(t) \in \mathbb{Z}$ and $\pi_n(t) \in T$. Assume that $\pi_0(t) = 0$ for each t and that the following properties hold: First, the values of $j_n(t)$ and $\pi_n(t)$ depend only on $A_n(t)$,*

$$\forall s, t \in T \;,\; \forall n \ge 0 \;;\; s \in A_n(t) \Rightarrow j_n(s) = j_n(t) \;;\; \pi_n(s) = \pi_n(t) \;. \tag{9.7}$$

The sequence $(j_n(t))_{n \ge 1}$ is non-decreasing:

$$\forall t \in T \;,\; \forall n \ge 0 \;,\; j_{n+1}(t) \ge j_n(t) \;. \tag{9.8}$$

When going from n to $n+1$ the value of $\pi_n(t)$ can change only when the value of $j_n(t)$ increases:

$$\forall t \in T \;,\; \forall n \ge 0 \;,\; j_n(t) = j_{n+1}(t) \Rightarrow \pi_n(t) = \pi_{n+1}(t) \;. \tag{9.9}$$

When going from n to $n+1$, if the value of $j_n(t)$ increases, then $\pi_{n+1}(t) \in A_n(t)$:

$$\forall t \in T \;,\; \forall n \ge 0 \;,\; j_{n+1}(t) > j_n(t) \Rightarrow \pi_{n+1}(t) \in A_n(t) \;. \tag{9.10}$$

For $t \in T$ and $n \ge 0$ we define $\Omega_n(t) \subset \Omega$ as $\Omega_0(t) = \Omega$ if $n = 0$ and

$$\Omega_n(t) = \left\{\omega \in \Omega \;;\; 0 \le q < n \Rightarrow |\pi_{q+1}(t)(\omega) - \pi_q(t)(\omega)| \le r^{-j_q(t)}\right\} . \tag{9.11}$$

[3] I spent a lot of time on this problem. This does not mean that the proof does not exist, simply that I did not look in the right direction.

Let us consider a parameter $u \geq 1$ and assume that

$$\forall t \in T\ ,\ \forall n \geq 0\ ,\ \int_{\Omega_n(t)} |r^{j_n(t)}(t(\omega) - \pi_n(t)(\omega))|^2 \wedge 1 d\nu(\omega) \leq u 2^n\ . \tag{9.12}$$

Then we can write $T \subset T_1 + T_2 + T_3$, where $0 \in T_1$, with

$$\gamma_2(T_1, d_2) \leq L\sqrt{u} \sup_{t \in T} \sum_{n \geq 0} 2^n r^{-j_n(t)}\ , \tag{9.13}$$

$$\gamma_1(T_1, d_\infty) \leq L \sup_{t \in T} \sum_{n \geq 0} 2^n r^{-j_n(t)}\ , \tag{9.14}$$

$$\forall t \in T_2\ ,\ \|t\|_1 \leq Lu \sup_{t \in T} \sum_{n \geq 0} 2^n r^{-j_n(t)}\ , \tag{9.15}$$

and where moreover

$$\forall t \in T_3\ ,\ \exists s \in T\ ,\ |t| \leq 5|s|\mathbf{1}_{\{2|s| \geq r^{-j_0(t)}\}}\ . \tag{9.16}$$

Proof of Theorem 9.2.1 We deduce this result from Theorem 9.2.4. We set $j_n(t) = \max_{0 \leq k \leq n} j_k(A_k(t))$, and we define

$$p(n, t) = \inf\{p \geq 0\ ;\ j_p(t) = j_n(t)\},$$

so that $p(n, t) \leq n$ and thus $A_{p(n,t)}(t) \supset A_n(t)$. Also, by definition of $j_n(t)$ for $p = p(n, t)$, we have

$$j_n(t) = j_p(t) = j_p(A_p(t))\ . \tag{9.17}$$

For each $t \in T$ we define $t_{0,T} = 0$. For $A \in \mathcal{A}_n\ ,\ n \geq 1$, we choose an arbitrary point $t_{n,A}$ in A. We define

$$\pi_n(t) = t_{p(n,t),B} \quad \text{where } B = A_{p(n,t)}(t)\ ,$$

and we note that $\pi_0(t) = 0$. When $s \in A_n(t)$ we have $A_p(s) = A_p(t)$ for $p \leq n$ and thus $p(n, s) = p(n, t)$ so that $\pi_n(s) = \pi_n(t)$. Also, if $j_{n+1}(t) = j_n(t)$ we have $p(n, t) = p(n + 1, t)$, so that $\pi_n(t) = \pi_{n+1}(t)$. This proves that (9.7) to (9.9) hold. Moreover, when $j_n(t) > j_{n-1}(t)$ we have $p(n, t) = n$ so that $\pi_n(t) = t_{n,A}$ for $A = A_n(t)$, and thus $\pi_n(t) \in A_n(t) \subset A_{n-1}(t)$, and this proves (9.10). Finally, (9.1) used for $p = p(n, t)$ and $A_p = A_p(t) = A_{p(n,t)}(t)$ reads

$$\forall s, s' \in B\ ,\ \int r^{2j_p(A_p)}|s - s'|^2 \wedge 1 d\nu \leq u 2^p$$

and this implies (9.12) since by (9.17) $j_p(A_p) = j_n(t)$ and $\pi_n(t) = t_{n,B} \in B$. The proof is complete. □

We turn to the proof of Theorem 9.2.4. The principle of the proof is, given $t \in T$, to produce a decomposition $t(\omega) = t^1(\omega) + t^2(\omega) + t^3(\omega)$ where one defines the values $t^1(\omega), t^2(\omega), t^3(\omega)$ from the values $\pi_n(t)(\omega)$ for $n \geq 1$ and to then check that the required conditions are satisfied. Despite considerable efforts, the proof is not really intuitive. Maybe it is unavoidable that the proof is not very simple. Theorem 9.4.1 is an immediate consequence of Theorem 9.2.1, and it has sweeping consequences.

Our strategy will be to define $t^1(\omega)$ as $\pi_{n(\omega)}(t)(\omega)$ for a cleverly chosen value of $n(\omega)$.[4] To prepare for the construction, we may assume that

$$\sup_{t \in T} \sum_{n \geq 0} 2^n r^{-j_n(t)} < \infty \,, \tag{9.18}$$

and in particular that

$$\forall t \in T \,, \quad \lim_{n \to \infty} j_n(t) = \infty \,. \tag{9.19}$$

For $t \in T$ and $\omega \in \Omega$, we define

$$m(t, \omega) = \inf \left\{ n \geq 0 \,;\; |\pi_{n+1}(t)(\omega) - \pi_n(t)(\omega)| > r^{-j_n(t)} \right\}$$

if the set on the right is not empty and $m(t, \omega) = \infty$ otherwise. In words, this is the first place at which $\pi_n(\omega)$ and $\pi_{n+1}(\omega)$ differ significantly. We note from the definition (9.11) of $\Omega_n(t)$ that

$$\Omega_n(t) = \{\omega \in \Omega \,;\; m(t, \omega) \geq n\} \,. \tag{9.20}$$

Lemma 9.2.5 *Under the assumptions of Theorem 9.2.4, if $n < m(t, \omega)$ then*

$$\sum_{n \leq m < m(t,\omega)} |\pi_{m+1}(t)(\omega) - \pi_m(t)(\omega)| \leq 2r^{-j_n(t)} \,. \tag{9.21}$$

Proof By definition of $m(t, \omega)$, we have

$$m < m(t, \omega) \Rightarrow |\pi_{m+1}(t)(\omega) - \pi_m(t)(\omega)| \leq r^{-j_m(t)} \,. \tag{9.22}$$

---

[4] One could also attempt to proceed as in the proof of Theorem 6.7.2: to write the chaining identity $t = \sum_{n \geq 1}(\pi_n(t) - \pi_{n-1}(t))$ and to use Lemma 6.7.1 for each of the increments $\pi_n(t) - \pi_{n-1}(t)$, with a suitable value of $u = u(t, n)$. This does not seem to make the proof any easier.

From (9.9), when $j_{m+1}(t) = j_m(t)$ we have $\pi_{m+1}(t) = \pi_m(t)$. Consequently for $m < m(t,\omega)$ we have

$$|\pi_{m+1}(t)(\omega) - \pi_m(t)(\omega)| \le r^{-j_m(t)} \mathbf{1}_{\{j_{m+1}(t) > j_m(t)\}} \,. \tag{9.23}$$

Therefore

$$\sum_{n \le m < m(t,\omega)} |\pi_{m+1}(t)(\omega) - \pi_m(t)(\omega)| \le \sum_{m \ge n} r^{-j_m(t)} \mathbf{1}_{\{j_{m+1}(t) > j_m(t)\}} \,.$$

The sum on the right is a sum of terms r^{-j} where the values of j are all different. Since $r \ge 2$, this sum is at most twice its largest term. □

When $m(t,\omega) = \infty$ it follows from (9.21) and (9.19) that the sequence $(\pi_n(t)(\omega))$ is a Cauchy sequence. Consequently $\lim_{n\to\infty} \pi_n(t)(\omega)$ exists. Let us then define

$$t^1(\omega) = \pi_{m(t,\omega)}(t)(\omega)$$

when $m(t,\omega) < \infty$ and

$$t^1(\omega) = \lim_{n\to\infty} \pi_n(t)(\omega)$$

if $m(t,\omega) = \infty$. It follows from (9.21) that for $n \le m < m(t,\omega)$ we have $|\pi_{m+1}(t)(\omega) - \pi_n(t)(\omega)| \le 2r^{-j_n(t)}$, so that (and since $t^1(\omega) - \pi_n(t)(\omega) = 0$ for $n = m(t,\omega)$)

$$n \le m(t,\omega) \Rightarrow |t^1(\omega) - \pi_n(t)(\omega)| \le 2r^{-j_n(t)} \,. \tag{9.24}$$

According to (9.7) the value of $j_0(t)$ is independent of t:

$$\forall t \in T \,,\; j_0(t) = j_0 \,. \tag{9.25}$$

Since $\pi_0(t) = 0$, (9.24) implies

$$|t^1(\omega)| \le 2r^{-j_0(t)} = 2r^{-j_0} \,. \tag{9.26}$$

For $t \in T$, we define

$$\Xi(t) = \{\omega \in \Omega \;;\; |t(\omega)| \le r^{-j_0}/2\}$$

and

$$t^2 := (t - t^1)\mathbf{1}_{\Xi(t)} \,,\; t^3 := (t - t^1)\mathbf{1}_{\Xi(t)^c} \,.$$

We define

$$T_1 = \{t^1 \ ; \ t \in T\} \ ; \ T_2 = \{t^2 \ ; \ t \in T\} \ ; \ T_3 = \{t^3 \ ; \ t \in T\} \ .$$

For $\omega \in \Xi(t)^c$ we have $|t(\omega)| \geq r^{-j_0}/2$, whereas $|t^1(\omega)| \leq 2r^{-j_0}$ by (9.26). Thus for such ω we have $|t^1(\omega)| \leq 4|t(\omega)|$. Therefore $|t^3(\omega)| = |t(\omega) - t^1(\omega)| \leq 5|t(\omega)|$. We have proved that $|t^3| \leq 5|t|\mathbf{1}_{\Xi(t)^c}$, so that the set T_3 satisfies (9.16).

We start the study of T_1. The reader would do well to review the proof of Theorem 6.7.2, since the method we use here is exactly the same as was used there for the control of T_1. The goal is to define a sequence (U_n) of sets with card $U_n \leq N_n$ which are approximations of T_1 both in the L^2 and L^∞ norm, after which we will obtain (9.13) from Proposition 2.9.7 and (9.14) through a similar principle. For $n \geq 0$, we define t_n^1 by

$$t_n^1(\omega) = \pi_{n \wedge m(t,\omega)}(t)(\omega) \ ,$$

so that

$$\forall\, \omega \in \Omega \ , \ t^1(\omega) = \lim_{n \to \infty} t_n^1(\omega) \ . \tag{9.27}$$

We define our approximating sets

$$U_n := \{t_n^1 \ ; \ t \in T\} \ .$$

We will first control the cardinality of U_n in the next lemma and then show that these sets are good approximations of T_1 both in the L^2 and the L^∞ norms.

Lemma 9.2.6 *We have* card $U_n \leq N_n$.

Proof We prove that when $s \in A_n(t)$ then $t_n^1 = s_n^1$. In other words, for $A \in \mathcal{A}_n$ all the elements t_n^1 for $t \in A$ are the same. Thus card $U_n \leq$ card $\mathcal{A}_n \leq N_n$. Consider $s \in A_n(t)$. Then $A_q(s) = A_q(t)$ for $q \leq n$, so that $\pi_q(s) = \pi_q(t)$ by (9.7). The definition of $m(t,\omega)$ shows that for any n', the points $\pi_q(t)$ for $0 \leq q \leq n'$ entirely determine whether or not it is true that $m(t,\omega) < n'$. Consequently, when $s \in A_n(t)$ we have $n \wedge m(t,\omega) = n \wedge m(s,\omega)$ for each ω, so that $t_n^1 = s_n^1$. □

Now we prove that the sets U_n approximate T_1 for the L^∞ norm. We note that $t^1(\omega) - t_n^1(\omega) = 0$ if $n \geq m(t,\omega)$, and by (9.21) that if $n < m(t,\omega)$, then

$$|t^1(\omega) - t_n^1(\omega)| \leq \sum_{n \leq m < m(t,\omega)} |\pi_{m+1}(t)(\omega) - \pi_m(t)(\omega)| \leq 2r^{-j_n(t)} \ .$$

Thus $\|t^1 - t_n^1\|_\infty \leq 2r^{-j_n(t)}$, and hence $d_\infty(t^1, U_n) \leq 2r^{-j_n(t)}$. Thus (9.14) follows from the analog of Proposition 2.9.7 for γ_1 rather then γ_2.

Before we continue, let us explain how we will use (9.12) and (9.10).

Lemma 9.2.7 *We have*

$$\int_{\Omega_n(t)} |r^{j_n(t)}(\pi_{n+1}(t)(\omega) - \pi_n(t)(\omega))|^2 \wedge 1 \mathrm{d}\nu(\omega) \le u2^n \tag{9.28}$$

and

$$\nu(\Omega_n(t) \setminus \Omega_{n+1}(t)) \le u2^n \ . \tag{9.29}$$

Proof To prove (9.28) we may assume that $\pi_n(t) \neq \pi_{n+1}(t)$ so that $j_{n+1}(t) > j_n(t)$ by (9.9). Then (9.10) shows that $s := \pi_{n+1}(t) \in A_n(t)$. Using (9.12) for s rather than t, and since $\pi_n(s) = \pi_n(t)$ and $j_n(s) = j_n(t)$ because $s \in A_n(t)$, we obtain (9.28). Then (9.29) follows since $r^{j_n(t)}|\pi_{n+1}(t) - \pi_n(t)| \ge 1$ on $\Omega_n(t) \setminus \Omega_{n+1}(t)$. □

We turn to the proof of (9.13). For this we will show that U_n approximates T_1 for the L^2 norm.

Lemma 9.2.8 *We have*

$$\|t^1_{n+1} - t^1_n\|_2 \le \sqrt{u}2^{n/2}r^{-j_n(t)} \ . \tag{9.30}$$

Proof First we observe that

$$t^1_{n+1} - t^1_n = (\pi_{n+1}(t) - \pi_n(t))\mathbf{1}_{\Omega_{n+1}(t)} \ .$$

Indeed, if $\omega \in \Omega_{n+1}(t) = \{m(t,\cdot) \ge n+1\}$, then $t^1_n(\omega) = \pi_n(t)(\omega)$ and $t^1_{n+1}(\omega) = \pi_{n+1}(t)(\omega)$, while if $\omega \notin \Omega_{n+1}(t)$, then $m(t,\omega) \le n$ and $t^1_n(\omega) = t^1_{n+1}(\omega) = \pi_{m(t,\omega)}(\omega)$.

By definition of $m(t,\omega)$ we have $|\pi_{n+1}(t) - \pi_n(t)| \le r^{-j_n(t)}$ whenever $m(t,\cdot) \ge n+1$, i.e., on $\Omega_{n+1}(t)$ by (9.20). Therefore,

$$\begin{aligned}\|t^1_{n+1} - t^1_n\|_2^2 &= \int_{\Omega_{n+1}(t)} |\pi_{n+1}(t)(\omega) - \pi_n(t)(\omega)|^2 \mathrm{d}\nu(\omega)\\ &\le r^{-2j_n(t)} \int_{\Omega_{n+1}(t)} |r^{j_n(t)}(\pi_{n+1}(t)(\omega) - \pi_n(t)(\omega))|^2 \wedge 1 \mathrm{d}\nu(\omega) \ ,\end{aligned}$$

so that (9.30) follows from (9.28). □

Proof of (9.13) Combining (9.30) and (9.18) implies that the sequence (t^1_q) is a Cauchy sequence in L^2, so that it converges to its limit, which is t^1 from (9.27), and hence $\lim_{q\to\infty} \|t^1 - t^1_q\|_2 = 0$, so that $\|t^1 - t^1_n\|_2 = \lim_{q\to\infty} \|t^1_q - t^1_n\|_2$.

Consequently

$$d_2(t^1, U_n) \le \|t^1 - t_n^1\|_2 = \lim_{q\to\infty} \|t_q^1 - t_n^1\|_2$$
$$\le \sum_{m\ge n} \|t_{m+1}^1 - t_m^1\|_2 \le \sqrt{u} \sum_{m\ge n} 2^{m/2} r^{-j_m(t)} . \tag{9.31}$$

Since

$$\sum_{n\ge 0} 2^{n/2} \sum_{m\ge n} 2^{m/2} r^{-j_m(t)} = \sum_{m\ge 0} 2^{m/2} r^{-j_m(t)} \sum_{n\le m} 2^{n/2} \le L \sum_{m\ge 0} 2^m r^{-j_m(t)} ,$$

we conclude by Proposition 2.9.7 again that (9.13) holds. □

We turn to the proof of (9.15). This is where there are new arguments compared to the case of Theorem 6.7.2. We define

$$r(t,\omega) = \inf\left\{n \ge 0\ ;\ |\pi_{n+1}(t)(\omega) - t(\omega)| \ge \frac{1}{2} r^{-j_{n+1}(t)}\right\}$$

if the set on the right is not empty and $r(t,\omega) = \infty$ otherwise.

Lemma 9.2.9 *Let us define* $t_n^2 = (t - t^1)\mathbf{1}_{\{r(t,\cdot)=n\}\cap \Xi(t)}$. *Then*

$$t^2 = \sum_{n\ge 0} t_n^2 \tag{9.32}$$

and

$$\|t_n^2\|_1 \le 3 r^{-j_n(t)} \nu(\{\omega \in \Omega\ ;\ r(t,\omega) = n\} \cap \Xi(t)) . \tag{9.33}$$

Proof Let us fix $\omega \in \Xi(t)$. Then

$$|\pi_0(t)(\omega) - t(\omega)| = |t(\omega)| \le r^{-j_0}/2 . \tag{9.34}$$

By definition of $r(t,\omega)$ we have

$$n < r(t,\omega) \Rightarrow |\pi_{n+1}(t)(\omega) - t(\omega)| < \frac{1}{2} r^{-j_{n+1}(t)} . \tag{9.35}$$

Consequently, for $0 \le n < r(t,\omega)$,

$$|\pi_{n+1}(t)(\omega) - \pi_n(t)(\omega)| \le |\pi_{n+1}(t)(\omega) - t(\omega)| + |\pi_n(t)(\omega) - t(\omega)|$$
$$\le \frac{1}{2}(r^{-j_{n+1}(t)} + r^{-j_n(t)}) \le r^{-j_n(t)} , \tag{9.36}$$

where for $n > 0$ we use (9.35) for n and $n-1$ and for $n = 0$ we use also (9.34). Consequently $r(t,\omega) \le m(t,\omega)$. When $r(t,\omega) = \infty$ then $m(t,\omega) = \infty$ so that, recalling (9.19), by (9.35) we have $t(\omega) = \lim_{n\to\infty} \pi_n(t)(\omega) = t^1(\omega)$ and $t^2(\omega) = t(\omega) - t^1(\omega) = 0$. Therefore we have proved that

$$\begin{aligned} t^2(\omega) = t(\omega) - t^1(\omega) &= (t(\omega) - t^1(\omega))\mathbf{1}_{\{r(t,\omega)<\infty\}} \\ &= \sum_{n\ge 0} (t(\omega) - t^1(\omega))\mathbf{1}_{\{r(t,\omega)=n\}} \, . \end{aligned} \tag{9.37}$$

Since this holds for each $\omega \in \Xi(t)$, we have proved (9.32). Now, when $n = r(t,\omega)$, we have $m(t,\omega) \ge n$ and, using (9.24), $|\pi_n(t)(\omega) - t^1(\omega)| \le 2r^{-j_n(t)}$. Now, if $n > 0$, using (9.35) for $n-1$, we have

$$|t(\omega) - \pi_n(t)(\omega)| \le \frac{1}{2} r^{-j_n(t)}$$

and if $n = 0$ this holds by (9.34). Consequently

$$|t(\omega) - t^1(\omega)| \le |t(\omega) - \pi_n(t)(\omega)| + |\pi_n(t)(\omega) - t^1(\omega)| \le 3r^{-j_n(t)} \, , \tag{9.38}$$

and this proves (9.33). □

Lemma 9.2.10 *Under the assumptions of Theorem 9.2.4, it holds that*

$$\nu(\{\omega \in \Omega \; ; \; r(t,\omega) = n\} \cap \Xi(t)) \le Lu2^n \, . \tag{9.39}$$

Proof Since for $\omega \in \Xi(t)$ we have $r(t,\omega) \le m(t,\omega)$, using (9.20) we get $\{\omega; r(t,\omega) = n\} \cap \Xi(t) \subset \Omega_n(t)$ and therefore

$$\begin{aligned} \nu(\{\omega \in \Omega \; ; \; r(t,\omega) = n\} \cap \Xi(t)) &\le \nu(\{\omega \in \Omega \; ; \; r(t,\omega) = n\} \cap \Omega_{n+1}(t)) \\ &\quad + \nu(\Omega_n(t) \setminus \Omega_{n+1}(t)) \, . \end{aligned} \tag{9.40}$$

Now, since $|\pi_{n+1}(t)(\omega) - t(\omega)| \ge r^{-j_{n+1}(t)}/2$ when $r(t,\omega) = n$, we have

$$\begin{aligned} &\frac{1}{4}\nu(\{\omega \in \Omega \; ; \; r(t,\omega) = n\} \cap \Omega_{n+1}(t)) \\ &\quad \le \int_{\Omega_{n+1}(t)} |r^{j_{n+1}(t)}(\pi_{n+1}(t)(\omega) - t(\omega))|^2 \wedge 1 d\nu(\omega) \le u2^{n+1} \, , \end{aligned}$$

using (9.12) for $n+1$ rather than n in the last inequality. Combining with (9.29) completes the proof. □

Combining (9.33) with (9.39) proves that $\|t_n^2\|_2 \le Lu2^n r^{-j_n(t)}$. Combining with (9.32), we have proved (9.15) and completed the proof of Theorem 9.2.4.

9.3 Philosophy

Let us stress how simple is the decomposition of Theorem 9.2.4. To explain this we focus on the case where $\|t\|_\infty \le r^{-j_0(t)}/2$ for each t (recalling that $j_0(t)$ does not depend on t). Then, recalling (9.11) (and that $\pi_0(t) = 0$), we have the formulas

$$t^1 = \sum_{n\ge 1}(\pi_n(t) - \pi_{n-1}(t))\mathbf{1}_{\Omega_n(t)} \; ; \; t^2 = \sum_{n\ge 0}(t - \pi_n(t))\mathbf{1}_{\Omega_n(t)\setminus\Omega_{n+1}(t)} \, . \tag{9.41}$$

The main difficulty in the proof of Theorem 9.2.4 is that it does not seem true that we can easily control $\sum_{n\ge 0}\|(t - \pi_n(t))\mathbf{1}_{\Omega_n(t)\setminus\Omega_{n+1}(t)}\|_1$ (in sharp contrast with the second proof of Theorem 6.7.2).

It is hard at that stage to really see the power of Theorem 9.2.4, which will come into full use later. Let us make some comments on this, although the reader may not fully understand them until she studies Chap. 10. Let us think that we are actually trying to construct the partitions $(\mathcal{A}_n)$ and the corresponding objects of Theorem 9.2.4. We have already constructed these objects up to level n, and we try to construct the next level. We have to ensure the constraint (9.10), but short of that we are pretty much free to choose $\pi_{n+1}(t)$ to our liking. The magic is that *whatever our choice*, we will drop the part $\Omega_n(t)\setminus\Omega_{n+1}(t)$ of Ω where $\pi_{n+1}(t)$ is too different from $\pi_n(t)$. The reason we can do that is that *on this set* the decomposition is *finished*: recalling that we assume for clarity that $\Xi(t) = \Omega$ for t and recalling that $\Omega_n(t) = \{m(t,\omega) \ge n\}$, on the set $\Omega_n(t) \setminus \Omega_{n+1}(t)$ we have $m(t,\omega) = n$ so that $t^1 = \pi_n(t)$ and $t^2 = t - \pi_n(t)$. We have already decided the values of t^1 and t^2! In particular for the $(n+1)$-st step of the construction, we are concerned only with points $\omega \in \Omega_n(t)$, and for these points the sequence $(\pi_q(t)(\omega))_{q\le n}$ does not have big jumps. We will take great advantage of this feature in the proof of Theorem 6.2.8.[5]

Another point which is quite obvious but needs to be stressed is that in performing a recursive construction of the $(\mathcal{A}_n)$, we only need to care about controlling the $\pi_n(t)$, and we do not have to worry "that we might loose information about t". If you need to visualize this fact, you may argue as follows. As we explained, if we know that $m(t,\omega) \le n$, we already know what are $t^1(\omega)$ and $t^2(\omega)$ so that we no longer have to worry about this value of ω in the sequel of the construction. When $m(t,\omega) \ge n$ we may write the peaky part $t^2(\omega)$ of t at ω as

$$\begin{aligned} t^2(\omega) &= t(\omega) - \pi_{m(t,\omega)}(t)(\omega) \\ &= t(\omega) - \pi_n(t)(\omega) + (\pi_n(t)(\omega) - \pi_{m(t,\omega)}(t)(\omega)) \, , \end{aligned}$$

[5] Despite the fact that the proof of Theorem 9.2.4 is identical to the proof of Theorem 9.2.1, the formulation of Theorem 9.2.1, which is due to W. Bednorz and R. Latała, is a great step forward, as it exactly identifies the essence of Theorem 9.2.1.

and, in a sense "we have attributed at stage n the part $t(\omega) - \pi_n(t)(\omega)$ of $t(\omega)$ to the peaky part $t^2(\omega)$ of t". (So that we no longer need not think about t itself.)

These considerations may look mysterious now, but they explain why certain constructions which we perform in Sect. 10.3 keep sufficient information to succeed.

9.4 Chaining for Bernoulli Processes

Our first result is a simple consequence of Theorem 9.2.1. It is a generalization of the generic chaining bound (2.59) to Bernoulli processes. The sweeping effectiveness of this result will be demonstrated soon. We consider a number $r \geq 2$ and we recall the quantity $b^*(T)$ of (6.9).

Theorem 9.4.1 *Consider a subset T of ℓ^2, and assume that $0 \in T$. Consider an admissible sequence of partitions $(\mathcal{A}_n)$ of T, and for $A \in \mathcal{A}_n$ consider a number $j_n(A) \in \mathbb{Z}$ with the following properties, where $u \geq 1$ is a parameter:*

$$\forall n \geq 0\,,\ \forall x, y \in A \in \mathcal{A}_n\,,\ \sum_{i \geq 1} |r^{j_n(A)}(x_i - y_i)|^2 \wedge 1 \leq u 2^n\,, \tag{9.42}$$

where $x \wedge y = \min(x, y)$. Then

$$b^*(T) \leq L\Big(u \sup_{x \in T} \sum_{n \geq 0} 2^n r^{-j_n(A_n(x))} + \sup_{x \in T} \sum_{i \geq 1} |x_i| \mathbf{1}_{\{2|x_i| \geq r^{-j_0(T)}\}}\Big)\,. \tag{9.43}$$

Moreover if ε_i are independent Bernoulli r.v.s, for any $p \geq 1$, we have

$$\Big(\mathsf{E} \sup_{x \in T} \Big|\sum_{i \geq 1} x_i \varepsilon_i\Big|^p\Big)^{1/p} \leq K(p) u \sup_{x \in T} \sum_{n \geq 0} 2^n r^{-j_n(A_n(x))}$$
$$+ L \sup_{x \in T} \sum_{i \geq 1} |x_i| \mathbf{1}_{\{2|x_i| \geq r^{-j_0(T)}\}}\,. \tag{9.44}$$

Conditions (9.42) is simply (9.1) in the case where the measure space is $\mathbb{N}$ with the counting measure. Condition (9.42) may also be written as

$$\forall x, y \in A \in \mathcal{A}_n\,,\ \sum_{i \geq 1} |x_i - y_i|^2 \wedge r^{-2j_n(A)} \leq u 2^n r^{-2j_n(A)}\,. \tag{9.45}$$

The point of writing (9.42) rather than (9.45) is simply that this is more in line with the generalizations of this statement that we shall study later.

If, instead of condition (9.45), we had the stronger condition

$$\forall x, y \in A \in \mathcal{A}_n\ , \ \sum_{i \ge 1} |x_i - y_i|^2 \le u 2^n r^{-2 j_n(A)}\ , \tag{9.46}$$

this would simply mean that $\Delta(A) \le \sqrt{u} 2^{n/2} r^{-j_n(A)}$ so that then $\gamma_2(T) \le \sqrt{u} \sup_{x \in T} \sum_{n \ge 0} 2^n r^{-j_n(A_n(x))}$ and we would prove nothing more than the generic chaining bound (2.59). The point of Theorem 9.4.1 is that (9.42) is significantly weaker than condition (9.46), because it requires a much weaker control on the large values of $x_i - y_i$. It may be difficult at this stage to really understand that this is a considerable gain, but some comments that might help may be found on page 435.

The proof of Theorem 9.4.1 relies on the following:

Proposition 9.4.2 *Under the conditions of Theorem 9.4.1, we can write $T \subset T_1 + T_2 + T_3$ where $0 \in T_1$ and*

$$\gamma_2(T_1, d_2) \le L\sqrt{u} \sup_{x \in T} \sum_{n \ge 0} 2^n r^{-j_n(A_n(x))}\ , \tag{9.47}$$

$$\gamma_1(T_1, d_\infty) \le L \sup_{x \in T} \sum_{n \ge 0} 2^n r^{-j_n(A_n(x))}\ , \tag{9.48}$$

$$\forall x \in T_2\ , \ \|x\|_1 \le L u \sup_{x \in T} \sum_{n \ge 0} 2^n r^{-j_n(A_n(x))}\ , \tag{9.49}$$

and

$$\forall x \in T_3\ , \ \exists y \in T\ , \ \forall i \ge 1\ , \ |x_i| \le 5 |y_i| \mathbf{1}_{\{2|y_i| \ge r^{-j_0(T)}\}}\ . \tag{9.50}$$

Proof This follows from Theorem 9.2.1 in the case $\Omega = \mathbb{N}^*$ and where ν is the counting measure. □

Proof of Theorem 9.4.1 To prove (9.43) we use that $T \subset T_1 + (T_2 + T_3)$, so that by definition of $b^*(T)$ we have

$$b^*(T) \le \gamma_2(T_1, d_2) + \sup_{x \in T_2 + T_3} \|x\|_1$$

and we use (9.47), (9.49), and (9.50). To prove (9.44) we show that for $j = 1, 2, 3$ the quantity $(\mathsf{E} \sup_{x \in T_j} |\sum_{i \ge 1} x_i \varepsilon_i|^p)^{1/p}$ is bounded by the right-hand side of (9.44). For $j = 1$ this follows from (2.66) (using also that $0 \in T_1$), and for $j = 2, 3$ this follows from the bound $(\mathsf{E} \sup_{x \in T_j} |\sum_{i \ge 1} \varepsilon_i x_i|^p)^{1/p} \le \sup_{x \in T_j} \|x\|_1$. The reason we have a factor u (rather than $\sqrt{u}$) in the right-hand side of (9.43) is that we have a factor u in (9.4) and hence in (9.49). As we will see this factor does not create problems and there is plenty of room. □

The following is a simple consequence of Theorem 9.4.1:

Corollary 9.4.3 *Assume that moreover*

$$\forall x \in T\ ,\ \|x\|_\infty < r^{-j_0(T)}/2\ . \tag{9.51}$$

Then

$$b^*(T) \le Lu \sup_{x\in T} \sum_{n\ge 0} 2^n r^{-j_n(A_n(x))}\ . \tag{9.52}$$

Proof In (9.43) the second term in the right-hand side is identically zero. □

Corollary 9.4.3 is in some sense optimal as the following shows:

Proposition 9.4.4 *Assume that* $0 \in T \subset \ell^2$. *Then we can find a sequence* $(\mathcal{A}_n)$ *of admissible partitions of* T *and for* $A \in \mathcal{A}_n$ *a number* $j_n(A)$ *such that conditions (9.51) and (9.42) are satisfied for* $u = 1$ *and moreover*

$$\sup_{x\in T} \sum_{n\ge 0} 2^n r^{-j_n(A_n(x))} \le K(r) b^*(T)\ . \tag{9.53}$$

This will be proved in Sect. 10.15. The situation here is the same as for the generic chaining bound for Gaussian processes. There is no magic wand to discover the proper choice of the partitions $\mathcal{A}_n$, and in specific situations this can be done only by understanding the "geometry of T", typically a very challenging problem.

We should point out one of the (psychological) difficulties in discovering the proof of Theorem 6.2.8 (the Latała-Bednorz theorem). Even though it turns out from Proposition 9.4.4 that one can find the partitions $\mathcal{A}_n$ such that (9.42) holds, when proving Theorem 6.2.8, it seems necessary to use partitions with a weaker property, which replaces the summation over all values of i in (9.42) by the summation over the values in an appropriate subset $\Omega_n(t)$ of $\mathbb{N}^*$, as in (9.12) above.

Efficient bounds on random Fourier series as presented in Sect. 7.8.5 were first discovered as an application of Theorem 9.4.1. The advantage of this method is that it bypasses the magic of Theorem 7.8.1 while following a conceptually transparent scheme. It is sketched in the following exercises (the solution of which is given in the briefest of manners).

Exercise 9.4.5 We consider complex numbers a_i and characters χ_i. We fix a number $r \ge 4$ and we define $\psi_j(s,t) = \sum_i |r^j a_i(\chi_i(s) - \chi_i(t))|^2 \wedge 1$. Consider a parameter $w \ge 1$ and for $n \ge 0$ set $D_n = \{s \in T;\, \psi_j(s,0) \le w2^n\}$. Assume that $\mu(D_0) \ge 3/4$ and that $\mu(D_n) \ge N_n^{-1}$ for $n \ge 1$. Prove that there is an admissible sequence of partitions $(\mathcal{A}_n)$ of T and integers j'_n with $j'_0 = j_0$ such that $\sum_{n\ge 0} 2^n r^{-j'_n} \le L \sum_{n\ge 0} 2^n r^{-j_n}$ and that $\psi_{j'_n}(s,t) \le Lw2^n$ for $s,t \in A \in \mathcal{A}_n$.

Exercise 9.4.6 Under the conditions of the previous exercise show that for each $p \geq 1$ it holds that

$$\Big(\mathsf{E}\sup_{s\in T}\Big|\sum_i \varepsilon_i a_i(\chi_i(s) - \chi_i(0))\Big|^p\Big)^{1/p} \leq K(r,p)w\sum_{n\geq 0} 2^n r^{-j_n} + K(r)\sum_i |a_i|\mathbf{1}_{\{4|a_i|\geq r^{-j_0}\}} \,. \tag{9.54}$$

Key Ideas to Remember

- We have introduced a method to measure the size of a set of measurable functions using a family of distances. Control of this size allows structural information of the set.
- Control of the size of a set of functions on $\mathbb{N}$ allows sharp bounds on the corresponding Bernoulli process.

9.5 Notes and Comments

Theorem 9.4.1 can be thought as an abstract version of Ossiander's bracketing theorem (that we shall prove in Sect. 14.1). The author proved Theorem 9.2.1 (in an essentially equivalent form) as early as [115] and in the exact form presented here in [129], but did not understand then its potential as a chaining theorem. The version of this work at the time the author received [16] contained only Theorem 9.2.1, with a proof very similar to the proof of Theorem 9.2.4 which we present here.

Chapter 10
Proof of the Bernoulli Conjecture

The present chapter will use to the fullest a number of the previous ideas, and the reader should have fully mastered Chaps. 2 and 6 as well as the statement of Theorem 9.2.4.

The overall strategy to prove the Bernoulli conjecture is somewhat similar to the one we used to prove Theorem 2.9.1: we recursively construct increasing families of partitions, and we measure the size of the elements of the partitions through certain functionals. Once this appropriate sequence has been constructed, the required decomposition will be provided by Theorem 9.2.4.

Just as in the case of the Gaussian case, Theorem 2.10.1, a main tool is the Sudakov minoration, which now refers to (6.21). In contrast with the Sudakov minoration in the Gaussian case, (2.116), we cannot use this result unless we control the supremum norm.

The basic tool to control the supremum norm is the method of "chopping maps" of Sect. 10.3. In this method to each element $t \in \ell^2$ we associate a new element $t' \in \ell^2$, of which we control the supremum norm. The difficulty is that this operation decreases the distance, $d(t', u') \leq d(t, u)$, and some of the information we had accumulated on the metric space (T, d) is lost in that step. The method may "change the set T" (as well as the set of underlying Bernoulli r.v.s) at every step of the construction, and consequently we will need to change the functional at every step of the construction. Furthermore, this functional will depend on the element of the partition we try to split further, a new technical feature.

A main difference between the proof of the Bernoulli conjecture and that of Theorem 2.9.1 is that instead of having two really different possibilities in partitioning a set, as in as in Lemma 2.9.4, there will now be three different possibilities (as expressed in Lemma 10.7.3). A radically new idea occurs here, which has no equivalent in the previous results, and we start with it in Sect. 10.1.

As the reader will soon realize, the proof of the Bernoulli conjecture is rather demanding. One reason probably is that there are missing pieces to our understanding. Finding a more transparent proof is certainly a worthy research project. The

M. Talagrand, *Upper and Lower Bounds for Stochastic Processes*, Ergebnisse der Mathematik und ihrer Grenzgebiete. 3. Folge / A Series of Modern Surveys in Mathematics 60, https://doi.org/10.1007/978-3-030-82595-9_10

good news however is that understanding the details of this proof is certain not required to continue reading this book, since the ideas on which the proof relies will not be met any more. It is however essential to understand well the meaning of the results of Sect. 10.15, which are the basis of the fundamental results of the next chapter.

I will not hesitate to state that the Latała-Bednorz theorem is the most magnificent piece of mathematics I ever came across[1] and that I find that it is well worth making some effort to understand it. Not only several beautiful ideas are required, but the way they are knitted together is simply breathtaking. I am greatly indebted to Kevin Tanguy for his help in making this chapter more accessible.

10.1 Latała's Principle

The principle proved in this section is a key to the entire chapter.[2] It was proved first by Rafał Latała in [49], but it was not obvious at the time how important this is.[3]

Consider a subset J of $\mathbb{N}^* := \mathbb{N} \setminus \{0\}$ and a subset T of ℓ^2. Our goal is to compare the processes $(\sum_{i\ge 1} \varepsilon_i t_i)_{t\in T}$ and $(\sum_{i\in J} \varepsilon_i t_i)_{t\in T}$.[4] We define $b_J(T) := \mathsf{E} \sup_{t\in T} \sum_{i\in J} \varepsilon_i t_i$. Thus $b_J(T) \le b(T)$. It may be the case that $b_J(T) = b(T)$ as, for example, when $t_i = 0$ for $i \notin J$ and $t \in T$. Note that in that case the diameter of T for the canonical distance d is the same as its diameter for the smaller distance d_J given by $d_J^2(s,t) = \sum_{i\in J}(s_i - t_i)^2$. This brings us to consider two typical (nonexclusive) situations:

- First, the diameter of T for d is about the same as its diameter for d_J.
- Second, $b_J(T)$ is significantly smaller than $b(T)$.

Latała's principle states that if one also controls the supremum norm, the set T can be decomposed into not too many pieces which satisfy one of these two conditions.

Proposition 10.1.1 (Latała's Principle) *There exists a constant L_1 with the following property. Consider a subset T of ℓ^2 and a subset J of $\mathbb{N}^*$. Assume that for certain numbers $c, \sigma > 0$ and an integer $m \ge 2$ the following holds:*

$$\forall s,t \in T\ ,\ \sum_{i\in J}(s_i - t_i)^2 \le c^2\ , \tag{10.1}$$

---

[1] A statement that has to be qualified by the fact that I do not read anything!

[2] It is in homage to this extraordinary result that I have decided to violate alphabetical order and to call Theorem 6.2.8 the Latała-Bednorz theorem. This is my personal decision, and impressive contributions of Witold Bednorz to this area of probability theory are brought to light in particular in Chap. 16.

[3] The simpler proof presented here comes from [17].

[4] The resulting information will allow us to *remove* some of the Bernoulli r.v.s at certain steps of our construction.

$$t \in T \Rightarrow \|t\|_\infty < \frac{8\sigma}{\sqrt{\log m}} \,. \tag{10.2}$$

Then provided

$$c \le \frac{\sigma}{L_1} \,, \tag{10.3}$$

we can find $m' \le m+1$ and a partition $(A_\ell)_{\ell \le m'}$ of T such that for each $\ell \le m'$ we have **either**[5]

$$\exists t^\ell \in T \,,\; A_\ell \subset B_2(t^\ell, \sigma) \,, \tag{10.4}$$

or else[6]

$$b_J(A_\ell) \le b(T) - \frac{\sigma}{L}\sqrt{\log m} \,. \tag{10.5}$$

In (10.1) we assume that the diameter of T is small for the *small* distance d_J. We then produce these sets A_ℓ on which extra information has been gained: they are either of small diameter for the *large* distance $d = d_{\mathbb{N}^*}$ as in (10.4), or they satisfy (10.5). The information of (10.4) and (10.5) are of a different nature. It is instructive to compare this statement with Lemma 2.9.4. We also note that we require control in supremum norm through (10.2). Indeed, this control is required to use the Sudakov minoration for Bernoulli processes (Theorem 6.4.1) which is a main ingredient of the proof.

Proof Let us fix a point t_0 of T and replace T by $T - t_0$. We then have

$$\forall t \in T \,,\; \sum_{i \in J} t_i^2 \le c^2 \tag{10.6}$$

and

$$t \in T \Rightarrow \|t\|_\infty < \frac{16\sigma}{\sqrt{\log m}} \,. \tag{10.7}$$

For $t \in T$ set

$$Y_t = \sum_{i \in J} \varepsilon_i t_i \;;\; Z_t = \sum_{i \notin J} \varepsilon_i t_i \,, \tag{10.8}$$

[5] As usual the ball $B(t^\ell, \sigma)$ below is the ball for the distance d.

[6] In fact, all the sets A_ℓ satisfy (10.4) except at most one which satisfies (10.5).

so that

$$b(T) = \mathsf{E} \sup_{t \in T} (Y_t + Z_t) \ . \tag{10.9}$$

We may assume that T cannot be covered by m balls of the type $B(t, \sigma)$, for the result is obvious otherwise. It thus makes sense to define

$$\alpha = \inf_{F \subset T, \operatorname{card} F \le m} \mathsf{E} \sup_{t \in T \setminus \bigcup_{s \in F} B(s,\sigma)} Y_t \ .$$

To prove the theorem, we shall prove that provided the constant L_1 of (10.3) is large enough, we have

$$\alpha \le b(T) - \frac{\sigma}{L} \sqrt{\log m} \ . \tag{10.10}$$

Indeed, consider a set $F = \{t^1, \dots, t^m\} \subset T$ such that

$$\mathsf{E} \sup_{t \in T \setminus \bigcup_{s \in F} B(s,\sigma)} Y_t = \mathsf{E} \sup_{t \in T \setminus \cup_{\ell \le m} B(t^\ell,\sigma)} Y_t \le b(T) - \sigma \sqrt{\log m}/L \ .$$

The required partition is then obtained by choosing $A_\ell \subset B(t^\ell, \sigma)$ for $\ell \le m$ and $A_{m+1} = T \setminus \cup_{\ell \le m} B(t^\ell, \sigma)$.

We turn to the proof of (10.10). By definition of α, given $F \subset T$ with $\operatorname{card} F \le m$, the r.v.

$$W := \sup_{t \in T \setminus \cup_{s \in F} B(s,\sigma)} Y_t \ \text{ satisfies } \ \mathsf{E} W \ge \alpha \ . \tag{10.11}$$

Moreover, using (10.6) and (6.14) with $a(t) = 0$, we obtain

$$\forall u > 0 \ , \ \mathsf{P}(|W - \mathsf{E}W| \ge u) \le L \exp\Big(- \frac{u^2}{Lc^2} \Big) \ . \tag{10.12}$$

Let us consider independent copies $(Y_t^k)_{t \in T}$ of the process $(Y_t)_{t \in T}$ (which are also independent of the r.v.s $(\varepsilon_i)_{i \ge 1}$) and a small number $\epsilon > 0$. First, we consider $W_1 := \sup_{t \in T} Y_t^1$ and we select a point $t^1 \in T$ (depending on the r.v.s Y_t^1) with

$$Y_{t^1}^1 \ge W_1 - \epsilon \ . \tag{10.13}$$

Next, we let $W_2 = \sup_{t \in T \setminus B(t^1,\sigma)} Y_t^2$ and we find $t^2 \notin B(t^1, \sigma)$ such that

$$Y_{t^2}^2 \ge W_2 - \epsilon \ . \tag{10.14}$$

We proceed in this manner, constructing points t^k with $Y^k_{t^k} \ge W_k - \varepsilon$ (where $W_k = \sup_{t \notin \sup_{\ell<k} B(t^\ell,\sigma)} Y^k_t$) and $t^k \notin \bigcup_{\ell<k} B(t^\ell, \sigma)$, until we construct a last point t^m.

The proof of (10.10) will follow from appropriate upper and lower bounds for the quantity

$$S := \mathsf{E} \max_{k \le m} (Y^k_{t^k} + Z_{t^k}) . \tag{10.15}$$

These bound are themselves obtained in the most natural manner using concentration of measure and Sudakov's minoration. To find a lower bound, we write

$$\max_{k \le m} (Y^k_{t^k} + Z_{t^k}) \ge \max_{k \le m} (W_k + Z_{t^k}) - \epsilon \ge \min_{k \le m} W_k + \max_{k \le m} Z_{t^k} - \epsilon , \tag{10.16}$$

and we proceed to evaluate the expected value of the right-hand side. First, fixing a value of $k \le m$ and using (10.11) given the points $t^1, \ldots, t^{k-1}$ implies that $\mathsf{E} W_k \ge \alpha$, because the process (Y^k_t) is independent of $t^1, \ldots, t^{k-1}$. Using (10.12) given $Y^1, \ldots, Y^{k-1}$ (and $t^1, \ldots, t^{k-1}$), we obtain that for $u > 0$ we have $\mathsf{P}(|W_k - \mathsf{E} W_k| \ge u) \le L \exp(-u^2/(Lc^2))$, and proceeding as in (2.123) we get $\mathsf{E} V \le Lc\sqrt{\log m}$ where $V = \max_{k \le m} |W_k - \mathsf{E} W_k|$. Since $W_k \ge \mathsf{E} W_k - V \ge \alpha - V$, we obtain

$$\mathsf{E} \min_{k \le m} W_k \ge \alpha - Lc\sqrt{\log m} . \tag{10.17}$$

Next, denoting by E_{J^c} expectation in the r.v.s $(\varepsilon_i)_{i \in J^c}$ only, we prove that

$$\mathsf{E}_{J^c} \max_{k \le m} Z_{t^k} \ge \frac{1}{L} \sigma \sqrt{\log m} . \tag{10.18}$$

For this we observe that for $s, t \in T$ with $d(s,t) = \|s - t\|_2 \ge \sigma$ then, using (10.1), and assuming without loss of generality $L_1 \ge 2$ in (10.3),

$$d_{J^c}(s,t)^2 = \sum_{i \notin J} (s_i - t_i)^2 = \sum_{i \ge 1} (s_i - t_i)^2 - \sum_{i \in J} (s_i - t_i)^2 \ge \sigma^2 - c^2 \ge (\sigma/2)^2 .$$

Now, by construction, for $k, \ell \le m$, $k \ne \ell$ we have $d(t^k, t^\ell) \ge \sigma$. We then apply Theorem 6.4.1 (the Sudakov minoration for Bernoulli processes) to J^c, and so, using also (10.7), (10.18) follows from (6.21). Taking expectation in (10.18) with respect to the r.v.s ε_i, $i \in J$ yields

$$\mathsf{E} \max_{k \le m} Z_{t^k} \ge \frac{1}{L} \sigma \sqrt{\log m} .$$

Taking expectation in (10.16) and letting $\epsilon \to 0$ we have proved that

$$S \ge \alpha + \Big(\frac{\sigma}{L} - Lc\Big) \sqrt{\log m} . \tag{10.19}$$

The next goal is to bound S from above. We first observe that

$$S \le \mathsf{E} \max_{k \le m} \sup_{t \in T} (Y_t^k + Z_t) \; . \tag{10.20}$$

Consider then some numbers $(a(t))_{t \in T}$. Using (6.14) and (10.6) we obtain

$$\forall u > 0 \, , \; \mathsf{P}\Big(\Big| \sup_{t \in T} (Y_t + a(t)) - \mathsf{E} \sup_{t \in T} (Y_t + a(t)) \Big| \ge u \Big) \le L \exp\Big(- \frac{u^2}{Lc^2} \Big) \, .$$

Proceeding as in (2.123) we obtain

$$\mathsf{E} \max_{k \le m} | \sup_{t \in T} (Y_t^k + a(t)) - \mathsf{E} \sup_{t \in T} (Y_t + a(t))| \le Lc\sqrt{\log m} \; ,$$

and finally

$$\mathsf{E} \max_{k \le m} \sup_{t \in T} (Y_t^k + a(t)) \le \mathsf{E} \sup_{t \in T} (Y_t + a(t)) + Lc\sqrt{\log m} \; . \tag{10.21}$$

Let us recall that Y_t^k does not depend on the r.v.s $(\varepsilon_i)_{i \in J^c}$, but only on the r.v.s $(\varepsilon_i)_{i \in J}$. Thus denoting E_J expectation in the r.v.s $(\varepsilon_i)_{i \in J}$ only (given the r.v.s $(\varepsilon_i)_{i \in J^c}$), we may rewrite (10.21) as

$$\mathsf{E}_J \max_{k \le m} \sup_{t \in T} (Y_t^k + a(t)) \le \mathsf{E}_J \sup_{t \in T} (Y_t + a(t)) + Lc\sqrt{\log m} \; . \tag{10.22}$$

Since Z_t depends only on the r.v.s $(\varepsilon_i)_{i \in J^c}$, so that given these r.v.s Z_t is just a number $a(t)$. Thus (10.22) implies

$$\mathsf{E}_J \max_{k \le m} \sup_{t \in T} (Y_t^k + Z_t) \le \mathsf{E}_J \sup_{t \in T} (Y_t + Z_t) + Lc\sqrt{\log m} \; .$$

Taking expectation and using (10.9) yields

$$\mathsf{E} \max_{k \le m} \sup_{t \in T} (Y_t^k + Z_t) \le b(T) + Lc\sqrt{\log m} \; . \tag{10.23}$$

Combining with (10.20) and (10.19), we obtain

$$\alpha + \Big(\frac{\sigma}{L} - Lc \Big) \sqrt{\log m} \le b(T) + Lc\sqrt{\log m} \; ,$$

so that $\alpha \le b(T) - (\sigma/L_2 - L_2 c)\sqrt{\log m}$ and indeed (10.10) holds true provided $c \le \sigma/(2L_2^2)$. □

10.2 Philosophy, I

Let us look at a high level at the work of the previous section. Given a subset J of $\mathbb{N}^*$, it is a completely natural question to ask when it is true that $b(T)$ is significantly larger than $b_J(T)$. One obvious way to guarantee this is as follows: for the typical value of $(\varepsilon_i)_{i\in J}$, the set T' of $t \in T$ for which $\sum_{i\in J}\varepsilon_i t_i \simeq \sup_{t\in T}\sum_{i\in J}\varepsilon_i t_i$ has to be such that $\mathsf{E}\sup_{t\in T'}\sum_{i\notin J}\varepsilon_i t_i$ is not very small. This is not the case in general. For example, it may happen that all the $(t_i)_{i\notin J}$ are the same and then $b(T) = b_J(T)$. More generally, it may happen that the sequence $(t_i)_{i\notin J}$ takes only a few values, and then $b(T)$ and $b_J(T)$ will be very close to each other. In some precise sense, Latała's principle states that this phenomenon just described is the only possibility for $b(T)$ to be about $b_J(T)$. Under (10.1) to (10.3) the set T can be decomposed into a few pieces A for which either $b_J(A)$ is significantly less than $b(A)$ or such that the set of $(t_i)_{i\notin J}$ is of small diameter (which, under the condition (10.1), takes the form that A itself is of small diameter).

10.3 Chopping Maps and Functionals

10.3.1 Chopping Maps

One of the most successful ideas about Bernoulli processes is that of chopping maps. The basic idea of chopping maps is to replace the individual r.v.s $\varepsilon_i x_i$ by a sum $\sum_j \varepsilon_{i,j}x_{i,j}$ where $\varepsilon_{i,j}$ are independent Bernoulli r.v.s and where $x_{i,j}$ are "small pieces of x_i". It is then easier to control the ℓ^∞ norm of the new process. This control is fundamental to be able to apply the Sudakov minoration (6.21) and its consequence Corollary 6.4.10 which are key elements of the proof of the Bernoulli conjecture.

Given $u \le v \in \mathbb{R}$ we define the function $\varphi_{u,v}$ as the unique continuous function for which $\varphi_{u,v}(0) = 0$, which is constant for $x \le u$ and $x \ge v$ and has slope 1 between these values, see Fig. 10.1. Thus

$$\varphi_{u,v}(x) = \min(v, \max(x,u)) - \min(v, \max(0,u)) \ . \tag{10.24}$$

Consequently

$$|\varphi_{u,v}(x)| \le v - u \ . \tag{10.25}$$

and

$$|\varphi_{u,v}(x) - \varphi_{u,v}(y)| \le |x-y| \tag{10.26}$$

with equality when $u \le x, y \le v$.

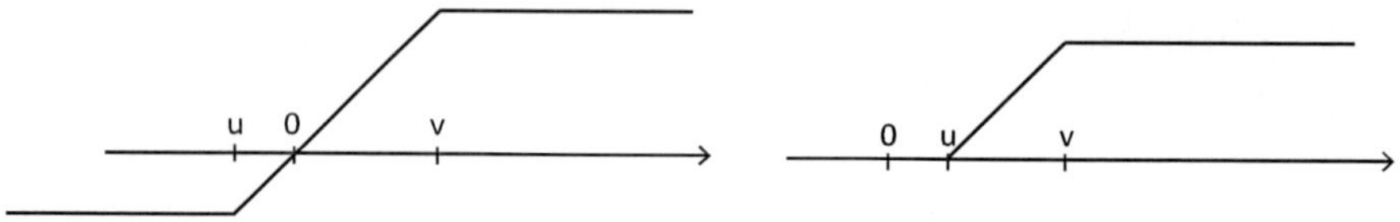

Fig. 10.1 The graph of $\varphi_{u,v}$, to the left when $u < 0 < v$, to the right when $0 < u < v$

It is very useful to note that if $u_1 \le u_2 \le \cdots \le u_k$, then

$$\varphi_{u_1,u_k}(x) = \sum_{1 \le \ell < k} \varphi_{u_\ell,u_{\ell+1}}(x) . \tag{10.27}$$

This is simply because both the left-hand side and the right-hand side are continuous, constant for $x \le u_1$ and $x \ge u_k$, have slope 1 between these values, and take the value 0 at 0.

Given a finite subset G of $\mathbb{R}$ we define

$$G^- := \{u \in G \; ; \; \exists v \in G \; , \; u < v\} ,$$

and for $u \in G^-$ we define $u^+ = \min\{v \in G \; ; \; u < v\}$, which we will call the *successor* of u. It will always be implicitly assumed that $\operatorname{card} G \ge 2$ so that $G^- \ne \emptyset$. In other words, if we enumerate $G = \{u_1, \dots, u_n\}$ where the sequence $(u_k)_{1 \le k \le n}$ is increasing, we have $G^- = \{u_1, \dots, u_{n-1}\}$ and for $k < n$ we have $u_k^+ = u_{k+1}$. To form a mental picture let us say that sets G consisting of a few evenly spaced points, say points of the type pa for $p \in \mathbb{Z}$, $p_0 \le p \le p_1$ (with $a \in \mathbb{R}^+$, $p_0, p_1 \in \mathbb{Z}$) will be essential although some slightly different situations will also be considered. A simple idea is that the family of numbers $\varphi_{u,u^+}(x)$ for $u \in G^-$ gives us good control of the values of x such that $\min G \le x \le \max G$. Different sets G_i will be considered to control each of the values of the different coordinates t_i of t in an appropriate range. As a consequence of (10.27), we obtain

$$\varphi_{\min G, \max G} = \sum_{u \in G^-} \varphi_{u,u^+} . \tag{10.28}$$

Lemma 10.3.1 *For each $x, y \in \mathbb{R}$ and each finite set G, we have*

$$\sum_{u \in G^-} |\varphi_{u,u^+}(x) - \varphi_{u,u^+}(y)| \le |x - y| . \tag{10.29}$$

Moreover there is equality if $\min G \le x, y \le \max G$.

Proof It suffices to prove (10.29) when $x > y$. As a consequence of the fact that φ_{u,u^+} is non-decreasing, the absolute values may be removed in the left-hand side, and (10.29) is then a consequence of (10.28) and (10.26). □

In particular, since $\varphi_{u,u^+}(0) = 0$, we have

$$\sum_{u \in G^-} |\varphi_{u,u^+}(x)| \le |x| \,. \tag{10.30}$$

In the remainder of this chapter we consider independent Bernoulli r.v.s $\varepsilon_{x,i}$ for $x \in \mathbb{R}$ and $i \in \mathbb{N}^*$. These are also assumed to be independent of all other Bernoulli r.v.s considered, in particular the ε_i.

Consider now for $i \ge 1$ a finite set $G_i \subset \mathbb{R}$. For $t \in \ell^2$ we consider the r.v.

$$X_t(G_i, i) := \sum_{u \in G_i^-} \varepsilon_{u,i}\varphi_{u,u^+}(t_i) \,. \tag{10.31}$$

That is, the value t_i is "chopped" into the potentially smaller pieces $\varphi_{u,u^+}(t_i)$.

Exercise 10.3.2 Consider $t, t' \in \ell^2$. Show that if $t_i, t'_i \ge \max G_i$ or if $t_i, t'_i \le \min G_i$ then $X_t(G_i, i) = X_{t'}(G_i, i)$.

We chop the value of t_i for all values of i. We write $\mathcal{G} = (G_i)_{i \ge 1}$ and we consider the r.v.

$$X_t(\mathcal{G}) := \sum_{i \ge 1} X_t(G_i, i) = \sum_{i \ge 1} \sum_{u \in G_i^-} \varepsilon_{u,i}\varphi_{u,u^+}(t_i) \,. \tag{10.32}$$

Combining (10.30) with the inequality $\sum a_k^2 \le (\sum |a_k|)^2$, we obtain that for $t \in \ell^2$ and $i \ge 1$, we have $\sum_{u \in G_i^-} \varphi_{u,u^+}(t_i)^2 \le t_i^2$ so that $X_t(\mathcal{G}) \in \ell^2$. In this manner to a Bernoulli process $(X_t)_{t \in T}$ we associate a new Bernoulli process $(X_t(\mathcal{G}))_{t \in T}$. Again the idea is that each value t_i is chopped into little pieces, and we should think of $\mathcal{G}$ as a parameter giving us the recipe to do that. Another way to look at this is as follows: Consider the index set

$$J^* = \{(i, u) \; ; \; i \in \mathbb{N}^*, u \in G_i^-\} \,, \tag{10.33}$$

and the map $\Phi : \ell^2(\mathbb{N}^*) \to \ell^2(J^*)$ given by

$$\Phi(t) = (\Phi(t)_j)_{j \in J^*} \text{ where } \Phi(t)_j = \varphi_{u,u^+}(t_i) \text{ when } j = (i, u) \,. \tag{10.34}$$

Then, replacing the process $(X_t)_{t \in T}$ by the process $(X_t(\mathcal{G}))_{t \in T}$ amounts to replacing the set T by the set $\Phi(T)$. The gain is that we now control $\|t\|_\infty$ for $t \in \Phi(T)$. We state this crucial fact explicitly.

Lemma 10.3.3 *For $t \in \Phi(T)$ we have*

$$\|t\|_\infty \le \max\{u^+ - u \; ; \; i \ge 1, u \in G_i^-\} \,.$$

This is an obvious consequence of (10.25). In order to take advantage of this bound, it is efficient to consider sets G_i consisting of evenly spaced points, as we will do. Lemma 10.3.3 is the purpose of the whole construction: controlling the supremum norm allows for use of the Sudakov minoration. Another important idea is that according to (10.29) and the inequality $\sum a_k^2 \le (\sum |a_k|)^2$, the canonical distance $d_{\mathcal{G}}$ associated with the new process satisfies

$$d_{\mathcal{G}}(s,t)^2 := \sum_{i\ge 1, u\in G_i^-} (\varphi_{u,u^+}(s_i) - \varphi_{u,u^+}(t_i))^2 \le \sum_{i\ge 1}(s_i - t_i)^2 = d(s,t)^2 \,. \tag{10.35}$$

The problem is that the reverse inequality is by no means true and that a set can very well be of small diameter for $d_{\mathcal{G}}$ but not for d. This is in a sense the main difficulty in using chopping maps. We shall discover soon how brilliantly Bednorz and Latała bypassed this difficulty using Proposition 10.1.1.

The following is fundamental. It asserts that in a sense the size of the process $X_t(\mathcal{G})$ is smaller than the size of the original process X_t.

Proposition 10.3.4 *For any family $\mathcal{G} = (G_i)_{i\ge 1}$ of finite subsets $G_i \subset \mathbb{R}$ and any finite set $T \subset \ell^2$, we have*

$$\mathsf{E} \sup_{t\in T} X_t(\mathcal{G}) \le b(T) = \mathsf{E} \sup_{t\in T} \sum_{i\ge 1} \varepsilon_i t_i \,. \tag{10.36}$$

Proof The families $(\varepsilon_{x,i})$ and $(\varepsilon_i \varepsilon_{x,i})$ have the same distribution, so that

$$\begin{aligned}\mathsf{E} \sup_{t\in T} X_t(\mathcal{G}) &= \mathsf{E} \sup_{t\in T} \sum_{i\ge 1, u\in G_i^-} \varepsilon_{u,i}\varphi_{u,u^+}(t_i) \\ &= \mathsf{E} \sup_{t\in T} \sum_{i\ge 1, u\in G_i^-} \varepsilon_i \varepsilon_{u,i}\varphi_{u,u^+}(t_i) \\ &= \mathsf{E}\Big(\mathsf{E}_\varepsilon \sup_{t\in T} \sum_{i\ge 1} \varepsilon_i \theta_i(t_i)\Big) \,,\end{aligned} \tag{10.37}$$

where the function θ_i is defined by $\theta_i(x) = \sum_{u\in G_i^-} \varepsilon_{u,i}\varphi_{u,u^+}(x)$ and where E_ε means expectation only in $(\varepsilon_i)_{i\ge 1}$. We note that θ_i is a contraction, since

$$|\theta_i(x) - \theta_i(y)| \le \sum_{u\in G_i^-} |\varphi_{u,u^+}(x) - \varphi_{u,u^+}(y)| \le |x-y|$$

by (10.29). The key point is (6.42) which implies

$$\mathsf{E}_\varepsilon \sup_{t\in T} \sum_{i\ge 1} \varepsilon_i \theta_i(t_i) \le \mathsf{E} \sup_{t\in T} \sum_{i\ge 1} \varepsilon_i t_i = b(T) \,.$$

Combining with (10.37) finishes the proof. □

The following exercise helps explaining the nice behavior of chopping maps with respect to the ℓ^2 and ℓ^1 norms, which will be used in the next result. There is no reason to believe that the constants are optimal; they are just a reasonable choice.

Exercise 10.3.5 Prove that for $x, y \in \mathbb{R}$ and $c \in \mathbb{R}^+$, we have

$$|x-y|^2 \mathbf{1}_{\{|x-y|<c\}} + c|x-y|\mathbf{1}_{\{|x-y|\geq c\}} \leq 3\sum_{\ell\in\mathbb{Z}} |\varphi_{c\ell,c(\ell+1)}(x) - \varphi_{c\ell,c(\ell+1)}(y)|^2 . \tag{10.38}$$

and

$$\sum_{\ell\in\mathbb{Z}} |\varphi_{c\ell,c(\ell+1)}(x) - \varphi_{c\ell,c(\ell+1)}(y)|^2 \leq |x-y|^2 \mathbf{1}_{\{|x-y|<c\}} + c|x-y|\mathbf{1}_{\{|x-y|\geq c\}} . \tag{10.39}$$

I invented chopping maps to prove the following version of Sudakov minoration which illustrates well their power. Compared with the Gaussian version (2.117) of Sudakov minoration, the ball ϵB_2 are enlarged into $\epsilon B_2 + Lb(T)B_1$ where B_1 denotes the unit ball of ℓ^1.

Proposition 10.3.6 *There exists a constant L such that for each subset T of ℓ^2 we have, for $\epsilon > 0$*

$$\epsilon\sqrt{\log N(T, \epsilon B_2 + Lb(T)B_1)} \leq Lb(T) ,$$

where $N(T, C)$ is the smallest number of translates of C that can cover T.

Exercise 10.3.7 will help you understand the formulation of this result and the need for the term $Lb(T)B_1$.

Proof Consider $c > 0$, and the map $\Psi_c : \ell^2 = \ell^2(\mathbb{N}^*) \to \ell^2(\mathbb{N}^* \times \mathbb{Z})$ given by $\Psi_c(t) = ((\varphi_{\ell c,(\ell+1)c}(t_i))_{(i,\ell)})$, and recall that according to (10.25) we have $\|t\|_\infty \leq c$ for $t \in \Psi_c(T)$. It then follows from Sudakov's minoration for Bernoulli processes (6.23) that

$$b(\Psi_c(T)) \geq \frac{1}{L}\min\left(\epsilon\sqrt{\log N(\Psi_c(T), \epsilon B_2)}, \frac{\epsilon^2}{c}\right) . \tag{10.40}$$

An obvious adaptation of Proposition 10.3.4 implies that $b(T) \geq b(\Psi_c(T))$. Combining with (10.40) for the value $c = \epsilon^2/(2Lb(T))$ where L is as in (10.40) we get

$$b(T) \geq \min\left(\frac{1}{L}\epsilon\sqrt{\log N(\Psi_c(T), \epsilon B_2)}, 2b(T)\right) ,$$

which implies that

$$Lb(T) \geq \epsilon\sqrt{\log N(\Psi_c(T), \epsilon B_2)}\ . \tag{10.41}$$

To conclude the proof it suffices to show that

$$N(\Psi_c(T), \epsilon B_2) \geq N(T, 4\epsilon B_2 + Lb(T)B_1)\ . \tag{10.42}$$

Letting $N = N(\Psi_c(T), \epsilon B_2)$, the set $\Psi_c(T)$ is covered by N balls of the type $z + \epsilon B_2$. It is therefore covered by N balls of the type $z + 2\epsilon B_2$ where now $z \in \Psi_c(T)$, i.e., it is covered by N balls of the type $\Psi_c(y) + 2\epsilon B_2$. Thus T is covered by N sets of the type $\Psi_c^{-1}(\Psi_c(y) + 2\epsilon B_2)$. Keeping in mind the value of c, it is enough to show that

$$\Psi_c^{-1}(\Psi_c(y) + 2\epsilon B_2) \subset y + 4\epsilon B_2 + \frac{12\epsilon^2}{c} B_1\ . \tag{10.43}$$

Consider x with $\Psi_c(x) \in \Psi_c(y) + 2\epsilon B_2$. Then the right-hand side of (10.38) is $\leq 12\epsilon^2$, and this inequality implies that $(x - y)\mathbf{1}_{\{|x-y|<c\}} \in 4\epsilon B_2$ and $(x - y)\mathbf{1}_{\{|x-y|\geq c\}} \in 12\epsilon^2 B_1/c$. Writing $x - y = (x - y)\mathbf{1}_{\{|x-y|<c\}} + (x - y)\mathbf{1}_{\{|x-y|\geq c\}}$ proves (10.43). □

Exercise 10.3.7 Take $T = B_1$, so that $b(T) = 1$. Prove that if $\epsilon + a < 1$ it is not possible to cover T by finitely many translates of the set $\epsilon B_2 + aB_1$.

Exercise 10.3.8 Deduce Proposition 10.3.6 from Theorem 6.2.8.

10.3.2 Basic Facts

For $i \geq 1$ we consider finite sets $G_i \subset G'_i$. Letting $\mathcal{G} = (G_i)_{i\geq 1}$ and $\mathcal{G}' = (G'_i)_{i\geq 1}$, we now want to compare the processes $(X_t(\mathcal{G}))_t$ and $(X_t(\mathcal{G}'))_t$. We start by comparing the associated distances. We recall the formula (10.35) for the distance $d_{\mathcal{G}}$.

Proposition 10.3.9

(a) Assume that for a certain integer q

$$\forall i \in \mathbb{N}^*\ ,\ \forall u \in G_i^-\ ,\ \mathrm{card}\left([u, u^+[\cap G'_i\right) \leq q\ , \tag{10.44}$$

where u^+ is the successor of u in G_i. Then

$$d_{\mathcal{G}} \leq \sqrt{q}\, d_{\mathcal{G}'}\ . \tag{10.45}$$

(b) Assume that

$$\forall i \in \mathbb{N}^* \; , \; \min G_i = \min G'_i \; , \; \max G_i = \max G'_i \; . \tag{10.46}$$

Then

$$d_{\mathcal{G}'} \le d_{\mathcal{G}} \; . \tag{10.47}$$

Proof Throughout the proof we write u an element of G_i^- and u^+ its successor in G_i; and v an element of $G_i'^-$ and v^+ its successor in G'_i. Thus, for $s, t \in T$ we have

$$d_{\mathcal{G}}(s,t)^2 = \sum_{i\ge 1}\sum_{u\in G_i^-} (\varphi_{u,u^+}(s_i) - \varphi_{u,u^+}(t_i))^2 \tag{10.48}$$

and

$$d_{\mathcal{G}'}(s,t)^2 = \sum_{i\ge 1}\sum_{v\in G_i'^-} (\varphi_{v,v^+}(s_i) - \varphi_{v,v^+}(t_i))^2 \; . \tag{10.49}$$

Given $i \in \mathbb{N}^*$ and $u \in G_i^-$ let us define the set $G_{i,u} = G_i'^- \cap [u, u^+[$. The sets $(G_{i,u})_{u\in G_i^-}$ are disjoint subsets of $G_i'^-$. The union of these sets is $G_i'^-$ exactly when $\min G_i = \min G'_i$ and $\max G_i = \max G'_i$.

Next, consider $u \in G_i^- \subset G'_i$ and the largest element v of $G_{i,u}$. Since $G_{i,u} \subset [u, u^+[$, we have $v < u^+ \in G_i \subset G'_i$. Thus v^+, the smallest element of G'_i which is $> v$ satisfies $v^+ \le u^+$. But then $v^+ = u^+$ for otherwise v would not be the largest element of $G_{i,u}$. It then follows from (10.27) that

$$|\varphi_{u,u^+}(s_i) - \varphi_{u,u^+}(t_i)| = \sum_{v\in G_{i,u}} |\varphi_{v,v^+}(s_i) - \varphi_{v,v^+}(t_i)| \; . \tag{10.50}$$

Thus, using the inequality $(\sum_{k\le q} a_k)^2 \le q\sum_{k\le q} a_k^2$, and since under condition (10.44) we have $\operatorname{card} G_{i,u} \le q$, we get then

$$(\varphi_{u,u^+}(s_i) - \varphi_{u,u^+}(t_i))^2 \le q \sum_{v\in G_{i,u}} (\varphi_{v,v^+}(s_i) - \varphi_{v,v^+}(t_i))^2 \; ,$$

and plugging into (10.48) we obtain

$$d_{\mathcal{G}}(s,t)^2 \le q \sum_{i\ge 1}\sum_{u\in G_i^-}\sum_{v\in G_{i,u}} (\varphi_{v,v^+}(s_i) - \varphi_{v,v^+}(t_i))^2 \; . \tag{10.51}$$

Now

$$\sum_{u \in G_i^-} \sum_{v \in G_{i,u}} (\varphi_{v,v^+}(s_i) - \varphi_{v,v^+}(t_i))^2 \le \sum_{v \in G_i'^-} (\varphi_{v,v^+}(s_i) - \varphi_{v,v^+}(t_i))^2$$

because each term in the double sum on the left is a term of the sum on the right. Using this in (10.51) and recalling (10.49), we have proved (10.45). Next, using again (10.50) as well as the inequality $(\sum_k |a_k|)^2 \ge \sum a_k^2$, we obtain

$$d_{\mathcal{G}}(s,t)^2 \ge \sum_{i \ge 1} \sum_{u \in G_i^-} \sum_{v \in G_{i,u}} (\varphi_{v,v^+}(s_i) - \varphi_{v,v^+}(t_i))^2 ,$$

and we have observed that under (10.46) the union of the sets $G_{i,u}$ for $u \in G_i^-$ is exactly $G_i'^-$ so that then right-hand side is exactly $d_{\mathcal{G}'}(s,t)^2$, so that we have proved (10.47) as well. □

Proposition 10.3.10 *For $i \ge 1$, consider finite sets $G_i \subset G_i'$ and let $\mathcal{G} = (G_i)_{i \ge 1}$ and $\mathcal{G}' = (G_i')_{i \ge 1}$. Assuming (10.46) we have*

$$\mathsf{E} \sup_{t \in T} X_t(\mathcal{G}') \le \mathsf{E} \sup_{t \in T} X_t(\mathcal{G}) . \tag{10.52}$$

This is a consequence of Proposition 10.3.4. It is quite intuitive, and the only difficulty is in the notation. We urge the reader to skip that tedious pensum until she has found motivation.

Proof Consider the set $J^* = \{(i,u) \; ; \; i \in \mathbb{N}^*, u \in G_i^-\}$ of (10.33) and the map Φ of (10.34). Let $T' = \Phi(T)$, so that

$$\mathsf{E} \sup_{t \in T} X_t(\mathcal{G}) = \mathsf{E} \sup_{s \in T'} \sum_{j \in J^*} \varepsilon_j s_j ,$$

where $\varepsilon_j = \varepsilon_{u,i}$ for $j = (i,u) \in J^*$. For each $j \in J^*$ consider a finite subset H_j of $\mathbb{R}$. Denoting by $(\varepsilon^*_{x,j})_{x \in \mathbb{R}, j \in J^*}$ a new sequence of independent Bernoulli r.v.s, it follows from Proposition 10.3.4 that

$$\mathsf{E} \sup_{s \in T'} \sum_{j \in J^*} \sum_{v \in H_j^-} \varepsilon^*_{v,j} \varphi_{v,v^+}(s_j) \le \mathsf{E} \sup_{s \in T'} \sum_{j \in J^*} \varepsilon_j s_j = \mathsf{E} \sup_{t \in T} X_t(\mathcal{G}) . \tag{10.53}$$

We will show that we can choose the sets H_j so that the left-hand side is $\mathsf{E} \sup_{t \in T} X_t(\mathcal{G}')$, and this will conclude the proof.

Let us start with a simple observation. To lighten notation, for $u < u'$ we set $\theta(u,u') = -\min(u', \max(0,u))$. We prove that

$$u \le v \le v' \le u' \Rightarrow \varphi_{v,v'}(x) = \varphi_{v+\theta(u,u'),v'+\theta(u,u')}(\varphi_{u,u'}(x)) . \tag{10.54}$$

First, we observe that (10.24) means that $\varphi_{u,u'}(x) = \min(u', \max(x, u)) + \theta(u, u')$. Thus, as x increases, the function $\varphi_{u,u'}(x)$ increases until it reaches the value $v + \theta(u, u')$ for $x = v$ and then the value $v' + \theta(u, u')$ for $x = v'$. Next, the function on the right-hand side of (10.54) is constant until $\varphi_{u,u'}(x)$ reaches the value $v+\theta(u, u')$, i.e., until $x = v$, then has a slope 1, and then is constant again after $\varphi_{u,u'}(x)$ passes the value $v' + \theta(u, u')$, i.e., after $x = v'$. It is also 0 for $x = 0$. Thus this function is $\varphi_{v,v'}$, which is characterized by these properties, and we have proved (10.54).

For $j = (i, u) \in J^*$, let us define the set

$$H_j = (G'_i \cap [u, u^+]) + \theta(u, u^+) ,$$

so that, recalling the sets $G_{i,u} = G'_i \cap [u, u^+[$, we have

$$H_j^- = G_{i,u} + \theta(u, u^+) \tag{10.55}$$

when $j = (i, u)$. Using the definition of J^* the left-hand side of (10.53) is then

$$\begin{aligned}
&\mathsf{E} \sup_{s \in T'} \sum_{i \geq 1} \sum_{u \in G_i^-} \sum_{v \in H^-_{(i,u)}} \varepsilon^*_{v,(i,u)} \varphi_{v,v^+}(s_{(i,u)}) \\
&= \mathsf{E} \sup_{s \in T'} \sum_{i \geq 1} \sum_{u \in G_i^-} \sum_{v \in G_{i,u}} \varepsilon^*_{v,(i,u)} \varphi_{v+\theta(u,u^+),v^++\theta(u,u^+)}(s_{(i,u)}) \\
&= \mathsf{E} \sup_{t \in T} \sum_{i \geq 1} \sum_{u \in G_i^-} \sum_{v \in G_{i,u}} \varepsilon^*_{v,(i,u)} \varphi_{v+\theta(u,u^+),v^++\theta(u,u^+)}(\varphi_{u,u^+}(t_i)) \\
&= \mathsf{E} \sup_{t \in T} \sum_{i \geq 1} \sum_{u \in G_i^-} \sum_{v \in G_{i,u}} \varepsilon^*_{v,(i,u)} \varphi_{v,v^+}(t_i) \ .
\end{aligned} \tag{10.56}$$

Here, we use (10.55) in the second line. In the third line, we use the definition of T':

$$T' = \{(\Phi(t)_j)_{j \in J^*} \ ; \ t \in T\} = \{(\varphi_{u,u^+}(t_i))_{i \in \mathbb{N}^*, u \in G_i^-} \ ; \ t \in T\}. \tag{10.57}$$

Finally we use (10.54) in the fourth line. Next, (10.46) ensures that $G_i'^- = \cup_{u \in G_i^-} G_{i,u}$. Thus the sequence

$$(\varepsilon^*_{v,(i,u)})_{i \in \mathbb{N}^*, u \in G_i^-, v \in G_{i,u}}$$

is simply a copy of an independent sequence $(\varepsilon_{v,i})_{i \in \mathbb{N}^*, v \in G_i'^-}$, and the expression on the last line of (10.56) equals $\mathsf{E} \sup_{t \in T} \sum_{i \geq 1} \sum_{v \in G_i'^-} \varepsilon_{v,i} \varphi_{v,v^+}(t_i)$. Since this last quantity is $\mathsf{E} \sup_{t \in T} X_t(\mathcal{G}')$, this concludes the proof. □

10.3.3 Functionals

We are now ready to define the functionals which we will use to prove Theorem 6.2.8, but the motivation for these definitions will become only gradually clear. These functionals depend on four parameters, two integers $k \le h \in \mathbb{Z}$, (yes, h denotes an **integer**), a point $w \in \ell^2$, and a subset I of $\mathbb{N}^*$. We fix an integer $r \ge 2$, which will be chosen later on. First, for $x \in \mathbb{R}$ and $k \in \mathbb{Z}$, we define the set

$$G(x,k) = \{pr^{-k}\ ;\ p \in \mathbb{Z}\ ,\ |pr^{-k} - x| \le 4r^{-k}\}\ . \tag{10.58}$$

If pr^{-k} and $p'r^{-k} \in G(x,k)$, then $|p - p'| \le 8$ so that $\operatorname{card} G(x,k) \le 9$. We also observe that (see Fig. 10.2)

$$x - 4r^{-k} \le \min G(x,k) \le x - 3r^{-k} \le x \le x + 3r^{-k} \le \max G(x,k) \le x + 4r^{-k}\ . \tag{10.59}$$

Next, given $k \le h \in \mathbb{Z}$ and $x \in \mathbb{R}$, we define the set

$$G(x,k,h) = \{pr^{-h}\ ;\ p \in \mathbb{Z}\ ,\ \min G(x,k) \le pr^{-h} \le \max G(x,k)\}\ . \tag{10.60}$$

In words, $G(x,k,h)$ consists of about $9 \cdot r^{h-k}$ points evenly spaced (with a spacing of r^{-h}) roughly centered on the point x. We should think of k as a scale parameter: the length of $G(x,k,h)$ is of order r^{-k}. We should think of h as a "granularity parameter": the distance between consecutive points of $G(x,k,h)$ is r^{-h}. We should think of x as a location parameter: $G(x,k,h)$ is roughly centered at x.

Let us note that $G(x,k) = G(x,k,k)$ (so that $\operatorname{card} G(x,k,k) \le 9$), that

$$\min G(x,k,h) = \min G(x,k)\ ;\ \max G(x,k,h) = \max G(x,k)\ . \tag{10.61}$$

Furthermore $G(x,k,h)$ increases with h, in the sense that $G(x,k,h) \subset G(x,k,h')$ if $h \le h'$, and decreases with k in the sense that if $k \le k'$ then $G(x,k',h) \subset G(x,k,h)$.

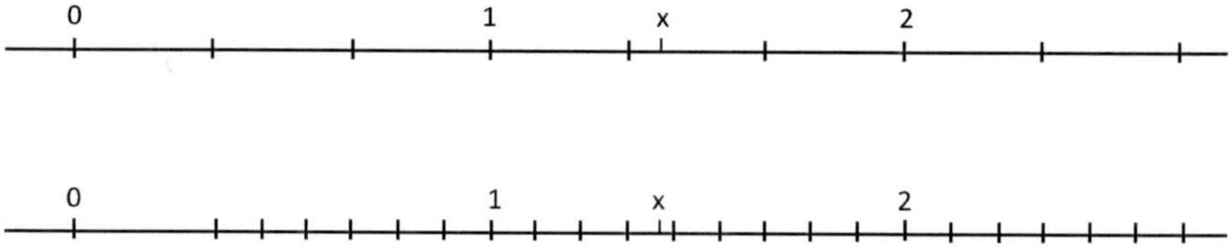

Fig. 10.2 The set $G(x, 1)$ on top. The spacing between the points is 1/3. The set $G(x, 1, 2)$ on bottom. The spacing between the points is 1/9. Here $r = 3$ and $4/3 < x < 5/3$

Definition 10.3.11 For a set $T \subset \ell^2$, integers $k \le h$, a point $w \in \ell^2$, and a subset I of $\mathbb{N}^*$, we define

$$F(T, I, w, k, h) = \mathsf{E} \sup_{t \in T} \sum_{i \in I} \sum_{u \in G(w_i, k, h)^-} \varepsilon_{u,i} \varphi_{u,u^+}(t_i) . \tag{10.62}$$

We denote by $d_{I,w,k,h}$ the corresponding distance

$$d_{I,w,k,h}(s,t)^2 = \sum_{i \in I} \sum_{u \in G(w_i, k, h)^-} (\varphi_{u,u^+}(s_i) - \varphi_{u,u^+}(t_i))^2 , \tag{10.63}$$

and $\Delta(T, I, w, k, h)$ the diameter of T for this distance

$$\Delta(T, I, w, k, h) = \sup_{s,t \in T} d_{I,w,k,h}(s,t) . \tag{10.64}$$

When $I = \mathbb{N}^*$ the distance $d_{I,w,k,h}$ is simply the distance $d_{\mathcal{G}}$ of (10.35) when $G_i = G(w_i, k, h)$. The effect of the parameter $w \in \ell^2$ is that the set $G(w_i, k, h)$ is roughly centered around w_i.

Even if we forget to mention it again, when writing these expressions, it is always assumed that $h \ge k$.

Let us look at the summation (10.62): decreasing I and increasing k decreases the number of terms in it. This opens the door to the use of Proposition 10.1.1 (Latała's principle).

Let us first point out some regularity properties of these functionals.

Lemma 10.3.12 *If $I' \subset I \subset \mathbb{N}^*$, $k' \ge k$ and $h' \ge h$ then*

$$F(T, I', w, k', h') \le F(T, I, w, k, h) \tag{10.65}$$

and

$$\Delta(T, I', w, k', h') \le \Delta(T, I, w, k, h) . \tag{10.66}$$

Proof That $F(T, I, w, k, h)$ is an increasing function of I follows from Jensen's inequality, by moving now the expectation over the r.v.s $\varepsilon_{u,i}$ for $i \in I' \setminus I$ inside the supremum rather than outside. Next if $k \le k' \le h$ then the inequality $F(T, I, w, k', h) \le F(T, I, w, k, h)$ follows similarly since $G(w_i, k', h)^- \subset G(w_i, k, h)^-$, by moving inside the supremum expectation with respect to the r.v.s $\varepsilon_{u,i}$ for $u \in G(w_i, k, h)^- \setminus G(w_i, k', h)^-$. That $F(T, I, w, k, h)$ is a decreasing function of h follows from Proposition 10.3.10 and (10.61). The statements concerning $\Delta(T, I, w, k, h)$ are easier, using now (10.47). □

Another key idea is that the distances (10.63) associated with the functionals (10.62) relate well to the distance considered in (9.12) in Theorem 9.2.4 (in the case where ν is the counting measure). Our next lemmas provide the main step in this direction.

Lemma 10.3.13 *Consider $x, y, z \in \mathbb{R}$ and assume that $|y - x| \le 2r^{-k}$. Then*

$$|y - z|^2 \wedge r^{-2h} \le 2 \sum_{u \in G(x,k,h)^-} (\varphi_{u,u^+}(y) - \varphi_{u,u^+}(z))^2 . \tag{10.67}$$

Proof First we reduce to the case where $|y - z| \le r^{-h}$. To do this, we replace z by the closest point to z in the interval $[y - r^{-h}, y + r^{-h}]$. This does not change the left-hand side of (10.67) and decreases the right-hand side because the functions φ_{u,u^+} are non-decreasing. So we assume now $|y - z| \le r^{-h}$. Since $|y - x| \le 2r^{-h}$ we have $x - 3r^{-h} \le y, z \le x + 3r^{-h}$ so that (10.59) implies that $\min G(x, k, h) \le y, z \le \max G(x, k, h)$. Then by the equality case of Lemma 10.3.1, we have

$$|y - z| = \sum_{u \in G(x,k,h)^-} |\varphi_{u,u^+}(y) - \varphi_{u,u^+}(z)| .$$

Now, since $|y - z| \le r^{-h}$, there are at most two non-zero terms in the right-hand side, and $(a + b)^2 \le 2(a^2 + b^2)$. □

Lemma 10.3.14 *Consider $s, t, w \in \ell^2$. Consider a set I of integers and assume that*

$$\forall i \in I \; , \; |s_i - w_i| \le 2r^{-k} . \tag{10.68}$$

Then

$$\sum_{i \in I} |r^h(t_i - s_i)|^2 \wedge 1 \le 2r^{2h} d_{I,w,k,h}(t, s)^2 . \tag{10.69}$$

Proof We use (10.67) for $x = w_i$, $y = s_i$, $z = t_i$ to obtain

$$|t_i - s_i|^2 \wedge r^{-2h} \le 2 \sum_{u \in G(w,k,h)^-} (\varphi_{u,u^+}(t_i) - \varphi_{u,u^+}(s_i))^2 .$$

We then sum over $i \in I$ and we use (10.3.1). □

10.4 Philosophy, II

In this section we try to shed some light on the construction we have started. Let us recall our goal: starting with $T \subset \ell^2$, we try to decompose each $t \in T$ as $t = t^1 + t^2$ is such a way that $\{t^1 ; t \in T\}$ is well behaved and that $\|t^2\|_1 \le Lb(T)$. For the purpose of the philosophical discussions, we will call t^2 the peaky part of t, even though the name is not really appropriate.

To prove the required decomposition, we will recursively construct an increasing sequence of partitions of T, and then the decomposition of T will be provided by Theorem 9.2.4 (used for the counting measure). At each level, for each element A of the partition, we will control a certain diameter $\Delta(A, I, w, k, h)$ and the corresponding functional $F(A, I, w, k, h)$. The first thought is that in order that the functional (10.62) really bear on A, "A should be chosen close to w". The precise meaning of this will be understood later, but for the time being, we keep in mind the idea that w provides information about the "location" of A. A second thought coming to mind will be that information seems lost when there are coordinates i and elements $t \in A$ with $|w_i - t_i| \ge 4r^{-k}$; see Exercise 10.3.2. This looks like a serious problem: the Latała-Bednorz theorem is absolutely sharp; we can never allow any essential information to be lost. The solution to that riddle was given at the end of Sect. 9.3: what we really need to keep track of are the values of $\pi_n(t)_i$, and this we really do as we explain below.

The details of what happens will be given as the proof develops, but we start to reveal some secrets. In order to be able to apply Theorem 9.2.4 (when ν is the counting measure), we need to have the crucial condition

$$\forall t \in T \;,\; \forall n \ge 0 \;,\; \sum_{i \in \Omega_n(t)} |r^{j_n(t)}(t_i - \pi_n(t)_i)|^2 \wedge 1 \le u 2^n \;, \tag{10.70}$$

where[7]

$$\Omega_n(t) = \big\{ i \;;\; 0 \le q < n \Rightarrow |\pi_{q+1}(t)_i - \pi_q(t)_i| \le r^{-j_q(t)} \big\} \;. \tag{10.71}$$

Not being precise about $\Omega_n(t)$ yet, we see that our best shot is to deduce (10.70) from (10.69) used for $s = \pi_n(t)$. We then guess that the number h relates to $j_n(t)$. We also realize the importance of the condition

$$\forall i \in I \;,\; |\pi_n(t)_i - w_i| \le 2r^{-k} \;. \tag{10.72}$$

The value of k will be decided later, but we should form the following picture: The value of k tells us that the range around w_i where we are getting some information

[7] In words, the points of $\Omega_n(t)$ are those for which the sequence $(\pi_q(t))_{q \le n+1}$ does not have big jumps.

on the i-th coordinate of the points of A is about r^{-k}. Most importantly, the value of $\pi_n(t)_i$ falls well within this range. It is necessary to keep this range, which might be much larger than $r^{-j_n(t)}$, because we do not have better information on the location of $\pi_n(t)$ than (10.72).

10.5 Latała's Step

We now state and prove the key new step in the proof of Theorem 6.2.8 (compared with the Gaussian case Theorem 2.10.1).

Proposition 10.5.1 *There exists a constant L_2 with the following property. Consider $w, w' \in \ell^2$, a set $I \subset \mathbb{N}^*$, and integers $k \le h$. Consider a subset T of ℓ^2 such that*

$$\Delta(T, I, w, k, h+2) \le c \,. \tag{10.73}$$

Assume that for a certain number σ

$$c \le \frac{\sigma}{L_2} \; ; \; r^{-h-1}\sqrt{\log m} \le \sigma \,. \tag{10.74}$$

Let[8]

$$I' = \{i \in I \; ; \; |w_i - w'_i| \le 2r^{-k}\} \,. \tag{10.75}$$

Then we can find $m' \le m+1$ and a partition $(A_\ell)_{\ell \le m'}$ of T such that for each $\ell \le m'$ we have **either**

$$\Delta(A_\ell, I, w, k, h+1) \le \sigma \tag{10.76}$$

or else

$$F(A_\ell, I', w', h+2, h+2) \le F(T, I, w, k, h+1) - \frac{\sigma}{L}\sqrt{\log m} \,, \tag{10.77}$$

$$\Delta(A_\ell, I', w', h+2, h+2) \le c \,. \tag{10.78}$$

In words, the proposition states that each piece produced by the previous decomposition is either such that its diameter for the large distance is small (when (10.76) holds), or else its size measured by the proper functionals has decreased (when (10.77) holds).

[8] The reader should notice that the set I' is not constructed, but is known once w and w' are known.

The fundamental point of this result is that the hypothesis on T involves a control of $\Delta(T, I, w, k, h + \mathbf{2})$, not of the larger quantity $\Delta(T, I, w, k, h + \mathbf{1})$, whereas the size of the pieces A_ℓ in (10.76) involves a control of the larger quantity $\Delta(A_\ell, I, w, k, h + \mathbf{1})$, not of the smaller quantity $\Delta(A_\ell, I, w, k, h + \mathbf{2})$.

Proof The proof relies on Latała's principle, Proposition 10.1.1, but requires some skills. There is no loss of generality to assume for notational convenience that $I = \mathbb{N}^*$. For $i \in \mathbb{N}^*$ consider the sets

$$G_i = G(w_i, k, h + 1) \ ,$$

and $\mathcal{G} = (G_i)_{i \geq 1}$. For $i \in \mathbb{N}^* \setminus I'$ let $G'_i = G_i$ and for $i \in I'$, define $G'_i = G_i \cup G(w'_i, h + 2, h + 2)$ and define $\mathcal{G}' = (G'_i)_{i \geq 1}$. The central object of the proof is the process $(X_t(\mathcal{G}'))_{t \in T}$.

First we observe that since $r \geq 2$ and $h \geq k$, and using (10.59) and (10.61),

$$G(w'_i, h + 2, h + 2) \subset [w'_i - 4r^{-h-2}, w'_i + 4r^{-h-2}] \subset [w'_i - r^{-k}, w'_i + r^{-k}]$$

and since $|w_i - w'_i| \leq 2r^{-k}$ for $i \in I'$, it follows then that

$$G(w'_i, h + 2, h + 2) \subset [w_i - 3r^{-k}, w_i + 3r^{-k}] \ .$$

Consequently from (10.59), we have

$$\begin{aligned} \max G(w'_i, h + 2, h + 2) \leq w_i + 3r^{-k} &\leq \max G(w_i, k, k) \\ &= \max G(w_i, k, h + 1) = \max G'_i \ . \end{aligned}$$

Proceeding similarly for the min shows that the sets G_i and G'_i satisfy (10.46).

Therefore by (10.52) we have

$$\mathsf{E} \sup_{t \in T} X_t(\mathcal{G}') \leq \mathsf{E} \sup_{t \in T} X_t(\mathcal{G}) = F(T, \mathbb{N}^*, w, k, h + 1) \ . \tag{10.79}$$

Next, since $\operatorname{card} G(w'_i, h + 2, h + 2) \leq 9$ and $G'_i = G_i \cup G(w'_i, h + 2, h + 2)$, for each i and $u \in G'_i$, we have $\operatorname{card}([u, u^+[\cap G'_i) \leq \operatorname{card}([u, u^+[\cap G_i) +$

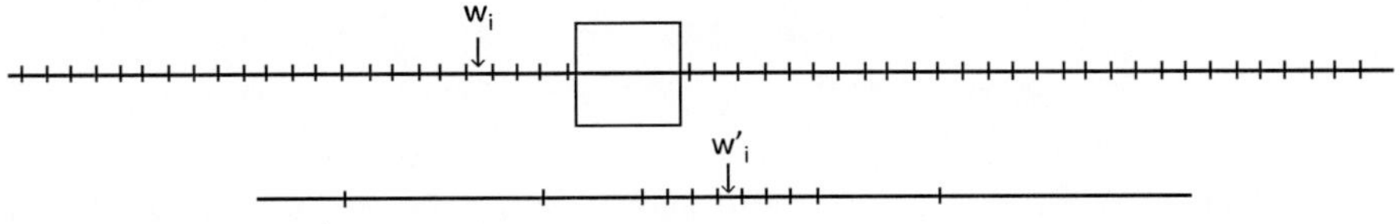

Fig. 10.3 The sets G'_i when $r = 8$ and $h = k + 1$. The bottom of the figure represents the part of the set contained in the box after magnification by a factor 8

card $G(w'_i, h+2, h+2) \le 1+9 = 10$. Thus the sets G_i and G'_i satisfy

$$\forall i \in \mathbb{N}^* \ , \ \forall u \in G_i^- \ , \ \mathrm{card}\left([u, u^+[\cap G'_i\right) \le 10 \ ,$$

and Proposition 10.3.9 implies that

$$d_{\mathcal{G}} \le 4 d_{\mathcal{G}'} \ . \tag{10.80}$$

We can appreciate the magic of this proof: neither the process $(X_t(\mathcal{G}'))_{t\in T}$ nor the distance it induces is exactly what we need, but they are related to the quantities of interest through the inequalities (10.79) and (10.80) which turn out to be in the right direction.

For $i \in I'$ we have $|w_i - w'_i| \le 2r^{-k}$ so that (using that $h \ge k$ and $r \ge 2$)

$$|pr^{-h-2} - w'_i| \le 4r^{-h-2} \Rightarrow |pr^{-h-2} - w_i| \le 2r^{-k} + 4r^{-h-2} \le 3r^{-k}$$

so hence by (10.59) again $G(w'_i, h+2, h+2) \subset G(w_i, k, h+2)$. In fact the points of the left-hand set are consecutive points of the right-hand set. Using (10.45) for $q = 1$ in the inequality (with I' instead of $\mathbb{N}^*$), we obtain

$$d_{I',w',h+2,h+2} \le d_{I',w',k,h+2} \ , \tag{10.81}$$

and this proves (10.78). Consequently[9]

$$\Delta(T, I', w', h+2, h+2) \le \Delta(T, I', w, k, h+2) \le \Delta(T, \mathbb{N}^*, w, k, h+2) \le c \ . \tag{10.82}$$

Let us consider the set J^* as in (10.33), where the family $\mathcal{G}$ has been replaced by the family $\mathcal{G}'$ so that $J^* = \{(i,u); i \in \mathbb{N}^*, u \in G_i'^-\}$. Let us consider the corresponding map Φ as in (10.34). Thus

$$s, t \in T \Rightarrow d_{\mathcal{G}'}(s,t) = \|\Phi(s) - \Phi(t)\|_2 \ , \tag{10.83}$$

where the norm on the right is in $\ell^2(J^*)$, and, using (10.79) in the second inequality,

$$b(\Phi(T)) = \mathsf{E} \sup_{t\in T} X_t(\mathcal{G}') \le F(T, \mathbb{N}^*, w, k, h+1) \ . \tag{10.84}$$

[9] Please observe what is happening here. The information (10.78) is a consequence of (10.73), not of the fact that we have partitioned the set T. This is coherent with our proof of Latała's principle. There is only one piece what satisfies (10.77) and (10.78). This piece is what is left of T after we have removed some parts which satisfy (10.76), and we took no action to decrease its diameter in any sense.

Consider the set $J \subset J^*$ given by

$$J = \{(i,u) \ ; \ i \in I', u \in G(w'_i, h+2, h+2)^-\} . \tag{10.85}$$

We will use Proposition 10.1.1, replacing the countable set $\mathbb{N}^*$ by J^*. With the notation b_J of Proposition 10.1.1, for $A \subset T$,

$$b_J(\Phi(A)) = F(A, I', w', h+2, h+2) . \tag{10.86}$$

The goal is to apply Proposition 10.1.1 to the set $\Phi(T)$ with $\sigma' = \sigma/8$ instead of σ and with set of indices J^* instead of $\mathbb{N}^*$. For this we check (10.1) to (10.3). First, (10.82) implies (10.1). Next, (10.2) holds since $\|t\|_\infty \le r^{-h-1}$ for $t \in \Phi(T)$ and since $r^{-h-1}\sqrt{\log m} \le \sigma = 8\sigma'$ by (10.74). Finally (10.3), i.e., the condition $c \le \sigma'/L_1$, follows from $c \le \sigma/L_2$ provided that $L_2 = 8L_1$. Thus we can apply Proposition 10.1.1. We then find a partition $(B_\ell)_{\ell \le m'}$ of $\Phi(T)$ such that for each $\ell \le m'$, we have either

$$\exists t^\ell \in \Phi(T) \ , \ B_\ell \subset B(t^\ell, \sigma/8) \ , \tag{10.87}$$

or else

$$b_J(B_\ell) \le b(\Phi(T)) - \frac{\sigma}{L}\sqrt{\log m} \ . \tag{10.88}$$

We then set $A_\ell = \Phi^{-1}(B_\ell)$. When B_ℓ satisfies (10.87), (10.83) implies that the diameter of A_ℓ for the distance $d_{\mathcal{G}'}$ is $\le \sigma/4$, and since $d_{\mathcal{G}} \le 4d_{\mathcal{G}'}$ by (10.80), its diameter for the distance $d_{\mathcal{G}}$ is $\le \sigma$. This distance is exactly the distance used in computing the diameter in (10.76). When B_ℓ satisfies (10.88) then A_ℓ satisfies (10.77) as follows from (10.84) and (10.86), and (10.78) follows from (10.82). □

10.6 Philosophy, III

There is a simple idea behind the occurrence of w and w'. We have to think of w as providing information about the "location" of T. As we keep splitting the sets of our partitions, we keep improving the information about their "location". The change from w to w' in Proposition 10.5.1 reflects the fact we now have a more accurate information than the one we used in the previous steps of the construction:

the location of the set T is actually better described using a slightly different point w' than using w.[10]

An intriguing feature is that when we are in the case (10.77), we will obviously need to replace I by I'. Why then crucial information is not lost? The answer to that lies in the mechanism explained in Sect. 9.3: at the time we will drop the coordinates in $I \setminus I'$, the decomposition will already have been determined on these coordinates. There will be a clever device to ensure that, which we will explain later.

Another intriguing feature of (10.77) is that on the left we have replaced the value k by the potentially much larger value $h+2$. It is absolutely essential to be able to do that to ensure that card $G(w'_i, h+2, h+2)$ remains bounded (in fact ≤ 9) because this is how we obtain the inequality (10.80) which is essential to obtain (10.76). But why can we afford to do that without losing critical information? At a high level this answer is that it is because we have a much better idea of the "localization" of the piece A_ℓ than of T (for the simple reason that A_ℓ is a "small part" of T), but the true mechanism is related to the fact that what really matters is the condition (10.72), and you have to wait a few more pages until Sect. 10.13 to have it explained in words.

Finally it cannot hurt to stress again the magic of this proposition, which lies in the use of the set (10.85). It is the use of this set which allows to use the weak hypothesis (10.73) while reaching the strong conclusion (10.76).

10.7 A Decomposition Lemma

Besides Proposition 10.5.1, we need another decomposition principle, very similar to what we did in Lemma 2.9.4 in the Gaussian case, which is just a reformulation of Lemma 6.6.4 (with $a = c/2$). Here Δ denotes the diameter for the ℓ^2 distance.

Lemma 10.7.1 *There exists a universal constant L_3 with the following property. Consider a set $T \subset \ell^2$ and $b, c > 0$. Assume that $\|t\|_\infty \leq b$ for all $t \in T$. Consider $m \geq 2$ with $b\sqrt{\log m} \leq c$. Then we can find $m' \leq m$ and a partition $(B_\ell)_{\ell \leq m'}$ of T such that for each $\ell \leq m'$ we have* **either**

$$\forall D \subset B_\ell \; ; \; \Delta(D) \leq \frac{c}{L_3} \Rightarrow b(D) \leq b(T) - \frac{c}{L}\sqrt{\log m} \tag{10.89}$$

or else

$$\Delta(B_\ell) \leq c \; . \tag{10.90}$$

We will need the following special case for our construction:

[10] You can try to visualize things that way: we use one single point w to describe the "position" of T. When we break T into small pieces, this gets easier. Furthermore, the position of some of the small pieces is better described by an other point than w.

Corollary 10.7.2 *Consider a set $T \subset \ell^2$ and $w \in \ell^2$. Consider $I \subset \mathbb{N}^*$, $c > 0$, and integers $k \le h$. Assume that $r^{-h}\sqrt{\log m} \le c$. Then we can find $m' \le m$ and a partition $(A_\ell)_{\ell \le m'}$ of T such that for each $\ell \le m'$ we have* **either**

$$\forall D \subset A_\ell \; ; \; \Delta(D, I, w, k, h) \le \frac{c}{L_3} \Rightarrow$$

$$F(D, I, w, k, h) \le F(T, I, w, k, h) - \frac{c}{L}\sqrt{\log m} \tag{10.91}$$

or else

$$\Delta(A_\ell, I, w, k, h) \le c\,. \tag{10.92}$$

Proof For notational convenience we assume $I = \mathbb{N}^*$. Set $G_i = G(w_i, k, h)$ and consider the set J^* and the map Φ as in (10.33) and (10.34). Then for a subset A of T, we have $b(\Phi(A)) = F(A, I, w, k, h)$. We construct a partition $(B_\ell)_{\ell \le m'}$ of $\Phi(T)$ using Lemma 10.7.1 with $b = r^{-h}$ and we set $A_\ell = \Phi^{-1}(B_\ell)$. □

We can now state and prove the basic tool to construct partitions.

Lemma 10.7.3 *Assuming that r is large enough, $r \ge L$, the following holds. Consider an integer $n \ge 2$. Consider a set $T \subset \ell^2$, a point $w \in \ell^2$, a subset $I \subset \mathbb{N}^*$, and integers $k \le j$. Then we can find $m \le N_n$ and a partition $(A_\ell)_{\ell \le m}$ such that for each $\ell \le m$, we have* **either** *of the following three properties:*

(a) We have

$$D \subset A_\ell \; ; \; \Delta(D, I, w, k, j+2) \le \frac{1}{L_4} 2^{(n+1)/2} r^{-j-1} \Rightarrow$$

$$F(D, I, w, k, j+2) \le F(T, I, w, k, j+2) - \frac{1}{L} 2^n r^{-j-1}\,, \tag{10.93}$$

or

(b)

$$\Delta(A_\ell, I, w, k, j+1) \le 2^{n/2} r^{-j-1}\,. \tag{10.94}$$

or else

(c) There exists $w' \in T$ such that for $I' = \{i \in I; |w_i - w'_i| \le 2r^{-k}\}$ we have

$$F(A_\ell, I', w', j+2, j+2) \le F(T, I, w, k, j+1) - \frac{1}{L} 2^n r^{-j-1}\,, \tag{10.95}$$

$$\Delta(A_\ell, I', w', j+2, j+2) \le 2^{n/2} r^{-j-1}\,. \tag{10.96}$$

At a high level, this lemma partitions T into not too many sets on which we have additional information. In the case (a) the new information is that the subsets of A_ℓ of a small diameter have a smaller size (as measured by the appropriate functional). In the case (b) it is the set A_ℓ itself which has a small diameter. These two cases are very similar to what happens in Lemma 2.9.4. This was to be expected since these cases are produced by the application of Lemma 10.7.1 which is very similar to Lemma 2.9.4. The really new feature is case (c), where again the size of the set A_ℓ has decreased as measured by an appropriate functional, while at the same time we control the diameter of A_ℓ (but for a much smaller distance than in case (b)). What is harder to visualize (but is absolutely essential) is the precise choice of the parameters in the distances and the functionals involved. It is absolutely essential that the condition on D in (10.93) bears on $\Delta(D, I, w, k, j+\mathbf{2})$, whereas the condition on A_ℓ in (10.94) bears on $\Delta(A_\ell, I, w, k, j+\mathbf{1})$. Let us also note that in the case (c) we have in particular, using (10.65) (i.e., the monotonicity in h and k)

$$F(A_\ell, I', w', j+2, j+2) \le F(T, I, w, k, j) - \frac{1}{L} 2^n r^{-j-1} , \tag{10.97}$$

Proof The principle of the proof is to apply first Corollary 10.7.2 and then to split again the resulting pieces of small diameter using Proposition 10.5.1.

Let us define $m = N_{n-1} - 1$. Since we assume $n \ge 2$, we have $2^{n/2}/L \le \sqrt{\log m} \le 2^{n/2}$. Let us set $c = 2^{n/2} r^{-j-1}/L_2$ so that $c\sqrt{\log m} \ge 2^n r^{-j-1}/L$ and $r^{-j-2}\sqrt{\log m} \le r^{-j-2}2^{n/2} \le L_2 c/r$. Assuming $r \ge L_2$ we then have $r^{-j-2}\sqrt{\log m} \le c$.

Let us recall the constant L_3 of Corollary 10.7.2. We then apply this corollary with these values of k, c, m, and with $h = j+2$.[11] This produces pieces $(C_\ell)_{\ell \le m'}$ with $m' \le m$ which satisfy either

$$\forall D \subset C_\ell \; ; \; \Delta(D, I, w, k, j+2) \le \frac{c}{L_3} \Rightarrow$$
$$F(D, I, w, k, j+2) \le F(T, I, w, k, j+2) - \frac{c}{L}\sqrt{\log m} \tag{10.98}$$

or else

$$\Delta(C_\ell, I, w, k, j+2) \le c \, . \tag{10.99}$$

Let us set $L_4 = \sqrt{2} L_2 L_3$, so that

$$\frac{c}{L_3} = \frac{1}{L_4} 2^{(n+1)/2} r^{-j-1} \, . \tag{10.100}$$

[11] The real reason to use $j+2$ rather than $j+1$ will become apparent only later. In short, it is because in (10.98), we absolutely need to have a condition bearing on $\Delta(D, I, w, k, j+2)$, not on the larger quantity $\Delta(D, I, w, k, j+1)$.

Thus the pieces C_ℓ of the partition which satisfy (10.98) also satisfy (10.93). We are done with these pieces.

The other pieces C_ℓ of the partition satisfy (10.99), that is, they satisfy (10.73) for $h = j$. Let us fix $w' \in T$. We split again these pieces C_ℓ into pieces $(C_{\ell,\ell'})_{\ell' \le m+1}$ using Proposition 10.5.1, with these values of k, m, with $h = j$, $\sigma = 2^{n/2} r^{-j-1}$, $c = 2^{n/2} r^{-j-1}/L_2 = \sigma/L_2$. Each of the resulting pieces $C_{\ell,\ell'}$ satisfies either

$$\Delta(C_{\ell,\ell'}, I, w, k, j+1) \le \sigma = 2^{n/2} r^{-j-1} , \tag{10.101}$$

and then we are in the case (b), or else (using that $\sigma\sqrt{\log m} \ge 2^n r^{-j-1}/L$) they satisfy

$$F(C_{\ell.\ell'}, I', w', j+2, j+2) \le F(T, I, w, k, j+1) - \frac{1}{L} 2^n r^{-j-1} , \tag{10.102}$$

$$\Delta(C_{\ell,\ell'}, I', w', j+2, j+2) \le c \le 2^n r^{-j-1} , \tag{10.103}$$

and then we are in the case (c).

Finally, the total number of pieces produced is $\le m(m+1) \le N_{n-1}^2 = N_n$. □

10.8 Building the Partitions

We will prove the basic partitioning result by iterating Lemma 10.7.3. A remarkable new feature of this construction is that the functionals we use depend on the set we partition. We recall the constant L_4 of Lemma 10.7.3. We fix an integer $\kappa \ge 3$ with $2^{\kappa/2} \ge 2L_4$, and we set $r = 2^\kappa$ (so that r *is now a universal constant* ≥ 8).

Consider a set $T \subset \ell^2$ with $0 \in T$. We plan to construct by induction over $n \ge 0$ an increasing sequence $(\mathcal{A}_n)$ of partitions of T, with $\operatorname{card} \mathcal{A}_n \le N_n$. To each $A \in \mathcal{A}_n$ we will attach a set $I_n(A) \subset \mathbb{N}^*$, a point $w_n(A) \in \ell^2$, and integers $k_n(A) \le j_n(A) \in \mathbb{Z}$, $0 \le p_n(A) \le 4\kappa - 1$. We are going to explain soon the meaning of these quantities and in particular of the integer $p_n(A)$. Let us right away introduce some basic notation. For $n \ge 0$, $A \in \mathcal{A}_n$, and $D \subset T$, we define

$$\Delta_{n,A}(D) := \Delta(D, I_n(A), w_n(A), k_n(A), j_n(A)) . \tag{10.104}$$

We will use this quantity to measure the size of the subsets of A. The following is obvious from the definition (10.104):

Lemma 10.8.1 *Assume that $B \in \mathcal{A}_n$ and $A \in \mathcal{A}_{n+1}$ and $I_{n+1}(A) = I_n(B)$, $w_{n+1}(A) = w_n(B)$, $k_{n+1}(A) = k_n(B)$, $j_{n+1}(A) = j_n(B)$. Then for $D \subset T$ we have $\Delta_{n,B}(D) = \Delta_{n+1,A}(D)$.*

To start the construction, we set $n_0 = 2$. For $n \le n_0 := 2$ we set $\mathcal{A}_n = \{T\}$, $I_n(T) = \mathbb{N}^*$, $w_n(T) = 0$, $p_n(T) = 0$ and $k_n(T) = j_n(T) = j_0$, where j_0 satisfies $\Delta(T) \le r^{-j_0}$.

For $n \ge n_0 = 2$ we will spell out rules by which we split an element $B \in \mathcal{A}_n$ into elements of $\mathcal{A}_{n+1}$ and how we attach the various quantities above to each newly formed element of $\mathcal{A}_{n+1}$. We will also show that certain relations are inductively satisfied. Two such conditions are absolutely central and bear on a certain diameter of A:

$$\forall A \in \mathcal{A}_n\ ,\ p_n(A) = 0 \Rightarrow \Delta_{n,A}(A) \le 2^{n/2} r^{-j_n(A)}\ . \tag{10.105}$$

$$\forall A \in \mathcal{A}_n\ ,\ p_n(A) > 0 \Rightarrow \Delta_{n,A}(A) \le 2^{(n-p_n(A))/2} r^{-j_n(A)+2}\ . \tag{10.106}$$

Let us observe that (10.106) gets more restrictive as $p_n(A)$ increases and that for the small values of $p_n(A)$ (e.g., $p_n(A) = 1$), this condition is very much weaker than (10.105) because of the extra factor r^2.

When $p_n(B) \ge 1$, observe first that from (10.106) we have

$$\Delta_{n,B}(B) \le 2^{(n-p_n(B))/2} r^{-j_n(B)+2}\ . \tag{10.107}$$

The rule for splitting B in that case is simple: we don't. We decide that $B \in \mathcal{A}_{n+1}$, and we set $I_{n+1}(B) = I_n(B)$, $w_{n+1}(B) = w_n(B)$, $k_{n+1}(B) = k_n(B)$, $j_{n+1}(B) = j_n(B)$. For further reference, let us state

$$p_n(B) > 0 \Rightarrow B \in \mathcal{A}_{n+1},\ j_{n+1}(B) = j_n(B)\ . \tag{10.108}$$

To define $p_{n+1}(B)$, we proceed as follows:

- If $p_n(B) < 4\kappa - 1$ we set $p_{n+1}(B) = p_n(B) + 1$.
- If $p_n(B) = 4\kappa - 1$ we set $p_{n+1}(B) = 0$.

When $p_{n+1}(B) = p_n(B) + 1 > 0$, we have to prove that $B \in \mathcal{A}_{n+1}$ satisfies (10.106), that is

$$\Delta_{n+1,B}(B) \le 2^{(n+1-p_{n+1}(B))/2} r^{-j_{n+1}(B)+2}\ .$$

This follows from (10.107), Lemma 10.8.1, and the fact that $(n+1) - p_{n+1}(B) = n - p_n(B)$.

When $p_{n+1}(B) = 0$ we have to prove that $B \in \mathcal{A}_{n+1}$ satisfies (10.105). Using Lemma 10.8.1 in the equality and (10.107) in the inequality, recalling that $p_n(B) = 4\kappa - 1$,

$$\Delta_{n+1,B}(B) = \Delta_{n,B}(B) \le 2^{(n-4\kappa+1)/2} r^{-j_n(B)+2} = 2^{(n+1)/2} r^{-j_{n+1}(B)}\ , \tag{10.109}$$

since $2^{-2\kappa} = r^{-2}$.

The integer $p_n(B)$ is a kind of counter. When $p_n(B) > 0$, this tells us that we are not permitted to split B, and we increment the counter, $p_{n+1}(B) = p_n(B) + 1$ *unless* $p_n(B) = 4\kappa - 1$, in which case we set $p_{n+1}(B) = 0$, which means we will split the set at the next step. More generally, the value of the counter tells us in how many steps we will split B: we will split B in $4\kappa - p_n(B)$ steps.

Let us now examine the main case, $p_n(B) = 0$. In that case we split B in at most N_n pieces using Lemma 10.7.3, with $I = I_n(B)$, $w = w_n(B)$, $j = j_n(B)$, and $k = k_n(B)$. There are three cases to consider.

(a) We are in case (a) of Lemma 10.7.3; the piece A produced has property (10.93). We define $p_{n+1}(A) = 0$. We then set

$$I_{n+1}(A) = I_n(B)\ ,\ \ w_{n+1}(A) = w_n(B)\ ,$$
$$j_{n+1}(A) = j_n(B)\ ,\ \ k_{n+1}(A) = k_n(B)\ . \tag{10.110}$$

(b) We are in case (b) of Lemma 10.7.3, and the piece A we produce has property (10.94). We then set $p_{n+1}(A) = 0$, $j_{n+1}(A) = j_n(B) + 1$, and we define

$$I_{n+1}(A) = I_n(B)\ ,\ \ w_{n+1}(A) = w_n(B)\ ,\ \ k_{n+1}(A) = k_n(B)\ . \tag{10.111}$$

(c) We are in case (c) of Lemma 10.7.3; the piece A produced has properties (10.95) and (10.96). We set $p_{n+1}(A) = 1$, and we define

$$j_{n+1}(A) = k_{n+1}(A) = j_n(B) + 2\ .$$

We define $w_{n+1}(A) = w' \in B$ and

$$I_{n+1}(A) = \left\{i \in I_n(B)\ ;\ |w_{n+1}(A)_i - w_n(B)_i| \le 2r^{-k_n(B)}\right\}\ , \tag{10.112}$$

so that in particular $I_{n+1}(A) \subset I_n(B)$.

In order to try to make sense of this, let us start with some very simple observations. We consider $B \in \mathcal{A}_n$ with $n \ge 0$ and $A \in \mathcal{A}_{n+1}$, $A \subset B$.

- In cases (a) and (b), we do not change the value of the counter: $p_{n+1}(A) = p_n(B) = 0$. Only in case (c) do we change this value, by setting $p_{n+1}(A) = 1$. This has the effect that the piece A will not be split in the next $4\kappa - 1$ steps, but will be split again exactly 4κ steps from now.
- There is a simple relation between $j_{n+1}(A)$ and $j_n(B)$. It should be obvious from our construction that the following conditions hold:

$$j_n(B) \le j_{n+1}(A) \le j_n(B) + 2\ . \tag{10.113}$$

$$p_{n+1}(A) = 0 \Rightarrow j_{n+1}(A) \le j_n(B) + 1\ . \tag{10.114}$$

$$p_{n+1}(A) = 1 \Rightarrow j_{n+1}(A) = j_n(B) + 2\ . \tag{10.115}$$

- It is also obvious by construction that "k, I, w did not change from step n to step $n+1$ unless $p_{n+1}(A) = 1$":

$$p_{n+1}(A) \neq 1 \Rightarrow k_{n+1}(A) = k_n(B) \; ; \\ I_{n+1}(A) = I_n(B) \; ; \; w_{n+1}(A) = w_n(B) \; . \qquad (10.116)$$

- The possibility that $p_{n+1}(A) \geq 2$ only arises from the case $p_n(B) \geq 1$, so we then have by construction

$$p_{n+1}(A) \geq 2 \Rightarrow p_n(B) = p_n(A) - 1 \; . \qquad (10.117)$$

Next we show that our construction satisfies the crucial conditions (10.105) and (10.106).

Lemma 10.8.2 *Conditions (10.105) and (10.106) hold for each n.*

Proof The proof goes by induction over n. We perform the induction step from n to $n+1$, keeping the notation $A \subset B$, $B \in \mathcal{A}_n$, $A \in \mathcal{A}_{n+1}$. We distinguish cases.

- We are in case (a) of Lemma 10.7.3; the piece A produced has property (10.93) and $p_{n+1}(A) = 0$. Using (10.110) and Lemma 10.8.1 and since $A \subset B$ we have $\Delta_{n+1,A}(A) = \Delta_{n,B}(A) \leq \Delta_{n,B}(B)$, so that (10.105) is satisfied for A and $n+1$ because it was satisfied for B and n.
- We are in case (b) of Lemma 10.7.3, the piece A we produce has property (10.94) and $p_{n+1}(A) = 0$. Since $j_{n+1}(A) = j_n(B) + 1$ the condition (10.94) means exactly that $\Delta_{n+1,A}(A) \leq 2^{n/2} r^{-j_{n+1}(A)}$ so that A satisfies (10.105).
- We are in case (c) of Lemma 10.7.3, the piece A produced has properties (10.95) and (10.96) and $p_{n+1}(A) = 1$. Then (10.96) means that $\Delta_{n+1,A}(A) \leq 2^{n/2} r^{-j_n(B)}$, so that

$$\Delta_{n+1,A}(A) \leq 2^{n/2} r^{-j_n(B)} = 2^{((n+1)-p_{n+1}(A))/2} r^{-j_{n+1}(A)+2}$$

since $j_{n+1}(A) = j_n(B) + 2$. Thus condition (10.106) holds for A. □

Let us explore more properties of the construction. For $n \geq 0$ and $B \in \mathcal{A}_n$ and $D \subset T$, let us define

$$F_{n,B}(D) := F(D, I_n(B), w_n(B), k_n(B), j_n(B)) \; . \qquad (10.118)$$

Lemma 10.8.3 *For any $n \geq 0$ when $p_{n+1}(A) = 1$ we have*

$$F_{n+1,A}(A) \leq F_{n,B}(B) - \frac{1}{L} 2^{n+1} r^{-j_{n+1}(A)} \; , \qquad (10.119)$$

$$w_{n+1}(A) \in B \; , \qquad (10.120)$$

$$I_{n+1}(A) = \left\{ i \in I_n(B) \;;\; |w_{n+1}(A)_i - w_n(B)_i| \le 2r^{-k_n(B)} \right\} . \tag{10.121}$$

Proof The only possibility that $p_{n+1}(A) = 1$ is when we are in the case (c) above, i.e., A is created by the case (c) of Lemma 10.7.3, and then A has property (10.97) which translates as (10.119). The other two properties hold by construction. □

Let us now introduce new notation. For $n \ge 1$, $B \in \mathcal{A}_n$, $D \subset T$ we define

$$\Delta^*_{n,B}(D) := \Delta(D, I_n(B), w_n(B), k_n(B), j_n(B) + 2) , \tag{10.122}$$

$$F^*_{n,B}(D) := F(D, I_n(B), w_n(B), k_n(B), j_n(B) + 2) , \tag{10.123}$$

and we learn to distinguish these quantities from those occurring in (10.104) and (10.118): here we have $j_n(B) + 2$ rather than $j_n(B)$.

Lemma 10.8.4 *Consider $B \in \mathcal{A}_n$ and $A \in \mathcal{A}_{n+1}$, $A \subset B$. If $n \ge 2$ and if $p_{n+1}(A) = 0$,* **either** *we have $p_n(B) = 4\kappa - 1$* **or** *$j_{n+1}(A) = j_n(B) + 1$* **or else** *we have*

$$D \subset A \ , \ \Delta^*_{n+1,A}(D) \le \frac{1}{L_4} 2^{(n+1)/2} r^{-j_{n+1}(A)-1} \Rightarrow$$
$$F^*_{n+1,A}(D) \le F^*_{n+1,A}(A) - \frac{1}{L} 2^n r^{-j_{n+1}(A)-1} . \tag{10.124}$$

Proof We may assume that $p_{n+1}(A) = 0$, $p_n(B) \ne 4\kappa - 1$, and $j_{n+1}(A) \ne j_n(B) + 1$. The set A has been produced by splitting B. There are three possibilities, as described on page 353. The possibility (b) is ruled out because $j_{n+1}(A) \ne j_n(B) + 1$. The possibility (c) is ruled out because $p_{n+1}(A) = 0$. So there remains only possibility (a), that is, A has been created by the case (a) of Lemma 10.7.3, and then (10.93) implies (10.124). □

Let us also observe another important property of the previous construction. If $B \in \mathcal{A}_n$, $A \in \mathcal{A}_{n+1}$, $A \subset B$, then

$$F_{n+1,A}(A) \le F_{n,B}(B) . \tag{10.125}$$

Indeed, if $p_{n+1}(A) \ne 1$ this follows from (10.113), (10.116), and (10.65), and if $p_{n+1}(A) = 1$, this is a consequence of (10.119).

10.9 Philosophy, IV

Let us stress some features of the previous construction. At a high level, cases (a) and (b) are just as in the Gaussian case. In these cases we do not change $I_n(B)$, $w_n(B)$, $k_n(B)$ when going from n to $n+1$. We split B into sets A which either have the property the a small D subset of A has a small functional (as is precisely stated in (10.94)) or which are such that "A is of small diameter". But the devil is in the fine print. "A is of small diameter" is not what you would obtain directly from Lemma 10.7.1, a control of $\Delta(A, I_{n+1}(A), w_{n+1}(A), k_{n+1}(A), j_{n+1}(A) + \mathbf{2})$. It is the much stronger control of $\Delta(A, I_{n+1}(A), w_{n+1}(A), k_{n+1}(A), j_{n+1}(A) + \mathbf{1})$. This stronger control of the diameter of A is essential to make the proof work and is permitted by a further splitting using Latała's principle.

The cost of using Latała's principle is that now we get a new case, (c). I like to think of this case as a really new start. We reset the values of $k_{n+1}(A)$ and $w_{n+1}(A)$, and we therefore lose a lot of the information we had gathered before. But the fundamental thing which happens in that case is that we have decreased the size of the set, as expressed by (10.119).

The counter $p_n(A)$ is not important; it is an artifact of the proof, just a way to slow down matters after we have been in case (c) so that they move at the same speed as in the other cases (instead of introducing more complicated notation).

10.10 The Key Inequality

Given $t \in T$ and $n \geq 0$, define then

$$j(n) := j_n(A_n(t)) \,,$$

where as usual $A_n(t)$ is the element of $\mathcal{A}_n$ which contains t. The fundamental property of the previous construction is as follows. It opens the door to the use of Theorem 9.2.4, since it controls the main quantity occurring there.

Proposition 10.10.1 *We have*

$$\forall t \in T \,, \ \sum_{n \geq 0} 2^n r^{-j_n(t)} \leq L(r^{-j_0} + b(T)) \,. \tag{10.126}$$

We set

$$a(n) = 2^n r^{-j(n)} = 2^n r^{-j_n(A_n(t))} \,.$$

Let us first observe that since $j(n) = j(0)$ for $n \leq 2$ we have $\sum_{n \leq 2} a(n) \leq L r^{-j(0)}$ so that it suffices to bound $\sum_{n \geq 2} a(n)$. Let us define

$$F(n) := F_{n, A_n(t)}(A_n(t)) \geq 0 \,,$$

where the functional $F_{n,A}$ has been defined in (10.118). As a consequence of (10.125) the sequence $(F(n))_{n\geq 0}$ is non-increasing, and of course $F(0) \leq b(T)$.[12]

Let us recall the definition $n_0 = 2$ and set

$$J_0 = \{n_0\} \cup \{n > n_0 \; ; \; j(n+1) > j(n)\} ,$$

which we enumerate as $J_0 = \{n_0, n_1, \ldots\}$. Since $n_k \in J_0$ we have $j(n_k + 1) > j(n_k)$. By (10.113) we have $j(n_k + 1) \leq j(n_k) + 2$. Also, for $n_k + 1 \leq n < n_{k+1}$ we have $j(n + 1) = j(n)$, so that

$$n_k + 1 \leq n \leq n_{k+1} \Rightarrow j(n) = j(n_k + 1) . \tag{10.127}$$

Taking for granted that the sequence $(a(n))$ is bounded (which we will show at the very end of the proof), and observing that $a(n + 1) = 2a(n)$ for $n \notin J_0$, Lemma 2.9.5 used for $\alpha = 2$ implies that $\sum_{n\geq n_0} a(n) \leq L \sum_{n\in J_0} a(n) = L \sum_{k\geq 0} a(n_k)$. Let us set

$$C^* := \left\{k \geq 0 \; ; \; \forall k' \geq 0 \; , \; a(n_k) \geq 2^{-|k-k'|} a(n_{k'})\right\} .$$

Using the Lemma2.9.5 again implies that

$$\sum_{k\geq 0} a(n_k) \leq L \sum_{k\in C^*} a(n_k) . \tag{10.128}$$

Thus, our task is to bound this later sum. In the next section, you will find some words trying to explain the structure of the proof, although they may not make sense before one had at least a cursory look at the arguments.

A good part of the argument is contained in the following fact, where we use the notation $p(n) := p_n(A_n(t))$:

Lemma 10.10.2 *Consider $k \in C^*$ with $k \geq 1$ and assume that*

$$n_k - 1 \leq m \leq n^* := n_{k+1} + 1 \Rightarrow p(m) = 0 . \tag{10.129}$$

Then

$$a(n_k) \leq L(F(n_k) - F(n_{k+2})) . \tag{10.130}$$

Proof The proof is very close to the proof of (2.94) which should be reviewed now. It will be deduced from the key property (10.124). A crucial fact here is that in the definition (10.122) of $\Delta^*_{n+1,A}(A)$, we have $j_{n+1}(A) + 2$ (and not $j_{n+1}(A)$).

[12] Actually we have $F(0) = b(T)$.

Let us fix k and set $n = n_k - 1$. The reader must keep this notation in mind at all time, and remember in particular that $n+1 = n_k$. We first prove that for $A = A_{n+1}(t) = A_{n_k}(t)$, (10.124) holds, i.e.

$$D \subset A \,,\; \Delta^*_{n+1,A}(D) \le \frac{1}{L_4} 2^{(n+1)/2} r^{-j_{n+1}(A)-1} \Rightarrow$$

$$F^*_{n+1,A}(D) \le F^*_{n+1,A}(A) - \frac{1}{L} 2^n r^{-j_{n+1}(A)-1} \,. \tag{10.131}$$

Since $p(n+1) = 0$, Lemma 10.8.4 states that there are three possibilities: either $p(n) = 4\kappa - 1$ or else $j(n) < j(n+1)$ or else (10.124) holds. We now rule out the first two possibilities. The first one is ruled out by (10.129), which asserts in particular that $p(n_k - 1) = p(n) = 0$. To rule out the second one, we assume for contradiction that $j(n) = j(n_k - 1) < j(n_k) = j(n+1)$. Then $n_k - 1 \in J_0$ so that $n_k - 1 = n_{k-1}$. But since $k \in C^*$, we have $a(n_{k-1}) \le 2a(n_k)$, i.e., $r^{j(n_k)-j(n_{k-1})} \le 2^{n_k+1-n_{k-1}} = 4$. Since $j(n_k) - j(n_{k-1}) = 1$, this is a contradiction since $r \ge 8$, which proves that $j(n) = j(n+1)$.

Thus we have proved (10.131), which is the crux of the proof.

It follows from (10.114) that for any $m \in J_0$, we have $j(m+1) = j(m) + 1$ when $p(m+1) = 0$. In particular (10.129) implies that this is the case for $m = n_k$ and $m = n_{k+1}$ so that, using also (10.127) in the third equality,

$$j(n^*) = j(n_{k+1} + 1) = j(n_{k+1}) + 1 = j(n_k + 1) + 1 = j(n_k) + 2 \,, \tag{10.132}$$

i.e.

$$j_{n^*}(A_{n^*}(t)) = j_{n_k}(A_{n_k}(t)) + 2 \,. \tag{10.133}$$

Furthermore (10.116) implies

$$w_{n^*}(A_{n^*}(t)) = w_{n_k}(A_{n_k}(t)) \;;\; I_{n^*}(A_{n^*}(t)) = I_{n_k}(A_{n_k}(t)) \;;$$
$$k_{n^*}(A_{n^*}(t)) = k_{n_k}(A_{n_k}(t)) \,. \tag{10.134}$$

We will prove later that

$$\Delta^*_{n+1,A}(A_{n^*}(t)) \le \frac{1}{L_4} 2^{(n+1)/2} r^{-j_{n+1}(A_{n+1}(t))-1} = \frac{1}{L_4} 2^{n_k/2} r^{-j_{n_k}(A_{n_k}(t))-1} \,. \tag{10.135}$$

For the time being, we assume that (10.135) holds, and we show how to conclude the proof of the lemma. Recalling that $n_k = n+1$, so that $j_{n+1}(A) = j_{n_k}(A) = j_{n_k}(A_{n_k}(t)) = j(n_k)$ we use (10.131) to obtain

$$F^*_{n_k,A}(A_{n^*}(t)) \le F^*_{n_k,A}(A_{n_k}(t)) - \frac{1}{L} 2^{n_k} r^{-j(n_k)-1} \,. \tag{10.136}$$

Using the monotonicity of (10.65) of F in the parameter j in the inequality, we obtain

$$
\begin{aligned}
&F^*_{n_k,A}(A_{n_k}(t)) &(10.137)\\
&= F(A_{n_k}(t), I_{n_k}(A_{n_k}(t)), w_{n_k}(A_{n_k}(t)), k_{n_k}(A_{n_k}(t)), j(n_k)+2)\\
&\le F(A_{n_k}(t), I_{n_k}(A_{n_k}(t)), w_{n_k}(A_{n_k}(t)), k_{n_k}(A_{n_k}(t)), j(n_k)) = F(n_k)\ .
\end{aligned}
$$

Using (10.134) and in an *absolutely crucial manner* that $j(n^*) = j(n_k) + 2$ by (10.132), we obtain

$$
\begin{aligned}
&F^*_{n_k,A}(A_{n*}(t)) &(10.138)\\
&= F(A_{n^*}(t), I_{n_k}(A_{n_k}(t)), w_{n_k}(A_{n_k}(t)), k_{n_k}(A_{n_k}(t)), j(n_k)+2)\\
&= F(A_{n^*}(t), I_{n^*}(A_{n^*}(t)), w_{n^*}(A_{n^*}(t)), k_{n^*}(A_{n^*}(t)), j(n^*)) = F(n^*)\ .
\end{aligned}
$$

Thus (10.136) implies

$$F(n^*) \le F(n_k) - \frac{1}{L} 2^{n_k} r^{-j(n_k)-1}\ ,$$

i.e., $a(n_k) \le L(F(n_k) - F(n^*)) \le L(F(n_k) - F(n_{k+2}))$ by (10.125), and this concludes the proof of the lemma.

We turn to the proof of (10.135). Using (10.132) and (10.134), we first obtain

$$
\begin{aligned}
&\Delta^*_{n+1,A}(A_{n^*}(t))\\
&= \Delta(A_{n^*}(t), I_{n_k}(A_{n_k}(t)), w_{n_k}(A_{n_k}(t)), k_{n_k}(A_{n_k}(t)), j(n_k)+2)\\
&= \Delta\big(A_{n^*}(t), I_{n^*}(A_{n^*}(t)), w_{n^*}(A_{n^*}(t)), k_{n^*}(A_{n^*}(t)), j(n^*)\big)\\
&= \Delta_{n^*, A_{n^*}(t)}(A_{n^*}(t))\ . &(10.139)
\end{aligned}
$$

Here also it is crucial that $j(n^*) = j(n_k) + 2$. Then we use (10.105) for n^* instead of n and $A = A_{n^*}(t)$ to obtain

$$\Delta_{n^*, A_{n^*}(t)}(A_{n^*}(t)) \le 2^{n^*/2} r^{-j(n^*)}\ . \tag{10.140}$$

By the definition of C^*, we have $a(n_k) \ge a(n_{k+1})/2$ i.e.

$$2^{n_k} r^{-j(n_k)} \ge 2^{n_{k+1}} r^{-j(n_{k+1})}/2\ ,$$

and thus, using again that $j(n_{k+1}) = j(n_k) + 1$,

$$2^{n_{k+1}-n_k} \le 2r = 2^{\kappa+1}$$

and therefore $n_{k+1} - n_k \le \kappa + 1$, so that $n^* = n_k + 1 \le n_k + \kappa + 2$, and since $j(n^*) = j(n_k) + 2$, we have $2^{n^*/2} r^{-j(n^*)} \le 2^{(2+\kappa)/2} 2^{n_k/2} r^{-j(n_k)-2}$. Using that $2^{\kappa/2} \ge 2L_4$ and $r = 2^\kappa$, so that $2^{(2+\kappa)/2} r^{-1} = 2^{1-\kappa/2} \le 1/L_4$, we get

$$2^{n^*/2} r^{-j(n^*)} \le 2^{(2+\kappa)/2} 2^{n_k/2} r^{-j(n_k)-2} \le \frac{1}{L_4} 2^{n_k/2} r^{-j(n_k)-1} . \tag{10.141}$$

Comparing with (10.140) and (10.139) yields the desired inequality (10.135). □

Corollary 10.10.3 *Consider the subset $\tilde{C}$ of C^* consisting of the integers $k \ge 1$, $k \in C^*$ for which (10.129) holds. Then*

$$\sum_{k \in \tilde{C}} a(n_k) \le Lb(T) . \tag{10.142}$$

Proof This follows from the usual "telescoping sum" argument, together with the fact that $F(n) \le Lb(0)$. □

So, we have now reduced the task to controlling $a(n_k)$ when $k \in C^* \setminus \tilde{C}$. We start by a simple observation.

Lemma 10.10.4 *We have*

$$p(m) > 0 \Rightarrow j(m) = j(m+1) \tag{10.143}$$

$$\forall k \; ; \; p(n_k) = 0 . \tag{10.144}$$

Proof Condition (10.143) holds by construction; see (10.108). Condition (10.144) is a corollary, since $j(n_k) < j(n_k + 1)$. □

The following lemma gives us precise information on these integers $k \in C^* \setminus \tilde{C}$:

Lemma 10.10.5 *If for a certain $k > 0$ we have $p(n_k - 1) = p(n_k + 1) = p(n_{k+1} + 1) = 0$ then $k \in \tilde{C}$.*

Proof We have to prove that $p(m) = 0$ for $n_k - 1 \le m \le n_{k+1} + 1$. Assume for contradiction that there exists $n_k - 1 \le m \le n_{k+1} + 1$ with $p(m) > 0$, and consider the smallest such m. Then certainly $n_k + 1 < m < n_{k+1}$ since $p(n_k - 1) = p(n_k + 1) = 0$ by hypothesis and since $p(n_k) = p(n_{k+1}) = 0$ by (10.108). Next we prove that $p(m) = 1$. Indeed otherwise $p(m) \ge 2$ and by (10.117) we have $p(m-1) = p(m) - 1 \ge 1$ which contradicts the minimality of m. Thus $p(m) = 1$. But by construction when $p(m) = 1$ then $j(m) = j(m-1) + 2$ (see (10.115)) so that $m - 1 \in J_0$. But since $n_k < m - 1 < n_{k+1}$, this contradicts the definition of n_{k+1} which is the smallest element of J_0 larger than n_k. □

Corollary 10.10.6 *For $k \in C^* \setminus \tilde{C}$, we have either $k = 0$ or $p(n_k - 1) > 0$ or $p(n_k + 1) = 1$ or $p(n_{k+1} + 1) = 1$.*

Proof According to Lemma 10.10.5 if $k \geq 1$ does not belong to $\tilde{C}$ then on of the quantities $p(n_k - 1), p(n_k + 1), p(n_{k+1} + 1)$ is > 0. Now, since $p(n_k) = 0$ and $p(n+1) \leq p(n) + 1$ we have $p(n_k + 1) \leq 1$. □

The goal now is to produce specific arguments to control $a(n_k)$ in the various situations which can happen when $k \in C^* \setminus \tilde{C}$, as brought forward by the previous corollary

Lemma 10.10.7 *Let* $J_1 = \{n \geq 0\ ;\ p(n+1) = 1\}$. *Then*

$$\sum_{n \in J_1} a(n) \leq Lb(T) . \tag{10.145}$$

Proof Indeed (10.119) implies that for $n \in J_1$,

$$a(n) \leq L(F(n) - F(n+1)) ,$$

and the telescopic sum is bounded by $LF(0) \leq Lb(T)$. □

Corollary 10.10.8 *Let* $C_1 = \{k \in C^*\ ;\ p(n_k + 1) = 1\} = \{k \in C^*\ ;\ n_k \in J_1\}$. *Then*

$$\sum_{k \in C_1} a(n_k) \leq Lb(T) . \tag{10.146}$$

Lemma 10.10.9 *Let* $C_2 := \{k \geq 0; n_{k+1} \in J_1\}$. *Then*

$$\sum_{k \in C_2} a(n_k) \leq Lb(T) . \tag{10.147}$$

Proof We have

$$a(n_k) = 2^{n_k} r^{-j(n_k)} \leq r^2 2^{n_{k+1}} r^{-j(n_{k+1})} = r^2 a(n_{k+1}) ,$$

where we have used in the first inequality that $j(n_{k+1}) \leq j(n_k) + 2$ by (10.113), and therefore (10.146) implies the result. □

Lemma 10.10.10 *Let* $C_3 := \{k \geq 0; p(n_k - 1) > 0\}$. *Then*

$$\sum_{k \in C_3} a(n_k) \leq Lb(T) . \tag{10.148}$$

Proof Let us recall that by construction $p(n+1) = p(n) + 1$ when $1 \leq p(n) \leq 4\kappa - 2$. Consequently the only possibility to have $p(n) > 0$ and $p(n+1) = 0$ is to have $p(n) = 4\kappa - 1$. Also, since $j(n_k + 1) > j(n_k)$, by construction we have $p(n_k) = 0$. Thus for $k \in C_3$ we have $p(n_k) = 0$ and $p(n_k - 1) > 0$ so that $p(n_k - 1) = 4\kappa - 1$, and then $p(n_k - 4\kappa + 1) = 1$, i.e., $n_k - 4\kappa \in J_1$.

Iteration of the relation $a(n) \le 2a(n-1)$ shows that $a(n_k) \le 2^{4\kappa} a(n_k - 4\kappa)$. Thus $\sum_{k\in C_3} a(n_k) \le L \sum_{k\in C_3} a(n_k - 4\kappa) \le \sum_{n\in J_1} a(n)$, and the result follows from (10.145). □

The following is a consequence of Corollary 10.10.6 and the subsequent lemmas:

Proposition 10.10.11 *We have*

$$\sum_{k\in C^*} a(n_k) \le L(r^{-j_0} + b(T)) . \tag{10.149}$$

Proof of Proposition 10.10.1 It follows by combining (10.128) and (10.149). But it remains to prove that the sequence $(a(n))$ is bounded. Using a much simplified version of the previous arguments, we prove that in fact

$$\forall n \,,\; a(n) \le L(r^{-j_0} + b(T)) . \tag{10.150}$$

By (10.145) this is true if $n \in J_1$. Next we recall that if $p(n) > 2$ then $p(n-1) = p(n) - 1$. Consequently $n \in J_2 := \{n; p(n-1) + p(n) > 0\}$, there exists $n' \in J_1$ with $n' \ge n - 4\kappa$. Also, since $a(m+1) \le 2a(m)$, we have $a(n) \le La(n')$, and we have shown that (10.150) holds for $n \in J_2$. Next we show that it also holds for $n \in J_3 := \{n \ge 3; j(n-1) = j(n), p(n-1) = p(n) = 0\}$. Indeed in that case, we use Lemma 10.8.4 for $n-1$ rather than n. Since $p(n) = p(n-1) = 0$, we are in the third case of the lemma, and (10.124) holds. Taking D reduced to a single point proves the required equality $a(n) \le LF(n_0)$. Now, if $n \ge 3$ and $n \notin J_2 \cup J_3$, we have $j(n) > j(n-1)$, and since $r > 4$, we have $a(n-1) > a(n)$. So to prove (10.150), we may replace n by $n-1$. Iteration of the procedure until we reach a point of $J_2 \cup J_3 \cup \{1, 2\}$ concludes the argument. □

10.11 Philosophy, V

In this section we try to describe at a high level some features of the previous proof. When in that proof $j(n_k + 1) > j(n_k)$, according to (10.114) and (10.115), there are two cases. First, it might happen that $p(n_k + 1) = 1$ and $j(n_k + 1) = j(n_k) + 2$. Second, it might happen that $p(n_k + 1) = 0$ and $j(n_k + 1) = j(n_k) + 1$. Let us think of the first case as an exceptional case. This exceptional case is a good thing because we then have no problem to control $a(n_k)$ thanks to (10.145). As expressed in (10.147) and (10.148), for semi-trivial reasons, we have no problem either to control $a(n_k)$ when $p(n_k - 1) = 1$ or $p(n_{k+1} + 1) = 1$, so that, so to speak, the problem is to control the $a(n_k)$ in the special case where there is no exceptional value k' near k. This is what Lemma 10.10.2 does. In that lemma we get by construction the information (10.131) on the small subsets D of $A = A_{n_k}(t)$. The idea is to use that information on a set $D = A_{n'}(t)$. For this we need to control the diameter of $A_{n'}(t)$. We should think of this diameter as governed by $2^{n'/2} r^{-j(n')}$.

For this to be small enough, we need $j(n') \geq j(n_k) + 2$. The smallest value of n' for which this happens is $n^* = n_{k+1} + 1$. An important feature of the argument is that our bound on the size of $A_{n_{k+1}+1}$ is smaller by a factor $2/r$ than the bound we had on the size of $A_{n_{k+1}}$ (which itself is not so much larger than the bound we had on the size of A_{n_k} thanks to the use of Lemma 2.9.5). In this manner we obtain the control $a(n_k) \leq L(F(n_k) - F(n^*))$.

10.12 Proof of the Latała-Bednorz Theorem

Without loss of generality, we assume $0 \in T$. First we use Lemma 6.3.6 to find j_0 such that $\Delta(T, d_2) \leq 2r^{-j_0} \leq Lb(T)$ so that in particular $|t_i| < r^{-j_0}/2$ for $t \in T$ and $i \in \mathbb{N}$. We then build a sequence of partitions as in Sect. 10.8, using this value of j_0. Then (10.126) yields

$$\sup_{t \in T} \sum_{n \geq 0} 2^n r^{-j_n(t)} \leq Lb(T) . \tag{10.151}$$

The plan is to produce the required decomposition of T using Theorem 9.2.4 for $\Omega = \mathbb{N}$ and μ the counting measure (and using also (10.151)). To apply this theorem, given n and given $A \in \mathcal{A}_n$, we will define an elements $\pi_n(A) \in T$. We will then define $\pi_n(t) = \pi_n(A_n(t))$. However, in contrast with what happened in many previous proofs, we will not require that $\pi_n(A) \in A$. It could be helpful to recall (9.9) and (9.10), which are the most stringent of the conditions we require on $\pi_n(t)$:

$$\forall t \in T \ , \ \forall n \geq 0 \ , \ j_n(t) = j_{n+1}(t) \Rightarrow \pi_n(t) = \pi_{n+1}(t) \ , \tag{9.9}$$

$$\forall t \in T \ , \ \forall n \geq 0 \ , \ j_{n+1}(t) > j_n(t) \Rightarrow \pi_{n+1}(t) \in A_n(t) \ . \tag{9.10}$$

The construction of the points $\pi_n(A)$ proceeds as follows. We choose $\pi_0(T) = 0$. Consider $A \in \mathcal{A}_{n+1}$ and $A \subset B$, $B \in \mathcal{A}_n$. According to (10.113) there are three disjoint cases which cover all possibilities:

- $j_{n+1}(A) = j_n(B)$. We then set $\pi_{n+1}(A) = \pi_n(B)$.
- $j_{n+1}(A) = j_n(B) + 1$. We take for $\pi_{n+1}(A)$ any point of B.
- $j_{n+1}(A) = j_n(B) + 2$. According to (10.114) we then have $p_{n+1}(A) = 1$, so that we are in the case (c) considered on page 353. We set $\pi_{n+1}(A) = w_{n+1}(A)$ so that $\pi_{n+1}(A) \in B$ using (10.120).

The important property (which obviously holds by construction) is that

$$j_{n+1}(A) = j_n(B) \Rightarrow \pi_{n+1}(A) = \pi_n(B) \ , \tag{10.152}$$

$$j_{n+1}(A) > j_n(B) \Rightarrow \pi_{n+1}(A) \in B \ . \tag{10.153}$$

Defining $\pi_n(t) = \pi_n(A_n(t))$, this implies that (9.9) and (9.10) hold, while (9.7) is obvious by construction. Also, according to (10.114) and (10.115), we have $j_{n+1}(A) = 1$ if and only if $j_{n+1}(A) = j_n(B)$, so that

$$p_{n+1}(A) = 1 \Rightarrow \pi_{n+1}(A) = w_{n+1}(A) . \tag{10.154}$$

Let us consider the set

$$\Omega_n(t) = \left\{ i \in \mathbb{N}^* \; ; \; \forall q < n \; , \; |\pi_q(t)_i - \pi_{q+1}(t)_i| \le r^{-j_q(t)} \right\} . \tag{10.155}$$

The key of the argument is to establish the inequality

$$\forall t \in T \; , \; \forall n \ge 0 \; , \; \sum_{i \in \Omega_n(t)} |r^{j_n(t)}(t_i - \pi_n(t)_i)|^2 \wedge 1 \le L2^n . \tag{10.156}$$

This inequality means that (9.12) holds for $u = L$. We can then apply Theorem 9.2.4 to obtain the required decomposition of T, since $T_3 = \{0\}$ by the choice of j_0.

We start the preparations for the proof of (10.156). Let us define

$$k_n(t) := k_n(A_n(t)) \; ; \; w_n(t) := w_n(A_n(t)) \; ; \; p_n(t) := p_n(A_n(t)) .$$

Then by (10.116) we have

$$p_{q+1}(t) \ne 1 \Rightarrow w_{q+1}(t) = w_q(t) \; ; \; k_{q+1}(t) = k_q(t) \; , \tag{10.157}$$

and (10.154) implies

$$p_{q+1}(t) = 1 \Rightarrow \pi_{q+1}(t) = w_{q+1}(t) . \tag{10.158}$$

Also, since $k_n(A) \le j_n(A)$ for $A \in \mathcal{A}_n$, we have

$$k_n(t) \le j_n(t) . \tag{10.159}$$

Our next goal is to prove the inequality

$$i \in \Omega_{n+1}(t) \Rightarrow |\pi_{n+1}(t)_i - w_n(t)_i| \le 2r^{-k_n(t)} . \tag{10.160}$$

To prepare for the proof, let $J' = \{0\} \cup \{n' ; \, p_{n'}(t) = 1\}$. Given $n \ge 0$ let us consider the largest $n' \in J'$ with $n' \le n$. Then by definition of n' for $n' \le q < n$ we have $p_{q+1}(t) \ne 1$, so that by (10.157) we have $w_{q+1}(t) = w_q(t)$ and $k_{q+1}(t) = k_q(t)$. Consequently we have

$$w_n(t) = w_{n'}(t) \; ; \; k_n(t) = k_{n'}(t) . \tag{10.161}$$

We prove next that $\pi_{n'}(t) = w_{n'}(t)$. If $n' = 0$ this is true because we defined $w_0(T) = \pi_0(T) = 0$. If $n' > 0$, by (10.158) we have $\pi_{n'}(t) = w_{n'}(t)$, and recalling (10.161) we have proved that $\pi_{n'}(t) = w_{n'}(t)$ as desired.

We observe next that by (10.159) we have $k'_{n'}(t) \le j_{n'}(t)$. Recalling (10.161), to prove (10.160) it suffices to prove that $|\pi_{n+1}(t)_i - \pi_{n'}(t)_i| \le 2r^{-j_{n'}(t)}$. We write

$$|\pi_{n+1}(t)_i - \pi_{n'}(t)_i| \le \sum_{n' \le q \le n} |\pi_{q+1}(t)_i - \pi_q(t)_i| = \sum_{i \in U} |\pi_{q+1}(t)_i - \pi_q(t)_i| \,, \tag{10.162}$$

where

$$U = \{q \; ; \; n' \le q \le n \,, \; \pi_{q+1}(t) - \pi_q(t) \ne 0\} \,.$$

Now, recalling (10.155), since $i \in \Omega_{n+1}(t)$ for $q \in U$ we have $|\pi_{q+1}(t)_i - \pi_q(t)_i| \le r^{-j_q(t)}$, so that by (10.162) we get

$$|\pi_{n+1}(t)_i - \pi_{n'}(t)_i| \le \sum_{q \in U} r^{-j_q(t)} \,. \tag{10.163}$$

Since the sequence $j_n(t)$ is non-decreasing, for $q \in U$ we have $j_q(t) \ge j_{n'}(t)$. Also, by the first part of (10.153) for $q \in U$, we have $j_{q+1}(t) \ne j_q(t)$ so that the numbers $j_q(t)$ for $q \in U$ are all different and the sum on the right-hand side of (10.163) is $\le \sum_{j \ge j_{n'}(t)} r^{-j} \le 2r^{-j_{n'}(t)}$ proving (10.160).

Next we prove by induction over n that $\Omega_n(t) \subset I_n(t) := I_n(A_n(t))$. This holds for $n = 0$ since $I_0(t) = \mathbb{N}^*$. The argument for the induction from n to $n+1$ depends on the value of $p_{n+1}(t)$. We start with the easy case is where $p_{n+1}(t) \ne 1$. Then $\Omega_{n+1}(t) \subset \Omega_n(t) \subset I_n(t)$ and $I_{n+1}(t) = I_n(t)$ by (10.116), concluding the argument. Let us now assume that $p_{n+1}(t) = 1$. We first note that according to (10.112) we then have

$$I_{n+1}(t) = \{i \in I_n(t) \; ; \; |w_{n+1}(t)_i - w_n(t)_i| \le 2r^{-k_n(t)}\} \,. \tag{10.164}$$

Also, by construction of $\pi_{n+1}(A)$, we have $\pi_{n+1}(t) = w_{n+1}(t)$, and then (10.160) implies that for $i \in \Omega_{n+1}(t)$ we have $|w_{n+1}(t)_i - w_n(t)_i| \le 2r^{-k_n(t)}$. Combining with the induction hypothesis, (10.164) concludes the proof that $\Omega_{n+1}(t) \subset I_{n+1}(t)$ and the induction.

Using (10.160) for $n-1$ rather than n, we obtain

$$i \in \Omega_n(t) \Rightarrow |\pi_n(t)_i - w_{n-1}(t)_i| \le 2r^{-k_{n-1}(t)} \,. \tag{10.165}$$

Since $\Omega_n(t) \subset I_n(t) \subset I_{n-1}(t)$, it follows from (10.69) that

$$\sum_{i \in \Omega_n(t)} |r^{j_n(t)}(t_i - \pi_n(t)_i)|^2 \wedge 1 \le 2r^{2j_n(t)} d_{I_{n-1}(t), w_{n-1}(t), k_{n-1}(t), j_n(t)}(t, \pi_n(t))^2$$

$$\le 2r^{2j_n(t)} d_{I_{n-1}(t), w_{n-1}(t), k_{n-1}(t), j_{n-1}(t)}(t, \pi_n(t))^2 , \qquad (10.166)$$

where in the second line we use that $j_{n-1}(t) \le j_n(t)$ and that $d_{I,w,k,j}$ decreases when j increases.

Finally we are ready to prove the main inequality (10.156). Let us assume first that $j_{n-1}(t) < j_n(t)$. In that case by (10.153) we have $\pi_n(t) \in A_{n-1}(t)$ so that the right-hand side of (10.166) is bounded by

$$2r^{2j_n(t)} \Delta(A_{n-1}(t), I_{n-1}(t), w_{n-1}(t), k_{n-1}(t), j_{n-1}(t))^2$$
$$= 2r^{2j_n(t)} \Delta_{n-1, A_{n-1}(t)}(A_{n-1}(t))^2 \le 2^n r^{2j_n(t) - 2j_{n-1}(t)} ,$$

where we have used (10.105) for $n-1$ rather than n in the inequality. This concludes the proof of (10.156) in that case since $j_{n-1}(t) \ge j_n(t) - 2$ and r is a universal constant.

To prove (10.156) in general, we proceed by induction over n. The inequality holds for $n = 0$ by our choice of j_0. For the induction step, according to the previous result, it suffices to consider the case where $j_n(t) = j_{n-1}(t)$. Then, according to (10.152) we have $\pi_n(t) = \pi_{n-1}(t)$, so the induction step is immediate since $\Omega_n(t) \subset \Omega_{n-1}(t)$. □

10.13 Philosophy, VI

Maybe we should stress how devilishly clever the previous argument is. On the one hand, we have the information (10.153) on $\pi_n(t)$, which, together with (10.105), allows us to control the right-hand side of (10.166). But we do not care about the right-hand side of this inequality; we care about the left-hand side. In order to be able to relate them using (10.69), we need to control the difference $|\pi_n(t)_i - w_{n-1}(t)_i|$ for many coordinates i. The coordinates for which we can achieve that depend on t.

Let us try to take the mystery out of the interplay between the sets $\Omega_n(t)$ and $I_n(t)$. The magic is already in the definition of the sets $\Omega_n(t)$: when going from $\Omega_n(t)$ to $\Omega_{n+1}(t)$, we drop the coordinates i for which $\pi_{n+1}(t)_i$ is significantly different from $\pi_n(t)_i$. Another important feature is that $w_n(t) = w_{n+1}(t)$ unless $p_{n+1}(t) = 1$, i.e., unless we are in the case (c) of Lemma 10.7.3. We then use a marvelously simple device. Each time we have just changed the value of $w_n(t)$ (i.e., we are at stage $n+1$ with $p_{n+1}(t) = 1$), we ensure that $\pi_{n+1}(t) = w_{n+1}(t)$. The point of doing this is that for the coordinates i, we have kept (i.e., those belonging to the set $\Omega_{n+1}(t)$) the value of $\pi_{n+1}(t)_i$ is nearly the same as the value $w_{n'}(t)_i$, where

n' is the last time we changed the value of $w_q(t)$. This is true *whatever our choice* for $\pi_{n+1}(t)$ and in particular for $\pi_{n+1}(t) = w_{n+1}(t)$. Thus we are *automatically* assured that for the coordinates i we keep $w_{n+1}(t)_i$ is nearly $w_{n'}(t)_i$ (i.e., that $\Omega_{n+1}(t) \subset I_{n+1}(t)$).

The purpose of (10.120) is precisely to be able to set $\pi_{n+1}(A) = w_{n+1}(A)$ when $p_{n+1}(A) = 1$ while respecting the crucial condition (10.153).

10.14 A Geometric Characterization of $b(T)$

The majorizing measure theorem (Theorem 2.10.1) asserts that for a subset T of ℓ^2, the "probabilistic" quantity $g(T) = \mathsf{E}\sup_{t\in T}\sum_{i\geq 1} g_i t_i$ is of the same order as the "geometric" quantity $\gamma_2(T,d)$. In this section we prove a similar result for the probabilistic quantity $b(T)$. The corresponding geometric quantity will use a familiar "family of distances" which we recall now. We fix $r \geq 8$, and for $j \in \mathbb{Z}$ and $s, t \in \ell^2$, we define (as we have done through Chap. 7)

$$\varphi_j(s,t) = \sum_{i\geq 1}(r^{2j}|s_i - t_i|^2) \wedge 1 \,. \tag{10.167}$$

Let us then consider the following "geometric measure of size of a subset T of ℓ^2":

Definition 10.14.1 Given a subset T of ℓ^2, we denote by $\tilde{b}(T)$ the infimum of the numbers S for which there exists an admissible sequence $(\mathcal{A}_n)$ of partitions of T, and for $A \in \mathcal{A}_n$ an integer $j_n(A)$ with the following properties:

$$s, t \in A \Rightarrow \varphi_{j_n(A)}(s,t) \leq 2^n \,, \tag{10.168}$$

$$\Delta(T, d_2) \leq r^{-j_0(T)} \,, \tag{10.169}$$

$$S = \sup_{t\in T}\sum_{n\geq 0} 2^n r^{-j_n(A_n(t))} \,. \tag{10.170}$$

We recall the quantity $b^*(T)$ from Definition 6.2.6.

Theorem 10.14.2 *For a subset T of ℓ^2 one has*

$$\tilde{b}(T) \leq L r^2 b^*(T) \,. \tag{10.171}$$

This theorem is a kind of converse to Theorem 2.10.1. Together with Corollary 9.4.3, it shows that the measures of size $b^*(T)$ and $\tilde{b}(T)$ are equivalent. The Latała-Bednorz theorem shows that the measures of size $b(T)$ and $b^*(T)$ are equivalent. Thus all three measures of size $b(T)$, $b^*(T)$, and $\tilde{b}(T)$ are equivalent. The equivalence of $b(T)$ and $\tilde{b}(T)$ parallels, for Bernoulli processes, the equivalence of $g(T)$ and $\gamma_2(T,d)$ for Gaussian processes. Observe that the proof of the equivalence of $\tilde{b}(T)$ and $b(T)$ is somewhat indirect, since we show that both quantities are equivalent to $b^*(T)$. There laid a considerable difficulty in discovering the proof of the Latała-Bednorz theorem: even though $b(T)$ and $\tilde{b}(T)$ are equivalent, it does not seem possible to directly construct the partition witnessing that $\tilde{b}(T) \le Lb(T)$.

Theorem 10.14.2 is a consequence of the Latała-Bednorz result and of the following facts:

Proposition 10.14.3 *Consider $a > 0$. The set $B_a = \{t \in \ell^2; \sum_{i\ge 1} |t_i| \le a\}$ satisfies $\tilde{b}(B_a) \le Lra$.*

Proposition 10.14.4 *For a subset T of ℓ^2 one has $\tilde{b}(T) \le Lr\gamma_2(T)$.*

Proposition 10.14.5 *Recalling that $T + T'$ denotes the Minkowski sum of T and T', for $T, T' \subset \ell^2$ one has*

$$\tilde{b}(T + T') \le Lr(\tilde{b}(T) + \tilde{b}(T')) . \tag{10.172}$$

The proof of Proposition 10.14.3 is delayed until Sect. 19.2.1 because this result has little to do with probability and bears on the geometry of B_a. The proof of Proposition 10.14.4 should be obvious to the reader having reached this point. If it is not, please review the discussion around (9.46), and if this does not suffice, try to figure it out by solving the next exercise.

Exercise 10.14.6 Write the proof of Proposition 10.14.4 in complete detail.

Proof of Proposition 10.14.5 We first observe that for a translation-invariant distance d, we have

$$d(s+s', t+t') \le d(s+s', t+s') + d(t+s', t'+s') \le d(s,t) + d(s',t') ,$$

so that since φ_j is the square of a translation-invariant distance,

$$\varphi_j(s+s', t+t') \le 2(\varphi_j(s,t) + \varphi_j(s',t')) . \tag{10.173}$$

For each $t \in T + T'$ let us pick in an arbitrary manner $u(t) \in T$ and $u'(t) \in T'$ with $t = u(t) + u'(t')$. For $A \subset T$, $A' \subset T'$ let us define

$$A * A' = \{t \in T \; ; \; u(t) \in A \; , \; u'(t) \in A'\} .$$

According to the definition of $\tilde{b}$, there exist admissible sequences $(\mathcal{A}_n)$ and $(\mathcal{A}'_n)$ on T and T', respectively, and for $A \in \mathcal{A}_n$ and $A' \in \mathcal{A}'_n$ corresponding integers $j_n(A)$ and $j'_n(A')$ as in (10.168) and (10.169) with

$$\sum_{n\ge 0} 2^n r^{-j_n(t)} \le 2\tilde{b}(T) \; ; \; \sum_{n\ge 0} 2^n r^{-j'_n(t)} \le 2\tilde{b}(T') . \tag{10.174}$$

Consider the family of subsets $\mathcal{B}_n$ of $T + T'$ consisting of the sets of the type $A * A'$ for $A \in \mathcal{A}_n$ and $A' \in \mathcal{A}'_n$. It obviously forms a partition of $T + T'$. The sequence $(\mathcal{B}_n)$ of partitions is increasing and card $\mathcal{B}_n \le N_n^2 \le N_{n+1}$. Also, for $t = u(t) + u'(t) \in T + T'$, we have $t \in A_n(u(t)) * A'_n(u'(t))$, so $A_n(u(t)) * A'_n(u'(t))$ is the element $B_n(t)$ of $\mathcal{B}_n$ containing t. For $B = A * A' \in \mathcal{B}_n$, we set

$$b_n(B) = \min(j_n(A), j'_n(A')) \ . \tag{10.175}$$

Thus for $s, t \in B$, we have, using (10.173) and also (10.168) in the last inequality,

$$\begin{aligned} \varphi_{b_n(B)}(s,t) &= \varphi_{b_n(B)}(u(s) + u'(s), u(t) + u'(t)) \\ &\le 2\varphi_{b_n(B)}(u(s), u(t)) + 2\varphi_{b_n(B)}(u'(s), u'(t)) \le 2^{n+2} \ . \end{aligned} \tag{10.176}$$

Let us then define a sequence $(\mathcal{C}_n)$ of partitions of $T + T'$ by setting $\mathcal{C}_n = \mathcal{B}_{n-2}$ for $n \ge 3$ and $\mathcal{C}_n = \{T + T'\}$ for $n \le 2$. Obviously, this sequence of partitions is admissible. Let us further define $k_n(B) = b_{n-2}(B)$ for $B \in \mathcal{C}_n = \mathcal{B}_{n-2}$ with $n \ge 3$ and $k_n(T + T') = \min(j_0(T), j'_0(T')) - 1$ for $n \le 2$. We will now check that the admissible sequence of partitions $(\mathcal{C}_n)$ together with the associated numbers $k_n(C)$ witness that $\tilde{b}(T + T') \le Lr(\tilde{b}(T) + \tilde{b}'(T))$. First, for any $t \in T + T'$ we have, using (10.174) in the third line,

$$\begin{aligned} \sum_{n \ge 0} 2^n r^{-k_n(C_n(t))} &\le Lr^{-\min(j_0(T), j'_0(T'))+1} + \sum_{n \ge 3} 2^n r^{-b_{n-2}(B_{n-2}(t))} \\ &\le Lr^{-\min(j_0(T), j'_0(T'))+1} + L \sum_{n \ge 1} 2^n r^{-b_n(B_n(t))} \\ &\le Lr \sum_{n \ge 0} 2^n r^{-\min(j_n(A_n(u(t))), j'_n(A'_n(u'(t))))} \\ &\le Lr(\tilde{b}(T) + \tilde{b}(T')) \ . \end{aligned} \tag{10.177}$$

Next, recalling (10.169), and using that $r \ge 2$ in the last inequality,

$$\begin{aligned} \Delta(T + T', d_2) &\le \Delta(T, d_2) + \Delta(T', d_2) \le r^{-j_0(T)} + r^{-j'_0(T')} \\ &\le 2r^{-\min(j_0(T) + j'_0(T'))} \le r^{-k_0(T+T')} \ , \end{aligned} \tag{10.178}$$

and we have checked that the sequence $(\mathcal{C}_n)$ satisfies (10.169). It remains to check (10.168). We use the inequalities $\varphi_j(s,t) \le r^{2j} d(s,t)^2$ to obtain that for $s, t \in T + T'$ and $n \le 2$, we have

$$\varphi_{k_n(T+T')}(s,t) \le r^{2k_n(T+T')} \Delta(T + T', d_2)^2 \le 1 \le 2^n \ ,$$

since $k_n(T+T') = k_0(T+T')$ and using (10.178). That is, for $n \le 2$ and $s, t \in C \in \mathcal{C}_n$, we have $\varphi_{k_n(C)}(s,t) \le 2^n$. This is also true for $n \ge 3$ because for $C \in \mathcal{C}_n$, we have $C \in \mathcal{B}_{n-2}$ and $k_n(C) = b_{n-2}(C)$ and using (10.176). Thus we have also checked (10.168) and (10.172) follows from (10.177). □

10.15 Lower Bounds from Measures

At this stage we advise the reader to review Sect. 3.3, as the main result of the present section, Theorem 10.15.1, is closely connected to Proposition 3.3.1. Given a set $T \subset \ell^2$ and a probability measure μ on T, we are now going to provide a lower bound for $b(T)$ using μ. This will be very useful later, in the study of certain random series in Chap. 11. We define j_0 to be the largest integer j such that

$$\forall s, t \in T\ ,\ \varphi_j(s,t) \le 1\ . \tag{10.179}$$

Thus we can find $s, t \in T$ with $\varphi_{j_0+1}(s,t) > 1$, and since $\varphi_{j_0+1}(s,t) \le r^{2(j_0+1)} d_2(s,t)^2$, we have $r^{j_0+1}\Delta(T, d_2) > 1$ and

$$r^{-j_0-1} < \Delta(T, d_2)\ . \tag{10.180}$$

Given $t \in T$ we define $\bar{j}_0(t) = j_0$, and for $n \ge 1$ we define

$$\bar{j}_n(t) = \sup\left\{j \in \mathbb{Z}\ ;\ \mu(\{s \in T\ ;\ \varphi_j(t,s) \le 2^n\}) \ge N_n^{-1}\right\}\ , \tag{10.181}$$

so that the sequence $(\bar{j}_n(t))_{n \ge 0}$ increases. We then set

$$I_\mu(t) = \sum_{n \ge 0} 2^n r^{-\bar{j}_n(t)}\ . \tag{10.182}$$

Theorem 10.15.1 *For any probability measure μ on T, we have*

$$\int_T I_\mu(t) \mathrm{d}\mu(t) \le L r^3 b(T)\ . \tag{10.183}$$

According to (10.171) it suffices to prove the following:

Proposition 10.15.2 *Given a probability measure μ on T, we have*

$$\int_T I_\mu(t) \mathrm{d}\mu(t) \le L r \tilde{b}(T)\ . \tag{10.184}$$

Proof Consider an admissible sequence $(\mathcal{A}_n)$ of partitions of T. For $A \in \mathcal{A}_n$ consider an integer $j_n(A)$ as in (10.168) and (10.169), and let $S =$

$\sup_{t\in T}\sum_{n\geq 0} 2^n r^{-j_n(A_n(t))}$. Comparing (10.169) and (10.180) yields $r^{-j_0-1} \leq r^{-j_0(T)} = r^{-j_0(A_0(t))} \leq S$. Consider $n \geq 1$ and $A \in \mathcal{A}_n$. Then for $t \in A$ we have

$$A \subset \left\{s \in T \; ; \; \varphi_{j_n(A)}(s,t) \leq 2^n\right\} \subset \left\{s \in T \; ; \; \varphi_{j_n(A)}(s,t) \leq 2^{n+1}\right\} .$$

Thus, by the definition (10.181) of $\bar{j}_{n+1}(t)$, if $\mu(A) \geq N_{n+1}^{-1}$ then $\bar{j}_{n+1}(t) \geq j_n(A)$ and thus

$$\int_A 2^{n+1} r^{-\bar{j}_{n+1}(t)} \mathrm{d}\mu(t) \leq 2 \int_A 2^n r^{-j_n(A_n(t))} \mathrm{d}\mu(t) . \tag{10.185}$$

On the other hand if $\mu(A) < N_{n+1}^{-1}$ then, since $\bar{j}_{n+1}(t) \geq \bar{j}_0(t) = j_0$,

$$\int_A 2^{n+1} r^{-\bar{j}_{n+1}(t)} \mathrm{d}\mu(t) \leq 2^{n+1} r^{-j_0} N_{n+1}^{-1} .$$

Summation over all $A \in \mathcal{A}_n$ implies (using in the second term that $\operatorname{card} \mathcal{A}_n \leq N_n$ and $N_n N_{n+1}^{-1} \leq N_n^{-1}$)

$$\int_T 2^{n+1} r^{-\bar{j}_{n+1}(t)} \mathrm{d}\mu(t) \leq 2 \int_T 2^n r^{-j_n(A_n(t))} \mathrm{d}\mu(t) + 2^{n+1} r^{-j_0} N_n^{-1} .$$

Summation over $n \geq 0$ implies

$$\int_T \sum_{n\geq 1} 2^n r^{-\bar{j}_n(t)} \mathrm{d}\mu(t) \leq LS + Lr^{-j_0} .$$

Since $2^n r^{-\bar{j}_n(t)} \leq Lr^{-j_0} \leq LrS$ for $n = 0$ or $n = 1$, we proved that $\int_T I_\mu(t)\mathrm{d}\mu(t) \leq LS$, which proves the result by definition of $\tilde{b}(T)$. □

One piece is missing to our understanding of Bernoulli processes. In the case of a metric space (T, d), one knows how to identify simple structures (trees), the presence of which provides a lower bound on $\gamma_2(T, d)$. One then can dream of identifying geometric structures inside a set $T \subset \ell^2$, which would provide lower bounds for $b(T)$ of the correct order. Maybe this dream is impossible to achieve, and not the least remarkable feature of the Latała-Bednorz proof of Theorem 6.2.8 is that it completely bypasses this problem. The following exercise gives an example of such a structure:

Exercise 10.15.3 For $n \geq 1$ let $I_n = \{0, 1\}^n$ and let $I = \bigcup_{n\geq 1} I_n$. For $\sigma \in \{0, 1\}^{\mathbb{N}^*}$ let us denote by $\sigma|n \in I_n$ the restriction of the sequence to its first n terms. Consider a sequence (α_n) with $\alpha_n > 0$ and $\sum_{n\geq 1} \alpha_n = 1$. Consider the set $T \subset \ell^1(I)$ consisting of the elements $x = (x_i)_{i\in I}$ such that there exists $\sigma \in \{0, 1\}^{\mathbb{N}^*}$ for which $x_i = \alpha_n$ if $i \in I_n$ and $i = \sigma|n$ and $x_i = 0$ otherwise. Prove that $\mathsf{E} \sup_{t\in T} \sum_{i\in I} \varepsilon_i t_i \geq 3/4$.

The following direction of investigation is related to Proposition 3.4.1:

Research Problem 10.15.4 Consider a $T \subset \ell^2$. Can we find a probability measure μ on T such that

$$b(T) \le L \inf_{t \in T} I_\mu(t) \ ?$$

Key Ideas to Remember

- The main difficulty in proving the Latała-Bednorz theorem is that we know little about Bernoulli processes unless we control their supremum norm. Such a control is required in particular to use the Sudakov minoration for these processes.
- To control the supremum norm, the main technical tool is chopping maps, which replace the process by a related process with a better control of the supremum norm.
- The strategy to prove the Latała-Bednorz theorem is to recursively construct increasing sequences of partitions of the index set. The size of each element of the partition is controlled by the value of a functional, which however depends on the element of the partition itself.
- Compared with the case of Gaussian processes, a fundamentally new partitioning principle is required, Latała's principle. Applying this partitioning principle requires changing the chopping map.
- The difficulty of the construction is to ensure that no essential information is lost at any stage.
- There is a natural geometric characteristic of a set $T \subset \ell^2$ equivalent to the size of the corresponding Bernoulli process. This geometric property involves the existence of an admissible sequences of partitions of the index set with precise smallness properties with respect to the canonical family of distances.
- The existence of such an admissible sequence of partitions in turn controls how "scattered" a probability measure on the index set can be. This property will be the key to using the Latała-Bednorz theorem.

10.16 Notes and Comments

I worked many years on the Bernoulli conjecture. The best I could prove is that if $p > 1$ for any set $T \subset \ell^2$, we can write $T \subset T_1 + T_2$ with $\gamma_2(T_1) \le K(p)$ and $T_2 \subset K(p)B_p$ where B_p is the unit ball of ℓ^p. This statement has non-trivial applications to Banach Space theory: it is sufficient to prove Theorem 19.1.5, but it is otherwise not very exciting. Nonetheless many results presented in Part II were results of efforts in this general direction.

Chapter 11
Random Series of Functions

As in the case of Chap. 7, the title of the chapter is somewhat misleading: our focus is not on the convergence of series, but on quantitative estimates bearing on sums of random functions.

11.1 Road Map

There are two fundamentally different reasons why a sum of random functions is not too large.[1]

- There is a lot of cancellation between the different terms.
- The sum of the absolute values of the functions is not too large.

One may of course also have mixtures of the previous two situations. Under rather general circumstances, we will prove the very striking fact that there are no other possibilities: *every situation is a mixture of the previous two situations.* Furthermore we will give an *exact quantitative description* of the cancellation, by exhibiting a chaining method which witnesses it.

Let us describe more precisely one of our central results in this direction, which is a vast generalization of Theorem 6.8.3.[2] (*The* central result of this chapter, Theorem 11.10.3, is an abstract version of Theorem 11.1.1 which is conceptually very close.) Consider independent r.v.s $(X_i)_{i\le N}$ valued in a measurable space Ω, and denote by λ_i the distribution of X_i. Set $\nu = \sum_{i\le N} \lambda_i$. We consider a set T of

[1] We have already seen this idea in Sect. 6.8 in the setting of empirical processes.

[2] Not only the proof of this generalization is identical to the proof of Theorem 6.8.3, but the generalization is powerful, as we will experience in Sect. 11.12.

M. Talagrand, *Upper and Lower Bounds for Stochastic Processes*, Ergebnisse der Mathematik und ihrer Grenzgebiete. 3. Folge / A Series of Modern Surveys in Mathematics 60, https://doi.org/10.1007/978-3-030-82595-9_11

functions on Ω, and we denote $s, t, \ldots$ the elements of T. We denote by d_2 and d_∞ the distances on T corresponding to the L^2 and L^∞ norm for ν.

Theorem 11.1.1 (The Decomposition Theorem for Empirical Processes) *There is a decomposition $T \subset T_1 + T_2$ such that*

$$\gamma_2(T_1, d_2) + \gamma_1(T_1, d_\infty) \le L \mathsf{E} \sup_{t \in T} \sum_{i \le N} \varepsilon_i t(X_i) \tag{11.1}$$

and

$$\mathsf{E} \sup_{t \in T_2} \sum_{i \le N} |t(X_i)| \le L \mathsf{E} \sup_{t \in T} \sum_{i \le N} \varepsilon_i t(X_i) . \tag{11.2}$$

To explain why this result fits in the previous conceptual framework, let us lighten notation by setting

$$S(T) = \mathsf{E} \sup_{t \in T} \sum_{i \le N} \varepsilon_i t(X_i) ,$$

and let us observe that when $T \subset T_1 + T_2$, we have $S(T) \le S(T_1) + S(T_2)$. When $T \subset T_1 + T_2$ we may think of T as a mixture of the sets T_1 and T_2, and to control $S(T)$ it suffices to control $S(T_1)$ and $S(T_2)$. There is a very clear reason why for $t \in T_2$ the sums $\sum_{i \le N} \varepsilon_i t(X_i)$ are not too large: it is because already the sums $\sum_{i \le N} |t(X_i)|$ are not too large. This is the content of (11.2).

To understand what happens for T_1, we recall the following fundamental bound:

Lemma 11.1.2 *We have $S(T) \le L(\gamma_2(T, d_2) + \gamma_1(T, d_\infty))$.*

Proof This follows from Bernstein's inequality (4.44) and Theorem 4.5.13 just as in the case of Theorem 4.5.16. □

Thus the information $\gamma_2(T_1, d_2) + \gamma_1(T_1, d_\infty) \le LS(T)$ of (11.1) is exactly what we need to prove that chaining controls $S(T_1)$.

Since this chaining is obtained through Bernstein's inequality, and since no cancellation is needed to explain the size of $S(T_2)$, we may picturesquely formulate Theorem 11.1.1 as *chaining using Bernstein's inequality captures all the cancellation.*

Despite the fact that Theorem 11.1.1 is of obvious theoretical importance, one must keep realistic expectations. The theorem does not contain a practical recipe to find the decomposition. In practical situations, such as those studied in Chap. 14, it is the part without cancellations which is difficult to control, and the cancellations are easily controlled through chaining (just as expected from Theorem 11.1.1).

The reader should review Sect. 6.8 now as well as Chap. 7, at least up to Sect. 7.7.

11.2 Random Series of Functions: General Setting

The setting in which we will work is more general than the setting of Theorem 11.1.1, and we describe it first. We consider an index set T and a random sequence $(Z_i)_{i\geq 1}$ of functions on T. We do not assume that this sequence is independent.[3] Consider an independent Bernoulli sequence $(\varepsilon_i)_{i\geq 1}$, which is independent of the sequence $(Z_i)_{i\geq 1}$. We are interested in the random sums $\sum_{i\geq 1} \varepsilon_i Z_i(t)$. We will measure their "size" by the quantity

$$S := \mathsf{E} \sup_{t\in T} \sum_{i\geq 1} \varepsilon_i Z_i(t) \; . \tag{11.3}$$

The crucial technical feature here is that given the randomness of the Z_i, we are considering a Bernoulli process. The most important case is when the sum in (11.3) is a finite sum (i.e., $Z_i = 0$ for i large enough). In the next chapter we will also consider situations where the sum in (11.3) is infinite, and we will then consider actual series of functions. To make sure that in this case the series (11.3) converges a.s., we assume

$$\forall t \in T \; , \; \sum_{i\geq 1} \mathsf{E}(|Z_i|^2 \wedge 1) < \infty \; , \tag{11.4}$$

a condition which is automatically satisfied when $Z_i = 0$ for i large enough.

Our main technical tool is that to each random sequence of functions, we can naturally associate a natural family of distances, and we explain this now. We fix a number $r \geq 4$ a number $j \in \mathbb{Z}$ and a given realization of the sequence $(Z_i)_{i\geq 1}$. We then consider the quantities

$$\psi_{j,\omega}(s,t) := \sum_{i\geq 1} (|r^j(Z_i(s) - Z_i(t))|^2 \wedge 1) \; . \tag{11.5}$$

In this notation ω symbolizes the randomness of the sequence $(Z_i)_{i\geq 1}$. We also define (please compare to (7.63))

$$\varphi_j(s,t) := \mathsf{E}\psi_{j,\omega}(s,t) = \sum_{i\geq 1} \mathsf{E}(|r^j(Z_i(s) - Z_i(t))|^2 \wedge 1) \; , \tag{11.6}$$

which is finite for each j, s, t. This is the "family of distances" we will use to control the size of T.

[3] Rather, we assume condition (11.8), which, as Lemma 11.2.1 shows, holds when the sequence (Z_i) is independent, but also in other cases which will be essential in Chap. 12.

It follows from Lebesgue's convergence theorem and (11.4) that

$$\forall s, t \in T, \lim_{j \to -\infty} \varphi_j(s,t) = 0 . \tag{11.7}$$

We will make the following (essential) additional hypothesis:

$$\forall j \in \mathbb{Z}\,,\ \forall s, t \in T\,,\ \mathsf{P}\big(\psi_{j,\omega}(s,t) \le \varphi_j(s,t)/4\big) \le \exp(-\varphi_j(s,t)/4)\,. \tag{11.8}$$

Lemma 11.2.1 *The condition (11.8) is satisfied when the r.v.s Z_i are independent.*

Proof This follows from Lemma 7.7.2(a), used for $W_i = |r^j(Z_i(s) - Z_i(t))|^2 \wedge 1$ and $A = \varphi_j(s,t)/4 = (1/4)\sum_{i\ge 1} \mathsf{E} W_i$. □

In Chap. 12 we will however meet slightly different situations where (11.8) is satisfied. Our main result will imply that under the previous conditions, an upper bound on S implies the existence of an admissible sequence of partitions of T whose size is suitably controlled by the φ_j.

11.3 Organization of the Chapter

The key result of the chapter, Theorem 11.7.1, states that a control of $S = \mathsf{E}\sup_{t\in T}\sum_{i\ge 1}\varepsilon_i Z_i(t)$ from above implies a kind of smallness of the index space T, in the form of the existence of a family of admissible partitions which is suitably small with respect to the family of distances (φ_j). It can be viewed as a lower bound for S. It can be seen as a generalization of Theorem 7.5.1 (or more precisely of (7.72)) to the case where one no longer has translation invariance.[4] It can also be seen as a generalization of Theorem 5.2.1. The main motivation of the author for presenting separately the results of Chap. 5 is actually to prepare the reader for the scheme of proof of Theorem 11.7.1, so the reader should review that chapter.

As in the case of Theorem 5.2.1, and in contrast with the situation of Chap. 7, the lower bound of Theorem 11.7.1 is by no means an upper bound. One should however refrain from the conclusion that this lower bound is "weak". In a precise sense, it is optimal. In the setting of Theorem 11.1.1, it contains the exact information needed to obtain (11.1) so that one could say that it contains the exact information needed to perform chaining witnessing whatever cancellation occurs between the terms of the random sum $\sum_{i\ge 1}\varepsilon_i Z_i$. What makes the situation subtle is that it may very well happen that such cancellation does not really play a role in the size of the random sum, that is, the sum $\sum_{i\ge 1}\varepsilon_i Z_i$ is not large simply because the larger sum $\sum_{i\ge 1}|Z_i|$ is not large. In this case the lower bound of Theorem 11.7.1 brings no

[4] So sequences of admissible partitions replace the "entropy numbers" implicitly used in (7.72) as explained in Exercise 7.5.4.

information, but we have another precious piece of information, namely, that the sum $\sum_{i\geq 1}|Z_i|$ is not large.

The main tool of the proof of Theorem 11.7.1 is the Latała-Bednorz theorem (Theorem 6.2.8). There is no question that this proof is rather difficult, so the reader may like, after understanding the statement of the theorem, to study as a motivation Sect. 11.8, where one learns to gain control of sums such as $\sum_{i\geq 1}|Z_i|$ and Sect. 11.9 where one proves Theorem 11.1.1.

The next three sections each contain a step of the proof of Theorem 11.7.1. Each of these steps corresponds quite closely to a step in the proof of Theorem 5.2.1, and this should help the reader to perceive the structure of the proof. In Sect. 11.10 we prove a decomposition theorem for random series which extends Theorem 11.1.1, and in the final section , we provide a spectacular application.

11.4 The Main Lemma

The reader should review the proof of Lemma 5.4.2 and of (5.17). We consider a random series of functions as in the previous section. We assume that T is finite and we keep (11.7) in mind. We define[5]

$$j_0 = \sup\left\{j \in \mathbb{Z}\,;\ \forall s,t \in T, \varphi_j(s,t) \leq 4\right\} \in \mathbb{Z} \cup \{\infty\}\,. \tag{11.9}$$

We denote by $\mathcal{M}^+$ the set of probability measures on T such that $\mu(\{t\}) > 0$ for each t in T. Given a $\mu \in \mathcal{M}^+$ and $t \in T$, we define

$$\bar{j}_0(t) = j_0 \tag{11.10}$$

and for $n \geq 1$ we define

$$\bar{j}_n(t) = \sup\left\{j \in \mathbb{Z}\,;\ \mu(B_j(t,2^n)) \geq N_n^{-1}\right\} \in \mathbb{Z} \cup \{\infty\}\,, \tag{11.11}$$

where as usual $B_j(t,2^n) = \{s \in T; \varphi_j(s,t) \leq 2^n\}$. This should be compared to (7.70). Thus the sequence $(\bar{j}_n(t))_{n\geq 0}$ increases, and $\bar{j}_n(t) = \infty$ whenever $\mu(\{t\}) \geq N_n^{-1}$, and in particular for n large enough. We define[6]

$$J_\mu(t) := \sum_{n\geq 0} 2^n r^{-\bar{j}_n(t)}\,. \tag{11.12}$$

Our goal in this section is to prove the following, where we recall (11.3):

[5] It may well happen that $\varphi_j(s,t) \leq 4$ for all j, for example, if $Z_i = 0$ for $i \geq 5$.

[6] The point of assuming $\mu(\{t\}) > 0$ is to ensure that $J_\mu(t) < \infty$.

Lemma 11.4.1 *For each probability measure $\mu \in \mathcal{M}^+$, we have*

$$\int J_\mu(t)\mathrm{d}\mu(t) \le KS\,. \tag{11.13}$$

Here and below, K depends on r only. To prove Lemma 11.4.1, we define

$$j_{0,\omega} = \sup\left\{j \in \mathbb{Z}\,;\ \forall s,t \in T\,,\ \psi_{j,\omega}(s,t) \le 1\right\} \in \mathbb{Z} \cup \{\infty\}\,,$$

and we set $j_{0,\omega}(t) = j_{0,\omega}$. For $n \ge 1$ we define

$$j_{n,\omega}(t) = \sup\left\{j \in \mathbb{Z}\,;\ \mu\left(\left\{s \in T\,;\ \psi_{j,\omega}(t,s) \le 2^n\right\}\right) \ge N_n^{-1}\right\} \in \mathbb{Z} \cup \{\infty\}\,.$$

Then the sequence $(j_{n,\omega}(t))_{n\ge 0}$ increases. We define

$$I_{\mu,\omega}(t) := \sum_{n\ge 0} 2^n r^{-j_{n,\omega}(t)}\,. \tag{11.14}$$

Given ω (i.e., given the sequence $(Z_i)_{i\ge 1}$) the process $X_t := \sum_{i\ge 1}\varepsilon_i Z_i(t)$ is a Bernoulli process, so that using Theorem 10.15.1 at a given ω, and denoting by E_ε expectation in the ε_i only, we obtain

$$\int I_{\mu,\omega}(t)\mathrm{d}\mu(t) \le K\mathsf{E}_\varepsilon \sup_{t\in T}\sum_{i\ge 1}\varepsilon_i Z_i(t)$$

and taking expectation yields

$$\mathsf{E}\int I_{\mu,\omega}(t)\mathrm{d}\mu(t) \le KS\,. \tag{11.15}$$

Lemma 11.4.2 *For each t we have $J_\mu(t) \le L\mathsf{E}I_{\mu,\omega}(t)$.*

Proof The proof is very similar to the proof of Lemma 7.7.3 (but is easier as we prove a weaker result). We will prove that

$$\mathsf{P}(j_{0,\omega}(t) \le \bar{j}_0(t)) = \mathsf{P}(j_{0,\omega} \le j_0) \ge 1/L\,, \tag{11.16}$$

$$n \ge 3 \Rightarrow \mathsf{P}(j_{n-3,\omega}(t) \le \bar{j}_n(t)) \ge 1/2\,. \tag{11.17}$$

These relations imply respectively that $\mathsf{E}r^{-j_{0,\omega}(t)} \ge r^{-j_0}/L$ and that

$$n \ge 3 \Rightarrow \mathsf{E}2^{n-3}r^{-j_{n-3,\omega}(t)} \ge 2^n r^{-\bar{j}_n(t)}/L\,.$$

Summing these relations implies $\sum_{n\geq 3} 2^n r^{-\bar{j}_n(t)} \leq L\mathsf{E}\sum_{n\geq 0} 2^n r^{-j_{n,\omega}(t)} = L\mathsf{E}I_{\mu,\omega}$. Since the sequence $(\bar{j}_n(t))$ increases, for $n \leq 2$ we have $2^n r^{-\bar{j}_n(t)} \leq 4r^{-j_0} \leq 4\mathsf{E}r^{-j_{0,\omega}} \leq 4\mathsf{E}I_{\mu,\omega}$, and this completes the proof.

We first prove (11.16). There is nothing to prove if $j_0 = \infty$. Otherwise, by definition of j_0, there exist $s, t \in T$ with $\varphi_{j_0+1}(s,t) > 4$, and by (11.8) we have

$$\mathsf{P}(\psi_{j_0+1,\omega}(s,t) > 1) \geq 1 - 1/e\ .$$

When $\psi_{j_0+1,\omega}(s,t) > 1$, by definition of $j_{0,\omega}$, we have $j_{0,\omega} \leq j_0$ and we have proved that $\mathsf{P}(j_{0,\omega} \leq j_0) \geq 1 - 1/e$.

We turn to the proof of (11.17). There is nothing to prove if $\bar{j}_n(t) = \infty$. Otherwise, by definition of $\bar{j}_n(t)$ we have

$$\mu_1 := \mu(\{s \in T\ ;\ \varphi_{\bar{j}_n(t)+1}(t,s) \leq 2^n\}) \leq \frac{1}{N_n}\ . \tag{11.18}$$

For $n \geq 2$ it follows from (11.8) that

$$\varphi_{\bar{j}_n(t)+1}(s,t) \geq 2^n \Rightarrow \mathsf{P}(\psi_{\bar{j}_n(t)+1,\omega}(s,t) \leq 2^{n-2}) \leq \exp(-2^{n-2}) \leq N_{n-2}^{-1}\ ,$$

so that

$$\mathsf{E}\mu\big(\{s \in T\ ;\ \varphi_{\bar{j}_n(t)+1}(s,t) \geq 2^n\ ,\ \psi_{\bar{j}_n(t)+1,\omega}(s,t) \leq 2^{n-2}\}\big) \leq N_{n-2}^{-1}$$

and thus, by Markov's inequality, with probability $\geq 1/2$ we have

$$\mu_2 := \mu\big(\{s \in T\ ;\ \varphi_{\bar{j}_n(t)+1}(s,t) \geq 2^n\ ,\ \psi_{\bar{j}_n(t)+1,\omega}(s,t) \leq 2^{n-2}\}\big) \leq 2N_{n-2}^{-1}\ .$$

When this occurs, recalling (11.18) we obtain

$$\mu(\{s \in T\ ;\ \psi_{\bar{j}_n(t)+1,\omega}(s,t) \geq 2^{n-2}\}) \leq \mu_1 + \mu_2 \leq N_n^{-1} + 2N_{n-2}^{-1} < N_{n-3}^{-1}\ ,$$

so that $j_{n-3,\omega}(t) \leq \bar{j}_n(t)$. □

Proof of Lemma 11.4.1 Combine the previous lemma with (11.15). □

11.5 Construction of the Majorizing Measure Using Convexity

The reader should review Sect. 3.3.2. The goal of this section is to prove the following:

Theorem 11.5.1 *Assume that T is finite. Then there exists a probability measure μ on T with*

$$\sup_{t \in T} J_\mu(t) \le KS\,. \tag{11.19}$$

For historical reasons (which have been briefly explained in Chap. 3), we call the measure μ a majorizing measure, and this explains the title of the section. It will then be a simple technical task to use this measure for a fruitful application in Theorem 11.6.3.

It is interesting to compare (11.19) with (7.71), which basically asserts that

$$\mathsf{P}\Big(\Big\|\sum_i \varepsilon_i Z_i\Big\| \ge \frac{1}{K}\sup_{t \in T} J_\mu(t)\Big) \ge \alpha_0\,,$$

where μ is the Haar probability on T. The previous inequality implies that $\mathsf{E}\|\sum_i \varepsilon_i Z_i\| \ge \sup_{t\in T} J_\mu(t)/K$, and Theorem 11.5.1 can be seen as a generalization of this fact. To clarify the relationship further, it can be shown in the translation invariant setting that for any probability measure ν one has $\sup_{t\in T} J_\mu(t) \le L\sup_{t\in T} J_\nu(t)$ where μ is the Haar measure (which we may assume here to be a probability). Thus, when one probability measure satisfies (11.19), then the Haar measure also satisfies it.

Our approach to Theorem 11.5.1 is to combine Lemma 11.4.1 with 3.3.2.

Corollary 11.5.2 *Assume that T is finite. Then there exist an integer M, probability measures $(\mu_i)_{i\le M}$ and numbers $(\alpha_i)_{i\le M}$ with $\alpha_i \ge 0$ and $\sum_{i\le M}\alpha_i = 1$ such that*

$$\forall t \in T\,,\ \sum_{i\le M} \alpha_i J_{\mu_i}(t) \le KS\,.$$

Proof Consider the set $\mathcal{S}$ of functions of the type $f(t) = J_\mu(t)$ where $\mu \in \mathcal{M}^+$.[7] Consider a probability measure ν on T and $\mu \in \mathcal{M}^+$ with $\mu \ge \nu/2$ (e.g., $\mu = \nu/2 + \lambda/2$ where λ is uniform over T). Then $\int J_\mu(t)\mathrm{d}\nu(t) \le 2\int J_\mu(t)\mathrm{d}\mu(t) \le KS$, using (11.13) in the last inequality. Thus by Lemma 3.3.2 there is a convex combination of functions of $\mathcal{S}$ which is $\le KS$. □

Theorem 11.5.1 is then a consequence of Corollary 11.5.2 and the next result.

Lemma 11.5.3 *Assume that T is finite and consider probability measures $(\mu_i)_{i\le M}$ on T and numbers $(\alpha_i)_{i\le M}$ with $\alpha_i \ge 0$ and $\sum_{i\le M}\alpha_i = 1$. Then the probability*

[7] The reason for which we require $\mu \in \mathcal{M}^+$ is to ensure that $f(t) < \infty$ for each t so that f is true function, as is required by our version of Lemma 3.3.2.

measure $\mu := \sum_{i \le M} \alpha_i \mu_i$ *satisfies*

$$\forall t \in T\ ,\ J_\mu(t) \le L \sum_{i \le M} \alpha_i J_{\mu_i}(t)\ . \tag{11.20}$$

Proof Let us fix $t \in T$. With obvious notation, for $n \ge 1$ let us define

$$U_n := \Big\{ i \le M\ ;\ r^{-\bar{j}_{n,i}(t)} \le 2 \sum_{s \le M} \alpha_j r^{-\bar{j}_{n,s}(t)} \Big\}\ .$$

Consider the probability measure P on $\{1, \ldots, M\}$ such that $\mathsf{P}(\{i\}) = \alpha_i$ and the function f on $\{1, \ldots, M\}$ given by $f(i) = r^{-\bar{j}_{n,i}(t)}$. By Markov's inequality we have $\mathsf{P}(f \ge 2\int f \mathrm{d}P) \le 1/2$, so that $\mathsf{P}(f \le 2\int f \mathrm{d}P) \ge 1/2$, i.e., $\sum_{i \in U_n} \alpha_i \ge 1/2$. For $n \ge 1$, let us denote by j_n the smallest integer with $r^{-j_n} \le 2 \sum_{i \le M} \alpha_i r^{-\bar{j}_{n,i}(t)}$. Then $\bar{j}_{n,i}(t) \ge j_n$ for $i \in U_n$. Thus by definition of $\bar{j}_{n,i}(t)$ we have $\mu_i(B_{j_n}(t, 2^n)) \ge N_n^{-1}$ and consequently

$$\mu(B_{j_n}(t, 2^n)) \ge N_n^{-1} \sum_{i \in U_n} \alpha_i \ge \frac{1}{2N_n} \ge \frac{1}{N_{n+1}}\ ,$$

so that by definition $\bar{j}_{n+1}(t) \ge j_n$. Thus (using also that $\bar{j}_{0,i}(t) = j_0 = \bar{j}_0(t)$ where j_0 is given by (11.9))

$$\sum_{n \ge 0} 2^n r^{-\bar{j}_n(t)} \le L \sum_{n \ge 0} 2^n r^{-j_n} \le L \sum_{i \le M} \alpha_i \sum_{n \ge 0} 2^n r^{-\bar{j}_{n,i}(t)} = L \sum_{i \le M} \alpha_i J_{\mu_i}(t)\ . \quad \square$$

11.6 From Majorizing Measures to Partitions

In the setting of one single distance, we had the inequality (3.41) to go from majorizing measures to sequences of partitions. We do not know how to generalize the proof of (3.41) to the setting of a family of distances, and we give a direct argument to pass directly from the existence of a majorizing measure to the existence of an appropriate increasing sequence of partitions. The very same argument was given in the case of one single distance in Sect. 3.3.3. Let us consider again the functions φ_j as in (11.6), so they satisfy

$$\varphi_j : T \times T \to \mathbb{R}^+\ ,\ \varphi_{j+1} \ge \varphi_j \ge 0\ ,\ \varphi_j(s,t) = \varphi_j(t,s)\ .$$

Since the φ_j are squares of distances, they also satisfy the properties

$$\forall s, t, u \in T\ ,\ \varphi_j(s,t) \le 2(\varphi_j(s,u) + \varphi_j(u,t))\ , \tag{11.21}$$

and $\varphi_j(t,t) = 0$. We recall the notation $B_j(t,r) = \{s \in T; \varphi_j(t,s) \le r\}$. As a consequence of (11.21) we have the following:

Lemma 11.6.1 *If $\varphi_j(s,t) > 4a > 0$ the balls $B_j(s,a)$ and $B_j(t,a)$ are disjoint.*

We assume that T is finite and we fix a probability measure μ on T.[8] We assume that there is a $j_0 \in \mathbb{Z}$ with

$$\forall s,t \in T\ ,\ \varphi_{j_0}(s,t) \le 4\ . \tag{11.22}$$

We assume that for $t \in T$ and $n \ge 0$, we are given an integer $j_n(t) \in \mathbb{Z}$ with the following properties:

$$\forall t \in T\ ,\ j_0(t) = j_0\ , \tag{11.23}$$

$$\forall t \in T\ ,\ \forall n \ge 0\ ,\ j_n(t) \le j_{n+1}(t) \le j_n(t) + 1\ , \tag{11.24}$$

$$\forall t \in T\ ,\ \forall n \ge 1\ ,\ \mu(B_{j_n(t)}(t, 2^n)) \ge N_n^{-1}\ . \tag{11.25}$$

Let us observe that we do require that $j_n(t)$ is the largest possible which would satisfy (11.25). Rather, we require the technical condition (11.24). To understand that important point, we urge the reader to study the following exercise:

Exercise 11.6.2 Assume that T is a group, and μ is the Haar measure, a probability. Assume $r \ge 4$. Assume that j_0 satisfies (11.22) and for $n \ge 1$ consider the numbers j_n defined as in (7.70). Prove that for $t \in T$ there exist numbers $j_n(t)$ satisfying (11.23) to (11.25) and for which $\sup_{t\in T} \sum_{n\ge 0} 2^n r^{-j_n(t)} \le L \sum_{n \ge 0} 2^n r^{-j_n}$. Hint: $j_n(t) = \sup_{p \le n} n - p + j_p$.

Theorem 11.6.3 *Under the previous conditions there exists an admissible sequence of partitions $(\mathcal{A}_n)_{n\ge 0}$ of T and for $A \in \mathcal{A}_n$ an integer $j_n(A) \in \mathbb{Z}$ such that*

$$s,t \in A \in \mathcal{A}_n \Rightarrow \varphi_{j_n(A)}(s,t) \le 2^{n+2} \tag{11.26}$$

and that the following holds for any value of $r \ge 1$:

$$\forall t \in T\ ,\ \sum_{n\ge 0} 2^n r^{-j_n(A_n(t))} \le L \sum_{n\ge 0} 2^n r^{-j_n(t)}\ . \tag{11.27}$$

In particular if $S^* := \sup_{t\in T} \sum_{n\ge 0} 2^n r^{-j_n(t)}$ (11.27) implies

$$\sup_{t\in T} \sum_{n\ge 0} 2^n r^{-j_n(A_n(t))} \le LS^*\ .$$

[8] For reasons explained in Chap. 3, we think of μ as a "majorizing measure". This explains the title of this section.

We note that (11.27) does not require a specific relation between the value of r and the φ_j, but this result will be interesting only when the relation (11.25) brings a precise control of S^*.

The basic brick of the construction is the following partitioning lemma, in the spirit of (2.46):

Lemma 11.6.4 *Consider a set $A \subset T$, an integer $j \in \mathbb{Z}$, and $n \geq 1$. Assume that $\mu(B_j(t, 2^n)) \geq 1/N_n$ for each $t \in A$. Then we can find a partition $\mathcal{A}$ of A with $\operatorname{card} \mathcal{A} \leq N_n$ such that for each $B \in \mathcal{A}$,*

$$s, t \in B \Rightarrow \varphi_j(s, t) \leq 2^{n+4} . \tag{11.28}$$

Proof Consider a subset U of A such that $\varphi_j(s, t) > 2^{n+2}$ for each $s, t \in U$, $s \neq t$. According to Lemma 11.6.1, the balls $B_j(t, 2^n)$ for $t \in U$ are disjoint. These balls are of measure $\geq N_n^{-1}$, so their union has measure $\geq N_n^{-1} \operatorname{card} U$. Since the measure is a probability, this proves that $\operatorname{card} U \leq N_n$. If $\operatorname{card} U$ is taken as large as possible the balls $B_j(t, 2^{n+2})$ centered at the points of U cover A. It follows from (11.21) that these balls satisfy (11.28). And A can be partitioned in at most N_n pieces, each of which is contained in a ball $B_j(t, 2^{n+2})$ where $t \in U$. □

Proof of Theorem 11.6.3 We are going to construct the partitions $\mathcal{A}_n$ and for $A \in \mathcal{A}_n$ integers $j_n(A)$. Our construction will satisfy the following property: For $n \geq 2$, $A \in \mathcal{A}_n$, the integer $j_n(A)$ is such that

$$t \in A \Rightarrow j_{n-2}(t) = j_n(A) . \tag{11.29}$$

To start the construction, we set $\mathcal{A}_0 = \mathcal{A}_1 = \mathcal{A}_2 = \{T\}$, and for $n \leq 2$ and $A \in \mathcal{A}_n$ we set $j_n(A) = j_0$.[9] According to (11.23), for $t \in T$ we have $j_0(t) = j_0$ so that (11.29) holds for $n = 2$ because then $j_0(t) = j_0 = j_2(A)$.

The rest of the construction proceeds by recursion. Having constructed $\mathcal{A}_n$ for some $n \geq 2$ we proceed as follows. According to (11.29) we have $j_{n-2}(t) = j_n(A)$ for $t \in A$, so that according to (11.24) for $t \in A$ we have $j_{n-1}(t) \in \{j_n(A), j_n(A)+1\}$. We set

$$A_0 = \{t \in A \,;\, j_{n-1}(t) = j_n(A)\} \,;\; A_1 = \{t \in A \,;\, j_{n-1}(t) = j_n(A) + 1\} .$$

Recalling (11.25), we can then apply Lemma 11.6.4 with $n - 1$ rather than n and $j = j_n(A) + 1$ to partition the set A_1 into N_{n-1} pieces. According to (11.28), for each piece B of A_1 thus created, we have

$$s, t \in B \Rightarrow \varphi_j(s, t) \leq 2^{n+3} . \tag{11.30}$$

[9] Since the only element of $\mathcal{A}_n$ is T, this means that $j_n(T) = j_0$ for $n \leq 2$.

For each piece B of A_1 thus created, we set $j_{n+1}(B) = j_n(A) + 1$, so that this piece satisfies (11.26) (for $n + 1$ rather than n). For a piece B contained in A_1 we set $j_{n+1}(B) = j_n(A) + 1$. We do not partition A_0, and we set $j_{n+1}(A_0) = j_n(A_0)$. In this manner, we have partitioned A into at most $N_{n-1} + 1 \le N_n$ pieces. We apply this procedure to each element A of $\mathcal{A}_n$ to obtain $\mathcal{A}_{n+1}$. Then card $\mathcal{A}_{n+1} \le N_n^2 = N_{n+1}$. Condition (11.29) is satisfied for $n + 1$ by construction, and so is (11.26) (see (11.30)). As for (11.27), it follows from the fact that $j_n(A_n(t)) = j_0$ if $n \le 2$ and $j_n(A_n(t)) = j_{n-2}(t)$ if $n \ge 2$ (as follows from (11.29)). □

11.7 The General Lower Bound

Still in the setting of Sect. 11.2, we have the following, where T is now countable rather than finite:

Theorem 11.7.1 *Assume that T is countable. Then there exists an admissible sequence $(\mathcal{A}_n)$ of partitions of T and for $A \in \mathcal{A}_n$ an integer $j_n(A)$ such that the following holds:*

$$\forall t \in T\ , \ \sum_{n\ge 0} 2^n r^{-j_n(A_n(t))} \le KS\ , \tag{11.31}$$

$$A \in \mathcal{A}_n\ ,\ C \in \mathcal{A}_{n-1}\ ,\ A \subset C \Rightarrow j_{n-1}(C) \le j_n(A) \le j_{n-1}(C) + 1\ , \tag{11.32}$$

$$s, t \in A \in \mathcal{A}_n \Rightarrow \varphi_{j_n(A)}(s,t) \le 2^{n+2}\ . \tag{11.33}$$

Proof Assume first that T is finite. Consider the probability measure μ on T provided by Theorem 11.5.1 and the corresponding numbers $\bar{j}_n(t)$. Let us define $j_n(t) = \min_{0\le p\le n}(\bar{j}_p(t) + n - p)$, so that $j_0(t) = \bar{j}_0(t) = j_0$ and $j_n(t) \le j_{n+1}(t) \le j_n(t) + 1$. Since $j_n(t) \le \bar{j}_n(t)$, by definition (11.11) of $\bar{j}_n(t)$ we have $\mu(B_{j_n(t)}(t, 2^n)) \ge N_n^{-1}$. Also, $r^{-j_n(t)} \le \sum_{p\le n} r^{-\bar{j}_p(t)-n+p}$ so that, since $r \ge 4$,

$$\sum_{n\ge 0} 2^n r^{-j_n(t)} \le \sum_{n\ge 0} 2^n \sum_{p\le n} r^{-\bar{j}_p(t)-n+p} = \sum_{p\ge 0} 2^p r^{-\bar{j}_p(t)} \sum_{n\ge p}\Big(\frac{2}{r}\Big)^{n-p} \le 2J_\mu(t)\ .$$

The result then follows from Theorem 11.6.3.

Assume next that T is countable. We write T as the union of an increasing sequence $(T_k)_{k\ge 1}$ of finite sets. We apply the previous result to each T_k, obtaining an admissible sequence $(\mathcal{A}_{n,k})$ of partitions of T_k. We number in an arbitrary way the sets of $\mathcal{A}_{n,k}$ as $(A_{n,k,\ell})_{\ell\le N_n}$, and for $t \in T_k$ we denote by $\ell_{n,k}(t)$ the unique integer $\ell \le N_n$ such that $A_{n,k}(t) = A_{n,k,\ell}$. For each $t \in T$ and each $n \ge 0$, the integer $\ell_{n,k}(t)$ is defined for k large enough since then $t \in T_k$. We may then assume by taking a subsequence that this integer is eventually equal to a number $\ell_n(t)$. Also, since the sequence $(\mathcal{A}_{n,k})$ is admissible, for each $\ell \le N_n$, there exists $\bar{\ell} \le N_{n-1}$ such that $A_{n,k,\ell} \subset A_{n-1,k,\bar{\ell}}$. We denote this integer by $\bar{\ell}_{n,k}(\ell)$. We may also assume by taking a subsequence that $\bar{\ell}_{n,k}(\ell)$ is eventually equal to an integer $\bar{\ell}_n(\ell)$. For $\ell \le N_n$ define $A_{n,\ell} = \{t \in T; \ell_n(t) = \ell\}$. Obviously the sets $A_{n,\ell}$ for $\ell \le N_n$ define a partition $\mathcal{A}_n$ of T and card $\mathcal{A}_n \le N_n$. Next for $t \in A_{n,\ell}$, for large k we have $t \in A_{n,k,\ell} \subset A_{n-1,k,\bar{\ell}_{n,k}(\ell)} = A_{n,\bar{\ell}_n(\ell)}$. This proves that $A_{n,\ell} \subset A_{n-1,\bar{\ell}_n(\ell)}$, so that the sequence $(\mathcal{A}_n)$ is admissible. For $s, t \in T_k$ we have $\varphi_{j_0(T_k)}(s,t) \le 4$. If the sequence $(j_0(T_k))$ is not bounded, then for all $s, t \in T$ and each $j \in \mathbb{Z}$, we have $\varphi_j(s,t) \le 4$, and the result is trivial since (11.33) is automatically satisfied. Thus we may assume that the sequence $(j_0(T_k))_{k\ge1}$ stays bounded. Since $j_n(A_{n,k,\ell}) \le j_0(T_k) + n$ by (11.32), each sequence $(j_n(A_{n,k,\ell}))_{k\ge1}$ stays bounded. We may then assume that for each n and ℓ, this sequence is eventually equal to a number $j_n(A_{n,\ell})$. It is straightforward to check that these numbers satisfy the desired requirements. □

The hypothesis that T is countable is of course largely irrelevant, as the following exercise shows:

Exercise 11.7.2 Extend Theorem 11.7.1 to the case where T is provided with a metric such that T is separable[10] and each function $\varphi_j(s,t)$ is continuous for this metric.

11.8 The Giné-Zinn Inequalities

Before we proceed we must build our tool kit. In this section we change topic, and we investigate a number of simple but fundamental inequalities. The main inequality, (11.37), allows to gain a control of $\sup_{t\in T} \sum_{i\le N} |t(X_i)|$.

We consider a class T of functions on a measurable space Ω. The elements of T will be denoted $s, t, \ldots$. We consider independent r.v.s $(X_i)_{i\le N}$ valued in Ω.

We denote by $(\varepsilon_i)_{i\le N}$ an independent sequence of Bernoulli r.v.s, independent of the sequence $(X_i)_{i\le N}$. We lighten notation by writing

$$\bar{S}(T) = \mathsf{E} \sup_{t\in T} \Big| \sum_{i\le N} \varepsilon_i t(X_i) \Big| . \tag{11.34}$$

The reader should not be disturbed by the fact that there are absolute values in (11.34) but not in (11.3). This is a purely technical matter, and in the key situations below, the absolute values do not matter by Lemma 2.2.1.

[10] That is, there is countable subset of T which is dense in T.

Lemma 11.8.1 *We have*

$$\mathsf{E}\sup_{t\in T}\Big|\sum_{i\le N}(t(X_i)-\mathsf{E}t(X_i))\Big|\le 2\bar{S}(T)\,. \tag{11.35}$$

Proof Consider an independent copy $(X'_i)_{i\le N}$ of the sequence $(X_i)_{i\le N}$. Then, using Jensen's inequality (i.e., taking expectation in the randomness of the X'_i inside the supremum and the absolute value on the left and outside these on the right)

$$\mathsf{E}\sup_{t\in T}\Big|\sum_{i\le N}(t(X_i)-\mathsf{E}t(X_i))\Big|\le \mathsf{E}\sup_{t\in T}\Big|\sum_{i\le N}(t(X_i)-t(X'_i))\Big|\,.$$

Now, the processes $(t(X_i)-t(X'_i))_{i\le N}$ and $(\varepsilon_i(t(X_i)-t(X'_i)))_{i\le N}$ have the same distribution so that

$$\mathsf{E}\sup_{t\in T}\Big|\sum_{i\le N}(t(X_i)-t(X'_i))\Big|=\mathsf{E}\sup_{t\in T}\Big|\sum_{i\le N}\varepsilon_i(t(X_i)-t(X'_i))\Big|\,.$$

The conclusion then follows from the triangle inequality. □

Lemma 11.8.2 *We have*

$$\mathsf{E}\sup_{t\in T}\Big|\sum_{i\le N}\varepsilon_i|t(X_i)|\Big|\le 2\bar{S}(T)\,. \tag{11.36}$$

Proof Use Corollary 6.5.2 at a given value of the $(X_i)_{i\le N}$. □

Theorem 11.8.3 (The Giné-Zinn Theorem [35]) *We have*

$$\mathsf{E}\sup_{t\in T}\sum_{i\le N}|t(X_i)|\le \sup_{t\in T}\sum_{i\le N}\mathsf{E}|t(X_i)|+4\bar{S}(T)\,. \tag{11.37}$$

To better understand this bound, observe that by Jensen's inequality, we have $\sup_{t\in T}\sum_{i\le N}\mathsf{E}|t(X_i)|\le\mathsf{E}\sup_{t\in T}\sum_{i\le N}|t(X_i)|$. The Giné-Zinn theorem is a kind of converse of this simple inequality. Once we control $\bar{S}(T)$ (a sum which involves cancellations) and $\sup_{t\in T}\sum_{i\le N}\mathsf{E}|t(X_i)|$, we control the left-hand side of (11.37), a sum which does not involve cancellations (as all the terms are of the same sign).

Proof We have

$$\sum_{i\le N}|t(X_i)|\le\sum_{i\le N}\mathsf{E}|t(X_i)|+\Big|\sum_{i\le N}\big(|t(X_i)|-\mathsf{E}|t(X_i)|\big)\Big|\,,$$

so that

$$\mathsf{E}\sup_{t\in T}\sum_{i\le N}|t(X_i)|\le\sup_{t\in T}\sum_{i\le N}\mathsf{E}|t(X_i)|+\mathsf{E}\sup_{t\in T}\Big|\sum_{i\le N}\big(|t(X_i)|-\mathsf{E}|t(X_i)|\big)\Big|\,.$$

The first term to the right is the same as the first term to the right of (11.37). Applying Lemma 11.8.1 to the second term to the right (and replacing T by the class $\{|t|; t \in T\}$) and then using (11.36) concludes the proof. □

The following is also useful:

Lemma 11.8.4 *If* $\mathsf{E}t(X_i) = 0$ *for each* $t \in T$ *and each* $i \le N$ *then*

$$\bar{S}(T) \le 2\mathsf{E}\sup_{t\in T}\Big|\sum_{i\le N} t(X_i)\Big| . \tag{11.38}$$

Proof We work conditionally on the sequence $(\varepsilon_i)_{i\le N}$. Setting $I = \{i \le N; \varepsilon_i = 1\}$ and $J = \{i \le N; \varepsilon_i = -1\}$, we obtain

$$\mathsf{E}\sup_{t\in T}\Big|\sum_{i\le N}\varepsilon_i t(X_i)\Big| \le \mathsf{E}\sup_{t\in T}\Big|\sum_{i\in I} t(X_i)\Big| + \mathsf{E}\sup_{t\in T}\Big|\sum_{i\in J} t(X_i)\Big| .$$

Now, since $\mathsf{E}t(X_i) = 0$ for each $t \in T$ and each $i \le N$, denoting E_J expectation in the r.v.s X_i for $i \in J$, we have $\mathsf{E}_J X_i = X_i$ if $i \in I$ and $\mathsf{E}_J X_i = 0$ if $i \notin I$ so that Jensen's inequality implies

$$\sup_{t\in T}\Big|\sum_{i\in I} t(X_i)\Big| = \sup_{t\in T}\Big|\mathsf{E}_J\sum_{i\le N} t(X_i)\Big| \le \mathsf{E}_J\sup_{t\in T}\Big|\sum_{i\le N} t(X_i)\Big| ,$$

so that taking expectations,

$$\mathsf{E}\sup_{t\in T}\Big|\sum_{i\in I} t(X_i)\Big| \le \mathsf{E}\sup_{t\in T}\Big|\sum_{i\le N} t(X_i)\Big| .$$

□

11.9 Proof of the Decomposition Theorem for Empirical Processes

Consider independent r.v.s $(X_i)_{i\le N}$ valued in a measurable space Ω, and denote by λ_i the distribution of X_i. Set $\nu = \sum_{i\le N}\lambda_i$. We consider a set T of functions on Ω, we use the notation

$$S(T) = \mathsf{E}\sup_{t\in T}\sum_{i\le N}\varepsilon_i t(X_i)\ ;\ \ \bar{S}(T) = \mathsf{E}\sup_{t\in T}\Big|\sum_{i\le N}\varepsilon_i t(X_i)\Big| , \tag{11.39}$$

and we start the preparations for the proof of Theorem 11.1.1. First we observe that without loss of generality, we may assume $0 \in T$. Indeed, if we fix $t_0 \in T$, the set $T - t_0 := \{t - t_0; t \in T\}$ satisfies $S(T - t_0) = S(T)$, and if we have a decomposition

$T - t_0 \subset T^1 + T^2$, we have a decomposition $T \subset (T^1 + t_0) + T^2$. So we now assume that $0 \in T$ and according to Lemma 2.2.1 we have

$$\bar{S}(T) \le 2S(T) . \tag{11.40}$$

For $i \le N$ consider the random function Z_i on T given by $Z_i(t) = t(X_i)$. Define $Z_i(t) = 0$ for $i > N$, so that the functions $(Z_i)_{i\ge 1}$ are independent. Then $S = \mathsf{E}\sup_{t\in T} \sum_{i\ge 1} \varepsilon_i Z_i(t) = S(T)$. The expressions (11.5) and (11.6) take the form

$$\psi_{j,\omega}(s,t) = \sum_{i\le N} (|r^j(s(X_i)) - t(X_i)|^2 \wedge 1) , \tag{11.41}$$

$$\varphi_j(s,t) = \sum_{i\le N} \mathsf{E}(|r^j(s(X_i)) - t(X_i)|^2 \wedge 1) = \int |r^j(s-t)|^2 \wedge 1 \mathrm{d}\nu . \tag{11.42}$$

The main idea is to combine Theorem 11.7.1 and Theorem 9.2.1, but we need an extra piece of information. Let us denote by $j_0 = j_0(T)$ the integer provided by (11.33) so that $\varphi_{j_0}(s,t) \le 4$ for $s, t \in T$.

Lemma 11.9.1 *Given $t \in T$ we have*

$$\int |t| \mathbf{1}_{\{2|t|\ge r^{-j_0}\}} \mathrm{d}\nu \le L\bar{S}(T) \le LS(T) . \tag{11.43}$$

Proof Since $0 \in T$ and $\varphi_{j_0}(0,t) \le 4$ we have $\int |r^{j_0} t|^2 \wedge 1 \mathrm{d}\nu \le 4$. In particular if $U = \{2|t| \ge r^{-j_0}\}$ then $\nu(U) \le 16$, that is, $\sum_{i\le N} \lambda_i(U) \le 16$. Let $A = \{i \le N \,;\, \lambda_i(U) \ge 1/2\}$, so that card $A \le 32$. For $i \notin A$ we have $1 - \lambda_i(U) \ge \exp(-2\lambda_i(U))$, so that $\prod_{j\notin A}(1 - \lambda_j(U)) \ge \exp(-32)$. For $i \le N$ consider the event Ξ_i given by $X_i \in U$ and $X_j \notin U$ for $j \ne i$ and $j \notin A$. Then

$$\mathsf{P}(\Xi_i) = \lambda_i(U) \prod_{j\ne i, j\notin A} (1 - \lambda_j(U)) \ge \lambda_i(U)/L .$$

Given Ξ_i, the r.v. X_i is distributed according to the restriction of λ_i to U, so that

$$\frac{1}{\mathsf{P}(\Xi_i)} \mathsf{E}\mathbf{1}_{\Xi_i} |t(X_i)| = \frac{1}{\lambda_i(U)} \int_U |t| \mathrm{d}\lambda_i ,$$

and hence

$$\int_U |t| \mathrm{d}\lambda_i \le L\mathsf{E}\mathbf{1}_{\Xi_i} |t(X_i)| \le L\mathsf{E}\mathbf{1}_{\Xi_i} \Big| \sum_{j\le N} \varepsilon_j t(X_j) \Big| , \tag{11.44}$$

where the last inequality follows by Jensen's inequality, averaging in the r.v.s ε_j for $j \ne i$ outside the absolute values rather than inside. As the events Ξ_i for $i \notin A$ are disjoint, and since card $A \le 32$, we have

$$\sum_{i \le N} \mathbf{1}_{\Xi_i} \le \sum_{i \notin A} \mathbf{1}_{\Xi_i} + \sum_{i \in A} \mathbf{1}_{\Xi_i} \le 1 + 32 = 33 .$$

Summation of the inequalities (11.44) over $i \le N$ and using (11.40) yields the result. □

Proposition 11.9.2 *We can decompose $T \subset T_1 + T_2$ where the set T_1 satisfies $0 \in T_1$ and*

$$\gamma_2(T_1, d_2) + \gamma_1(T_1, d_\infty) \le LS(T) \tag{11.45}$$

and where

$$\forall t \in T_2 \,, \quad \int |t| \mathrm{d}\nu \le LS(T) \,. \tag{11.46}$$

Proof We apply Theorem 11.7.1 and then Theorem 9.2.1, calling T_2 what is called $T_2 + T_3$ there. Then (11.46) is a consequence of Lemma 11.9.1. □

Proof of Theorem 11.1.1 Combining (11.45) and Lemma 11.1.2 yields $S(T_1) \le LS(T)$, so that also $\bar{S}(T_1) \le LS(T)$ since $0 \in T_1$. We may assume that $T_2 \subset T - T_1$, simply by replacing T_2 by $T_2 \cap (T - T_1)$. Thus $\bar{S}(T_2) \le \bar{S}(T) + \bar{S}(T_1) \le LS(T)$. Combining with (11.46), Theorem 11.8.3 then implies $\mathsf{E} \sup_{t \in T_2} \sum_{i \le N} |t(X_i)| \le LS(T)$ and finishes the proof. □

Proof of Theorem 6.8.3 We apply Theorem 11.1.1 to the case where $\lambda_i = \lambda$ is independent of i so that $\nu = N\lambda$. Then, with obvious notation, $\gamma_2(T_1, d_{2,\nu}) = \sqrt{N}\gamma_2(T_1, d_{2,\lambda})$. We also use Lemma 11.8.4. □

11.10 The Decomposition Theorem for Random Series

We will now apply the previous result to random series, and we go back to that setting, as in Sect. 11.2. We assume that for some integer N, we have $Z_i = 0$ for $i > N$ and that $(Z_i)_{i \le N}$ are independent, but not necessarily identically distributed. We consider on T the following two distances:

$$d_2(s,t)^2 = \sum_{i \le N} \mathsf{E}|Z_i(s) - Z_i(t)|^2 \tag{11.47}$$

$$d_\infty(s,t) = \inf\left\{a \;;\; \forall i \le N \;;\; |Z_i(s) - Z_i(t)| \le a \text{ a.e.}\right\} , \tag{11.48}$$

and we assume that they are both finite. The following provides an upper bound for $S = \mathsf{E}\sup_{t\in T}\sum_{i\le N}\varepsilon_i Z_i(t)$:

Theorem 11.10.1 *We have*

$$S \le L\big(\gamma_2(T,d_2) + \gamma_1(T,d_\infty)\big) . \tag{11.49}$$

Proof As always, this follows from Bernstein's inequality (4.44) and Theorem 4.5.13. □

In another direction, the following trivial bound does not involve cancellation:

Proposition 11.10.2 *We have*

$$S \le \mathsf{E}\sup_{t\in T}\sum_{i\le N}|Z_i(t)| . \tag{11.50}$$

As our next result shows, these two methods are the only possible methods to bound S. In other words, every situation is a mixture of these. In loose words: *chaining using Bernstein's inequality explains all the part of the boundedness which is due to cancellation.*

Theorem 11.10.3 *For any independent sequence $(Z_i)_{i\le N}$ of random functions, we may find a decomposition $Z_i = Z_i^1 + Z_i^2$ such that each of the sequences $(Z_i^1)_{i\le N}$ and $(Z_i^2)_{i\le N}$ are independent, and the following hold: First,*

$$\gamma_2(T,d_2^1) + \gamma_1(T,d_\infty^1) \le LS , \tag{11.51}$$

where the distances d_2^1 and d_∞^1 are given by (11.47) and (11.48) where Z_i is replaced by Z_i^1. Second,

$$\mathsf{E}\sup_{t\in T}\sum_{i\le N}|Z_i^2(t)| \le LS . \tag{11.52}$$

Certainly it would be of interest to consider more precise situations, such as the case where T is a metric space and where Z_i is a continuous function on T. In that case however, it is not claimed that the previous decomposition consists of continuous functions. The possibility of this is better left for further research.

Proof Theorem 11.10.3 is a simple consequence of Theorem 11.1.1. The one difficulty lies in the high level of abstraction required. The independent sequence of random functions $(Z_i)_{i\le N}$ is just an independent sequence of random variables valued in the space $\Omega = \mathbb{R}^T$. We denote by λ_i the law of Z_i on $\Omega = \mathbb{R}^T$, and we set $\nu = \sum_{i\le N}\lambda_i$.

To each element $t \in T$ we associate the corresponding coordinate function $\theta(t)$ on $\Omega = \mathbb{R}^T$. That is, for $x = (x(s))_{s\in T} \in \Omega$ we have $\theta(t)(x) = x(t)$. Thus we have $\theta(t)(Z_i) = Z_i(t)$.[11] It should be obvious that

$$S = \mathsf{E} \sup_{t \in T} \sum_{i \le N} \varepsilon_i Z_i(t) = \mathsf{E} \sup_{t \in T} \sum_{i \le N} \varepsilon_i \theta(t)(Z_i) = S(\theta(T)) ,$$

where $\theta(T) = \{\theta(t); t \in T\}$ is a set of functions on Ω. We use the decomposition of $\theta(T)$ provided by Theorem 11.1.1. For each $t \in T$ we fix a decomposition $\theta(t) = \theta(t)^1 + \theta(t)^2$ where $\theta(t)^1 \in \theta(T)_1$ and $\theta(t)^2 \in \theta(T)_2$.

We define then the random functions Z_i^1 and Z_i^2 on T by

$$Z_i^1(t) = \theta(t)^1(Z_i) \; ; \; Z_i^2(t) = \theta(t)^2(Z_i) , \tag{11.53}$$

so that $Z_i(t) = Z_i^1(t) + Z_i^2(t)$. The definition of $Z_i^2(t)$ should make it obvious that (11.52) follows from (11.2). Next,

$$d_2^1(s,t)^2 = \sum_{i \le N} \mathsf{E}|Z_i^1(s) - Z_i^1(t)|^2 = \sum_{i \le N} \mathsf{E}|\theta(s)^1(Z_i) - \theta(t)^1(Z_i)|^2$$

$$= \int_C |\theta(s)^1 - \theta(t)^1|^2 \mathrm{d}\nu ,$$

so that, with obvious notation, $d_2^1(s,t) = \|\theta(s)^1 - \theta(t)^1\|_{2,\nu}$. That is, the map $t \mapsto \theta(t)^1$ witnesses that the metric space (T, d_2^1) is isometric to a subspace of the metric space $(\theta(T)_1, d_{2,\nu})$ and thus $\gamma_2(T, d_2^1) \le \gamma_2(\theta(T)_1, d_{2,\nu}) \le LS$, using (11.45) in the last inequality. The rest is similar. □

It is probably worth insisting on the highly non-trivial definition (11.53). For $j = 1, 2$ we may define a map $\Xi_j : \Omega \to \Omega = \mathbb{R}^T$ by $\Xi_j(x)(t) = \theta(t)^j(x)$. These are fairly complicated maps. The formula (11.53) reads as $Z_i^j = \Xi_j(Z_i)$. The next example stresses this point.

Example 11.10.4 We should stress a subtle point: we apply Theorem 11.1.1 to $\theta(T)$, a set of functions on $\mathcal{C}$. When T itself is naturally a set of functions on some other space, say on $[0, 1]$, it is not the same to decompose T as a set of functions on $[0, 1]$, or $\theta(T)$ as a set of functions on $\mathcal{C}$. To explain this, consider a set T of functions on $[0, 1]$ and independent r.v.s $(\xi_i, X_i)_{i \le N}$ valued in $\mathbb{R} \times [0, 1]$. To study the quantity $\sup_{t \in T} \sum_{i \le N} \varepsilon_i \xi_i t(X_i)$, we have to consider the functions $\theta(t)$ on $\mathbb{R} \times [0, 1]$, given for $(x, y) \in \mathbb{R} \times [0, 1]$ by $\theta(t)(x, y) = xt(y)$. It is *this* function which is decomposed. In particular, there is no reason why one should have $\theta(t)^1(x, y)$ of the type $xt'(y)$ for a certain function t' on $[0, 1]$.

[11] The best way to write the proof is to lighten notation by writing $t(Z_i)$ rather than $\theta(t)(Z_i)$ and to think of T as a set of functions on Ω. Please attempt this exercise after you understand the present argument.

11.11 Selector Processes and Why They Matter

Given a number $0 < \delta < 1$, we consider i.i.d. r.v.s $(\delta_i)_{i \le M}$ with

$$\mathsf{P}(\delta_i = 1) = \delta \; ; \; \mathsf{P}(\delta_i = 0) = 1 - \delta \, . \tag{11.54}$$

We will assume that $\delta \le 1/2$, the most interesting case.[12] The r.v.s δ_i are often called selectors, because they allow to select a random subset J of $\{1, \dots, M\}$ of cardinality about $\mathsf{E}\,\mathrm{card}\, J = \delta M$, namely, the set $\{i \le M ; \delta_i = 1\}$. They will be used for this purpose in Sects. 19.3.1 and 19.3.2.

Selector processes are also important as that they provide a discrete approximation of the fundamental procedure of constructing independent random points $(X_j)_{j \le N}$ distributed according to μ in a probability space (Ω, μ). We will explain this is a very informal manner. Assuming for clarity that μ has no atoms, let us divide Ω into M small pieces Ω_i of equal measure, where M is much larger than N. Consider then selectors $(\delta_i)_{i \le M}$ where δ in (11.54) is given by $\delta = N/M$. When $\delta_i = 1$ let us choose a point Y_i in Ω_i. Since Ω_i is small, how we do this is not very important, but let us be perfectionist and choose Y_i according to the conditional probability that $Y_i \in \Omega_i$. Then the collection of points $\{Y_i ; \delta_i = 1\}$ resembles a collection $\{X_j ; j \le N'\}$ where the points $(X_j)_{j \le M'}$ are independent, distributed according to μ, and where $N' = \sum_{i \le M} \delta_i$. For M large, N' is nearly a Poisson r.v. of expectation $\delta M = N$. So a more precise statement is that selector processes approximate the operation of choosing a set of N' independent random points in a probability space, where N' is a Poisson r.v.[13] Many problems are "equivalent", where one considers this random number of points $(X_j)_{j \le N'}$, or a fixed number of points $(X_j)_{j \le N}$ (the so-called Poissonization procedure).

We will call a family of r.v.s of the type $\sum_{i \le M} t_i(\delta_i - \delta)$ where t varies over a set T of sequences a "selector process", and we set

$$\delta(T) := \mathsf{E} \sup_{t \in T} \Big| \sum_{i \le M} t_i(\delta_i - \delta) \Big| \, . \tag{11.55}$$

According to the previous discussion, we expect for selector processes a result similar to Theorem 6.8.3. This will be proved in the next section as a consequence of Theorem 11.1.1.

[12] The case $\delta \ge 1/2$ actually follows by the transformation $\delta_i \to 1 - \delta_i$ and $\delta \to 1 - \delta$.

[13] If you want to emulate an independent sequence $(X_j)_{j \le N'}$ by this procedure, you first consider the collection $\{Y_i ; \delta_i = 1\}$ and you number those Y_i in a random order.

11.12 Proving the Generalized Bernoulli Conjecture

Since $\mathsf{E}(\delta_i - \delta)^2 = \delta(1-\delta) \le 1/4$, in the present case Bernstein's inequality (4.44) yields

$$\mathsf{P}\Big(\Big|\sum_{i\le M} t_i(\delta_i - \delta)\Big| \ge v\Big) \le 2\exp\Big(-\min\Big(\frac{v^2}{\delta\sum_{i\le M} t_i^2}, \frac{v}{2\max_{i\le M}|t_i|}\Big)\Big) . \tag{11.56}$$

Combining with Theorem 4.5.13, (11.56) implies a first bound on selector processes. If T is a set of sequences with $0 \in T$, then (recalling the quantity $\delta(T)$ of (11.55)),

$$\delta(T) \le L\big(\sqrt{\delta}\gamma_2(T, d_2) + \gamma_1(T, d_\infty)\big) . \tag{11.57}$$

The following shows that the chaining argument of (11.57) takes care of all "the part of boundedness which comes from cancellation":

Theorem 11.12.1 *Given a set T of sequences we can write $T \subset T_1 + T_2$ with*

$$\gamma_2(T_1, d_2) \le \frac{L\delta(T)}{\sqrt{\delta}} \; ; \; \gamma_1(T_1, d_\infty) \le L\delta(T) \tag{11.58}$$

and

$$\mathsf{E}\sup_{t\in T_2}\sum_{i\le M}|t_i|\delta_i \le L\delta(T) . \tag{11.59}$$

Proof We will show that this is a special case of Theorem 11.1.1. Consider the space $\Omega = \{0, 1, \dots, M\}$, and for an element $t = (t_i)_{i\le M} \in T$, consider the real-valued function $\tilde{t}$ on Ω given by $\tilde{t}(0) = 0$ and $\tilde{t}(i) = t_i$ for $1 \le i \le M$. Conversely, for a real-valued function u on Ω, denote by $P(u)$ the sequence $(u(i))_{1\le i\le M}$, and note that $P(\tilde{t}) = t = (t_i)_{i\le M}$. For $1 \le i \le M$ consider the r.v. X_i valued in Ω given by $X_i = 0$ if $\delta_i = 0$ and $X_i = i$ if $\delta_i = 1$. Then $\tilde{t}(X_i) = \delta_i t_i$ and if $\tilde{T} = \{\tilde{t}; t \in T\}$ then

$$S(\tilde{T}) := \mathsf{E}\sup_{\tilde{t}\in\tilde{T}}\sum_{i\le M}\varepsilon_i\tilde{t}(X_i) = \mathsf{E}\sup_{t\in T}\sum_{i\le M}\varepsilon_i\delta_i t_i . \tag{11.60}$$

We will prove later that

$$\mathsf{E}\sup_{t\in T}\Big|\sum_{i\le M}\varepsilon_i\delta_i t_i\Big| \le 4\delta(T) , \tag{11.61}$$

so that in particular

$$S(\tilde{T}) \le L\delta(T) . \qquad (11.62)$$

The law λ_i of X_i is such that $\lambda_i(\{i\}) = \mathsf{P}(X_i = i) = \delta$ and $\lambda_i(\{0\}) = \mathsf{P}(X_i = 0) = 1 - \delta$. The measure $\nu = \sum_{i \le M} \lambda_i$ is then such that for any function u on Ω, we have

$$\delta \|P(u)\|^2 = \delta \sum_{i \le M} (u(i))^2 \le \int u^2 \mathrm{d}\nu = \|u\|^2 ,$$

where the norms are, respectively, in $\ell^2(M)$ and in $L^2(\nu)$. Consequently,

$$\sqrt{\delta}\|P(u)\|_2 \le \|u\|_2 \; ; \; \|P(u)\|_\infty \le \|u\|_\infty . \qquad (11.63)$$

Consider the decomposition $\tilde{T} \subset U_1 + U_2$ provided by Theorem 11.1.1, so that, using (11.62) in the last inequality,

$$\gamma_2(U_1, d_2) + \gamma_1(U_1, d_\infty) \le LS(\tilde{T}) \le L\delta(T) \qquad (11.64)$$

and

$$\mathsf{E} \sup_{u \in U_2} \sum_{i \le M} |u(X_i)| \le LS(\tilde{T}) \le L\delta(T) . \qquad (11.65)$$

Since $T = P(\tilde{T})$, this provides a decomposition $T \subset T_1 + T_2$ where $T_j = P(U_j)$. It follows from (11.63) that

$$\gamma_2(T_1, d_2) \le \frac{1}{\sqrt{\delta}} \gamma_2(U_1, d_2) \; ; \; \gamma_1(T_1, d_\infty) \le \gamma_1(U_1, d_\infty) ,$$

and then (11.64) implies that (11.58) holds. Furthermore since $|u(X_i)| \ge \delta_i |u(i)|$ (11.65) implies (11.59).

It remains only to prove (11.61). For this let us denote by $(\delta'_i)_{i \le M}$ a copy of the sequence $(\delta_i)_{i \le M}$, which is independent of all the other r.v.s previously used. We first note that by the triangle inequality we have

$$\mathsf{E} \sup_{t \in T} \Big| \sum_{i \le M} (\delta_i - \delta'_i) t_i \Big| \le 2\delta(T) . \qquad (11.66)$$

Next, the sequences $(\varepsilon_i|\delta_i - \delta_i'|)_{i\le M}$, $(\varepsilon_i(\delta_i - \delta_i'))_{i\le M}$ and $(\delta_i - \delta_i')_{i\le M}$ have the same distribution, so that

$$\mathsf{E}\sup_{t\in T}\Big|\sum_{i\le M}\varepsilon_i|\delta_i-\delta_i'|t_i\Big| = \mathsf{E}\sup_{t\in T}\Big|\sum_{i\le M}\varepsilon_i(\delta_i-\delta_i')t_i\Big| = \mathsf{E}\sup_{t\in T}\Big|\sum_{i\le M}(\delta_i-\delta_i')t_i\Big| . \tag{11.67}$$

Now, using Jensen's inequality, and since $\mathsf{E}|\delta_i - \delta_i'| = 2\delta(1-\delta) \ge \delta$,

$$\delta\mathsf{E}\sup_{t\in T}\Big|\sum_{i\le M}\varepsilon_i t_i\Big| \le \mathsf{E}\sup_{t\in T}\Big|\sum_{i\le M}\varepsilon_i|\delta_i-\delta_i'|t_i\Big| \le 2\delta(T) , \tag{11.68}$$

where we use (11.67) and (11.66) in the last inequality. Next, we write

$$\mathsf{E}\sup_{t\in T}\Big|\sum_{i\le M}\varepsilon_i\delta_i t_i\Big| \le \mathsf{E}\sup_{t\in T}\Big|\sum_{i\le M}\varepsilon_i(\delta_i-\delta)t_i\Big| + \delta\mathsf{E}\sup_{t\in T}\Big|\sum_{i\le M}\varepsilon_i t_i\Big| . \tag{11.69}$$

Using Jensen's inequality and (11.66) to obtain

$$\mathsf{E}\sup_{t\in T}\Big|\sum_{i\le M}\varepsilon_i(\delta_i-\delta)t_i\Big| \le \mathsf{E}\sup_{t\in T}\Big|\sum_{i\le M}\varepsilon_i(\delta_i-\delta_i')t_i\Big| \le 2\delta(T) ,$$

and combining (11.68) and (11.69) we obtain (11.61). □

Exercise 11.12.2 Consider a set T of sequences and assume that for a certain number $A > 0$,

$$\gamma_2(T, d_2) \le \frac{A}{\sqrt{\delta}} \; ; \; \gamma_1(T, d_\infty) \le A .$$

(a) Prove that then $\operatorname{conv} T \subset T_1 + T_2$ where $\mathsf{E}\sup_{t\in T_2}\sum_{i\le M}|t_i|\delta_i \le LA$ and $\gamma_2(T_1, d_2) \le LA/\sqrt{\delta}$; $\gamma_1(T_1, d_\infty) \le LA$.
(b) Prove that it is *not* always true that

$$\gamma_2(\operatorname{conv} T, d_2) \le \frac{LA}{\sqrt{\delta}} \; ; \; \gamma_1(\operatorname{conv} T, d_\infty) \le LA .$$

Hint: Use Exercise 8.3.7 and choose δ appropriately small.

Theorem 11.12.1 shows that chaining, when performed using Bernstein's inequality, already captures all the possible cancellation. This is remarkable because Bernstein's inequality is not always sharp. We can see that by comparing it with the following simple lemma:

Lemma 11.12.3 *Consider a fixed set I. If $u \geq 6\delta \operatorname{card} I$ we have*

$$\mathsf{P}\Big(\sum_{i\in I}\delta_i \geq u\Big) \leq \exp\Big(-\frac{u}{2}\log\frac{u}{2\delta \operatorname{card} I}\Big)\,. \tag{11.70}$$

Proof We are dealing here with the tails of the binomial law and (11.70) follows from the Chernov bounds. For a direct proof, considering $\lambda > 0$ we write

$$\mathsf{E}\exp\lambda\delta_i \leq 1 + \delta e^{\lambda} \leq \exp(\delta e^{\lambda})$$

so that we have

$$\mathsf{E}\exp\lambda\sum_{i\in I}\delta_i \leq \exp(\delta e^{\lambda}\operatorname{card} I)$$

and

$$\mathsf{P}\Big(\sum_{i\in I}\delta_i \geq u\Big) \leq \exp(\delta e^{\lambda}\operatorname{card} I - \lambda u)\,.$$

We then take $\lambda = \log(u/(2\delta \operatorname{card} I))$, so that $\lambda \geq 1$ since $u \geq 6\delta \operatorname{card} I$ and $\delta e^{\lambda}\operatorname{card} I = u/2 \leq \lambda u/2$. □

Exercise 11.12.4 Prove that the use of (11.56) to bound the left-hand side of (11.70) is sharp (in the sense that the logarithm of the bound it provides is of the correct order) only for u of order $\delta \operatorname{card} I$.

Key Ideas to Remember

- A natural way to bound the discrepancy $\mathsf{E}\sup_{f\in\mathcal{F}}|\sum_{i\leq N}\varepsilon_i f(X_i)|$ of a class of functions is to use chaining and Bernstein's inequality.
- An alternate way to control this discrepancy is to give up on possible cancellations and to use the inequality

$$\mathsf{E}\sup_{f\in\mathcal{F}}\Big|\sum_{i\leq N}\varepsilon_i f(X_i)\Big| \leq \mathsf{E}\sup_{f\in\mathcal{F}}\sum_{i\leq N}|f(X_i)|\,.$$

- Amazingly, there is never a better method than interpolating between the previous two methods: all possible cancellation can be witnessed by Bernstein's inequality.
- The previous result can be interpreted in terms of certain random series of functions. The uniform convergence of these can either be proved from Bernstein's inequality, or without using cancellation, or by a mixture of these two methods.

11.13 Notes and Comments

Even though Theorem 11.6.3 uses only ideas that are also used elsewhere in the book, I formulated it only after reading the paper [18] by Witold Bednorz and Rafał Martynek, where the authors prove in a more complicated way that chaining can be performed from a majorizing measure. In fact, the possibility of performing chaining from such a measure goes back at least to [112]. The contents of Sect. 11.5 and in particular Lemma 11.5.3 are taken from [18]. The use of convexity to construct majorizing measures goes back to Fernique (see Sect. 3.3.2) but is used for the first time in [18] in the context of families of distances.

Chapter 12
Infinitely Divisible Processes

The secret of the present chapter can easily be revealed. Infinite divisible processes can be seen as sums of random series $\sum_{i\geq 1} \varepsilon_i Z_i$ of functions where the sequence $(\varepsilon_i)_{i\geq 1}$ is an independent sequence of Bernoulli r.v.s, which is independent of the sequence (Z_i), and where the sequence of functions $(Z_i)_{i\geq 1}$ shares enough properties with an independent sequence to make all results of the previous chapter go through. Moreover, when Z_i is a multiple of a character, the process behaves just as a random Fourier series, and the results on these extend to this case.

The main result of this chapter is a decomposition theorem for infinitely divisible processes, Theorem 12.3.5 in the spirit of Theorem 11.1.1. It is a consequence of Theorem 11.10.3, our main result on random series of functions.

12.1 Poisson r.v.s and Poisson Point Processes

We start by recalling some classical facts. A reader needing more details may refer to her favorite textbook.

A Poisson r.v. X of expectation a is a r.v. such that

$$\forall n \geq 0\ ;\ \mathsf{P}(X = n) = \frac{a^n}{n!}\exp(-a)\ , \tag{12.1}$$

and indeed $\mathsf{E}X = a$. Then, for any $b \in \mathbb{C}$,

$$\mathsf{E}b^X = \exp(-a)\sum_{n\geq 0} b^n \frac{a^n}{n!} = \exp(a(b-1))\ , \tag{12.2}$$

M. Talagrand, *Upper and Lower Bounds for Stochastic Processes*, Ergebnisse der Mathematik und ihrer Grenzgebiete. 3. Folge / A Series of Modern Surveys in Mathematics 60, https://doi.org/10.1007/978-3-030-82595-9_12

and in particular

$$\mathsf{E}\exp(\lambda X) = \exp\big(a(\exp\lambda - 1)\big)\,. \tag{12.3}$$

Consequently, the sum of two independent Poisson r.v.s is Poisson.

Consider a σ-finite measure ν on a measurable space Ω. A Poisson point process of intensity ν is a random subset Π (at most countable) with the following properties: For any measurable set, A of finite measure,

$$\mathrm{card}(A \cap \Pi) \text{ is a Poisson r.v. of expectation } \nu(A)\,, \tag{12.4}$$

and moreover

$$\begin{aligned}&\text{If } A_1, \ldots, A_k \text{ are disjoint measurable sets, the r.v.s}\\ &(\mathrm{card}(A_\ell \cap \Pi))_{\ell \le k} \text{ are independent}\,.\end{aligned} \tag{12.5}$$

A very important result (which we do not prove) is as follows:

Lemma 12.1.1 *Consider a Poisson point process of intensity ν and a set A with $0 < \nu(A) < \infty$. Given* $\mathrm{card}(\Pi \cap A) = N$, *the set $\Pi \cap A$ has the same distribution as a set $\{X_1, \ldots, X_N\}$, where the variables X_i are independent and distributed according to the probability λ on A given by $\lambda(B) = \nu(A \cap B)/\nu(A)$ for $B \subset A$.*

The purpose of the next exercise is to provide a proof of the previous result and give you a chance to really understand it.[1]

Exercise 12.1.2 Assuming that $\nu(\Omega) < \infty$ consider a subset Π of Ω generated by the following procedure. First, consider a Poisson r.v. M with $\mathsf{E}M = \nu(\Omega)$. Second, given M, consider i.i.d. points $Y_1, \ldots, Y_M$ distributed according to the probability $\mathsf{P}(A)$ on Ω given by $\mathsf{P}(A) = \nu(A)/\nu(\Omega)$, and set $\Pi = \{Y_1, \ldots, Y_M\}$. Prove that (12.4) holds for a subset A of Ω and that the property of Lemma 12.1.1 holds too. When $\nu(\Omega)$ is not finite but Ω is σ-finite, show how to actually construct a Poisson point process on it.

We will enumerate all the points of the set Π as $(Z_i)_{i \ge 1}$.[2] We observe first that as a consequence of (12.4) for any set A

$$\sum_{i \ge 1} \mathsf{P}(Z_i \in A) = \mathsf{E}\,\mathrm{card}(\Pi \cap A) = \nu(A)\,. \tag{12.6}$$

---

[1] If this does not suffices, you may look into [46] Proposition 3.8.

[2] Here we implicitly assume that there are infinitely many such points, i.e., ν has infinite mass, which is the case of interest. One should also keep in mind that there are many possible way to enumerate the points, and one should be careful to write only formulas with make sense independently of the way these points are enumerated.

Consequently, if f is an integrable function on Ω we have

$$\mathsf{E}\sum_{i\geq 1} f(Z_i) = \int f(\beta)\mathrm{d}\nu(\beta)\ , \tag{12.7}$$

as is seen by approximation by functions taking only finitely many values. If A is a measurable set of finite measure, $f = c\mathbf{1}_A$ and $X = \operatorname{card} A\cap\Pi$ then $\sum_{i\geq 1} f(Z_i) = cX$ so that

$$\mathsf{E}\exp\lambda\sum_{i\geq 1} f(Z_i) = \mathsf{E}\exp(\lambda cX) = \exp(\nu(A)(\exp\lambda c - 1))\ .$$

where we use that X is a Poisson r.v. of expectation $\nu(A)$ and (12.3) in the second equality. When f is a step function, $f = \sum_{\ell\leq k} c_\ell \mathbf{1}_{A_\ell} := \sum_{\ell\leq k} f_\ell$ for disjoint sets A_ℓ, the previous formula combined with (12.5) implies

$$\begin{aligned}\mathsf{E}\exp\lambda\sum_{i\geq 1} f(Z_i) &= \prod_{\ell\leq k}\mathsf{E}\exp\lambda\sum_{i\geq 1} f_\ell(Z_i) = \exp\Big(\sum_{\ell\leq k}\nu(A_\ell)(\exp\lambda c_\ell - 1)\Big)\\ &= \exp\Big(\int(\exp(\lambda f(\beta)) - 1)\mathrm{d}\nu(\beta)\Big)\ ,\end{aligned} \tag{12.8}$$

a formula which also holds by approximation under the condition that the exponent in the right-hand side is well defined. This is in particular the case if f is bounded above and if $\int |f|\wedge 1\mathrm{d}\nu < \infty$, where we recall the notation $a\wedge b = \min(a,b)$. This formula will let us obtain bounds on the quantities $\sum_{i\geq 1} f(Z_i)$ pretty much as if the Z_i were independent r.v.s. It contains almost all that we need to know about Poisson point processes. Let us state right away the basic lemma.

Lemma 12.1.3 *Consider* $0\leq f\leq 1$. *Then:*

(a) If $4A\leq\int f\mathrm{d}\nu$ *we have*

$$\mathsf{P}\Big(\sum_{i\geq 1} f(Z_i)\leq A\Big)\leq\exp(-A)\ . \tag{12.9}$$

(b) If $A\geq 4\int f\mathrm{d}\nu$ *we have*

$$\mathsf{P}\Big(\sum_{i\geq 1} f(Z_i)\geq A\Big)\leq\exp\Big(-\frac{A}{2}\Big)\ . \tag{12.10}$$

Proof Using (12.8) the proof is nearly identical to that of Lemma 7.7.2. □

12.2 A Shortcut to Infinitely Divisible Processes

Infinitely divisible r.v.s are standard fare of probability theory. Intuitively they are r.v.s which are sums of infinitely small independent r.v.s. We will call an infinitely divisible random process a family (X_t) of r.v.s such that every finite linear combination of them is infinitely divisible.

Let us not be misled by our terminology. What is called in the literature an infinitely divisible process is usually a very much more special structure.[3] The "infinitely divisible processes" of the literature *are to the processes we study what Brownian motion is to general Gaussian processes.*

The beautiful classical body of knowledge about infinitely divisible r.v.s (such as the so-called Lévy-Kintchin representation of their characteristic functional) bears little on our study because what matters here is a certain representation of infinitely divisible processes as sums of random series which are conditionally Bernoulli processes. For this reason we will directly define infinitely divisible processes as sums of certain random series, and we postpone to Appendix C the task of relating this definition to the classical one.

Consider an index set T and the measurable space $\mathcal{C} = \mathbb{C}^T$, provided with the σ-algebra generated by the coordinate functions.[4] We consider a σ-finite measure ν on $\mathcal{C}$, and we make the fundamental hypothesis that

$$\forall t \in T \; ; \quad \int_{\mathcal{C}} |\beta(t)|^2 \wedge 1 d\nu(\beta) < \infty \; . \tag{12.11}$$

A Poisson process of intensity measure ν generates a sequence $(Z_i)_{i\ge 1}$ of points of $\mathcal{C}$, that is, a sequence of functions on T. Under (12.11), given $t \in T$ it follows form the formula (12.7) (applied to the function $\beta \mapsto |\beta(t)|^2 \wedge 1$) that $\mathsf{E}\sum_{i\ge 1} |Z_i(t)|^2 \wedge 1 = \int_{\mathcal{C}} |\beta(t)|^2 \wedge 1 d\nu(\beta) < \infty$, so that $\sum_{i\ge 1} |Z_i(t)|^2 \wedge 1 < \infty$ a.s., and hence also $\sum_{i\ge 1} |Z_i(t)|^2 < \infty$. Consider an independent Bernoulli sequence $(\varepsilon_i)_{i\ge 1}$, independent of the process (Z_i). Then the series $X_t = \sum_{i\ge 1} \varepsilon_i Z_i(t)$ converges a.s.

Definition 12.2.1 An infinitely divisible (symmetric and without Gaussian component) process is a collection $(X_t)_{t\in T}$ as above where

$$X_t = \sum_{i\ge 1} \varepsilon_i Z_i(t) \; . \tag{12.12}$$

The measure ν is called the *Lévy measure* of the process.

[3] That is, a process on $\mathbb{R}$ with stationary increments

[4] As usual, we will not care about measure-theoretic details because when considering processes $(X_t)_{t\in T}$ are only interested in the joint distribution of finite collections of the r.v.s X_t.

Thus, given the randomness of the (Z_i), an infinitely divisible process is a Bernoulli process. This will be used in a fundamental way. Typically, the convergence of the series $\sum_{i\geq 1} \varepsilon_i Z_i(t)$ is permitted by cancellation between the terms.

There is no reason why X_t should have an expectation. (It can be shown that this is the case if and only if $\int_C |\beta(t)| \mathrm{d}\nu(\beta) < \infty$.) When studying infinitely divisible processes, medians are a better measure of their size than expectation. To keep the statements simple, we have however decided to stick to expectations.

It is an important fact that p-stable processes (in the sense of Sect. 5.1) are infinitely divisible processes in the sense of Definition 12.2.1. This is explained in the next section.

12.3 Overview of Results

Throughout this section $(X_t)_{t\in T}$ denotes an infinitely divisible process, as in Definition 12.2.1, of which we keep the notation. Following our general philosophy, our goal is to relate the size of the r.v. $\sup_{t\in T} X_t$ with a proper measure of "the size of T". Given a number $r \geq 4$, we will use the "family of distances" on $T \times T$ given by

$$\varphi_j(s,t) = \int_C |r^j(\beta(s) - \beta(t))|^2 \wedge 1 \mathrm{d}\nu(\beta) \tag{12.13}$$

(where $j \in \mathbb{Z}$) to measure the size of T, where of course ν is the Lévy measure of the process.

12.3.1 The Main Lower Bound

Let us stress that for the time being, we consider only real-valued processes.

In words our main lower bound shows that the boundedness of an infinitely divisible process implies a certain smallness condition of the index set T.The level of abstraction reached here makes it difficult to understand the power of this result, which will become apparent in the next section, which relies on it.

Theorem 12.3.1 *Consider an infinitely divisible process* $(X_t)_{t\in T}$, *as in Definition 12.2.1, and assume that* T *is countable. Then there exists an admissible sequence* $(\mathcal{A}_n)$ *of partitions of* T *and for* $A \in \mathcal{A}_n$ *an integer* $j_n(A)$ *such that the following holds:*

$$\forall t \in T\ , \sum_{n\geq 0} 2^n r^{-j_n(A_n(t))} \leq K\mathsf{E}\sup_{t\in T} X_t\ , \tag{12.14}$$

$$s,t \in A \in \mathcal{A}_n \Rightarrow \varphi_{j_n(A)}(s,t) \leq 2^{n+2}\ , \tag{12.15}$$

where $\varphi_j(s,t)$ *is given by 12.13.*

Proof This is a special instance of Theorem 11.7.1. The condition (11.8) is a consequence of Lemma 12.1.3. □

Exercise 12.3.2 Analyze the proof to convince yourself that in the right-hand side of (12.14), one could write instead KM where M is a median of $\mathsf{E}_\varepsilon \sup_{t\in T}\sum_{i\geq 1}\varepsilon_i Z_i(t)$.

We explain now why Theorem 12.3.1 can be seen as a considerable extension of Theorem 5.2.1. As is explained in Sect. C.4, a p-stable process is infinitely divisible, and moreover its Lévy measure ν is obtained by the following construction. Denoting by λ Lebesgue's measure on $\mathbb{R}^+$, there exists a finite positive measure m on $\mathcal{C} = \mathbb{C}^T$ such that ν is the image of the measure $\mu \otimes \lambda$ on $\mathbb{R}^+ \times \mathcal{C}$ under the map $(x,\beta) \mapsto x^{-1/p}\beta$. By change of variable, it is obvious that

$$\int_{\mathbb{R}^+} (|ax^{-1/p}|^2 \wedge 1)\mathrm{d}x = C_1(p)|a|^p\ . \tag{12.16}$$

Consequently

$$\begin{aligned}\varphi_j(s,t) &= \int_{\mathcal{C}}\int_{\mathbb{R}^+} \big(|x^{-1/p} r^j(\omega(s)-\omega(t))|^2 \wedge 1\big)\mathrm{d}m(\omega)\mathrm{d}x\\ &= C_1(p) r^{jp}\int_{\mathcal{C}} |\omega(s)-\omega(t)|^p \mathrm{d}m(\omega) = C_1(p) r^{jp}\bar{d}(s,t)^p\ ,\end{aligned}$$

where $\bar{d}(s,t)^p = \int_{\mathcal{C}} |\omega(s)-\omega(t)|^p\mathrm{d}m(\omega)$. It is possible to show that the distance d associated with the p-stable process (X_t) as in (5.4) is a multiple of $\bar{d}$ (see Appendix C). Then (12.15) implies $\Delta(A,d) \leq K2^{n/p}r^{-j_n(A)}$, and (12.14) yields

$$\sum_{n\geq 0} 2^{n/q}\Delta(A_n(t),d) \leq K\mathsf{E}\sup_{t\in T} X_{\tilde{t}}\ ,$$

where $1/q = 1 - 1/p$ and thus $\gamma_q(T,d) \leq K\mathsf{E}\sup_{t\in T} X_t$, which is the content of Theorem 5.2.1.

12.3.2 The Decomposition Theorem for Infinitely Divisible Processes

Let us now turn to the difficult problem of bounding infinitely divisible processes. Let us first show how to bound infinitely divisible processes using chaining. On T we consider the distances

$$d_2^2(s,t) = \int_C |\beta(s) - \beta(t)|^2 \mathrm{d}\nu(\beta) , \tag{12.17}$$

$$d_\infty(s,t) = \inf\{a > 0 \; ; \; |\beta(s) - \beta(t)| \le a \; \nu\text{-a.e.}\} . \tag{12.18}$$

The distance d_∞ is simply the distance induced by the norm of $L^\infty(\nu)$ when one considers a point t of T as the functions on $\mathcal{C}$ given by the map $\beta \mapsto \beta(t)$, and similarly the distance d_2 is the distance induced by the norm of $L^2(\nu)$. We will prove a suitable version of Bernstein's inequality which will make the next result appear as a chaining bound (4.56).

Theorem 12.3.3 *We have*

$$\mathsf{E} \sup_{t \in T} X_t \le L\big(\gamma_2(T, d_2) + \gamma_1(T, d_\infty)\big) . \tag{12.19}$$

There is however a method very different from chaining to control the size of an infinitely divisible process, a method which owes nothing to cancellation, using the inequality $|X_t| = |\sum_{i\ge1} \varepsilon_i Z_i(t)| \le \sum_{i\ge1} |Z_i(t)|$. This motivates the following definition:

Definition 12.3.4 Consider a set T, a σ-finite measure ν on $\mathcal{C} = \mathbb{C}^T$ and assume that $\int |\beta(t)| \wedge 1 \mathrm{d}\nu(\beta) < \infty$ for each $t \in T$. Then we define the process $(|X|_t)_{t\in T}$ by

$$|X|_t = \sum_{i \ge 1} |Z_i(t)| . \tag{12.20}$$

When we control the supremum of the process $(|X|_t)_{t\in T}$, we may say that the boundedness of the process $(X_t)_{t\in T}$ owes nothing to cancellation.

We have described two very different reasons why an infinitely divisible process $(X_t)_{t\in T}$ may be bounded.

- The boundedness may be witnessed by chaining as in (12.19).
- It may happen that the process $(|X|_t)_{t\in T}$ is already bounded, and then the boundedness of (X_t) owes nothing to cancellation.

The main result of this chapter, the decomposition theorem for infinitely divisible processes below states that there is *no other possible reason*: every bounded infinitely divisible process is a mixture of the previous two situations.

Theorem 12.3.5 (The Decomposition Theorem for Infinitely Divisible Processes) *Consider an infinitely divisible process* $(X_t)_{t\in T}$*, as in Definition 12.2.1, and assume that* T *is countable. Let* $S = \mathsf{E}\sup_{t\in T} X_t$*. Then we can write* $X_t = X_t^1 + X_t^2$ *in a manner that each of the processes* $(X_t^1)_{t\in T}$ *and* $(X_t^2)_{t\in T}$ *is infinitely divisible and that*

$$\gamma_2(T, d_2^1) + \gamma_1(T, d_\infty^1) \le KS\ , \tag{12.21}$$

where the distances d_2^1 *and* d_∞^1 *are given by (12.18) and (12.17) (for the process* (X_t^1) *rather than* (X_t)*) whereas*

$$\mathsf{E}\sup_{t\in T} |X^2|_t \le KS\ . \tag{12.22}$$

This decomposition witnesses the size of $\mathsf{E}\sup_{t\in T} X_t$. Indeed $\mathsf{E}\sup_{t\in T} X_t \le \mathsf{E}\sup_{t\in T} X_t^1 + \mathsf{E}\sup_{t\in T} X_t^2$. The first term on the right is bounded through chaining; see (12.19). The second term is bounded because $\mathsf{E}\sup_{t\in T_2} |X_t| \le \mathsf{E}\sup_{T\in T_2} |X|_t$ is already bounded by (12.22).

The decomposition theorem is a close cousin of Theorem 11.10.3. In words it can be formulated as follows:

> *Chaining using Bernstein's inequality captures* exactly *the part of the boundedness of an infinitely divisible process that is due to cancellation.*

Exercise 12.3.6 Learn about the Lévy measure of a p-stable process in Sect. C.4 (which was described at the end of Sect. 12.3.1). Show that if such a process is not zero, the process $(|X|_t)_{t\in T}$ is not defined. Conclude that when applying the decomposition theorem to a p-stable process $(X_t)_{t\in T}$, it is not possible to take the pieces $(X_t^1)_{t\in T}$ and $(X_t^2)_{t\in T}$ both p-stable.

12.3.3 Upper Bounds Through Bracketing

Our result is called a "bracketing theorem" because for each $A \in \mathcal{A}_n$, we control the size of the "brackets" $h_A(\omega) = [\inf_{t\in A}\omega(t), \sup_{t\in A}\omega(t)] = \sup_{s,t\in A}|\omega(s) - \omega(t)|$.

Theorem 12.3.7 *Consider an admissible sequence* $(\mathcal{A}_n)$ *of* T*, and for* $A \in \mathcal{A}_n$ *and* $\omega \in \mathcal{C} = \mathbb{R}^T$ *consider* $h_A(\omega) = \sup_{s,t\in A}|\omega(s) - \omega(t)|$*. Assume that for* $A \in \mathcal{A}_n$ *we are given* $j_n(A) \in \mathbb{Z}$ *satisfying*

$$A \in \mathcal{A}_n\ ,\ C \in \mathcal{A}_{n-1}\ ,\ A \subset C \Rightarrow j_n(A) \ge j_{n-1}(C)\ .$$

Assume that for some numbers $r \geq 2$ and $S > 0$ we have

$$\forall A \in \mathcal{A}_n\,, \int \big(r^{2j_n(A)} h_A^2 \wedge 1\big) \mathrm{d}\nu \leq 2^n\,, \tag{12.23}$$

$$\int h_T \mathbf{1}_{\{2h_T \geq r^{-j_0(T)}\}} \mathrm{d}\nu \leq S\,, \tag{12.24}$$

and

$$\forall t \in T\,, \sum_{n \geq 0} 2^n r^{-j_n(A_n(t))} \leq S\,. \tag{12.25}$$

Then $\mathsf{E} \sup_{t \in T} |X_t| \leq KS$.

The principle of Theorem 12.3.7 goes back at least to [112], but its power does not seem to have been understood.

12.3.4 Harmonizable Infinitely Divisible Processes

Motivated by Chap. 7, we may expect that when "there is stationarity" (in the sense that there is a kind of translation invariant structure), it will be much easier to find upper bounds for infinitely divisible processes. In this section, in contrast with the rest of the chapter, infinitely divisible processes are permitted to be complex-valued.

Consider (for simplicity) a metrizable compact group T, and its dual G, the set of continuous characters on T. We denote by $\mathbb{C}G$ the set of functions on T which are of the type $\alpha\chi$ where $\alpha \in \mathbb{C}$ and χ is a character.

Definition 12.3.8 If T is a metrizable compact group, an infinitely divisible process $(X_t)_{t \in T}$ as in Definition 12.2.1 is called *harmonizable* if its Lévy measure is supported by $\mathbb{C}G$.

Special classes of such processes were extensively studied by M. Marcus and G. Pisier [62] and later again by M. Marcus [59]. Although it would be hard to argue that these processes are intrinsically important, our results exemplify the amount of progress permitted by the idea of families of distances.[5] To bring forward that the study of these processes is closely related to that of random Fourier series, we state four results which parallel Lemmas 7.10.4 to 7.10.7 and provide a complete understanding of when these processes are bounded a.s.[6] Here μ denotes the Haar

[5] For example, Marcus [59] obtains necessary and sufficient conditions for boundedness only in the case of harmonizable p-stable processes considered in Sect. 12.3.5.

[6] Since the proofs of these results are, in a high level sense, mere translations of the proofs of the results for the random Fourier series, we will not provide them.

measure of G and $(X_t)_{t\in T}$ an infinitely divisible harmonizable process. The first lemma is obvious.

Lemma 12.3.9 *If the process $(X_t)_{t\in T}$ is a.s. uniformly bounded, given $\alpha > 0$ there exists M such that $\mathsf{P}(\sup_{t\in T}|X_t| \geq M) \leq \alpha$.*

Theorem 12.3.10 *There exists a number $\alpha_1 > 0$ with the following property. Assume that for a certain number M we have*

$$\mathsf{P}\Big(\sup_{t\in T}|X_t| \geq M\Big) \leq \alpha_1 . \tag{12.26}$$

Then there exists an integer j_0 such that

$$\forall s,t\in T\ ,\ \varphi_{j_0}(s,t) \leq 1\ , \tag{12.27}$$

and for $n \geq 1$ an integer j_n with

$$\mu(\{s\in T\ ;\ \varphi_{j_n}(s,0) \leq 2^n\}) \geq N_n^{-1}\ , \tag{12.28}$$

for which

$$\sum_{n\geq 0} 2^n r^{-j_n} \leq KM . \tag{12.29}$$

Theorem 12.3.11 *Consider a harmonizable infinitely divisible process and integers $j_n \in \mathbb{Z}$, $n \geq 0$ that satisfy the conditions (12.27) and (12.28). Then we can split the Lévy measure in three parts ν^1, ν^2, ν^3, such that ν^1, the restriction of ν to the set $\{\beta;\ |\beta(0)| \geq 2r^{-j_0}\}$, is such that its total mass $|\nu^1|$ satisfies $|\nu^1| \leq L$ and that*

$$\int |\beta(0)|\mathrm{d}\nu^2(\beta) \leq K\sum_{n\geq 0} 2^n r^{-j_n} . \tag{12.30}$$

$$\gamma_2(T,d) \leq K\sum_{n\geq 0} 2^n r^{-j_n}\ , \tag{12.31}$$

where the distance d is given by

$$d(s,t)^2 = \int |\beta(s)-\beta(t)|^2\mathrm{d}\mu(\beta) . \tag{12.32}$$

Theorem 12.3.12 *When the Lévy measure is as in Theorem 12.3.11, the process $(X_t)_{t\in T}$ is almost surely bounded.*

Keeping in mind the representation (12.12), this is proved by considering separately the case of ν^ℓ for $\ell = 1,2,3$. For $\ell = 1$, a.s. there are only finitely many Z_i since ν^1 is of finite mass. For $\ell = 2$ we have $\mathsf{E}\sum_i |Z_i(0)| < \infty$, and $|\sum_i Z_i(t)| \le \sum_i |Z_i(0)|$ since $Z_i \in \mathbb{C}G$. For $\ell = 3$ we use a suitable version of (7.23). Let us stress the content of the previous results.

Theorem 12.3.13 *For a harmonizable infinitely divisible process $(X_t)_{t\in T}$ we have $\sup_{t\in T}|X_t| < \infty$ a.s. if and only if there exist integers j_0, $(j_n)_{n\ge 1}$ in $\mathbb{Z}$ satisfying (12.27) and (12.28) for which $\sum_{n\ge 0} 2^n r^{-j_n} < \infty$.*

This theorem should be compared to Theorem 7.5.16.

12.3.5 Example: Harmonizable p-Stable Processes

Let us illustrate Theorem 12.3.13 in the simpler case of "harmonizable p-stable processes", where $1 < p < 2$. By definition such a process is infinitely divisible such that its Lévy measure ν is obtained by the following construction: Starting with a finite measure m on G, ν is the image on $\mathbb{C}G$ of the measure $\mu \otimes m$ on $\mathbb{R}^+ \times G$ under the map $(x, \chi) \to x\chi$, where μ has density x^{-p-1} with respect to Lebesgue's measure on $\mathbb{R}^+$. In that case for a certain constant C_p, we have $\varphi_j(s,t) = C_p r^{jp} d(s,t)^p$ for a certain distance d on T. We explore the situation through a sequence of exercises.[7]

Exercise 12.3.14

(a) When $\varphi_j(s,t) = C_p r^{jp} d(s,t)^p$ for $p > 1$ prove that there exists a sequence (j_n) satisfying the conditions (12.27) and (12.28) as well as $\sum_{n\ge 0} 2^n r^{-j_n} < \infty$ if and only if $\gamma_q(T,d) < \infty$, where q is the conjugate exponent of p. Hint: Basically because $\varphi_j(s,t) \le 2^n$ if and only if $d(s,t) \le 2^{n/p} r^{-j}$.
(b) When $p = 1$ prove that this is the case if and only if there exists a sequence $(\epsilon_n)_{n\ge 0}$ such that $\sum_{n\ge 0} \epsilon_n < \infty$ and

$$\mu(\{s \in T\ ;\ d(s,t) \le \epsilon_n\}) \ge N_n^{-1}\ . \tag{12.33}$$

Exercise 12.3.15 In the case $p = 1$, prove that the condition $\sum_n \epsilon_n < \infty$ is equivalent to the condition $\gamma_\infty(T,d) < \infty$ where the quantity $\gamma_\infty(T,d)$ is defined in (5.20).

Exercise 12.3.16 Prove that the Lévy measure ν of a harmonizable p-stable process satisfies $\nu(\{\beta;\ |\beta(0)| \ge u\}) \le Cu^{-p}$ for a constant C independent of u.

Exercise 12.3.17 Prove that if $1 < p < 2$ then $\mathsf{E}\sup_{t\in T}|X_t| < \infty$ if and only if $\gamma_q(T,d) < \infty$. Prove that if $p = 1$, then $\sup_{t\in T}|X_t| < \infty$ a.s. if and only if there exists a sequence (ϵ_n) such that $\sum_n \epsilon_n < \infty$ and (12.33) holds. Hint: Use the previous exercise.

[7] The sketch of proofs of which are especially concise.

Exercise 12.3.18 In the case of harmonizable 1-stable processes, prove the estimate $\mathsf{P}(\sup_{t\in T}|X_t| \geq u) \leq C/u$ for a number C independent of u. Hint: Use Exercise 12.3.16.

12.4 Proofs: The Bracketing Theorem

Proof of Theorem 12.3.7 The plan is to use Theorem 9.4.1. Given $u \geq 1$, Lemma 12.1.3 (b) implies that for each n and each $A \in \mathcal{A}_n$ we have

$$\mathsf{P}\Big(\sum_{i\geq 1}(r^{2j_n(A)}h_A(Z_i)^2 \wedge 1) \leq u2^{n+2}\Big) \geq 1 - \exp(-u2^{n+1}) .$$

Consequently, the r.v. U defined as

$$U = \sup\big\{2^{-n-2}(r^{2j_n(A)}h_A(Z_i)^2 \wedge 1) \; ; \; n \geq 0, A \in \mathcal{A}_n\big\}$$

satisfies $\mathsf{P}(U \geq u) \leq \sum_{n\geq 0}\exp(-u2^{n+1}) \leq L\exp(-u)$ for $u \geq L$. In particular $\mathsf{E}U \leq L$. We observe the fundamental fact: if $s, t \in A$ then

$$\sum_{i\geq 1}|r^{j_n(A)}(Z_i(s) - Z_i(t))|^2 \wedge 1 \leq \sum_{i\geq 1}(r^{2j_n(A)}h_A(Z_i)^2 \wedge 1) \leq 2^{n+2}U ,$$

and therefore using (9.44) with $p = 1$ and $u = 4U$, we obtain

$$\mathsf{E}_\varepsilon \sup_{t\in T}\Big|\sum_{i\geq 1}\varepsilon_i Z_i(t)\Big| \leq KUS + K\sum_{i\geq 1}h_T(Z_i)\mathbf{1}_{\{2h_T(Z_i)\geq r^{-j_0(T)}\}} , \tag{12.34}$$

where K depends on r only. Since $\mathsf{E}U \leq L$, taking expectation yields

$$\mathsf{E}\sup_{t\in T}\Big|\sum_{i\geq 1}\varepsilon_i Z_i(t)\Big| \leq KS + K\mathsf{E}\sum_{i\geq 1}h_T(Z_i)\mathbf{1}_{\{2h_T(Z_i)\geq r^{-j_0(T)}\}} . \tag{12.35}$$

Now (12.7) yields

$$\mathsf{E}\sum_{i\geq 1}h_T(Z_i)\mathbf{1}_{\{2h_T(Z_i)\geq r^{-j_0(T)}\}} = \int h_T(\beta)\mathbf{1}_{\{2h_T(\beta)\geq r^{-j_0(T)}\}}d\nu(\beta) ,$$

and (12.24) proves that this quantity is $\leq S$. Combining with (12.35) proves that $\mathsf{E}\sup_{t\in T}|X_t| \leq KS$. □

We have imposed condition (12.24) in order to get a clean statement. Its use is simply to control the size of the last term in (12.34). This hypothesis is absolutely inessential: this term is a.s. finite because the sum contains only finitely many non-zero terms. Its size can then be controlled in specific cases through specific methods.

We refer the reader to [132] where it is shown how to deduce recent results of [63] from Theorem 12.3.7.

12.5 Proofs: The Decomposition Theorem for Infinitely Divisible Processes

The decomposition theorem for infinitely divisible processes is a close cousin of Theorem 11.10.3 and also ultimately relies on Theorem 9.2.1. As in the proof of Theorem 11.10.3, a significant level of abstraction is required, so that before we get into the details, it could be worth to give an outline of proof. The main idea is to consider the elements of T as functions on $\mathcal{C} = \mathbb{C}^T$, that is, to each $t \in T$ we associate the function $\theta(t)$ on $\mathcal{C}$ given by $\theta(t)(\beta) = \beta(t)$. We will then suitably decompose each function $\theta(t)$ as a sum $\theta(t) = \theta^1(t) + \theta^2(t)$ of two functions on $\mathcal{C}$, and for $j = 1, 2$ we will define the process X_t^j as $\sum_{i\geq 1} \varepsilon_i \theta^j(t)(Z_i)$. To describe these processes in the language of Definition 12.2.1, for $j = 1, 2$ let us define a map $\Xi^j : \mathcal{C} \to \mathcal{C} = \mathbb{C}^T$ by the formula $\Xi^j(\beta)(t) = \theta^j(t)(\beta)$, so that $\theta^j(t)(Z_i) = \Xi^j(Z_i)(t)$. Define then the positive measure ν^j on $\mathcal{C}$ as the image of ν under the map Ξ^j. It is simple to see that the points $\Xi^j(Z_i)$ arise from a Poisson point process of intensity measure ν^j, the image of ν under the map Ξ^j, so that ν^j is the Lévy measure of the process $(X_t^j)_{t\in T}$.

This having been spelled out, to lighten notation we consider T as space of functions on $\mathcal{C}$ by simply identifying an element $t \in T$ with the function $\beta \mapsto \beta(t)$ on $\mathcal{C}$, so that we write $t(Z_i)$ rather than $Z_i(t)$.

We first prove a suitable version of Bernstein's inequality.

Lemma 12.5.1 *Consider a function u on $\mathcal{C}$. Assume that $\|u\|_2 < \infty$ where $\|u\|_2^2 = \int |u(\beta)|^2 d\nu(\beta)$ and that $\|u\|_\infty = \sup_{\beta\in\Omega} |u(\beta)| < \infty$. Then for $v \geq 0$ we have*

$$\mathsf{P}\Big(\Big|\sum_{i\geq 1} \varepsilon_i u(Z_i)\Big| \geq v\Big) \leq 2\exp\Big(-\frac{1}{L}\min\Big(\frac{v^2}{\|u\|^2}, \frac{v}{\|u\|_\infty}\Big)\Big) . \tag{12.36}$$

Proof Leaving some convergence details to the reader[8], we get

$$\mathsf{E}_\varepsilon \exp\lambda \sum_{i\geq 1} \varepsilon_i u(Z_i) = \prod_{i\geq 1} \cosh \lambda u(Z_i) = \exp \sum_{i\geq 1} \log\cosh \lambda u(Z_i) ,$$

[8] It might be a good idea here to review Exercise (6.1.2).

and taking expectation and using (12.8), we obtain

$$\mathsf{E}\exp\lambda\sum_{i\geq 1}\varepsilon_i u(Z_i) = \exp\int(\cosh\lambda u(\beta)-1)\mathrm{d}\nu(\beta)\ .$$

Since $\cosh x \leq 1 + Lx^2$ for $|x| \leq 1$ it follows that for $\lambda\|u\|_\infty \leq 1$ we have $\cosh\lambda u(\beta) - 1 \leq L\lambda^2 u(\beta)^2$ and then

$$\mathsf{E}\exp\lambda\sum_{i\geq 1}\varepsilon_i u(Z_i) \leq \exp\Big(L\lambda^2\int u(\beta)^2\mathrm{d}\nu(\beta)\Big)\ ,$$

and as in the proof of Bernstein's inequality (Lemma 4.5.6), this implies (12.36). □

Proof of Theorem 12.3.3 It follows from (12.36) that for $s, t \in T$ and $v > 0$, we have

$$\mathsf{P}\Big(\Big|\sum_{i\geq 1}\varepsilon_i(s(Z_i)-t(Z_i))\Big| \geq v\Big) \leq \exp\Big(-\frac{1}{L}\min\Big(\frac{v^2}{d_2(s,t)^2}, \frac{v}{d_\infty(s,t)}\Big)\Big)\ ,$$

where the distance d_2 and d_∞ are those defined in (12.17) and (12.18), so that (12.19) follows from Theorem 4.5.13. □

The next result is in the spirit of the Giné-Zinn theorem (Sect. 11.8.)

Theorem 12.5.2 *We have*

$$\mathsf{E}\sup_{t\in T}|X|_t \leq \sup_{t\in T}\int|t(\beta)|\mathrm{d}\nu(\beta) + 4\mathsf{E}\sup_{t\in T}|X_t|\ .$$

Proof Consider a subset $A \subset \mathcal{C}$ with $\nu(A) < \infty$. Consider an independent sequence $(Y_i)_{i\leq N}$ of r.v.s which are distributed according to the probability P on A given by $\mathsf{P}(B) = \nu(B\cap A)/\nu(A)$. Consider an independent sequence $(\varepsilon_i)_{i\geq 1}$ of Bernoulli r.v.s which is independent of the sequence (Y_i). We apply (11.37) so that $\mathsf{E}|t(Y_i)| = \nu(A)^{-1}\int_A|t(\beta)|\mathrm{d}(\beta)\nu$, and we obtain

$$\mathsf{E}\sup_{t\in T}\sum_{i\leq N}|t(Y_i)| \leq \frac{N}{\nu(A)}\sup_{t\in T}\int_A|t(\beta)|\mathrm{d}\nu(\beta) + 4\mathsf{E}\sup_{t\in T}\Big|\sum_{i\leq N}\varepsilon_i t(Y_i)\Big|\ . \tag{12.37}$$

Consider then a Poisson point process (Z_i) of intensity measure ν and let $N = \mathrm{card}\{i \geq 1;\ Z_i \in A\}$. Given N, according to Lemma 12.1.1, the r.v.s $Z_i\mathbf{1}_A(Z_i)$ are distributed like an independent sequence $(Y_i)_{i\leq N}$, where Y_i is distributed according to the probability P above. We use (12.37) given $N = \mathrm{card}\{i \geq 1;\ Z_i \in A\}$ for the sequence $(Y_i)_{i\le N}$, and we take expectation to obtain (and since $\mathsf{E}N = \nu(A)$):

$$\mathsf{E}\sup_{t\in T}\sum_{i\ge1}|t(Z_i)|\mathbf{1}_A(Z_i) \le \sup_{t\in T}\int_A |t(\beta)|\mathrm{d}\nu(\beta) + 4\mathsf{E}\sup_{t\in T}\Big|\sum_{i\le N}\varepsilon_i t(Z_i)\mathbf{1}_A(Z_i)\Big|$$

$$\le \sup_{t\in T}\int |t(\beta)|\mathrm{d}\nu(\beta) + 4\mathsf{E}\sup_{t\in T}\Big|\sum_{i\le N}\varepsilon_i t(Z_i)\Big| \,, \quad (12.38)$$

by using Jensen's inequality in the second line (i.e., taking expectation in the r.v.s ε_i for which $\mathbf{1}_A(Z_i) = 0$ outside the supremum and the absolute value rather than inside). The result follows since A is arbitrary. □

Let us now prepare for the proof of Theorem 12.3.5. Without loss of generality, we may assume that $0 \in T$, so that $\mathsf{E}\sup_{t\in T}|X_t| \le 2S$ by Lemma 2.2.1.

Lemma 12.5.3 *Let $j_0 = j_0(T)$ be as in Theorem 12.3.1. Then we have*

$$\forall t \in T\ ;\ \int_\Omega |t|\mathbf{1}_{\{2|t|\ge r^{-j_0}\}}\mathrm{d}\nu \le LS\,. \quad (12.39)$$

Proof Using (12.15) for $n = 0$ and since $0 \in T$ we have

$$\int_\Omega r^{2j_0}|t^2| \wedge 1\mathrm{d}\nu \le 4$$

so that by Markov's inequality $U := \{2|t| \ge r^{-j_0}\}$ satisfies $\nu(U) \le 16$ and

$$\int_\Omega |t|\mathbf{1}_{\{2|t|\ge r^{-j_0}\}}\mathrm{d}\nu = \int_U |t|\mathrm{d}\nu\,. \quad (12.40)$$

Consider the event Ξ given by $\mathrm{card}\{i \ge 1;\ Z_i \in U\} = 1$. We lighten notation by assuming that the r.v.s Z_i are numbered in such a way that $Z_1 \in U$ when Ξ occurs. According to Lemma 12.1.1, conditionally on Ξ the r.v. Z_1 is uniformly distributed on U, so that

$$\frac{1}{\mathsf{P}(\Xi)}\mathsf{E}\mathbf{1}_\Xi|t(Z_1)| = \frac{1}{\nu(U)}\int_U |t|\mathrm{d}\nu\,.$$

Furthermore, since (Z_i) is a Poisson point process of intensity ν, we have $\mathsf{P}(\Xi) = \nu(U)\exp(-\nu(U)) \ge \nu(U)/L$ so that

$$\int_U |t|\mathrm{d}\nu \le L\mathsf{E}\mathbf{1}_\Xi|t(Z_1)|\,. \quad (12.41)$$

Now, denoting by E_ε expectation in the r.v.s. ε_i only, we have, using Jensen's inequality in the first inequality,

$$\mathbf{1}_\Xi|t(Z_1)| = \mathsf{E}_\varepsilon \mathbf{1}_\Xi|\varepsilon_1 t(Z_1)| \le \mathsf{E}_\varepsilon \mathbf{1}_\Xi\Big|\sum_{i\ge 1}\varepsilon_i t(Z_i)\Big| = \mathsf{E}_\varepsilon \mathbf{1}_\Xi|X_t| \le \mathsf{E}_\varepsilon|X_t| \,.$$

Taking expectation we obtain $\mathsf{E}\mathbf{1}_\Xi|t(Z_1)| \le \mathsf{E}|X_t| \le 2S$, and using (12.41) and (12.40) concludes the proof. □

Proposition 12.5.4 *There exists a decomposition $T \subset T_1 + T_2$, such that $0 \in T_1$ and*

$$\gamma_2(T_1, d_2) \le KS \,, \ \gamma_1(T_1, d_\infty) \le KS \,, \tag{12.42}$$

$$\sup_{t\in T_2} \int |t(\beta)|d\nu(\beta) \le KS \,. \tag{12.43}$$

Here d_2 and d_∞ are as always the distances induced by the L^2 and the L^∞ norm when T is seen as a space of functions on the measured space $(\mathcal{C}, \nu)$. These are the same distances as in (12.17) and (12.18).

Proof We combine Theorem 12.3.1 with Theorem 9.2.1, calling T_2 what is called $T_2 + T_3$ there. Lemma 12.5.3 asserts that $\int |t(\beta)|d\nu(\beta) \le KS$ for $t \in T_3$. □

Proof of Theorem 12.3.5 Consider the decomposition of T provided by Proposition 12.5.4. Combining (12.42) with Theorem 12.3.3 yields $\mathsf{E}\sup_{t\in T_1} X_t \le KS$. Since $0 \in T_1$, combining with Lemma 2.2.1 yields $\mathsf{E}\sup_{t\in T_1}|X_t| \le KS$. We may assume that $T_2 \subset T - T_1$, simply by replacing T_2 by $T_2 \cap (T - T_1)$. Thus

$$\mathsf{E}\sup_{t\in T_2}|X_t| \le \mathsf{E}\sup_{t\in T_1}|X_t| + \mathsf{E}\sup_{t\in T}|X_t| \le KS \,.$$

Combining with (12.43), Theorem 12.5.2 then implies $\mathsf{E}\sup_{t\in T_2}|X|_t \le KS$. Every element $t \in T$ has a decomposition $t = t^1 + t^2$ with $t^1 \in T_1$ and $t^2 \in T_2$. We set $X_t^1 = X_{t^1}$ and $X_t^2 = X_{t^2}$ to finish the proof since (12.21) is a consequence of (12.42). □

Key Ideas to Remember

- Infinitely divisible processes (symmetric, without Gaussian components) are standard fare of probability theory. They can be viewed as sum of certain random series of functions (the terms of which are not independent).
- Our general results about random series of functions apply to this setting and considerably clarify the question of boundedness of such processes. This boundedness can always be witnessed by a suitable use of Bernstein's inequality, or by forgetting about possible cancellations, or by a mixture of both methods.

12.6 Notes and Comments

Theorem 12.3.5 (the decomposition theorem) was first proved in [112] under an extra technical hypothesis. Basically the same proof was given in [132]. While preparing the present edition, I discovered a much simpler proof (still using the same extra technical hypothesis) based on the Latała-Bednorz theorem. Namely, I proved Lemma 11.4.1, and I used the functionals $F_n(A) = \sup_\mu \inf_{t \in A} J_\mu(t)$ (where the supremum is taken over all probability measures μ with $\mu(A) = 1$) and Theorem 8.1.2 to obtain Theorem 12.3.1. The technical condition was necessary to prove that these functionals satisfy the appropriate growth condition. Witold Bednorz and Rafał Martynek [18], who had the early version of this book, combined the method of Lemma 11.4.1 with the use of convexity[9] as in Lemma 3.3.2 to construct the majorizing measure of Theorem 11.5.1 and to show (in a somewhat complicated manner) that this majorizing measure can be used to perform the appropriate chaining. In this manner in [18] they proved Theorem 12.3.5 in the slightly weaker form where in (12.22) there is an extra term $K \sup_{t \in T} \int_\Omega |t| \mathbf{1}_{\{2|t| \geq r^{-j_0}\}}$. This extra term was removed here using the simple Lemma 12.5.3.

[9] This use of convexity goes back to Fernique [32].

Chapter 13
Unfulfilled Dreams

We have made much progress on several of the dreams which were born in Sect. 2.12 (which the reader should review now). Some of this progress is partial; in Theorems 6.8.3, 11.12.1, and 12.3.5, we have shown that "chaining explains all the boundedness due to cancellation". But what could we say about boundedness of processes where no cancellation occurs? In this chapter, we dream about this, in the simplest case of positive selector processes. Our goal does not vary: trying to show that when a process is bounded, this can be witnessed in a simple manner: using the union bound (through chaining), maybe taking convex hull, or some other simple idea (we will use positivity below). The most important material in this chapter is in Sect. 13.2, where the analysis leads us to a deep question of combinatorics. The author has spent considerable time studying it and offers a prize for its solution.

13.1 Positive Selector Processes

Theorem 11.12.1 reduces the study of the boundedness of selector processes to the study of the boundedness of positive selector processes. That is, we have to understand the quantity

$$\delta^+(T) := \mathsf{E} \sup_{t \in T} \sum_{i \le M} t_i \delta_i \tag{13.1}$$

where T is a set of sequences $t = (t_i)_{i \le M}$ with $t_i \ge 0$, and where the r.v.s δ_i are independent, $\mathsf{P}(\delta_i = 1) = \delta$, and $\mathsf{P}(\delta_i = 0) = 1 - \delta$. The study of positive selector processes is in any case fundamental, since, following the same steps as in Sect. 11.11, it is essentially the same problem as understanding the quantity $\mathsf{E} \sup_{f \in \mathcal{F}} \sum_{i \le N} f(X_i)$ when $\mathcal{F}$ is a class of non-negative functions.

M. Talagrand, *Upper and Lower Bounds for Stochastic Processes*, Ergebnisse der Mathematik und ihrer Grenzgebiete. 3. Folge / A Series of Modern Surveys in Mathematics 60, https://doi.org/10.1007/978-3-030-82595-9_13

It is difficult to imagine as a function of which geometrical characteristics of T one should evaluate the quantity (13.1). We shall discuss in Sect. 13.4 the case where T consists of indicator of sets, and in particular we shall give concrete examples which illustrate this point.

An important feature of positive selector processes is that we can use *positivity* to construct new processes from processes we already know how to bound. To implement the idea, given a set T, we denote by solid T its "solid convex hull", i.e., the set of sequences $(s_i)_{i\le M}$ for which there exists $t \in \operatorname{conv} T$ such that $s_i \le t_i$ for each $i \le M$. It should be obvious that

$$\sup_{t\in\operatorname{solid} T}\sum_{i\le M} t_i\delta_i = \sup_{t\in T}\sum_{i\le M} t_i\delta_i . \tag{13.2}$$

Taking expectation shows that $\delta^+(\operatorname{solid} T) = \delta^+(T)$. Thus, to bound $\delta^+(T)$ from above, it suffices to find a set T' for which $T \subset \operatorname{solid} T'$ and such that we control $\delta^+(T')$. Recalling (2.149), we define

$$S(T) = \inf\Big\{S > 0\,;\ \int_S^\infty \sum_{t\in T}\mathsf{P}\Big(\sum_{i\le M} t_i\delta_i \ge u\Big)du \le S\Big\} . \tag{13.3}$$

In Lemma 2.12.1, we proved that $\delta^+(T) \le 2S(T)$, so that if $T \subset \operatorname{solid} T'$, then $\delta^+(T) \le 2S(T')$. Wishful thinking, supplemented by a dreadful lack of imagination[1], leads to the following:

Research Problem 13.1.1 Does there exist a universal constant L such that for any set T of sequences $t = (t_i)$, $t_i \ge 0$, one can find a set T' with $S(T') \le L\delta^+(T)$ and $T \subset \operatorname{solid} T'$?

13.2 Explicitly Small Events

Our ultimate goal should be to give a complete description of the quantity $\delta^+(T)$ "as a function of the geometry of the metric space (T, d)". This motivated Research Problem 13.1.1. At present, we simply have no clue on how to approach this problem, and in the rest of the chapter, we explore different directions.

We proved that as a consequence of Theorem 2.11.9, for any Gaussian process, we can find a jointly Gaussian sequence (u_k) such that

$$\Big\{\sup_{t\in T}|X_t| \ge L\mathsf{E}\sup_{t\in T}|X_t|\Big\} \subset \bigcup_{k\ge 1}\{u_k \ge 1\} \tag{13.4}$$

[1] In other words, we could not think of any other way to bound $\delta^+(T)$.

and moreover

$$\sum_{k\geq 1} \mathsf{P}(u_k \geq 1) \leq \frac{1}{2} .$$

The sets $\{u_k \geq 1\}$ are *simple concrete witnesses* that the event on the left-hand side of (13.4) has a probability at most $1/2$.[2]

Let us explore the same idea for positive selector processes. Does there exist a universal constant L such that for each set T of sequences $t = (t_i)_{i\geq 1}$, $t_i \geq 0$, there exist "simple witnesses" that the event

$$\sup_{t\in T} \sum_{i\leq M} \delta_i t_i \geq L\delta^+(T) \tag{13.5}$$

has a probability at most $1/2$?

There is a simple and natural choice for these witnesses. For a finite subset I of $\{1, \ldots, M\}$, let us consider the event H_I defined by

$$H_I = \{\forall i \in I, \delta_i = 1\} , \tag{13.6}$$

so that $\mathsf{P}(H_I) = \delta^{\operatorname{card} I}$. The events H_I play the role that the half-spaces play for Gaussian processes in (13.4) (see (2.153)).

Definition 13.2.1 Given a positive number $\eta > 0$, an event Ω is η-small if we can find a family $\mathcal{G}$ of subsets I of $\{1, \ldots, M\}$ with

$$\sum_{I\in\mathcal{G}} \eta^{\operatorname{card} I} \leq 1/2 \tag{13.7}$$

and

$$\Omega \subset \bigcup_{I\in\mathcal{G}} H_I . \tag{13.8}$$

The choice of the constant $1/2$ in (13.7) is rather arbitrary. Since $\mathsf{P}(H_I) = \delta^{\operatorname{card} I}$, a δ-small event is of probability $\leq 1/2$, but it is such in an "explicit" way (hence the title of the section). The sets H_I as in (13.8) are "simple concrete witnesses" of that.

The first point to make is that there exist sets of small probability which do not look at all like δ-small sets. A typical example is as follows. Let us consider two integers k, r, and r disjoint subsets $I_1, \ldots, I_r$ of $\{1, \ldots, M\}$, each of cardinality k.

[2] The existence of these witnesses is a not as strong as the information provided by Theorem 2.10.1. It is easy to deduce it from Theorem 2.10.1, but it does not seem easy to go the other way around.

Let us consider the set

$$A = \{(\delta_i)_{i \le M} \; ; \; \forall \ell \le r \; , \; \exists i \in I_\ell \; , \; \delta_i = 1\} \; . \tag{13.9}$$

It is straightforward to see that $\mathsf{P}(A) = (1 - (1 - \delta)^k)^r$. In particular, given k, one can choose r large so that $\mathsf{P}(A)$ is small. We leave the following as a teaser to the reader:

Exercise 13.2.2 Prove that the set A is not $1/k$-small. Hint: A carries a probability measure ν such that $\nu(H_I) \le k^{-\operatorname{card} I}$ for each I.

The following asks if the event (13.5) is "explicitly small":

Research Problem 13.2.3 Is it true that we can find a universal constant L such that for any class of sequences T as in (13.5), the event

$$\Big\{ \sup_{t \in T} \sum_{i \le M} \delta_i t_i \ge L\delta^+(T) = L\mathsf{E} \sup_{t \in T} \sum_{i \le M} \delta_i t_i \Big\} \tag{13.10}$$

is δ-small?

Even proving that the set (13.10) is $\alpha\delta$-small, where α is some universal constant, would be of interest. The main result of Sect. 13.4 is a positive answer to this problem when T consists of indicators of sets.

Proposition 13.2.4 *If Problem 13.1.1 has a positive answer, then so does Problem 13.2.3.*

In view of (13.2), this proposition is an immediate consequence of the following, where $S(T)$ is defined in (13.3):

Proposition 13.2.5 *For any set T, the event*

$$\Big\{ \sup_{t \in T} \sum_{i \le M} \delta_i t_i \ge LS(T) \Big\}$$

is δ-small.

This result is very much weaker than a positive answer to Problem 13.2.3 because we expect $S(T)$ to by typically infinite or much larger than $\delta^+(T)$.[3] Thus, Proposition 13.2.4 achieves little more than checking that our conjectures are not blatantly wrong, and it seems better to refer to [132] for a proof.[4]

---

[3] It would be an astonishing fact if it were true that $S(T) \le L\delta^+(T)$, and proving it would be a sensational result.

[4] We do not reproduce this proof here because it uses the rather complicated Theorem 11.1 of [131], and we hope that a creative reader will invent a better argument.

13.3 My Lifetime Favorite Problem

Problem 13.2.3 motivates a more general question, which the author believes to be of fundamental importance.[5] It is considerably easier to explain this question if we identify $\{0,1\}^M$ with the class $\mathcal{M}$ of subsets of $M^* := \{1,\ldots,M\}$ in the obvious manner, identifying a point $(x_i)_{i\le M} \in \{0,1\}^M$ with the set $\{i \le M; x_i = 1\}$. We do this throughout this section.

We first explain the central combinatorial definition.

Definition 13.3.1 Given a class $D \subset \mathcal{M}$ and an integer q, we define the class $D^{(q)} \subset \mathcal{M}$ as the class of subsets of M^* which are not included in the union of any q subsets of M^* belonging to D.

It is useful to think of the points of $D^{(q)}$ as being "far from D". To make sure you understand the definition, convince yourself that if for an integer k we have

$$D = \{J \subset M^* \ ; \ \operatorname{card} J \le k\} , \tag{13.11}$$

then

$$D^{(q)} = \{J \subset M^* \ ; \ \operatorname{card} J \ge kq + 1\} . \tag{13.12}$$

Given $0 < \delta < 1$ and the corresponding independent sequence $(\delta_i)_{i\le m}$, let us denote by P_δ the law of the random set $\{i \le M; \delta_i = 1\} \in \mathcal{M}$.

Research Problem 13.3.2 Prove (or disprove) that there exists an integer q with the following property. Consider any value of δ, any value of M, and any subset D of $\mathcal{M}$ with $\mathsf{P}_\delta(D) \ge 1 - 1/q$. Then the set $D^{(q)}$ is δ-small.

In other words, we simply look for $\epsilon > 0$ small and q large such that $D^{(q)}$ is δ-small whenever $\mathsf{P}_\delta(D) \ge 1 - \epsilon$.

To understand this problem, it helps to analyze the example (13.11). Then the set H_I of (13.6) is now described by $H_I = \{J \in \mathcal{M}; I \subset J\}$. According to (13.12), we have

$$D^{(q)} \subset \bigcup_{I \in \mathcal{G}} H_I ,$$

where $\mathcal{G} = \{I \in \mathcal{M}; \operatorname{card} I = kq + 1\}$. Thus, using the elementary inequality

$$\binom{n}{k} \le \left(\frac{en}{k}\right)^k \tag{13.13}$$

[5] Far more so than Problem 13.2.3 itself

we obtain

$$\sum_{I \in \mathcal{G}} \delta^{\operatorname{card} I} = \binom{M}{kq+1} \delta^{kq+1} \le \left(\frac{eM\delta}{kq+1}\right)^{kq+1} . \tag{13.14}$$

It is elementary to show that when $\mathsf{P}_\delta(D) \ge 1/2$, one has $k \ge \delta M/L$. It then follows that if q is a large enough universal constant, the right-hand side of (13.14) is $\le 1/2$. That is, we have proved that sets D of the type (13.11) have the property described in Problem 13.3.2 that $D^{(q)}$ is δ-small.

To believe that Problem 13.3.2 has a positive solution, one has to believe that the simple case above is "extremal", i.e., "the worst possible". It might be possible to provide a negative solution to Problem 13.3.2 in a few lines: it "suffices" to invent a new type of set D to solve it negatively!

A solution to Problem 13.3.2 will be rewarded by a \$1000 prize, even if it applies only to sufficiently small values of δ. It seems probable that progress on this problem requires methods unrelated to those of this book. A simple positive result in the right direction is provided in the next section.

Proposition 13.3.3 *A positive solution to Problem 13.3.2 implies a positive solution to Problem 13.2.3.*

Proof Let q be as provided by the positive solution of Problem 13.3.2. It follows from Markov's inequality that the event $\{\sup_{t\in T} \sum_{i\le M} \delta_i t_i \le q\delta^+(T)\}$ has probability $\ge 1 - 1/q$ which in our current language means that $\mathsf{P}_\delta(D) \ge 1 - 1/q$ where $D = \{J \in \mathcal{M};\ \sup_{t\in T} \sum_{i\in J} t_i \le q\delta^+(T)\}$. Now, if $J^1, \dots, J^q \in D$ and $J = \cup_{\ell \le q} J^\ell$, it is obvious that $\sup_{t\in T} \sum_{i \in J} t_i \le q^2\delta^+(T)$. Consequently,

$$\Big\{J \in \mathcal{M}\ ;\ \sup_{t\in T} \sum_{i\in J} t_i > q^2 \delta^+(T)\Big\} \subset D^{(q)} \tag{13.15}$$

and the positive solution of Problem 13.3.2 asserts that this set is δ-small. □

The author has spent considerable energy on Problem 13.3.2. It would not be realistic to attempt to convey the depth of this problem in a few pages. A sequence of conjectures of increasing strength, of which a positive answer to (a weak version of) Problem 13.3.2 is the weakest, can be found in [131].

13.4 Classes of Sets

In this section, we consider positive selector processes in the simpler case where T consists of indicators of sets. Please be careful that the notation does not coincide with that of the previous section. It is the elements of T which are now identified to points of $\mathcal{M}$. Considering a class $\mathcal{J}$ of subsets of $M^* = \{1, \dots, M\}$, we try to bound the quantity

$$\delta(\mathcal{J}) := \mathsf{E} \sup_{J \in \mathcal{J}} \sum_{i \in J} \delta_i \ .$$

The main result of this section is Theorem 13.4.4. Before we come to it, we explore a few naive ways to bound $\delta(\mathcal{J})$.

Proposition 13.4.1 *Assume that for some number $S > 0$, we have*

$$\sum_{J \in \mathcal{J}} \Big(\frac{\delta \operatorname{card} J}{S}\Big)^S \le 1 \ . \tag{13.16}$$

Then $\delta(\mathcal{J}) \le LS$.

Proof We first observe that by (13.16), each term in the summation is ≤ 1, so that $\delta \operatorname{card} J \le S$ whenever $J \in \mathcal{J}$, and thus $u \ge 6\delta \operatorname{card} J$ whenever $u \ge 6S$. We then simply use Lemma 11.12.3 to obtain that for $u \ge 6S$, we have, using (11.70) in the second inequality,

$$\mathsf{P}\Big(\sup_{J \in \mathcal{J}} \sum_{i \in J} \delta_i \ge u\Big) \le \sum_{J \in \mathcal{J}} \mathsf{P}\Big(\sum_{i \in J} \delta_i \ge u\Big) \le \sum_{J \in \mathcal{J}} \Big(\frac{2\delta \operatorname{card} J}{u}\Big)^{u/2} \ .$$

To finish the proof, it is enough to integrate in u the previous inequality and to use (13.16) and simple estimates. □

For a class $\mathcal{J}$ of sets, let us define $S_\delta(\mathcal{J})$ as the infimum of the numbers S for which (13.16) holds. Thus, the inequality $\delta(\mathcal{J}) \le LS$ implies

$$\delta(\mathcal{J}) \le L S_\delta(\mathcal{J}) \ . \tag{13.17}$$

Exercise 13.4.2 Prove that the inequality (13.17) cannot be reversed. That is, given $A > 0$, construct a class $\mathcal{J}$ of sets for which $A\delta(\mathcal{J}) \le S_\delta(\mathcal{J})$. Hint: Consider many disjoint sets of the same cardinality.

Given a class $\mathcal{J}$ of sets and two integers n and m, let us define the class $\mathcal{J}(n, m)$ as follows:

$$\forall J \in \mathcal{J}(n, m) \ , \ \exists J_1, \dots, J_n \in \mathcal{J} \ ; \ \operatorname{card}\Big(J \setminus \bigcup_{\ell \le n} J_\ell\Big) \le m \ . \tag{13.18}$$

Then for each realization of the r.v.s (δ_i), one has

$$\sum_{i \in J} \delta_i \le m + \sum_{\ell \le n} \sum_{i \in J_\ell} \delta_i$$

and consequently

$$\delta(\mathcal{J}(n,m)) \le n\delta(\mathcal{J}) + m . \tag{13.19}$$

Combining (13.19) and (13.17), one obtains

$$\delta(\mathcal{J}(n,m)) \le LnS_\delta(\mathcal{J}) + m . \tag{13.20}$$

In particular, taking $n = 1$, for two classes $\mathcal{I}$ and $\mathcal{J}$ of sets, one has

$$\mathcal{I} \subset \mathcal{J}(1,m) \Rightarrow \delta(\mathcal{I}) \le LS_\delta(\mathcal{J}) + m ,$$

and thus

$$\delta(\mathcal{I}) \le L \inf\{S_\delta(\mathcal{J}) + m \; ; \; \mathcal{I} \subset \mathcal{J}(1,m)\} , \tag{13.21}$$

where the infimum is over all classes of sets $\mathcal{J}$ and all m for which $\mathcal{I} \subset \mathcal{J}(1,m)$. The following (very) challenging exercise disproves a most unfortunate conjecture stated in [130] and [131], which overlooked the possibility of taking $n \ge 2$ in (13.20):

Exercise 13.4.3 Using the case $n = 2$, $m = 0$ of (13.20), prove that the inequality (13.21) cannot be reversed. That is, given $A > 0$ (however large), construct a class of sets $\mathcal{I}$ such that $A\delta(\mathcal{I}) \le S_\delta(\mathcal{J}) + m$ for each class of sets $\mathcal{J}$ and each m for which $\mathcal{I} \subset \mathcal{J}(1,m)$.

In words, we can prove that (13.21) cannot be reversed because we have found a genuinely different way to bound $\delta(\mathcal{I})$, namely, (13.20) for $n = 2$.

In the same line as Exercise 13.4.3, it would seem worth investigating whether given a number A we can construct a class of sets $\mathcal{I}$ such that $A\delta(\mathcal{I}) \le nS_\delta(\mathcal{J}) + m$ whenever $\mathcal{I} \subset \mathcal{J}(n,m)$. This seems plausible, because we have a (seemingly) more general way to bound $\delta(\mathcal{I})$ than (13.19), namely, the "solid convex hull" method of Sect. 13.1.

In the remainder of this section, we prove the following:

Theorem 13.4.4 ([130]) *For any class $\mathcal{J}$ of subsets of M^*, the event (13.10)*

$$\Big\{ \sup_{J \in \mathcal{J}} \sum_{i \in J} \delta_i \ge L\delta(\mathcal{J}) \Big\}$$

is δ-small.

That is, Problem 13.2.3 has a positive solution when T consists of indicators of sets. This result is a simple consequence of the following:

Proposition 13.4.5 *Consider a class $\mathcal{J}$ of subsets of M^* and an integer n. If the event*

$$\Big\{\sup_{J\in\mathcal{J}}\sum_{i\in J}\delta_i \geq n\Big\} \tag{13.22}$$

is not δ-small, then

$$\delta(\mathcal{J}) \geq n/L_0 \;. \tag{13.23}$$

Proof of Theorem 13.4.4 Considering a class $\mathcal{J}$ of subsets of M^*, we will prove that the event

$$\Big\{\sup_{J\in\mathcal{J}}\sum_{i\in J}\delta_i \geq 3L_0\delta(\mathcal{J})\Big\} \tag{13.24}$$

is δ-small. Assume for contradiction that this is not the case. Then for $n \leq 3L_0\delta(\mathcal{J})$, the larger event (13.22) is not δ-small, and (13.23) shows that $n \leq L_0\delta(\mathcal{J})$. Thus, whenever $n \leq 3L_0\delta(\mathcal{J})$, we also have $n \leq L_0\delta(\mathcal{J})$. This means that there is no integer in the interval $]L_0\delta(\mathcal{J}), 3L_0\delta(\mathcal{J})]$, so that this interval is of length ≤ 1, i.e., $2L_0\delta(\mathcal{J}) \leq 1$. Thus, (13.23) fails for $n = 1$, so that by Proposition 13.4.5 again the event $\{\sup_{J\in\mathcal{J}}\sum_{i\in J}\delta_i \geq 1\} = \{\sup_{J\in\mathcal{J}}\sum_{i\in J}\delta_i > 0\}$ is δ-small, and the smaller event (13.24) is δ-small. This contradiction finishes the proof. □

We start the proof of Proposition 13.4.5. We fix n once and for all, and we define

$$\mathcal{J}' = \{J' \in \mathcal{M} \;;\; \operatorname{card} J' = n \;,\; \exists J \in \mathcal{J} \;,\; J' \subset J\} \;. \tag{13.25}$$

We observe that

$$\Big\{\sup_{J\in\mathcal{J}}\sum_{i\in J}\delta_i \geq n\Big\} = \Big\{\sup_{J\in\mathcal{J}'}\sum_{i\in J}\delta_i \geq n\Big\} \;. \tag{13.26}$$

For an integer $1 \leq k \leq n$, we set

$$d(k) = 2\Big(\frac{4en\delta}{k}\Big)^k \;. \tag{13.27}$$

Lemma 13.4.6 *Assume that the event (13.26) is not δ-small. Then there exists a probability measure ν on $\mathcal{J}'$ with the following property: For each set $A \subset M^*$ with $1 \leq \operatorname{card} A \leq n$, we have*

$$\nu(\{J \in \mathcal{J}' \;;\; A \subset J\}) \leq d(\operatorname{card} A) \;. \tag{13.28}$$

Proof For such a set A, consider the function f_A on $\mathcal{J}'$ given by

$$f_A(J) = \frac{1}{d(\mathrm{card}\, A)} \mathbf{1}_{\{A \subset J\}} \,.$$

The main argument is to prove that any convex combination of functions of the type f_A takes at least one value < 1 (we will then appeal to the Hahn-Banach theorem). Suppose, for contradiction, that this is not the case, so that there exist coefficients $\alpha_A \geq 0$ of sum 1 for which

$$\forall J \in \mathcal{J}' \,, \; \sum_A \alpha_A f_A(J) = \sum_{A \subset J} \frac{\alpha_A}{d(\mathrm{card}\, A)} \geq 1 \,. \tag{13.29}$$

For $1 \leq k \leq n$, let $\mathcal{G}_k$ be the collection of all the sets A for which $\mathrm{card}\, A = k$ and $\alpha_A \geq 2^{k+1}\delta^k$. Since $\sum_A \alpha_A = 1$, we have $\mathrm{card}\, \mathcal{G}_k \leq \delta^{-k} 2^{-k-1}$, and thus

$$\sum_{k \geq 1} \delta^k \,\mathrm{card}\, \mathcal{G}_k \leq \frac{1}{2} \,. \tag{13.30}$$

We claim that

$$\forall J \in \mathcal{J}' \,; \; \exists k \leq n \,, \; \exists A \in \mathcal{G}_k \,; \; A \subset J \,. \tag{13.31}$$

Indeed, otherwise, we can find $J \in \mathcal{J}'$ for which

$$A \subset J \,, \;\; \mathrm{card}\, A = k \,, \;\; k \leq n \Rightarrow \alpha_A < 2^{k+1}\delta^k$$

and thus, using the definition of $d(k)$ and (13.13),

$$\sum_{A \subset J} \frac{\alpha_A}{d(\mathrm{card}\, A)} < \sum_{1 \leq k \leq n} \binom{n}{k} \frac{2^{k+1}\delta^k}{d(k)} \leq 1 \,.$$

This contradicts (13.29) and proves (13.31).

To conclude the argument, we consider $\mathcal{G} = \bigcup_{1 \leq k \leq n} \mathcal{G}_k$. Consider $(\delta_i)_{i \leq M}$ such that $\sum_{i \in J} \delta_i \geq n$ for some $J \in \mathcal{J}'$. Then (13.31) proves that J contains a set $A \in \mathcal{G}$, so that $(\delta_i)_{i \leq M} \in H_A$, and we have shown that the event $\{\sup_{J \in \mathcal{J}'} \sum_{i \in J} \delta_i \geq n\}$ in the right of (13.26) is contained in $\cup_{A \in \mathcal{G}} H_A$. Now $\sum_{A \in \mathcal{G}} \delta^{\mathrm{card}\, A} \leq 1/2$ from (13.30). Thus, this event is δ-small. Using (13.26), we obtain that the event (13.22) is δ-small, a contradiction which proves that (13.29) is impossible.

We have proved that the convex hull $\mathcal{C}$ of the functions of the type f_A is disjoint from the set $\mathcal{U}$ of functions which are everywhere > 1. The set $\mathcal{U}$ is open and convex. The Hahn-Banach theorem asserts that we can separate the convex sets $\mathcal{C}$ and $\mathcal{U}$ by a linear functional. That is, there exist such a functional φ on $\mathbb{R}^{\mathcal{J}'}$ and a number a

such that $\varphi(f) \leq a \leq \varphi(g)$ for $f \in \mathcal{C}$ and $g \in \mathcal{U}$. For $g \in \mathcal{U}$ and $h \geq 0$, we have $g + \lambda h \in \mathcal{U}$ for each $\lambda > 0$. Thus, $\varphi(g + \lambda h) \geq a$, and hence, $\varphi(h) \geq 0$. Thus, φ is positive. We can then assume that φ is given by a probability measure ν on $\mathcal{J}'$, $\varphi(f) = \int f(J) \mathrm{d}\nu(J)$. Since $a \leq \varphi(g)$ whenever g is a constant > 1, we get $a \leq 1$. Thus, $\int f(J) \mathrm{d}\nu(J) \leq 1$ for each $f \in \mathcal{C}$ and in particular for f of the type f_A. We have proved (13.28). □

Lemma 13.4.7 *Assume that the event (13.26) is not δ-small. Then this event has a probability* $\geq \exp(-Ln)$.

Proof Consider the probability ν on the set $\mathcal{J}'$ of (13.25) as in (13.28) and the r.v. (depending on the random input $(\delta_i)_{i \leq M}$)

$$Y = \nu(\{J \in \mathcal{J}' \; ; \; \forall i \in J \; , \; \delta_i = 1\}) = \nu(\{J \; ; \; (\delta_i) \in H_J\})$$
$$= \int \mathbf{1}_{\{(\delta_i) \in H_J\}} \mathrm{d}\nu(J) \; .$$

Obviously, the event (13.26) contains the event $Y > 0$. The plan is to use the Paley-Zygmund inequality in the weak form

$$\mathsf{P}(Y > 0) \geq \frac{(\mathsf{E}Y)^2}{\mathsf{E}Y^2} \; , \tag{13.32}$$

which is a simple consequence of the Cauchy-Schwarz inequality. First,

$$\mathsf{E}Y = \mathsf{E} \int \mathbf{1}_{\{(\delta_i) \in H_J\}} \mathrm{d}\nu(J) = \int \mathsf{P}(H_J) \mathrm{d}\nu(J) = \delta^n \; , \tag{13.33}$$

since ν is supported by $\mathcal{J}'$ and card $J = n$ for $J \in \mathcal{J}'$. Next,

$$Y^2 = \nu^{\otimes 2}(\{(J, J') \; ; \; (\delta_i) \in H_J, (\delta_i) \in H_{J'}\})$$
$$= \nu^{\otimes 2}(\{(J, J') \; ; \; (\delta_i) \in H_J \cap H_{J'}\}) \; ,$$

so that, proceeding as in (13.33), and since $\mathsf{P}((\delta_i) \in H_J \cap H_{J'}) = \delta^{\mathrm{card}(J \cup J')}$,

$$\mathsf{E}Y^2 = \int \delta^{\mathrm{card}(J \cup J')} \mathrm{d}\nu(J) \mathrm{d}\nu(J') \; . \tag{13.34}$$

Now, the choice $A = J \cap J'$ shows that

$$\delta^{\mathrm{card}(J \cup J')} \leq \sum_{A \subset J} \delta^{2n - \mathrm{card}\, A} \mathbf{1}_{\{A \subset J'\}}$$

and therefore, using (13.28) in the second line and again (13.13) in the last line,

$$\begin{aligned}\int \delta^{\mathrm{card}(J\cup J')}\mathrm{d}\nu(J') &\le \sum_{A\subset J}\delta^{2n-\mathrm{card}\,A}\nu(\{J'\,;\,A\subset J'\})\\ &\le \sum_{0\le k\le n}\binom{n}{k}\delta^{2n-k}d(k)\\ &\le 2\delta^{2n}\sum_{0\le k\le n}\Bigl(\frac{2en}{k}\Bigr)^{2k}\,. \end{aligned} \tag{13.35}$$

An elementary computation shows that the last term dominates in the sum, so that the right-hand side of (13.35) is less than $\le \delta^{2n}\exp Ln$, and recalling (13.34), this proves that $\mathsf{E}Y^2 \le \exp(Ln)(\mathsf{E}Y)^2$ and completes the proof using (13.32). □

Proof of Proposition 13.4.5 Consider the r.v. $X = \sup_{J\in\mathcal{J}}\sum_{i\in J}\delta_i$. We assume that the event $\{X \ge n\}$ is not δ-small. Lemma 13.4.7 implies that

$$\mathsf{P}(X \ge n) \ge \exp(-L_1 n)\,. \tag{13.36}$$

From this fact alone, we shall bound from below $\delta(\mathcal{J}) = \mathsf{E}X$. Using Markov's inequality, we know that $\mathsf{P}(D) \ge 1/2$, where $D = \{X \le 2\delta(\mathcal{J})\}$. Recalling the set $D^{(q)}$ of Definition 13.3.1, given two integers q and $k \ge 0$, we define similarly $D^{(q,k)}$ as the set of subsets J of M^* which have the property that whenever one considers $J^1,\dots,J^q \in D$, then

$$\mathrm{card}\bigl(J\setminus \cup_{\ell\le q}J^\ell\bigr) \ge k+1\,.$$

Thus, $D^{(q,0)} = D^{(q)}$, and as in (13.15), one proves that

$$\{X \ge 2q\delta(\mathcal{J}) + k\} \subset D^{(q,k)}\,. \tag{13.37}$$

The heart of the matter is Theorem 3.1.1 of [121] which asserts that

$$\mathsf{P}(D^{(q,k)}) \le \frac{2^q}{q^k}\,.$$

Comparing with (13.36) and (13.37) then yields

$$2q\delta(\mathcal{J}) + k \le n \Rightarrow \exp(-L_1 n) \le \frac{2^q}{q^k}\,.$$

Let us fix q with $q \ge \exp(2L_1)$, so that q is now a universal constant. If $2q\delta(\mathcal{J}) \ge n$, we have proved that $\delta(\mathcal{J}) \ge n/L$ so we may assume that $2q\delta(\mathcal{J}) < n$. Let us consider $k \ge 0$ with $2q\delta(\mathcal{J}) + k \le n$. Then $\exp(-L_1 n) \le 2^q/q^k$ so that

$\exp(-L_1 n) \le 2^q \exp(-2L_1 k)$, and thus $2k - n \le L$, so that $k \le n/2 + L$. So, we have proved that $k \le n/2 + L$ whenever $k \le n - 2q\delta(\mathcal{J})$. Consequently $n - 2q\delta(\mathcal{J}) \le n/2 + L + 1$, and thus $\delta(\mathcal{J}) \ge (n - L_0)/L_0$. We have proved that $\delta(\mathcal{J}) \ge n/L$ when $n \ge 2L_0$. This finishes the proof in that case.

Finally, we finish the proof when $n \le 2L_0$. Since our assumption is that the event $X \ge n$ is not δ-small, the larger event $\{X \ge 1\}$ is not δ-small. Consider the union I of all the elements of $\mathcal{J}$, and let $m = \operatorname{card} I$. Then $\{X \ge 1\} \subset \cup_{i \in I} H_{\{i\}}$ so that since this event is not small, we have $\delta \operatorname{card} I \ge 1/2$. So if $\Xi = \{\exists i \in I, \delta_i = 1\}$, then $\mathsf{P}(\Xi) = 1 - (1 - \delta)^m \ge 1/L$. Now $X \ge \mathbf{1}_\Xi$ so that taking expectation, we get $\delta(\mathcal{J}) \ge \mathsf{P}(\Xi) \ge 1/L$. □

Part III
Practicing

Chapter 14
Empirical Processes, II

The reader should review Sect. 6.8 where we started the study of empirical processes. Empirical processes are a vast topic, but here our goal is pretty limited. In Sect. 14.1, we prove a "bracketing theorem" to illustrate again the power of the methods of Sect. 9.4. In Sects. 14.2 and 14.3, we prove two specific results, which illustrate in particular that Proposition 6.8.2 performs no miracle: it is the part "without cancellation" which requires work and for which one must use a specific method in each case.

We denote by (X_i) an i.i.d. sequence of r.v.s valued in a measure space (Ω, μ), μ being the common law of the (X_i).

14.1 Bracketing

Theorem 14.1.1 *Consider a countable class $\mathcal{F}$ of functions in $L^2(\mu)$ with $0 \in \mathcal{F}$. Consider an admissible sequence $(\mathcal{A}_n)$ of partitions of $\mathcal{F}$. For $A \in \mathcal{A}_n$, define the function h_A by*

$$h_A(\omega) = \sup_{f, f' \in A} |f(\omega) - f'(\omega)| . \tag{14.1}$$

Consider an integer $N \geq 1$. Assume that for a certain $j_0 = j_0(\mathcal{F}) \in \mathbb{Z}$, we have

$$\|h_{\mathcal{F}}\|_2 \leq \frac{2^{-j_0}}{\sqrt{N}} . \tag{14.2}$$

M. Talagrand, *Upper and Lower Bounds for Stochastic Processes*, Ergebnisse der Mathematik und ihrer Grenzgebiete. 3. Folge / A Series of Modern Surveys in Mathematics 60, https://doi.org/10.1007/978-3-030-82595-9_14

Assume that for each $n \geq 1$ and each $A \in \mathcal{A}_n$, we are given a number $j_n(A) \in \mathbb{Z}$ with

$$\int (2^{2j_n(A)} h_A^2) \wedge 1 \mathrm{d}\mu \leq \frac{2^n}{N} \tag{14.3}$$

and let

$$S = \sup_{f \in \mathcal{F}} \sum_{n \geq 0} 2^{n - j_n(A_n(f))} , \tag{14.4}$$

where $A_n(f)$ denotes the unique element of $\mathcal{A}_n$ which contains f. Then

$$S_N(\mathcal{F}) := \mathsf{E} \sup_{f \in \mathcal{F}} \Big| \sum_{i \leq N} (f(X_i) - \mu(f)) \Big| \leq LS . \tag{14.5}$$

It is instructive to rewrite (14.2) as $\int 2^{2j_0(\mathcal{F})} h_{\mathcal{F}}^2 \mathrm{d}\mu \leq 1/N$ in order to compare it with (14.3). The normalization used in this theorem is not intuitive, but should be clearer after you study the proof of the next corollary.

Corollary 14.1.2 *With the notation of Theorem 14.1.1, define now*

$$S^* = \sup_{f \in \mathcal{F}} \sum_{n \geq 0} 2^{n/2} \| h_{A_n(f)} \|_2 . \tag{14.6}$$

Then

$$S_N(\mathcal{F}) \leq L\sqrt{N} S^* . \tag{14.7}$$

Since $\Delta(A) \leq \|h_A\|_2$, we have $\gamma_2(\mathcal{F}, d_2) \leq S^*$; but it is not true in general that $S_N(\mathcal{F}) \leq L\sqrt{N}\gamma_2(\mathcal{F}, d_2)$.

Proof This is routine. Define $j_n(A)$ as the largest integer j for which $\|h_A\|_2 \leq 2^{n/2-j}/\sqrt{N}$, so that $2^{n/2-j_n(A)} \leq 2\sqrt{N}\|h_A\|_2$, and consequently,

$$\sum_{n \geq 0} 2^{n - j_n(A_n(f))} \leq 2\sqrt{N} \sum_{n \geq 0} 2^{n/2} \| h_{A_n(f)} \|_2 .$$

Since $\|h_A\|_2 \leq 2^{n/2-j_n(A)}/\sqrt{N}$, (14.3) holds, and the result follows from Theorem 14.1.1. □

Exercise 14.1.3 Given two (measurable) functions $f_1 \leq f_2$, define the bracket $[f_1, f_2]$ as the set of functions $\{f; f_1 \leq f \leq f_2\}$. Given a class $\mathcal{F}$ of functions and $\epsilon > 0$, define $N_{[\,]}(\mathcal{F}, \epsilon)$ as the smallest number of brackets $[f_1, f_2]$ with

$\|f_2 - f_1\|_2 \le \epsilon$ which can cover $\mathcal{F}$. Use Corollary 14.1.2 to prove that

$$S_N(\mathcal{F}) \le L\sqrt{N} \int_0^\infty \sqrt{\log N_{[\,]}(\mathcal{F}, \epsilon)} d\epsilon \,. \tag{14.8}$$

Inequality (14.8) is known as Ossiander's bracketing theorem [77], and (14.7) is simply the "generic chaining version" of it. The proof of Ossiander's bracketing theorem requires a tricky idea beyond the ideas of Dudley's bound. In our approach, Ossiander's bracketing theorem is a straightforward consequence of Theorem 14.1.1, itself a straightforward consequence of Theorem 9.4.1. None of the simple arguments there involves chaining. All the work involving chaining has already been performed in Theorem 9.4.1. As suggested in Sect. 6.9, in some sense, Theorem 9.4.1 succeeds in extending Ossiander's bracketing theorem to a considerably more general setting.

Proof of Theorem 14.1.1 Let us fix $A \in \mathcal{A}_n$, and consider the r.v.s $W_i = (2^{2j_n(A)} h_A(X_i)^2) \wedge 1$, so that by (14.3), we have $\sum_{i\le N} \mathsf{E} W_i \le 2^n$. Consider a parameter $u \ge 1$. Then Lemma 7.7.2 (b) yields

$$\mathsf{P}\Big(\sum_{i\le N} W_i \ge u2^{n+2}\Big) \le \exp(-u2^{n+1}) \,. \tag{14.9}$$

Consider the event $\Omega(u)$ defined by

$$\forall n \ge 0 \,,\ \forall A \in \mathcal{A}_n \,,\ \sum_{i\le N} 2^{2j_n(A)} h_A(X_i)^2 \wedge 1 \le u2^{n+2} \,, \tag{14.10}$$

so that (14.9) and the union bound yield $\mathsf{P}(\Omega(u)) \ge 1 - L\exp(-u)$. Let us consider independent Bernoulli r.v.s ε_i, which are independent of the X_i, and let us recall that E_ε denotes expectation in the r.v.s ε_i only. Given the r.v.s X_i, we consider the set T of all sequences of the type $(x_i)_{1\le i\le N} = (f(X_i))_{1\le i\le N}$ for $f \in \mathcal{F}$. To bound $\mathsf{E}\sup_{x\in T} |\sum_{i\le N} \varepsilon_i x_i|$, we appeal to Theorem 9.4.1. Also, since $|f(X_i) - f'(X_i)| \le h_A(X_i)$ for $f, f' \in A$, (9.42) (with $4u$ rather than u) follows from (14.10). Finally, for $f \in \mathcal{F}$, we have $|f(X_i)| \le h_{\mathcal{F}}(X_i)$, so that

$$|f(X_i)|\mathbf{1}_{\{2|f(X_i)|\ge 2^{-j_0(\mathcal{F})}\}} \le h_{\mathcal{F}}(X_i)\mathbf{1}_{\{2h_{\mathcal{F}}(X_i)\ge 2^{-j_0(\mathcal{F})}\}} \,.$$

We then use (9.44) with $p = 1$ to obtain

$$\mathsf{E}_\varepsilon \sup_{f\in\mathcal{F}} \Big|\sum_{i\le N} \varepsilon_i f(X_i)\Big| \le Lu \sup_{f\in\mathcal{F}} \sum_{n\ge 0} 2^{n-j_n(A_n(f))} + L\sum_{i\le N} h_{\mathcal{F}}(X_i)\mathbf{1}_{\{2h_{\mathcal{F}}(X_i)\ge 2^{-j_0(\mathcal{F})}\}} \,. \tag{14.11}$$

The expectation of the last term is $LN \int h_{\mathcal{F}} \mathbf{1}_{\{2h_{\mathcal{F}} \ge 2^{-j_0(\mathcal{F})}\}} \mathrm{d}\mu$. Now, since $h\mathbf{1}_{\{h \ge v\}} \le h^2/v$, and using (14.2) in the last inequality,

$$N \int h_{\mathcal{F}} \mathbf{1}_{\{2h_{\mathcal{F}} \ge 2^{-j_0(\mathcal{F})}\}} \mathrm{d}\mu \le N 2^{j_0(\mathcal{F})+1} \int h_{\mathcal{F}}^2 \mathrm{d}\mu \le 2^{-j_0(\mathcal{F})+1} .$$

Consequently, taking expectation in (14.11) and using that $\mathsf{P}(\Omega(u)) \ge 1 - L\exp(-u)$, we obtain

$$\mathsf{E} \sup_{f \in \mathcal{F}} \Big| \sum_{i \le N} \varepsilon_i f(X_i) \Big| \le L \sup_{f \in \mathcal{F}} \sum_{n \ge 0} 2^{n - j_n(A_n(f))} = LS ,$$

and we conclude the proof using Lemma 11.8.4. □

14.2 The Class of Squares of a Given Class

It is beyond the scope of this book to cover the theory of empirical processes (even restricted to its applications to Analysis and Banach Space theory). In the rest of this chapter, we give two sample results, which are facets of the following problem. Consider independent r.v.s X_i valued in $\mathbb{R}^m$. Denoting by $\langle \cdot, \cdot \rangle$ the canonical duality of $\mathbb{R}^m$ with itself, and T a subset of $\mathbb{R}^m$, we are interested in bounding the quantity

$$\sup_{t \in T} \Big| \sum_{i \le N} (\langle X_i, t \rangle^2 - \mathsf{E}\langle X_i, t \rangle^2) \Big| . \tag{14.12}$$

As a warm-up, we recommend that the reader studies the following exercise. The results there are often needed.

Exercise 14.2.1 Given a probability μ, for a measurable function f, we define the following two norms (Orlicz norms)

$$\|f\|_{\psi_1} = \inf\Big\{A > 0 \,;\, \int \exp\Big(\frac{|f|}{A}\Big) \mathrm{d}\mu \le 2\Big\} \tag{14.13}$$

and

$$\|f\|_{\psi_2} = \inf\Big\{A > 0 \,;\, \int \exp\Big(\frac{f^2}{A^2}\Big) \mathrm{d}\mu \le 2\Big\} . \tag{14.14}$$

(a) Prove that if $k \ge 1$

$$\int \exp|f| \mathrm{d}\mu \le 2^k \Rightarrow \|f\|_{\psi_1} \le k . \tag{14.15}$$

Hint: Use Hölder's inequality.

(b) Prove that

$$\forall u > 0\ ,\ \mathsf{P}(|f| \geq u) \leq 2\exp(-u) \Rightarrow \|f\|_{\psi_1} \leq L$$

and

$$\forall u > 0\ ,\ \mathsf{P}(|f| \geq u) \leq 2\exp(-u^2) \Rightarrow \|f\|_{\psi_2} \leq L\ .$$

(c) Prove that

$$\|f\|_{\psi_1} \leq L\|f\|_{\psi_2} \tag{14.16}$$

and

$$\|f_1 f_2\|_{\psi_1} \leq \|f_1\|_{\psi_2}\|f_2\|_{\psi_2}\ . \tag{14.17}$$

(d) On a rainy day, obtain a completely uninteresting and useless result by computing the exact value of $\|g\|_{\psi_2}$ where g is a standard Gaussian r.v.
(e) If (ε_i) denote independent Bernoulli r.v.s and (a_i) denote real numbers, prove that

$$\Big\|\sum_i a_i\varepsilon_i\Big\|_{\psi_2} \leq L\Big(\sum_i a_i^2\Big)^{1/2}\ . \tag{14.18}$$

Hint: Use the sub-Gaussian inequality (6.1.1).
(f) Prove that if the r.v.s Y_i are independent and centered, then for $v > 0$, it holds that

$$\mathsf{P}\Big(\sum_{i\geq 1} Y_i \geq v\Big) \leq \exp\Big(-\frac{1}{L}\min\Big(\frac{v^2}{\sum_{i\leq N}\|Y_i\|_{\psi_1}^2}, \frac{v}{\max_{i\leq N}\|Y_i\|_{\psi_1}}\Big)\Big)\ . \tag{14.19}$$

Hint: Prove that for $|\lambda|\|Y\|_{\psi_1} \leq 1/2$, we have $\mathsf{E}\exp\lambda Y \leq \exp(\lambda^2\|Y\|_{\psi_1}^2/L)$, and copy the proof of Bernstein's inequality.
(g) Prove that if the r.v.s Y_i are independent and centered, then for $v > 0$, it holds

$$\mathsf{P}\Big(\sum_{i\geq 1} Y_i \geq v\Big) \leq \exp\Big(-\frac{v^2}{L\sum_{i\leq N}\|Y_i\|_{\psi_2}^2}\Big)\ . \tag{14.20}$$

The tail inequalities (14.19) and (14.20) motivate the use of the distances d_{ψ_1} and d_{ψ_2} associated with the norms $\|\cdot\|_{\psi_1}$ and $\|\cdot\|_{\psi_2}$.

As in the previous section, we consider a probability space (Ω, μ), and we denote by $(X_i)_{i\leq N}$ r.v.s valued in Ω of law μ. We recall the norm $\|\cdot\|_{\psi_1}$ of (14.13) and the associated distance d_{ψ_1}. Before we come to our main result, we prove a simpler but connected fact. We consider independent Bernoulli r.v.s ε_i independent of the r.v.s X_i.

Theorem 14.2.2 *Consider a (countable) class of functions $\mathcal{F}$ with $0 \in \mathcal{F}$. Then for any integer N,*

$$\mathsf{E} \sup_{f \in \mathcal{F}} \Big| \sum_{i \le N} \varepsilon_i f(X_i) \Big| \le L\sqrt{N}\gamma_2(\mathcal{F}, d_{\psi_1}) + L\gamma_1(\mathcal{F}, d_{\psi_1}) . \tag{14.21}$$

To understand this statement, we observe that $\gamma_2(\mathcal{F}, d_{\psi_1}) \le \gamma_1(\mathcal{F}, d_{\psi_1})$, but that the factor $\sqrt{N}$ is in front of the smaller term $\gamma_2(\mathcal{F}, d_{\psi_1})$.

Exercise 14.2.3 After studying the proof of Theorem 14.2.2, produce a decomposition of $\mathcal{F}$ as in Theorem 6.8.3, where $S_N(\mathcal{F})$ is replaced by $L\sqrt{N}\gamma_2(\mathcal{F}, d_{\psi_1}) + L\gamma_1(\mathcal{F}, d_{\psi_1})$.

We start the proof of Theorem 14.2.2 with a simple fact.

Lemma 14.2.4 *If $u \ge 1$, then*

$$\mathsf{P}\Big(\sum_{i \le N} |f(X_i)| \ge 2uN\|f\|_{\psi_1}\Big) \le \exp(-uN) . \tag{14.22}$$

Proof Replacing f by $f/\|f\|_{\psi_1}$, we may assume $\|f\|_{\psi_1} = 1$. Then

$$\mathsf{E}\exp \sum_{i \le N} |f(X_i)| \le 2^N \le e^N .$$

Using the inequality $\mathsf{P}(X \ge v) \le \exp(-v)\mathsf{E}\exp X$ yields $\mathsf{P}(\sum_{i \le N} |f(X_i)| \ge N(u+1)) \le \exp(-uN)$. □

We consider an admissible sequence $(\mathcal{A}_n)$ of partitions of $\mathcal{F}$ such that

$$\forall f \in \mathcal{F} , \sum_{n \ge 0} 2^{n/2} \Delta(A_n(f), d_{\psi_1}) \le L\gamma_2(\mathcal{F}, d_{\psi_1}) \tag{14.23}$$

$$\forall f \in \mathcal{F} , \sum_{n \ge 0} 2^{n} \Delta(A_n(f), d_{\psi_1}) \le L\gamma_1(\mathcal{F}, d_{\psi_1}) . \tag{14.24}$$

For each $A \in \mathcal{A}_n$, we choose a point $f_{n,A} \in A$. We will lighten notation by writing f_A rather than $f_{n,A}$. For $A \in \mathcal{A}_n$ with $n \ge 1$, we denote by A' the unique element of $\mathcal{A}_{n-1}$ that contains A. This defines as usual a chaining in $\mathcal{F}$, by choosing $\pi_n(f) = f_A$ where $A = A_n(f)$.

We denote by n_1 the largest integer with $2^{n_1} \le N$, so that $N \le 2^{n_1+1}$.

Lemma 14.2.5 *Consider a parameter* $u \geq 1$ *and the event* $\Omega(u)$ *defined by the following conditions:*

$$\forall n\,,\ 1 \leq n \leq n_1\,,\ \forall A \in \mathcal{A}_n\,,$$
$$\Big|\sum_{i\leq N} \varepsilon_i (f_A(X_i) - f_{A'}(X_i))\Big| \leq Lu2^{n/2}\sqrt{N}\Delta(A', d_{\psi_1})\,. \tag{14.25}$$

$$\forall n > n_1\,,\ \forall A \in \mathcal{A}_n\,,\ \sum_{i\leq N} |f_A(X_i) - f_{A'}(X_i)| \leq Lu2^n\Delta(A', d_{\psi_1})\,. \tag{14.26}$$

Then

$$\mathsf{P}(\Omega(u)) \geq 1 - L\exp(-u)\,. \tag{14.27}$$

Proof The r.v. $Y_i = \varepsilon_i(f_A(X_i) - f_{A'}(X_i))$ is centered, and since f_A and $f_{A'}$ belong to A', we have $\|Y_i\|_{\psi_1} \leq \Delta(A', d_{\psi_1})$. We use (14.19) to obtain that for any $v > 0$,

$$\mathsf{P}\Big(\Big|\sum_{i\leq N} Y_i\Big| \geq v\Delta(A', d_{\psi_1})\Big) \leq 2\exp\Big(-\frac{1}{L}\min\Big(\frac{v^2}{N}, v\Big)\Big)\,. \tag{14.28}$$

Since $n \leq n_1$, we have $\sqrt{N} \geq 2^{n/2}$. For $u \geq 1$, setting $v = Lu2^{n/2}\sqrt{N}$, then $v^2/N \geq Lu2^n$ and $v \geq Lu2^n$. Thus, (14.28) implies that the inequality in (14.25) occurs with probability $\geq 1 - L\exp(-2u2^n)$.

Next, since $\|f_A - f_{A'}\|_{\psi_1} \leq \Delta(A', d_{\psi_1})$ and $u2^n/N \geq 1$ for $n > n_1$, using Lemma 14.2.4 for $2u2^n/N$ rather than u implies as desired that the right-hand side of (14.26) occurs with probability $\geq 1 - L\exp(-2u2^n)$.

Since $\operatorname{card}\mathcal{A}_n \leq N_n = 2^{2^n}$ and since $\sum_{n\geq 0} 2^{2^n}\exp(-2u2^n) \leq L\exp(-u)$, (14.27) follows from the union bound. □

Proof of Theorem 14.2.2 We prove that on the event $\Omega(u)$ of Lemma 14.2.5, we have

$$\sup_{f\in\mathcal{F}} \Big|\sum_{i\leq N} \varepsilon_i f(X_i)\Big| \leq Lu(\sqrt{N}\gamma_2(\mathcal{F}, d_{\psi_1}) + \gamma_1(\mathcal{F}, d_{\psi_1}))\,, \tag{14.29}$$

which by taking expectation and using (14.27) implies the result. To prove (14.29), since $0 \in \mathcal{F}$, we may assume that $\pi_0(f) = 0$. We deduce from (14.25) that for each n with $1 \leq n \leq n_1$, one has

$$\Big|\sum_{i\leq N} \varepsilon_i(\pi_n(f)(X_i) - \pi_{n-1}(f)(X_i))\Big| \leq Lu2^{n/2}\sqrt{N}\Delta(A_{n-1}(f), d_{\psi_1})\,. \tag{14.30}$$

For $n > n_1$, we use that from (14.26), we have

$$\Big|\sum_{i\le N}\varepsilon_i(\pi_n(f)(X_i)-\pi_{n-1}(f)(X_i))\Big| \le \sum_{i\le N}\big|\pi_n(f)(X_i)-\pi_{n-1}(f)(X_i)\big|$$
$$\le u2^n\Delta(A_{n-1}(f),d_{\psi_1})\,. \tag{14.31}$$

Summation of these inequalities together with (14.23) and (14.24) concludes the proof. □

We now come to the main result of this section. We recall the norm $\|\cdot\|_{\psi_2}$ of (14.14). We denote by d_{ψ_2} the associated distance.

Theorem 14.2.6 ([41,73]) *Consider a (countable) class of functions $\mathcal{F}$ with $0\in\mathcal{F}$. Assume that*

$$\forall f\in\mathcal{F}\,,\ \|f\|_{\psi_2}\le\Delta^*\,. \tag{14.32}$$

Then for any integer N,

$$\mathsf{E}\sup_{f\in\mathcal{F}}\Big|\sum_{i\le N}(f(X_i)^2-\mathsf{E}f^2)\Big|\le L\sqrt{N}\Delta^*\gamma_2(\mathcal{F},d_{\psi_2})+L\gamma_2(\mathcal{F},d_{\psi_2})^2\,. \tag{14.33}$$

The point of the theorem is that we use information on the class $\mathcal{F}$ to bound the empirical process on the class $\mathcal{F}^2=\{f^2;\,f\in\mathcal{F}\}$. This theorem does not follow from Theorem 14.2.2 applied to the class $\mathcal{F}^2=\{f^2;\,f\in\mathcal{F}\}$. As we will show, it is true that $\gamma_2(\mathcal{F}^2,d_{\psi_1})\le\Delta^*\gamma_2(\mathcal{F},d_{\psi_2})$, so that the first term in the right-hand side of (14.33) is really the same as in the right-hand side of (14.21) but the functional γ_1 no longer occurs in (14.33).

As an example of relevant situation, let us consider the case where $\Omega=\mathbb{R}^m$ and where μ is the canonical Gaussian measure on $\mathbb{R}^m$, i.e., the law of an independent sequence $(g_i)_{i\le m}$ of standard Gaussian r.v.s. Recalling that $\langle\cdot,\cdot\rangle$ denotes the canonical duality between $\mathbb{R}^m$ and itself, for any $t\in\mathbb{R}^m$, we have

$$\int\langle t,x\rangle^2\mathrm{d}\mu(x)=\|t\|_2^2, \tag{14.34}$$

where $\|t\|_2$ denotes the Euclidean norm of t. In words, μ is "isotropic". Thus, if X_i has law μ, then $\mathsf{E}\langle t,X_i\rangle^2=\|t\|_2^2$. Consider a subset T of $\mathbb{R}^m$, which is seen as a set $\mathcal{F}$ of functions on Ω through the canonical duality $\langle\cdot,\cdot\rangle$. The left-hand side of (14.33) is then simply

$$\mathsf{E}\sup_{t\in T}\Big|\sum_{i\le N}(\langle t,X_i\rangle^2-\|t\|_2^2)\Big|\,. \tag{14.35}$$

A bound for this quantity is relevant in particular to the problem of signal reconstruction, i.e., of (approximately) finding the transmitted signal $t \in T$ when observing only the data $(\langle t, X_i \rangle)_{i \le N}$; see [73] for details.[1] In these applications, one does not like to have $0 \in \mathcal{F}$, but one assumes instead that $\mathcal{F}$ is symmetric (i.e., $-f \in \mathcal{F}$ if $f \in \mathcal{F}$). It is simple to show that (14.33) still holds in this case. (Let us also observe that (14.33) does not hold when $\mathcal{F}$ is reduced to a single non-zero function.)

Now, what is a possible strategy to prove Theorem 14.2.6? First, rather than the left-hand side of (14.33), we shall bound $\mathsf{E} \sup_{f \in \mathcal{F}} |\sum_{i \le N} \varepsilon_i f(X_i)^2|$, where (ε_i) are independent Bernoulli r.v.s, independent of the r.v.s (X_i) and use Lemma 11.8.4. We have to bound the empirical process on the class $\mathcal{F}^2 = \{f^2; f \in \mathcal{F}\}$. There is a natural chaining $(\pi_n(f))$ on $\mathcal{F}$, witnessing the value of $\gamma_2(\mathcal{F}, d_{\psi_2})$. There simply seems to be no other way than to use the chaining $(\pi_n(f)^2)$ on $\mathcal{F}^2$ and to control the "increments along the chain":

$$\sum_{i \le N} \varepsilon_i (\pi_n(f)(X_i)^2 - \pi_{n-1}(f)(X_i)^2) .$$

It seems unavoidable that we will have to control some of the quantities $|\pi_n(f)(X_i)^2 - \pi_{n-1}(f)(X_i)^2|$. Our hypotheses on $\mathcal{F}$ do not yield naturally a control of these quantities, but rather of the quantities $|\pi_n(f)(X_i) - \pi_{n-1}(f)(X_i)|$. Since

$$\pi_n(f)(X_i)^2 - \pi_{n-1}(f)(X_i)^2 = (\pi_n(f)(X_i) - \pi_{n-1}(f)(X_i))(\pi_n(f)(X_i) + \pi_{n-1}(f)(X_i)) ,$$

it seems impossible to achieve anything unless we have some additional control of the sequence $(\pi_n(f)(X_i) + \pi_{n-1}(f)(X_i))_{i \le N}$, which most likely means that we must gain some control of the sequence $(f(X_i))_{i \le N}$ for all $f \in \mathcal{F}$. Indeed, a key step of the proof will be to show that

$$\mathsf{E} \sup_{f \in \mathcal{F}} \Big(\sum_{i \le N} f(X_i)^2 \Big)^{1/2} \le L(\sqrt{N}\Delta^* + \gamma_2(\mathcal{F}, d_{\psi_2})) . \tag{14.36}$$

We now prepare the proof of Theorem 14.2.6. We consider an admissible sequence $(\mathcal{A}_n)$ of partitions of $\mathcal{F}$ such that

$$\forall f \in \mathcal{F} , \sum_{n \ge 0} 2^{n/2} \Delta(A_n(f), d_{\psi_2}) \le 2\gamma_2(\mathcal{F}, d_{\psi_2}) . \tag{14.37}$$

[1] A few words about this general direction may be found in Sect. D.6.

For each $A \in \mathcal{A}_n$, we choose a point $f_{n,A} \in A$. We will lighten notation by writing f_A rather than $f_{n,A}$. For $A \in \mathcal{A}_n$ with $n \geq 1$, we denote by A' the unique element of $\mathcal{A}_{n-1}$ that contains A. This defines as usual a chaining in $\mathcal{F}$, by choosing $\pi_n(f) = f_A$ where $A = A_n(f)$.

We consider Bernoulli r.v.s (ε_i) independent of the r.v.s (X_i). We denote by n_1 the largest integer with $2^{n_1} \leq N$, so that $N \leq 2^{n_1+1}$.

Lemma 14.2.7 *Consider a parameter $u \geq 1$ and the event $\Omega(u)$ defined by the following conditions:*

$$\forall n\ ,\ 1 \leq n \leq n_1\ ,\ \forall A \in \mathcal{A}_n\ ,$$

$$\Big|\sum_{i \leq N} \varepsilon_i (f_A(X_i)^2 - f_{A'}(X_i)^2)\Big| \leq Lu 2^{n/2} \sqrt{N} \Delta^* \Delta(A', d_{\psi_2})\ . \tag{14.38}$$

$$\forall n > n_1\ ,\ \forall A \in \mathcal{A}_n\ ,\ \sum_{i \leq N} (f_A(X_i) - f_{A'}(X_i))^2 \leq Lu 2^n \Delta(A', d_{\psi_2})^2\ . \tag{14.39}$$

$$\forall A \in \mathcal{A}_{n_1}\ ,\ \sum_{i \leq N} f_A(X_i)^2 \leq LuN\Delta^{*2}\ . \tag{14.40}$$

Then

$$\mathsf{P}(\Omega(u)) \geq 1 - L\exp(-u)\ . \tag{14.41}$$

Proof Let us first study (14.38). By (14.17), we have

$$\|f_A^2 - f_{A'}^2\|_{\psi_1} \leq \|f_A - f_{A'}\|_{\psi_2} \|f_A + f_{A'}\|_{\psi_2} \leq \Delta(A', d_{\psi_2}) \times 2\Delta^*\ .$$

Consequently, the r.v. $Y_i = \varepsilon_i (f_A(X_i)^2 - f_{A'}(X_i)^2)$ is centered and $\|Y_i\|_{\psi_1} \leq 2\Delta^* \Delta(A', d_{\psi_2})$. We prove that the inequality in (14.38) occurs with probability $\geq 1 - L\exp(-2u2^n)$ just as in the case of (14.25).

Let us turn to the study of (14.39). It is obvious from the definition (or from (14.17)) that $\|f^2\|_{\psi_1} \leq \|f\|_{\psi_2}^2$, so the function $f = (f_A - f_{A'})^2$ satisfies $\|f\|_{\psi_1} \leq \|f_A - f_{A'}\|_{\psi_2}^2 \leq \Delta(A', d_{\psi_2})^2$. Also $u2^n/N \geq 1$ for $n > n_1$. Using Lemma 14.2.4 for $2u2^n/N$ rather than u implies as desired that the right-hand side of (14.39) occurs with probability $\geq 1 - L\exp(-2u2^n)$.

Using again Lemma 14.2.4, and since $\|f_A^2\|_{\psi_1} \leq \|f_A\|_{\psi_2}^2 \leq \Delta^{*2}$ by (14.32), we obtain that for any $A \in \mathcal{A}_{n_1}$, inequality (14.40) holds with probability $\geq 1 - L\exp(-2Nu)$.

Finally, we use the union bound. Since $\operatorname{card} \mathcal{A}_n \leq N_n = 2^{2^n}$ and in particular $\operatorname{card} \mathcal{A}_{n_1} \leq N_{n_1} \leq 2^N$, and since $\sum_{n \geq 0} 2^{2^n} \exp(-2u2^n) \leq L\exp(-u)$, the result follows. □

We consider the random norm $W(f)$ given by

$$W(f) = \Big(\sum_{i\le N} f(X_i)^2\Big)^{1/2} . \tag{14.42}$$

Lemma 14.2.8 *On the event $\Omega(u)$, we have*

$$\forall f \in \mathcal{F}\ ,\ W(f) \le L\sqrt{u}(\sqrt{N}\Delta^* + \gamma_2(\mathcal{F}, d_{\psi_2})) . \tag{14.43}$$

Proof Given $f \in \mathcal{F}$, we denote by $\pi_n(f)$ the element f_A where $A = A_n(f)$. We also observe that $A_{n-1}(f)$ is the unique element A' in $\mathcal{A}_{n-1}$ which contains A. Writing $f = \pi_{n_1}(f) + \sum_{n>n_1}(\pi_n(f) - \pi_{n-1}(f))$, using the triangle inequality for W implies

$$W(f) \le W(\pi_{n_1}(f)) + \sum_{n>n_1} W\big(\pi_n(f) - \pi_{n-1}(f)\big) .$$

Next, (14.40) implies $W(\pi_{n_1}(f)) \le L\sqrt{Nu}\Delta^*$, and (14.39) implies that for $n > n_1$, one has

$$W\big(\pi_n(f) - \pi_{n-1}(f)\big) \le L2^{n/2}\sqrt{u}\Delta(A_{n-1}(f), d_{\psi_2}) . \tag{14.44}$$

We conclude the proof with (14.37). □

Proof of Theorem 14.2.6 Let us recall the event $\Omega(u)$ of Lemma 14.2.7. Our goal is to prove that when this event occurs, then

$$\sup_{f\in\mathcal{F}} \Big|\sum_{i\le N} \varepsilon_i f(X_i)^2\Big| \le Lu\gamma_2(\mathcal{F}, d_{\psi_2})(\sqrt{N}\Delta^* + \gamma_2(\mathcal{F}, d_{\psi_2})) , \tag{14.45}$$

which by taking expectation and using (14.41) implies

$$\mathsf{E} \sup_{f\in\mathcal{F}} \Big|\sum_{i\le N} \varepsilon_i f(X_i)^2\Big| \le L\big(\sqrt{N}\Delta^*\gamma_2(\mathcal{F}, d_{\psi_2}) + \gamma_2(\mathcal{F}, d_{\psi_2})^2\big) .$$

Using (11.34) and (11.35), this proves (14.33) and finishes the proof.

To prove (14.45), since $0 \in \mathcal{F}$, we may assume that $\pi_0(f) = 0$. For each n with $1 \le n \le n_1$, (14.38) means that one has

$$\Big|\sum_{i\le N} \varepsilon_i(\pi_n(f)(X_i)^2 - \pi_{n-1}(f)(X_i)^2)\Big| \le Lu2^{n/2}\sqrt{N}\Delta^*\Delta(A_{n-1}(f), d_{\psi_2}) . \tag{14.46}$$

For $n > n_1$, we write

$$\Big|\sum_{i\le N}\varepsilon_i(\pi_n(f)(X_i)^2-\pi_{n-1}(f)(X_i)^2)\Big| \le \sum_{i\le N}\big|\pi_n(f)(X_i)^2-\pi_{n-1}(f)(X_i)^2\big| . \tag{14.47}$$

Recalling the random norm $W(f)$ of (14.42), and since $a^2-b^2=(a-b)(a+b)$, using the Cauchy-Schwarz inequality, the right-hand side of (14.47) is at most

$$\begin{aligned} &W\big(\pi_n(f)-\pi_{n-1}(f)\big)W\big(\pi_n(f)+\pi_{n-1}(f)\big)\\ &\qquad\le W\big(\pi_n(f)-\pi_{n-1}(f)\big)\big(W(\pi_n(f))+W(\pi_{n-1}(f))\big) , \end{aligned} \tag{14.48}$$

where we have used the triangle inequality for W, and from (14.44) and (14.43), this is at most

$$Lu2^{n/2}\Delta(A_{n-1}(f),d_{\psi_2})(\gamma_2(\mathcal{F},d_{\psi_2})+\sqrt{N}\Delta^*) .$$

Combining with (14.46) and summation over n using (14.37) proves (14.45) and concludes the proof of Theorem 14.2.6. □

A statement similar to Theorem 14.2.6, but with considerably weaker hypothesis, was proved by S. Mendelson and G. Paouris [72]. We present here a key step of the proof of this result (the full proof is too technical to belong here), corresponding to (14.36): the control of the quantities $\sum_{i\le N} f(X_i)^2$. We follow the recent observation of W. Bednorz [15] that Fernique's argument of Theorem 7.13.1 works well here.

Theorem 14.2.9 ([72]) *Consider a (countable) class of functions $\mathcal{F}$ with $0\in\mathcal{F}$. Consider two distances d_1 and d_2 on $\mathcal{F}$. Assume that given $f, f'\in\mathcal{F}$, then*[2]

$$\forall u>0\ ,\ \mu(\{|f-f'|\ge u\})\le 2\exp\Big(-\min\Big(\frac{u^2}{d_2(f,f')^2},\frac{u}{d_1(f,f')}\Big)\Big) . \tag{14.49}$$

Let $S=\gamma_2(\mathcal{F},d_2)+\gamma_1(\mathcal{F},d_1)$. Then

$$\mathsf{E}\sup_{f\in\mathcal{F}}\Big(\sum_{i\le N} f(X_i)^2\Big)^{1/2}\le L\big(S+\sqrt{N}\Delta(\mathcal{F},d_2)+\sqrt{N}\Delta(\mathcal{F},d_1)\big) . \tag{14.50}$$

[2] In the paper [72], the assumption that one controls the diameter of $\mathcal{F}$ for d_1 and d_2 is relaxed into the much weaker condition that for a certain number $q>4$ and a number C, we have $\forall f\in\mathcal{F}$, $\forall u>0$, $\mu(\{|f|\ge u\})\le (C/u)^q$.

The control on the size of $\mathcal{F}$ here is considerably weaker than the control of $\gamma_2(\mathcal{F}, \psi_2)$ assumed in Theorem 14.2.6 since (14.49) holds for $d_1 = 0$ and d_2 the distance d_{ψ_2} associated with the norm $\|\cdot\|_{\psi_2}$.

Before we start the proof, we must understand the tail behavior of sums $\sum_{i \ge 1} a_i Y_i$ where a_i are numbers and where the independent r.v.s Y_i satisfy the tail condition (14.49). We proved much more general results in Sect. 8.2, but the reader wanting a direct proof should try the following exercise:

Exercise 14.2.10 Consider a centered r.v. Y, and assume that for two numbers $0 \le B \le A$, we have

$$\forall u > 0 \,,\ \mathsf{P}(|Y| \ge u) \le 2 \exp\Big(- \min\Big(\frac{u^2}{A^2}, \frac{u}{B}\Big)\Big) \,. \tag{14.51}$$

Then

$$0 \le \lambda \le 1/(2B) \Rightarrow \mathsf{E} \exp \lambda Y \le \exp(L\lambda^2 A^2) \,. \tag{14.52}$$

As in the proof of Bernstein's inequality (4.44), we deduce the following:

Lemma 14.2.11 *Consider i.i.d. copies $(Y_i)_{i \le k}$ of a centered r.v. Y which satisfies the condition (14.51). Then for numbers $(a_i)_{i \le k}$ and any $u > 0$, we have*

$$\mathsf{P}\Big(\Big|\sum_{i \le k} a_i Y_i\Big| \ge u\Big) \le L \exp\Big(- \frac{1}{L} \min\Big(\frac{u^2}{A^2 \sum_{i \le k} a_i^2}, \frac{u}{B \max_{i \le k} |a_i|}\Big)\Big) \,. \tag{14.53}$$

A convenient way to use (14.53) is the following, which is now obvious:

Lemma 14.2.12 *Consider i.i.d. copies $(Y_i)_{i \le k}$ of a centered r.v. Y which satisfies the condition (14.51). If $w > 0$ and*

$$v = LA \sqrt{w \sum_{i \le k} a_i^2} + LBw \max_{i \le k} |a_i| \,, \tag{14.54}$$

then

$$\mathsf{P}\Big(\Big|\sum_{i \le k} a_i Y_i\Big| \ge v\Big) \le L \exp(-w) \,. \tag{14.55}$$

We will use this result for the r.v.s $Y_i = \varepsilon_i (f(X_i) - f'(X_i))$ for $f, f' \in \mathcal{F}$. According to (14.49) in (14.54), we may then take $A = d_2(f, f')$ and $B = d_1(f, f')$.

We denote by $B_{N,2}$ the unit ball of $\ell^2(N)$. We write $\Delta_1 := \Delta(\mathcal{F}, d_1)$ and $\Delta_2 := \Delta(\mathcal{F}, d_2)$. Consider an independent sequence of Bernoulli r.v.s. Theorem 14.2.9 is an immediate consequence of the following:

Theorem 14.2.13 *Given a number $u > 0$, we have*

$$\sup_{\alpha \in B_{N,2}} \sup_{f \in \mathcal{F}} \Big|\sum_{i \le N} \varepsilon_i \alpha_i f(X_i)\Big| \le L\Big(\sqrt{u}\gamma_2(\mathcal{F}, d_2) + u\gamma_1(\mathcal{F}, d_1) + u\sqrt{N}\Delta_1 + \sqrt{u}\sqrt{N}\Delta_2\Big) \quad (14.56)$$

with probability $\ge 1 - L\exp(-Lu)$.

Indeed, taking the supremum over α in the left-hand side of (14.56) shows that this left-hand side[3] is $\sup_{f \in \mathcal{F}}(\sum_{i \le N} f(X_i)^2)^{1/2}$, and the conclusion follows by the general method described at the end of Sect. 2.4.

Before we prove Theorem 14.2.13, we need a simple lemma. We think of N as fixed, and we denote by $e_{n,2}$ and $e_{n,\infty}$ the entropy numbers of $B_{N,2}$ for the ℓ_2 and ℓ_∞ distances (see Sect. 2.5).

Lemma 14.2.14 *We have*

$$\sum_{n \ge 0} 2^{n/2} e_{n,2} \le L\sqrt{N}\ ; \ \sum_{n \ge 0} 2^n e_{n,\infty} \le L\sqrt{N}\ . \quad (14.57)$$

Proof of Theorem 14.2.13 We can find an increasing sequence of partitions $(\mathcal{A}_n)_{n \ge 0}$ of $\mathcal{F}$ with card $\mathcal{A}_0 = 1$ and card $\mathcal{A}_n \le N_{n+1}$ such that for each $f \in \mathcal{F}$,

$$\sum_{n \ge 0}(2^n \Delta(A_n(f), d_1) + 2^{n/2}\Delta(A_n(f), d_2)) \le L(\gamma_1(\mathcal{F}, d_1) + \gamma_2(\mathcal{F}, d_2))\ . \quad (14.58)$$

We can find an increasing sequence of partitions $\mathcal{B}_n$ of $B_{N,2}$ such that card $\mathcal{B}_0 = 1$ and card $\mathcal{B}_n \le N_{n+1}$ such that

$$B \in \mathcal{B}_n \Rightarrow \Delta(B, d_2) \le e_{n,2}\ ; \ \Delta(B, d_\infty) \le e_{n,\infty}\ . \quad (14.59)$$

We will perform chaining on $B_{N,2} \times \mathcal{F}$ using the increasing sequence of partitions $\mathcal{C}_n$ consisting of the sets $B \times A$ for $B \in \mathcal{B}_n$ and $A \in \mathcal{A}_n$. The chaining is routine once we prove the following inequality: If $f, f' \in A \in \mathcal{A}_n$ and $\alpha, \alpha' \in B \in \mathcal{B}_n$, then with probability $\ge 1 - L\exp(-u2^n)$

$$\Big|\sum_{i \le N} \varepsilon_i \alpha_i f(X_i) - \sum_{i \le N} \varepsilon_i \alpha'_i f'(X_i)\Big| \le L\Big(u2^n\Delta(A, d_1) + \sqrt{u2^n}\Delta(A, d_2) + u2^n e_{n,\infty}\Delta_1 + \sqrt{u2^n} e_{n,2}\Delta_2\Big)\ .$$

[3] So, this left-hand side actually does not depend on the values of the ε_i, but nonetheless these are required to be permitted to use (14.55).

To prove this, we start with the triangle inequality

$$\Big|\sum_{i\le N} \varepsilon_i \alpha_i f(X_i) - \sum_{i\le N} \varepsilon_i \alpha'_i f'(X_i)\Big|$$
$$\le \Big|\sum_{i\le N} \varepsilon_i(\alpha_i - \alpha'_i) f(X_i)\Big| + \Big|\sum_{i\le N} \varepsilon_i \alpha'_i (f(X_i) - f'(X_i))\Big| .$$

We use (14.55) with $Y_i = \varepsilon_i f(X_i)$, $a_i = \alpha_i - \alpha'_i$, $A = \Delta_2$, and $B = \Delta_1$ to obtain that the event

$$\Big|\sum_{i\le N} \varepsilon_i(\alpha_i - \alpha'_i) f(X_i)\Big| \ge L\Big(\sqrt{u2^n} e_{n,2}\Delta_2 + u2^n e_{n,\infty}\Delta_1\Big)$$

occurs with probability $\le L\exp(-Lu2^n)$ by (14.55). We use again (14.55), with now $Y_i = \varepsilon_i(f(X_i) - f'(X_i))$ and $a_i = \alpha'_i$, to obtain that the event

$$\Big|\sum_{i\le N} \varepsilon_i \alpha'_i (f(X_i) - f'(X_i))\Big| \ge L\Big(\sqrt{u2^n}\Delta(A, d_2) + u2^n \Delta(A, d_1)\Big)$$

also has probability $\le L\exp(-Lu^2 2^n)$ (using now that $\sum_{i\le N}(\alpha'_i)^2 \le 1$). □

Let us recall the formula

$$\binom{N}{k} \le \Big(\frac{eN}{k}\Big)^k = \exp(k\log(eN/k)) . \tag{14.60}$$

Proof of Lemma 14.2.14 We prove only the statement related to $e_{n,\infty}$ since the other statement is much easier. Let us denote by $(e_i)_{i\le N}$ the canonical basis of $\ell^2(N)$. Consider the largest integer k_1 such that $2^{2k_1} \le N$. For $1 \le k \le k_1$, let us denote by D_k the set of vectors of the type $2^{-k}\sum_{i\in I}\eta_i e_i$ where card $I = 2^{2k}$ and $\eta_j \in \{-1, 0, 1\}$. It should be clear that a point of $B_{N,2}$ is within supremum distance 2^{-k} of a point of $D_1 + \ldots + D_k$. Consequently,

$$\log N(B_{N,2}, d_\infty, 2^{-k}) \le \sum_{j\le k} \log \operatorname{card} D_j . \tag{14.61}$$

A vector in D_j is determined by the choice of the set I of cardinality 2^{2j} (for which there are at most $\binom{N}{2^{2j}}$ possibilities), and once this set is chosen, we have another $3^{2^{2j}}$ possibilities for the choice of the η_i for $i \in I$. Thus, $\operatorname{card} D_j \le 3^{2^{2j}}\binom{N}{2^{2j}}$ so we have $\log\operatorname{card} D_j \le 2^{2j}\log(3eN/2^{2j})$ by (14.60). There is now plenty of room in the estimates, although the computations are messy. Fixing $\beta = 1/4$ and since $\log x \le x^\beta$ for $x \ge 1$, we then have $\log\operatorname{card} D_j \le L2^{2j(1-\beta)}N^\beta$, and (14.61) implies that for $k \le k_1$, we have $\log N(B_{N,2}, d_\infty, 2^{-k}) \le L2^{2k(1-\beta)}N^\beta$. Thus, for $n \le 2k_1$, we have $e_{n,\infty} \le 2^{-k}$ when $L2^{2k(1-\beta)}N^{\beta} \le 2^n$ which occurs for $2^{-k} \simeq LN^{1/(2(1-\beta))}2^{-n/(2(1-\beta))}$. Since

$$\sum_{n\le 2k_1} 2^n N^{1/(2(1-\beta))}2^{-n/(2(1-\beta))} \le L\sqrt{N}\,,$$

this proves that $\sum_{n\le 2k_1} 2^n e_{n,\infty} \le L\sqrt{N}$. We leave to the reader the much easier proof that $\sum_{n>2k_1} 2^n e_{n,\infty} \le L\sqrt{N}$, using the fact that the quantity $e_{n,\infty}$ decreases very fast as $n > 2k_1$ increases. □

14.3 When Not to Use Chaining

In this section, we work in the space $\mathbb{R}^n$ provided with the Euclidean norm $\|\cdot\|$. We denote by $\langle\cdot,\cdot\rangle$ the canonical duality of $\mathbb{R}^n$ with itself. We consider a sequence $(X_i)_{i\le N}$ of independent $\mathbb{R}^n$-valued random vectors, and we assume that

$$\|x\| \le 1 \Rightarrow \mathsf{E}\exp|\langle x, X_i\rangle| \le 2 \tag{14.62}$$

and

$$\max_{i\le N}\|X_i\| \le (Nn)^{1/4}\,. \tag{14.63}$$

Theorem 14.3.1 ([1, 2]) *We have*

$$\sup_{\|x\|\le 1}\Big|\sum_{i\le N}(\langle x, X_i\rangle^2 - \mathsf{E}\langle x, X_i\rangle^2)\Big| \le L\sqrt{nN}\,, \tag{14.64}$$

with probability $\ge 1 - L\exp(-(Nn)^{1/4}) - L\exp(-n)$.

A particularly striking application of this theorem is to the case where the X_i are i.i.d. with law μ where μ is isotropic (see (14.34)) and log-concave[4] It is known in that case (see [1] for the details) that for each $x \in \mathbb{R}^n$, we have $\|\langle x, X_i\rangle\|_{\psi_1} \le L\|x\|$ and the hypothesis (14.62) appears now completely natural. We may then write (14.64) as

$$\sup_{\|x\|\le 1}\Big|\frac{1}{N}\sum_{i\le N}\langle x, X_i\rangle^2 - \|x\|^2\Big| \le L\sqrt{\frac{n}{N}}\,.$$

[4] For example, when μ is the uniform measure on a convex set, it is log-concave, as follows from the Brunn-Minkowski inequality.

Many of the ideas of the proof go back to a seminal paper of J. Bourgain [23].[5] In the present case, rather than chaining, it is simpler to use the following elementary fact:

Lemma 14.3.2 *In $\mathbb{R}^k$, there is a set U with $\operatorname{card} U \le 5^k$ consisting of vectors of norm ≤ 1, with the property that $x \in 2\operatorname{conv} U = \operatorname{conv} 2U$ whenever $\|x\| \le 1$. Consequently,*

$$\forall x \in \mathbb{R}^k\ ,\ \exists a \in U\ ,\ \sum_{i\le k} a_i x_i \ge \frac{1}{2}\Big(\sum_{i\le k} x_i^2\Big)^{1/2}\ . \tag{14.65}$$

Proof It follows from (2.47) that there exists a subset U of the unit ball of $\mathbb{R}^k$ with $\operatorname{card} U \le 5^k$ such that every point of this ball is within distance $\le 1/2$ of a point of U. Given a point x of the unit ball, we can inductively pick points u_ℓ in U such that $\|x - \sum_{1\le \ell \le n} 2^{\ell-1} u_\ell\| \le 2^{-n}$, and this proves that $x \in 2\operatorname{conv} U$. Given $x \in \mathbb{R}^k$ and using that $x/\|x\| \in 2\operatorname{conv} U$, we obtain that $\|x\|^2 = \langle x, x\rangle = \|x\|\langle x, x/\|x\|\rangle \le 2\|x\| \sup_{a\in U}\langle x, a\rangle$ which proves (14.65). □

For $k \le 2N$, we use the notation

$$\varphi(k) = k\log(eN/k)\ ,$$

so that (14.60) becomes

$$\binom{N}{k} \le \Big(\frac{eN}{k}\Big)^k = \exp\varphi(k)\ . \tag{14.66}$$

Thus, $\varphi(k) \ge k$, the sequence $(\varphi(k))$ increases, and $\varphi(k) \ge \varphi(1) = 1 + \log N$.

The difficult part of the proof of Theorem 14.3.1 is the control of the random quantities

$$A_k := \sup_{\|x\|\le 1}\ \sup_{\operatorname{card} I \le k}\Big(\sum_{i\in I}\langle x, X_i\rangle^2\Big)^{1/2}\ . \tag{14.67}$$

It will be achieved through the following:

Proposition 14.3.3 *For $u > 0$, with probability $\ge 1 - L\exp(-u)$, we have*

$$\forall k \ge 1\ ,\ A_k \le L\Big(u + \varphi(k)/\sqrt{k} + \max_{i\le N}\|X_i\|\Big)\ . \tag{14.68}$$

[5] It would be nice if one could deduce Theorem 14.3.1 from a general principle such as Theorem 14.2.9, but unfortunately we do not know how to do this, even when the sequence (X_i) is i.i.d.

Corollary 14.3.4 *If $N \le n$, then with probability $\ge 1 - L\exp(-(nN)^{1/4})$, we have*

$$\sup_{\|x\|\le 1}\sum_{i\le N}\langle x, X_i\rangle^2 \le L\sqrt{Nn}\ .$$

Proof Since $\varphi(N) = N$, we may use (14.68) for $k = N$ and $u = (nN)^{1/4}$ and (14.63) to obtain $A_N \le L(Nn)^{1/4} + L\sqrt{N} \le L(Nn)^{1/4}$. □

We start the proof of Proposition 14.3.3 with the identity

$$A_k = \sup_{\|x\|\le 1}\ \sup_{\operatorname{card} I\le k}\ \sup_{\sum_{i\in I}a_i^2\le 1}\sum_{i\in I}a_i\langle x, X_i\rangle = \sup_{\operatorname{card} I\le k}\ \sup_{\sum_{i\in I}a_i^2\le 1}\Big\|\sum_{i\in I}a_iX_i\Big\|\ . \tag{14.69}$$

The proof will require several steps to progressively gain control.

Lemma 14.3.5 *Consider $x \in \mathbb{R}^n$ with $\|x\| \le 1$ and an integer $1 \le k \le N$. Then for $u > 0$, with probability $\ge 1 - L\exp(-u - 3\varphi(k))$, the following occurs: For each set $I \subset \{1,\dots,N\}$ with $\operatorname{card} I = m \ge k$, we have*

$$\sum_{i\in I}|\langle x, X_i\rangle| \le 6\varphi(m) + u\ . \tag{14.70}$$

Proof Given a set I with $\operatorname{card} I = m$, (14.62) implies $\mathsf{E}\exp\sum_{i\in I}|\langle x, X_i\rangle| \le 2^m \le \exp m \le \exp\varphi(m)$, and thus by Markov's inequality,

$$\mathsf{P}\Big(\sum_{i\in I}|\langle x, X_i\rangle| \ge 6\varphi(m) + u\Big) \le \exp(-5\varphi(m))\exp(-u)\ .$$

Since there are at most $\exp\varphi(m)$ choices for I by (14.66), the union bound implies

$$\sum_{\operatorname{card} I\ge k}\mathsf{P}\Big(\sum_{i\in I}|\langle x, X_i\rangle| \ge 6\varphi(m) + u\Big) \le \sum_{m\ge k}\exp(-4\varphi(m))\exp(-u)\ .$$

Now we observe that $\varphi(m) \ge \varphi(k)$ for $m \ge k$ and that $\varphi(m) \ge m$, so that

$$\sum_{m\ge k}\exp(-4\varphi(m)) \le \exp(-3\varphi(k))\sum_{m\ge 1}\exp(-\varphi(m)) \le L\exp(-3\varphi(k))\ .$$ □

Using Lemma 14.3.2, for each $1 \le k \le N$, and each subset I of $\{1,\dots,N\}$ of cardinality k, we construct a subset $\mathcal{S}_{k,I}$ of the unit ball of $\mathbb{R}^I$ with $\operatorname{card}\mathcal{S}_{k,I} \le 5^k$, such that $\operatorname{conv} 2\mathcal{S}_{k,I}$ contains this unit ball. Consequently,

$$x \in \mathbb{R}^I \Rightarrow \sup_{a\in\mathcal{S}_{k,I}}\sum_{i\in I}a_ix_i \ge \frac{1}{2}\Big(\sum_{i\in I}x_i^2\Big)^{1/2}\ . \tag{14.71}$$

Lemma 14.3.6 *With probability* $\geq 1 - L\exp(-u)$*, the following occurs. Consider disjoint subsets* I, J *of* $\{1, \dots, N\}$ *with* $\operatorname{card} I = m \geq \operatorname{card} J = k$*, and consider any* $a \in \mathcal{S}_{k,J}$*. Then*

$$\sum_{i \in I} \left|\left\langle X_i, \sum_{j \in J} a_j X_j \right\rangle\right| \leq (6\varphi(m) + u) \left\| \sum_{j \in J} a_j X_j \right\| . \tag{14.72}$$

Proof First, we prove that given J and $a \in \mathcal{S}_{k,J}$, the probability that (14.72) occurs for each choice of I of cardinality m and disjoint of J is at least $1 - L(k/eN)^{3k}\exp(-u)$. To prove this, we show that this is the case given the r.v.s X_j for $j \in J$ by using Lemma 14.3.5 for $x = y/\|y\|$, $y = \sum_{j \in J} a_j X_j$. Next, there are at most $\exp\varphi(k)$ choices of J of cardinality k, and for each such J, there are at most 5^k choices for a. Moreover, since $\varphi(k) \geq k$,

$$\sum_{k \leq N} \exp(-3\varphi(k)) 5^k \exp\varphi(k) = \sum_{k \leq N} \exp(-2\varphi(k)) 5^k \leq \sum_{k \geq 1} e^{-2k} 5^k \leq L .$$

The result then follows from the union bound. □

Corollary 14.3.7 *For* $u > 0$*, with probability* $\geq 1 - L\exp(-u)$*, the following occurs. Consider disjoint subsets* I, J *of* $\{1, \dots, N\}$ *with* $\operatorname{card} I = m \geq \operatorname{card} J = k$*, and consider any sequence* $(b_i)_{i \in J}$ *with* $\sum_{i \in J} b_i^2 \leq 1$*. Then*

$$\sum_{i \in I} \left|\left\langle X_i, \sum_{j \in J} b_j X_j \right\rangle\right| \leq L(\varphi(m) + u) A_k . \tag{14.73}$$

Proof With probability $\geq 1 - L\exp(-u)$, (14.72) occurs for every choice of $a \in \mathcal{S}_{k,J}$. We prove that then (14.73) holds. Since $\sum_{j \in J} a_j^2 \leq 1$ for $a \in \mathcal{S}_{k,J}$, for each such sequence, we then have $\|\sum_{j \in J} a_j X_j\| \leq A_k$, and then (14.72) implies

$$\sum_{i \in I} \left|\left\langle X_i, \sum_{j \in J} a_j X_j \right\rangle\right| \leq (6\varphi(m) + u) A_k .$$

Now, each sequence $(b_j)_{j \in J}$ with $\sum_{j \in J} b_j^2 \leq 1$ is in the convex hull of $2\mathcal{S}_{k,J}$, and this proves (14.73). □

We will need the following elementary property of the function φ:

Lemma 14.3.8 *For* $1 \leq x \leq y \leq N$*, we have*

$$\varphi(x) \leq L(x/y)^{3/4} \varphi(y) , \tag{14.74}$$

Proof The function $x \mapsto x^{1/4}\log(eN/x)$ increases for $1 \leq x \leq Ne^{-3}$, so that (14.74) holds with $L = 1$ for $y \leq Ne^{-3}$, and $\varphi(y)$ is about $\varphi(N)$ for $Ne^{-3} \leq y \leq N$. □

We are now ready for the main step of the proof of Proposition 14.3.3.

Proposition 14.3.9 *When the event of Corollary 14.3.7 occurs, for any two disjoint sets I, J of cardinality $\le k$ and any sequence $(a_i)_{i\le N}$ with $\sum_{i\le N} a_i^2 \le 1$, we have*

$$\Big|\Big\langle \sum_{i\in I} a_i X_i, \sum_{j\in J} a_j X_j\Big\rangle\Big| \le L(u + \varphi(k)/\sqrt{k})A_k \,. \tag{14.75}$$

Proof The idea is to suitably group the terms depending on the values of the coefficients (a_i). Let $\kappa = \operatorname{card} I$, and let us enumerate $I = \{i_1, \dots, i_\kappa\}$ in such a way that the sequence $(|a_{i_s}|)_{1\le s\le\kappa}$ is non-increasing. Let us define a partition $(I'_\ell)_{0\le\ell\le\ell_1}$ of I as follows. Consider the largest integer ℓ_1 with $2^{\ell_1} \le 2\operatorname{card} I = 2\kappa$. For $0 \le \ell < \ell_1$, let $I_\ell = \{i_1, \dots, i_{2^\ell}\}$, and let $I_{\ell_1} = I$. We set $I'_0 = I_0$, and for $1 \le \ell \le \ell_1$, we set $I'_\ell = I_\ell \setminus I_{\ell-1}$, so that the sets I'_ℓ for $0 \le \ell \le \ell_1$ form a partition of I. For $0 \le \ell \le \ell_1$, we set $y_\ell = \sum_{i\in I'_\ell} a_i X_i$, so that

$$\sum_{i\in I} a_i X_i = \sum_{0\le\ell\le\ell_1} y_\ell \,.$$

Let us then define similarly for $0 \le \ell \le \ell_2$ sets $J_\ell \subset J$ with $\operatorname{card} J_\ell = 2^\ell$ for $\ell < \ell_2$, sets J'_ℓ and elements $z_\ell = \sum_{j\in J'_\ell} a_j X_j$ so that $\sum_{j\in J} a_j X_j = \sum_{0\le\ell\le\ell_2} z_\ell$. Without loss of generality, we assume $\operatorname{card} I \ge \operatorname{card} J$, so that $\ell_1 \ge \ell_2$. We write

$$\Big\langle \sum_{i\in I} a_i X_i, \sum_{j\in J} a_j X_j\Big\rangle = \Big\langle \sum_{0\le\ell\le\ell_1} y_\ell, \sum_{0\le\ell'\le\ell_2} z_{\ell'}\Big\rangle = \mathrm{I} + \mathrm{II} \,, \tag{14.76}$$

where

$$\mathrm{I} = \sum_{0\le\ell\le\ell_1} \Big\langle y_\ell, \sum_{0\le\ell'\le\min(\ell,\ell_2)} z_{\ell'}\Big\rangle \,;\ \mathrm{II} = \sum_{0\le\ell'\le\ell_2} \Big\langle \sum_{0<\ell<\ell',\ell\le\ell_1} y_\ell, z_{\ell'}\Big\rangle \,.$$

This identity is obvious if we observe that I is the sum of the quantities $\langle y_\ell, z_{\ell'}\rangle$ over the set $\{0 \le \ell \le \ell_1, 0 \le \ell' \le \ell_2, \ell' \le \ell\}$, whereas II is the sum of these quantities over the set $\{0 \le \ell \le \ell_1, 0 \le \ell' \le \ell_2, \ell' > \ell\}$.

We bound I. Since the sequence $(|a_i|)$ is non-increasing, we have $s|a_{i_s}|^2 \le \sum_{i\le\kappa} |a_i|^2 \le 1$ so that $|a_{i_s}| \le 1/\sqrt{s}$ and in particular $|a_i| \le 2^{-\ell/2+1}$ for $i \in I'_\ell$. Next, for each vector x and each $0 \le \ell \le \ell_1$, we have

$$|\langle y_\ell, x\rangle| = \Big|\Big\langle \sum_{i\in I'_\ell} a_i X_i, x\Big\rangle\Big| \le \sum_{i\in I'_\ell} |a_i||\langle X_i, x\rangle| \le 2^{-\ell/2+1} \sum_{i\in I'_\ell} |\langle X_i, x\rangle| \,. \tag{14.77}$$

Using this inequality for $x = \sum_{0\le \ell' \le \ell} z_{\ell'} = \sum_{j\in J_\ell} a_j X_j$ and since $I'_\ell \subset I_\ell$, we get

$$\Big|\Big\langle y_\ell, \sum_{0\le \ell'\le \ell} z_{\ell'}\Big\rangle\Big| \le 2^{-\ell/2+1} \sum_{i\in I'_\ell} \Big|\Big\langle X_i, \sum_{j\in J_\ell} a_j X_j\Big\rangle\Big|$$
$$\le 2^{-\ell/2+1} \sum_{i\in I_\ell} \Big|\Big\langle X_i, \sum_{j\in J_\ell} a_j X_j\Big\rangle\Big| .$$

Thus, we may use (14.73) for $I = I_\ell$ and $J = J_\ell$, and

$$\mathrm{card}\, I_\ell = \min(2^\ell, \mathrm{card}\, I) \ge \min(2^\ell, \mathrm{card}\, J) = \mathrm{card}\, J_\ell$$

to obtain, since $A_{\mathrm{card}\, J} \le A_k$,

$$\sum_{i\in I_\ell} \Big|\Big\langle X_i, \sum_{0\le\ell'\le\ell} z_{\ell'}\Big\rangle\Big| \le L(u + \varphi(\mathrm{card}\, I_\ell))A_k$$
$$\le L(u + (\mathrm{card}\, I_\ell)^{3/4} k^{-3/4}\varphi(k))A_k ,$$

using (14.74) in the second inequality. Thus, we have shown that

$$\mathrm{I} \le \sum_{0\le\ell\le\ell_1} L2^{-\ell/2}\big(u + (\mathrm{card}\, I_\ell)^{3/4} k^{-3/4}\varphi(k)\big)A_k .$$

Since $\mathrm{card}\, I_\ell \le 2^\ell$, we have $\sum_{0\le\ell\le\ell_1} 2^{-\ell/2}(\mathrm{card}\, I_\ell)^{3/4} \le Lk^{1/4}$ so that $\mathrm{I} \le L(u + \varphi(k)/\sqrt{k})A_k$. The same argument proves that this bound also holds for II (using now that $\mathrm{card}\, I_{\ell'-1} \le 2^{\ell'-1} \le \mathrm{card}\, J_{\ell'}$ if $\ell' \le \ell_2$). □

Proposition 14.3.10 *When the event of Corollary 14.3.7 occurs, we have*

$$\forall k \ge 1 , \; A_k^2 \le \max_{i\le N} \|X_i\|^2 + L(u + \varphi(k)/\sqrt{k})A_k . \tag{14.78}$$

Proof We fix once and for all an integer k. Consider a subset W of $\{1, \dots, N\}$ with $\mathrm{card}\, W = k$. Consider $(a_i)_{i\in W}$ with $\sum_{i\in W} a_i^2 \le 1$. Then

$$\Big\|\sum_{i\in W} a_i X_i\Big\|^2 = \sum_{i\in W} a_i^2 \|X_i\|^2 + \sum_{i,j\in W, i\ne j} \langle a_i X_i, a_j X_j\rangle . \tag{14.79}$$

We use the obvious bound for the first term:

$$\sum_{i\in W} a_i^2 \|X_i\|^2 \le \max_{i\le N} \|X_i\|^2 . \tag{14.80}$$

For the second term, we use a standard "decoupling device". Consider independent Bernoulli r.v.s ε_i, and observe that for $i \ne j$, we have $\mathsf{E}(1 - \varepsilon_i)(1 + \varepsilon_j) = 1$, so that by linearity of expectation, and denoting by E_ε expectation in the r.v.s ε_i only,

$$\sum_{i,j\in W,i\neq j} \langle a_i X_i, a_j X_j\rangle = \mathsf{E}_\varepsilon \sum_{i,j\in W,i\neq j} (1+\varepsilon_i)(1-\varepsilon_j)\langle a_i X_i, a_j X_j\rangle \,.$$

Given (ε_i), observe that if $I = \{i \in W;\, \varepsilon_i = 1\}$ and $J = W \setminus I$,

$$\begin{aligned}\frac{1}{4}\sum_{i,j\in W,i\neq j} (1+\varepsilon_i)(1-\varepsilon_j)\langle a_i X_i, a_j X_j\rangle &= \sum_{i\in I,j\in J} \langle a_i X_i, a_j X_j\rangle \\ &= \Big\langle \sum_{i\in I} a_i X_i, \sum_{j\in J} a_j X_j\Big\rangle \,. \qquad (14.81)\end{aligned}$$

The bound (14.75) completes the proof. □

Proof of Proposition 14.3.3 We use that $BA_k \le (B^2 + A_k^2)/2$ with $B = L(u + \varphi(k)/\sqrt{k})$ to deduce from (14.78) that $A_k^2 \le L(\max_{i\le N}\|X_i\|^2 + (u+\varphi(k)/\sqrt{k})^2)$. This implies (14.68). □

Proof of Theorem 14.3.1. According to Corollary 14.3.4, we may assume $N \ge n$. According to Lemma 14.3.2, there exist a subset U of $\mathbb{R}^n$ with $\operatorname{card} U \le 5^n$, consisting of elements of norm ≤ 2 and such that its convex hull contains the unit ball of $\mathbb{R}^n$. Thus (considering that one may take $y = x$ to obtain the first inequality),

$$\begin{aligned}&\sup_{\|x\|\le 1}\Big|\sum_{i\le N}(\langle x, X_i\rangle^2 - \mathsf{E}\langle x, X_i\rangle^2)\Big| \\ &\quad\le \sup_{\|x\|,\|y\|\le 1}\Big|\sum_{i\le N}(\langle x, X_i\rangle\langle y, X_i\rangle - \mathsf{E}\langle x, X_i\rangle\langle y, X_i\rangle)\Big| \\ &\quad\le \sup_{x,y\in U}\Big|\sum_{i\le N}(\langle x, X_i\rangle\langle y, X_i\rangle - \mathsf{E}\langle x, X_i\rangle\langle y, X_i\rangle)\Big| \,. \qquad (14.82)\end{aligned}$$

The plan is to assume that

$$\forall k\ge 1\,,\; A_k \le L(\sqrt{k}\log(eN/k) + (Nn)^{1/4})\,, \qquad (14.83)$$

and to prove that then with probability $\ge 1 - \exp(-n)$, the right-hand side of (14.82) is $\le L\sqrt{Nn}$. This completes the proof of Theorem 14.3.1 because Proposition 14.3.3 and (14.63) show that (14.83) occurs with probability $\ge 1 - L\exp(-(Nn)^{1/4})$.

Consider a truncation level $B \ge 0$ which we will determine later (depending only on N), and define

$$Z_i(x,y) = \langle x, X_i\rangle\langle y, X_i\rangle \mathbf{1}_{\{|\langle x, X_i\rangle\langle y, X_i\rangle|\le B\}}$$

and

$$Y_i(x,y) = \langle x, X_i\rangle\langle y, X_i\rangle \mathbf{1}_{\{|\langle x,X_i\rangle\langle y,X_i\rangle|>B\}} ,$$

so that $\langle x, X_i\rangle\langle y, X_i\rangle = Z_i(x,y) + Y_i(x,y)$ and $Y_i(x,y) \neq 0 \Rightarrow |Y_i(x,y)| \geq B$. This argument is yet another instance of a decomposition in a "spread out part" and a "peaky part". The peaky part $\sum_i Y_i(x,y)$ will be controlled as usual without using cancellations, i.e., we will control $\sum_i |Y_i(x,y)|$.[6] We bound the right-hand side of (14.82) by I + II + III, where

$$\mathrm{I} = \sup_{x,y\in U} \Big|\sum_{i\leq N}(Z_i(x,y) - \mathsf{E}Z_i(x,y))\Big| , \tag{14.84}$$

$$\mathrm{II} = \sup_{x,y\in U} \sum_{i\leq N} |Y_i(x,y)| , \tag{14.85}$$

$$\mathrm{III} = \sup_{x,y\in U} \sum_{i\leq N} |\mathsf{E}Y_i(x,y)| . \tag{14.86}$$

The fun is to bound II. We prove that when (14.83) occur, then $\mathrm{II} \leq L\sqrt{Nn}$. For this, let us fix $x, y \in U$ and set

$$I = \big\{i \leq N \,;\, |Y_i(x,y)| > B\} = \{i \leq N \,;\, |Y_i(x,y)| \neq 0\big\} .$$

Defining $m = \operatorname{card} I$ we have, using the Cauchy-Schwarz inequality in the second inequality, and recalling the definition (14.67) of A_m,

$$mB \leq \sum_{i\in I} |Y_i(x,y)| \leq \Big(\sum_{i\in I}\langle x, X_i\rangle^2\Big)^{1/2}\Big(\sum_{i\in I}\langle y, X_i\rangle^2\Big)^{1/2} \leq 4A_m^2 , \tag{14.87}$$

and thus from (14.83),

$$mB \leq L_1(m(\log(eN/m))^2 + \sqrt{Nn}) . \tag{14.88}$$

Without loss of generality, we assume from Corollary 14.3.4 that $N > n$ and then $N > \sqrt{Nn}$. Thus, we may consider the smallest integer $k_0 \leq N$ such that $k_0(\log(eN/k_0))^2 > \sqrt{Nn}$.

---

[6] And as usual this control is far harder than the control of the cancellations.

Let us now choose $B = 2L_1(\log(eN/k_0))^2$. Assuming if possible that $m \geq k_0$, then (14.88) implies

$$2L_1 m(\log(eN/k_0))^2 = mB \leq L_1(m(\log(eN/m))^2 + \sqrt{Nn}) \\ \leq L_1(m(\log(eN/k_0))^2 + \sqrt{Nn}) ,$$

so that $m(\log(eN/k_0))^2 \leq \sqrt{Nn}$ and thus

$$k_0(\log(eN/k_0))^2 \leq m(\log(eN/k_0))^2 \leq \sqrt{Nn} .$$

This is impossible by the definition of k_0, so that we have proved that $m < k_0$. By definition of k_0, we then have $m(\log(eN/m))^2 \leq \sqrt{Nn}$, i.e., $\sqrt{m}\log(eN/m) \leq (Nn)^{1/4}$. Thus, by (14.83), we have $A_m \leq L(Nn)^{1/4}$, and finally by (14.87) and since $\sum_{i \leq N} |Y_i(x, y)| = \sum_{i \in I} |Y_i(x, y)|$ that $\mathrm{II} \leq L\sqrt{Nn}$.

Next, let us control III. Recalling (14.63), we have $\mathrm{II} \leq \sum_{i \leq N} \|X_i\|^2 \leq N\sqrt{Nn}$. Moreover, we have just shown that when (14.83) occurs, i.e., with probability $\geq 1 - L\exp(-(Nn)^{1/4})$, we have in fact $\mathrm{II} \leq L\sqrt{Nn}$. Thus, $\mathrm{III} \leq \mathsf{E}\,\mathrm{II} \leq L\sqrt{Nn} + L\exp(-(Nn)^{1/4})N\sqrt{Nn} \leq L\sqrt{Nn}$.

It remains to bound I. Since $(\log x)^2 \leq L\sqrt{x}$ for $x \geq e$,

$$\sqrt{Nn} < k_0(\log(eN/k_0))^2 \leq Lk_0\sqrt{N/k_0} ,$$

and thus $k_0 \geq n/L$. Therefore, with huge room to spare,

$$B = 2L_1(\log(eN/k_0))^2 \leq L\sqrt{N/n} .$$

Since $|Z_i(x, y)| \leq B$ and $\mathsf{E}Z_i(x, y)^2 \leq L$ (using the Cauchy-Schwarz inequality and (14.62)), it follows from Bernstein's inequality (4.44) that

$$\mathsf{P}\Big(\Big|\sum_{i \leq N}(Z_i(x, y) - \mathsf{E}Z_i(x, y))\Big| \geq t\Big) \leq 2\exp\Big(-\min\Big(\frac{t^2}{LN}, \frac{t}{L\sqrt{N/n}}\Big)\Big) .$$

The right-hand side is $\leq 5^{-3n}$ for $t = L\sqrt{Nn}$. There are at most 5^{2n} choices for the pair $(x, y) \in U^2$, so that by the union bound $\mathrm{I} \leq L\sqrt{Nn}$ with probability $1 - 5^{-n} \geq 1 - \exp(-n)$. This completes the proof of Theorem 14.3.1. □

Chapter 15
Gaussian Chaos

Gaussian chaos are simply polynomials in Gaussian r.v.s. In this chapter, we investigate two questions related to chaos. Since we understand the boundedness of Gaussian processes well, we might hope that this could be the case of "chaos processes". Unfortunately, even in the simplest case of order 2 chaos, this is far from being the case, as we will explain in Sect. 15.1. It is striking that there exist apparently rather different methods to bound a chaos process, and it remains unclear how (if at all possible) we may describe the supremum of a chaos process in terms of geometric characteristics of the index set. Section 15.2 investigates a different topic, the size of the tails of a single chaos, a deep result of R. Latała.

15.1 Order 2 Gaussian Chaos

15.1.1 Basic Facts

Consider independent standard Gaussian sequences (g_i), (g'_j), $i, j \geq 1$. Given a double sequence $t = (t_{i,j})_{i,j\geq 1}$, we consider the r.v.

$$X_t = \sum_{i,j\geq 1} t_{i,j} g_i g'_j \ . \tag{15.1}$$

The series converges in L^2 as soon as $\sum_{i,j\geq 1} t_{i,j}^2 < \infty$, but for the present purpose of proving inequalities, we may as well assume that only finitely many coefficients $t_{i,j}$ are not 0. This random variable is called a (decoupled) order 2 Gaussian chaos. There is also a theory of non-decoupled chaos, $\sum_{i>j\geq 1} t_{i,j} g_i g_j$. For the present purposes of finding upper bounds, this theory reduces to the decoupled case using well-understood arguments such as the following:

M. Talagrand, *Upper and Lower Bounds for Stochastic Processes*, Ergebnisse der Mathematik und ihrer Grenzgebiete. 3. Folge / A Series of Modern Surveys in Mathematics 60, https://doi.org/10.1007/978-3-030-82595-9_15

Lemma 15.1.1 *We have*

$$\mathsf{E}\sup_{t\in T}\Big|\sum_{i\neq j} t_{i,j}g_ig_j+\sum_{i\geq 1} t_{i,i}(g_i^2-1)\Big|\leq 2\mathsf{E}\sup_{t\in T}\Big|\sum_{i,j\geq 1} t_{i,j}g_ig'_j\Big|\,. \tag{15.2}$$

Proof It is obvious by Jensen's inequality (taking the expectation in the r.v.s g'_i inside rather than outside the supremum and the absolute values) that

$$\mathsf{E}\sup_{t\in T}\Big|\sum_{i\neq j} t_{i,j}g_ig_j+\sum_{i\geq 1} t_{i,i}(g_i^2-1)\Big|\leq \mathsf{E}\sup_{t\in T}\Big|\sum_{i,j} t_{i,j}(g_i+g'_i)(g_j-g'_j)\Big|\,,$$

and the right-hand side is just the right-hand side of (15.2) because the families $(g_i+g'_i)/\sqrt{2}$ and $(g_i-g'_i)/\sqrt{2}$ are independent sequences of standard Gaussian r.v.s independent of each other. □

Given a finite family T of double sequences $t=(t_{i,j})$, we would like to find upper and lower bounds for the quantity

$$S(T)=\mathsf{E}\sup_{t\in T} X_t\,. \tag{15.3}$$

We would like in fact to understand the value of $S(T)$ as a function of "the geometry of T" as we did in the case of Gaussian processes. Surprisingly, the difficulty of this problem is of an entirely different magnitude from the Gaussian case, and we have only limited results to offer at this point.

Let us start with the basics. We denote by B the unit ball of $\ell^2(\mathbb{N}^*)$, $B=\{\alpha=(\alpha_j)_{j\geq 1};\sum_{j\geq 1}\alpha_j^2\leq 1\}$, and we note the following fundamental fact: For real numbers $(x_j)_{j\geq 1}$, we have

$$\Big(\sum_{j\geq 1}x_j^2\Big)^{1/2}=\sup_{\alpha\in B}\sum_{j\geq 1}\alpha_jx_j\,. \tag{15.4}$$

Given an array $t=(t_{i,j})$, we define

$$\begin{aligned}\|t\|&=\sup_{\alpha}\Big(\sum_{i\geq 1}\Big(\sum_{j\geq 1}\alpha_jt_{i,j}\Big)^2\Big)^{1/2}\\ &=\sup\Big\{\sum_{i,j\geq 1}\alpha_j\beta_it_{i,j}\,;\,\sum_{j\geq 1}\alpha_j^2\leq 1\,,\,\sum_{i\geq 1}\beta_i^2\leq 1\Big\}\,.\end{aligned}$$

If we think of t as a matrix, $\|t\|$ is the operator norm of t from ℓ^2 to ℓ^2. We will also need the Hilbert-Schmidt norm of this matrix, given by

$$\|t\|_{HS} = \Big(\sum_{i,j\geq 1} t_{i,j}^2\Big)^{1/2} .$$

Thus, $\|t\| \leq \|t\|_{HS}$ by the Cauchy-Schwarz inequality.

We find it convenient to assume that the underlying probability space is a product $(\Omega \times \Omega', \mathsf{P} = \mathsf{P}_0 \otimes \mathsf{P}')$, so that

$$X_t(\omega, \omega') = \sum_{i,j} t_{i,j} g_i(\omega) g'_j(\omega') .$$

Conditionally on ω, X_t is a Gaussian r.v. Denoting by E' integration in ω' only (i.e., conditional expectation given ω), we have

$$\mathsf{E}' X_t^2 = \sum_{j\geq 1}\Big(\sum_{i\geq 1} t_{i,j} g_i(\omega)\Big)^2 . \tag{15.5}$$

Consider the r.v.

$$\sigma_t = \sigma_t(\omega) = (\mathsf{E}' X_t^2)^{1/2} ,$$

and note that $\mathsf{E}\sigma_t^2 = \mathsf{E}X_t^2$. The importance of this r.v. is made clear by the fact that the random distance d_ω associated with the Gaussian process X_t (at given ω) is

$$d_\omega(s,t) = \sigma_{s-t}(\omega) . \tag{15.6}$$

Thus, $d_\omega(s,t)^2 = \sum_{j\geq 1}(\sum_{i\geq 1}(s_{i,j} - t_{i,j}) g_i(\omega))^2$. A fundamental difference with the situation of Chap. 11 is that there is no reason why the probability that $d_\omega(s,t)$ is small should be small. A particularly striking case of this is when there is only one non-zero term in the sum $\sum_j$. (The condition (11.8) plays a fundamental role in Chap. 11.)

Lemma 15.1.2 *We have*

$$\mathsf{P}(|\sigma_t - \|t\|_{HS}| \geq v + L\|t\|) \leq 2\exp\Big(-\frac{v^2}{2\|t\|^2}\Big) . \tag{15.7}$$

Proof Given $\alpha \in B$, we consider the r.v.

$$g_{\alpha,t} := \sum_{i\geq 1} g_i(\omega)\Big(\sum_{j\geq 1}\alpha_j t_{i,j}\Big) = \sum_{j\geq 1}\alpha_j\Big(\sum_{i\geq 1} t_{i,j} g_i(\omega)\Big) , \tag{15.8}$$

so that from (15.4)

$$\sup_{\alpha \in B} g_{\alpha,t} = \sigma_t \ , \tag{15.9}$$

and also

$$(\mathsf{E} g_{\alpha,t}^2)^{1/2} = \Big(\sum_{i \ge 1} \Big(\sum_{j \ge 1} \alpha_j t_{i,j}\Big)^2\Big)^{1/2} \le \|t\| \ .$$

Following (15.9), (2.118) implies that for $v > 0$,

$$\mathsf{P}(|\sigma_t - \mathsf{E}\sigma_t| \ge v) \le 2 \exp\Big(-\frac{v^2}{2\|t\|^2}\Big) \ . \tag{15.10}$$

In particular, we have $\|\sigma_t - \mathsf{E}\sigma_t\|_2 \le L\|t\|$ where $\|\cdot\|_2$ denotes the norm in $L^2(\Omega)$. Using the general inequality $|\|X\| - \|Y\|| \le \|X - Y\|$ yields $|\|\sigma_t\|_2 - |\mathsf{E}\sigma_t|| \le L\|t\|$, and since $\mathsf{E}\sigma_t \ge 0$, we obtain $|\|\sigma_t\|_2 - \mathsf{E}\sigma_t| \le L\|t\|$. Now

$$\|\sigma_t\|_2 = (\mathsf{E}\sigma_t^2)^{1/2} = (\mathsf{E}X_t^2)^{1/2} = \|t\|_{HS} \ , \tag{15.11}$$

so that $|\mathsf{E}\sigma_t - \|t\|_{HS}| \le L\|t\|$ and (15.10) implies (15.7). □

We are now ready to prove a simple classical fact (first obtained in [38]).

Lemma 15.1.3 *For $v \ge 0$, we have*

$$\mathsf{P}(|X_t| \ge v) \le L \exp\Big(-\frac{1}{L}\min\Big(\frac{v^2}{\|t\|_{HS}^2}, \frac{v}{\|t\|}\Big)\Big) \ . \tag{15.12}$$

Proof Given ω, the r.v. X_t is Gaussian so that

$$\mathsf{P}'(|X_t| \ge v) \le 2 \exp\Big(-\frac{v^2}{2\sigma_t^2}\Big) \ ,$$

and, given $a > 0$,

$$\begin{aligned} \mathsf{P}(|X_t| \ge v) = \mathsf{E}\mathsf{P}'(|X_t| \ge v) &\le 2\mathsf{E} \exp\Big(-\frac{v^2}{2\sigma_t^2}\Big) \\ &\le 2 \exp\Big(-\frac{v^2}{2a^2}\Big) + 2\mathsf{P}(\sigma_t \ge a) \ . \end{aligned}$$

We now estimate from above the last term of the previous inequality. It follows from (15.7) that $\mathsf{P}(\sigma_t \geq v + \|t\|_{HS} + L\|t\|) \leq L\exp(-v^2/2\|t\|^2)$. Since $\|t\| \leq \|t\|_{HS}$, we have in particular that $\mathsf{P}(\sigma_t \geq v + L_0\|t\|_{HS}) \leq L\exp(-v^2/2\|t\|^2)$, and thus, when $a \geq 2L_0\|t\|_{HS}$,

$$\mathsf{P}(\sigma_t \geq a) \leq \mathsf{P}\Big(\sigma_t \geq \frac{a}{2} + L_0\|t\|_{HS}\Big) \leq L\exp\Big(-\frac{a^2}{L\|t\|^2}\Big) .$$

Consequently, when $a \geq 2L_0\|t\|_{HS}$,

$$\mathsf{P}(|X_t| \geq v) \leq 2\exp\Big(-\frac{v^2}{2a^2}\Big) + L\exp\Big(-\frac{a^2}{L\|t\|^2}\Big) . \tag{15.13}$$

To finish the proof, we take $a = \max\big(L\|t\|_{HS}, \sqrt{v\|t\|}\big)$. The last term in (15.13) is always at most $L\exp(-v/(L\|t\|))$, and the first term is always at most $L\exp(-v/(L\|t\|)) + L\exp(-v^2/L\|t\|_{HS})$. □

Consider the two distances on T defined by

$$d_\infty(s,t) = \|t - s\| \;,\; d_2(s,t) = \|t - s\|_{HS} . \tag{15.14}$$

As a consequence of (15.12), we have

$$\mathsf{P}(|X_s - X_t| \geq v) \leq L\exp\Big(-\frac{1}{L}\min\Big(\frac{v^2}{d_2^2(s,t)}, \frac{v}{d_\infty(s,t)}\Big)\Big) \tag{15.15}$$

and Theorem 4.5.13 implies the following:

Theorem 15.1.4 *For a set T of sequences $(t_{i,j})$, we have*

$$S(T) = \mathsf{E}\sup_{t\in T} X_t \leq L\big(\gamma_1(T, d_\infty) + \gamma_2(T, d_2)\big) . \tag{15.16}$$

We analyze now a very interesting example of set T, which will show in particular that (15.16) cannot be reversed. Given an integer n, we consider

$$T = \{t \;;\; \|t\| \leq 1 \,,\, t_{i,j} \neq 0 \Rightarrow i \,,\, j \leq n\} . \tag{15.17}$$

Since

$$\sum_{i,j} t_{ij} g_i g'_j \leq \Big(\sum_{i\leq n} g_i^2\Big)^{1/2}\Big(\sum_{j\leq n} g_j'^2\Big)^{1/2}\|t\| ,$$

taking supremum over $t \in T$ and expectation (and using the Cauchy-Schwarz inequality) implies that $S(T) \le n$. Volume arguments show that $\log N(T, d_\infty, 1/4) \ge n^2/L$, so that $\gamma_1(T, d_\infty) \ge n^2/L$. It is also simple to prove that (see [53])

$$\log N(T, d_2, \sqrt{n}/L) \ge n^2/L$$

and that $S(T)$ is about n, $\gamma_1(T, d_\infty)$ is about n^2, and $\gamma_2(T, d_2)$ is about $n^{3/2}$. In this case, (15.16) is not sharp, which means that there is no hope of reversing this inequality in general. This is so despite the fact that we have used a competent chaining method and that the bounds (15.15) are essentially optimal (as follows, e.g., from the left-hand side of (15.69)). It can also be shown that in the case where the elements t of T satisfy $t_{i,j} = 0$ for $i \ne j$, the bound (15.16) can be reversed. This is essentially proved in Theorem 8.3.3.

15.1.2 When T Is Small for the Distance d_∞

We continue the study of general chaos processes. When T is "small for the distance d_∞", it follows from (15.15) that the process $(X_t)_{t\in T}$ resembles a Gaussian process, so that there should be a close relationship between $S(T) = \mathsf{E}\sup_{t\in T} X_t$ and $\gamma_2(T, d_2)$. The next result is a step in this direction. It should be compared with Theorem 6.6.1.

Theorem 15.1.5 *We have*

$$\gamma_2(T, d_2) \le L\big(S(T) + \sqrt{S(T)\gamma_1(T, d_\infty)}\big) . \tag{15.18}$$

The example (15.17) provides a situation where this inequality is sharp, since then both the left-hand and the right-hand sides are of order $n^{3/2}$. Combining with Theorem 15.1.4, this implies the following:

Corollary 15.1.6 *Defining*

$$R = \frac{\gamma_1(T, d_\infty)}{\gamma_2(T, d_2)} ,$$

we have

$$\frac{1}{L(1+R)}\gamma_2(T, d_2) \le S(T) \le L(1+R)\gamma_2(T, d_2) . \tag{15.19}$$

In particular, $S(T)$ is of order $\gamma_2(T, d_2)$ when R is of order 1 or smaller.

Proof The right-hand side is obvious from (15.16). To obtain the left-hand side, we simply write in (15.18) that, since $\sqrt{ab} \le (a+b)/2$,

$$\sqrt{S(T)\gamma_1(T,d_\infty)} = \sqrt{S(T)R\gamma_2(T,d_2)} \le \frac{1}{2}\Big(\frac{1}{L}\gamma_2(T,d_2) + LS(T)R\Big)$$

where L is as in (15.18), and together with (15.18), this yields

$$\gamma_2(T,d_2) \le LS(T) + \frac{1}{2}\gamma_2(T,d_2) + LS(T)R \, . \qquad \square$$

In the examples of interest, $\gamma_1(T,d_\infty)$ has a tendency to be large, and the previous results are not sharp. Nonetheless, we will prove Theorem 15.1.5 as an exercise in using functionals in a slightly new way, and the reader who is not interested in this aspect is invited to jump to Proposition 15.1.13. We recall the random distance d_ω of (15.6).

Lemma 15.1.7 *We have*

$$\mathsf{P}\Big(d_\omega(s,t) \le \frac{1}{2}d_2(s,t)\Big) \le L\exp\Big(-\frac{d_2^2(s,t)}{Ld_\infty^2(s,t)}\Big) \, . \tag{15.20}$$

Proof Taking $v = \|t\|_{HS}/4$ in (15.7), we obtain

$$\mathsf{P}\Big(|\sigma_t - \|t\|_{HS}| \ge \frac{\|t\|_{HS}}{4} + L_1\|t\|\Big) \le 2\exp\Big(-\frac{\|t\|_{HS}^2}{L\|t\|^2}\Big) \, .$$

When $L_1\|t\| \le \|t\|_{HS}/4$, this gives

$$\mathsf{P}\Big(\sigma_t \le \frac{\|t\|_{HS}}{2}\Big) \le L\exp\Big(-\frac{\|t\|_{HS}^2}{L\|t\|^2}\Big) \, , \tag{15.21}$$

whereas when $L_1\|t\| \ge \|t\|_{HS}/4$, (15.21) holds automatically if the constant in front of the exponential is large enough. $\square$

We will deduce Theorem 15.1.5 from the following general abstract result:

Theorem 15.1.8 *Consider a finite set T, provided with two distances d and d_1. Consider a random distance d_ω on T and a number $0 < \alpha \le 1/2$. Assume that*

$$\forall s,t \in T \, , \; \mathsf{P}\big(d_\omega(s,t) \ge \alpha d(s,t)\big) \ge \alpha \tag{15.22}$$

$$\forall s,t \in T \, , \; \mathsf{P}\big(d_\omega(s,t) \le \alpha d(s,t)\big) \le \frac{1}{\alpha}\exp\Big(-\alpha\frac{d^2(s,t)}{d_1^2(s,t)}\Big) \, . \tag{15.23}$$

Consider a number M such that

$$\mathsf{P}(\gamma_2(T, d_\omega) \le M) \ge 1 - \alpha/2 . \tag{15.24}$$

Then

$$\gamma_2(T, d) \le K(\alpha)\big(M + \sqrt{M\gamma_1(T, d_1)}\big) , \tag{15.25}$$

where $K(\alpha)$ depends on α only.

Proof of Theorem 15.1.5 We first prove that the pair of distances $d_1 = d_\infty$ and $d = d_2$ of (15.14) and the random distance $d_\omega(s,t) = \sigma_{s-t}(\omega)$ of (15.6) satisfy (15.22) and (15.23) whenever α is small enough. For (15.23), this is a consequence of (15.20). Next, the formula (15.9) makes σ_t, and hence σ_{s-t}, appear as the supremum of a Gaussian process. Applying the Paley-Zygmund inequality (6.15) to this process yields $\mathsf{P}(\sigma_{s-t} \ge (\mathsf{E}\sigma_{s-t}^2)^{1/2}/L) \ge 1/L$. Since $\mathsf{E}\sigma_{s-t}^2 = \|s-t\|_{HS}^2 = d_2(s,t)^2$, (15.22) holds whenever α is small enough.

Next, we prove that (15.24) holds for $M = LS(T)/\alpha$. Since $\mathsf{E}\mathsf{E}' \sup_{t\in T} X_t = S(T)$, and since $\mathsf{E}' \sup_{t\in T} X_t \ge 0$, Markov inequality implies

$$\mathsf{P}\Big(\mathsf{E}' \sup_{t\in T} X_t \le 2S(T)/\alpha\Big) \ge 1 - \alpha/2 .$$

Since $L\mathsf{E}' \sup_{t\in T} X_t \ge \gamma_2(T, d_\omega)$ by Theorem 2.10.1, this proves that (15.24) holds for $M = LS(T)/\alpha$. Thus, (15.25) holds by Theorem 15.1.8, and it implies (15.18). □

It would be nice to have a proof of Theorem 15.1.8 which falls into our general scheme of proof. We do not know how to do that. The next exercise shows how to obtain a weaker result in the same direction following this scheme of proof.

Exercise 15.1.9

(a) Consider an admissible sequence $(\mathcal{D}_n)$ of partitions of T and a probability measure μ on T. For each $t \in T$, we define $\eta_{0,\omega}(t) = \Delta(T, d_\omega)$, and for $n \ge 1$, we define $\eta_{n,\omega}(t) = \Delta(T, d_\omega)$ if $\mu(D_n(t)) \le 2N_{n+2}^{-1}$, and otherwise, we define

$$\eta_{n,\omega}(t) = \inf\big\{\epsilon > 0 \; ; \; \mu(D_{n+1}(t) \cap B_{d_\omega}(t, \epsilon)) \ge 2N_{n+2}^{-1}\big\} . \tag{15.26}$$

Prove that

$$\int_T \sum_{n\ge 0} 2^{n/2} \eta_{n,\omega}(t) \mathrm{d}\mu(t) \le L\gamma_2(T, d_\omega) . \tag{15.27}$$

(b) Set $\epsilon_n(t) = \inf\{\epsilon > 0; \mu(B(t,\epsilon)) \geq N_{n+2}^{-1}\} \leq \theta_n(t)$. Prove that

$$\epsilon_n(t) \leq K 2^{n/2} \Delta(D_n(t), d_1) + K \mathsf{E}\eta_{n,\omega}(t) . \tag{15.28}$$

and conclude. Hint: Review Proposition 3.3.1 for (a) and Theorem 5.4.1 for (b).

As a warm-up for the proof of Theorem 15.1.8, we recommend that the reader masters Exercise 3.4.2. Our proof of Theorem 15.1.8 will use the following functional, related to that exercise:

Definition 15.1.10 Consider a number M as in (15.24). For any set $H \subset T$ and any probability measure μ on T with $\mu(H) = 1$, we define

$$F(\mu, H) = \mathsf{E}\mathbf{1}_U \inf_{(\mathcal{A}_n)} \int \sum_{n \geq 0} 2^{n/2} \min\big(\Delta(A_n(t), d_\omega), \Delta(A_n(t), d)\big) \mathrm{d}\mu(t) , \tag{15.29}$$

where U is the set $\{\gamma_2(T, d_\omega) \leq M\}$ and where the infimum is computed over all sequences of admissible partitions *of* H. For any set $H \subset T$, we then define

$$F(H) = \sup\{F(\mu, H) \; ; \; \mu(H) = 1\} .$$

Lemma 15.1.11 *We have*

$$\Delta(T, d) \leq K F(T) . \tag{15.30}$$

Proof Considering just the term for $n = 0$ in (15.29), and since $\mathcal{A}_0 = \{T\}$, we get that for any measure μ,

$$F(\mu, T) \geq \mathsf{E}\mathbf{1}_U \min(\Delta(T, d_\omega), \Delta(T, d)) .$$

Using (15.22) for s, t with $d(s, t) \geq \Delta(T, d)/2$, we obtain that $\mathsf{P}(\Delta(T, d_\omega) \geq \alpha\Delta(T, d)/2) \geq \alpha$. Since $\mathsf{P}(U) \geq 1 - \alpha/2$ by (15.24), it follows that $\mathsf{P}(U \cap \{\Delta(T, d_\omega) \geq \alpha\Delta(T, d)/2\}) \geq \alpha/2$ so that $F(\mu, T) \geq \Delta(T, d)/K$ by (2.7). □

The following lemma provides the appropriate growth condition for the functional F:

Lemma 15.1.12 *Assume the conditions of Theorem 15.1.8. There exists a constant $K_0 = K_0(\alpha)$ with the following property. Consider an integer $m \geq 2$. Consider a set $D \subset T$ with $\Delta(D, d_1) \leq 2a/(K_0\sqrt{\log m})$, and for $\ell \leq m$, consider points $t_\ell \in D$ that satisfy $d(t_\ell, t_{\ell'}) \geq a$ for $\ell \neq \ell'$. Consider moreover for $\ell \leq m$ sets $H_\ell \subset B(t_\ell, a/K_0)$. Then*

$$F\Big(\bigcup_{\ell \leq m} H_\ell\Big) \geq \frac{a}{K}\sqrt{\log m} + \min_{\ell \leq m} F(H_\ell) . \tag{15.31}$$

Proof of Theorem 15.1.8 We repeat the proof of Theorem 6.6.1, using the functional $F(H)$ rather than the functional $b(H)$ and the distance $K_0 d_1$ rather than the distance d_∞. The only two properties of the functional $b(H)$ which were used in the proof of Theorem 6.6.1 are those proved in Lemmas 15.1.11 and 15.1.12. Thus, we obtain that $\gamma_2(T,d) \le K(\alpha)(F(T) + \sqrt{F(T)\gamma_1(T,d_1)})$. But it is obvious that $F(T) \le \mathsf{E}\mathbf{1}_U \gamma_2(T,d_\omega) \le M$. □

Proof of Lemma 15.1.12 Let us fix for each $\ell \le m$ a probability measure μ_ℓ with $\mu_\ell(H_\ell) = 1$, and let $\mu = m^{-1}\sum_{\ell\le m}\mu_\ell$. For $t \in H := \bigcup_{\ell\le m} H_\ell$, let us define $\ell(t) \le m$ by $t \in H_{\ell(t)}$. Let us assume first that $m \ge 32$. Let us then consider the largest integer n_0 with $N_{n_0} \le \sqrt{m/32}$, so that $2^{n_0/2} \ge \sqrt{\log m}/K$. Considering a given ω and a given admissible sequence $(\mathcal{A}_n)$ of partitions of H, for $t \in H$, let us define

$$f(t,\omega,(\mathcal{A}_n)) = \min\big(\Delta(A_{n_0}(t),d_\omega),\Delta(A_{n_0}(t),d)\big) - \min\Big(\Delta(A_{n_0}(t)\cap H_{\ell(t)},d_\omega),\Delta(A_{n_0}(t)\cap H_{\ell(t)},d)\Big)\,. \tag{15.32}$$

Thus, given the admissible sequence $(\mathcal{A}_n)$, we have

$$\sum_{n\ge0} 2^{n/2}\min\big(\Delta(A_n(t),d_\omega),\Delta(A_n(t),d)\big) \ge 2^{n_0/2} f(t,\omega,(\mathcal{A}_n)) + \sum_{n\ge0} 2^{n/2}\min\big(\Delta(A_n(t)\cap H_{\ell(t)},d_\omega),\Delta(A_n(t)\cap H_{\ell(t)},d)\big)\,. \tag{15.33}$$

Now, for each ℓ, the sets $A\cap H_\ell$ for $A\in\mathcal{A}_n$ form a partition $\mathcal{A}_{n,\ell}$ of H_ℓ, and the sequence $(\mathcal{A}_{n,\ell})_{n\ge0}$ is an admissible sequence of partitions of H_ℓ, so that

$$\begin{aligned}
&\int \sum_{n\ge0} 2^{n/2}\min\big(\Delta(A_n(t)\cap H_{\ell(t)},d_\omega),\Delta(A_n(t)\cap H_{\ell(t)},d)\big)\mathrm{d}\mu(t)\\
&= \frac{1}{m}\sum_{\ell\le m}\int\sum_{n\ge0}2^{n/2}\min\big(\Delta(A_n(t)\cap H_\ell,d_\omega),\Delta(A_n(t)\cap H_\ell,d)\big)\mathrm{d}\mu_\ell(t)\\
&\ge \frac{1}{m}\sum_{\ell\le m}\inf_{\mathcal{B}_n}\int\sum_{n\ge0}2^{n/2}\min\big(\Delta(B_n(t),d_\omega),\Delta(B_n(t),d)\big)\mathrm{d}\mu_\ell(t)\,,
\end{aligned} \tag{15.34}$$

where the last infimum is taken over all admissible sequences of partitions $(\mathcal{B}_n(t))$ of H_ℓ. Integrating (15.33) with respect to $\mathrm{d}\mu$, combining with (15.34), taking the infimum over the choice of the sequence $(\mathcal{A}_n)$, multiplying by $\mathbf{1}_U$, and taking expectation, we obtain

$$F(\mu,H) \ge 2^{n_0/2}\mathsf{E}\mathbf{1}_U \inf_{(\mathcal{A}_n)}\int f(t,\omega,(\mathcal{A}_n))\mathrm{d}\mu(t) + \frac{1}{m}\sum_{\ell\le m}F(\mu_\ell,H_\ell)\,. \tag{15.35}$$

The goal now is to bound from below the first term on the right-hand side of (15.35). Consider the set

$$B = \{(s,t) \in H \times H \; ; \; d(s,t) \geq a/2\} = H \times H \setminus \Big(\bigcup_{\ell \leq m} H_\ell \times H_\ell \Big) \,.$$

Since the sets $H_\ell \times H_\ell$ are disjoint and satisfy $\mu \otimes \mu(H_\ell \times H_\ell) = 1/m^2$, we have $\mu \otimes \mu(B) = 1 - 1/m$. Given $(s,t) \in B$, it follows from (15.23) (and since $d_1(s,t) \leq 2a/(K_0\sqrt{\log m})$) that

$$\mathsf{P}(d_\omega(s,t) \geq \alpha a/2) \leq \frac{1}{\alpha} \exp\Big(- \frac{K_0^2 \log m}{K} \Big) \,.$$

If K_0 is large enough, the right-hand side is $\leq 1/4m$, so that then

$$\mathsf{E}\mu \otimes \mu(\{(s,t) \in B \; ; \; d_\omega(s,t) \geq \frac{\alpha a}{2}\}) \leq \frac{1}{4m} \,.$$

It then follows from Markov's inequality that the event U_0 defined by

$$\mu \otimes \mu(\{(s,t) \in B \; ; \; d_\omega(s,t) \leq \frac{\alpha a}{2}\}) \leq \frac{1}{m} \,,$$

has probability $\geq 3/4$. Since

$$\{(s,t) \in H \times H \; ; \; d_\omega(s,t) \leq \frac{\alpha a}{2}\} \subset \{(s,t) \in B \; ; \; d_\omega(s,t) \leq \frac{\alpha a}{2}\} \cup (H \times H \setminus B) \,,$$

and since $\mu \otimes \mu(H \times H \setminus B) \leq 1/m$, when U_0 occurs, we have

$$\mu \otimes \mu(\{(s,t) \in H \times H \; ; \; d_\omega(s,t) \leq \frac{\alpha a}{2}\}) \leq \frac{2}{m} \,. \tag{15.36}$$

Assuming that (15.36) holds and that K_0 is large enough, we prove that

$$\inf_{(\mathcal{A}_n)} \int f(t, \omega, (\mathcal{A}_n)) \mathrm{d}\mu(t) \geq \frac{a}{K} \,. \tag{15.37}$$

Combining with (15.35), using that $\mathsf{P}(U \cap U_0) \geq 1/2$, and taking the supremum over the choice of the measures, μ_ℓ completes the proof of (15.31).

We start the proof of (15.37). Let

$$B_1 = \bigcup \{A \in \mathcal{A}_{n_0} ; \; \Delta(A, d_\omega) \leq \alpha a/2\} \; ; \; B_2 = \bigcup \{A \in \mathcal{A}_{n_0} ; \; \Delta(A, d) \leq a/2\} \,. \tag{15.38}$$

If $A \in \mathcal{A}_{n_0}$ satisfies $A \subset B_1$ then $\Delta(A, d_\omega) \le \alpha a/2$ and $A \times A \subset \{(s,t)\ ;\ d_\omega(s,t) \le \alpha a/2\}$. Thus, by (15.36), it holds that $\mu \otimes \mu(A \times A) \le 2/m$, so that $\mu(A) \le \sqrt{2/m}$. Since B_1 is the union of at most $N_{n_0} \le \sqrt{m/32}$ such sets, we have $\mu(B_1) \le N_{n_0}\sqrt{2/m} \le 1/4$. Next, if $A \in \mathcal{A}_{n_0}$ satisfies $A \subset B_2$, then $\Delta(A, d) \le a/2$. Since $d(t_\ell, t_{\ell'}) \ge a$ for $\ell \ne \ell'$, A is entirely contained in a set H_ℓ, so that $\mu(A) \le 1/m$. Since B_2 is the union of at most N_{n_0} such sets, we have $\mu(B_2) \le N_{n_0}/m \le 1/4$.

Denoting by C the complement of $B_1 \cup B_2$, we have shown that $\mu(C) \ge 1/2$. Now, by definition of B_1 and B_2, if $t \in C$, then

$$\Delta(A_{n_0}(t), d_\omega) \ge \alpha a/2\ ;\ \Delta(A_{n_0}(t), d) \ge a/2$$

so that since $\Delta(H_\ell, d) \le 2a/K_0$, we obtain for such t that

$$f(t, \omega, (\mathcal{A}_n)) \ge a(\min(\alpha/2, 1/2) - 2/K_0) \ge a/K\ , \tag{15.39}$$

if K_0 has been taken large enough. Thus, $f(t, \omega, (\mathcal{A}_n)) \ge a/K$ on a set of measure $\ge 1/2$, and this completes the proof of (15.37) and concludes the argument when $m \ge 32$.

Let us now consider the case where $m \le 32$. Then we set $n_0 = 0$, and we proceed in a similar but much simpler manner. We define $f(t, \omega, (\mathcal{A}_n))$ as in (15.32) so that

$$f(t, \omega, (\mathcal{A}_n)) \ge \min(\Delta(H, d_\omega), \Delta(H, d)) - \Delta(H_{\ell(t)}, d)\ .$$

and it suffices to use (15.23) to prove that with high probability, we have $\Delta(H, d_\omega) \ge a/K$ to conclude as previously. □

15.1.3 Covering Numbers

Let us give a simple consequence of Theorem 15.1.5. We recall the covering numbers $N(T, d, \epsilon)$ of Sect. 1.4. We recall that $S(T) = \mathsf{E}\sup_{t\in T} X_t$.

Proposition 15.1.13 *There exists a constant L with the following property:*

$$\epsilon \ge L\sqrt{\Delta(T, d_\infty)S(T)} \Rightarrow \epsilon\sqrt{\log N(T, d_2, \epsilon)} \le LS(T)\ . \tag{15.40}$$

A remarkable feature of (15.40) is that, as we shall now prove, the right-hand side need not hold if $\epsilon \le \sqrt{\Delta(T, d_\infty)S(T)}/L$ (see however (15.43)). To see this, let us consider the example (15.17). For $\epsilon = \sqrt{n}/L$, we have $\epsilon\sqrt{\log(N(T, d_2, \epsilon))} \ge n^{3/2}/L$, while $S(T) \le Ln$, so that the right-hand side of (15.40) does not hold. Moreover, since $\Delta(T, d_\infty) = 2$, $\epsilon \ge \sqrt{\Delta(T, d_\infty)S(T)}/L$. This shows that the condition $\epsilon \ge L\sqrt{\Delta(T, d_\infty)S(T)}$ in (15.40) is rather precise.

Proof of Proposition 15.1.13 To lighten notation, we set $\Delta = \Delta(T, d_\infty)$. Consider $\epsilon > 0$ and a finite set $T' \subset T$ such that

$$\forall s, t \in T' \,,\; s \neq t \,,\; d_2(s,t) = \|t - s\|_{HS} \geq \epsilon \,. \tag{15.41}$$

Let $m = \operatorname{card} T'$. Thus, $N(T', d, \epsilon/2) \geq m$ so that by Exercise 2.7.8(b), we have $\gamma_2(T', d_2) \geq \epsilon\sqrt{\log m}/L$. Next, we have $\gamma_1(T', d_\infty) \leq L\Delta \log m$. This is witnessed by an admissible sequence $(\mathcal{A}_n)$ such that for $N_n \geq m$, then each set $A \in \mathcal{A}_n$ contains exactly one point (see Exercise 2.7.5). Now (15.18) implies

$$\begin{aligned}\frac{\epsilon}{L}\sqrt{\log m} \leq \gamma_2(T', d_2) &\leq L\big(S(T') + \sqrt{S(T')\gamma_1(T', d_\infty)}\big)\\ &\leq L\big(S(T) + \sqrt{S(T)\Delta \log m}\big) \,.\end{aligned} \tag{15.42}$$

Let us denote by L_2 the constant in the previous inequality. Now, if $\epsilon \geq L_3\sqrt{\Delta S(T)}$ where $L_3 = 2(L_2)^2$, we have $\sqrt{S(T)\Delta \log m} \leq \epsilon\sqrt{\log m}/L_3$, so that (15.42) implies

$$\frac{\epsilon}{L_2}\sqrt{\log m} \leq L_2 S(T) + \frac{1}{2L_2}\epsilon\sqrt{\log m}$$

and therefore $\epsilon\sqrt{\log m} \leq LS(T)$. Assuming $m = \operatorname{card} T'$ as large as possible, the balls centered at the points of T' of radius ϵ cover T and $N(T, d_2, \epsilon) \leq m$. □

The proof of Proposition 15.1.13 does not use the full strength of Theorem 15.1.8, and we propose the following as a very challenging exercise:

Exercise 15.1.14 Find a direct proof that under the conditions of Theorem 15.1.8, one has

$$\epsilon \geq L\sqrt{M\Delta(T, d_1)} \Rightarrow \epsilon\sqrt{\log N(T, d, \epsilon)} \leq LM \,,$$

and use this result to find a more direct proof of Proposition 15.1.13.

For completeness, let us mention the following, which should be compared with (15.40):

Proposition 15.1.15 *For each* $\epsilon > 0$, *we have*

$$\epsilon(\log N(T, d_2, \epsilon))^{1/4} \leq LS(T) \,. \tag{15.43}$$

In the previous example (15.17), both sides are of order n for $\epsilon = \sqrt{n}/L$.

Research Problem 15.1.16 Is it true that

$$\epsilon\sqrt{\log N(T,d_\infty,\epsilon)} \le LS(T)\ ? \tag{15.44}$$

For a partial result, and a proof of Proposition 15.1.15, see [109].

Exercise 15.1.17 Prove that (15.43) is true if (15.44) always hold.

15.1.4 *Another Way to Bound* $S(T)$

Next, we describe a way to control $S(T)$ from above, which is really different from the method of Theorem 15.1.4. Given a convex balanced subset U of ℓ^2 (i.e., $\lambda U \subset U$ for $|\lambda| \le 1$, or, equivalently, $U = -U$), we define

$$g(U) = \mathsf{E} \sup_{(u_i)\in U} \sum_{i\ge 1} u_i g_i$$

$$\sigma(U) = \sup_{(u_i)\in U} \Big(\sum_{i\ge 1} u_i^2\Big)^{1/2} .$$

Given convex balanced subsets U and V of ℓ^2, we define

$$T_{U,V} = \Big\{ t = (t_{i,j})\ ;\ \forall (x_i)_{i\ge 1}\,,\ \forall (y_j)_{j\ge 1}\,,$$
$$\sum t_{i,j} x_i y_j \le \sup_{(u_i)\in U} \sum_{i\ge 1} x_i u_i \sup_{(v_j)\in V} \sum_{j\ge 1} y_j v_j \Big\} .$$

This is a generalization of the example (15.17) to other norms than the Euclidean norm. It follows from (2.118) that, if $w > 0$,

$$\mathsf{P}\Big(\sup_{(u_i)\in U} \sum_{i\ge 1} g_i u_i \ge g(U) + w\sigma(U)\Big) \le 2\exp\Big(-\frac{w^2}{2}\Big) ,$$

so that (using that for positive numbers, when $ab > cd$, we have either $a > c$ or $b > d$)

$$\mathsf{P}\Big(\sup_{(u_i)\in U} \sum_{i\ge 1} g_i u_i \sup_{(v_j)\in V} \sum_{j\ge 1} g'_j v_j \ge (g(U) + w\sigma(U))(g(V) + w\sigma(V))\Big)$$
$$\le 4\exp\Big(-\frac{w^2}{2}\Big) .$$

Now,

$$\sup_{t \in T_{U,V}} X_t \le \sup_{(u_i) \in U} \sum_{i \ge 1} u_i g_i \sup_{(v_j) \in V} \sum_{j \ge 1} v_j g'_j ,$$

so that, whenever $g(U), g(V) \le 1$ and $\sigma(U), \sigma(V) \le 2^{-n/2}$, we obtain

$$\mathsf{P}\Big(\sup_{t \in T_{U,V}} X_t \ge (1 + 2^{-n/2} w)^2 \Big) \le 4 \exp\Big(-\frac{w^2}{2} \Big) .$$

Changing w into $2^{n/2} w$, this yields

$$\mathsf{P}\Big(\sup_{t \in T_{U,V}} X_t \ge (1 + w)^2 \Big) \le 4 \exp(-2^{n-1} w^2) . \tag{15.45}$$

Proposition 15.1.18 *Consider for $n \ge 0$ a family $\mathcal{C}_n$ of pairs (U, V) of convex balanced subsets of ℓ^2. Assume that* $\operatorname{card} \mathcal{C}_n \le N_n$ *and that*

$$\forall (U, V) \in \mathcal{C}_n , \; g(U) , \; g(V) \le 1 \; ; \; \sigma(U) , \; \sigma(V) \le 2^{-n/2} .$$

Then, the set

$$T = \operatorname{conv} \Big\{ \bigcup_n \bigcup_{(U,V) \in \mathcal{C}_n} T_{U,V} \Big\}$$

satisfies $S(T) \le L$.

Proof It follows from (15.45) that for $w \ge 2$,

$$\begin{aligned} \mathsf{P}\Big(\sup_T X_t \ge (1 + w)^2 \Big) &\le \sum_{n \ge 0} \sum_{(U,V) \in \mathcal{C}_n} \mathsf{P}\Big(\sup_{t \in T_{U,V}} X_t \ge (1 + w)^2 \Big) \\ &\le \sum_{n \ge 0} N_n \exp(-2^{n-1} w^2) \le L \exp(-w^2/4) , \end{aligned}$$

and the conclusion by (2.6) as usual. □

15.1.5 Yet Another Way to Bound $S(T)$

We end this section by a result involving a very special class of chaos, which we will bound by a method which is apparently different both from the method of Theorem 15.1.4 and from the method used for the set (15.17). To lighten notation, we denote by tg the sequence $(\sum_{j\geq 1} t_{i,j}g_j)_{i\geq 1}$, by $\langle\cdot,\cdot\rangle$ the dot product in ℓ^2, and by $\|\cdot\|_2$ the corresponding norm. For $t = (t_{i,j})$, let us write

$$Y_t^* := \sum_{i\geq 1}\Big(\sum_{j\geq 1} t_{i,j}g_j\Big)^2 = \|tg\|_2^2 = \langle tg, tg\rangle = \sum_{i\geq 1}\sum_{j,k\geq 1} t_{i,j}t_{i,k}g_jg_k \tag{15.46}$$

and

$$Y_t := Y_t^* - \mathsf{E}Y_t^* = \sum_{i\geq 1}\sum_{j\neq k} t_{i,j}t_{i,k}g_jg_k + \sum_{i\geq 1}\sum_{j\geq 1} t_{i,j}^2(g_j^2 - 1)\;, \tag{15.47}$$

which is a chaos of order 2.

Theorem 15.1.19 ([44]) *For any set T with $0 \in T$, we have*

$$\mathsf{E}\sup_{t\in T}|Y_t| \leq L\gamma_2(T, d_\infty)\Big(\gamma_2(T, d_\infty) + \sup_{t\in T}\|t\|_{HS}\Big)\;. \tag{15.48}$$

Let us define, with obvious notation,

$$Z_t = \sum_{i,j,k\geq 1} t_{i,j}t_{i,k}g_jg_k' = \langle tg, tg'\rangle\;.$$

The main step of the proof of Theorem 15.1.19 is as follows.

Proposition 15.1.20 *Let $U^2 := \mathsf{E}\sup_{t\in T}\|tg\|_2^2$. Then*

$$\mathsf{E}\sup_{t\in T}|Z_t| \leq LU\gamma_2(T, d_\infty)\;. \tag{15.49}$$

Proposition 15.1.21 *We set $V = \sup_{t\in T}\|t\|_{HS}$. Then*

$$U \leq L(V + \gamma_2(T, d_\infty))\;. \tag{15.50}$$

Proof We have $V^2 = \sup_{t\in T}\|t\|_{HS}^2 = \sup_{t\in T}\sum_{i,j\geq 1} t_{i,j}^2 = \sup_{t\in T}\mathsf{E}Y_t^*$. For $t \in T$, we have $\|tg\|_2^2 = Y_t^* = Y_t + \mathsf{E}Y_t^* \leq Y_t + V^2$, and thus,

$$U^2 \leq V^2 + \mathsf{E}\sup_{t\in T}|Y_t|\;. \tag{15.51}$$

Now, combining (15.2) and (15.47), we have

$$\mathsf{E}\sup_{t\in T}|Y_t| \leq 2\mathsf{E}\sup_{t\in T}|Z_t|\;. \tag{15.52}$$

Combining with (15.51) and (15.49), we obtain $U^2 \leq V^2 + LU\gamma_2(T, d_\infty)$, and this proves (15.50). □

Research Problem 15.1.22 The statement (15.50) is basically a bound on $\mathsf{E}\sup_{t\in T,\|x\|\le 1}\langle x, tg\rangle$, the supremum of a Gaussian process. Provide a direct construction of an admissible sequence on the index set which witnesses this bound. Recalling that t is viewed as an operator on ℓ^2, denote by t^* its adjoint, so that$\langle x, tg\rangle = \langle t^*x, g\rangle$. Prove that $\gamma_2(H, d_2) \le L(V + \gamma_2(T, d_\infty))$ where $H = \{t^*x; t \in T, \|x\| \le 1\}$.

Proof of Theorem 15.1.19 Plug (15.50) in (15.49). □

Proof of Proposition 15.1.20 Without loss of generality, we assume that T is finite. Consider an admissible sequence $(\mathcal{A}_n)$ with

$$\sup_{t\in T}\sum_{n\ge 0} 2^{n/2}\Delta(A_n(t)) \le 2\gamma_2(T, d_\infty)\ ,$$

where the diameter Δ is for the distance d_∞. For $A \in \mathcal{A}_n$, consider an element $t_{A,n} \in A$, and define as usual a chaining by $\pi_n(t) = t_{A_n(t),n}$. Since $0 \in T$, without loss of generality, we may assume that $\pi_0(t) = 0$. We observe that

$$Z_{\pi_n(t)} - Z_{\pi_{n-1}(t)} = \langle(\pi_n(t) - \pi_{n-1}(t))g, \pi_n(t)g'\rangle \\ + \langle\pi_{n-1}(t)g, (\pi_n(t) - \pi_{n-1}(t))g'\rangle\ . \tag{15.53}$$

Recalling that we think of each t as an operator on ℓ^2, let us denote by t^* its adjoint. Thus,

$$\langle(\pi_n(t) - \pi_{n-1}(t))g, \pi_n(t)g'\rangle = \langle g, (\pi_n(t) - \pi_{n-1}(t))^*\pi_n(t)g'\rangle\ . \tag{15.54}$$

Here, $(\pi_n(t) - \pi_{n-1}(t))^*\pi_n(t)g'$ is the element of ℓ^2 obtained by applying the operator $(\pi_n(t) - \pi_{n-1}(t))^*$ to the vector $\pi_n(t)g'$. Let us now consider the r.v.s $W = \sup_{t\in T}\|tg\|_2$ and $W' = \sup_{t\in T}\|tg'\|_2$. Then

$$\|(\pi_n(t) - \pi_{n-1}(t))^*\pi_n(t)g'\|_2 \le \|(\pi_n(t) - \pi_{n-1}(t))^*\|\|\pi_n(t)g'\|_2 \\ \le \Delta(A_{n-1}(t))W'.$$

It then follows from (15.54) that, conditionally on g', the quantity $\langle(\pi_n(t) - \pi_{n-1}(t))g, \pi_n(t)g'\rangle$ is a Gaussian r.v. G with $(\mathsf{E}G^2)^{1/2} \le \Delta(A_{n-1}(t))W'$. Thus, we obtain that for $u \ge 1$

$$\mathsf{P}\Big(|\langle(\pi_n(t) - \pi_{n-1}(t))g, \pi_n(t)g'\rangle| \ge 2^{n/2}u\Delta(A_{n-1}(t))W'\Big) \le \exp(-u^2 2^n/2)\ .$$

Proceeding in a similar fashion for the second term in (15.53), we get

$$\mathsf{P}(|Z_{\pi_n(t)} - Z_{\pi_{n-1}(t)}| \ge 2^{n/2}u\Delta(A_{n-1}(t))(W + W')) \le 2\exp(-u^2 2^n/2)\ .$$

Using that $Z_{\pi_0(t)} = 0$, and proceeding just as in the proof of the generic chaining bound (2.33), we obtain that for $u \geq L$,

$$\mathsf{P}\Big(\sup_{t \in T} |Z_t| \geq Lu\gamma_2(T, d_\infty)(W + W')\Big) \leq L \exp(-u^2) .$$

In particular, the function $R = \sup_{t\in T} |Z_t|/(W+W')$ satisfies $\mathsf{E}R^2 \leq L\gamma_2(T, d_\infty)^2$. Since $\sup_{t\in T} |Z_t| = R(W + W')$ and $\mathsf{E}W^2 = \mathsf{E}W'^2 = U^2$, the Cauchy-Schwarz inequality yields (15.49). □

Having found three distinct ways, (15.16), Proposition 15.1.18, and (15.48) of controlling $S(T)$, one should certainly ask whether there are more. It simply seems difficult to even make a sensible conjecture about what might be the "most general way to bound a chaos process".

15.2 Tails of Multiple-Order Gaussian Chaos

In this section, we consider a single-order d (decoupled) Gaussian chaos, that is, a r.v. X of the type

$$X = \sum_{i_1,\ldots,i_d} a_{i_1,\ldots,i_d} g^1_{i_1} \cdots g^d_{i_d} , \tag{15.55}$$

where $a_{i_1,\ldots,i_d}$ are numbers and g^j_i are independent standard Gaussian r.v.s. The sum is finite; each index i_ℓ runs from 1 to m. Our purpose is to estimate the higher moments of the r.v. X as a function of certain characteristics of

$$A := (a_{i_1,\ldots,i_d})_{i_1,\ldots,i_d \leq m} . \tag{15.56}$$

Estimating the higher moments of the r.v. X amounts to estimate its tails, and it is self-evident that this is a natural question. This topic runs into genuine notational difficulties. One may choose to avoid considering tensors, in which case one faces heavy multi-index notation. Or one may entirely avoid multi-index notation using tensors, but one gets dizzy from the height of the abstraction. We shall not try for elegance in the presentation, but rather to minimize the amount of notation the reader has to assimilate. Our approach will use a dash of tensor vocabulary, but does not require any knowledge of what these are. In any case for the really difficult arguments, we shall focus on the case $d = 3$.

Let us start with the case $d = 2$ that we considered at length in the previous section. In that case, one may think of A as a linear functional on $\mathbb{R}^{m^2}$ by the formula

$$A(x) = \sum_{i,j} a_{i,j} x_{i,j} , \tag{15.57}$$

where $x = (x_{i,j})_{i,j \le m}$ is the generic element of $\mathbb{R}^{m^2}$. It is understood that in (15.57), the sum runs over $i, j \le m$. When we provide $\mathbb{R}^{m^2}$ with the canonical Euclidean structure, the norm of A viewed as a linear functional on $\mathbb{R}^{m^2}$ is simply

$$\|A\|_{\{1,2\}} := \Big(\sum_{i,j} a_{i,j}^2\Big)^{1/2} . \tag{15.58}$$

This quantity was denoted $\|A\|_{HS}$ in the previous section, but here we need new notation. We may also think of A as a bilinear functional on $\mathbb{R}^m \times \mathbb{R}^m$ by the formula

$$A(x, y) = \sum_{i,j} a_{i,j} x_i y_j , \tag{15.59}$$

where $x = (x_i)_{i \le m}$ and $y = (y_i)_{i \le m}$. In that case, if we provide both copies of $\mathbb{R}^m$ with the canonical Euclidean structure, the corresponding norm of A is

$$\|A\|_{\{1\}\{2\}} := \sup\Big\{\Big|\sum_{i,j} a_{i,j} x_i y_j\Big| \; ; \; \sum x_i^2 \le 1, \sum y_j^2 \le 1\Big\} , \tag{15.60}$$

which is also the operator norm when one sees A as a matrix, i.e., an operator from $\mathbb{R}^m$ to $\mathbb{R}^m$. One observes the inequality $\|A\|_{\{1\}\{2\}} \le \|A\|_{\{1,2\}}$.

Let us now turn to the case $d = 3$. One may think of A as a linear functional on $\mathbb{R}^{m^3}$, obtaining the norm

$$\|A\|_{\{1,2,3\}} := \Big(\sum_{i,j,k} a_{i,j,k}^2\Big)^{1/2} , \tag{15.61}$$

or think of A as a trilinear functional on $(\mathbb{R}^m)^3$, obtaining the norm

$$\|A\|_{\{1\}\{2\}\{3\}} := \sup\Big\{\Big|\sum_{i,j,k} a_{i,j,k} x_i y_j z_k\Big| \; ; \; \sum x_i^2 \le 1, \sum y_j^2 \le 1, \sum z_k^2 \le 1\Big\} . \tag{15.62}$$

One may also view A as a bilinear function on $\mathbb{R}^{m^2} \times \mathbb{R}^m$ by the formula

$$A(x, y) = \sum_{i,j,k} a_{i,j,k} x_{i,j} y_k , \tag{15.63}$$

for $x = (x_{i,j})_{i,j} \in \mathbb{R}^{m^2}$ and $(y_k) \in \mathbb{R}^m$. One then obtains the norm

$$\|A\|_{\{1,2\}\{3\}} := \sup\Big\{\Big|\sum_{i,j,k} a_{i,j,k} x_{i,j} y_k\Big| \; ; \; \sum x_{i,j}^2 \le 1, \sum y_k^2 \le 1\Big\} . \tag{15.64}$$

We observe the inequality

$$\|A\|_{\{1\}\{2\}\{3\}} \le \|A\|_{\{1,2\}\{3\}} \le \|A\|_{\{1,2,3\}} . \tag{15.65}$$

More generally, given a partition $\mathcal{P} = \{I_1, \dots, I_k\}$ of $\{1, \dots, d\}$, we may define the norm

$$\|A\|_{\mathcal{P}} = \|A\|_{I_1,\dots,I_k} \tag{15.66}$$

by viewing A as a k-linear form C on $F_1 \times \cdots \times F_k$ where $F_\ell = \mathbb{R}^{m^{\operatorname{card} I_\ell}}$ and defining

$$\|A\|_{I_1,\dots,I_k} = \|C\|_{\{1\}\{2\}\cdots\{k\}} , \tag{15.67}$$

where the right-hand side is defined as in (15.62). When the partition $\mathcal{P}'$ is finer than the partition $\mathcal{P}$, then

$$\|A\|_{\mathcal{P}'} \le \|A\|_{\mathcal{P}} . \tag{15.68}$$

The moments of the r.v. X of (15.55) are then evaluated by the following formula:

Theorem 15.2.1 (R. Latała [48]) *For $p \ge 1$, we have*

$$\frac{1}{K(d)} \sum_{\mathcal{P}} p^{\operatorname{card} \mathcal{P}/2} \|A\|_{\mathcal{P}} \le \|X\|_p \le K(d) \sum_{\mathcal{P}} p^{\operatorname{card} \mathcal{P}/2} \|A\|_{\mathcal{P}} , \tag{15.69}$$

where $\mathcal{P}$ runs over all partitions of $\{1, \dots, d\}$.

A multidimensional array as in (15.56) will be called a *tensor of order d* (the value of m may depend on the context).

Exercise 15.2.2 Generalize Theorem 15.1.4 to a set T of tensors of order d using the upper bound of (15.69). Hint: This assumes that you know how to transform (15.69) into a tail estimate.

The proof of the lower bound in (15.69) is significantly easier than the proof of the upper bound, and we start with it.

Proof of the Lower Bound in Theorem 15.2.1 We shall prove this lower bound only for $p \ge 2$. First, we observe that for $d = 1$, this simply reflects the fact that for a standard Gaussian r.v. g, one has $(\mathsf{E}|g|^p)^{1/p} \ge \sqrt{p}/L$. (No, this has not been proved anywhere in this book, but see Exercise 2.3.9.) Next, we prove by induction on d that for each d, one has

$$(\mathsf{E}|X|^p)^{1/p} \ge \frac{\sqrt{p}}{K} \|A\|_{\{1,2,\dots,d\}} = \frac{\sqrt{p}}{K} \Big(\sum_{i_1,\dots,i_d} a_{i_1,\dots,i_d}^2 \Big)^{1/2} . \tag{15.70}$$

For this, we consider the random tensor B of order $d-1$ given by

$$b_{i_1,\ldots,i_{d-1}} = \sum_{i\le m} a_{i_1,\ldots,i_{d-1},i}\, g_i^d \,.$$

Applying the induction hypothesis to B given the r.v.s g_i^d, and denoting by E' expectation given these variables, we obtain

$$(\mathsf{E}'|X|^p)^{1/p} \ge \frac{\sqrt{p}}{K}\Big(\sum_{i_1,\ldots,i_{d-1}} b^2_{i_1,\ldots,i_{d-1}}\Big)^{1/2} .$$

We compute the norm in L^p of both sides, using that for $p \ge 2$ one has $(\mathsf{E}|Y|^p)^{1/p} \ge (\mathsf{E}Y^2)^{1/2}$ to obtain

$$(\mathsf{E}|X|^p)^{1/p} \ge \frac{\sqrt{p}}{K}\Big(\mathsf{E}\sum_{i_1,\ldots,i_{d-1}} b^2_{i_1,\ldots,i_{d-1}}\Big)^{1/2} ,$$

which yields (15.70) for d. (It is only at this place that a tiny extra effort is required if $p \le 2$.)

Let us now prove by induction over k that

$$(\mathsf{E}|X|^p)^{1/p} \ge \frac{p^{k/2}}{K}\|A\|_{I_1,\ldots,I_k} \,. \tag{15.71}$$

The case $k=1$ is (15.70). For the induction from $k-1$ to k, let us assume without loss of generality that $I_k = \{r+1,\ldots,d\}$, and let us define an order $d-r$ random tensor C by

$$c_{i_{r+1},\ldots,i_d} = \sum_{i_1,\ldots,i_r} a_{i_1,\ldots,i_d}\, g^1_{i_1}\cdots g^r_{i_r} \,,$$

so that

$$X = \sum_{i_{r+1},\ldots,i_d} c_{i_{r+1},\ldots,i_d}\, g^{r+1}_{i_{r+1}}\cdots g^d_{i_d} \,.$$

Denoting now by $\mathsf{E}^{\sim}$ expectation only in the r.v.s $g^\ell_{i_\ell}$ for $r+1 \le \ell \le d$, we use (15.70) to obtain

$$(\mathsf{E}^{\sim}|X|^p)^{1/p} \ge \frac{\sqrt{p}}{K}\Big(\sum_{i_{r+1},\ldots,i_d} c^2_{i_{r+1},\ldots,i_d}\Big)^{1/2} .$$

Consequently, if $x_{i_{r+1},\dots,i_d}$ are numbers with $\sum_{i_{r+1},\dots,i_d} x^2_{i_{r+1},\dots,i_d} \le 1$, one gets

$$
\begin{aligned}
(\mathsf{E}^{\sim}|X|^p)^{1/p} &\ge \frac{\sqrt{p}}{K}\Big|\sum_{i_{r+1},\dots,i_d} c_{i_{r+1},\dots,i_d} x_{i_{r+1},\dots,i_d}\Big| \\
&= \frac{\sqrt{p}}{K}\Big|\sum_{i_1,\dots,i_r} d_{i_1,\dots,i_r} g^1_{i_1}\cdots g^r_{i_r}\Big| ,
\end{aligned}
\tag{15.72}
$$

where

$$
d_{i_1,\dots,i_r} = \sum_{i_{r+1},\dots,i_d} a_{i_1,\dots,i_d} x_{i_{r+1},\dots,i_d} .
$$

We now compute the L^p norm of both sides of (15.72), using the induction hypothesis to obtain

$$
(\mathsf{E}|X|^p)^{1/p} \ge \frac{p^{k/2}}{K}\|D\|_{I_1,\dots,I_{k-1}} ,
$$

where D is the tensor $(d_{i_1,\dots,i_r})$. The supremum of the norms on the right-hand side over the choices of $(x_{i_{r+1},\dots,i_d})$ with $\sum_{i_{r+1},\dots,i_d} x^2_{i_{r+1},\dots,i_d} \le 1$ is $\|A\|_{I_1,\dots,I_k}$. (A formal definition of these norms by induction over k would be based exactly on this property.) □

Let us denote by $E_1,\dots,E_d$ copies of $\mathbb{R}^m$. The idea is that E_k is the copy that corresponds to the k-th index of A. Given a vector $x \in E_d$, we may then define the contraction $\langle A, x\rangle$ as the tensor $(b_{i_1,\dots,i_{d-1}})$ of order $d-1$ given by

$$
b_{i_1,\dots,i_{d-1}} = \sum_{i\le m} a_{i_1,\dots,i_{d-1},i} x_i .
$$

The summation here is on the d-th index, consistent with the fact that $x \in E_d$.

If G is a standard Gaussian random vector valued in E_d, i.e., $G = (g_i)_{i\le m}$, where g_i are independent standard r.v.s, then $\langle A, G\rangle$ is a random tensor of order $d-1$. We shall deduce Theorem 15.2.1 from the following fact, of independent interest:

Theorem 15.2.3 *Consider $d \ge 2$. Then for all $\tau \ge 1$, we have*

$$
\mathsf{E}\|\langle A, G\rangle\|_{\{1\}\cdots\{d-1\}} \le K\sum_{\mathcal{P}} \tau^{\operatorname{card}\mathcal{P}-d+1}\|A\|_{\mathcal{P}} ,
\tag{15.73}
$$

where $\mathcal{P}$ runs over all the partitions of $\{1,\dots,d\}$.

Here, as well as in the rest of this section, K denotes a number that depends only on the order of the tensor considered and certainly not on τ.

The bound (15.73) has the mind-boggling feature that the powers of τ in the right-hand side may have different signs. This feature will actually appear very naturally in the course of the proof. It will be used only through Corollary 15.2.4.

If we think of A as a d-linear form on $E_1 \times \cdots \times E_d$, then

$$Y := \|\langle A, G\rangle\|_{\{1\}\cdots\{d-1\}} = \sup A(x^1, \ldots, x^{d-1}, G) , \tag{15.74}$$

where the supremum is over all choices of x^ℓ with $\|x^\ell\| \le 1$. Therefore, the issue to prove (15.73) is to bound the supremum of a certain complicated Gaussian process.

Corollary 15.2.4 *For all $p \ge 1$, one has*

$$\big(\mathsf{E}\|\langle A, G\rangle\|^p_{\{1\}\cdots\{d-1\}}\big)^{1/p} \le K \sum_{\mathcal{P}} p^{(\operatorname{card}\mathcal{P}-d+1)/2}\|A\|_{\mathcal{P}} . \tag{15.75}$$

Proof As witnessed by (15.74), the r.v. $Y = \|\langle A, G\rangle\|_{\{1\}\cdots\{d-1\}}$ is the supremum of Gaussian r.v.s of the type $Z = A(x^1, \ldots, x^{d-1}, G)$, where in this formula we view A as a d-linear map on $E_1 \times \cdots \times E_d$ and where x^ℓ is a vector of norm ≤ 1. Now, the Gaussian r.v. Z is of the type $Z = \sum_i a_i g_i$, and the formula

$$\Big(\mathsf{E}\Big(\sum_i a_i g_i\Big)^2\Big)^{1/2} = \Big(\sum_i a_i^2\Big)^{1/2} = \sup\Big\{\sum_i a_i x_i \; ; \; \sum_i x_i^2 \le 1\Big\}$$

implies

$$(\mathsf{E}Z^2)^{1/2} = \sup_{\|x\|\le 1} |A(x^1, \ldots, x^{d-1}, x)| \le \sigma := \|A\|_{\{1\}\cdots\{d-1\}\{d\}} .$$

It then follows from (2.118) that for $u > 0$, the r.v. Y satisfies

$$\mathsf{P}(|Y - \mathsf{E}Y| \ge u) \le 2\exp\Big(-\frac{u^2}{2\sigma^2}\Big) .$$

Then (2.24) implies

$$(\mathsf{E}|Y - \mathsf{E}Y|^p)^{1/p} \le L\sqrt{p}\sigma ,$$

and since $(\mathsf{E}|Y|^p)^{1/p} \le \mathsf{E}|Y| + (\mathsf{E}|Y - \mathsf{E}Y|^p)^{1/p} \le \mathsf{E}|Y| + L\sqrt{p}\sigma$, using (15.73) for $\tau = p^{1/2}$ to bound $\mathsf{E}|Y|$, the result follows. □

Proof of the Upper Bound in Theorem 15.2.1 We proceed by induction over d, using also (15.75). For $d = 1$, (15.69) reflects the growth of the moments of a single Gaussian r.v. as captured by (2.24). Assuming that the result has been proved for $d-1$, we prove it for d. We consider the Gaussian random vector $G=(g_i^d)$ and the order $d-1$ random tensor

$$B=\langle A,G\rangle=(b_{i_1,\ldots,i_{d-1}})\ ,$$

where

$$b_{i_1,\ldots,i_{d-1}}=\sum_{i\le m}a_{i_1,\ldots,i_{d-1},i}\,g_i^d\ .$$

Thus,

$$X=\sum_{i_1,\ldots,i_d}a_{i_1,\ldots,i_d}g_{i_1}^1\cdots g_{i_d}^d=\sum_{i_1,\ldots,i_{d-1}}b_{i_1,\ldots,i_{d-1}}g_{i_1}^1\cdots g_{i_{d-1}}^{d-1}\ .$$

Let us denote by E' expectation given G. Then the induction hypothesis applied to B implies

$$(\mathsf{E}'|X|^p)^{1/p}\le K\sum_{\mathcal{Q}}p^{\operatorname{card}\mathcal{Q}/2}\|B\|_{\mathcal{Q}}\ , \tag{15.76}$$

where the sum runs over all partitions $\mathcal{Q}$ of $\{1,\ldots,d-1\}$. We now compute the L^p norm of both sides, using the triangle inequality in L^p to obtain

$$(\mathsf{E}|X|^p)^{1/p}\le K\sum_{\mathcal{Q}}p^{\operatorname{card}\mathcal{Q}/2}(\mathsf{E}\|B\|_{\mathcal{Q}}^p)^{1/p}\ . \tag{15.77}$$

Our next goal is to prove that

$$\big(\mathsf{E}\|B\|_{\mathcal{Q}}^p\big)^{1/p}\le Kp^{-\operatorname{card}\mathcal{Q}/2}\sum_{\mathcal{P}}p^{\operatorname{card}\mathcal{P}/2}\|A\|_{\mathcal{P}}\ , \tag{15.78}$$

providing the same bound $K\sum_{\mathcal{P}}p^{\operatorname{card}\mathcal{P}/2}\|A\|_{\mathcal{P}}$ for each term in the summation of (15.77) and finishing the proof of (15.69).

To prove (15.78), we denote by $I_1,\ldots,I_k$ the elements of $\mathcal{Q}$, so that

$$\big(\mathsf{E}\|B\|_{\mathcal{Q}}^p\big)^{1/p}=\big(\mathsf{E}\|\langle A,G\rangle\|_{I_1,\ldots,I_k}^p\big)^{1/p}\ . \tag{15.79}$$

For $\ell\le k$, we define $F_\ell=\mathbb{R}^{m^{\operatorname{card}I_\ell}}$, and we define $F_{k+1}=\mathbb{R}^m$. Let us view A as a $(k+1)$-linear form C on the space $F_1\times\cdots\times F_{k+1}$. Thus, from (15.67), we have $\|\langle A,G\rangle\|_{I_1,\ldots,I_k}=\|\langle C,G\rangle\|_{\{1\},\ldots,\{k\}}$. Applying (15.75) to C (with $d=k+1$), we get

$$\big(\mathsf{E}\|\langle C,G\rangle\|_{\{1\},\ldots,\{k\}}^p\big)^{1/p}\le Kp^{-k/2}\sum_{\mathcal{R}}p^{\operatorname{card}\mathcal{R}/2}\|C\|_{\mathcal{R}}\ ,$$

where $\mathcal{R}$ ranges over all partitions of $\{1, \ldots, k+1\}$. Combining with (15.79), we have obtained a stronger form of (15.78), where the summation in the right-hand side is restricted to the partitions $\mathcal{P}$ which are coarser than the partition $I_1, \ldots, I_k, \{d\}$ of $\{1, \ldots, d\}$. □

The rest of this section in devoted to the proof of Theorem 15.2.3. The proof is by induction over d. The case $d = 2$ is very simple. We simply write

$$\|\langle A, G\rangle\|_{\{1\}} = \sup_{\|x\|\le 1} A(x, G) = \Big(\sum_i \Big(\sum_j a_{i,j} g_j\Big)^2\Big)^{1/2},$$

and use of the Cauchy-Schwarz inequality proves that

$$\mathsf{E}\|\langle A, G\rangle\|_{\{1\}} \le \Big(\sum_{i,j} a_{i,j}^2\Big)^{1/2} = \|A\|_{\{1,2\}}, \tag{15.80}$$

while for $\operatorname{card}\mathcal{P} = 1$, one has $\operatorname{card}\mathcal{P} - d + 1 = 0$ and $\tau^{\operatorname{card}\mathcal{P}-d+1} = 1$.

In order to help the reader penetrate the very deep ideas involved in the proof of Theorem 15.2.3, we will now assume that $d = 3$. The proof of Theorem 15.2.3 for the general value of d does not require any essentially new idea, but it is more complicated to write. We refer to Latała's paper for this. We start with some tools (of fundamental importance).

Lemma 15.2.5 *Denoting by μ the canonical Gaussian measure on $\mathbb{R}^m$, then for each closed symmetric set V of $\mathbb{R}^m$, one has*

$$\mu(V + x) \ge \mu(V)\exp\Big(-\frac{\|x\|^2}{2}\Big). \tag{15.81}$$

Proof Let us denote λ Lebesgue's measure on $\mathbb{R}^m$. Then, using symmetry in the third line, the parallelogram identity and convexity of the exponential in the fourth line, and setting $c = (2\pi)^{-m/2}$,

$$\begin{aligned}
\mu(x + V) &= c\int_{x+V} \exp\Big(-\frac{\|y\|^2}{2}\Big)\mathrm{d}\lambda(y)\\
&= c\int_V \exp\Big(-\frac{\|x+y\|^2}{2}\Big)\mathrm{d}\lambda(y)\\
&= c\int_V \frac{1}{2}\Big(\exp\Big(-\frac{\|x+y\|^2}{2}\Big) + \exp\Big(-\frac{\|x-y\|^2}{2}\Big)\Big)\mathrm{d}\lambda(y)\\
&\ge c\int_V \exp\Big(-\frac{\|x\|^2+\|y\|^2}{2}\Big)\mathrm{d}\lambda(y)\\
&= \exp\Big(-\frac{\|x\|^2}{2}\Big)\mu(V).
\end{aligned}$$

□

Lemma 15.2.6 *Consider a standard Gaussian vector G valued in $\mathbb{R}^m$ and x in $\mathbb{R}^m$. Consider a semi-norm α on $\mathbb{R}^m$. Then*

$$\mathsf{P}\big(\alpha(G-x)\le 4\mathsf{E}\alpha(G)\big)\ge\frac{1}{2}\exp\Big(-\frac{\|x\|^2}{2}\Big)\,. \tag{15.82}$$

Proof We consider the set $V=\{y\in\mathbb{R}^m\ ;\ \alpha(y)\le 4\mathsf{E}\alpha(G)\}$, so that by Markov's inequality, $\mu(V^c)\le 1/4$ and consequently $\mu(V)\ge 3/4$. Then (15.81) implies

$$\mathsf{P}(\alpha(G-x)\le 4\mathsf{E}\alpha(G))=\mu(V+x)\ge\frac{3}{4}\exp\Big(-\frac{\|x\|^2}{2}\Big)\,,$$

where we have used in the equality that μ is the law of G and (15.81) in the last inequality. □

We recall the entropy numbers $e_n(T,d)$ of (2.36). The next consequence of Lemma 15.2.6 is called "the dual Sudakov inequality" and is *extremely useful* to estimate these entropy numbers.

Lemma 15.2.7 *Consider a semi-norm α on $\mathbb{R}^m$ and a standard Gaussian r.v. G valued in $\mathbb{R}^m$. Then if d_α is the (quasi)-distance associated with α, the Euclidean unit ball B of $\mathbb{R}^m$ satisfies*

$$e_n(B,d_\alpha)\le L2^{-n/2}\mathsf{E}\alpha(G)\,. \tag{15.83}$$

Proof From (15.82), we get

$$\mathsf{P}(\alpha(G-x)\le 4\mathsf{E}\alpha(G))\ge\frac{1}{2}\exp\Big(-\frac{\|x\|^2}{2}\Big)\,,$$

and, by homogeneity, for $x\in B$ and $\tau>0$,

$$\mathsf{P}(\alpha(\tau G-x)\le 4\tau\mathsf{E}\alpha(G))\ge\frac{1}{2}\exp\Big(-\frac{1}{2\tau^2}\Big)\,. \tag{15.84}$$

The proof is really similar to the argument of Exercise 2.5.9. We repeat this argument for the convenience of the reader. Consider $\epsilon=4\tau\mathsf{E}\alpha(G)$ and a subset U of B such that any two points of U are at mutual distances $\ge 3\epsilon$. Then the closed[1] balls for d_α of radius ϵ centered at the points of U are disjoint. Now, (15.84) asserts that the probability that G belongs to any such ball is $\ge\exp(-1/(2\tau^2)/2$, so that $\operatorname{card}U\le 2\exp(1/2\tau^2)$. Taking U as large as possible, the balls centered at U of radius $3\epsilon=12\tau\mathsf{E}\alpha(G)$ cover B. Taking τ such that $2\exp(1/2\tau^2)=2^{2^n}$ (so that $\tau\le L2^{-n/2}$), we have covered B by at most 2^{2^n} ball of radius $L2^{-n/2}\mathsf{E}\alpha(G)$. □

[1] We take the centers of the balls at mutual distance $\ge 3\epsilon$ to ensure that the closed balls centered at these points are disjoint.

In the next page or so, we make a detour from our main story to provide a proof of the Sudakov minoration.

To understand the difference between the Sudakov minoration and the dual Sudakov minoration, consider the set T which is the polar set of the unit ball for α. That is, denoting by $\langle x, y\rangle$ the duality of $\mathbb{R}^m$ with itself,

$$T = \{x \in \mathbb{R}^m \; ; \; \forall y \in \mathbb{R}^m, \alpha(y) \le 1 \Rightarrow \langle x, y\rangle \le 1\} \, .$$

Then $\alpha(G) = \sup_{x\in T}\langle x, G\rangle$. According to (2.117), the Sudakov minoration states that $e_n(T, d) \le L2^{-n/2}\mathsf{E}\alpha(G)$, where d is the Euclidean distance.

Exercise 15.2.8 Consider a symmetric subset A of $\mathbb{R}^M$ (i.e., $x \in A \Rightarrow -x \in A$) and the semi-norm α on $\mathbb{R}^m$ given by $\alpha(y) = \sup\{|\langle x, y\rangle|; x \in A\}$. Prove that $\mathsf{E}\alpha(G) = g(A)$.

The following is a simple yet fundamental fact about entropy numbers:

Lemma 15.2.9 *Consider two distances d_1 and d_2 on $\mathbb{R}^m$ that arise from semi-norms, of unit balls U_1 and U_2, respectively. Then for any set $T \subset \mathbb{R}^m$, one has*

$$e_{n+1}(T, d_2) \le 2e_n(T, d_1)e_n(U_1, d_2) \, . \tag{15.85}$$

Proof Consider $a > e_n(T, d_1)$ so that we can find points $(t_\ell)_{\ell\le N_n}$ of T such $T \subset \cup_{\ell\le N_n}(t_\ell + aU_1)$. Consider $b > e_n(U_1, d_2)$, so that we can find points $(u_\ell)_{\ell\le N_n}$ for which $U_1 \subset \cup_{\ell\le N_n}(u_\ell + bU_2)$. Then

$$T \subset \bigcup_{\ell,\ell'\le N_n} (t_\ell + au_{\ell'} + abU_2) \, .$$

Let

$$I = \{(\ell, \ell') \; ; \; \ell, \ell' \le N_n \, , \; (t_\ell + au_{\ell'} + abU_2) \cap T \ne \emptyset\} \, ,$$

so that $\operatorname{card} I \le N_n^2 = N_{n+1}$. For $(\ell, \ell') \in I$, let $v_{\ell,\ell'} \in (t_\ell + au_{\ell'} + abU_2) \cap T$. Then

$$T \subset \bigcup_{(\ell,\ell')\in I} (v_{\ell,\ell'} + 2abU_2) \, ,$$

so that $e_{n+1}(T, d_2) \le 2ab$. □

The following very nice result is due to N. Tomczak-Jaegermann:

Lemma 15.2.10 *Consider on $\mathbb{R}^m$ a distance d_V induced by a norm of unit ball V, and let V° be the polar set of V as in (8.55). Denote by B the Euclidean ball of $\mathbb{R}^m$ and by d_2 the Euclidean distance. Assume that for some numbers $\alpha \geq 1$, A and n^*, we have*

$$0 \leq n \leq n^* \Rightarrow e_n(B, d_V) \leq 2^{-n/\alpha} A \,. \tag{15.86}$$

Then

$$0 \leq n \leq n^* \Rightarrow e_n(V^\circ, d_2) \leq 16 \cdot 2^{-n/\alpha} A \,. \tag{15.87}$$

Proof Consider $n \leq n^*$. Using (15.85) in the first inequality and (15.86) in the second one, we obtain

$$e_{n+1}(V^\circ, d_V) \leq 2e_n(V^\circ, d_2)e_n(B, d_V) \leq 2^{-n/\alpha+1} Ae_n(V^\circ, d_2) \,. \tag{15.88}$$

Let us now denote by $\langle \cdot, \cdot \rangle$ the canonical duality of $\mathbb{R}^m$ with itself, so that if $y \in V$ and $z \in V^\circ$, we have $\langle y, z \rangle \leq 1$. Consider $x, t \in V^\circ$, and $a = d_V(x, t)$. Then $x - t \in 2V^\circ$ and $x - t \in aV$, so that

$$\|x - t\|_2^2 = \langle x - t, x - t \rangle \leq 2a \,,$$

and thus $d_2(x, t)^2 \leq 2d_V(x, t)$. Consequently, $e_{n+1}(V^\circ, d_2)^2 \leq 2e_{n+1}(V^\circ, d_V)$. Combining with (15.88),

$$e_{n+1}(V^\circ, d_2)^2 \leq 2^{-n/\alpha+2} Ae_n(V^\circ, d_2) \,,$$

from which (15.87) follows by induction over n. □

Proof of Sudakov Minoration (Lemma 2.10.2) We keep the previous notations. Consider a finite subset T of $\mathbb{R}^m$, and define the semi-norm α on $\mathbb{R}^m$ by $\alpha(x) = \sup_{t \in T} |\langle x, t \rangle|$, so that $\mathsf{E}\alpha(G) = \mathsf{E} \sup_{t \in T} |X_t|$ where $X_t = \langle G, t \rangle$, and by (15.83), we have $e_n(B, d_\alpha) \leq L2^{-n/2}\mathsf{E}|\alpha(G)|$, where d_α is the distance associated with the semi-norm α. Denoting by V the unit ball of the semi-norm d_α, the set T is a subset of V°, and Lemma 15.2.10 implies $e_n(T, d_2) \leq L2^{-n/2}\mathsf{E}|\alpha(G)| = L2^{-n/2}\mathsf{E} \sup_{t \in T} |X_t|$. To evaluate $e_n(T, d_2)$, we may by translation assume that $0 \in T$, and by (2.3), we obtain $e_n(T, d_2) \leq L2^{-n/2}\mathsf{E} \sup_{t \in T} X_t$, which is the desired result by (2.117). □

We go back to our main story. Our next goal is to prove special extensions of Lemmas 15.2.6 and 15.2.7 "to the two-dimensional case". We consider two copies E_1 and E_2 of $\mathbb{R}^m$, and for vectors $y^\ell \in E_\ell$, $y^\ell = (y_i^\ell)_{i \leq m}$, we define their tensor product $y^1 \otimes y^2$ as the vector (z_{i_1,i_2}) in $\mathbb{R}^{m^2}$ given by $z_{i_1,i_2} = y_{i_1}^1 y_{i_2}^2$. Let us consider for $\ell \le 2$ independent standard Gaussian vectors G^ℓ valued in E_ℓ, and let us fix vectors $x^\ell \in E_\ell$. We define

$$U_\emptyset = x^1 \otimes x^2 \; ; \; U_{\{1\}} = G^1 \otimes x^2 \; ; \; U_{\{2\}} = x^1 \otimes G^2 \; ; \; U_{\{1,2\}} = G^1 \otimes G^2 \; .$$

We denote by $\|x\|$ the Euclidean norm of a vector x of E_ℓ.

Lemma 15.2.11 *Set* $\mathcal{I} = \{\{1\}, \{2\}, \{1,2\}\}$*, and consider a semi-norm* α *on* $\mathbb{R}^{m^2}$*. Then*

$$\mathsf{P}\Big(\alpha(U_{\{1,2\}} - U_\emptyset) \le \sum_{I \in \mathcal{I}} 4^{\operatorname{card} I} \mathsf{E}\alpha(U_I)\Big) \ge \frac{1}{4} \exp\Big(-\frac{1}{2}(\|x^1\|^2 + \|x^2\|^2)\Big) \,. \tag{15.89}$$

Proof We will deduce the result from (15.82). We consider the quantities

$$S = 4\alpha(G^1 \otimes G^2) = 4\alpha(U_{\{1,2\}}) \; ; \; T = 4\alpha(G^1 \otimes x^2) = 4\alpha(U_{\{1\}} \; .$$

We denote by E^2 conditional expectation given G^2, and we consider the events

$$\Omega_1 = \{\alpha(U_{\{2\}} - U_\emptyset) \le 4\mathsf{E}\alpha(U_{\{2\}})\} \,,$$

$$\Omega_2 = \{\alpha(U_{\{1,2\}} - U_{\{2\}}) \le \mathsf{E}^2 S\} \,,$$

and

$$\Omega_3 = \{\mathsf{E}^2 S \le 4\mathsf{E}S + \mathsf{E}T\} \,.$$

When these three events occur simultaneously, we have

$$\begin{aligned} \alpha(U_{\{1,2\}} - U_\emptyset) &\le \alpha(U_{\{1,2\}} - U_{\{2\}}) + \alpha(U_{\{2\}} - U_\emptyset) \\ &\le \mathsf{E}^2 S + 4\mathsf{E}\alpha(U_{\{2\}}) \\ &\le 4\mathsf{E}S + \mathsf{E}T + 4\mathsf{E}\alpha(U_{\{2\}}) \\ &\le \sum_{I \in \mathcal{I}} 4^{\operatorname{card} I} \mathsf{E}\alpha(U_I) \,, \end{aligned} \tag{15.90}$$

where we have used that $\mathsf{E}T = 4\mathsf{E}\alpha(U_{\{1\}})$ and $\mathsf{E}S = 4\mathsf{E}\alpha(U_{\{1,2\}})$. Next, we prove that

$$\mathsf{P}(\Omega_1 \cap \Omega_3) \ge \frac{1}{2} \exp\Big(-\frac{\|x^2\|^2}{2}\Big) \,. \tag{15.91}$$

For this, we consider on E_2 the semi-norms

$$\alpha_1(y) := \alpha(x^1 \otimes y)\ ;\ \alpha_2(y) := 4\mathsf{E}\alpha(G^1 \otimes y)\ .$$

Thus, $\mathsf{E}^2 S = \alpha_2(G^2)$ and $\mathsf{E}T = \alpha_2(x^2)$. Since $U_{\{2\}} - U_\emptyset = x^1 \otimes (G^2 - x^2)$, we have

$$\Omega_1 = \{\alpha_1(G^2 - x^2) \le 4\mathsf{E}\alpha_1(G^2)\}\ ,$$

$$\Omega_3 = \{\alpha_2(G^2) \le 4\mathsf{E}\alpha_2(G^2) + \alpha_2(x^2)\}\ .$$

Consider the convex symmetric set

$$V = \left\{y \in E_2\ ;\ \alpha_1(y) \le 4\mathsf{E}\alpha_1(G^2)\ ,\ \alpha_2(y) \le 4\mathsf{E}\alpha_2(G^2)\right\}\ .$$

Then Markov's inequality implies that $\mathsf{P}(G^2 \in V) \ge 1/2$, so that (15.81) yields

$$\mathsf{P}(G^2 \in V + x^2) \ge \frac{1}{2}\exp\Big(-\frac{\|x^2\|^2}{2}\Big)\ . \tag{15.92}$$

The definition of V and the inequality $\alpha_2(G^2) \le \alpha_2(G^2 - x^2) + \alpha_2(x^2)$ imply that

$$\{G^2 \in V + x^2\} \subset \Omega_1 \cap \Omega_3\ ,$$

so (15.92) implies (15.91).

Finally, we prove that if P^2 denotes probability given G^2, then

$$\mathsf{P}^2(\Omega_2) \ge \frac{1}{2}\exp\Big(-\frac{\|x^1\|^2}{2}\Big)\ . \tag{15.93}$$

For this, we may think of G^2 as a given deterministic vector of E_2. We then consider on $\mathbb{R}^m$ the semi-norm α' given by $\alpha'(y) = \alpha(y \otimes G^2)$. Since $U_{\{1\},\{2\}} - U_{\{2\}} = (G^1 - x^1) \otimes G^2$, we have

$$\Omega_2 = \left\{\alpha'(G^1 - x^1) \le \mathsf{E}^2 S = 4\mathsf{E}^2\alpha'(G_1)\right\}\ ,$$

so that (15.93) follows from (15.82).

Now the events Ω_1 and Ω_3 depend on G^2 so that (15.93) implies that $\mathsf{P}(\Omega_1 \cap \Omega_3 \cap \Omega_2) \ge \exp(-\|x^1\|^2/2)\mathsf{P}(\Omega_1 \cap \Omega_3)/2$ and (15.91) proves that the probability that Ω_1, Ω_2 and Ω_3 occur simultaneously is at least $2^{-2}\exp(-(\|x^1\|^2 + \|x^2\|^2)/2)$. Combining with (15.90) completes the proof. □

Through the remainder of the section, we write

$$B = \{x = (x^1, x^2) \in E_1 \times E_2 \; ; \; \|x^1\| \le 1, \|x^2\| \le 1\} , \tag{15.94}$$

and we first draw some consequences of Lemma 15.2.11.

Lemma 15.2.12 *Consider a subset T of $2B$ and a semi-norm α on $\mathbb{R}^{m^2}$. Consider the distance d_α on T defined for $x = (x^1, x^2)$ and $y = (y^1, y^2)$ by*

$$d_\alpha(x, y) = \alpha(x^1 \otimes x^2 - y^1 \otimes y^2) . \tag{15.95}$$

Let us define

$$\bar{\alpha}(T) = \sup_{x \in T} \left(\mathsf{E}\alpha(x^1 \otimes G^2) + \mathsf{E}\alpha(G^1 \otimes x^2)\right) . \tag{15.96}$$

Then

$$e_n(T, d_\alpha) \le L(2^{-n/2}\bar{\alpha}(T) + 2^{-n}\mathsf{E}\alpha(G^1 \otimes G^2)) . \tag{15.97}$$

Proof For any $\tau > 0$, using (15.89) for $\bar{x}^\ell := x^\ell/\tau$ rather than x^ℓ, we obtain

$$\mathsf{P}\Big(\alpha(\tau^2 G^1 \otimes G^2 - x^1 \otimes x^2) \le W\Big) \ge \frac{1}{4} \exp\Big(-\frac{1}{2\tau^2}\sum_{\ell \le 2} \|x^\ell\|^2\Big) ,$$

where

$$W = 4\tau(\mathsf{E}\alpha(x^1 \otimes G^2) + \mathsf{E}\alpha(G^1 \otimes x^2)) + 16\tau^2\mathsf{E}\alpha(G^1 \otimes G^2) .$$

When $x = (x^1, x^2) \in T \subset 2B$, one has $\|x^1\|^2 + \|x^2\|^2 \le 8$ and (recalling (15.96)) $W \le 4\tau\bar{\alpha}(T) + 16\tau^2\mathsf{E}\alpha(G^1 \otimes G^2)$, so that

$$\mathsf{P}\Big(\alpha(\tau^2 G^1 \otimes G^2 - x^1 \otimes x^2) \le 4\tau\bar{\alpha}(T) + 16\tau^2\mathsf{E}\alpha(G^1 \otimes G^2)\Big) \ge \frac{1}{4} \exp\Big(-\frac{4}{\tau^2}\Big) . \tag{15.98}$$

Let

$$\epsilon = 4\tau\bar{\alpha}(T) + 16\tau^2\mathsf{E}\alpha(G^1 \otimes G^2) ,$$

and consider a subset U of T such that any two points of U are at mutual distances $\ge 3\epsilon$ for d_α. Then the sets $\{z \in \mathbb{R}^{m^2} ; \alpha(z - x^1 \otimes x^2) \le \epsilon\}$ for $x \in U$ are disjoint, so that (15.98) implies

$$\operatorname{card} U \le 4\exp(4\tau^{-2}) . \tag{15.99}$$

Taking U maximal for the inclusion proves that $N(T, d_\alpha, 3\epsilon) \le 4\exp(4\tau^{-2})$. Choosing τ so that this quantity is 2^{2^n} finishes the proof. □

We are now ready to start the proof of Theorem 15.2.3 for $d = 3$. For a subset T of $E_1 \times E_2$, we define

$$F(T) := \mathsf{E} \sup_{x \in T} A(x^1, x^2, G) . \tag{15.100}$$

Since all our spaces are finite dimensional, this quantity is finite whenever T is bounded. The goal is to bound

$$F(B) = \mathsf{E}\|\langle A, G\rangle\|_{\{1\}\{2\}} . \tag{15.101}$$

We consider the semi-norm α on $\mathbb{R}^{m^2}$ given for $z = (z_{i,j})_{i,j \le m}$ by

$$\alpha(z) = \Big(\sum_k \Big(\sum_{i,j} a_{i,j,k} z_{i,j}\Big)^2\Big)^{1/2} . \tag{15.102}$$

Then the corresponding distance d_α on $E_1 \times E_2$ given by (15.95) is the canonical distance associated with the Gaussian process $X_x = A(x^1, x^2, G)$.[2] In particular, we have

$$\mathsf{E}A(x_1, x_2, G)^2 = d_\alpha(0, x)^2 . \tag{15.103}$$

Lemma 15.2.13 *We have*

$$\mathsf{E}\alpha(G^1 \otimes x^2) \le \|\langle A, x^2\rangle\|_{\{1,3\}} \tag{15.104}$$

$$\mathsf{E}\alpha(x^1 \otimes G^2) \le \|\langle A, x^1\rangle\|_{\{2,3\}} \tag{15.105}$$

$$\mathsf{E}\alpha(G^1 \otimes G^2) \le \|A\|_{\{1,2,3\}} . \tag{15.106}$$

Proof Here, if $A = (a_{i,j,k})$ and $x^2 = (x^2_j)$, then $\langle A, x^2\rangle$ is the matrix $(b_{i,k})$ where $b_{i,k} = \sum_j a_{i,j,k} x^2_j$, and $\|\langle A, x^2\rangle\|_{\{1,3\}} = (\sum_{i,k} b^2_{i,k})^{1/2}$. To prove (15.104), we simply observe that $\alpha(G^1 \otimes x^2) = (\sum_k (\sum_i b_{i,k} g^1_i)^2)^{1/2}$, so that $\mathsf{E}\alpha(G^1 \otimes x^2) \le (\sum_{i,k} b^2_{i,k})^{1/2} = \|\langle A, x^2\rangle\|_{\{1,3\}}$. The rest is similar. □

Lemma 15.2.14 *For $u = (u^1, u^2) \in E_1 \times E_2$ and $T \subset 2B$, one has*

$$F(u + T) \le F(T) + 2\|\langle A, u^1\rangle\|_{\{2,3\}} + 2\|\langle A, u^2\rangle\|_{\{1,3\}} . \tag{15.107}$$

[2] This semi-norm will be used until the end of the proof.

Proof The proof starts with the identity

$$A(x^1+u^1, x^2+u^2, G) = A(x^1, x^2, G) + A(u^1, x^2, G) + A(x^1, u^2, G) + A(u^1, u^2, G) \;. \tag{15.108}$$

We take the supremum over $x \in T$ and then expectation to obtain (using that $\mathsf{E}A(u^1, u^2, G) = 0$)

$$F(T+u) \le F(T) + C_1 + C_2 \;,$$

where

$$C_1 = \mathsf{E} \sup_{\|x^2\| \le 2} A(u^1, x^2, G) \;;\; C_2 = \mathsf{E} \sup_{\|x^1\| \le 2} A(x^1, u^2, G) \;.$$

We then apply (15.80) to the tensor $\langle A, u^1 \rangle$ to obtain $C_1 \le 2\|\langle A, u^1\rangle\|_{\{2,3\}}$ and similarly for C_2. □

This result motivates the introduction on $E_1 \times E_2$ of the semi-norm

$$\alpha^*(x) = \|\langle A, x^1\rangle\|_{\{2,3\}} + \|\langle A, x^2\rangle\|_{\{1,3\}} \;. \tag{15.109}$$

We may then rewrite (15.107) as

$$F(u+T) \le F(T) + 2\alpha^*(u) \;. \tag{15.110}$$

We denote by d_{α^*} the distance on $E_1 \times E_2$ associated with the semi-norm α^*.

The semi-norm α^* has another use: the quantity $\bar{\alpha}(T)$ defined in (15.96) satisfies

$$\bar{\alpha}(T) \le \sup\{\alpha^*(x) \;;\; x \in T\} \;, \tag{15.111}$$

as follows from (15.104) and (15.105).

Lemma 15.2.15 *We have*

$$e_n(2B, d_{\alpha^*}) \le L2^{-n/2}\|A\|_{\{1,2,3\}} \;. \tag{15.112}$$

Proof A standard Gaussian random vector valued in the space $E_1 \times E_2$ is of the type (G^1, G^2) where G^1 and G^2 are independent standard Gaussian random vectors. Proceeding as in (15.80), we get

$$\mathsf{E}\|\langle A, G^1\rangle\|_{\{2,3\}} \le \|A\|_{\{1,2,3\}} \;, \tag{15.113}$$

and similarly $\mathsf{E}\|\langle A, G^2\rangle\|_{\{1,3\}} \le \|A\|_{\{1,2,3\}}$, so that

$$\mathsf{E}\alpha^*(G^1, G^2) \le 2\|A\|_{\{1,2,3\}} .$$

Lemma 15.2.7 then implies the result. □

Exercise 15.2.16 Write the proof of (15.113) in detail.

Given a point $y \in B$ and $a, b > 0$, we define

$$C(y, a, b) = \big\{x \in B - y \; ; \; d_\alpha(0, x) \le a \; , \; d_{\alpha^*}(0, x) \le b\big\} . \tag{15.114}$$

We further define

$$W(a, b) = \sup\{F(C(y, a, b)) \; ; \; y \in B\} . \tag{15.115}$$

Since $C(y, a, b) \subset C(y, a', b')$ for $a \le a', b \le b'$, it follows that $W(a, b)$ is monotone increasing in both a and b. This will be used many times without further mention. The center of the argument is as follows, where we lighten notation by setting

$$S_1 = \|A\|_{\{1,2,3\}} . \tag{15.116}$$

Lemma 15.2.17 *For all values of $a, b > 0$ and $n \ge 0$, we have*

$$W(a, b) \le L2^{n/2}a + Lb + W(L2^{-n/2}b + L2^{-n}S_1, L2^{-n/2}S_1) . \tag{15.117}$$

Proof Consider $y \in B$ so that $B - y \subset 2B$ and $T := C(y, a, b) \subset 2B$. Since $d_{\alpha^*}(0, x) = \alpha^*(x)$, it follows from (15.111) and (15.114) that $\bar{\alpha}(T) \le b$. Combining (15.97) and (15.106), we obtain that $e_n(T, d_\alpha) \le \delta := L(2^{-n/2}b + 2^{-n}S_1)$. Using also (15.112), we find a partition of $T = C(y, a, b)$ into $N_{n+1} = 2^{2^{n+1}}$ sets which are of diameter $\le \delta$ for d_α and of diameter $\le \delta^* := L2^{-n/2}S_1$ for d_{α^*}. Thus, we can find points $y_i \in C(y, a, b)$ for $i \le N_{n+1}$ such that

$$C(y, a, b) = \bigcup_{i \le N_{n+1}} T_i , \tag{15.118}$$

where

$$T_i = \big\{x \in E_1 \times E_2 \; ; \; x \in C(y, a, b) \; , \; d_\alpha(y_i, x) \le \delta \; , \; d_{\alpha^*}(y_i, x) \le \delta^*\big\} .$$

For $x \in B - y$, we have $x - y_i \in B - (y + y_i)$, so that

$$T_i - y_i \subset C(y + y_i, \delta, \delta^*) \tag{15.119}$$

and thus

$$T_i \subset y_i + C(y + y_i, \delta, \delta^*) . \tag{15.120}$$

Also, since $y_i \in B - y$, we have $y + y_i \in B$, so that

$$F(C(y + y_i, \delta, \delta^*)) \leq W(\delta, \delta^*) ,$$

and combining with (15.120) and (15.110), and since $\alpha^*(y_i) = d_{\alpha^*}(y_i, 0) \leq b$ because $y_i \in C(y, a, b)$, we obtain

$$F(T_i) \leq W(\delta, \delta^*) + 2b . \tag{15.121}$$

Furthermore, since $T_i \subset C(y, a, b)$, for $x \in T_i$, we have $d_\alpha(0, x) \leq a$. Recalling (15.100) and (15.103), it follows from (2.124) that

$$F\Big(\bigcup_{i \leq N_{n+1}} T_i\Big) \leq La\sqrt{\log N_{n+1}} + \max_{i \leq N_{n+1}} F(T_i) .$$

Combining with (15.121) and (15.118) implies

$$F(C(y, a, b)) \leq La2^{n/2} + 2b + W(\delta, \delta^*) ,$$

which is the desired conclusion. □

Proposition 15.2.18 *For $n \geq 0$, we have*

$$W(a, b) \leq L(2^{n/2}a + b + 2^{-n/2}S_1) . \tag{15.122}$$

Proof of Theorem 15.2.3 for $d = 3$ We set

$$S_3 = \|A\|_{\{1\}\{2\}\{3\}}$$
$$S_2 = \|A\|_{\{1\}\{2,3\}} + \|A\|_{\{2\}\{1,3\}} + \|A\|_{\{3\}\{1,2\}} .$$

Since $\alpha(x^1 \otimes x^2) = \sup\{A(x^1, x^2, x^3); \|x^3\| \leq 1\}$, we have $d_\alpha(x, 0) \leq S_3$ for $x \in B$. Also by (15.109), we have $d_{\alpha^*}(0, x) = \alpha^*(x) \leq S_2$ for $x \in B$. Therefore, $B \subset C(0, S_3, S_2)$ so that using (15.122),

$$F(B) \leq W(S_3, S_2) \leq L(2^{n/2}S_3 + S_2 + 2^{-n/2}S_1) .$$

Recalling (15.101) and choosing n so that $2^{n/2}$ is about τ proves (15.73). □

Proof of Proposition 15.2.18 Using the monotonicity of W in a and b, one may assume that all the constants in (15.117) are the same. Denoting by L_0 this constant, (15.117) implies in particular that for $n, n_0 \geq 0$, one has

$$W(a,b)$$
$$\leq L_0 2^{(n+n_0)/2} a + L_0 b + W(L_0 2^{-(n+n_0)/2} b + L_0 2^{-n-n_0} S_1, L_0 2^{-(n+n_0)/2} S_1) .$$

Fixing n_0 a universal constant such that $L_0 2^{-n_0/2} \leq 2^{-2}$ implies that for $n \geq 0$, one has

$$W(a,b) \leq L 2^{n/2} a + Lb + W(2^{-n/2-2} b + 2^{-n-2} S_1, 2^{-(n+1)/2} S_1) . \tag{15.123}$$

Using this for $a = 2^{-n} S_1$ and $b = 2^{-n/2} S_1$, we obtain

$$W(2^{-n} S_1, 2^{-n/2} S_1) \leq L 2^{-n/2} S_1 + W(2^{-n-1} S_1, 2^{-(n+1)/2} S_1) .$$

Given $r \geq 0$, summation of these relations for $n \geq r$ implies

$$W(2^{-r} S_1, 2^{-r/2} S_1) \leq L 2^{-r/2} S_1 . \tag{15.124}$$

Using this relation, we then deduce from (15.123) that

$$W(a, 2^{-n/2} S_1) \leq L 2^{n/2} a + L 2^{-n/2} S_1 ,$$

and bounding the last term of (15.123) using this inequality yields (15.122). □

We strongly encourage the reader to carry out the proof in the case $d = 4$, using (15.122) and the induction hypothesis.

15.3 Notes and Comments

Our exposition of Latała's result in Sect. 15.2 brings no new idea whatsoever compared to his original paper [48]. (Improving the mathematics of Rafał Latała seems extremely challenging.) Whatever part of the exposition might be better than in the original paper draws heavily on J. Lehec's paper [54]. This author found [48] very difficult to read and included Sect. 15.2 in an effort to make these beautiful ideas more accessible. It seems most probable that Latała started his work with the case $d = 3$, but one has to do significant reverse engineering to get this less technical case out of his paper.

Chapter 16
Convergence of Orthogonal Series: Majorizing Measures

16.1 A Kind of Prologue: Chaining in a Metric Space and Pisier's Bound

As will become apparent soon, the questions considered in this chapter have a lot to do with the control of stochastic processes which satisfy the condition (1.17) for the function $\varphi(x) = x^2$. Before we get into the sophisticated considerations of the next few sections, it could be helpful to learn a simpler and rather robust method (even though it is not directly connected to most of the subsequent material). The present section is a continuation of Sect. 1.4 which the reader should review now.

Definition 16.1.1 We say that a function $\varphi : \mathbb{R} \to \mathbb{R}$ is a Young function if $\varphi(0) = 0$, $\varphi(-x) = \varphi(x)$, φ is convex, and $\varphi \not\equiv 0$.

On a metric space (T, d), we consider processes $(X_t)_{t\in T}$ that satisfy the condition

$$\forall s, t \in T\ ,\ \mathsf{E}\varphi\Big(\frac{X_s - X_t}{d(s,t)}\Big) \le 1\ . \tag{16.1}$$

Condition (16.1) is quite natural, because given the process $(X_t)_{t\in T}$, it is simple to show that the quantity

$$d(s,t) = \inf\Big\{u > 0\ ;\ \mathsf{E}\varphi\Big(\frac{X_s - X_t}{u}\Big) \le 1\Big\} \tag{16.2}$$

is a quasi-distance on T, for which (16.1) is satisfied.[1]

[1] It would also be natural to consider processes where the size of the "increments" $X_s - X_t$ is controlled by a distance d in a different manner, e.g., for all $u > 0$, $\mathsf{P}(|X_s - X_t| \ge ud(s,t)) \le$

M. Talagrand, *Upper and Lower Bounds for Stochastic Processes*, Ergebnisse der Mathematik und ihrer Grenzgebiete. 3. Folge / A Series of Modern Surveys in Mathematics 60, https://doi.org/10.1007/978-3-030-82595-9_16

We are interested in bounding processes which satisfy the condition (16.1). We have already established the bound (1.18). In this bound occurs the term $N(T,d,\epsilon)^2$ rather than $N(T,d,\epsilon)$. This does not matter if $\varphi(x) = \exp(x^2/4) - 1$ (so that $\varphi^{-1}(N^2) = 2\sqrt{\log(N^2+1)} \le 2\varphi^{-1}(N)$), but it does matter if, say, $\varphi(x) = |x|^p$. We really do not have the right integral on the right-hand side. In this section, we show how to correct this, illustrating again that even in a structure as general as a metric space, not all arguments are trivial. The same topic will be addressed again in Sect. 16.8 at a higher level of sophistication.

To improve the brutal chaining argument leading to (1.18), without loss of generality, we assume that T is finite. For $n \ge n_0$, we consider a map $\theta_n : T_{n+1} \to T_n$ such that $d(\theta_n(t), t) \le 2^{-n}$ for each $t \in T_{n+1}$. Since we assume that T is finite, we have $T = T_m$ when m is large enough. We fix such an m, and we define $\pi_n(t) = t$ for each $t \in T$ and each $n \ge m$. Starting with $n = m$, we then define recursively $\pi_n(t) = \theta_n(\pi_{n+1}(t))$ for $n \ge n_0$. The point of this construction is that $\pi_{n+1}(t)$ determines $\pi_n(t)$ so that there are at most $N(T,d,2^{-n-1})$ pairs $(\pi_{n+1}(t), \theta_n(\pi_{n+1}(t))) = (\pi_{n+1}(t), \pi_n(t))$, and the bound (1.13) implies

$$\mathsf{E} \sup_{t\in T} |X_{\pi_{n+1}(t)} - X_{\pi_n(t)}| \le 2^{-n}\varphi^{-1}(N(T,d,2^{-n-1})) \,. \tag{16.3}$$

Using the chaining identity

$$X_t - X_{\pi_n(t)} = \sum_{k\ge n} X_{\pi_{k+1}(t)} - X_{\pi_k(t)} \,,$$

we have proved the following:

Lemma 16.1.2 *We have*

$$\mathsf{E} \sup_{t\in T} |X_t - X_{\pi_n(t)}| \le \sum_{k\ge n} 2^{-k}\varphi^{-1}(N(T,d,2^{-k-1})) \,. \tag{16.4}$$

Taking $n = n_0$, this yields the following clean result (due to G. Pisier):

Theorem 16.1.3 (G. Pisier) *We have*

$$\mathsf{E} \sup_{s,t\in T} |X_s - X_t| \le L \int_0^{\Delta(T)} \varphi^{-1}(N(T,d,\epsilon))d\epsilon \,. \tag{16.5}$$

In this chapter, we will learn how to go beyond the bound (16.5) when the function $\varphi(x)$ has a much weaker rate of growth than $\exp x^2 - 1$, and first of all, $\varphi(x) = x^2$.

$\psi(u)$, for a given function ψ, see [143]. This question has received considerably less attention than the condition (16.1).

In contrast with (1.19), Theorem 16.1.3 does not provide a uniform modulus of continuity. We investigate this side story in the rest of this section. A clever twist is required in the argument.

Theorem 16.1.4 *For any $\delta > 0$, $n \geq n_0$, we have*

$$\mathsf{E} \sup_{d(s,t)<\delta} |X_s - X_t| \leq \delta\varphi^{-1}(N(T,d,2^{-n})^2) + 4\sum_{k\geq n} 2^{-k}\varphi^{-1}(N(T,d,2^{-k-1})) . \tag{16.6}$$

To ensure that the right-hand side is small, we fix n large enough so that the sum is small, and then we take δ small enough that the first term of the right-hand side is small.

Proof We fix n and we set $Z = \sup_{t\in T} |X_t - X_{\pi_n(t)}|$. We define

$$V = \{(\pi_n(s), \pi_n(t)) \; ; \; d(s,t) < \delta\} \subset T_n \times T_n .$$

Given $s, t \in T$ with $d(s,t) < \delta$, we have $|X_s - X_{\pi_n(s)}| \leq Z$ and $|X_t - X_{\pi_n(t)}| \leq Z$ and so that $|X_s - X_t| \leq |X_{\pi_n(s)} - X_{\pi_n(t)}| + 2Z$ and thus

$$\sup_{d(s,t)<\delta} |X_s - X_t| \leq \sup_{(a,b)\in V} |X_a - X_b| + 2Z . \tag{16.7}$$

For $(a,b) \in V$, we choose $(s(a,b), t(a,b)) \in T \times T$ such that $d(s(a,b), t(a,b)) < \delta$ and $a = \pi_n(s(a,b))$, $b = \pi_n(t(a,b))$. Thus, $|X_a - X_{s(a,b)}| \leq Z$ and $|X_b - X_{t(a,b)}| \leq Z$, so that $|X_a - X_b| \leq |X_{s(a,b)} - X_{t(a,b)}| + 2Z$. Combining with (16.7),

$$\sup_{d(s,t)<\delta} |X_s - X_t| \leq \sup_{(a,b)\in V} |X_{s(a,b)} - X_{t(a,b)}| + 4Z . \tag{16.8}$$

Using (1.13), we have

$$\mathsf{E} \sup_{(a,b)\in V} |X_{s(a,b)} - X_{t(a,b)}| \leq \delta\varphi^{-1}(N(T,d,2^{-n})^2) ,$$

so that taking expectation in (16.8) and using (16.4) completes the proof. □

16.2 Introduction to Orthogonal Series: Paszkiewicz's Theorem

An orthonormal sequence $(\varphi_m)_{m\geq 1}$ on a probability space (Ω, P) is a sequence such that $\mathsf{E}\varphi_m^2 = 1$ for each n and $\mathsf{E}\varphi_m\varphi_n = 0$ for $m \neq n$. A classical question asks which are the sequences (a_m) for which the series

$$\sum_m a_m\varphi_m \tag{16.9}$$

converges a.s. whatever the choice of the orthonormal sequence (φ_m) and of the probability space. (See Sect. 16.10 for comments on this question.) Since the series $\sum_{m\geq 1} a_m \varepsilon_m$ must converge a.s., where ε_m are independent Bernoulli r.v.s, we have $\sum_{m\geq 1} a_m^2 < \infty$ (see Exercise 6.3.4). As we shall see, the condition $\sum_{m\geq 1} a_m^2 < \infty$ is however far from sufficient: there exist an orthonormal sequence (φ_m) and coefficients a_m such that $\sum_{m\geq 1} a_m^2 < \infty$ and the series $\sum_{m\geq 1} a_m \varphi_m$ diverges everywhere.

Let us consider the set

$$T = \Big\{ \sum_{m\leq n} a_m^2 \; ; \; n \geq 1 \Big\} . \tag{16.10}$$

Since $\sum_{m\geq 1} a_m^2 < \infty$, we may assume without loss of generality that $T \subset]0,1]$. We may also assume that $a_m \neq 0$ for each m. Let us denote by $\mathcal{I}_n$ the family of the 2^n dyadic intervals $](i-1)2^{-n}, i2^{-n}]$ for $1 \leq i \leq 2^n$. For a point $t \in]0,1]$, we denote by $I_n(t)$ the unique interval of $\mathcal{I}_n$ that contains t.

Theorem 16.2.1 (A. Paszkiewicz [81]) *Given the sequence (a_m), and hence the set T, the following are equivalent:*

(a) The series (16.9) converges a.s. for every choice of the orthonormal sequence (φ_n).

(b) There exists a probability measure μ on T such that

$$\sup_{t\in T} \sum_{n\geq 0} \frac{1}{\sqrt{2^n \mu(I_n(t))}} < \infty . \tag{16.11}$$

(c) There exists a number B such that for every probability measure μ on T, one has

$$\sum_{n\geq 0} \sum_{I\in\mathcal{I}_n} \sqrt{2^{-n}\mu(I)} \leq B . \tag{16.12}$$

(d) There exists a number B' such that for each process $(X_t)_{t\in T}$ which satisfies

$$\forall s,t \in T \; , \; \mathsf{E}(X_s - X_t)^2 \leq |s-t| \; , \tag{16.13}$$

we have

$$\mathsf{E} \sup_{s,t\in T} |X_s - X_t| \leq B' . \tag{16.14}$$

(e) For each process $(X_t)_{t\in T}$ which satisfies (16.13), $\lim_{k\to\infty} X_{t_k}$ exists a.s. where $t_k = \sum_{m\leq k} a_m^2$.

At this stage, this theorem should look completely mysterious. It will take the work of several sections to clarify the underlying issues. Let us start by some simple observations. The conditions (b) to (e) do not involve orthonormal series, but only the set T. This set T is just the set of points of the increasing sequence $t_n = \sum_{m \le n} a_m^2$. Such a set has the notable property that its closure is the set $T \cap \{t^*\}$, where $t^* = \lim_{n \to \infty} t_n = \sum_{m \ge 1} a_m^2$. The condition 16.13 is the special case of Condition (16.1) where $\varphi(x) = x^2$ and $d(s,t) = \sqrt{|s-t|}$. Thus, Pisier's bound (16.5) ensures that (d) holds when the integral $\int_0^{\Delta(T,d)} \sqrt{N(T,d,\epsilon)} d\epsilon$ is finite or, equivalently, when $\sum_{n \ge 0} \sqrt{2^{-n} N(T,d,2^{-n/2})} < \infty$. However, the necessary and sufficient conditions (b) and (c) are somewhat weaker than this condition (although this is not obvious yet). In Sect. 16.8, we will provide considerable generalizations of the equivalence of (b) and (c) on the one hand and (d) on the other hand, but we will first provide specific proofs of this fact in the context of Paszkiewicz's theorem.

Exercise 16.2.2 Prove that the condition $\sum_{n \ge 0} \sqrt{2^{-n} N(T, d^2, 2^{-n})} < \infty$ is equivalent to the condition

$$\sum_{n \ge 0} \sqrt{2^{-n} \operatorname{card}\{I \in \mathcal{I}_n \; ; \; I \cap T \ne \emptyset\}} < \infty \,. \tag{16.15}$$

Prove that under this condition, (c) and (d) of Theorem 16.2.1 hold.

16.3 Recovering the Classical Results

In this section, we recover two classical results from Theorem 16.2.1. This will help us get a feeling for the conditions of this theorem. On a less positive note, it will also illustrate that working with these conditions is not as easy as what one would like it to be.

Corollary 16.3.1 (Rademacher-Menchov [66, 90]) *If*

$$\sum_{m \ge 1} a_m^2 (\log m)^2 < \infty \,, \tag{16.16}$$

then for each choice of the orthonormal sequence (φ_m), *the series* $\sum_m a_m \varphi_m$ *converges a.s.*

Proof We shall prove that (c) is satisfied. We consider a probability measure μ on T, and we aim to bound the left-hand side of (16.12). The plan is for each n to split the sum $\sum_{I \in \mathcal{I}_n} \sqrt{2^{-n} \mu(I)}$ into several suitable pieces and to bound each of them using the Cauchy-Schwarz inequality in the following form: For $\mathcal{J} \subset \mathcal{I}_n$, then

$$\sum_{I\in\mathcal{J}} \sqrt{2^{-n}\mu(I)} = 2^{-n/2} \sum_{I\in\mathcal{J}} \sqrt{\mu(I)} \le 2^{-n/2}\sqrt{\operatorname{card}\mathcal{J}}\sqrt{\mu(\cup_{I\in\mathcal{J}} I)}\,, \tag{16.17}$$

where we have used that for a disjoint family $\mathcal{I}$ of intervals, $\sum_{I\in\mathcal{I}} \mu(I) = \mu(\cup_{I\in\mathcal{I}} I)$. But, first, we must reformulate (16.16). For $n \ge 1$, let $t_n = \sum_{1\le m\le n} a_m^2$. For $k \ge 0$, let $u_k = t_{2^{2^k}}$, so that

$$\sum_{k\ge0} 2^{2k}(u_{k+1} - u_k) = \sum_{k\ge0} \sum_{2^{2^k} < m \le 2^{2^{k+1}}} 2^{2k} a_m^2 \le L \sum_{m\ge2} a_m^2 (\log m)^2 < \infty\,, \tag{16.18}$$

using that $2^k \le L\log m$ for $m \ge 2^{2^k}$. In particular, we have $u_{k+1} - u_k \le C2^{-2k}$ so that if $t^* = \sum_{m\ge1} a_m^2$, then $t^* - u_k = \sum_{r\ge k}(u_{r+1} - u_r) \le C2^{-2k}$.

We now fix k and consider $2^k \le n < 2^{k+1}$ and turn to the task of splitting the sum $\sum_{I\in\mathcal{I}_n} \sqrt{2^{-n}\mu(I)}$ into suitable pieces. Consider $I \in \mathcal{I}_n$ with $\mu(I) > 0$, so that $I \cap T \ne \emptyset$. We claim that at least one of the following four cases must occur:

- I contains a point u_p for $k-1 \le p \le 2k$ or contains t^*
- $I \subset]0, u_{k-1}]$
- $I \subset]u_\ell, u_{\ell+1}]$ for some $k-1 \le \ell \le 2k$
- $I \subset]u_{2k}, t^*]$.

To see this, we simply observe that if the interval I does not contain either the point t^* or one of the points u_p for $k-1 \le p \le 2k$, then it must be contained in one the subintervals of $[0, 1]$ created when removing these points from $[0, 1]$. However, since $T \subset [0, t^*]$, it cannot be contained in the interval $]t^*, 1]$, so that it is contained in one of the other intervals left, which is exactly what the last three bullets state.

Consequently, for $2^k \le n < 2^{k+1}$, we may write

$$\sum_{I\in\mathcal{I}_n} \sqrt{2^{-n}\mu(I)} = \mathrm{I} + \mathrm{II} + \mathrm{III} + \sum_{k-1\le\ell\le2k} V(\ell)\,, \tag{16.19}$$

where

- I is the sum over $I \subset]0, u_{k-1}]$. This sum has at most $2^{2^{k-1}}$ non-zero terms, because when $T \cap I \ne \emptyset$, I must contain at least one point t_m with $m \le 2^{2^{k-1}}$. Then (16.17) implies that the sum is $\le 2^{-n/2}2^{2^{k-2}} \le 2^{-2^{k-2}}$.
- II is the sum over the intervals I that contain a point u_p for $k-1 \le p \le 2k$ or that contain the point t^*. This sum has at most $k+2$ terms and is bounded as above.

- III is the sum over the family of intervals I contained in $]u_{2k}, t^*]$. Here we use that, if $u \le v$, and $\mathcal{J}$ denotes the family of intervals $I \in \mathcal{I}_n$, $I \subset]u, v]$, then

$$\sum_{I \in \mathcal{J}} \sqrt{2^{-n}\mu(I)} \le \sqrt{v-u}\sqrt{\mu(]u,v])} . \tag{16.20}$$

  This follows from (16.17) because card $\mathcal{J} \le 2^n(v-u)$ since the intervals I of $\mathcal{I}_n$ have length 2^{-n}. Thus, III $\le \sqrt{t^* - u_{2k}} \le C2^{-2k}$.
- $V(\ell)$ is the sum over the intervals $I \subset]u_\ell, u_{\ell+1}]$, which, as witnessed by (16.20), is bounded by $\sqrt{u_{\ell+1} - u_\ell}\sqrt{\mu(]u_{\ell+1}, u_\ell])}$.

Summation of the inequalities (16.19) over n with $2^k \le n < 2^{k+1}$ and then over k yields that for a certain number C'

$$\sum_{n \ge 0}\sum_{I \in \mathcal{I}_n} \sqrt{2^{-n}\mu(I)} \le C' + \sum_{k \ge 1} 2^k \sum_{k-1 \le \ell \le 2k} V(\ell) .$$

We want to prove that this quantity is finite. First,

$$\sum_{k \ge 1} 2^k \sum_{k-1 \le \ell \le 2k} V(\ell) \le \sum_{\ell \ge 0} V(\ell) \sum_{k-1 \le \ell} 2^k \le 4 \sum_{\ell \ge 0} 2^\ell V(\ell) ,$$

and then (recalling that $V(\ell) \le \sqrt{u_{\ell+1} - u_\ell}\sqrt{\mu(]u_{\ell+1}, u_\ell])}$)

$$\sum_{\ell \ge 0} 2^\ell V(\ell) \le \sum_{\ell \ge 0} 2^\ell \sqrt{u_{\ell+1} - u_\ell}\sqrt{\mu(]u_{\ell+1}, u_\ell])} < \infty$$

using the Cauchy-Schwarz inequality and (16.18). □

Corollary 16.3.2 (Tandori [135]) *If for each choice of the orthonormal sequence* (φ_m) *the series* $\sum_m a_m \varphi_m$ *converges a.s., then*

$$\sum_{m \ge 1} a_m^2 (\log |a_m|)^2 < \infty . \tag{16.21}$$

Before we start the preparation for the proof, let us introduce some notation that will be used throughout the book. We will write in a standard way expressions such as $\sum_{i \le n} a_i$ or $\sum_{i \in I} a_i$. However, when we want in the same line to describe both the summation and the set over which the summation occurs, we will write $\sum\{a_i; i \in I\}$, with the expression $i \in I$ replaced if necessary by the description of the set, as is done in (16.22). We will use the convention not only for sums but also for inf, sup, max, and min (where it is more standard). Using this convention for $k \ge 0$, we write

$$b_k = \sum \left\{a_m^2 \; ; \; 2^{-2^{k+1}} < a_m^2 \le 2^{-2^k}\right\} , \tag{16.22}$$

which as explained means $b_k = \sum_{m\in U_k} a_m^2$ where

$$U_k = \{m\ ;\ 2^{-2^{k+1}} < a_m^2 \le 2^{-2^k}\}\,. \tag{16.23}$$

Lemma 16.3.3 *For $n \ge 1$, let $t_n = \sum_{m\le n} a_m^2$. Fix k with $b_k > 0$. Consider the probability measure μ_k on T given by $\mu_k(\{t_n\}) = a_n^2/b_k$ if $n \in U_k$ and $\mu_k(\{t_n\}) = 0$ if $n \notin U_k$. Consider n with $2^{k-1} \le n < 2^k$. Then*

$$\sum_{I\in\mathcal{I}_n} \sqrt{2^{-n}\mu_k(I)} \ge \sqrt{\frac{b_k}{2}}\,. \tag{16.24}$$

Proof To prove (16.24), it suffices to prove that if $2^{k-1} \le n < 2^k$, then for each $I \in \mathcal{I}_n$

$$\mu_k(I) \le \frac{2^{-n+1}}{b_k}\,, \tag{16.25}$$

because then $\sqrt{2^{-n}\mu_k(I)} \ge \mu_k(I)\sqrt{b_k/2}$, from which (16.24) follows by summation over $I \in \mathcal{I}_n$. By definition of μ_k, it suffices the prove that $\sum\{a_m^2;\, m \in U_k,\, t_m \in I\} \le 2^{-n+1}$. Denoting $I =]a,b]$, consider the interval $I' =]a - 2^{-2^k}, b]$, so that the length of I' is $2^{-n} + 2^{-2^k} \le 2^{-n+1}$. When $t_m \in I$, the interval $]t_{m-1}, t_m]$ is entirely contained in I', and its length is exactly a_m^2. As these intervals are disjoint as m varies, this shows that $\sum\{a_m^2;\, m \in U_k,\, t_m \in I\}$ is at most the length of I'. □

Proof of Corollary 16.3.2 Since $\log|a_m| \le L2^k$ for $m \in U_k$ and $b_k = \sum_{m\in U_k} a_m^2$, it suffices to prove that $\sum_{k\ge 0} 2^{2k} b_k = \sum_{k\in J} 2^{2k} b_k < \infty$.

By Theorem 16.2.1, we know that (16.12) holds. We will apply this condition to a suitable probability measure μ, which we construct now. Consider numbers $(\alpha_k)_{k\in J}$ with $\alpha_k \ge 0$ and $\sum_k \alpha_k = 1$. Consider the probability measure $\mu = \sum_{k\in J} \alpha_k \mu_k$, where μ_k was described in Lemma 16.3.3. Consider n with $2^{k-1} \le n < 2^k$. Using (16.24) in the second inequality, we obtain

$$\sum_{I\in\mathcal{I}_n} \sqrt{2^{-n}\mu(I)} \ge \sqrt{\alpha_k} \sum_{I\in\mathcal{I}_n} \sqrt{2^{-n}\mu_k(I)} \ge \sqrt{\frac{\alpha_k b_k}{2}}\,. \tag{16.26}$$

We then sum (16.26) over $2^{k-1} \le n < 2^k$ and then over k to obtain, using also (16.12) in the first inequality,

$$B \ge \sum_{n\ge 0} \sum_{I\in\mathcal{I}_n} \sqrt{2^{-n}\mu(I)} \ge \frac{1}{2\sqrt{2}} \sum_{k\in J} \sqrt{\alpha_k} 2^k \sqrt{b_k}\,,$$

and since the sequence α_k is arbitrary with $\sum \alpha_k = 1$, optimization over this sequence yields $\sum_{k \in J} 2^{2k} b_k \le 8B^2$. □

The necessary condition (16.21) is by no means sufficient for the convergence of each series $\sum_{m \ge 1} a_m \varphi_m$. This is obvious from Theorem 16.7.1.

Exercise 16.3.4 Prove that the conditions (16.21) and (16.16) are equivalent when the sequence (a_m) is non-increasing. Consequently, (16.21) is a necessary and sufficient condition so that one can find a permutation π such that the series $\sum_m a_{\pi(m)} \varphi(m)$ converges a.s. for each orthonormal sequence (φ_m).

Exercise 16.3.5 For a finite subset T of $]0,1]$, consider the following quantity $M(T)$. If $\operatorname{card} T = 1$, we set $M(T) = 0$. Otherwise, let $n(T)$ be the largest integer such that there exists $I \in \mathcal{I}_{n(T)}$ for which $T \subset I$. Call this interval I_T. Define then

$$M(T) = \inf \sup_{t \in T} \sum_{n \ge n(T)} \frac{1}{\sqrt{2^n \mu(I_n(t))}} ,$$

where the infimum is computed over all choices of probability measures on T. Now, I_T is the union of two intervals I_1 and I_2 of $\mathcal{I}_{n(T)+1}$. Explain how to compute $M(T)$ when you know $M(T \cap I_j)$ for $j = 1, 2$. In this manner, the quantity $M(T)$ can be "computed recursively".

16.4 Approach to Paszkiewicz's Theorem: Bednorz's Theorem

We now describe our approach to Theorem 16.2.1. The following is an obvious consequence of orthonormality:

Lemma 16.4.1 *For $t = \sum_{m \le n} a_m^2 \in T$, let us define*

$$X_t = \sum_{m \le n} a_m \varphi_m . \tag{16.27}$$

Then

$$\forall s, t \in T \ , \ \mathsf{E}(X_s - X_t)^2 = |s - t| . \tag{16.28}$$

This makes it obvious that (e) implies (a) in Theorem 16.2.1. It also motivates the following:

Definition 16.4.2 If T is a subset of $[0,1]$, we say that the process $(X_t)_{t \in T}$ is *orthonormal* if it satisfies (16.28) and if moreover $\mathsf{E} X_t = 0$ for each t.

The main ingredient in the proof of Theorem 16.2.1 is the following result:

Theorem 16.4.3 (W. Bednorz [13]) *Consider a finite subset T of $[0, 1]$, and define*

$$F^*(T) = \sup \mathsf{E} \sup_{t \in T} X_t \;, \tag{16.29}$$

where the first supremum is taken over all orthonormal processes indexed by T. Then for each probability measure μ on T, we have

$$\sum_{n \ge 0} \sum_{I \in \mathcal{I}_n} \sqrt{2^{-n} \mu(I)} < L(1 + F^*(T)) \;. \tag{16.30}$$

Our first task is to make the link between Theorems 16.2.1 and 16.4.3.

Lemma 16.4.4 *If the process $(X_t)_{t \in T}$ is orthonormal, then*

$$t_1 \le t_2 \le t_3 \le t_4 \in T \Rightarrow \mathsf{E}(X_{t_4} - X_{t_3})(X_{t_2} - X_{t_1}) = 0 \;. \tag{16.31}$$

Proof Consider $t_1 \le t_2 \le t_3 \in T$. Then

$$\begin{aligned} t_3 - t_1 &= \mathsf{E}(X_{t_3} - X_{t_1})^2 \\ &= \mathsf{E}(X_{t_3} - X_{t_2})^2 + \mathsf{E}(X_{t_2} - X_{t_1})^2 + 2\mathsf{E}(X_{t_3} - X_{t_2})(X_{t_2} - X_{t_1}) \\ &= t_3 - t_2 + t_2 - t_1 + 2\mathsf{E}(X_{t_3} - X_{t_2})(X_{t_2} - X_{t_1}) \;, \end{aligned}$$

so that we have proved

$$t_1 \le t_2 \le t_3 \in T \Rightarrow \mathsf{E}(X_{t_3} - X_{t_2})(X_{t_2} - X_{t_1}) = 0 \;. \tag{16.32}$$

We use (16.32) to write

$$\begin{aligned} 0 &= \mathsf{E}(X_{t_4} - X_{t_3})(X_{t_3} - X_{t_1}) \\ &= \mathsf{E}(X_{t_4} - X_{t_3})(X_{t_2} - X_{t_1}) + \mathsf{E}(X_{t_4} - X_{t_3})(X_{t_3} - X_{t_2}) \\ &= \mathsf{E}(X_{t_4} - X_{t_3})(X_{t_2} - X_{t_1}) \;, \end{aligned}$$

using again (16.32) in the third inequality. □

We will also need a classical result of Tandori [136]. This lemma really brings out the strength of the statement "for every orthonormal sequence…".

Lemma 16.4.5 *Assume that the sequence (a_n) has the property that for every orthonormal sequence (φ_n), the series $\sum_{m \ge 1} a_m \varphi_m$ converges a.s. Then there exists a number A such that for each orthonormal sequence (φ_n), we have*

$$\mathsf{E} \sup_{n \ge 1} \Big(\sum_{1 \le m \le n} a_m \varphi_m \Big)^2 \le A \;. \tag{16.33}$$

Proof For $1 \le p \le q$, let us define

$$V(p,q) = \mathsf{E} \sup_{p \le n \le q, (\varphi_n)} \Big(\sum_{p \le m \le n} a_m \varphi_m \Big)^2 , \tag{16.34}$$

where the supremum is also over all orthonormal sequences (φ_n). Let us assume for contradiction that (16.33) fails, i.e., that $\lim_{q \to \infty} V(1,q) = \infty$. The inequality $(a+b)^2 \le 2a^2 + 2b^2$ shows that $2(\sum_{p \le m \le n} a_m \varphi_m)^2 \ge (\sum_{1 \le m \le n} a_m \varphi_m)^2 - 2(\sum_{1 \le m < p} a_m \varphi_m)^2$. Consequently, for each p, we have $\lim_{q \to \infty} V(p,q)^2 = \infty$, and therefore, we can find an increasing sequence (p_k) such that $V(p_k, p_{k+1}) > 1$ for each k. By definition of $V(p,q)$, we can then find an orthonormal sequence $(\varphi_{m,k})_{m \ge 1}$ for which

$$W(k) := \max_{p_k \le n \le p_{k+1}} \Big| \sum_{p_k \le m \le n} a_m \varphi_{m,k} \Big|$$

satisfies $\mathsf{E} W(k)^2 \ge 1$. Let us define the function

$$\theta_k = \frac{W(k)^2}{\mathsf{E} W(k)^2} ,$$

so that $\mathsf{E}\theta_k = 1$. For $p_k < m \le p_{k+1}$, we have $|a_m \varphi_{m,k}| \le 2W(k)$ since $a_m \varphi_{m,k} = \sum_{p_k \le s \le m} a_s \varphi_{s,k} - \sum_{p(k) \le s \le m-1} a_s \varphi_{s,k}$. Consequently, $\varphi_{m,k} = 0$ when $\theta_k = 0$. For $p_k < m \le p_{k+1}$, we may then define $\varphi'_m = \varphi_{m,k}/\sqrt{\theta_k}$. Since

$$\int \varphi'_m \varphi'_{m'} \theta_k \mathrm{d}\mathsf{P} = \int \varphi_{m,k} \varphi_{m',k} \mathrm{d}\mathsf{P} ,$$

the functions $(\varphi'_n)_{p_k \le n < p_{k+1}}$ still form an orthonormal sequence for the probability measure $\mathsf{P}' = \theta_k \mathsf{P}$ and satisfy

$$\max_{p_k \le n \le p_{k+1}} \Big| \sum_{p_k \le m \le n} a_m \varphi'_m \Big| = \frac{W(k)}{\sqrt{\theta_k}} = (\mathsf{E} W(k)^2)^{1/2} \ge 1 . \tag{16.35}$$

We can moreover assume that $\mathsf{E}\varphi'_m = 0$. To see this, we replace the sequence $(\varphi'_m)_{p_k \le m < p_{k+1}}$ by the sequence $(\varepsilon \varphi'_m)_{p_k \le m < p_{k+1}}$ where ε is a Bernoulli r.v. independent of all the r.v.s φ'_m.

We construct a sequence $(\psi_m)_{m \ge 1}$ using independent blocks: for each $k \ge 1$, the sequence $(\psi_m)_{p_k \le m < p_{k+1}}$ is a copy of the sequence $(\varphi'_m)_{p_k \le m < p_{k+1}}$, and these sequences are globally independent as k varies. The sequence $(\psi_m)_{m \ge p(1)}$ is orthonormal. Indeed, if $p_k \le m, m' \le p_{k+1}$ for some k, then $\mathsf{E}\psi_m \psi_{m'} = \mathsf{E}\varphi'_m \varphi'_{m'} = 0$, while if this is not the case, ψ_m and $\psi_{m'}$ are independent and of expectation 0. We complete in any way we like in an orthonormal sequence $(\psi_m)_{m \ge 1}$ (e.g., we set $\psi_m = \varepsilon_m$ for $1 \le m < p(1)$ where (ε_m) is an independent

Bernoulli sequence). According to (16.35), the series $\sum_{m\geq 1} a_m \psi_m$ diverges a.s. This contradiction shows that (16.33) must hold and concludes the proof. □

Corollary 16.4.6 *Under the hypothesis of Lemma 16.4.5 for each orthonormal process* $(X_t)_{t\in T}$, *one has*

$$\mathsf{E} \sup_{s,t\in T} |X_s - X_t| \leq 2\sqrt{A} . \tag{16.36}$$

Proof We recall that $T = \{t_1, t_2, \ldots\}$ where $t_m = \sum_{1\leq k\leq m} a_k^2$. It follows from Lemma 16.4.4 that the sequence $(\varphi_m)_{m\geq 2}$ given by

$$\varphi_m = a_m^{-1}(X_{t_m} - X_{t_{m-1}})$$

is an orthonormal sequence. If $p \leq q$, then $X_{t_q} - X_{t_p} = \sum_{p<m\leq q} a_m \varphi_m$, so that

$$\sup_{s,t\in T_k} |X_s - X_t| \leq \sup_{p,q} \Big| \sum_{p<m\leq q} a_m \varphi_m \Big| \leq 2 \sup_n \Big| \sum_{2\leq m\leq n} a_m \varphi_m \Big|$$

and (16.33) implies (16.36). □

Assuming Theorem 16.4.3, we are now ready to prove the "hard part" of Theorem 16.2.1.

Theorem 16.4.7 *If the series (16.9) converges a.s. for each choice of the orthonormal sequence* (φ_n) *or if (d) of Theorem 16.2.1 holds, then there exists a number* B *such that for each probability measure* μ *on* T, *one has*

$$\sum_{n\geq 0} \sum_{I\in \mathcal{I}_n} \sqrt{2^{-n}\mu(I)} \leq B . \tag{16.37}$$

Proof Assuming first that the series (16.9) converges a.s. for each choice of the orthonormal sequence (φ_n), it follows from (16.36) that the quantity $F^*(T)$ of (16.29) satisfies $F^*(T) \leq 2\sqrt{A}$.

Consequently, Theorem 16.4.3 implies that for each probability measure μ on a finite subset of T, one has

$$\sum_{n\geq 0} \sum_{I\in \mathcal{I}_n} \sqrt{2^{-n}\mu(I)} \leq B := L(1+\sqrt{A}) .$$

It then should be obvious that this implies the same inequality for each probability measure μ on T.

Assuming now that (d) holds, one has $F^*(T) \leq B'$ and the proof is the same. □

We shall also use the following, which is a variation on Lemma 3.3.3:

Lemma 16.4.8 *Consider a finite set $T \subset [0, 1]$, and assume that for each probability measure μ on T, one has*

$$\sum_{n\geq 0}\sum_{I\in\mathcal{I}_n}\sqrt{2^{-n}\mu(I)} \leq B\,. \tag{16.38}$$

Then there is a probability measure μ on T for which[2]

$$\sup_{t\in T}\sum_{n\geq 0}\frac{1}{\sqrt{2^n\mu(I_n(t))}} \leq 2B\,. \tag{16.39}$$

Proof The key argument is the Hahn-Banach theorem, in the form of Lemma 3.3.2. Let us denote by $\mathcal{M}(T)$ the set of probability measures on T, and for $\mu \in \mathcal{M}(T)$, let us consider the function

$$f_\mu(t) := \sum_{n\geq 0}\frac{1}{\sqrt{2^n\mu(I_n(t))}}\,.$$

Since $I_n(t) = I$ for $t \in I \in \mathcal{I}_n$, we have $\int_I (\mu(I_n(t)))^{-1/2}\mathrm{d}\mu(t) = \sqrt{\mu(I)}$ so that, using (16.38) in the inequality,

$$\int f_\mu(t)\mathrm{d}\mu(t) = \sum_{n\geq 0}\sum_{I\in\mathcal{I}_n}\sqrt{2^{-n}\mu(I)} \leq B\,. \tag{16.40}$$

Since the function $x \mapsto 1/\sqrt{x}$ is convex, the map $\mu \mapsto f_\mu$ is convex. Consequently, the class $\mathcal{C}$ of functions f on T that satisfy

$$\exists\mu \in \mathcal{M}(T)\;;\;\forall t \in T\;,\; f_\mu(t) \leq f(t)$$

is convex. For each probability measure μ on T, (16.40) shows that there exists f in $\mathcal{C}$ (namely, $f = f_\mu$!) with $\int f\mathrm{d}\mu \leq B$. Consequently, by Lemma 3.3.2 (used for $\epsilon = B$), there exists $f \in \mathcal{C}$ such that $f \leq 2B$. □

Proposition 16.4.9 *If condition (c) of Theorem 16.2.1 holds, so does condition (b).*

Proof Let $t_n = \sum_{1\leq m\leq n} a_m^2$, $t^* = \lim_{n\to\infty} t_n = \sum_{m\geq 1} a_m^2$ so that $T^* = T \cup \{t^*\}$ is compact. Consider $T_k = \{t_n, n \leq k\}$. Combining (c) with Lemma 16.4.8, we obtain a probability measure μ_k on T_k for which

$$\sup_{t\in T_k}\sum_{n\geq 0}\frac{1}{\sqrt{2^n\mu_k(I_n(t))}} \leq 2B\,. \tag{16.41}$$

[2] If one works a tad harder, one may get B rather than $2B$ in the next inequality.

From here, the proof is basically a compactness argument. Taking a subsequence if necessary, we may assume that the sequence (μ_k) converges weakly as $k \to \infty$ to a probability measure μ' on T^*. Then for each compact subset K of T, we have $\mu'(K) \ge \liminf_k \mu_k(K)$. This applies in particular to the sets $I \cap T^*$ for $I \in \mathcal{I}_n$, which are compact. It then follows from (16.41) that

$$\sup_{t\in T} \sum_{n\ge 0} \frac{1}{\sqrt{2^n \mu'(I_n(t))}} \le LB \; . \tag{16.42}$$

The problem is that it might happen that $\mu'(\{t^*\}) > 0$, and then μ' is not supported by T. We modify μ' to take care of this problem. We consider a probability measure μ of the form $\mu = \mu'/2 + \mu_1$ where μ_1 is a positive measure on T of mass $1/2$ which for each n gives mass $\ge 2^{-n/2}/L$ to the interval $I \in \mathcal{I}_n$ which contains t^*. This is possible because this interval is of the type $]u, v]$ so that it meets T.

Then, for $I \in \mathcal{I}_n$, we have $\mu(I) \ge \mu'(I)/2$ if $t^* \notin I$, while if $t^* \in I$, then $\mu(I) \ge 2^{-n/2}/L$. It is then immediate to check that μ satisfies (16.11). □

16.5 Chaining, I

We need one more ingredient to complete the proof of Theorem 16.2.1 (still assuming Theorem 16.4.3). We need to control the supremum of a stochastic process under condition (16.13). Theorem 16.1.3 does not suffice for this purpose. In this section, we develop a more efficient chaining scheme. We consider a general finite metric space (T, d), and we try to bound a process $(X_t)_{t\in T}$ which satisfies

$$\forall s, t \in T \; , \; \mathsf{E}(X_s - X_t)^2 \le d(s,t)^2 \; . \tag{16.43}$$

When T is a subset of the unit interval and when $d(s,t) = \sqrt{|s-t|}$, this covers the case of the processes satisfying (16.13).

Consider a sequence $(T_n)_{n\ge 0}$ of subsets of T. We assume that $\operatorname{card} T_0 = 1$, and we denote by t_0 the unique element of T_0. We assume that for each $n \ge 1$, we are given a map $\theta_n : T_n \to T_{n-1}$. As in the proof of Theorem 16.1.3, this map will help us to build a proper chaining. Since we assume that T is finite, it is not much of a restriction to assume that $T_m = T$ for a certain (large) integer m. We define $\pi_m(t) = t$ for each t and recursively $\pi_{n-1}(t) = \theta_n(\pi_n(t))$. First, as usual, we write

$$|X_t - X_{t_0}| \le \sum_{1\le n\le m} |X_{\pi_n(t)} - X_{\pi_{n-1}(t)}| \; . \tag{16.44}$$

Using the inequality $xy \le x^2 + y^2$, it is rather natural to write that, for $s \in T_n$, and introducing a parameter $c_n(s)$,

$$|X_s - X_{\theta_n(s)}| \le \frac{d(s,\theta_n(s))}{c_n(s)} + d(s,\theta_n(s))c_n(s)\left(\frac{X_s - X_{\theta_n(s)}}{d(s,\theta_n(s))}\right)^2 .$$

Let us assume for simplicity that for numbers $\epsilon_n > 0$, we have

$$\forall s \in T_n \ , \ d(s,\theta_n(s)) \le \epsilon_n \ . \tag{16.45}$$

Then

$$|X_s - X_{\theta_n(s)}| \le \frac{\epsilon_n}{c_n(s)} + \epsilon_n c_n(s)\left(\frac{X_s - X_{\theta_n(s)}}{d(s,\theta_n(s))}\right)^2 .$$

Using this for $s = \pi_n(t)$, and recalling that $\pi_{n-1}(t) = \theta_n(\pi_n(t))$, we obtain (using a crude bound on the last term to make it independent of t)

$$|X_{\pi_{n-1}(t)} - X_{\pi_n(t)}| \le \frac{\epsilon_n}{c_n(\pi_n(t))} + \sum_{s\in T_n} \epsilon_n c_n(s)\left(\frac{X_s - X_{\theta_n(s)}}{d(s,\theta_n(s))}\right)^2 .$$

We then deduce from (16.44)

$$|X_t - X_{t_0}| \le \sum_{1\le n\le m} \frac{\epsilon_n}{c_n(\pi_n(t))} + \sum_{1\le n\le m}\sum_{s\in T_n} \epsilon_n c_n(s)\left(\frac{X_s - X_{\theta_n(s)}}{d(s,\theta_n(s))}\right)^2 . \tag{16.46}$$

Let us now set

$$S = \sup_{t\in T} \sum_{1\le n\le m} \frac{\epsilon_n}{c_n(\pi_n(t))} \ , \tag{16.47}$$

$$S^* = \sum_{1\le n\le m}\sum_{s\in T_n} \epsilon_n c_n(s) \ . \tag{16.48}$$

Then (16.46) yields

$$\sup_{t\in T} |X_t - X_{t_0}| \le S + \sum_{1\le n\le m}\sum_{s\in T_n} \epsilon_n c_n(s)\left(\frac{X_s - X_{\theta_n(s)}}{d(s,\theta_n(s))}\right)^2 . \tag{16.49}$$

Taking expectation and using (16.43), we obtain the following important relation:

Lemma 16.5.1 *Recalling (16.47) and (16.48), we have*

$$\mathsf{E}\sup_{t\in T}|X_t - X_{t_0}| \le S + S^* . \tag{16.50}$$

Corollary 16.5.2 *We have*

$$\mathsf{E}\sup_{t\in T}|X_t - X_{t_0}| \le L\sum_{n\ge 1}\epsilon_n\sqrt{\operatorname{card} T_n} . \tag{16.51}$$

Proof Choose $c_n(t) = 1/\sqrt{\operatorname{card} T_n}$ for $t \in T_n$. □

Exercise 16.5.3 Prove that (16.51) implies Pisier's bound (16.5) in the case $\varphi(x) = x^2$.

We recall the notation $I_n(t)$ of Theorem 16.2.1.

Corollary 16.5.4 *Consider a countable subset T of $[0,1]$. Assume that for a certain integer $n_0 \ge 0$ and a certain $I_0 \in \mathcal{I}_{n_0}$, we have $T \subset I_0$. Consider a probability measure μ on T such that*

$$A := \sup_{t\in T}\sum_{n\ge n_0}\frac{1}{\sqrt{2^n\mu(I_n(t))}} < \infty .$$

Then for each process $(X_t)_{t\in T}$ that satisfies (16.13), we have

$$\mathsf{E}\sup_{s,t\in T}|X_s - X_t| \le LA. \tag{16.52}$$

Proof Since the process satisfies (16.13), it satisfies (16.43) for $d(s,t) = \sqrt{|s-t|}$. The plan is to use (16.50), and we construct the relevant chaining. We construct inductively for $n \ge n_0$ a set $T_n \subset T$ such that $\operatorname{card} T_n \cap I = 1$ whenever $I \in \mathcal{I}_n$ and $I \cap T \ne \emptyset$ and such that moreover $T_{n-1} \subset T_n$. When s is the unique point of $T_n \cap I$, let us then set

$$c_n(s) = \sqrt{\mu(I)} .$$

Let us moreover define the map $\theta_n : T_n \to T_{n-1}$ in the canonical manner. That is, if s is the unique point of $T_n \cap I$ where $I \in \mathcal{I}_n$, there are a unique $I' \in \mathcal{I}_{n-1}$ with $I \subset I'$ and a unique point s' in $T_{n-1} \cap I'$. We then set $\theta_n(s) = s'$. We have $|s - \theta_n(s)| = |s - s'| \le 2^{-(n-1)}$, so that $d(s,\theta_n(s)) = \sqrt{|s-\theta_n(s)|} \le \epsilon_n := 2^{-(n-1)/2}$, i.e., (16.45) holds for this value of ϵ_n.

Considering an arbitrary integer m, we now use the bound (16.49) for T_m rather than T. Then, for $t \in T_m$, we have

$$\sum_{1\le n\le m} \frac{\epsilon_n}{c_n(\pi_n(t))} = \sum_{1\le n\le m} \frac{2^{-(n-1)/2}}{\sqrt{\mu(I_n(t))}} \le 2A ,$$

so that $S \le 2A$. Also, integrating the inequality

$$\forall t \in T \,, \sum_{n\ge n_0} \frac{1}{\sqrt{2^n \mu(I_n(t))}} \le A$$

with respect to μ, we obtain (using as always that $I(t) = I$ for $t \in I$)

$$\sum_{n\ge n_0} \sum_{I\in\mathcal{I}_n} 2^{-n/2}\sqrt{\mu(I)} \le A .$$

This means that $S^* \le LA$. Consequently, the bound (16.50) implies

$$\mathsf{E} \sup_{s,t\in T_m} |X_s - X_t| \le LA ,$$

and since m is arbitrary, this proves (16.52). □

Proof of Theorem 16.2.1 We proved that (c) implies (b) in Proposition 16.4.9. We proved that (b) implies (d) in Corollary 16.5.4. We proved that (d) implies (c) in Theorem 16.4.7. Thus, (b), (c), and (d) are equivalent.

We proved that (e) implies (a) in Lemma 16.4.1. We proved that (a) implies (c) in Theorem 16.4.7.

We now prove that (b) implies (e), completing the proof of Theorem 16.2.1. Let us consider the point $t^* = \sum_{m\ge 1} a_m^2 = \lim_{k\to\infty} t_k$, the supremum of T. Let us consider an integer n_0 and the unique $I_0 \in \mathcal{I}_{n_0}$ with $t^* \in I_0$. Consider the set $T' = T \cap I_0$, so that $t_k \in T'$ for k large enough. Then

$$\sup_{t\in T'} \sum_{n\ge 0} \frac{1}{\sqrt{2^n \mu(I_n(t))}} \le A^* := \sup_{t\in T} \sum_{n\ge 0} \frac{1}{\sqrt{2^n \mu(I_n(t))}} .$$

Consequently, the probability measure μ' on T' given for $B \subset T'$ by $\mu'(B) = \mu(B \cap T')/\mu(T') = \mu(B \cap I_0)/\mu(I_0)$ satisfies

$$\sup_{t\in T'} \sum_{n\ge n_0} \frac{1}{\sqrt{2^n \mu'(I_n(t))}} \le A^*\sqrt{\mu(I_0)} .$$

The bound (16.52) used for T' and μ' then implies that each process $(X_t)_{t\in T}$ which satisfies (16.13) also satisfies

$$\mathsf{E} \sup_{s,t\in T'} |X_s - X_t| \le LA^* \sqrt{\mu(I_0)} \; .$$

Now for n_0 large enough $\mu(I_0)$ is arbitrarily small since $\bigcap_{\epsilon>0}(T \cap]t^* - \epsilon, t^*]) = \emptyset$. Consequently,

$$\lim_{n\to\infty} \mathsf{E} \sup_{k,\ell\ge n} |X_{t_k} - X_{t_\ell}| = 0 \; .$$

This concludes the proof. □

16.6 Proof of Bednorz's Theorem

The main step in the proof of Bednorz's theorem is, given a finite subset T of $]0, 1]$, to relate the "size" of T with the size of the four sets $T_j = T \cap I_j$ where for $1 \le j \le 4$, I_j is the interval $](j-1)/4, j/4]$. It is performed in Proposition 16.6.3. The reason why we use 4-adic partitions is that we are certain that "T_1 is far apart from T_3"(etc.), whereas one cannot say the same about, say, T_1 and T_2 since T_1 might be located to the very right of I_1 and T_2 might be located to the very left of I_2. (This is why dyadic partitions would not work.)

Definition 16.6.1 Consider an interval $J =]c, d] \subset [0, 1]$ and $\overline{J} = [c, d]$. We say that the process $(X_t)_{t\in\overline{J}}$ is *normalized* if $\mathsf{E}X_t = 0$, $X_c = X_d = 0$ and

$$\forall s, t \in \overline{J} \; , \; s < t \; , \; \mathsf{E}(X_s - X_t)^2 = t - s - (d-c)^{-1}(t-s)^2 \; . \tag{16.53}$$

The reason behind the formula in the right-hand side of (16.53) will be explained soon. We fix the finite set $T \subset]0, 1]$ once and for all. For an interval $J =]c, d] \subset [0, 1]$, we consider the quantity

$$F(J) = \sup \mathsf{E} \sup_{t\in T\cap J} X_t \; , \tag{16.54}$$

where the first supremum is taken over all normalized processes indexed by $\overline{J} = [c, d]$. Although $X_c = 0$ is defined, in (16.54), the supremum is only over $T \cap J$, not over $T \cap \overline{J}$. We define $F(J) = 0$ when $T \cap J = \emptyset$.

The quantity $F(J)$ will be our "measure of size of J". We first relate it to the quantity $F^*(T)$ of (16.29).

Lemma 16.6.2 *We have* $F(]0, 1]) \le F^*(T)$.

Proof Consider a normalized process $(X_t)_{t\in[0,1]}$. Consider a centered r.v. Z, independent of this process, and such that $\mathsf{E}Z^2 = (d-c)^{-1}$. Then the process $Y_t = X_t + tZ$ is orthonormal. Using the definition of F^* in the first inequality and using Jensen's inequality (taking the expectation in Z inside the supremum) in the second inequality yields

$$F^*(T) \ge \mathsf{E}\sup_{t\in T} Y_t \ge \mathsf{E}\sup_{t\in T} X_t ,$$

and since the normalized process (X_t) is arbitrary, this proves that $F(]0,1]) \le F^*(T)$. □

We state now the main step in the proof of Bednorz's theorem.

Proposition 16.6.3 *Consider* $I =]0,1]$, *and for* $j = 1,2,3,4$, *consider* $I_j =](j-1)/4, j/4]$ *and numbers* $\alpha_j \ge 0$ *such that* $\sum_{j\le 4}\alpha_j = 1$. *Then*

$$F(I) \ge \sum_{1\le j\le 4} \sqrt{\alpha_j} F(I_j) . \tag{16.55}$$

Moreover, if for each $1 \le j \le 4$ *we have* $\alpha_j \ge 1/400$ *and* $T \cap I_j \ne \emptyset$, *then*

$$F(I) \ge \sum_{1\le j\le 4} \sqrt{\alpha_j} F(I_j) + \frac{1}{80} . \tag{16.56}$$

The inequality (16.56) is a kind of growth condition. The constants 80 and 400 are just convenient choices and do not carry special meaning.[3] There is no magic: to prove this result, given normalized processes on the intervals I_j, we have to construct a normalized process witnessing (16.55), and this will require hard work. This work will be performed at the end of the section, and we first turn to the comparatively easier task of deducing Bednorz's theorem from this result. We first state a kind of rescaling of Proposition 16.6.3, where, instead of starting with $I =]0,1]$, we start with any dyadic interval.

Corollary 16.6.4 *Consider* $I \in \mathcal{I}_m$ *and the four intervals* I_j *of* $\mathcal{I}_{m+2}$ *for* $j = 1,2,3,4$ *which it contains. Consider numbers* $\alpha_j \ge 0$ *such that* $\sum_{j\le 4}\alpha_j = 1$. *Then*

$$F(I) \ge \sum_{1\le j\le 4} \sqrt{\alpha_j} F(I_j) . \tag{16.57}$$

[3] The condition $\alpha_j \ge 1/400$ simply ensures that α_j stays away from 0.

Moreover, if for each $1 \le j \le 4$ *we have* $\alpha_j \ge 1/400$ *and* $T \cap I_j \neq \emptyset$, *then*

$$F(I) \ge \sum_{1 \le j \le 4} \sqrt{\alpha_j} F(I_j) + 2^{-m/2} \frac{1}{80} \,. \tag{16.58}$$

Proof Denoting by $F_T(J)$ the quantity (16.54) to indicate the dependence in T, it suffices to prove that for $a > 0$ and $b \in \mathbb{R}$, we have, with obvious notation, $F_{aT+b}(aJ + b) = \sqrt{a} F_T(J)$. This follows from the fact that if the process $(X_t)_{t \in aJ+b}$ is normalized on the interval $aJ+b$, then the process $Y_t = a^{1/2} X_{(t-b)/a}$ is normalized on the interval J. □

We recall that $\mathcal{I}_n$ denotes the family of 2^n dyadic intervals of length 2^{-n}. Theorem 16.4.3 is a consequence of Lemma 16.6.2 and the following:

Proposition 16.6.5 *Consider a finite set* $T \subset [0, 1]$. *Then, given a probability measure* μ *on* T, *we have*

$$\sum_{n \ge 0} \sum_{I \in \mathcal{I}_n} 2^{-n/2} \sqrt{\mu(I)} \le L(1 + F(]0, 1])) \,. \tag{16.59}$$

As a preparation for the proof of this result, we fix a probability measure μ on T, and for $n \ge 0$, we define $\mathcal{I}^*_{2n}$ as the collection of intervals $I \in \mathcal{I}_{2n}$ that have the following property:

$$I' \in \mathcal{I}_{2n+2} \,,\; I' \subset I \Rightarrow \mu(I') \ge \mu(I)/400 \,. \tag{16.60}$$

We then define

$$M_n = 2^{-n} \sum_{I \in \mathcal{I}^*_{2n}} \sqrt{\mu(I)} \,. \tag{16.61}$$

Lemma 16.6.6 *We have*

$$\sum_{n \ge 0} M_n \le 80 F(]0, 1]) \,. \tag{16.62}$$

Proof We recall that $F(J) = 0$ when $J \cap T = \emptyset$. We prove that for each $n \ge 0$, we have

$$\sum_{I \in \mathcal{I}_{2n}} \sqrt{\mu(I)} F(I) \ge \frac{1}{80} M_n + \sum_{I \in \mathcal{I}_{2n+2}} \sqrt{\mu(I)} F(I) \,. \tag{16.63}$$

Given $I \in \mathcal{I}_{2n}$, let us denote by I_1, I_2, I_3, and I_4 the intervals of $\mathcal{I}_{2n+2}$ which are contained in I. Then, using (16.57) for $\alpha_j = \mu(I_j)/\mu(I)$, we obtain

$$\sqrt{\mu(I)}F(I) \geq \sum_{j\leq 4}\sqrt{\mu(I_j)}F(I_j) = \sum_{I'\subset I, I'\in\mathcal{I}_{2n+2}} \sqrt{\mu(I')}F(I') \,. \tag{16.64}$$

If moreover $I \in \mathcal{I}^*_{2n}$, we can now use (16.58) with the same choice of α_i and $m = 2n$ (so that $2^{-n} = 2^{-m/2}$) to obtain the better inequality

$$\sqrt{\mu(I)}F(I) \geq \sum_{I'\subset I, I'\in\mathcal{I}_{2n+2}} \sqrt{\mu(I')}F(I') + \frac{1}{80}2^{-n}\sqrt{\mu(I)} \,. \tag{16.65}$$

Summation of the inequalities (16.64) and (16.65) over $I \in \mathcal{I}_{2n}$ completes the proof of (16.63). Rewriting (16.63) as $M_n \leq 80(S_n - S_{n+1})$ where $S_n = \sum_{I\in\mathcal{I}_{2n}} \sqrt{\mu(I)}F(I)$ and summing over n completes the proof. □

Lemma 16.6.7 *Consider numbers* $\alpha_j \geq 0$ *for* $j = 1, 2, 3, 4$ *such that* $\sum_{1\leq j\leq 4}\alpha_j = 1$. *Then*

$$\min_{1\leq j\leq 4} \alpha_j \leq \frac{1}{400} \Rightarrow \frac{1}{2}\sum_{1\leq j\leq 4}\sqrt{\alpha_j} \leq \frac{9}{10} \,. \tag{16.66}$$

Proof Assume, for example, that $\alpha_1 \leq 1/400$. Concavity of the function $x \mapsto \sqrt{x}$ shows that $\sqrt{\alpha_2} + \sqrt{\alpha_3} + \sqrt{\alpha_4} \leq \sqrt{3(\alpha_2 + \alpha_3 + \alpha_4)} \leq \sqrt{3}$, so that

$$\frac{1}{2}\sum_{1\leq j\leq 4}\sqrt{\alpha_j} \leq \frac{1}{2}\Big(\frac{1}{20} + \sqrt{3}\Big) \leq \frac{9}{10} \,.$$ □

The following will complete the proof of Proposition 16.6.5:

Lemma 16.6.8 *We have*

$$\sum_{n\geq 0}\sum_{I\in\mathcal{I}_n} 2^{-n/2}\sqrt{\mu(I)} \leq 2S(\mu) := 2\sum_{n\geq 0}\sum_{I\in\mathcal{I}_{2n}} 2^{-n}\sqrt{\mu(I)} \,, \tag{16.67}$$

and

$$S(\mu) \leq 10 + 10\sum_{n\geq 0} M_n \,. \tag{16.68}$$

Proof An interval $I \in \mathcal{I}_n$ is the union of two intervals I' and I'' of $\mathcal{I}_{n+1}$ and $\mu(I) = \mu(I') + \mu(I'')$. The inequality $\sqrt{a} + \sqrt{b} \leq \sqrt{2}\sqrt{a+b}$ implies that $2^{-(n+1)/2}(\sqrt{\mu(I')} + \sqrt{\mu(I'')}) \leq 2^{-n/2}\sqrt{\mu(I)}$. Consequently, the sequence $c_n := \sum_{I\in\mathcal{I}_n} 2^{-n/2}\sqrt{\mu(I)}$ is non-increasing. Thus, $\sum_{n\geq 0} c_{2n} \leq \sum_{n\geq 0} c_n \leq 2\sum_{n\geq 0} c_{2n}$ which is (16.67).

For $I \in \mathcal{I}_{2n}$, let

$$w(I) = 2^{-n-1} \sum_{J \subset I, J \in \mathcal{I}_{2n+2}} \sqrt{\mu(J)} \,.$$

The equality

$$S(\mu) = 1 + \sum_{n \geq 0} \sum_{I \in \mathcal{I}_{2n}} w(I) \tag{16.69}$$

holds because all the terms in the summation that defines $S(\mu)$ occur in one of the terms $w(I)$, except the term for $n = 0$, which is equal to 1 since μ is a probability. Given $n \geq 0$,

$$\sum_{I \in \mathcal{I}_{2n}} w(I) = \sum_{I \in \mathcal{I}_{2n}^*} w(I) + \sum_{I \notin \mathcal{I}_{2n}^*} w(I) \,. \tag{16.70}$$

Consider $I \notin \mathcal{I}_{2n}^*$, and denote by I_j, $j = 1, 2, 3, 4$, the four intervals of $\mathcal{I}_{2n+2}$ contained in I. Define $\alpha_j = \mu(I_j)/\mu(I)$. By definition of $\mathcal{I}_{2n}$, the smallest of these four numbers is $< 1/400$ so that Lemma 16.6.7 implies that

$$w(I) < \frac{9}{10} 2^{-n} \sqrt{\mu(I)} \,. \tag{16.71}$$

Summation of the relations (16.71) yields

$$\sum_{n \geq 0} \sum_{I \notin \mathcal{I}_{2n}^*} w(I) \leq \frac{9}{10} \sum_{n \geq 0} \sum_{I \notin \mathcal{I}_{2n}^*} \leq 2^{-n} \sqrt{\mu(I)} \leq \frac{9}{10} S(\mu) \,.$$

Combining with (16.69) and (16.70) and recalling (16.61), we thus obtain

$$S(\mu) \leq 1 + \frac{9}{10} S(\mu) + \sum_{n \geq 0} \sum_{I \in \mathcal{I}_{2n}^*} 2^{-n} \sqrt{\mu(I)} \leq 1 + \frac{9}{10} S(\mu) + \sum_{n \geq 0} M_n \,.$$

This completes the proof. □

It remains only to prove Proposition 16.6.3. This proof occupies the rest of the present section. To prove the proposition, starting with normalized processes[4] $(Y_t^j)_{t \in \overline{I}_j}$ for $1 \leq j \leq 4$, we shall construct a suitable normalized process on $I = [0, 1]$. Let us set

$$\varphi(x) = x - 4x^2 \,, \tag{16.72}$$

[4] We remind the reader that in particular $\mathsf{E} Y_t^j = 0$.

so that saying that the process Y_t^j is normalized means exactly that for $s, t \in I_j$, $s < t$, we have

$$\mathsf{E}(Y_s^j - Y_t^j)^2 = \varphi(t - s) .$$

We start with some preparations. For $t \in I$ and $1 \le j \le 4$, let us define

$$t^j = \max(\min(t, j/4), (j-1)/4) \in \overline{I}_j = [(j-1)/4, j/4] .$$

In words, $t_j = (j-1)/$ if $t \le j/4$, $t_j = t$ if $(j-1)/4 \le t \le j/4$ and $t^j = j/4$ if $t \ge j/4$.

For $0 \le s \le t \le 1$, the interval $]s, t]$ is the disjoint union of the intervals $]s^j, t^j]$ for $1 \le j \le 4$. In particular, we have

$$\mathbf{1}_{]s,t]} = \sum_{j \le 4} \mathbf{1}_{]s^j, t^j]} . \tag{16.73}$$

Consider the probability space $[0, 1]$ provided with Lebesgue's measure. (Thus, E refers simply to integration with respect to this measure.) The archetypical example of a normalized process on $[0, 1]$ is given by the formula

$$W_t = \mathbf{1}_{[0,t]} - t . \tag{16.74}$$

Our first goal is to play with this process to discover the useful algebraic identity (16.76). Consider the algebra $\mathcal{S}$ of subsets of $[0, 1]$ generated by the intervals I_j for $1 \le j \le 4$, and denote by $\mathsf{E}_{\mathcal{S}}$ conditional expectation with respect to this algebra. We define

$$V_t = \mathsf{E}_{\mathcal{S}} W_t = \mathsf{E}_{\mathcal{S}}(\mathbf{1}_{[0,t]} - t) . \tag{16.75}$$

Lemma 16.6.9 *We have the identity*

$$t - s - (t-s)^2 = \sum_{1 \le j \le 4} \varphi(s^j - t^j) + \mathsf{E}(V_s - V_t)^2 . \tag{16.76}$$

Proof We define $V_t' = W_t - \mathsf{E}_{\mathcal{S}} W_t$, so that $W_t = V_t' + V_t$ and V_t is $\mathcal{S}$-measurable, while $\mathsf{E}_{\mathcal{S}}(V_t') = 0$. Given two function f, f' with $\mathsf{E}_{\mathcal{S}} f = 0$ and f' $\mathcal{S}$-measurable, then $\mathsf{E} f f' = \mathsf{E}\mathsf{E}_{\mathcal{S}} f f' = \mathsf{E} f' \mathsf{E}_{\mathcal{S}} f = 0$, so that $\mathsf{E}(f + f')^2 = \mathsf{E} f^2 + \mathsf{E} f'^2$. Consequently, for $s \le t$,

$$t - s - (t-s)^2 = \mathsf{E}(W_s - W_t)^2 = \mathsf{E}(V_s' - V_t')^2 + \mathsf{E}(V_s - V_t)^2 . \tag{16.77}$$

Keeping in mind that $\lambda(I_j) = 1/4$, we obtain from (16.73) that

$$\mathsf{E}_{\mathcal{S}}\mathbf{1}_{[s,t]} = 4\sum_{j\le 4}(t^j - s^j)\mathbf{1}_{I_j}\ . \tag{16.78}$$

Since $\mathsf{E}\mathbf{1}_{I_j} = 1/4$, we obtain

$$\mathsf{E}(\mathsf{E}_{\mathcal{S}}\mathbf{1}_{[s,t]})^2 = 4\sum_{j\le 4}(t^j - s^j)^2\ . \tag{16.79}$$

For $s \le t$, we have

$$\mathsf{E}(V'_s - V'_t)^2 = \mathsf{E}(\mathbf{1}_{[s,t]} - \mathsf{E}_{\mathcal{S}}\mathbf{1}_{[s,t]})^2 = \mathsf{E}\mathbf{1}_{[s,t]} - \mathsf{E}(\mathsf{E}_{\mathcal{S}}\mathbf{1}_{[s,t]})^2\ .$$

Using (16.79) and since $\mathsf{E}(\mathbf{1}_{[s,t]} = t - s = \sum_{j\le 4} t^j - s^j$, we obtain

$$\mathsf{E}(V'_s - V'_t)^2 = \sum_{1\le j\le 4}\varphi(t^j - s^j)\ . \tag{16.80}$$

We then conclude from (16.77). □

We go back to our main construction. It involves an auxiliary process $(Z_t)_{t\in T}$ and a r.v. $\tau \in \{1,2,3,4\}$. Throughout the proof, we assume the following:

$$\text{The processes } Y^j_t \text{ are independent of each other and of the r.v.s } Z_t \text{ and } \tau\ , \tag{16.81}$$

$$\forall j \le 4\ ;\ \mathsf{P}(\tau = j) = \alpha_j\ , \tag{16.82}$$

$$\mathsf{E}Z_t = 0\ ;\ \mathsf{E}(Z_s - Z_t)^2 = \mathsf{E}(V_s - V_t)^2\ . \tag{16.83}$$

We do *not* assume that Z_t and τ are independent. When $\alpha_j = 0$, for $t \in I$, let us define $U^j_t = Y^j_{t^j}$. Otherwise, we define

$$U^j_t = \frac{1}{\sqrt{\alpha_j}}\mathbf{1}_{\{\tau=j\}}Y^j_{t^j}\ , \tag{16.84}$$

and we set

$$S_t = \sum_{1\le j\le 4} U^j_t\ . \tag{16.85}$$

We then transform the process (S_t) into a normalized process by adding (Z_t) as follows:

Lemma 16.6.10 *The process*

$$X_t = S_t + Z_t \tag{16.86}$$

is normalized.

Proof Assume $s < t$. Using the independence of τ and Y_t^j in the second equality, and that the process Y^j is normalized in the third one, we obtain

$$\mathsf{E}(U_s^j - U_t^j)^2 = \frac{1}{\alpha_j}\mathsf{E}\mathbf{1}_{\{\tau=j\}}(Y_{t^j}^j - Y_{s^j}^j)^2 = \mathsf{E}(Y_{t^j}^j - Y_{s^j}^j)^2 = \varphi(t^j - s^j)\,. \tag{16.87}$$

This formula remains true even when $\alpha_j = 0$, since then $U_t^j = Y_{t^j}^j$. It follows from (16.81) and the fact that Y^j and $Y^{j'}$ are independent for $j \neq j'$ that then $\mathsf{E}U_s^j U_t^{j'} = 0$, so that

$$\mathsf{E}(S_s - S_t)^2 = \sum_{1\le j\le 4} \mathsf{E}(U_s^j - U_t^j)^2 = \sum_{1\le j\le 4} \varphi(t^j - s^j)\,. \tag{16.88}$$

It follows from (16.81) that $\mathsf{E}S_s Z_t = 0$, so that

$$\mathsf{E}(X_s - X_t)^2 = \mathsf{E}(S_s - S_t)^2 + \mathsf{E}(Z_s - Z_t)^2\,,$$

and the result follows from (16.83), (16.88), and (16.76). □

Lemma 16.6.11 *Assume that $T \cap I_j \neq \emptyset$ for each $j \le 4$. Then*

$$\mathsf{E}\sup_{t\in T} X_t \ge \sum_{1\le j\le 4}\Big(\sqrt{\alpha_j}\,\mathsf{E}\sup_{t\in T\cap I_j} Y_t^j + \inf_{t\in T\cap I_j} \mathsf{E}\mathbf{1}_{\{\tau=j\}} Z_t\Big)\,. \tag{16.89}$$

Proof First, we observe that, using that $T \cap I_j \neq \emptyset$ in the last inequality,

$$\begin{aligned}\mathsf{E}\sup_{t\in T} X_t &= \sum_{1\le j\le 4} \mathsf{E}\mathbf{1}_{\{\tau=j\}} \sup_{t\in T} X_t\\ &= \sum_{1\le j\le 4} \mathsf{E}\sup_{t\in T} \mathbf{1}_{\{\tau=j\}} X_t\\ &\ge \sum_{1\le j\le 4} \mathsf{E}\sup_{t\in T\cap I_j} \mathbf{1}_{\{\tau=j\}} X_t\,.\end{aligned} \tag{16.90}$$

If $\alpha_j = 0$, we use that trivially $\mathsf{E}\sup_{t\in T\cap I_j} \mathbf{1}_{\{\tau=j\}} X_t \ge \inf_{t\in T\cap I_j} \mathsf{E}\mathbf{1}_{\{\tau=j\}} Z_t$ because $\mathsf{E}\mathbf{1}_{\{\tau=j\}} S_t = 0$. Let us fix $j \le 4$ with $\alpha_j \neq 0$ and denote by E^j conditional expectation given the r.v.s Y_t^j. Then Jensen's inequality implies

$$\mathsf{E} \sup_{t\in T\cap I_j} \mathbf{1}_{\{\tau=j\}} X_t \geq \mathsf{E} \sup_{t\in T\cap I_j} \mathsf{E}^j \mathbf{1}_{\{\tau=j\}} X_t \ . \tag{16.91}$$

Let us fix $t \in I_j$, so that then $t^j = t$. Since $\mathbf{1}_{\{\tau=j\}}\mathbf{1}_{\{\tau=j'\}} = 0$ for $j' \neq j$, we have by definition of X_t

$$\mathbf{1}_{\{\tau=j\}} X_t = \frac{1}{\sqrt{\alpha_j}} \mathbf{1}_{\{\tau=j\}} Y_t^j + \mathbf{1}_{\{\tau=j\}} Z_t \ .$$

Using the independence of τ and Y_t^j in the first equality, and the independence of Y_t^j from τ and Z in the second equality, we get

$$\mathsf{E}^j \mathbf{1}_{\{\tau=j\}} X_t = \sqrt{\alpha_j} Y_t^j + \mathsf{E}^j \mathbf{1}_{\{\tau=j\}} Z_t = \sqrt{\alpha_j} Y_t^j + \mathsf{E}\mathbf{1}_{\{\tau=j\}} Z_t \ . \tag{16.92}$$

To conclude, we simply use that $\sup_t (y_t + z_t) \geq \sup_t y_t + \inf_t z_t$, and thus

$$\mathsf{E} \sup_{t\in T\cap I_j} \mathsf{E}^j \mathbf{1}_{\{\tau=j\}} X_t \geq \sqrt{\alpha_j} \mathsf{E} \sup_{t\in T\cap I_j} Y_t^j + \inf_{t\in T\cap I_j} \mathsf{E}\mathbf{1}_{\{\tau=j\}} Z_t \ . \qquad \square$$

Lemma 16.6.12 *Even when $T \cap I_j = \emptyset$ for some $j \leq 4$, if the process (Z_t) is independent of τ, we have*

$$\mathsf{E} \sup_{t\in T} X_t \geq \sum_{j\in J} \sqrt{\alpha_j} \mathsf{E} \sup_{t\in T\cap I_j} Y_t^j \ , \tag{16.93}$$

where $J = \{j \leq 4; I_j \neq \emptyset\}$.

Proof As in (16.90), we obtain

$$\mathsf{E} \sup_{t\in T} X_t = \sum_{1\leq j\leq 4} \mathsf{E} \sup_{t\in T} \mathbf{1}_{\{\tau=j\}} X_t \geq \sum_{j\in J} \mathsf{E} \sup_{t\in T\cap I_j} \mathbf{1}_{\{\tau=j\}} X_t \ ,$$

where we use that $\mathsf{E}\sup_{t\in T} \mathbf{1}_{\{\tau=j\}} X_t \geq \mathsf{E}\sup_{t\in T\cap I_j} \mathbf{1}_{\{\tau=j\}} X_t$ if $T \cap I_j \neq \emptyset$ and $\mathsf{E}\sup_{t\in T} \mathbf{1}_{\{\tau=j\}} X_t \geq 0$ because $\mathsf{E}\mathbf{1}_{\{\tau=j\}} X_t = 0$ for each t. Since τ and Z_t are independent, and since $\mathsf{E} Z_t = 0$, then (16.92) implies $\mathsf{E}^j \mathbf{1}_{\{\tau=j\}} X_t = \sqrt{\alpha_j} Y_t^j$ and the conclusion from (16.91). $\square$

We need one more ingredient, which is a consequence of the definition $V_t = \mathsf{E}_{\mathcal{S}}(\mathbf{1}_{[0,t]} - t)$.

Lemma 16.6.13 *We have*

$$\inf_{t\in I_1} \mathsf{E}(V_t \mathbf{1}_{I_4}) \geq -\frac{1}{16} \;;\; \inf_{t\in I_2} \mathsf{E}(V_t \mathbf{1}_{I_1}) \geq \frac{1}{8}$$

$$\inf_{t\in I_3} \mathsf{E}(V_t \mathbf{1}_{I_2}) \geq \frac{1}{16} \;;\; \inf_{t\in I_4} \mathsf{E}(V_t \mathbf{1}_{I_3}) \geq 0\,. \tag{16.94}$$

Proof Indeed, for $t \in I_1$ and $x \in I_4$, we have $V_t(x) = -t \geq -1/4$; for $x \in I_1$ and $t \in I_2$, we have $V_t(x) = 1 - t \geq 1/2$; etc. □

Proof of Proposition 16.6.3 To prove (16.55), we simply choose Z_t independent of τ (e.g., a copy of the process (V_t) which is independent of all the other processes considered), and we use (16.93) since by definition $F(I) = 0$ when $I \cap T = \emptyset$. It remains only to prove (16.56). We shall use (16.89) with an appropriate choice of the process (Z_t) (which will no longer be independent of τ). This appropriate choice will make the quantity $\sum_{1\leq j\leq 4} \inf_{t\in T\cap I_j} \mathsf{E}\mathbf{1}_{\{\tau=j\}} Z_t$ large.

For a subset A of $[0, 1]$, we denote by $A/100$ the set $\{x/100; x \in A\}$. Thus, $I/100 =]0, 1/100]$ is the union of the four intervals $I_j/100$, each of length $1/400$. For each $j \leq 4$, we have $P(\tau = j) = \alpha_j \geq 1/400$. Without loss of generality, we may then assume that the underlying probability space is $[0, 1]$ provided with Lebesgue's measure and that for $j \leq 4$,

$$(I/100) \cap \{\tau = j\} = I_{n(j)}/100\,, \tag{16.95}$$

where $n(1) = 4, n(2) = 1, n(3) = 2$, and $n(4) = 3$. This will greatly simplify notation. Let us then define $Z_t(x) \equiv 0$ for $x > 1/100$, and for $x \leq 1/100$, let us define $Z_t(x) = 10V_t(100x)$, where V_t is defined in (16.75). Using change of variable, we see that (16.83) holds.

The fundamental relation is, recalling that $I_j =](j-1)/4, j/4]$,

$$\mathsf{E}\mathbf{1}_{\{\tau=j\}} Z_t = \frac{1}{10} \mathsf{E}\mathbf{1}_{I_{n(j)}} V_t\,.$$

The proof is straightforward by a change of variable:

$$\begin{aligned}\mathsf{E}\mathbf{1}_{\{\tau=j\}} Z_t &= 10 \int_{\{\tau=j\}\cap]0,1/100]} V_t(100x)\mathrm{d}x = 10 \int_{I_{n(j)}/100} V_t(100x)\mathrm{d}x \\ &= \frac{1}{10} \int_{I_{n(j)}} V_t(x)\mathrm{d}x = \frac{1}{10}\mathsf{E}\mathbf{1}_{I_{n(j)}} V_t\,.\end{aligned} \tag{16.96}$$

It then follows from Lemma 16.6.13 that

$$\sum_{1\leq j\leq 4} \inf_{t\in I_j} \mathsf{E}\mathbf{1}_{\{\tau=j\}} Z_t = \frac{1}{10} \sum_{1\leq j\leq 4} \inf_{t\in I_j} \mathsf{E}\mathbf{1}_{I_{n(j)}} V_t \geq \frac{1}{10}\Big(-\frac{1}{16} + \frac{1}{8} + \frac{1}{16}\Big) = \frac{1}{80}\,,$$

and combining with (16.89), this completes the proof. □

16.7 Permutations

One may also ask the following question: What are the sequences (a_m) such that for any permutation π and any orthonormal sequence (φ_m), the series $\sum_m a_{\pi(m)}\varphi_m$ converges a.s.? The answer to this question was also discovered by A. Paszkiewicz and is announced in [81]. Given the sequence (a_m) and the permutation π of $\mathbb{N}$, we define the set

$$T_\pi = \Big\{ \sum_{1\le m\le n} a^2_{\pi(m)} \; ; \; n \ge 1 \Big\} . \tag{16.97}$$

We also consider the numbers

$$b_k := \sum \Big\{ a_m^2 \; ; \; 2^{-2^{k+1}} < a_m^2 \le 2^{-2^k} \Big\} . \tag{16.98}$$

Without loss of generality, we assume that $\sum_m a_m^2 \le 1/2$, so that $\sum_k b_k \le 1/2$.

Theorem 16.7.1 *For a sequence (a_m), the following are equivalent:*

(f) For every permutation π and every orthonormal sequence (φ_m), the series $\sum_m a_{\pi(m)}\varphi_m$ converges a.s.
(g) We have

$$\sum_{k\ge 1} 2^k \sqrt{b_k} < \infty . \tag{16.99}$$

(h) There exists π such that

$$\sum_{n\ge 0} \sqrt{2^{-n}\,\mathrm{card}\{I \in \mathcal{I}_n \; ; \; I \cap T_\pi \ne \emptyset\}} < \infty . \tag{16.100}$$

(i) Condition (16.100) holds for each π.

This should be compared with Corollary 16.3.2, which asserts that when the series $\sum_m a_m\varphi_m$ converges a.s. whatever the choice of the orthonormal sequence φ_m, then $\sum_{k\ge 1} 2^{2k} b_k < \infty$. The stronger hypothesis of Theorem 16.7.1 implies the stronger conclusion (16.99).

It is not very difficult to prove the equivalence of (g) to (i), and this is what we will do first. We will then deduce the rest of Theorem 16.7.1 from Theorem 16.2.1. Condition (16.100) is the natural "covering number condition" adapted to processes that satisfy (16.13) (see Exercise 16.2.2). The deepest idea of the proof is that the more sophisticated conditions of Theorem 16.2.1 are equivalent to this natural covering condition "when the set T is homogeneous", and to prove that (f) implies (h), we will construct π so that T_π is "as homogeneous as possible".

Let us prove "the easy part" of Theorem 16.7.1.

Proposition 16.7.2 *Conditions (g) to (i) of Theorem 16.7.1 are equivalent.*

Proof We first prove that conditions (16.99) and (16.100) are equivalent when π is the identity. We first prove in that case that (16.100) implies (16.99). For $n \geq 1$, we define $\mathcal{J}_n$ as the set of dyadic intervals $I \in \mathcal{I}_n$ for which $T \cap I \neq \emptyset$.

In Lemma 16.3.3, we constructed a probability measure μ_k on T which satisfies (16.24) for $2^{k-1} \leq n < 2^k$, i.e., the first inequality in Eq. (16.101) below. Thus,

$$\frac{1}{2}\sqrt{b_k} \leq \sum_{I \in \mathcal{I}_n} \sqrt{2^{-n}\mu_k(I)} = \sum_{I \in \mathcal{J}_n} \sqrt{2^{-n}\mu_k(I)} \leq \sqrt{2^{-n}\operatorname{card}\mathcal{J}_n}\,, \qquad (16.101)$$

where we have used the Cauchy-Schwarz inequality as in (16.17) in the last inequality. Summing over n with $2^{k-1} \leq n < 2^k$ yields

$$2^{k-2}\sqrt{b_k} \leq \sum_{2^{k-1} \leq n < 2^k} \sqrt{2^{-n}\operatorname{card}\mathcal{J}_n}\,.$$

Summing then over k proves that (16.100) for the identity implies (16.99).

We next prove that (16.99) implies (16.100) when π is the identity. We shall prove that

$$\sum_{n \geq 0} \sqrt{2^{-n}\operatorname{card}\mathcal{J}_n} < \infty\,. \qquad (16.102)$$

Let us as usual enumerate T as a sequence $t_n = \sum_{1 \leq m \leq n} a_m^2$, and let $t^* = \sum_{m \geq 1} a_m^2$. Define

$$W_k = \{t_n\,;\ \max(a_n^2, a_{n+1}^2) > 2^{-2^k}\} \cup \{t_1, t^*\}\,,$$

$$V_k = \bigcup \big\{[t_n, t_{n+1}]\,;\ a_{n+1}^2 = t_{n+1} - t_n \leq 2^{-2^k}\big\} \subset [0, 1]\,.$$

Denoting Lebesgue's measure by λ, we deduce from (16.98) that

$$\lambda(V_k) = \sum \big\{a_{n+1}^2\,;\ a_{n+1}^2 \leq 2^{-2^k}\big\} \leq \sum_{r \geq k} b_r\,. \qquad (16.103)$$

Also,

$$\operatorname{card} W_k \leq 2 + \operatorname{card}\{n;\, \max(a_n^2, a_{n+1}^2) > 2^{-2^k}\} \leq 2 + 2\operatorname{card}\{n;\, a_n^2 > 2^{-2^k}\}\,,$$

and since $\operatorname{card}\{n; a_n^2 > 2^{-2^k}\} \le 2^{2^k} \sum_{m\ge1} a_m^2 \le 2^{2^k}$, we obtain

$$\operatorname{card} W_k \le 2 + 2 \cdot 2^{2^k} . \tag{16.104}$$

Consider $2^{k+1} \le \ell < 2^{k+2}$ and an interval $I \in \mathcal{J}_\ell$. We will prove that one of the following occurs: Either

$$I \cap W_k \ne \emptyset \tag{16.105}$$

or else

$$I \subset V_k . \tag{16.106}$$

To prove this, we assume $I \cap W_k = \emptyset$, and we prove (16.106). First, since $I \in \mathcal{J}_\ell$, then I meets $T \subset [t_1, t^*]$ by definition, so that either $t_1 \in I$, or $t^* \in I$, or else

$$I \subset]t_1, t^*[\subset \bigcup_{n\ge1} [t_n, t_{n+1}] . \tag{16.107}$$

Since we assume that $I \cap W_k = \emptyset$, we have in particular that $t_1, t^* \notin I$ and thus (16.107) holds. Consider an interval $[t_n, t_{n+1}]$ which meets I. Then either t_n or t_{n+1} belongs to I, for otherwise $I \subset]t_n, t_{n+1}[$, contradicting the fact that $I \cap T \ne \emptyset$. Since $I \cap W_k = \emptyset$, it cannot happen that both t_n and t_{n+1} belong to W_k. The definition of W_k shows that $a_{n+1}^2 \le 2^{-2^k}$, so that $[t_n, t_{n+1}] \subset V_k$ by definition of V_k. We have shown that every interval $[t_n, t_{n+1}]$ which meets I is a subset of V_k. Since (16.107) holds, we have proved (16.106), finishing the proof that either (16.105) or (16.106) holds.

There are at most $\operatorname{card} W_k$ intervals $I \in \mathcal{I}_\ell$ which satisfy (16.105). Since $\lambda(I) = 2^{-\ell}$ for $I \in \mathcal{I}_\ell$, there are at most $2^\ell \lambda(V_\ell)$ such intervals contained in V_k. Since every interval $I \in \mathcal{I}_\ell$ satisfies either (16.105) or (16.106), recalling (16.103),

$$\operatorname{card} \mathcal{J}_\ell \le \operatorname{card} W_k + 2^\ell \lambda(V_k) \le 2 + 2 \cdot 2^{2^k} + 2^\ell \sum_{r\ge k} b_r ,$$

so that (using the inequality $\sqrt{a+b} \le \sqrt{a} + \sqrt{b}$)

$$\sqrt{\operatorname{card} \mathcal{J}_\ell} \le L 2^{2^{k-1}} + 2^{\ell/2} \sqrt{\sum_{r\ge k} b_r} .$$

This holds whenever $2^{k+1} \le \ell < 2^{k+2}$ and therefore

$$\sum_{2^{k+1}\le\ell<2^{k+2}} 2^{-\ell/2} \sqrt{\operatorname{card} \mathcal{J}_\ell} \le L 2^{-2^k} 2^{2^{k-1}} + L 2^k \sqrt{\sum_{r\ge k} b_r} .$$

Summation of these inequalities over $k \geq 1$, use of the inequality $\sqrt{\sum_{r \geq k} b_r} \leq \sum_{r \leq k} \sqrt{b_r}$ and of (16.99) implies (16.102). This concludes the proof that (16.99) implies (16.100) when π is the identity.

The case of general π follows, because the value of b_k does not depend on the order in which we consider the elements a_m. So we simply replace the numbers a_m by the numbers $a_{\pi(m)}$ to obtain the equivalence of (16.99) and (16.100). □

The next goal is to complete the proof of Theorem 16.7.1. First, we prove that (i) implies (f). This is because each set T_π satisfies (16.100) so that it satisfies condition (c) of Theorem 16.2.1, as follows from the Cauchy-Schwarz inequality (as in (16.17) again):

$$\sum_{I \in \mathcal{I}_n} \sqrt{2^{-n} \mu(I)} \leq \sqrt{2^{-n} \operatorname{card}\{I \in \mathcal{I}_n \,;\, I \cap T_\pi \neq \emptyset\}} \,.$$

Now we come to the main argument: the proof that (f) implies (g). Let us first introduce some notations. Given a finite set U, let us denote by $\mathcal{T}(U)$ the collection of sets of the type

$$T = \big\{0, u_1, u_1 + u_2, \ldots, u_1 + u_2 + \cdots + u_q\big\}$$

where $q = \operatorname{card} U$ and $(u_\ell)_{\ell \leq q}$ is an enumeration of the elements of U. In other words, to construct $T \in \mathcal{T}(U)$, we start with 0, and, having constructed an element of T, we construct the next largest element by adding an element of U, in such a manner that each element of U is used exactly once for this purpose. For such a set T, we denote by $t^*(T)$ its largest element, so that $t^*(T)$ is the sum of the elements of U. The center of the proof is the following:

Proposition 16.7.3 *There exists a universal constant L^*, with the following property. Consider a finite set $J \subset \{10, 11, \ldots\}$ with the property that*

$$k < k' \in J \Rightarrow k' - k \geq 5 \,. \tag{16.108}$$

For $k \in J$, consider a finite set $U_k \subset \mathbb{R}^+$, assume that

$$u \in U_k \Rightarrow 2^{-2^{k+1}} < u \leq 2^{-2^k} \,, \tag{16.109}$$

and set

$$b_k = \sum_{u \in U_k} u \,. \tag{16.110}$$

Assume also that $\sum_{k\in J} b_k \le 1/2$, *and denoting by* k^* *the smallest element of* J, *assume that*

$$\forall k \in J\ ,\ b_k \ge 2^{-4k} 2^{-2^{k^*-3}}\ . \tag{16.111}$$

Let $U = \cup_{k\in J} U_k$. *Then there exist a set* $T \in \mathcal{T}(U)$ *(so that* $t^*(T) = \sum_{k\in J} b_k$*) and a probability measure* μ *on* T *such that* $\mu(\{0\}) = 0$ *with the following property. Consider* $x \in [0, 1]$ *with* $x + t^*(T) \le 1$. *Then for each* $k \in J$, *we have*

$$\sum_{2^{k-2}\le n\le 2^{k-1}} \sum_{I\in\mathcal{I}_n, I-x\subset[0,t^*(T)]} \sqrt{2^{-n}\mu(I-x)} \ge \frac{2^k\sqrt{b_k}}{L^*}\ , \tag{16.112}$$

so that in particular

$$\sum_{n\ge 0} \sum_{I\in\mathcal{I}_n, I-x\subset[0,t^*(T)]} \sqrt{2^{-n}\mu(I-x)} \ge \sum_{k\in J} \frac{2^k\sqrt{b_k}}{L^*}\ . \tag{16.113}$$

The number 5 in (16.108) is simply a convenient choice whose relevance will became apparent gradually. The condition (16.111) is purely technical, and its relevance will become apparent when we prove the proposition by induction over card J. The important part is condition (16.112). To understand it better, we note that $I - x \subset [0, t^*(T)]$ if and only if $I \subset [x, x + t^*(T)]$ and that for $x + t^*(T) \le 1$, the interval $[x, x + t^*(T)]$ is a subinterval of $[0, 1]$, so that it has a good chance to contain plenty of intervals $I \in \mathcal{I}_n$ which will contribute to making the left-hand side of (16.113) large (this would be less the case if $x + t^*(T)$ was ≥ 1).

We will discuss and prove Proposition 16.7.3 later, but our first goal is to complete the proof of Theorem 16.7.1, i.e., to prove that (f) implies (g).

Lemma 16.7.4 *Assume that the sequence* $(b_k)_{k\ge 1}$ *satisfies* $\sum_{k\ge 1} 2^k\sqrt{b_k} = \infty$. *Then given* $\tilde{k}$ *and* $A > 0$, *we can find a finite set* $J \subset \{\tilde{k}, \tilde{k}+1, \ldots\}$ *which satisfies (16.108) and* $\sum_{k\in J} 2^k\sqrt{b_k} \ge A$ *and for which* $b_k \ge 2^{-4k}$ *for all* $k \in J$.

Proof For $0 \le j \le 4$, consider the set I_j of integers $\ge \tilde{k}$ which are equal to j modulo 5, so that each such set satisfies (16.108). There exist $0 \le j \le 4$ such that $\sum_{k\in I_j} 2^k\sqrt{b_k} = \infty$. Define now $I = \{k \in I_j; b_k \ge 2^{-4k}\}$. Since $\sum_{b_k\le 4^{-k}} 2^k\sqrt{b_k} < \infty$, we have $\sum_{k\in I} 2^k\sqrt{b_k} = \infty$. Then there is a finite subset J of I with $\sum_{k\in J} 2^k\sqrt{b_k} > A$. □

Proof that (f) implies (g) We argue by contradiction, and we assume that (g) fails, i.e., $\sum_{k\ge 1} 2^k\sqrt{b_k} = \infty$. Consider the set $V = \{a_m^2; m \ge 1\}$. We construct by induction finite sets $V_s \subset V$ with $\max V_{s+1} < \min V_s$, sets $T_s \in \mathcal{T}(V_s)$, and probability measures μ_s on T_s with $\mu_s(\{0\}) = 0$ and the following property.

Consider $x \in [0, 1]$ with $x + t^*(T_s) \le 1$. Then

$$\sum_{n\ge 0}\ \sum_{I\in\mathcal{I}_n, I-x\subset[0,t^*(T_s)]} \sqrt{2^{-n}\mu_s(I-x)} \ge 2^s\ . \tag{16.114}$$

The construction is inductive. Having constructed V_s, consider k_s such that $2^{-2^{k_s}} < u$ for all $u \in V_s$. We construct a set J by applying Lemma 16.7.4 with $\bar{k} = k_s + 1$. We then apply Proposition 16.7.3 with $U_k = \{a_m^2; 2^{-2^{k+1}} < a_m^2 \le 2^{-2^k}\}$ for $k \in J$ to find V_{s+1}, completing the induction.

By construction for $s \ge 1$, we have $T_s = \{0\} \cup \{\sum_{1\le m\le n} a^2_{r_{m,s}}; n \le q_s\}$ where the integers $r_{m,s}$ for $s \ge 1$ and $m \le q_s$ are all distinct. Consider then a permutation π with the property that for each s, the integers $r_{m,s}$, $1 \le m \le q_s$ occur as $\pi(j_s+1), \ldots, \pi(j_s+q_s)$ for consecutive integers $j_s+1, \ldots, j_s+q_s$. Let us set $x_s = \sum_{m\le j_s} a^2_{\pi(m)}$. Then for each $n \le q_s$, we have $x_s + \sum_{1\le m\le n} a^2_{r_{m,s}} = \sum_{1\le m\le j_s+n} a^2_{\pi(m)}$ so that $x_s + T_s \subset T_\pi$. In particular, $x_s + t^*(T_s) \le 1$, so that we can use (16.114) for $x = x_s$ to obtain

$$\sum_{n\ge 0}\sum_{I\in\mathcal{I}_n} \sqrt{2^{-n}\mu_s(I-x_s)} \ge \sum_{n\ge 0}\ \sum_{I\in\mathcal{I}_n, I-x_s\subset[0,t^*(T_s)]} \sqrt{2^{-n}\mu_s(I-x_s)} \ge 2^s\ .$$

This proves that the probability measure ν on T_π given by $\nu(C) = \mu_s(C - x_s)$ satisfies

$$\sum_{n\ge 0}\sum_{I\in\mathcal{I}_n} \sqrt{2^{-n}\nu(I)} \ge 2^s\ .$$

Thus, condition (c) of Theorem 16.2.1 is not satisfied. Thus, there exists an orthonormal sequence (φ_m) such that the series $\sum_m a_{\pi(m)}\varphi_m$ does not converge a.s. so that (f) fails and the proof of Theorem 16.7.1 is complete. □

We turn to the discussion of Proposition 16.7.3. The main idea is very simple (but unfortunately the details are quite taxing). For each k in J, the following happens:

- If $n \le 2^{k-1}$, at scale 2^{-n}, the measure μ looks very much like the uniform measure on a set S_k with $\lambda(S_k) \ge b_k$.
- The set S_k is a union of intervals. If $n \ge 2^{k-2}$, all these intervals are very much longer than 2^{-n}.

Assuming $2^{k-2} \le n \le 2^{k-1}$, we can then pretend that μ is the uniform measure on S_k to estimate the term $\sum_{I\in\mathcal{I}_n, I-x\subset[0,t^*(T)]}\sqrt{2^{-n}\mu(I-x)}$. Let us first estimate how many $I \in \mathcal{I}_n$ are such that $I - x \subset S_k$. We have $I - x \subset S_k$ if and only if $I \subset x + S_k$, and when $x < 1 - t^*(T)$, we have $x + S_k \subset [0, 1]$ because $S_k \subset [0, t^*(T)]$. Since S_k is a union of intervals, each of which is of length much larger than 2^{-n}, for $x < 1 - t^*(T)$, there are about $2^n\lambda(S_k)$ sets $I \in \mathcal{I}_n$ such that $I - x \subset S_k$. For each of these sets, let us estimate $\mu(I - x)$. Since we pretend that μ is the uniform measure on S_k, it has density $1/\lambda(S_k)$ on S_k, so that if $I - x \subset S_k$, then

$\mu(I-x)$ is about $2^{-n}/\lambda(S_k)$, and hence $\sqrt{2^{-n}\mu(I-x)}$ is about $2^{-n}/\sqrt{\lambda(S_k)}$. Thus, the sum $\sum_{I\in\mathcal{I}_n, I-x\subset[0,t^*(T)]}\sqrt{2^{-n}\mu(I-x)}$ has about $2^n\lambda(S_k)$ terms each about $2^{-n}/\sqrt{\lambda(S_k)}$ so it is $\geq\sqrt{\lambda(S_k)}/L\geq\sqrt{b_k}/L$, and then

$$\sum_{2^{k-2}\leq n\leq 2^{k-1}}\ \sum_{I\in\mathcal{I}_n, I-x\subset[0,t^*(T)]}\sqrt{2^{-n}\mu(I-x)}\geq\frac{2^k\sqrt{b_k}}{L}\,, \tag{16.115}$$

which is the desired inequality (16.112).

How is the situation of Proposition 16.7.3 possible? The important idea is *separation of scales*. If $k'>k+1$, the elements of $U_{k'}$ are in a sense "infinitely smaller" than those of U_k. The use of (16.108) is to ensure separation of scales between the different values of $k\in J$. Let us try to visualize the measure μ by blowing the picture by a factor 2 every second [31]. At the beginning, the measure μ appears uniform on a given interval $[0,b]$. After a while however, this no longer appears to be the case, gaps appear, and μ seems to be carried by a union of many very small intervals (which can be of rather widely different sizes). It really looks like the uniform measure on the union of these intervals. Waiting quite longer, this appears to be no longer the case. Gaps appear. Each of the very small intervals breaks into extremely small intervals. These can also be widely different sizes. However, the longer of them are still very much shorter than the shortest of the previous very small intervals. And μ looks like the uniform measure on the union of these intervals. The amount of time it takes to go through one step of the process increases as a geometric series.

The main ingredient of the construction is the following principle. Consider two probability measures on a set $S\subset[0,1]$ which is a union of very small intervals. If these two measures give the same (small) mass to each of these intervals, at a large scale, they are nearly identical. The principle will be used when one of the probability measures is the uniform probability on S.

It could help the reader to start with the case card $J=1$.

Proof of Proposition 16.7.3 when card $J=1$ Writing $k=k^*$ for simplicity, we have $J=\{k\}$, and all the elements u of $U=U_k$ satisfy $2^{-2^{k+1}}<u\leq 2^{-2^k}$. We enumerate $U=\{u_1,\ldots,u_q\}$, so that $b:=b_k=\sum_{\ell\leq q}u_\ell\geq 2^{-4k}2^{-2^{k-3}}$ by (16.111). For $\ell\leq q$, let $t_\ell=\sum_{m\leq\ell}u_m$, and let $T=\{t_0=0,t_1,\ldots,t_q\}\in\mathcal{T}(U)$. Consider the probability measure μ on T such that $\mu(\{t_\ell\})=u_\ell/b=t_\ell-t_{\ell_1}$ for $\ell\leq q$. One way to visualize this measure is to start with the uniform measure on the interval $[0,b]$, which is the union of the intervals $]t_{\ell-1},t_\ell]$. The small interval $]t_{\ell-1},t_\ell]$ has total mass u_ℓ/b. This mass is then swept to the right of this interval. What matters is that the small interval $]t_{\ell-1},t_\ell]$ has the same mass for the uniform measure on the interval $[0,b]$ and for μ.

For $n\leq 2^{k-1}$, at the scale 2^{-n}, the probability μ looks like the uniform measure on the interval $[0,b]$ because the distance between two consecutive elements of T is smaller than 2^{-2^k}, which is very much smaller than 2^{-n}. For $I\in\mathcal{I}_n$ with $I-x\subset[0,b]$, the measure of I for the uniform measure on $[0,b]$ is $2^n/b$ so that

$\mu(I-x)$ will have nearly the same value, and in particular, we will have $\mu(I-x) \geq 2^{-n}/Lb$. Consequently, $\sqrt{2^{-n}\mu(I-x)} \geq 2^{-n}/(L\sqrt{b})$. Next, if $2^{k-2} \leq n$, then $b2^n \geq 2^n 2^{-4k} 2^{-2^{k-3}}$ is much larger than 1. Thus, given $0 \leq x \leq 1 - t^*(T)$, there are then about $2^n b$ sets $I \in \mathcal{I}_n$ for which $I \subset [x, x + t^*(T)] = [x, x+b] \subset [0,1]$, i.e., there are about $2^n b$ sets $I \in \mathcal{I}_n$ such that $I - x \subset [0, t^*(T)] = [0, b]$ so that $\sum_{I \in \mathcal{I}_n, I - x \subset [0, t^*(T)]} \sqrt{2^{-n}\mu(I-x)}$ is at least of order $2^n b \times 2^{-n}/L\sqrt{b}$, i.e., $\sqrt{b}/L$. As this is true for each $2^{k-2} \leq n \leq 2^{k-1}$, the left-hand side of (16.112) is at least of order $2^k\sqrt{b}/L$ as desired.

Proving that things happen the way we described them requires no skill whatsoever because there is all the room in the world for the estimates. This is better left to the reader, as any attempt to write these estimates makes the proof unreadable.[5] □

Proof of Proposition 16.7.3 The proof is by induction over card J. We denote by k^* the smallest element of J, and we set $J' = J \setminus \{k^*\}$. We enumerate U_{k^*} as $u_1, \ldots, u_q$. The sum of these elements is at most 1, and each of them is $\geq 2^{-2^{k^*+1}}$. Thus, $q \leq 2^{2^{k^*+1}}$. For $\ell \leq q$, we set $\beta_\ell = u_\ell / b_{k^*}$, so that $\sum_{\ell \leq q} \beta_\ell = 1$.

The proof will require using the induction hypothesis for each $\ell \leq q$. The first step of the proof is to partition each set U_k ($k \in J'$) into q disjoint sets $(U_{k,\ell})_{\ell \leq q}$, so that the elements of $U_{k,\ell}$ will be used for the construction associated with ℓ. This partitioning is done in such a way that the proportion of U_k attributed to $U_{k,\ell}$ is about β_ℓ, that is,

$$b_{k,\ell} := \sum_{u \in U_{k,\ell}} u \simeq \beta_\ell b_k = \beta_\ell \sum_{u \in U_k} u \,. \tag{16.116}$$

To prove that this is possible, it suffices to show that for $k \in J'$, the elements of U_k are very small compared to $\beta_\ell b_k$. This is because since $\beta_\ell = u_\ell / b_{k^*} \geq u_\ell \geq 2^{-2^{k^*+1}}$ by (16.111), we have

$$\beta_\ell b_k \geq 2^{-4k} 2^{-2^{k^*+1}} 2^{-2^{k^*-3}} \geq 2 \times 2^{-4k} 2^{-2^{k^*+2}} \,.$$

The smallest element $\hat{k}$ of J' satisfies $\hat{k} \geq k^* + 5$ so that since $b_{k,\ell} \simeq \beta_\ell b_k$, we will have $b_{k,\ell} \geq 2^{-4k} 2^{-2^{\hat{k}-3}}$, i.e., the condition (16.111) is satisfied for the sets $U_{k,\ell}$ for $k \in J'$. Given ℓ, we may then use the induction hypothesis for these sets. We then obtain sets $T_\ell \in \mathcal{T}(\cup_{k \in J'} U_{k,\ell})$, with

$$t^*(T_\ell) = \sum_{k \in J'} b_{k,\ell} \simeq \beta_\ell \sum_{k \in J'} b_k \,, \tag{16.117}$$

[5] Again, no skill whatsoever is required there.

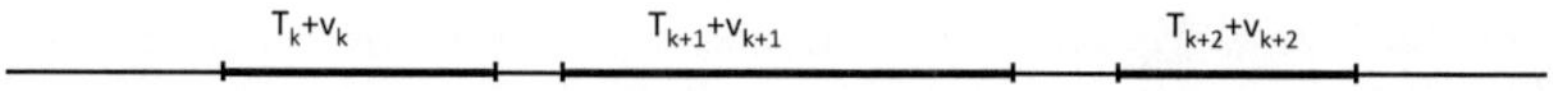

Fig. 16.1 View of the set T at a scale where the intervals between the sets $T_k + v_k$ can be seen, but not yet the individual structure of these sets

and probability measures μ_ℓ on T_ℓ with the following property. Consider $x \in [0, 1]$ with $x + t^*(T_\ell) \le 1$. Then for each $k \in J'$, we have

$$\sum_{2^{k-2} \le n \le 2^{k-1}} \; \sum_{I \in \mathcal{I}_n, I - x \subset [0, t^*(T_\ell)]} \sqrt{2^{-n} \mu_\ell(I - x)} \ge \frac{2^k \sqrt{b_{k,\ell}}}{L^*} . \tag{16.118}$$

□

The set $T \setminus \{0\}$ will be the union over $\ell \le q$ of translates $v_\ell + T_\ell$ of the sets T_ℓ. The numbers v_ℓ are recursively determined as follows, $v_1 = u_1$, and, once v_ℓ has been constructed, $v_{\ell+1}$ is such that the interval between the largest element w_ℓ of $v_\ell + T_\ell$ and $v_{\ell+1}$ is $u_{\ell+1}$, i.e., $v_{\ell+1} = w_{\ell+1} + u_{\ell+1}$.[6] It should be obvious that

$$t^*(T) = \sum_{\ell \le q} u_\ell + \sum_{\ell \le q} t^*(T_\ell) = \sum_{k \in J} b_k .$$

Next, we claim that $T \in \mathcal{T}(U)$. Recall that a set in $\mathcal{T}(U)$ is a set such that to go from one element of this set to the next largest one, one adds an element of U, in such a way that each element of U is used exactly once in this manner. The elements u_ℓ of U_{k^*} is used to go from $w_{\ell-1}$ to v_ℓ. The elements of $\cup_{k \in J'} U_{k,\ell}$ are used when going from one element of $v_k + T_k$ to the next (since $T_k \in \mathcal{T}(\cup_{k \in J'} U_{k,\ell})$). Conversely, going from one element $t \in v_\ell + T_\ell \subset T$ to the next element of T requires adding an element of $\cup_{k \in J'} U_{k,\ell}$, unless $x = w_\ell$ in which case this requires adding $u_{\ell+1}$ (Fig. 16.1).

The probability measure μ on T is defined as $\sum_{\ell \le q} \beta_\ell \mu'_\ell$ where μ'_ℓ is the translation by v_ℓ of the probability μ_ℓ. The main idea behind this construction is that at the scale 2^{-n} for $n < 2^{k^*-1}$, the measure μ will look uniform on the interval $[0, t^*(T)]$. The reason for this is very simple. Observe that since $\mu_\ell(\{0\}) = 0$, the probability μ'_ℓ is supported by the interval $]v_\ell, v_{\ell+1}]$, so that $\mu(]v_\ell, v_{\ell+1}]) = \beta_\ell$. However, recalling (16.117), we have

$$v_{\ell+1} - v_\ell = u_\ell + t^*(T_\ell) \simeq \beta_\ell b_{k^*} + \beta_\ell \sum_{k \in J'} b_k = \beta_\ell \sum_{k \in J} b_k = \beta_\ell t^*(T) .$$

[6] And, as we look at the structure at increasingly finer scale, these are the first gaps which will appear, and the gaps inside each block are much smaller.

Thus, the measures of the very small intervals $]v_\ell, v_{\ell+1}]$ are almost exactly proportional to their lengths, so that μ looks uniform on the interval $[0, t^*(T)]$ at the scale 2^{-n} for $n < 2^{k^*-1}$, and as we explained, this implies (16.112) for $k = k^*$ (and since $t^*(T) \geq b_{k^*}$).

Now, we have to prove (16.112) for $k \in J'$. Consider $I \in \mathcal{I}_n$ with $I - x \subset [v_\ell, v_\ell + t^*(T_\ell)]$, so that $\mu(I - x) \geq \beta_\ell \mu'_\ell(I - x) = \beta_\ell \mu_\ell(I - (x + v_\ell))$, using in the last equality that μ'_ℓ is the translation of μ_ℓ by v_ℓ. We have shown that

$$\sum_{2^{k-2} \leq n \leq 2^{k-1}} \sum_{I \in \mathcal{I}_n, I - x \subset [v_\ell, v_\ell + t^*(T_\ell)]} \sqrt{2^{-n}\mu(I - x)}$$
$$\geq \sqrt{\beta_\ell} \sum_{2^{k-2} \leq n \leq 2^{k-1}} \sum_{I \in \mathcal{I}_n, I - (x + v_\ell) \subset [0, t^*(T_\ell)]} \sqrt{2^{-n}\mu_\ell(I - (x + v_\ell))} \,. \tag{16.119}$$

Consider $0 \leq x \leq 1 - t^*(T)$, and observe that then for each $\ell \leq q$, we have $x + v_\ell \leq 1 - t^*(T_\ell)$. We can then use the induction hypothesis (16.118) (with $x + v_\ell$ instead of x) to obtain that the right-hand side of (16.119) is $\geq 2^k \sqrt{\beta_\ell b_{k,\ell}}/L^*$. Thus, we have shown that

$$\sum_{2^{k-2} \leq n \leq 2^{k-1}} \sum_{I \in \mathcal{I}_n, I - x \subset [v_\ell, v_\ell + t^*(T_\ell)]} \sqrt{2^{-n}\mu(I - x)} \geq \frac{2^k \sqrt{\beta_\ell b_{k,\ell}}}{L^*} \,. \tag{16.120}$$

Since the intervals $[v_\ell, v_\ell + t^*(T_\ell)]$ for $\ell \leq q$ are disjoint subintervals of $[0, t^*(T)]$, by summation of the inequalities (16.120) over $\ell \leq q$, we obtain

$$\sum_{2^{k-2} \leq n \leq 2^{k-1}} \sum_{I \in \mathcal{I}_n, I - x \subset [0, t^*(T)]} \sqrt{2^{-n}\mu(I - x)} \geq \sum_{\ell \leq q} \frac{2^k \sqrt{\beta_\ell b_{k,\ell}}}{L^*} \,.$$

Since $b_{k,\ell} \simeq \beta_\ell b_k$, the right-hand side is nearly $2^k \sqrt{b_k}/L^*$, and we almost obtain the required inequality (16.112). To make the proof complete, it suffices to quantify the errors made in the statement (16.116) and to show that they do not destroy the argument. Since there is plenty of room, this is better left to the reader.[7]

16.8 Chaining, II

For the special sets T of the type (16.10), the equivalence of (c) and (d) of Theorem 16.2.1 tells us for which sets T all the processes satisfying the increment condition $\mathsf{E}(X_s - X_t)^2 \leq |s - t|$ for $s, t \in T$ are bounded. Our goal is to investigate

[7] One may use in particular that since the elements of $U_{k,\ell}$ are $\leq 2^{-2^k}$, it should be obvious that one can achieve $b_{k,\ell} \geq \beta_\ell b_k - 2^{-2^k} \geq (1 - 2^{-k})\beta_\ell b_k$.

the same question for more general metric spaces under the more general increment conditions (16.1):

$$\forall s,t \in T\ ,\ \mathsf{E}\varphi\Big(\frac{X_s - X_t}{d(s,t)}\Big) \le 1\ , \tag{16.1}$$

where φ is a Young function as in Definition 16.1.1.

What are the weakest possible natural conditions that will ensure that we control the size of the process $(X_t)_{t\in T}$ under (16.1)? We consider this question in the remainder of this chapter.

The material of this section is self-contained, but the reader might do well to master first the simpler ideas of Sect. 16.1 to provide perspective. For simplicity, we consider only the case where T is finite. We will first develop a chaining scheme. This scheme is related, but different, from the scheme considered in Sect. 16.5 (which was well adapted to our limited goals there).

We say that a sequence $\mathcal{T} = (T_n)_{n\ge 0}$ of subsets of T is *admissible* if it satisfies

$$\operatorname{card} T_0 = 1 \tag{16.121}$$

and

$$\operatorname{card} T_n \le \varphi(4^n)\ . \tag{16.122}$$

We *do not* require the sequence (T_n) to be increasing. Let us consider the following quantities:

$$S_d(\mathcal{T}) = \sup_{t\in T} \sum_{n\ge 0} 4^n d(t, T_n) \tag{16.123}$$

and

$$S_d^*(\mathcal{T}) = \sum_{n\ge 1} \sum_{s\in T_n} \frac{4^n d(s, T_{n-1})}{\varphi(4^n)}\ . \tag{16.124}$$

In the case where $\varphi(x) = \exp(x^2) - 1$, which corresponds to Gaussian processes, we have $\operatorname{card} T_n \le \exp(4^{2n})$, and the quantity (16.123) is then basically the right-hand side of (2.34) (the difference is that we change n into $4n$). The new feature here is the quantity $S_d^*(\mathcal{T})$, which was not needed in the Gaussian case or more generally in the case where one has "exponential tails". The formulation of the following theorem is due again to W. Bednorz:

Theorem 16.8.1 *Consider a process that satisfies (16.1). Then, for each sequence* $\mathcal{T} = (T_n)_{n\geq 0}$ *of admissible sets, we have*

$$\mathsf{E} \sup_{s,t\in T} |X_s - X_t| \leq L(S_d(\mathcal{T}) + S_d^*(\mathcal{T})) . \tag{16.125}$$

It is not required here that $\mathsf{E}X_t = 0$.

Proof For $n \geq 1$, let us define a map $\theta_n : T_n \to T_{n-1}$ such that for $s \in T_n$, one has

$$d(s, \theta_n(s)) = d(s, T_{n-1}) . \tag{16.126}$$

We may assume that $S_d(\mathcal{T}) < \infty$ for otherwise there is nothing to prove. This implies that for large m, T_m is a good approximation of T, and in particular, since T is finite, there exists m with $T = T_m$. Let us consider such a value of m. For $t \in T$, we define $\pi_m(t) = t$, and we define recursively $\pi_{n-1}(t) = \theta_n(\pi_n(t))$, so that (16.126) implies

$$d(\pi_n(t), \pi_{n-1}(t)) = d(\pi_n(t), T_{n-1}) . \tag{16.127}$$

For $x, y > 0$, the inequality

$$\frac{y}{x} \leq 1 + \frac{\varphi(y)}{\varphi(x)} . \tag{16.128}$$

is obvious if $y \leq x$, and if $x \leq y$ follows from the fact that $\varphi(x) \leq x\varphi(y)/y$ by convexity of φ. We use (16.128) with $y = |X_s - X_{\theta_n(s)}|/d(s, \theta_n(s))$ and $x = 4^n$ to obtain (since $\varphi(y) = \varphi(|y|)$),

$$|X_s - X_{\theta_n(s)}| \leq 4^n d(s, \theta_n(s)) + \frac{4^n d(s, \theta_n(s))}{\varphi(4^n)} \varphi\Big(\frac{X_s - X_{\theta_n(s)}}{d(s, \theta_n(s))}\Big) . \tag{16.129}$$

Using this for $s = \pi_n(t)$ yields (using a crude bound to obtain a last term independent of t)

$$|X_{\pi_n(t)} - X_{\pi_{n-1}(t)}| \leq 4^n d(\pi_n(t), \pi_{n-1}(t)) + \sum_{s\in T_n} \frac{4^n d(s, \theta_n(s))}{\varphi(4^n)} \varphi\Big(\frac{X_s - X_{\theta_n(s)}}{d(s, \theta_n(s))}\Big).$$

Combining with (16.44) if $T_0 = \{t_0\}$, we obtain

$$|X_t - X_{t_0}| \leq \sum_{n\geq 1} 4^n d(\pi_{n-1}(t), \pi_n(t))$$
$$+ \sum_{n\geq 1} \sum_{s\in T_n} \frac{4^n d(s, \theta_n(s))}{\varphi(4^n)} \varphi\Big(\frac{X_s - X_{\theta_n(s)}}{d(s, \theta_n(s))}\Big) , \tag{16.130}$$

and consequently,

$$\sup_{t\in T}|X_t - X_{t_0}| \le \sup_{t\in T}\sum_{n\ge 1}4^n d(\pi_{n-1}(t),\pi_n(t)) + \sum_{n\ge 1}\sum_{s\in T_n}\frac{4^n d(s,\theta_n(s))}{\varphi(4^n)}\varphi\Big(\frac{X_s - X_{\theta_n(s)}}{d(s,\theta_n(s))}\Big) . \tag{16.131}$$

Taking expectation and using (16.1) yields

$$\mathsf{E}\sup_{t\in T}|X_t - X_{t_0}| \le \sup_{t\in T}\sum_{n\ge 1}4^n d(\pi_{n-1}(t),\pi_n(t)) + S_d^*(\mathcal{T}) . \tag{16.132}$$

Now, recalling (16.127),

$$\begin{aligned} d(\pi_{n-1}(t),\pi_n(t)) &= d(\pi_n(t),T_{n-1}) \\ &\le d(t,T_{n-1}) + d(t,\pi_n(t)) \\ &\le d(t,T_{n-1}) + \sum_{k\ge n} d(\pi_k(t),\pi_{k+1}(t)) . \end{aligned}$$

Thus, using that $\sum_{n\le k}4^n \le 4^{k+1}/2$ we get

$$\begin{aligned} \sum_{n\ge 1}4^n d(\pi_{n-1}(t),\pi_n(t)) &\le \sum_{n\ge 1}4^n d(t,T_{n-1}) + \sum_{n\ge 1}4^n\sum_{k\ge n} d(\pi_k(t),\pi_{k+1}(t)) \\ &= \sum_{n\ge 1}4^n d(t,T_{n-1}) + \sum_{k\ge 1}\Big(\sum_{n\le k}4^n\Big) d(\pi_k(t),\pi_{k+1}(t)) \\ &\le \sum_{n\ge 1}4^n d(t,T_{n-1}) + \frac{1}{2}\sum_{k\ge 1}4^{k+1} d(\pi_k(t),\pi_{k+1}(t)) \\ &\le \sum_{n\ge 1}4^n d(t,T_{n-1}) + \frac{1}{2}\sum_{n\ge 1}4^n d(\pi_{n-1}(t),\pi_n(t)) , \end{aligned}$$

so that recalling (16.123), we get

$$\sum_{n\ge 1}4^n d(\pi_{n-1}(t),\pi_n(t)) \le 2\sum_{n\ge 1}4^n d(t,T_{n-1}) = 8\sum_{n\ge 0}4^n d(t,T_n) \le 8S_d(\mathcal{T}) . \tag{16.133}$$

Combining with (16.132), this finishes the proof. □

Interestingly, the previous proof does not use (16.122)!

Corollary 16.8.2 *Define* $e_0^* = \Delta(T, d)$ *and for* $n \geq 1$*, define*

$$e_n^* = \inf\{\epsilon > 0 \; ; \; \exists U \subset T \; , \; \operatorname{card} U \leq \varphi(4^n) \; , \; \forall t \in T \; , \; d(t, U) \leq \epsilon\} \; . \tag{16.134}$$

Then

$$\mathsf{E} \sup_{s,t \in T} |X_s - X_t| \leq L \sum_{n \geq 0} 4^n e_n^* \; . \tag{16.135}$$

Proof Consider an arbitrary point t_0 of T, and set $T_0 = \{t_0\}$. For $n \geq 1$, consider a subset T_n of T with $\operatorname{card} T_n \leq \varphi(4^n)$ and $d(t, T_n) \leq 2e_n^*$ for each $t \in T$. It is then obvious that the quantities $S_d(\mathcal{T})$ and $S_d^*(\mathcal{T})$ of (16.123) and (16.124) satisfy

$$S_d(\mathcal{T}) \leq L \sum_{n \geq 0} 4^n e_n^* \; ; \; S_d^*(\mathcal{T}) \leq L \sum_{n \geq 0} 4^n e_n^* \; . \qquad \square$$

Exercise 16.8.3 Deduce Pisier's bound (16.5) from Corollary 16.8.2.

The bound of Theorem 16.8.1 raises two questions: How to construct admissible sequences? How sharp is this result?

Definition 16.8.4 For a metric space (T, d), let

$$S(T, d, \varphi) = \sup \left\{ \mathsf{E} \sup_{s,t \in T} |X_s - X_t| \right\} , \tag{16.136}$$

where the supremum is taken over all the processes which satisfy (16.1).

The reader will need this definition throughout the rest of this chapter. We reformulate (16.125) as

$$S(T, d, \varphi) \leq L(S_d(\mathcal{T}) + S_d^*(\mathcal{T})) \; , \tag{16.137}$$

and the question arises as to which extent this inequality is sharp for the best possible choice of $\mathcal{T}$. W. Bednorz has recently discovered a rather general setting where this is the case. To describe it, we need the following concept:

Definition 16.8.5 Consider $p > 1$. A distance d on a metric space is called p-concave if d^p is still a distance, i.e.,

$$d(s, t)^p \leq d(s, v)^p + d(v, t)^p \; . \tag{16.138}$$

This definition is well adapted to the study of the distance $d(s, t) = \sqrt{|s - t|}$, which is 2-concave. Unfortunately, the usual distance on $\mathbb{R}^n$ is not p-concave, and as we shall see later, this case is considerably more complex.

One of the results we will prove is that for a p-concave distance, the inequality (16.137) can be reversed. The proof is indirect. We will show that both sides of this inequality are equivalent to a new quantity, itself of independent interest.

Theorem 16.8.6 (W. Bednorz [8]) *Assume that*

$$\text{the function } x \mapsto \varphi^{-1}(1/x) \text{ is convex} . \tag{16.139}$$

Assume that the distance d is p-concave. Then there exists a probability measure μ on T for which

$$\sup_{t \in T} \int_0^{\Delta(T,d)} \varphi^{-1}\Big(\frac{1}{\mu(B(t,\epsilon))}\Big) d\epsilon \le K(p) S(T,d,\varphi) , \tag{16.140}$$

where $S(T,d,\varphi)$ is defined in (16.136).

Condition (16.139) is inessential and is imposed only for simplicity. It is the behavior of φ^{-1} at zero that matters.

Theorem 16.8.7 (W. Bednorz, [11]) *Consider a probability measure μ on T, and let*

$$B = \sup_{t \in T} \int_0^{\Delta(T,d)} \varphi^{-1}\Big(\frac{1}{\mu(B(t,\epsilon))}\Big) d\epsilon . \tag{16.141}$$

Then there is an admissible sequence $\mathcal{T}$ of subsets of T for which

$$S_d(\mathcal{T}) \le LB \; ; \; S_d^*(\mathcal{T}) \le LB . \tag{16.142}$$

Thus, through (16.137) and (16.142), any probability measure μ yields a bound on $\mathcal{S}(T,d,\varphi)$. In this context, such a probability μ on (T,d) is traditionally called a *majorizing measure*. The importance of majorizing measures seemed to decrease considerably with the invention of the generic chaining, as they seemed to have limited use in the context of Gaussian processes (see Sect. 3.1), but as we saw in Chap. 11, matters are more complicated than that.

Definition 16.8.8 For a metric space (T,d), we define

$$\mathcal{M}(T,d,\varphi) = \inf\Big\{ \sup_{t \in T} \int_0^{\Delta(T,d)} \varphi^{-1}\Big(\frac{1}{\mu(B(t,\epsilon))}\Big) d\epsilon \Big\} , \tag{16.143}$$

where the infimum is taken over all probability measures μ on T.

Combining Theorems 16.8.1, 16.8.6, and 16.8.7, we have proved the following:

Theorem 16.8.9 *Assuming (16.139), if the distance d is p-concave, then*

$$S(T, d, \varphi) \le L \inf_{\mathcal{T}} (S_d(\mathcal{T}) + S_d^*(\mathcal{T})) \le L\mathcal{M}(T, d, \varphi) \le K(p) S(T, d, \varphi) \,. \tag{16.144}$$

Thus, $S(T, d, \varphi)$ is of the same order as $\mathcal{M}(T, d, \varphi)$, but the determination of the quantity $\mathcal{M}(T, d, \varphi)$ is by no means easy.

Let us turn to the proof of Theorem 16.8.6. A p-concave distance satisfies the following improved version of the triangle inequality:

Lemma 16.8.10 *If the distance d is p-concave, then for* $s, t, v \in T$*, we have*

$$d(s, v) - d(t, v) \le d(s, t) \Big(\frac{d(s, t)}{d(t, v)} \Big)^{p-1} \,. \tag{16.145}$$

Proof We have

$$d(s, v)^p \le d(t, v)^p + d(s, t)^p = d(t, v)^p \Big(1 + \frac{d(s, t)^p}{d(t, v)^p} \Big) \,,$$

so that since (crudely) $(1 + x)^{1/p} \le 1 + x$ for $x \ge 0$ and $p \ge 1$,

$$d(s, v) \le d(t, v) \Big(1 + \frac{d(s, t)^p}{d(t, v)^p} \Big)^{1/p} \le d(t, v) + d(s, t) \Big(\frac{d(s, t)}{d(t, v)} \Big)^{p-1} \,. \qquad \square$$

Lemma 16.8.11 *Consider* $s, t \in T$*. Then for each probability measure* μ *on* T*, one has*

$$\int_T \mathrm{d}\mu(\omega) \int_{\min(d(s,\omega), d(t,\omega))}^{\max(d(s,\omega), d(t,\omega))} \frac{1}{\mu(B(\omega, 3\epsilon))} \mathrm{d}\epsilon \le K(p) d(s, t) \,. \tag{16.146}$$

Proof Let us consider the set $A = \{\omega \in T \,;\, d(t, \omega) \le d(s, \omega)\}$, so that for $\omega \in A$, the second integral in (16.146) is from $d(t, \omega)$ to $d(s, \omega)$. Since for $d(t, \omega) \le \epsilon$ we have $B(t, 2\epsilon) \subset B(\omega, 3\epsilon)$, it suffices to prove that

$$\int_A \mathrm{d}\mu(\omega) \int_{d(t,\omega)}^{d(s,\omega)} \frac{1}{\mu(B(t, 2\epsilon))} \mathrm{d}\epsilon \le K(p) d(s, t) \tag{16.147}$$

and to use the similar result where s and t are exchanged. Let

$$A_0 = \{\omega \in A \,;\, d(t, \omega) \le 2d(s, t)\} \,,$$

so that

$$\int_{A_0} \mathrm{d}\mu(\omega) \int_{d(t,\omega)}^{d(s,\omega)} \frac{1}{\mu(B(t,2\epsilon))} \mathrm{d}\epsilon$$
$$= \iint \mathbf{1}_{\{d(t,\omega)\le\epsilon\le d(s,\omega)\}} \mathbf{1}_{\{d(t,\omega)\le 2d(s,t)\}} \frac{1}{\mu(B(t,2\epsilon))} \mathrm{d}\epsilon \mathrm{d}\mu(\omega) .$$

Then, since $d(s,\omega) \le d(s,t) + d(t,\omega)$, for $\epsilon \le d(s,\omega)$ and $d(t,\omega) \le 2d(s,t)$, we have $\epsilon \le 3d(s,t)$ so that

$$\begin{aligned}
\int_{A_0} \mathrm{d}\mu(\omega) \int_{d(t,\omega)}^{d(s,\omega)} \frac{1}{\mu(B(t,2\epsilon))} \mathrm{d}\epsilon
&\le \iint \mathbf{1}_{\{d(t,\omega)\le\epsilon\}} \mathbf{1}_{\{\epsilon\le 3d(s,t)\}} \frac{1}{\mu(B(t,2\epsilon))} \mathrm{d}\epsilon \mathrm{d}\mu(\omega) \\
&= \int \mathbf{1}_{\{\epsilon\le 3d(s,t)\}} \frac{1}{\mu(B(t,2\epsilon))} \mathrm{d}\epsilon \int \mathbf{1}_{\{d(t,\omega)\le\epsilon\}} \mathrm{d}\mu(\omega) \\
&= \int \mathbf{1}_{\{\epsilon\le 3d(s,t)\}} \frac{\mu(B(t,\epsilon))}{\mu(B(t,2\epsilon))} \mathrm{d}\epsilon \\
&\le \int \mathbf{1}_{\{\epsilon\le 3d(s,t)\}} \mathrm{d}\epsilon = 3d(s,t) .
\end{aligned} \tag{16.148}$$

Next, for $n \ge 1$, let

$$A_n = \{\omega \in A \,;\, 2^n d(s,t) \le d(t,\omega) \le 2^{n+1} d(s,t)\} \subset B(t, 2^{n+1} d(s,t)) ,$$

so that the sets $(A_n)_{n\ge 0}$ cover A. It follows from (16.145) that for $\omega \in A_n$,

$$d(s,\omega) - d(t,\omega) \le 2^{-n(p-1)} d(s,t) .$$

Furthermore, for $\omega \in A_n$ and $d(t,\omega) \le \epsilon$, we have $2^{n+1} d(s,t) \le 2\epsilon$ so that $B(t, 2^{n+1} d(s,t)) \subset B(t, 2\epsilon)$ and

$$\int_{d(t,\omega)}^{d(s,\omega)} \frac{1}{\mu(B(t,2\epsilon))} \mathrm{d}\epsilon \le 2^{-n(p-1)} d(s,t) \frac{1}{\mu(B(t, 2^{n+1} d(s,t)))} .$$

Consequently, since $\mu(A_n) \le \mu(B(t, 2^{n+1} d(s,t)))$,

$$\int_{A_n} \mathrm{d}\mu(\omega) \int_{d(t,\omega)}^{d(s,\omega)} \frac{1}{\mu(B(t,2\epsilon))} \mathrm{d}\epsilon \le 2^{-n(p-1)} d(s,t) .$$

Then (16.147) follows by summation over $n \ge 1$ and combining with (16.148). □

Proposition 16.8.12 *If the distance d is p-concave, then for each probability measure μ on T, one has*

$$\int_T \mathrm{d}\mu(t) \int_0^{\Delta(T,d)} \varphi^{-1}\Big(\frac{1}{\mu(B(t,\epsilon))}\Big)\mathrm{d}\epsilon \le K(p) S(T,d,\varphi) , \tag{16.149}$$

where $S(T,d,\varphi)$ is defined in (16.136).

Proof On the probability space (T,μ), consider the process $(X_t)_{t\in T}$ given by

$$X_t(\omega) = c \int_{\min(\Delta(T,d)/2, d(t,\omega))}^{\Delta(T,d)/2} \varphi^{-1}\Big(\frac{1}{\mu(B(\omega,3\epsilon))}\Big)\mathrm{d}\epsilon , \tag{16.150}$$

where the constant $c \le 1$ will be determined later. Next, we claim that

$$\sup_{s,t\in T} |X_s(\omega) - X_t(\omega)| \ge c \int_0^{\Delta(T,d)/2} \varphi^{-1}\Big(\frac{1}{\mu(B(\omega,3\epsilon))}\Big)\mathrm{d}\epsilon .$$

To see this, we choose s such that $d(s,\omega) \ge \Delta(T,d)/2$, so that $X_s(\omega) = 0$, we choose $t = \omega$, and we compute $X_t(\omega)$ by the formula (16.150). Consequently,

$$\mathsf{E} \sup_{s,t\in T} |X_s - X_t| \ge c \int_T \mathrm{d}\mu(\omega) \int_0^{\Delta(T,d)/2} \varphi^{-1}\Big(\frac{1}{\mu(B(\omega,3\epsilon))}\Big)\mathrm{d}\epsilon . \tag{16.151}$$

Next, we set $a_0(\omega) = \min(d(s,\omega), d(t,\omega))$ and $b_0(\omega) = \max(d(s,\omega), d(t,\omega))$ and $a(\omega) = \min(\Delta(T,d)/2, a_0(\omega))$, $b(\omega) = \min(\Delta(T,d)/2, b_0(\omega))$. Then

$$|X_s(\omega) - X_t(\omega)| = c \int_{a(\omega)}^{b(\omega)} \varphi^{-1}\Big(\frac{1}{\mu(B(\omega,3\epsilon))}\Big)\mathrm{d}\epsilon .$$

Since $b(\omega) - a(\omega) \le d(s,t)$, we have $c(b(\omega) - a(\omega))/d(s,t) \le 1$. Using the convexity of φ in the first inequality, and Jensen's inequality in the second inequality,

$$\begin{aligned}
&\varphi\Big(\frac{X_s(\omega) - X_t(\omega)}{d(s,t)}\Big) \\
&= \varphi\Big(\frac{c(b(\omega) - a(\omega))}{d(s,t)} \frac{1}{b(\omega) - a(\omega)} \int_{a(\omega)}^{b(\omega)} \varphi^{-1}\Big(\frac{1}{\mu(B(\omega,3\epsilon))}\Big)\mathrm{d}\epsilon\Big) \\
&\le \frac{c(b(\omega) - a(\omega))}{d(s,t)} \varphi\Big(\frac{1}{b(\omega) - a(\omega)} \int_{a(\omega)}^{b(\omega)} \varphi^{-1}\Big(\frac{1}{\mu(B(\omega,3\epsilon))}\Big)\mathrm{d}\epsilon\Big) \\
&\le \frac{c}{d(s,t)} \int_{a(\omega)}^{b(\omega)} \frac{1}{\mu(B(\omega,3\epsilon))}\mathrm{d}\epsilon .
\end{aligned} \tag{16.152}$$

Now, if $a_0(\omega) \geq \Delta(T,d)/2$, then $b_0(\omega) \geq a_0(\omega) \geq \Delta(T,d)/2$, and then $a(\omega) = b(\omega) = \Delta(T,d)/2$, and the term in the last line of (16.152) is 0. If $a_0(\omega) \leq \Delta(T,d)/2$, then $a(\omega) = \alpha_0(\omega) \leq b(\omega) \leq b_0(\omega)$, and it follows from Lemma 16.8.11 that the term on the last line of (16.152) is $\leq cK(p)$. Consequently, we may choose $c = 1/K(p)$ depending on p only such (16.1) holds. Combining (16.151) with the definition of $S(T,d,\varphi)$, we then obtain

$$\sup_{\omega\in T}\int_0^{\Delta(T,d)/2} \varphi^{-1}\Big(\frac{1}{\mu(B(\omega,3\epsilon))}\Big)\mathrm{d}\epsilon \leq K(p)S(T,d,\varphi) ,$$

and a change of variable then completes the proof. □

Proof of Theorem 16.8.6 Combine Proposition 16.8.12 and Lemma 3.3.3 used for $\Phi(x) = \varphi^{-1}(1/x)$. □

We now turn to the proof of Theorem 16.8.7. We have to use a probability measure as in (16.141) to construct a suitable admissible sequence. There is a genuine difficulty in this construction, namely, that the measure of the balls $B(t,\epsilon)$ can greatly vary for a small variation of t. This difficulty has been bypassed in full generality by an argument of W. Bednorz, which we present now. This argument is so effective that the difficulty might no longer be noticed. Without loss of generality, we assume

$$\varphi(1) = 1 , \tag{16.153}$$

but (16.139) is not required.

The proof of Theorem 16.8.7 is based on the functions $\epsilon_n(t)$ defined for $n \geq 0$ as

$$\epsilon_n(t) = \inf\Big\{\epsilon > 0 \,;\; \mu(B(t,\epsilon)) \geq \frac{1}{\varphi(4^n)}\Big\} . \tag{16.154}$$

This quantity is well defined since $\varphi(4^n) \geq 1$ for $n \geq 0$.

Lemma 16.8.13 *We recall the quantity B of (16.141). We have*

$$\mu(B(t,\epsilon_n(t))) \geq \frac{1}{\varphi(4^n)} , \tag{16.155}$$

$$|\epsilon_n(s) - \epsilon_n(t)| \leq d(s,t) , \tag{16.156}$$

$$\forall t \in T , \sum_{n\geq 0} 4^n \epsilon_n(t) \leq 2B . \tag{16.157}$$

Proof First, (16.155) is obvious, and since $B(t,\epsilon) \subset B(s,\epsilon + d(s,t))$, $\epsilon_n(s) \le \epsilon_n(t) + d(s,t)$ and (16.156) follows. Next, since

$$\epsilon < \epsilon_n(t) \Rightarrow \varphi^{-1}\Big(\frac{1}{\mu(B(t,\epsilon))}\Big) > 4^n ,$$

we have

$$B \ge \sum_{n\ge 0} \int_{\epsilon_{n+1}(t)}^{\epsilon_n(t)} \varphi^{-1}\Big(\frac{1}{\mu(B(t,\epsilon))}\Big) \mathrm{d}\epsilon \ge \sum_{n\ge 0} 4^n(\epsilon_n(t) - \epsilon_{n+1}(t)) .$$

Now,

$$\sum_{n\ge 0} 4^n(\epsilon_n(t) - \epsilon_{n+1}(t)) = \sum_{n\ge 0} 4^n \epsilon_n(t) - \sum_{n\ge 1} 4^{n-1}\epsilon_n(t) \ge \frac{1}{2}\sum_{n\ge 0} 4^n \epsilon_n(t) . \qquad \square$$

Lemma 16.8.14 *For each $n \ge 0$, there exists a subset T_n of T that satisfies the following conditions:*

$$\operatorname{card} T_n \le \varphi(4^n) . \tag{16.158}$$

$$\textit{The balls } B(t,\epsilon_n(t)) \textit{ for } t \in T_n \textit{ are disjoint} . \tag{16.159}$$

$$\forall t \in T \ , \ d(t, T_n) \le 4\epsilon_n(t) . \tag{16.160}$$

$$\forall t \in T_n \ , \ \forall s \in B(t,\epsilon_n(t)) \ , \ \epsilon_n(s) \ge \frac{1}{2}\epsilon_n(t) . \tag{16.161}$$

Proof We define $D_0 = T$, and we choose $t_1 \in D_0$ such that $\epsilon_n(t_1)$ is as small as possible. Assuming that we have constructed $D_{k-1} \ne \emptyset$, we choose $t_k \in D_{k-1}$ such that $\epsilon_n(t_k)$ is as small as possible, and we define

$$D_k = \big\{t \in D_{k-1} \ ; \ d(t,t_k) \ge 2(\epsilon_n(t) + \epsilon_n(t_k))\big\} .$$

The construction continues as long as possible. It stops at the first integer p for which $D_p = \emptyset$. We define $T_n = \{t_1, t_2 \ldots, t_p\}$. Consider $t_k, t_{k'} \in T_n$ with $k < k'$. Then by construction, and since the sequence (D_k) decreases, $t_{k'} \in D_k$, so that

$$d(t_{k'}, t_k) \ge 2(\epsilon_n(t_{k'}) + \epsilon_n(t_k)) ,$$

and therefore the balls $B(t_k, \epsilon_n(t_k))$ and $B(t_{k'}, \epsilon_n(t_{k'}))$ are disjoint. This proves (16.159) and (16.155) imply (16.158). To prove (16.160), consider $t \in T$ and the largest $k \ge 1$ such that $t \in D_{k-1}$. Then by the choice of t_k, we have

$\epsilon_n(t) \geq \epsilon_n(t_k)$. Since by definition of k we have $t \notin D_k$, the definition of D_k shows that

$$d(t, t_k) < 2(\epsilon_n(t) + \epsilon_n(t_k)) \leq 4\epsilon_n(t) ,$$

and since $t_k \in T_n$, this proves (16.160).

Finally, consider t_k and $s \in B(t_k, \epsilon_n(t_k))$. If $s \in D_{k-1}$, then $\epsilon_n(s) \geq \epsilon_n(t_k)$, and (16.161) is proved. Otherwise, the unique k' such that $s \in D_{k'-1}$ and $s \notin D_{k'}$ satisfies $k' < k$. Since $s \in D_{k'-1}$ but $s \notin D_{k'}$, the definition of this set shows that

$$d(s, t_{k'}) \leq 2(\epsilon_n(s) + \epsilon_n(t_{k'})) ,$$

and since $d(s, t_k) \leq \epsilon_n(t_k)$, we get

$$d(t_k, t_{k'}) \leq d(s, t_k) + d(s, t_{k'}) \leq \epsilon_n(t_k) + 2(\epsilon_n(s) + \epsilon_n(t_{k'})) . \tag{16.162}$$

On the other hand, since $k' < k$, then $t_k \in D_{k-1} \subset D_{k'}$ so the definition of this set implies

$$d(t_k, t_{k'}) \geq 2(\epsilon_n(t_k) + \epsilon_n(t_{k'})) ,$$

and comparing with (16.162) completes the proof of (16.161). □

Proof of Theorem 16.8.7 For $n \geq 0$, we consider the set T_n provided by Lemma 16.8.14, so card $T_0 = 1$. Combining (16.157) and (16.160), we obtain

$$\sum_{n \geq 0} 4^n d(t, T_n) \leq 8B ,$$

and this proves that $S_d(\mathcal{T}) \leq 8B$.

Next, since $\mu(B(s, \epsilon_n(s)) \geq 1/\varphi(4^n)$ by (16.155) and since $d(s, T_{n-1}) \leq 4\epsilon_{n-1}(s)$ by (16.160), for $n \geq 1$, we have

$$\sum_{s \in T_n} \frac{d(s, T_{n-1})}{\varphi(4^n)} \leq 4 \sum_{s \in T_n} \int_{B(s, \epsilon_n(s))} \epsilon_{n-1}(s) \mathrm{d}\mu(t) .$$

It follows from (16.161) that for $t \in B(s, \epsilon_n(s))$, one has $\epsilon_n(s) \leq 2\epsilon_n(t)$. Combining with (16.156) implies

$$\epsilon_{n-1}(s) \leq \epsilon_{n-1}(t) + \epsilon_n(s) \leq \epsilon_{n-1}(t) + 2\epsilon_n(t) ,$$

and since the balls $B(s, \epsilon_n(s))$ are disjoint for $s \in T_n$, this yields

$$\sum_{s \in T_n} \frac{4^n d(s, T_{n-1})}{\varphi(4^n)} \le 4^{n+1} \int_T (\epsilon_{n-1}(t) + 2\epsilon_n(t)) \mathrm{d}\mu(t) \ .$$

Summation over $n \ge 1$ and use of (16.157) conclude the proof. □

When we do not assume that the distance is p-concave, the last inequality in (16.144) need not hold. This considerably more complex situation will be briefly discussed in the next section, and we end up the present section by discussing two more specialized questions. A striking feature of Theorem 16.2.1 is that even though we studied processes that satisfied $\mathsf{E}(X_s - X_t)^2 = d(s,t)$ where d is the usual distance on the unit interval, we ended up considering the sequence $\mathcal{I}_n$ of partitions of this unit interval and, implicitly, the distance δ given by $\delta(s,t) = 2^{-n}$ where n is the largest integer for which s, t belong to the same element of $\mathcal{I}_n$. This distance is *ultrametric*, i.e., it satisfies

$$\forall s, t, v \in T \ , \ \delta(s,t) \le \max(\delta(s,v), \delta(t,v)) \ . \tag{16.163}$$

In particular, a distance is ultrametric if and only if it is p-concave for all p. Ultrametric distances are intimately connected to increasing sequences of partitions, because the balls of a given radius form a partition in a ultrametric space. As the following shows, the implicit occurrence of an ultrametric structure is very frequent:

Theorem 16.8.15 (W. Bednorz [12]) *Let us assume that the Young function φ satisfies*

$$\forall k \ge 1 \ , \ \sum_{n > k} \frac{4^n}{\varphi(4^n)} \le C \frac{4^k}{\varphi(4^k)} \ . \tag{16.164}$$

Consider an admissible sequence $\mathcal{T}$ of subsets of (T, d). Then there exist an ultrametric distance $\delta \ge d$ and an admissible sequence $\mathcal{T}^*$ of subsets of (T, δ) such that

$$S_\delta(\mathcal{T}^*) + S^*_\delta(\mathcal{T}^*) \le K(C)(S_d(\mathcal{T}) + S^*_d(\mathcal{T})) \ , \tag{16.165}$$

where $K(C)$ depends on C only.

In words, this means that if the existence of an admissible sequence provides a bound for processes that satisfy the increment condition (16.1), then there is an ultrametric distance δ greater than d and such that the processes satisfying the increment condition (16.1) for this greater distance basically still satisfy the same bound.

Proof Let $\mathcal{T} = (T_n)_{n\ge 0}$. As a first step, we prove that we may assume that the sequence (T_n) increases. Define $T'_0 = T_0$, and for $n \ge 1$, define $T'_n = \cup_{k<n} T_k$. Thus,

$$\operatorname{card} T'_n \le \sum_{k<n} \varphi(4^k) \le \sum_{k<n} 4^{k-n}\varphi(4^n) \le \varphi(4^n)\,,$$

where in the second inequality we have used that $\varphi(x) \le x\varphi(y)/y$ by convexity of φ. Thus, the sequence $\mathcal{T}' = (T'_n)_{n\ge 1}$ is admissible. Since $d(t, T'_n) \le d(t, T_{n-1})$ for $n \ge 1$, it follows from the definition (16.123) that $S_d(\mathcal{T}') \le 4S_d(\mathcal{T})$. Next, we observe that for $n \ge 2$, and since $T'_n = \cup_{k<n} T_k$,

$$\sum_{s\in T'_n} d(s, T'_{n-1}) \le \sum_{k<n}\sum_{s\in T_k} d(s, T'_{n-1}) \le \sum_{k<n}\sum_{s\in T_k} d(s, T_{k-1})\,, \tag{16.166}$$

because $T_{k-1} \subset T'_{n-1}$ for $k < n$. For $n = 1$, $d(s, T'_{n-1}) = 0$ for $s \in T'_1 = T_0$. Thus, using (16.166) in the second line and (16.164) in the last line,

$$\begin{aligned} S^*_d(\mathcal{T}') &= \sum_{n\ge 1}\sum_{s\in T'_n} \frac{4^n d(s, T'_{n-1})}{\varphi(4^n)} \\ &\le \sum_{n\ge 1}\sum_{k<n}\sum_{s\in T_k} \frac{4^n d(s, T_{k-1})}{\varphi(4^n)} \\ &= \sum_{k\ge 1}\sum_{s\in T_k} d(s, T_{k-1}) \sum_{n>k} \frac{4^n}{\varphi(4^n)} \\ &\le C S^*_d(\mathcal{T})\,. \end{aligned}$$

In summary, the sequence $\mathcal{T}'$ is admissible and increasing and satisfies $S_d(\mathcal{T}') \le 4S_d(\mathcal{T})$ and $S^*_d(\mathcal{T}') \le CS^*_d(\mathcal{T})$. Therefore, replacing $\mathcal{T}$ by $\mathcal{T}'$, we now assume that the sequence (T_n) increases.

Let us consider the points $\pi_n(t)$ as in the proof of Theorem 16.8.1. Since the sequence (T_n) increases, we have $\pi_k(t) = t$ for $t \in T_n$ and $k \ge n$. Given $s, t \in T$, let us consider the largest integer m for which $\pi_m(s) = \pi_m(t)$ and define

$$\delta(s, t) = 2\max\Big(\sum_{k\ge m} d(\pi_k(t), \pi_{k+1}(t)), \sum_{k\ge m} d(\pi_k(s), \pi_{k+1}(s))\Big)\,. \tag{16.167}$$

It is straightforward to check that this defines an ultrametric distance. Moreover, we have $d(s,t) \le d(s, \pi_m(s)) + d(t, \pi_m(s)) = d(s, \pi_m(s)) + d(t, \pi_m(t))$, using the definition of m in the last equality. Furthermore, since $t = \pi_n(t)$ for n large enough, the triangle inequality implies $d(t, \pi_m(t)) \le \sum_{k\ge m} d(\pi_k(t), \pi_{k+1}(t))$, and the same inequality for s then yields that $d(s,t) \le \sum_{k\ge m} d(\pi_k(t),\pi_{k+1}(t)) + \sum_{k\ge m} d(\pi_k(s),\pi_{k+1}(s)) \le \delta(s,t)$.

Consider now $t \in T$ and $s = \pi_n(t) \in T_n$. Then $\pi_k(t) = \pi_k(s)$ for $k \le n$, and $\pi_k(s) = s$ for $k \ge n$. Consequently, the definition of δ shows that

$$\delta(t, T_n) \le 2 \sum_{k\ge n} d(\pi_k(t), \pi_{k+1}(t)) . \tag{16.168}$$

Interchanging as usual the sums over k and n yields

$$\sum_{n\ge 0} 4^n \delta(t, T_n) \le \sum_{k\ge 0} d(\pi_k(t), \pi_{k+1}(t)) \sum_{n\le k} 4^n \le 2 \sum_{k\ge 0} 4^k d(\pi_k(t), \pi_{k+1}(t)) .$$

Denoting by $\mathcal{T}^*$ the admissible sequence (T_n), then (16.133) proves that $S_\delta(\mathcal{T}^*) \le L S_d(\mathcal{T})$.

Now, if $t \in T_{n+1}$, we have $\pi_k(t) = t$ for $k \ge n+1$, and thus (16.168) and (16.127) yield $\delta(t, T_n) \le 2d(\pi_{n+1}(t), \pi_n(t)) = 2d(\pi_{n+1}(t), T_n) = 2d(t, T_n)$. This implies that $S^*_\delta(\mathcal{T}^*) \le 2S^*_d(\mathcal{T})$. □

The conclusion of Theorem 16.8.15 is not true without some kind of condition on φ such as (16.164). A counterexample is provided in [104] in the case $\varphi(x) = x$.

Finally, we briefly investigate the extent to which we can improve (16.125) by requiring a stronger integrability condition on $\sup_{s,t} |X_s - X_t|$. For a Young function φ, and a r.v. X, let us define

$$\|X\|_\varphi = \inf\left\{u > 0 \,;\ \mathsf{E}\varphi(X/u) \le 1\right\} , \tag{16.169}$$

so that the distance of (16.2) is simply $\|X_s - X_t\|_\varphi$. It would be nice if we could replace the left-hand side of (16.125) by

$$\Big\| \sup_{s,t\in T} |X_s - X_t| \Big\|_\varphi ,$$

but unfortunately this is not true. However, we have the following (which is a special case of a general principle, see [104]):

Proposition 16.8.16 *Assume that for a Young function ψ, we have*

$$x \ge \varphi^{-1}(1) = 1\ ,\ y \ge 1 \Rightarrow \varphi(xy) \ge \varphi(x)\psi(y) . \tag{16.170}$$

Then we may replace (16.125) by

$$\Big\| \sup_{s,t\in T} |X_s - X_t| \Big\|_\psi \le L(S_d(\mathcal{T}) + S^*_d(\mathcal{T})) . \tag{16.171}$$

In particular, we may improve the metric entropy bound (16.5) into

$$\| \sup_{s,t\in T} |X_s - X_t| \|_{\psi} \le L \int_0^{\Delta(T,d)} \varphi^{-1}(N(T,d,\epsilon))d\epsilon \,. \tag{16.172}$$

Proof Proceeding as in (16.130), we observe that for each number $a > 0$, we have

$$\sup_{t\in T} |X_t - X_{t_0}| \le a \sup_{t\in T} \sum_{n\ge 1} 4^n d(\pi_{n-1}(t), \pi_n(t)) + \sum_{n\ge 1}\sum_{s\in T_n} b(n,s)\varphi\Big(\frac{Y_{n,s}}{a}\Big) \,, \tag{16.173}$$

where we lighten notation by writing

$$b(n,s) = \frac{4^n d(s,\theta_n(s))}{\varphi(4^n)} \;;\; Y_{n,s} = \frac{|X_s - X_{\theta_n(s)}|}{d(s,\theta_n(s))} \,.$$

Let us define

$$h(\omega) = \inf\Big\{ a > 0 \,;\, \sum_{n\ge 1}\sum_{s\in T_n} b(n,s)\varphi\Big(\frac{Y_{n,s}(\omega)}{a}\Big) \le 2S_d^*(\mathcal{T}) \Big\} \,, \tag{16.174}$$

so that obviously $h(\omega) > 0$. Using (16.173) for any $a > h(\omega)$ implies

$$\sup_{t\in T} |X_t(\omega) - X_{t_0}(\omega)| \le h(\omega) \sup_{t\in T} \sum_{n\ge 1} 4^n d(\pi_{n-1}(t), \pi_n(t)) + 2S_d^*(\mathcal{T}) \,,$$

and recalling (16.133), it suffices to prove that $\|h\|_{\psi} \le L$. Let us consider $g(\omega) < h(\omega)$. We deduce from (16.174)

$$\sum_{n\ge 1}\sum_{s\in T_n} b(n,s)\varphi\Big(\frac{Y_{n,s}(\omega)}{g(\omega)}\Big) \ge 2S_d^*(\mathcal{T}) \,. \tag{16.175}$$

Recalling that $\varphi(1) = 1$ (so that $\varphi(x) \le 1$ for $|x| \le 1$) and that the sum of the coefficients $b(n,s)$ for $s \in T_n$ and $n \ge 1$ is $S_d^*(\mathcal{T})$,

$$\sum_{n\ge 1}\sum_{s\in T_n} b(n,s)\varphi\Big(\frac{Y_{n,s}(\omega)}{g(\omega)}\Big)\mathbf{1}_{\{|Y_{n,s}(\omega)|\le g(\omega)\}} \le S_d^*(\mathcal{T}) \,,$$

and combining with (16.175),

$$\sum_{n\ge 1}\sum_{s\in T_n} b(n,s)\varphi\Big(\frac{Y_{n,s}(\omega)}{g(\omega)}\Big)\mathbf{1}_{\{|Y_{n,s}(\omega)|> g(\omega)\}} \ge S_d^*(\mathcal{T}) \,. \tag{16.176}$$

Now, (16.170) implies that $\varphi(y/g(\omega)) \leq \varphi(y)/\psi(g(\omega))$ for $y \geq g(\omega)$ and $g(\omega) \geq 1$. Multiplying both sides of (16.176) by $\mathbf{1}_{\{g(\omega)\geq 1\}}$ and using that $\varphi(Y_{n,s}(\omega)/g(\omega))\mathbf{1}_{\{|Y_{n,s}(\omega)|>g(\omega)\}}\mathbf{1}_{\{g(\omega)\geq 1\}} \leq \varphi(Y_{n,s}(\omega))/\psi(g(\omega))$, we obtain

$$\mathbf{1}_{\{g(\omega)\geq 1\}}\psi(g(\omega))S_d^*(\mathcal{T}) \leq \sum_{n\geq 1}\sum_{s\in T_n} b(n,s)\varphi(Y_{n,s}(\omega)) .$$

Since the expected value of the right-hand side is $\leq S_d^*(\mathcal{T})$, taking expectation implies $\mathsf{E}\mathbf{1}_{\{g(\omega)\geq 1\}}\psi(g(\omega)) \leq 1$. Taking $x = y = 1$ in (16.170) proves that $\psi(1) \leq 1$, so that $\mathsf{E}\mathbf{1}_{\{g(\omega)\leq 1\}}\psi(g(\omega)) \leq 1$ and then $\mathsf{E}\psi(g) \leq 2$. Taking $g(\omega) = \alpha h(\omega)$ with $0 < \alpha < 1$ and letting $\alpha \to 1$ shows that $\mathsf{E}\psi(h) \leq 2$ so that $\mathsf{E}\psi(h/2) \leq 1$ and $\|h\|_\psi \leq 2$. □

Condition (16.170) is essentially optimal, as the following challenging exercise shows:

Exercise 16.8.17 Investigate the necessary conditions on the function ψ so that for any metric space and any process $(X_t)_{t\in T}$ that satisfies (16.1), one has

$$\Big\| \sup_{s,t\in T} |X_s - X_t| \Big\|_\psi \leq L \int_0^{\Delta(T,d)} \varphi^{-1}(N(T,d,\epsilon))\mathrm{d}\epsilon . \tag{16.177}$$

Hint: Consider N and the space T of cardinality N where any two distinct points are at distance 1. Consider $\epsilon < 1$, and consider disjoint events $(\Omega_t)_{t\in T}$ with $\mathsf{P}(\Omega_t) = \epsilon/N$. Apply (16.177) to the process $(X_t)_{t\in T}$ given by $X_t = \varphi^{-1}(N/2\epsilon)\mathbf{1}_{\Omega_t}$.

16.9 Chaining, III

We now briefly discuss the problem of the boundedness of processes that satisfy (16.1) in a general metric space, when the distance is not assumed to be p-concave (and in particular when d is the usual Euclidean distance on $\mathbb{R}^n$). In this case, there is a new phenomenon which takes place. In all the examples of chaining we have met up to now, the interpolation points $\pi_n(t)$ converge geometrically toward t, but this feature is not always optimal. To understand this, consider a toy example, the unit interval with the usual distance.

Proposition 16.9.1 *Consider a process $(X_t)_{t\in[0,1]}$ that satisfies*

$$\forall s,t \in [0,1] ,\ \mathsf{E}|X_s - X_t| \leq |s-t| . \tag{16.178}$$

Then

$$\mathsf{E} \sup_{0\leq s,t\leq 1} |X_s - X_t| \leq 1 . \tag{16.179}$$

Proof We have to show that if $F \subset [0,1]$ is finite, then $\mathsf{E}\sup_{s,t\in F}|X_s - X_t| \le 1$. Let $F = \{t_1, \ldots, t_n\}$ with $0 \le t_1 < \ldots < t_n \le 1$. Then

$$\mathsf{E}\sup_{\ell<\ell'}|X_{t_\ell} - X_{t_{\ell'}}| \le \mathsf{E}\sum_{1\le\ell<n}|X_{t_{\ell+1}} - X_{t_\ell}| \le \sum_{1\le\ell\le n} t_{\ell+1} - t_\ell \le 1 . \qquad \square$$

The following exercise shows that Proposition (16.9.1) cannot be deduced from Theorem 16.8.7:

Exercise 16.9.2 Prove that if μ is a probability measure on $[0,1]$, then

$$\int_0^1 \mathrm{d}t \int_0^1 \frac{1}{\mu(B(t,\epsilon))}\mathrm{d}\epsilon = \infty .$$

The next exercise shows that the result of Proposition 16.9.1 cannot be explained by the size of the covering numbers of $[0,1]$.

Exercise 16.9.3 Denote by $t = (t_i)_{i\ge1}$ the generic point of $T = \{0,1\}^{\mathbb{N}}$. On T, consider the ultrametric distance δ given by $\delta(s,t) = 2^{-i+1}$, where i is the smallest integer for which $s_i \ne t_i$. Construct an unbounded process (X_t) on (T,δ) that satisfies $\mathsf{E}|X_s - X_t| \le \delta(s,t)$ for each $s,t \in T$. Compare the covering numbers of (T,δ) and $([0,1],d)$ where d is the usual distance.

Exercise 16.9.4 Review the proof of Theorem 16.8.1 to show that when $\varphi(x) = |x|$, then one can improve (16.125) into

$$\mathsf{E}\sup_{s,t\in T}|X_s - X_t| \le LS_d^*(\mathcal{T}) .$$

We consider now the case of the processes on $T = [0,1]^p$, provided with the usual distance d. Which are the Young functions φ such that all processes satisfying (16.1), i.e.,

$$\forall s,t \in T \; ; \; \mathsf{E}\varphi(X_s - X_t) \le d(s,t)$$

are bounded? The covering numbers $N(T,d,\epsilon)$ behave like ϵ^{-p}, so that (16.5) implies that it suffices that

$$\sum_n 2^{-n}\varphi^{-1}(2^{np}) < \infty . \tag{16.180}$$

Theorem 16.9.5 ([104, Theorem 5.1]) *Processes which satisfy (16.1) are all bounded if and only if φ satisfies the condition*

$$\sum_n 2^{-n}(\varphi')^{-1}(2^{n(1-p)}) < \infty , \tag{16.181}$$

where φ' is the derivative of φ.

The difference between (16.180) and (16.181) can be seen clearly in the case $p = 1$ where (16.181) is automatically satisfied but (16.180) is not. What happens here is that as in Proposition 16.9.1, one can join two elements of T by a long chain of small steps (and this is not the case in the setting of Exercise 16.9.3). Theorem 16.9.5 is obviously more of theoretical than practical interest so we do not reproduce the specialized proof.

This is however not the end of the story. The reason why the weak condition (16.181) suffices for boundedness is a kind of "connectivity" in the structure of $[0, 1]^p$. This connectivity structure does not exist when the distance is ultrametric, as Exercise 16.9.3 shows. There are also "intermediate situations" where both aspects are present, e.g., if one takes a product of $[0, 1]^p$ with a ultrametric space. Complicated necessary and sufficient conditions are found in [104] in such a case. This probably indicates that no simple complete description of the metric spaces for which condition (16.1) implies boundedness can be found, even in the "homogeneous situation" where covering numbers suffice.

16.10 Notes and Comments

Il y a les questions qui se posent, et les questions que l'on se pose.[8] *Henri Poincaré*

Obviously, Poincaré had better mathematical taste than many subsequent mathematicians. My own view is that many of the problems considered in the chapter belong to the second category rather than the first, and I have included their solution in this book only because I find it excessively beautiful, in particular thanks to the work of W. Bednorz, who discovered a number of very clean and seemingly final arguments. A particularly important contribution of W. Bednorz is to have brought to light the technical importance of (16.12), after which everything becomes much easier.

I undertook a systematic study of boundedness of stochastic processes under increment conditions using majorizing measure in [104]. This paper contains near-optimal results, but several arguments have been greatly simplified by W. Bednorz. I undertook this project despite the fact that I felt the topic to be of marginal importance, because I thought that I had no chance of making progress in the Gaussian case without having first mastered the elusive notion of majorizing measure. This strategy was successful.

When the Young function φ as in Sect. 16.8 has "polynomial growth" rather than "exponential growth", it does not seem possible to characterize the size of T according to majorizing measures in terms of the size of the trees it contains, as we did in Sect. 3.1.

[8] I won't dare attempting a literal translation, but roughly this distinguishes between questions of self-evident importance and more arbitrary questions one may ask.

Chapter 17
Shor's Matching Theorem

17.1 Introduction

This chapter continues Chap. 4, which should be fresh in the reader's mind before attempting to penetrate the more difficult material presented here. In particular, the notion of "evenly spread" points is explained on page 121. The main result is as follows:

Theorem 17.1.1 (P. Shor) *Consider evenly spread points* $(Y_i)_{i\le N}$ *of* $[0,1]^2$. *Set* $Y_i = (Y_i^1, Y_i^2)$. *Consider i.i.d. points* $(X_i)_{i\le N}$ *uniform over* $[0,1]^2$, *and set* $X_i = (X_i^1, X_i^2)$. *Then with probability* $\ge 1 - LN^{-10}$, *there exists a matching* π *such that*

$$\sum_{i\le N} |X_i^1 - Y_{\pi(i)}^1| \le L\sqrt{N\log N} \tag{17.1}$$

$$\sup_{i\le N} |X_i^2 - Y_{\pi(i)}^2| \le L\sqrt{\frac{\log N}{N}}\,. \tag{17.2}$$

The power N^{-10} plays no special role. Theorem 17.1.1 improves upon Theorem 4.5.1 because when $|X_i^2 - Y_{\pi(i)}^2| \le L\sqrt{\log N/N}$ for each i, then $\sum_{i\le N}|X_i^2 - Y_{\pi(i)}^2| \le L\sqrt{N\log N}$.

A remarkable feature of Theorem 17.1.1 is that both coordinates do not play the same role. Following this idea, one may ask the following:

M. Talagrand, *Upper and Lower Bounds for Stochastic Processes*, Ergebnisse der Mathematik und ihrer Grenzgebiete. 3. Folge / A Series of Modern Surveys in Mathematics 60, https://doi.org/10.1007/978-3-030-82595-9_17

Research Problem 17.1.2 (The Ultimate Matching Conjecture) Prove or disprove the following. Consider $\alpha_1, \alpha_2 > 0$ with $1/\alpha_1 + 1/\alpha_2 = 1/2$. Then with high probability, we can find a matching π such that, for $j = 1, 2$, we have

$$\sum_{i \le N} \exp\left(\sqrt{\frac{N}{\log N}} \frac{|X_i^j - Y_{\pi(i)}^j|}{L}\right)^{\alpha_j} \le 2N \, . \tag{17.3}$$

In Chap. 18, we shall prove a suitable version of the ultimate matching conjecture in dimension $d \ge 3$. Noting that

$$\sum_{i \le N} \exp a_i^4 \le 2N \Rightarrow \max_{i \le N} |a_i| \le L(\log N)^{1/4}$$

and that $\exp a^4 \ge |a|$ shows that the case $\alpha_1 = \alpha_2 = 4$ would provide a very neat common generalization of Theorems 4.5.1 and 4.7.1. Another case of special interest is the case "$\alpha_1 = 2, \alpha_2 = \infty$" for which one may interpret (17.3) for $j = 2$ as meaning (17.2). Then (17.3) for $j = 1$ is very much stronger than (17.1). A partial result in the direction of this special case of Problem 17.1.2 is as follows:

Theorem 17.1.3 *Consider a number $0 < \alpha < 1/2$, an integer $N \ge 2$, and evenly spread points $(Y_i)_{i \le N}$ of $[0, 1]^2$. Set $Y_i = (Y_i^1, Y_i^2)$. Consider i.i.d. points $(X_i)_{i \le N}$ uniform over $[0, 1]^2$, and set $X_i = (X_i^1, X_i^2)$. Then with probability $\ge 1 - LN^{-10}$, there exists a matching π such that*

$$\sum_{i \le N} \exp\left(\sqrt{\frac{N}{\log N}} \frac{|X_i^1 - Y_{\pi(i)}^1|}{K(\alpha)}\right)^{\alpha} \le 2N \tag{17.4}$$

$$\sup_{i \le N} |X_i^2 - Y_{\pi(i)}^2| \le K(\alpha)\sqrt{\frac{\log N}{N}} \, . \tag{17.5}$$

Since $\exp |x|^\alpha \ge |x|/K(\alpha)$, (17.1) follows from (17.4), and thus Theorem 17.1.3 improves upon Theorem 17.1.1. One may show that when α increases, the conclusion of Theorem 17.1.3 becomes stronger. We do not know how to prove Theorem 17.1.3 for $\alpha \ge 1/2$.

Conjecture 17.1.4 Theorem 17.1.3 holds for $\alpha = 2$.

This is the special case "$\alpha_1 = 2, \alpha_2 = \infty$" of the ultimate matching conjecture and a nice research problem by itself.

A central difficulty in the proof of a matching theorem is how to relate it to a suitable discrepancy theorem (here Theorem 17.2.1), and the most instructive part of the present section is how we pass this difficulty. There is however a lethal weakness in our approach to the discrepancy theorem. It is explained in Sect. 17.3. Until one finds a way to correct this weakness, no amount of technical effort is going to succeed in reaching the value $\alpha = 2$. For this reason, we only outline the proof of Theorem 17.1.1, and we refer the reader to [129] for the far more technical proof of Theorem 17.1.3.

17.2 The Discrepancy Theorem

The proof of Theorem 17.1.1 relies again on Proposition 4.3.2 and a "discrepancy theorem" of the same nature as (4.38), but for a more complicated class of functions. This is Theorem 17.2.1. It requires some preparations to state this discrepancy theorem.

To prove Theorem 17.1.1, we may assume N large, and we do not care about what happens at a scale less than $\sqrt{\log N}/\sqrt{N}$. We will then replace $[0, 1]^2$ by a discrete approximation at that scale. More precisely, let us consider a universal constant L^* which we shall choose later. Consider the largest integer p with $2^{-p} \geq L^*\sqrt{\log N}/\sqrt{N}$, so that when N is large enough, $p \leq \log N$. This idea is to replace $[0, 1]^2$ by the set of points $(k2^{-p}, \ell 2^{-p})$ for $1 \leq k, \ell \leq 2^p$. It is however pointless to carry the factor 2^{-p} through our calculations, so that we re-scale this set: we consider the set $G = \{1, \ldots, 2^p\}^2$.[1] A generic point of G is denoted by τ. Since $G = \{1, \ldots, 2^p\}^2$, we may also denote a point of G by its coordinates (k, ℓ) which are two integers between 1 and 2^p. To each point, $\tau = (k, \ell)$ of G corresponds a little square $H_\tau =](k-1)2^{-p}, k2^{-p}] \times](\ell-1)2^{-p}, \ell 2^{-p}]$ with sides of length 2^{-p}, and these little squares form a partition of $]0, 1]^2$. We define "evenly spread" points $(Z_i)_{i \leq N}$ of G as follows: we set $Z_i = \tau$ if Y_i belongs to H_τ. Thus, denoting by Z_i^1 and Z_i^2 the components of Z_i, we have

$$|2^{-p}Z_i^1 - Y_i^1| \leq 2^{-p} \; ; \; |2^{-p}Z_i^2 - Y_i^2| \leq 2^{-p} \,. \tag{17.6}$$

We define

$$n(\tau) = \operatorname{card}\{i \leq N \; ; \; Z_i = \tau\} \,, \tag{17.7}$$

so that[2]

$$\sum_{\tau \in G} n(\tau) = N \,. \tag{17.8}$$

Since 2^{-p} is about $L^*\sqrt{\log N}/\sqrt{N}$, $N2^{-2p}$ is about $L^{*2} \log N$ and hence large. Each square H_τ contains a large number of points Y_i, and due to the fact that these

---

[1] The notation G *does not* have the same meaning as in Chap. 4. Now the "grid" G is not a subset of $[0, 1]^2$!

[2] We assume of course that the points Y_i belong to $]0, 1]^2$.

points are evenly spread, it should be obvious that this number of points is about the same for each square H_τ. That is, given N large, for a certain integer m_0, we have

$$\forall \tau \in G \; ; \; m_0 \le n(\tau) \le 2m_0 \; . \tag{17.9}$$

Summation of these inequalities over τ together with (17.8) implies

$$N2^{-2p-1} \le m_0 \le N2^{-2p} \; . \tag{17.10}$$

Since $2^{-2p}N$ is about $L^{*2} \log N$ while $p \le \log N$, when L^* is large, the ratio m_0/p is large. We will prove that our arguments work when this ratio is large enough, which can be achieved by taking L^* large enough.

In our discrete approximation, the points Z_i replace the points Y_i. Now we have to construct the random points U_i valued in G which replace the points X_i. The obvious procedure (to define $U_i = \tau$ when $X_i \in H_\tau$) is not the correct one,[3] which we describe now. By definition of an evenly spread family, each of the points Y_i belongs to a little rectangle of area $1/N$. We denote by K_τ the union of these little rectangles for which $Y_i \in H_\tau$ so that K_τ is of area $n(\tau)/N$. We consider the G-valued r.v.s U_i such that $U_i = \tau$ when $X_i \in K_\tau$. Thus, the r.v.s U_i are i.i.d. with law μ, where the probability measure μ on G is given by

$$\mu(\{\tau\}) = \frac{n(\tau)}{N} \; . \tag{17.11}$$

Also, when $U_i = \tau = (k, \ell)$, we have by definition that X_i belongs to the small rectangle associated with a point $Y_j \in H_\tau$. Then

$$|X_i^1 - Y_j^1| \le \frac{L}{\sqrt{N}} \; ; \; |X_i^2 - Y_j^2| \le \frac{L}{\sqrt{N}}$$

and

$$|2^{-p}U_i^1 - Y_j^1| \le 2^{-p} \; ; \; |2^{-p}U_i^2 - Y_j^2| \le 2^{-p} \; ,$$

and combining these, we have

$$|2^{-p}U_i^1 - X_i^1| \le L2^{-p} \; ; \; |2^{-p}U_i^2 - X_i^2| \le L2^{-p} \; . \tag{17.12}$$

For a function h on G, we have

$$\int h \mathrm{d}\mu = \frac{1}{N} \sum_{\tau \in G} n(\tau)h(\tau) \; . \tag{17.13}$$

[3] We do not want that $\mathsf{P}(U_i = \tau) = 2^{-2p}$ but rather $\mathsf{P}(U_i = \tau) = n(\tau)/N$.

We consider the class $\mathcal{H}$ of functions $h : G \to \mathbb{R}$ such that

$$\sum_{1 \le k \le 2^p \, , \, 1 \le \ell \le 2^p - 1} |h(k, \ell + 1) - h(k, \ell)| \le 2^{2p} \tag{17.14}$$

$$\forall k, \ell \, , \; |h(k + 1, \ell) - h(k, \ell)| \le 1 \, . \tag{17.15}$$

To lighten notation, we will write (17.14) as $\sum |h(k, \ell + 1) - h(k, \ell)| \le 2^{2p}$, and we will not mention any more that it is always understood that when a quantity such as $h(k, \ell + 1) - h(k, \ell)$ occurs in a summation, we consider only the values of ℓ with $\ell + 1 \le 2^p$. In a similar manner, when the quantity $|h(k + 1, \ell) - h(k, \ell)|$ occurs in a condition, it is always understood that we consider only the values of k for which $k + 1 \le 2^p$.

It will become clear only gradually that the class $\mathcal{H}$ of functions is related to a matching problem. Let us say however that the weak restriction (17.14) is related to the fact that we ask the strong condition (17.2) on the second coordinates, whereas the strong restriction (17.15) is related to the fact that we ask only the weak condition (17.1) on the first coordinates.

The central ingredient of our approach is the following:

Theorem 17.2.1 *Consider independent r.v.s U_i valued in G, of law μ. Then, with probability $\ge 1 - \exp(-46p)$, we have*

$$\forall h \in \mathcal{H} \, , \; \Big| \sum_{i \le N} (h(U_i) - \int h \mathrm{d}\mu) \Big| \le L\sqrt{pm_0}\, 2^{2p} \, . \tag{17.16}$$

We shall explain soon how to turn this type of result into a matching theorem. The larger the class $\mathcal{H}$ in (17.16), the better the matching theorem one gets. It is therefore a natural question to wonder for which classes of functions a result such as Theorem 17.2.1 might be true.

Research Problem 17.2.2 Consider two functions $\theta_1(x) \ge x, \theta_2(x) \ge x$. Consider the class $\mathcal{H}$ of functions $h : G \to \mathbb{R}$ such that

$$\sum \theta_1(|h(k+1, \ell) - h(k, \ell)|) + \sum \theta_2(|h(k, \ell+1) - h(k, \ell)|) \le 2^{2p} \, . \tag{17.17}$$

What are the conditions on θ_1 and θ_2 so that

$$\mathsf{E} \sup_{h \in \mathcal{H}} \Big| \sum_{i \le N} (h(U_i) - \int h \mathrm{d}\mu) \Big| \le K\sqrt{pm_0}\, 2^{2p} \tag{17.18}$$

for a constant K independent of p?

Of particular interest is the case $\theta_1(x) = x(\log(3+x))^{1/2}$ and $\theta_2(x) = x$. A positive answer (and significant extra work) would allow one to prove Conjecture 17.1.4.

We shall outline the proof of Theorem 17.2.1 in the next section, but we first state the matching theorem it implies. In the following statement, we denote by U_i^1 and U_i^2 the components of U_i and similarly for Z_i:

Theorem 17.2.3 *There exists a number L_0 with the following property. Recalling the number m_0 of (17.9), assume that*

$$p \le \frac{m_0}{L_0} \,. \tag{17.19}$$

Consider points $(Z_i)_{i\le N}$ in G, and assume that for each $\tau \in G$, we have $\mathrm{card}\{i \le N; Z_i = \tau\} = n(\tau) = N\mu(\{\tau\})$. *Consider points $(U_i)_{i\le N}$ in G, and assume that (17.16) holds. Then we can find a permutation π of $\{1,\ldots,N\}$ for which*

$$\sum_{i\le N} |U_i^1 - Z_{\pi(i)}^1| \le N \,, \tag{17.20}$$

$$\forall i \le N \,,\ |U_i^2 - Z_{\pi(i)}^2| \le 1 \,. \tag{17.21}$$

It is unimportant to have N rather than LN in (17.20).

Proof of Theorem 17.1.1. We consider the points Z_i of G given by $Z_i = \tau$ when $Y_i \in H_\tau$ and the random points U_i in G given by $U_i = \tau$ when $X_i \in K_\tau$ (the union of the little rectangles associated with the points $Y_i \in H_\tau$). As we have already observed just below (17.10), when L^* becomes large, the ratio m_0/p becomes large, so that if L^* is large enough, then (17.19) holds. Also since p is about $L^* \log N$, for L^* large, we have $\exp(-46p) \le N^{-10}$ so that according to Theorem 17.2.1, (17.16) holds with probability $\ge 1 - N^{10}$. When this is the case, and since $2^{-p} \le L\sqrt{\log N}/\sqrt{N}$, it follows from (17.6) and (17.12) that the permutation π which satisfies (17.20) and (17.21) also satisfies (17.1) and (17.2). □

Beginning of the Proof of Theorem 17.2.3. The first steps of the proof look canonical. Let us define

$$M_1 = \sup \sum_{i\le N} (w_i + w_i') \,, \tag{17.22}$$

where the supremum is taken over all families (w_i), (w_i') for which

$$\forall i, j \le N \,,\ |U_i^2 - Z_j^2| \le 1 \Rightarrow w_i + w_j' \le |U_i^1 - Z_j^1| \,. \tag{17.23}$$

We first claim that there is a permutation π which satisfies (17.21) and for which $\sum_{i\le N} |U_i^1 - Z_{\pi(i)}^1| \le M_1$. To see this, let us consider a number $c > 0$, and let us define $c_{i,j} = |U_i^1 - Z_j^1|$ when $|U_i^2 - Z_j^2| \le 1$ and $c_{i,j} = c$ otherwise. Consider then numbers $(w_i)_{i\le N}$ and $(w_i')_{i\le N}$ such that $w_i + w_j' \le c_{i,j}$ for each $i, j \le N$. Then (17.23) holds so that by definition of M_1, we have $\sum_{i\le N} w_i + w_i' \le$

M_1. It follows from Proposition 4.3.2 that there is a permutation π such that $\sum_{i\le N} c_{i,\pi(i)} \le M_1$. If one takes $c = 2M_1$, this shows in particular that no term in the sum equals c, so that for each term, we have $|U_i^2 - Z_{\pi(i)}^2| \le 1$ and $c_{i,\pi(i)} = |U_i^1 - Z_{\pi(i)}^1|$. This proves the claim.

The idea now is, considering a family (w_i'), to define the function h' on G given by

$$h'(k,\ell) = \min_j \bigl\{|k - Z_j^1| - w_j' \; ; \; |\ell - Z_j^2| \le 1\bigr\} .$$

Rewriting (17.23) as

$$\forall i, j \le N \, , \, |U_i^2 - Z_j^2| \le 1 \Rightarrow w_i \le |U_i^1 - Z_j^1| - w_j' \, ,$$

we have $h'(U_i) \ge w_i$, and thus (17.22) implies

$$M_1 \le \sum_{i\le N} (h'(U_i) + w_i') \, . \tag{17.24}$$

This construction is a bit clumsy, because given $\tau \in G$, there are quite a few values of i for which $Z_i = \tau$. A moment thinking shows that one can only increase the left-hand side of (17.24) if one replaces the corresponding values of w_i' by their average (over all the values of i for which $Z_i = \tau$). To pursue this idea, given numbers $(u(\tau))_{\tau\in G}$, we define the function h on G given by

$$h(k,\ell) = \min \bigl\{|k - r| + u(r,s) \; ; \; (r,s) \in G, |\ell - s| \le 1\bigr\} . \tag{17.25}$$

For $\tau = (k,\ell) \in G$, let us then choose

$$u(\tau) = -\frac{1}{n(\tau)} \sum \{w_i' \; ; \; Z_i = \tau\} \, , \tag{17.26}$$

so that

$$\sum_{i\le N, Z_i=\tau} w_i' = -\sum_{\tau\in G} n(\tau) u(\tau) \, . \tag{17.27}$$

The infimum of the numbers $-w_i'$ for $Z_i = \tau$ is less than their average so that given τ and j with $Z_j = \tau$, we can find j' with $Z_{j'} = Z_j = \tau$ and $-w_{j'}' \le u(\tau)$. Comparing the definitions of h and h' proves that $h' \le h$. Consequently, (17.24) and (17.27) imply

$$M_1 \le \sum_{i\le N} h(U_i) - \sum_{\tau\in G} n(\tau) u(\tau) \, . \tag{17.28}$$

Using (17.13) and (17.28), we get

$$M_1 \le \Big|\sum_{i\le N}(h(U_i) - \int h\mathrm{d}\mu)\Big| - \sum_{\tau\in G} n(\tau)(u(\tau) - h(\tau)) . \tag{17.29}$$

The hope now is that

$$\sum_{\tau\in G} n(\tau)(u(\tau) - h(\tau)) \text{ small} \Rightarrow \text{the function } h \text{ behaves well,} \tag{17.30}$$

so that we may have a chance that with high probability $\sup_h |\sum_{i\le N}(h(U_i) - \int h\mathrm{d}\mu)|$ (where the supremum is taken over all functions h arising in this manner) is small, and consequently the right-hand side of (17.29) has a chance to bounded. The difficulty (which is generic when deducing matching theorems from Proposition 4.3.2) is to find a usable way to express that "h behaves well". In the present case, this difficulty is solved by the following result:

Proposition 17.2.4 *Consider numbers $u(k,\ell)$ for $(k,\ell) \in G = \{1,\dots,2^p\}^2$, and consider the function h of (17.25), i.e.,*

$$h(k,\ell) = \inf\big\{u(r,s) + |k-r|\ ;\ (r,s)\in G\,,\ |\ell - s| \le 1\big\} . \tag{17.31}$$

Then

$$\forall k,\ell\,,\ \ |h(k+1,\ell) - h(k,\ell)| \le 1 \tag{17.32}$$

and, assuming (17.19),

$$m_0 \sum_{k,\ell} |h(k,\ell+1) - h(k,\ell)| \le L \sum_{\tau\in G} n(\tau)(u(\tau) - h(\tau)) . \tag{17.33}$$

So, when the left-hand side of (17.30) is small, h behaves well in the sense that (17.32) holds and that $m_0\sum_{k,\ell}|h(k,\ell+1) - h(k,\ell)|$ is also small. This is what motivated the introduction of the class $\mathcal{H}$ and of (17.14). The proof of Proposition 17.2.4 is elementary and is rather unrelated with the main ideas of this work. It is given in Sect. B.6.

End of the Proof of Theorem 17.2.3. We have to show that provided the constant L_0 of (17.19) is large enough, then the right-hand side of (17.29) is $\le N$. We define

$$B = 2^{-2p}\sum |h(k,\ell+1) - h(k,\ell)| \tag{17.34}$$

and $B' = B + 1$ so that $B' \ge 1$ and $h/B' \in \mathcal{H}$. Then (17.16) implies

$$\Big|\sum_{i\le N}(h(U_i) - \int h\mathrm{d}\mu)\Big| \le L\sqrt{pm_0}\,2^{2p}B' , \tag{17.35}$$

whereas (17.33) and (17.34) imply

$$\sum n(\tau)(u(\tau) - h(\tau)) \geq \frac{m_0}{L} \sum |h(k, \ell+1) - h(k, \ell)| = \frac{m_0 2^{2p}}{L} B .$$

Combining with (17.29) and (17.35), we get, since $B' = B + 1$,

$$\begin{aligned} M_1 &\leq L\sqrt{pm_0}\, 2^{2p} B' - \frac{m_0}{L} 2^{2p} B \\ &\leq B 2^{2p}\Big(L\sqrt{pm_0} - \frac{m_0}{L}\Big) + L\sqrt{pm_0} 2^{2p} \end{aligned} \tag{17.36}$$

using that $B' = B + 1$ in the last inequality. Consequently, if the constant L_0 in (17.19) is large enough, the first term is negative, so that (17.36) implies as desired that $M_1 \leq L\sqrt{pm_0} 2^{2p} \leq m_0 2^{2p} \leq N$ using (17.19) for an appropriate choice of L_0 and (17.10). □

17.3 Lethal Weakness of the Approach

We turn to the proof of Theorem 17.2.1. The main difficulty is to control $\gamma_2(\mathcal{H}, d_2)$, where d_2 is the distance in $\ell^2(G)$ or, equivalently, the Euclidean distance on $\mathbb{R}^G$. Ideally, this should be done by using a suitable functional and Theorem 2.9.1. However, the functional can be discovered only by understanding the underlying geometry, which the author does not.

How should one use condition (17.14)? To bypass this difficulty, we will replace this condition by the more familiar Lipschitz-type condition

$$|h(k, \ell+1) - h(k, \ell)| \leq 2^j . \tag{17.37}$$

More precisely, we consider the class $\mathcal{H}_1$ consisting of the functions $h : G \to \mathbb{R}$ such that

$$\forall k, \ell\,,\ |h(k+1, \ell) - h(k, \ell)| \leq 1\ ;\ |h(k, \ell+1) - h(k, \ell)| \leq 1\ . \tag{17.38}$$

Given an integer $j \geq 2$, for a number $V > 0$, we consider the class $\mathcal{H}_j(V)$ of functions $h : G \to \mathbb{R}$ such that

$$\forall k, \ell\,,\ |h(k+1, \ell) - h(k, \ell)| \leq 1\,,\ |h(k, \ell+1) - h(k, \ell)| \leq 2^j \tag{17.39}$$

$$\operatorname{card}\{(k, \ell) \in G\ ;\ h(k, \ell) \neq 0\} \leq V\ . \tag{17.40}$$

The key to our approach is the following, which is proved in Sect. B.5:

Proposition 17.3.1 *If $h \in \mathcal{H}$, we can decompose*

$$h = \sum_{j \ge 1} h_j \text{ where } h_1 \in L\mathcal{H}_1 \text{ and } h_j \in L\mathcal{H}_j(2^{2p-j}) \text{ for } j \ge 2 \,. \tag{17.41}$$

Before we analyze the classes $\mathcal{H}_j := \mathcal{H}_j(2^{2p-j})$, let us reveal the dirty secret. To prove (17.16), we will deduce from (17.41) the inequality

$$\sup_{h \in \mathcal{H}} \Big| \sum_{i \le N} h(U_i) - \int h d\mu \Big| \le L \sum_{j \ge 1} \sup_{h \in \mathcal{H}_j} \Big| \sum_{i \le N} h(U_i) - \int h d\mu \Big| \,. \tag{17.42}$$

That is (Heaven forbid!), we bound the supremum of a sum by the sum of the suprema.[4] This entails a loss which seems to prevent reaching the correct value $\alpha = 2$ in Theorem 17.1.3.

Let us now try to understand the size of the classes $\mathcal{H}_j$. In (17.39), the first and second coordinates play a different role. The continuous equivalent of this is the class of functions f on the unit square which satisfy $|f(x, y) - f(x', y)| \le |x - x'|$ and $|f(x, y') - f(x, y)| \le 2^j |y' - y|$. The function $T(f)$ given by $T(f)(x, y) = f(U(x, y))$ where $U(x, y) = (2^{j/2}x, 2^{-j/2}y)$ is basically $2^{j/2}$-Lipschitz, whereas U preserves Lebesgue's measure. Thus, one should think (17.39) means "the function is $2^{j/2}$-Lipschitz". On the other hand, it turns out (even though this is not obvious at this stage) that condition (17.40) creates a factor 2^{-j}. Thus, the "size of the class $\mathcal{H}_j$ should be $2^{-j/2}$ the size of the class $\mathcal{H}_1$". We thus expect that the right-hand side of (17.42) will converge as a geometric series, therefore requiring the level zero of sophistication.

The central step in the proof of Theorem 17.2.1 is as follows:

Proposition 17.3.2 *Consider $1 \le k_1 \le k_2 \le 2^p$, $1 \le \ell_1 \le \ell_2 \le 2^p$ and $R = \{k_1, \dots, k_2\} \times \{\ell_1, \dots, \ell_2\}$. Assume that*

$$\ell_2 - \ell_1 + 1 = 2^{-j}(k_2 - k_1 + 1) \,. \tag{17.43}$$

Consider independent r.v.s U_i valued in G, of law μ. Then, with probability at least $1 - L\exp(-50p)$, the following occurs. Consider any function $h : G \to \mathbb{R}$, and assume that

$$h(k, \ell) = 0 \text{ unless } (k, \ell) \in R \,. \tag{17.44}$$

$$(k, \ell)\,,\ (k+1, \ell) \in R \Rightarrow |h(k+1, \ell) - h(k, \ell)| \le 1 \tag{17.45}$$

$$(k, \ell)\,,\ (k, \ell+1) \in R \Rightarrow |h(k, \ell+1) - h(k, \ell)| \le 2^j \tag{17.46}$$

$$\forall (k, \ell) \in R\,,\ |h(k, \ell)| \le 2(k_2 - k_1) \,. \tag{17.47}$$

---

[4] Just as what happens in the traditional chaining based on entropy numbers.

Then

$$|\sum_{i\le N}(h(U_i) - \int h\mathrm{d}\mu)| \le L2^{j/2}\sqrt{pm_0}\ \mathrm{card}\,R\ , \tag{17.48}$$

where m_0 *is as in (17.9) and (17.10).*

In words, R is a rectangle of the appropriate shape (17.43), reflecting the fact that the two coordinates in (17.39) do not play the same role. The function h is zero outside R, and its restriction to R satisfies the conditions (17.45) and (17.46). This however does not suffice to obtain a control as in (17.48), as is shown by the case where h is constant on R, so that a mild control on the size of h is required as in (17.47).

Besides the fact that we do not assume that h is zero on the boundary of R (and require a stronger control than just in expectation), Proposition 17.3.2 is really a discrete version of Proposition 4.5.19. Therefore, as expected, no really new idea is required, only technical work, as, for example, replacing the use of the Fourier transform by a discrete version of it (a technique which is detailed in Sect. 4.5.2). For this reason, we have decided not to include the proof of Proposition 17.3.2.[5]

We will also use the following, in the same spirit as Proposition 17.3.2, but very much easier. It covers the case of "flat rectangles".

Proposition 17.3.3 *Consider* $1 \le k_1 \le k_2 \le 2^p$, $1 \le \ell_0 \le 2^p$ *and* $R = \{k_1, \ldots, k_2\} \times \{\ell_0\}$. *Assume that*

$$k_2 - k_1 + 1 \le 2^j\ . \tag{17.49}$$

Consider independent r.v.s U_i *valued in* G, *of law* μ. *Then, with probability at least* $1 - L\exp(-50p)$, *the following occurs. Consider any function* $h : G \to \mathbb{R}$, *and assume that*

$$h(k,\ell) = 0 \ \ \text{unless}\ \ (k,\ell) \in R\ , \tag{17.50}$$

$$(k,\ell)\ ,\ (k+1,\ell) \in R \Rightarrow |h(k+1,\ell) - h(k,\ell)| \le 1\ , \tag{17.51}$$

$$\forall (k,\ell) \in R\ ,\ |h(k,\ell)| \le 2(k_2 - k_1)\ . \tag{17.52}$$

Then

$$|\sum_{i\le N}(h(U_i) - \int h\mathrm{d}\mu)| \le L\sqrt{m_0}(k_2 - k_1 + 1)^{3/2} \le L2^{j/2}\sqrt{m_0}\ \mathrm{card}\,R\ . \tag{17.53}$$

[5] Most readers are likely to be satisfied with a global understanding of Shor's matching theorem and would not read these proofs. The exceptional reader willing to have a run at the ultimate matching conjecture should figure out the details herself as preparatory training. Finally, those really eager on mastering all the technical details can find them in [132].

Proof of Theorem 17.2.1. In Proposition 17.3.2, there are (crudely) at most 2^{4p} choices for the quadruplet $(k_1, k_2, \ell_1, \ell_2)$. Thus, with probability at least $1 - L\exp(-46p)$, the conclusions of Proposition 17.3.2 are true for all values of k_1, k_2, ℓ_1, and ℓ_2, and the conclusions of Proposition 17.3.3 hold for all values of k_1, k_2, and ℓ_0. We assume that this is the case in the remainder of the proof. Under these conditions, we show the following:

$$h \in \mathcal{H}_1 \Rightarrow \Big|\sum_{i\le N}(h(U_i) - \int h\mathrm{d}\mu)\Big| \le L\sqrt{pm_0}\,2^{2p}\,. \tag{17.54}$$

$$h \in \mathcal{H}_j = \mathcal{H}_j(2^{2p-j}) \Rightarrow \Big|\sum_{i\le N}(h(U_i) - \int h\mathrm{d}\mu)\Big| \le L\sqrt{pm_0}\,2^{2p-j/2}\,. \tag{17.55}$$

The conclusion then follows from the decomposition (17.41) of a function in $\mathcal{H}$ provided by Proposition 17.3.1.

The proof of (17.54) relies on the case $k_1 = \ell_1 = 1$ and $k_2 = \ell_2 = 2^p$ of Proposition 17.3.3. The function h satisfies (17.45) and (17.46), and hence $|h(k,\ell) - h(1,1)| \le 2^{p+1} - 2$ for each $(k,\ell) \in G$. Consequently, the function $h^*(k,\ell) = h(k,\ell) - h(1,1)$ satisfies (17.44), (17.45), (17.46), and (17.47). Therefore, h^* satisfies (17.54), and consequently this is also the case for h.

Next, we turn to the proof of (17.55), and we fix j once and for all. The method is to find a family $\mathcal{R}$ of rectangles with the following properties:

- The rectangles in $\mathcal{R}$ all have the appropriate shape to apply either Proposition 17.3.2 or Proposition 17.3.3.
- The family $\mathcal{R}$ is disjoint and covers the support of h.
- The sum of the cardinalities of the rectangles in $\mathcal{R}$ is at most eight times the cardinality of the support of h.
- Each of the functions $h\mathbf{1}_R$ for $R \in \mathcal{R}$ satisfies either (17.47) or (17.52).

We then write $h = \sum_{R\in\mathcal{R}} h\mathbf{1}_R$, and we apply either (17.48) or (17.53) to each term to obtain the desired result. To understand the idea of the construction of the family $\mathcal{R}$, one can start with an exercise. Denoting by λ Lebesgue's measure on the unit square, the exercise is to prove that a measurable subset A of the unit square with $\lambda(A) \le 1/8$ can be covered by a union of disjoint dyadic squares such that for each square C in this family, one has $\lambda(C)/8 \le \lambda(A\cap C) \le \lambda(C)/2$. To see this, one recursively removes (starting with the larger squares) the dyadic squares C for which $\lambda(C\cap A) \ge \lambda(C)/8$. The condition $\lambda(A\cap C) \le \lambda(C)/2$ is an automatic consequence of the fact that the dyadic square four times the size of C and containing C has not been previously removed. The point of this condition is to ensure that C is not contained in A. In the construction of the family $\mathcal{R}$, this will ensure that h takes the value zero for at least one point on each square R, so that since it is 1-Lipschitz, we may bound it as required by (17.47) or (17.52). The unsurprising details of the construction may be found in Sect. B.4.

Chapter 18
The Ultimate Matching Theorem in Dimension 3

18.1 Introduction

In this chapter, we continue the study of matchings, but in dimension $d = 3$ rather than 2.[1] We consider i.i.d. r.v.s $(X_i)_{i \le N}$ uniformly distributed over the set $[0, 1]^3$. We want to match these points to nonrandom "evenly spread" points $(Y_i)_{i \le N}$.[2] Here, we say that $(Y_i)_{i \le N}$ are *evenly spread* if one can cover $[0, 1]^3$ with N rectangular boxes with disjoint interiors, such that each box R has a three-dimensional volume $1/N$, contains exactly one point Y_i, and is such that $R \subset B(Y_i, 10N^{-1/3})$. Each point of $[0, 1]^3$ belongs to such a box R and is within distance $10N^{-1/3}$ of a point Y_i.

The plan is to prove that for as large as possible a function φ, with probability close to 1, there exists a permutation π of $\{1, \ldots, N\}$ such that

$$\frac{1}{N} \sum_{i \le N} \varphi\Big(\frac{X_i - Y_{\pi(i)}}{LN^{-1/3}}\Big) \le 2 \tag{18.1}$$

where $N^{-1/3}$ is the scaling factor which is appropriate to dimension 3. For example, one might consider a function such as $\varphi(X) = \exp d(X, 0)^\alpha$, where $d(X, 0)$ is the distance between X and 0. Then (18.1) implies that this distance between X_i and $Y_{\pi(i)}$ is typically about $N^{-1/3}$ and gives a precise control on the number of indexes i for which it is significantly larger.

Let us try to explain in words the difference between the situation in dimension 3 and in dimension 2. In dimension 2, there are irregularities at all scales in the distribution of a random sample $(X_i)_{i \le N}$ of $[0, 1]^2$, and these irregularities

[1] No new ideas are required to cover the case $d > 3$.

[2] The reader may review the beginning of Sect. 4.3 at this stage.

M. Talagrand, *Upper and Lower Bounds for Stochastic Processes*, Ergebnisse der Mathematik und ihrer Grenzgebiete. 3. Folge / A Series of Modern Surveys in Mathematics 60, https://doi.org/10.1007/978-3-030-82595-9_18

combine to create the mysterious fractional powers of $\log N$. In dimension 3, no such phenomenon occurs, but there are still irregularities at many different scales. Cubes of volume about A/N with a dramatic deficit of points X_i exist for A up to about $\log N$. The larger A, the fewer such cubes. The essential feature of dimension ≥ 3 is that, as we will detail below, irregularities at different scales *cannot combine*. Still there typically exists a cube of side about $(\log N/N)^{1/3}$ which contains no point X_i. A point Y_i close to the center of this cube has to be matched with a point X_i at distance about $(\log N/N)^{1/3}$. Thus, if φ satisfies (18.1) and is a function $\varphi(X) = f(d(X,0))$ of the distance of X to 0, the function f cannot grow faster than $\exp x^3$.

We may also consider functions φ for which the different coordinates in $\mathbb{R}^3$ play different roles. It is then the scarcity of points X_i inside certain rectangles which will provide obstacles to matchings.

Although this is not obvious at this stage, it turns out that the important characteristic of the function φ is the sequence of sets $(\{\varphi \leq N_n\})_{n\geq 1}$. We will assume that these sets are rectangles with sides parallel to the coordinate axes, and we change perspective; we use these rectangles (rather than φ) as the basic object. To describe such rectangles, for each $k \geq 0$, we consider three integers $n_j(k) \geq 0$, $1 \leq j \leq 3$ with the following properties: Each sequence $(n_j(k))_{k\geq 0}$ is non-decreasing and

$$\sum_{j\leq 3} n_j(k) = k\ . \tag{18.2}$$

We define

$$S_k = \prod_{j\leq 3}[-2^{n_j(k)}, 2^{n_j(k)}]\ , \tag{18.3}$$

so that, denoting by λ the volume measure, $\lambda(S_k) = 2^{k+3}$ by (18.2). Thus, to go from S_k to S_{k+1}, one doubles the size of one side, not necessarily the same at each step. We note for further use that

$$S_{k+1} \subset 2S_k\ . \tag{18.4}$$

Recalling our notation $N_0 = 1$, $N_k = 2^{2^k}$ for $k \geq 1$, let us then define a function φ by

$$\varphi(x) = \inf\{N_k\ ;\ x \in S_k\}\ , \tag{18.5}$$

so that $\varphi(x) = 1$ if $x \in S_0$, $\varphi(x) = \infty$ if $x \notin \bigcup_{k\geq 0} S_k$ and $\varphi(x) = N_k$ if $x \in S_k \setminus S_{k-1}$. Also (and this motivated the construction), we have

$$\{\varphi \leq N_k\} = S_k\ . \tag{18.6}$$

Thus, the function φ depends on the three sequences of integers $(n_j(k))_{k\geq 0}$, but the notation does not indicate this.

Theorem 18.1.1 *Consider the function φ as above. Then with probability $\geq 1 - LN^{-10}$, there exists a permutation π of $\{1,\ldots,N\}$ such that*

$$\frac{1}{N}\sum_{i\leq N}\varphi\Big(\frac{X_i - Y_{\pi(i)}}{LN^{-1/3}}\Big) \leq 2\,. \tag{18.7}$$

Before we discuss this result, we state an elementary fact, which will be used during this discussion.

Lemma 18.1.2 *Assume that $[0,1]^3$ is divided into sets of equal measure $\leq \log N/(2N)$. Then if N is large enough, with probability close to one (and certainly $\geq 3/4$), there is one of these sets which contains no point X_i.*

Informal Proof. The probability for any one of these sets not to contain a point X_i is at least $(1 - \log N/(2N))^N \simeq 1/\sqrt{N}$. There are more than $2N/\log N$ such sets and $1/\sqrt{N} \times N/\log(N) \gg 1$. The result follows if we pretend that these events are independent because for independent events $(\Omega_i)_{i\leq k}$, the probability that none of the events occurs is $\prod_{i\leq k}(1 - \mathsf{P}(\Omega_i)) \leq \exp(-\sum_{i\leq k}\mathsf{P}(\Omega_i))$. The assertion that the event is independent is rigorous for a Poisson point process, and it suffices to compare the actual process with such a process. □

Let us now argue that Theorem 18.1.1 is sharp. Denoting by λ Lebesgue's measure, the function φ satisfies $\lambda(\{\varphi \leq N_k\}) = \lambda(S_k) = 2^{k+3}$. This condition restricts the growth of the function φ. We will show that it is basically necessary.

Proposition 18.1.3 *Assume that φ is of the type (18.5), but without requiring that $\sum_{j\leq 3} n_j(k) = k$. Assume that for a certain number C and all N large enough with probability $\geq 1/2$, we can find a permutation such that*

$$\frac{1}{N}\sum_{i\leq N}\varphi\Big(\frac{X_i - Y_{\pi(i)}}{CN^{-1/3}}\Big) \leq 2\,. \tag{18.8}$$

Then for every k large enough, we have $\lambda(\{\varphi \leq N_k\}) \geq 2^k/LC^3$.

Proof Without loss of generality, we may assume that the number C in (18.8) equals 2^{n_0} for a certain $n_0 \geq 1$. Let us then fix k and $n \geq 2 + n_0 + \max_{j\leq 3} n_j(k)$, so that the set $2^{-n+n_0}S_k$ is a rectangle of the type $\prod_{j\leq 3}[-2^{-m_j}, 2^{-m_j}]$ where $m_j = n - n_0 - n_j(k) \geq 2$. Then we may divide $[0,1]^3$ into sets A_ℓ which are translates of $2^{-n+n_0}S_k$. The sets A_ℓ are of measure $a := \lambda(2^{-n+n_0}S_k) = 2^{-3n+3n_0}\lambda(S_k)$. Consider then $N = 2^{3n+3}$, so that $N^{-1/3} = 2^{-n-1}$. The sides of the rectangles A_ℓ are $2^{-m_j+1} = 2^{-n+n_0+n_j(k)+1} = N^{-1/3}2^{n_0+n_j(k)+2}$. If $(Y_i)_{i\leq N}$ are evenly spread, for each set A_ℓ, there is a point Y_ℓ within distance $10N^{-1/3}$ of its center, and this point is well inside of A_ℓ, say, if $A_\ell = c_\ell + 2^{-n+n_0}S_k$, then $Y_\ell \in c_\ell + 2^{-n+n_0-1}S_k$.[3]

Let us assume if possible that

$$a = 2^{-3n+3n_0}\lambda(S_k) \le \frac{\log N}{2N} . \tag{18.9}$$

Applying Lemma 18.1.2, then with probability > 0, there will at the same time exist a matching as in (18.8), and one of the sets A_ℓ will not contain any point X_j. The corresponding point $Y_\ell \in c_\ell + 2^{-n+n_0-1}S_k$ near the center of A_ℓ can only be matched to a point $X_i \notin A_\ell = c_\ell + 2^{-n+n_0}S_k$. Then $Y_\ell - X_i \notin 2^{-n+n_0-1}S_k = 2^{n_0}N^{-1/3}S_k$ so that $(Y_\ell - X_i)/(2^{n_0}N^{-1/3}) \notin S_k = \{\varphi \le N_k\}$ and thus $\varphi((Y_\ell - X_i)/2^{n_0}N^{-1/3}) \ge N_k$. On the other hand, since X_i and Y_ℓ are matched together, (18.8) implies that $\varphi((Y_\ell - X_i)/2^{n_0}N^{-1/3}) \le 2N$. In particular, we have shown that $N_k \le 2N$. Turning things around, when $2N < N_k$, (18.9) must fail, that is, we have $\log N/(2N) \le 2^{-3n+3n_0}\lambda(S_k) = 2^{3n_0+3}\lambda(S_k)/N$, and thus $\log N \le 2^{3n_0}L\lambda(S_k) \le LC^3\lambda(S_k)$. Choosing n as large as possible with $2N = 2^{3n+4} < N_k$ yields $N_k \le 4N$ and $2^k/L \le \log N \le LC^3\lambda(S_k)$ which is the desired result. □

Exercise 18.1.4 Consider a convex function $\psi \ge 0$ on $\mathbb{R}^3$, with $\psi(0) = 0$, which is allowed to take infinite values. Assume that it satisfies the following:

$$\forall u \ge 1 \;,\; \lambda(\{\psi \le u\}) \ge \log u \;, \tag{18.10}$$

$$\epsilon_1, \epsilon_2, \epsilon_3 = \pm 1 \Rightarrow \psi(\epsilon_1 x_1, \epsilon_2 x_2, \epsilon_3 x_3) = \psi(x_1, x_2, x_3) \;, \tag{18.11}$$

$$\psi(1,0,0) \le 1 \;;\; \psi(0,1,0) \le 1 \;;\; \psi(0,0,1) \le 1 \;. \tag{18.12}$$

Then there are a constant L and a function φ as in Theorem 18.1.1 with $\psi(x) \le \varphi(Lx)$. Hint: All it takes is to observe that a convex set invariant by the symmetries around the coordinate planes is basically a rectangular box with sides parallel to the coordinate axes.

As a consequence of this exercise, Theorem 18.1.1 also applies to such functions. Consider, for example, $\alpha_1, \alpha_2, \alpha_3 \in]0, \infty]$ with

$$\frac{1}{\alpha_1} + \frac{1}{\alpha_2} + \frac{1}{\alpha_3} = 1$$

and the function

$$\psi(x_1, x_2, x_3) = \exp \frac{1}{3}(|x_1|^{\alpha_1} + |x_2|^{\alpha_2} + |x_3|^{\alpha_3}) - 1 \;.$$

[3] Note that without loss of generality, we may assume $n_0 \ge 10$ to give us all the room we need.

Here, we define $|x|^\infty = 0$ if $|x| < 1$ and $|x|^\infty = \infty$ if $|x| \geq 1$. Then

$$\{\psi \leq u\} \supset \{(x_1, x_2, x_3) \; ; \; \forall j \leq 3 \; , \; |x_j| < (\log(1+u))^{1/\alpha_j}\} \; ,$$

and consequently,

$$\lambda(\{\psi \leq u\}) \geq \log(1+u) \; .$$

Thus, Theorem 18.1.1 proves in this setting the "ultimate matching conjecture" of Problem 17.1.2.

The special case $\alpha_1 = \alpha_2 = \alpha_3 = 3$ is essentially the case where $\psi(x) = \exp(\|x\|^3)$. It was proved earlier by J. Yukich using the so-called transportation method (unpublished), but the transportation method seems powerless to prove anything close to Theorem 18.1.1. This special case shows that with probability $\geq 1 - LN^{-10}$, we can find a matching for which

$$\sum_{i \leq N} \exp(Nd(X_i, Y_{\pi(i)})^3/L) \leq 2N \; ,$$

so that in particular $\sum_{i \leq N} d(X_i, Y_{\pi(i)})^3 \leq L$ (since $x \leq \exp x$) and for each i, $\exp(Nd(X_i, Y_{\pi(i)})^3/L) \leq 2N$, which implies $\max_{i \leq N} d(X_i, Y_{\pi(i)}) \leq LN^{-1/3}(\log N)^{1/3}$ (a result first obtained by J. Yukich and P. Shor in [97]).

Research Problem 18.1.5 Find a proof of Theorem 18.1.1 a few pages long. The current proof occupies the entire chapter.

18.2 Regularization of φ

For purely technical reasons, we will not be able to work directly with the function φ, so in this section, we construct a regularized version of it. Our goal is to prove the following:

Proposition 18.2.1 *There exists a function φ^* with $\varphi^*(0) = 0$ that satisfies the following properties:*

$$\forall k \geq 0 \; , \; 8S_k \subset \{\varphi^* \leq N_k\} \subset 16S_k \; , \tag{18.13}$$

$$\text{the set } \{\varphi^* \leq u\} \text{ is convex for each } u > 0 \; , \tag{18.14}$$

$$\forall x \; , \; \varphi^*(x) = \varphi^*(-x) \; , \tag{18.15}$$

$$u \geq N_5 \Rightarrow \frac{3}{4}\{\varphi^* \leq u\} \subset \{\varphi^* \leq \frac{u}{4}\} \; . \tag{18.16}$$

The crucial new property of φ^* compared to φ is (18.16). Please note that condition (18.14) does not say that φ^* is convex.

Let us start some auxiliary constructions. For each $j \le 3$, we have $n_j(k) \le n_j(k+1) \le n_j(k)+1$. Thus, the sequence $(n_j(k))_{k\ge 0}$ takes all the values $0 \le n \le n_j^* := \sup_k n_j(k) \in \mathbb{N} \cup \{\infty\}$. For $n \le n_j^*$, $n \in \mathbb{N}$, we define

$$k_j(n) = \inf\{k \; ; \; n_j(k) = n\} \, . \tag{18.17}$$

In particular, $k_j(0) = 0$, and, as will be of constant use, $k_j(n+1) \ge k_j(n)+1$. We then define a function $\theta_j : \mathbb{R}^+ \to \mathbb{R}^+$ by the following properties:

$$0 \le t \le 2^3 \Rightarrow \theta_j(t) = 0 \, . \tag{18.18}$$

$$n \le n_j^* \Rightarrow \theta_j(2^{n+3}) = \log N_{k_j(n)} \, . \tag{18.19}$$

$$\theta_j \text{ is linear between } 2^{n+3} \text{ and } 2^{n+4} \text{ for } 0 \le n < n_j^* \, . \tag{18.20}$$

$$t > 2^{n_j^*+3} \Rightarrow \theta_j(t) = \infty \, . \tag{18.21}$$

Obviously, this function is non-decreasing.

Lemma 18.2.2 *For $k \ge 0$, we have*

$$[0, 2^{n_j(k)+3}] \subset \{\theta_j \le \log N_k\} \subset [0, 2^{n_j(k)+4}] \, . \tag{18.22}$$

Proof Let $n = n_j(k)$ so that by definition of $k_j(n)$, we have $k_j(n) \le k < k_j(n+1)$, and by (18.19) $\theta_j(2^{n+3}) = \log N_{k_j(n)} \le \log N_k$ so that $[0, 2^{n_j(k)+3}] \subset \{\theta_j \le \log N_k\}$. Next, for $t > 2^{n+4}$, by (18.19) again, we have $\theta_j(t) \ge \log N_{k_j(n+1)} > \log N_k$ so that $\{\theta_j \le \log N_k\} \subset [0, 2^{n_j(k)+4}]$, and (18.22) is proved. □

Lemma 18.2.3 *For $t \in \mathbb{R}^+$, we have*

$$\theta_j(t) \ge \log N_4 + 2\log 2 \Rightarrow \theta_j(3t/4) \le \theta_j(t) - 2\log 2 \, . \tag{18.23}$$

Proof A first observation is that for $n \ge 1$, the slope of θ_j on the interval $[2^{n+3}, 2^{n+4}]$ is (recalling that $\log N_k = 2^k \log 2$ for $k \ge 2$)

$$2^{-n-3}(\log N_{k_j(n+1)} - \log N_{k_j(n)}) = 2^{-n-3}(2^{k_j(n+1)} - 2^{k_j(n)}) \log 2 \, .$$

A technical problem here is that there seems to be no reason why this quantity would increase with n. On the other hand, since $k_j(n) \le k_j(n+1) - 1$, this slope is at least

$$\chi(n) := \log 2 \times 2^{k_j(n+1)-n-4} \, , \tag{18.24}$$

which satisfies $\chi(n) \le \chi(n+1)$. As a consequence, the slope of θ_j on the interval $[2^{n+1}, \infty[$ is at least $\chi(n)$. To prove (18.23), we may assume $\theta_j(3t/4) > \log N_4$ for the proof is finished otherwise. Thus, $\{\theta_j \le \log N_4\} \subset [0, 3t/4]$, and from (18.18), we have $3t/4 > 2^3$. Consider the largest integer n^* with $2^{n^*+3} \le 3t/4$, so that $3t/4 \le 2^{n^*+4}$ and thus

$$\log N_4 < \theta_j(3t/4) \le \theta_j(2^{n^*+4}) = \log N_{k_j(n^*+1)} ,$$

and hence $k_j(n^* + 1) \ge 4$. Since $3t/4 \ge 2^{n^*+3}$, we have $t/4 > 2^{n^*+1}$, so that $t - 3t/4 = t/4 > 2^{n^*+1}$. As we noted, the slope of θ_j on the interval $[2^{n^*+3}, \infty]$ is everywhere at least $\chi(n^*)$. Thus, we have proved that

$$\theta_j(t) - \theta_j(3t/4) \ge 2^{n^*+1}\chi(n^*) = \log 2 \times 2^{k_j(n^*+1)-3} \ge 2\log 2 . \qquad \square$$

Proof of Proposition 18.2.1. We define $\psi_j(t) = |t|/8$ for $|t| \le 1$ and $\psi_j(t) = \exp\theta_j(|t|)$ for $|t| \ge 1$. We define

$$\varphi^*(x_1, x_2, x_3) = \max_{j\le 3} \psi_j(x_j) , \tag{18.25}$$

so that (18.13) follows from (18.22) and the equality $\{\varphi^* \le u\} = \prod_{j\le 3}\{\psi_j \le u\}$, which also proves (18.14). Certainly, (18.15) is obvious.

We turn to the proof of (18.16). Consider $u \ge N_5$. Given x with $\varphi^*(x) \le u$, we have to prove that $\varphi^*(3x/4) \le u/4$. Since $\varphi^*(x) \le u$, we have $\max_{j\le 3}\theta_j(|x_j|) \le \log u$ so that, since $\log u \ge \log N_5$, we have

$$\max(\log N_5, \max_{j\le 3}\theta_j(|x_j|)) \le \log u . \tag{18.26}$$

Now, since $\log N_5 = 2^5 \log 2 \ge 4\log 2 + 2^4 \log 2 = 4\log 2 + \log N_4$, when $\theta_j(|x_j|) \le \log N_4 + 2\log 2$, then

$$\theta_j(3|x_j|/4) \le \theta_j(|x_j|) \le \log N_5 - 2\log 2 ,$$

where we use that θ_j is increasing in the first inequality. Now, if $\theta_j(|x_j|) \ge \log N_4 + 2\log 2$, then (18.23) implies that $\theta_j(3|x_j|/4)) \le \theta_j(|x_j|) - 2\log 2$. Consequently,

$$\log\varphi^*(3x/4) = \max_{j\le 3}\theta_j(3|x_j|/4) \le \max(\log N_5, \max_{j\le 3}\theta_j(|x_j|)) - 2\log 2 ,$$

and using (18.26), we obtain $\log\varphi^*(3x/4) \le \log u - 2\log 2 = \log(u/4)$. $\square$

18.3 Discretization

We now think of N as fixed and large. Consider a universal constant L^* which will be determined later. We define p as

$$\text{the largest integer such that } L^* 2^{3p} \le N \ , \tag{18.27}$$

so that $p \ge 2$ for N large. Let $G = \{1, \dots, 2^p\}^3$. We denote by τ the generic element of G. To each $\tau = (\tau_j)_{j \le 3}$ corresponds the small cube $H_\tau := \prod_{j \le 3}]2^{-p}(\tau_j - 1), 2^{-p}\tau_j]$ of side 2^{-p} of which $2^{-p}\tau$ is a vertex. These cubes form a partition of $]0, 1]^3$. The idea is simply that since we are not interested in what happens at scales less than $N^{-1/3}$, we replace H_τ by the point $2^{-p}\tau$, and G is a discrete model for $[0, 1]^3$ (although we have to keep in mind the scaling factor 2^{-p}).

Recalling our evenly spread points Y_i, we define $Z_i = \tau$ where $\tau \in G$ is determined by $Y_i \in H_\tau$. We note that each coordinate of Y_i differs by at most 2^{-p} from the corresponding coordinate of $2^{-p} Z_i$, which for further use we express as (and recalling that $S_0 = [-1, 1]^3$ by (18.3))

$$2^p Y_i - Z_i \in S_0 \ . \tag{18.28}$$

We set

$$n(\tau) = \operatorname{card}\{i \le N \ ; \ Z_i = \tau\} = \operatorname{card}\{i \le N \ ; \ Y_i \in H_\tau\} \ . \tag{18.29}$$

Lemma 18.3.1 *If L^* is large enough, there exists an integer $m_0 \ge L^*/2$ such that*

$$\forall \tau \in G \ , \ m_0 \le n(\tau) \le 2m_0 \tag{18.30}$$

and

$$m_0 2^{3p} \le N \le 2m_0 2^{3p} \ . \tag{18.31}$$

Proof We first observe that (18.31) follows from summation of the inequalities (18.30) over $\tau \in G$.

Since the points Y_i are evenly spread, there exists a partition of $[0, 1]^3$ in rectangular boxes R_i with $Y_i \in R_i \subset B(Y_i, 10/N^{1/3})$. Each of these boxes has volume N^{-1}. Let us fix $\tau \in G$. Let W_1 be the union of the R_i such that $R_i \subset H_\tau$, and observe that $N\lambda(W_1)$ is just the number of these boxes $R_i \subset H_\tau$. When $R_i \subset W_1$, we have $Y_i \in H_\tau$ so that

$$N\lambda(W_1) \le \operatorname{card}\{i \le N; Y_i \in H_\tau\} = n(\tau) \ . \tag{18.32}$$

Let W_2 be the union of the R_i such that $R_i \cap H_\tau \neq \emptyset$, so that $W_2 \subset H_\tau$. When $Y_i \in H_\tau$, we have $R_i \cap H_\tau \neq \emptyset$ so that $R_i \subset W_2$ and

$$n(\tau) = \operatorname{card}\{i \leq N; Y_i \in H_\tau\} \leq N\lambda(W_2) . \tag{18.33}$$

On the other hand, we have $\lambda(W_1) \leq \lambda(H_\tau) = 2^{-3p} \leq \lambda(W_2)$ so that $N\lambda(W_1) \leq N2^{-3p} \leq N\lambda(W_2)$, and combining with (18.32) and (18.33), we obtain

$$|n(\tau) - N2^{-3p}| \leq N\lambda(W_2 \setminus W_1) , \tag{18.34}$$

where $W_2 \setminus W_1$ is the union of the boxes R_i for which $R_i \cap H_\tau \neq \emptyset$ and $R_i \cap H_\tau^c \neq \emptyset$. Since R_i is of diameter $\leq 20N^{-1/3}$, every point of $W_2 \setminus W_1$ is within distance $20N^{-1/3}$ of the boundary of H_τ. Since $L^*2^{3p} \leq N$, we have $N^{-1/3} \leq 2^{-p}(L^*)^{-1/3}$ so that when L^* is large, we have $20N^{-1/3} \ll 2^{-p}$. We should then picture $W_2 \setminus W_1$ as contained in a thin shell around the boundary of H_τ, and the volume of this shell is a small proportion of the volume of H_τ, say less than 1/3 of this volume, $\lambda(W_2 \setminus W_1) \leq \lambda(H_\tau)/3 = 2^{-3p}/3$, and then (18.34) yields

$$\frac{2}{3}N2^{-3p} \leq n(\tau) \leq \frac{4}{3}N2^{-3p} .$$

Since $N2^{-3p} \geq L^*$, the smallest integer $m_0 \geq 2N2^{-3p}/3$ satisfies $m_0 \geq L^*/2$ and $4N2^{-3p}/3 \leq 2m_0$ so that

$$\forall \tau \in G , \ m_0 \leq n(\tau) \leq 2m_0 . \qquad \square$$

We will now forget about L^* until the very end of the proof. We will prove results that hold when $m_0 \geq L$ for a large universal constant, a condition which can be achieved by taking L^* large enough.

18.4 Discrepancy Bound

Let us recall that each of the evenly spread points Y_j belongs to a little box R_j of volume $1/N$. For each $\tau \in G$, we define K_τ as the union of the boxes R_j for which the corresponding point $Y_j \in H_\tau$. We define the r.v.s U_i by $U_i = \tau$ where τ is the random point of G determined by $X_i \in K_\tau$. Since the boxes R_j have a diameter $\leq 10N^{-1/3}$ and since $N^{-1} \leq 2^{-3p}/L^*$, assuming $L^* \geq 10^3$, given any point of H_τ and any point of K_τ, their difference has coordinates $\leq 2^{-p+1}$. In particular, we have

$$2^p X_i - U_i \in 2S_0 = 2[-1, 1]^3 . \tag{18.35}$$

The r.v.s U_i are i.i.d. of law μ where μ is the probability measure on G given by

$$\forall \tau \in G \,,\; \mu(\{\tau\}) = \frac{n(\tau)}{N} \,, \tag{18.36}$$

so that according to (18.30),

$$\forall \tau \in G \,,\; \frac{m_0}{N} \le \mu(\{\tau\}) \le \frac{2m_0}{N} \,. \tag{18.37}$$

Thus, μ is nearly uniform on G. To each function $w : G \to \mathbb{R}$, we associate the function $h_w : G \to \mathbb{R}$ given by

$$h_w(\tau) = \inf\{w(\tau') + \varphi^*(\tau - \tau') \,;\; \tau' \in G\} \,. \tag{18.38}$$

Since $\varphi^*(0) = 0$, we have

$$h_w \le w \,, \tag{18.39}$$

and we define[4]

$$\Delta(w) = \int (w - h_w)\mathrm{d}\mu \ge 0 \,. \tag{18.40}$$

Since $\varphi^* \ge 0$ and G is finite, we have $\Delta(w) < \infty$. The crucial ingredient for Theorem 18.1.1 is the following discrepancy bound:

Theorem 18.4.1 *Consider an i.i.d. sequence of r.v.s $(U_i)_{i \le N}$ distributed like μ. Then with probability $\ge 1 - L\exp(-100p)$, the following occurs:*

$$\forall w : G \to \mathbb{R} \,,\; \Big|\sum_{i \le N} \Big(h_w(U_i) - \int h_w \mathrm{d}\mu\Big)\Big| \le L\sqrt{m_0}2^{3p}(\Delta(w) + 1) \,. \tag{18.41}$$

The essential difficulty in a statement of this type is to understand which kind of information on the function h_w we may obtain from the fact that $\Delta(w)$ is given. In very general terms, there is no choice: we must extract information showing that such functions "do not vary wildly" so that we may bound the left-hand side of (18.41) with overwhelming probability. In still rather general terms, we shall prove that control of $\Delta(w)$ implies a kind of local Lipschitz condition on h_w. This is the goal of Sect. 18.5. This local Lipschitz condition implies in turn a suitable control on the coefficients of a Haar basis expansion of h_w, and this will allow us to conclude. The proof does not explicitly use chaining, although it is in a similar

[4] The use of the notation Δ has nothing to do with the Laplacian and everything to do with the fact that $\Delta(w)$ "measure the size of the difference between w and h_w".

spirit. The formulation in an abstract setting of the principle behind this proof is a possible topic for further research.

In the remainder of this section, we first prove a matching theorem related to the bound (18.41), and we then use Theorem 18.4.1 to complete the proof of Theorem 18.1.1.

Theorem 18.4.2 *There exists a constant L_1 such that the following occurs. Assume that*

$$m_0 \geq L_1 \; . \tag{18.42}$$

Consider points $(U_i)_{i\leq N}$ as in (18.41). Then there exists a permutation π of $\{1, \ldots, N\}$ for which

$$\sum_{i\leq N} \varphi^*(U_i - Z_{\pi(i)}) \leq N \; . \tag{18.43}$$

Proof First we deduce from Proposition 4.3.2 that

$$\inf_{\pi} \sum_{i\leq N} \varphi^*(U_i - Z_{\pi(i)}) = \sup \sum_{i\leq N} (w_i + w'_i) \; , \tag{18.44}$$

where the supremum is over all families $(w_i)_{i\leq N}$ and $(w'_i)_{i\leq N}$ for which

$$\forall i, j \leq N \; , \; w_i + w'_j \leq \varphi^*(U_i - Z_j) \; . \tag{18.45}$$

Given such families (w_i) and (w'_i), for $\tau \in G$, let us then define

$$h(\tau) = \inf_{j\leq N} (-w'_j + \varphi^*(\tau - Z_j)) \; , \tag{18.46}$$

so that from (18.45), we obtain $w_i \leq h(U_i)$ and thus

$$\sum_{i\leq N} (w_i + w'_i) \leq \sum_{i\leq N} (h(U_i) + w'_i) \; . \tag{18.47}$$

For $\tau \in G$, we define

$$w(\tau) := \inf\{-w'_j \; ; \; Z_j = \tau\} \; , \tag{18.48}$$

so that, taking in (18.46) the infimum first at a given value τ' of Z_j, we obtain

$$h(\tau) = \inf\{w(\tau') + \varphi^*(\tau - \tau') \; ; \; \tau' \in G\} \; , \tag{18.49}$$

and consequently, recalling the notation (18.38),

$$h(\tau) = h_w(\tau) .$$

Also, (18.48) implies

$$-w(\tau) = \sup\{w'_j \; ; \; Z_j = \tau\} ,$$

so that, using (18.36),

$$\sum\{w'_j \; ; \; Z_j = \tau, j \le N\} \le \operatorname{card}\{j \; ; \; Z_j = \tau\} \sup\{w'_j \; ; \; Z_j = \tau\}$$
$$= -N\mu(\{\tau\})w(\tau) ,$$

and by summation of these inequalities over $\tau \in G$,

$$\sum_{i \le N} w'_i \le -N \int w \mathrm{d}\mu . \tag{18.50}$$

Consequently,

$$\begin{aligned}\sum_{i \le N} (h(U_i) + w'_i) &\le \sum_{i \le N} h(U_i) - N \int w \mathrm{d}\mu \\ &\le \sum_{i \le N} \Big(h(U_i) - \int h \mathrm{d}\mu\Big) - N \int (w - h) \mathrm{d}\mu \\ &= \sum_{i \le N} \Big(h(U_i) - \int h \mathrm{d}\mu\Big) - N\Delta(w) . \end{aligned} \tag{18.51}$$

Now (18.41) implies, since $h = h_w$,

$$\sum_{i \le N} \Big(h(U_i) - \int h \mathrm{d}\mu\Big) \le L\sqrt{m_0} 2^{3p} (\Delta(w) + 1) , \tag{18.52}$$

and combining with (18.47) and (18.51), we have proved that all families (w_i) and (w'_i) as in (18.45) satisfy

$$\sum_{i \le N} (w_i + w'_i) \le L\sqrt{m_0} 2^{3p} (\Delta(w) + 1) - N\Delta(w) .$$

Recalling that $N \ge m_0 2^{3p}$ by (18.31), we obtain that for $m_0 \ge L_1$, the right-hand side is $\le N$. □

The idea to prove Theorem 18.1.1 is the obvious one: when we match the points U_i and $Z_{\pi(i)}$ in the discretized version of the problem, we will match the points X_i and $Y_{\pi(i)}$ in the original problem, and the next result allows to use the information provided by (18.30).

Lemma 18.4.3 *For any $i, j \le N$, we have*

$$\varphi(2^{p-6}(X_i - Y_j)) \le 1 + \varphi^*(U_i - Z_j) . \tag{18.53}$$

Proof Assume first that $U_i - Z_j \in 16S_0$. Combining with (18.28) and (18.35), we obtain $2^p(X_i - Y_j) = 2^pX_i - U_i + U_i - Z_j + Z_j - 2^pY_j \in 19S_0$ so that in particular $2^{p-6}(X_i - Y_j) \in S_0$ and since $\varphi = 1$ on S_0, we get $\varphi(2^{p-6}(X_i - Y_j)) = 1$, proving (18.53).

Assume next that $U_i - Z_j \notin 16S_0$, and consider the smallest k for which $U_i - Z_j \in 16S_k$ (if no such k exists, there is nothing to prove since $\varphi^*(U_i - Z_j) = \infty$), so that $k \ge 1$. Since $U_i - Z_j \notin 16S_{k-1}$, we have $\varphi^*(U_i - Z_j) \ge N_{k-1}$ by (18.13). Since $S_k \subset 2S_{k-1}$ by (18.4), we have $U_i - Z_j \in 16S_k \subset 32S_{k-1}$. Combining with (18.28) and (18.35), we obtain $2^p(X_i - Y_j) \in 35S_{k-1}$ so that $2^{p-6}(X_i - Y_j) \in S_{k-1}$ and $\varphi(2^{p-6}(X_i - Y_j)) \le N_{k-1}$ by (18.6). We have proved (18.35). □

Corollary 18.4.4 *Consider a permutation π of $\{1, \dots, N\}$ such that (18.43) holds. Then (18.7) holds.*

Proof This is obvious from (18.53) because by definition of p, we have $L^*2^{3p+3} \ge N$ so that $2^{p-6} \ge N^{1/3}/L$. □

Proof of Theorem 18.1.1. According to Theorem 18.4.1, (18.41) occurs with probability $\ge 1 - L\exp(-100p) \ge 1 - LN^{-10}$. When (18.41) occurs, Theorem 18.4.2 produces a permutation π for which (18.43) holds, and Corollary 18.4.4 shows that (18.7) holds for the same permutation. □

18.5 Geometry

The real work toward the proof of Theorem 18.4.1 starts here. To lighten notation, *we assume from now on that $\varphi = \varphi^*$ satisfies (18.13) to (18.16).*

In this section, we carry out the task of finding some "regularity" of the functions h_w (defined in (18.38)) for which $\Delta(w)$ is not too large. In other words, we describe some of the underlying "geometry" of this class of functions.

We define

$$s_j(k) = \min(p, n_j(k)) \; ; \; s(k) = \sum_{j \le 3} s_j(k) . \tag{18.54}$$

It follows from (18.2) that $n_j(k) \le k$, so that

$$k \le p \Rightarrow s_j(k) = n_j(k) \tag{18.55}$$

and also using again (18.2),

$$k \le p \Rightarrow s(k) = k\ . \tag{18.56}$$

It is the nature of the problem that different scales must be used. They will appear in terms of the sets we define now. We consider the collection $\mathcal{P}_k$ of subsets of G of the form

$$\prod_{j\le 3}\{b_j 2^{s_j(k)} + 1, \ldots, (b_j+1)2^{s_j(k)}\} \tag{18.57}$$

for $b_j \in \mathbb{N}$, $0 \le b_j \le 2^{p-s_j(k)} - 1$. For lack of a better word, subsets of G which are product of three intervals will be called *rectangles*. There are $2^{3p-s(k)}$ rectangles of the previous type, which form a partition of G. Each of these rectangles has cardinality $2^{s(k)}$:

$$C \in \mathcal{P}_k \Rightarrow \operatorname{card} C = 2^{s(k)}\ . \tag{18.58}$$

We say that a subset A of G is $\mathcal{P}_k$*-measurable* if it is a union of rectangles belonging to $\mathcal{P}_k$.

Let us say that two rectangles of $\mathcal{P}_k$ are *adjacent* if for all $j \le 3$, the corresponding values of b_j differ by at most 1. Thus, a rectangle is adjacent to itself and to at most 26 other rectangles. Given an integer q, let us say that two rectangles of $\mathcal{P}_k$ are *q-adjacent* if for each $j \le 3$, the corresponding values of b_j differ by at most q. Thus, at most $(2q+1)^3$, rectangles of $\mathcal{P}_k$ are q-adjacent to a given rectangle. The elementary proof of the following is better left to the reader. We recall the definition (18.3) of the sets S_k.

Lemma 18.5.1

(a) *If C, C' in $\mathcal{P}_k$ are adjacent, and if $\tau \in C$, $\tau' \in C'$, then $\tau - \tau' \in 2S_k$.*
(b) *If $\tau \in C \in \mathcal{P}_k$, $A \subset \tau + qS_k$, $A \in \mathcal{P}_k$, then A and C are q-adjacent.*

Recalling the definition (18.38), given $\Delta \ge 0$, we define the class $\mathcal{S}(\Delta)$ of functions on G by

$$\mathcal{S}(\Delta) = \left\{h_w\ ;\ \Delta(w) = \int (w - h_w)\mathrm{d}\mu \le \Delta\right\}. \tag{18.59}$$

We next state the main result of this section. The essence of this result is that for $h \in \mathcal{S}(\Delta)$ at each point, there is a scale at which it is appropriate to control the variations of h.

Theorem 18.5.2 *Given $\Delta \geq 1$, for every function $h \in \mathcal{S}(\Delta)$, we can find a partition $(B_k)_{k\geq 4}$ of G such that B_k is $\mathcal{P}_k$-measurable and such that for each $C \in \mathcal{C}_k := \{C \in \mathcal{P}_k ; C \subset B_k\}$, we can find a number $z(C)$ such that the following properties hold:*

$$\sum_{k\geq 4}\sum_{C\in\mathcal{C}_k} 2^{s(k)} z(C) \leq L2^{3p}\Delta \,, \tag{18.60}$$

$$k \geq 4 \,,\; C \in \mathcal{C}_k \Rightarrow N_k \leq z(C) \leq N_{k+1} \,. \tag{18.61}$$

For every $k \geq 4$, if $C \in \mathcal{C}_k$ and if $C' \in \mathcal{P}_k$ is adjacent to C, then

$$\tau \in C \,,\; \tau' \in C' \Rightarrow |h(\tau) - h(\tau')| \leq z(C) \,. \tag{18.62}$$

Let us stress that (18.62) holds in particular for $C' = C$.

Any point τ of G belongs to some B_k. Close to τ, the relevant scale to control the variations of h is given by the partition $\mathcal{P}_k$. Condition (18.61) controls the size of the numbers $z(C)$, depending on the local scale at which the kind of Lipschitz condition (18.62) holds. The restriction $z(C) \leq N_{k+1}$ is essential; the lower bound $z(C) \geq N_k$ is purely technical. Finally, the global size of the weighted quantities $z(C)$ is controlled by (18.60).

In the remainder of this chapter, we shall use the information provided by Theorem 18.5.2 to prove Theorem 18.4.1.

We start the proof of Theorem 18.5.2, which will occupy us until the end of the present section. This proof is fortunately not as formidable as the statement of the theorem itself. Considering a function $h \in \mathcal{S}(\Delta)$, we have to construct the objects of Theorem 18.5.2. By definition of $\mathcal{S}(\Delta)$, we can find a function $w : G \to \mathbb{R}$ such that

$$\forall \tau \in G \,,\; h(\tau) = \inf\{w(\tau') + \varphi(\tau - \tau') \;;\; \tau' \in G\} \,, \tag{18.63}$$

while

$$\int (w - h)\mathrm{d}\mu \leq \Delta \,. \tag{18.64}$$

For each τ and τ' in G, we have

$$h(\tau) \leq w(\tau') + \varphi(\tau - \tau') \,,$$

so that

$$w(\tau') \geq h(\tau) - \varphi(\tau - \tau') \,.$$

Let us then define

$$\widehat{h}(\tau') = \sup\{h(\tau) - \varphi(\tau - \tau')\ ;\ \tau \in G\}\ , \tag{18.65}$$

so that

$$h \le \widehat{h} \le w\ . \tag{18.66}$$

One may think of $\widehat{h}$ as a regularized version of the function w. Moreover,

$$\int (\widehat{h} - h)\mathrm{d}\mu \le \int (w - h)\mathrm{d}\mu \le \Delta\ . \tag{18.67}$$

For C in $\mathcal{P}_k$, let us define

$$y(C) = \min_{\tau \in C} \widehat{h}(\tau) - \max_{\tau \in C} h(\tau)\ . \tag{18.68}$$

Thus, for $\tau \in C$, we have $y(C) \le \widehat{h}(\tau) - h(\tau)$, so that

$$\mu(C)y(C) \le \int_C (\widehat{h}(\tau) - h(\tau))\mathrm{d}\mu(\tau)\ .$$

Using (18.58) and (18.37), we have $\mu(C) \ge 2^{s(k)} m_0/N$. Using also that $m_0/N \ge 2^{-3p-1}$ by (18.31), we finally obtain

$$2^{s(k)} y(C) \le L 2^{3p} \int_C (\widehat{h}(\tau) - h(\tau))\mathrm{d}\mu(\tau)\ . \tag{18.69}$$

Lemma 18.5.3 *There exists a disjoint sequence $(L_k)_{k \ge 5}$ of $\mathcal{P}_k$-measurable sets with the following properties:*

(a) If $A \in \mathcal{P}_k$ and $A \subset L_k$, then $y(A) \ge N_k/2$.
(b) Consider $\ell \ge 5$ and $A \in \mathcal{P}_\ell$ with $y(A) \ge N_\ell/2$. Then there exist $\ell' \ge \ell$ and $A' \in \mathcal{P}_{\ell'}$ with $A \subset A'$ and $y(A') \ge N_{\ell'}/2$.

Proof If for each $k \ge 5$ and each $A \in \mathcal{P}_k$ we have $y(A) < N_k/2$, there is nothing to do. Otherwise, consider the largest k_0 for which there exists $A \in \mathcal{P}_{k_0}$ with $y(A) \ge N_{k_0}/2$. For $k > k_0$, we set $L_k = \emptyset$. Let L_{k_0} be the union of all such rectangles $A \in \mathcal{P}_{k_0}$ with $y(A) \ge N_{k_0}/2$. We then construct the sets L_k by decreasing induction over k. Having constructed L_ℓ for $\ell \ge k$, we define L_{k-1} as the union of all rectangles $A \in \mathcal{P}_{k-1}$ for which $A \not\subset \cup_{\ell \ge k} L_\ell$ and $y(A) \ge N_{k-1}/2$. It is obvious that this sequence has the required properties. □

From this point on, we set $q = 32$. The construction of the partition $(B_k)_{k \ge 4}$ is obtained in the next result, although it will take further work to prove that this partition has the right properties.

Proposition 18.5.4 *There exists a partition $(B_k)_{k\geq 4}$ of G consisting of $\mathcal{P}_k$-measurable sets with the following properties:*

(a) If $C \in \mathcal{P}_k$, $C \subset B_k$, there exists $A \in \mathcal{P}_k$, $A \subset L_k$ which is q-adjacent to C.
(b) Consider $C \in \mathcal{P}_k$, $C \subset B_k$, $k \geq 5$. Consider $\ell \geq k$. Consider the unique $D \in \mathcal{P}_\ell$ with $C \subset D$. If there exists $A \in \mathcal{P}_\ell$ with $y(A) \geq N_\ell/2$ which is q-adjacent to D, then $k = \ell$.

Proof If for each $k \geq 5$ and each $A \in \mathcal{P}_k$ we have $y(A) < N_k/2$, we set $B_4 = G$ and $B_k = \emptyset$ for $k > 4$. Otherwise, consider the sequence $(L_k)_{k\geq 5}$ constructed in the previous lemma. We denote by k_0 the largest integer for which there exists $A \in \mathcal{P}_{k_0}$ with $y(A) \geq N_{k_0}/2$. We construct the sequence (B_k) as follows. For $k > k_0$, we set $B_k = \emptyset$. We define B_{k_0} as the union of all the rectangles $C \in \mathcal{P}_{k_0}$ which are q-adjacent to a rectangle $A \in \mathcal{P}_{k_0}$, $A \subset L_{k_0}$. We then define the sequence $(B_k)_{k\geq 5}$ by decreasing induction over k. Having defined B_ℓ for $\ell \geq k$, we define B_{k-1} as the union of all the rectangles $C \in \mathcal{P}_{k-1}$ which are q-adjacent to an $A \in \mathcal{P}_{k-1}$ with $A \subset L_{k-1}$ but for which $C \not\subset \cup_{\ell\geq k} B_\ell$. Finally, B_4 is what is left after we have constructed $(B_k)_{k\geq 5}$.

The property (a) is obvious by construction. To prove (b), we note that by Lemma 18.5.3, there exists $\ell' \geq \ell$ for which the element $A' \in \mathcal{P}_{\ell'}$ with $A \subset A'$ satisfies $A' \subset L_{\ell'}$. Then the element $D' \in \mathcal{P}_{\ell'}$ such that $D \subset D'$ is q-adjacent to A'. By construction of $B_{\ell'}$, we have $D' \subset \cup_{\ell''\geq \ell'} B_{\ell''}$. Since $C \subset D'$ and $C \subset B_k$, and since the sets $(B_{\ell''})$ form a partition, we have $\ell' \leq k$ so that $\ell = \ell' = k$. □

The reader should keep in mind the following important notation which will be used until the end of the chapter. For $k \geq 4$, we write

$$\mathcal{C}_k = \{C \in \mathcal{P}_k, C \subset B_k\} . \tag{18.70}$$

For $C \in \mathcal{C}_k$, $k \geq 5$, we set

$$x(C) = \max\{y(C')\ ;\ C' \in \mathcal{P}_k\ ,\ C' \subset L_k\ ,\ C' \text{ is } q\text{-adjacent to } C\}\ , \tag{18.71}$$

using that such C' exists by Proposition 18.5.4, (a). Using Lemma 18.5.3, (a) we further have

$$x(C) \geq N_k/2 . \tag{18.72}$$

Our next goal is to prove the following, which is the main step toward (18.60):

Lemma 18.5.5 *We have*

$$\sum_{k\geq 5} \sum_{C\in\mathcal{C}_k} 2^{s(k)} x(C) \leq L2^{3p} \Delta . \tag{18.73}$$

Proof When $C \in \mathcal{C}_k$ by definition of $x(C)$, there exists $\bar{C} \in \mathcal{P}_k$ which is q-adjacent to C and such that $x(C) = y(\bar{C})$ and $\bar{C} \subset L_k$. Thus,

$$\sum_{k \geq 5} \sum_{C \in \mathcal{C}_k} 2^{s(k)} x(C) = \sum_{k \geq 5} \sum_{C \in \mathcal{C}_k} 2^{s(k)} y(\bar{C})$$

$$\leq (2q+1)^3 \sum_{k \geq 5} \sum_{C \in \mathcal{C}_k, C \subset L_k} 2^{s(k)} y(C) \,, \qquad (18.74)$$

because there are at most $(2q+1)^3$ sets $C' \in \mathcal{C}_k$ for which $\bar{C}'$ is a given $C \in \mathcal{C}_k$. Since the sets L_k are disjoint, it follows from (18.69) and (18.67) that the sum on the right-hand side is $\leq L 2^{3p} \Delta$. □

For $C \in \mathcal{C}_k$, $k \geq 5$, we set

$$z(C) = \min(2x(C), N_{k+1}) \geq N_k \,.$$

If $C \in \mathcal{C}_4$, we set $z(C) = N_5$. Thus, (18.61) holds. Moreover,

$$\sum_{C \in \mathcal{C}_4} 2^{s(4)} z(C) = N_5 2^{s(4)} \operatorname{card} \mathcal{C}_4 \leq N_5 2^{s(4)} \operatorname{card} \mathcal{P}_4 = N_5 2^{3p} \leq L 2^{3p} \Delta$$

since $\Delta \geq 1$. Therefore, (18.60) follows from (18.73).

We turn to the proof of (18.62), the core of Theorem 18.5.2.

Proposition 18.5.6 *If $k \geq 4$, $C \in \mathcal{C}_k$, $C' \in \mathcal{P}_k$ are adjacent, and $\tau \in C, \tau' \in C'$, then*

$$h(\tau') \leq h(\tau) + z(C) \,. \qquad (18.75)$$

Since $z(C) \geq N_{\max(k,5)}$, there is nothing to prove unless

$$h(\tau') - h(\tau) \geq N_{\max(k,5)} \,,$$

so we assume that this is the case in the rest of the argument. It follows from (18.63) that for some $\rho \in G$, we have

$$h(\tau) = w(\rho) + \varphi(\tau - \rho) \,. \qquad (18.76)$$

We fix such a ρ, and we define

$$u = \max(h(\tau') - h(\tau), \varphi(\tau - \rho)) \,, \qquad (18.77)$$

so that $u \geq N_{\max(k,5)}$. We set

$$U = \{\varphi \leq u\} \,.$$

According to (18.14), this is a convex set, and (18.15) implies that $U = -U$. Consider the largest integer $\ell \geq 1$ such that $u \geq N_\ell$. Since $u \geq N_{\max(k,5)}$, we have $\ell \geq \max(k, 5)$, and by the definition of ℓ, we have $u < N_{\ell+1}$. We then use (18.13) as well as $S_{\ell+1} \subset 2S_\ell$ to obtain

$$8S_\ell \subset U \subset 32S_\ell \,. \tag{18.78}$$

Lemma 18.5.7 *There exists $A \in \mathcal{P}_\ell$ with*

$$A \subset V := \frac{\rho + \tau'}{2} + S_\ell \subset \Big(\tau' + \frac{3U}{4}\Big) \cap \Big(\rho + \frac{3U}{4}\Big) \,. \tag{18.79}$$

Proof Since C and C' are adjacent, it follows from Lemma 18.5.1 (a) that

$$\tau' - \tau \in 2S_k \subset 2S_\ell \subset \frac{U}{4} \,.$$

Since $\varphi(\tau - \rho) \leq u$, we have $\tau - \rho \in U$, so that $\tau' - \rho = \tau' - \tau + \tau - \rho \in 5U/4$ and therefore $\rho - \tau' \in 5U/4$ since U is symmetric by (18.15). Consequently,

$$\frac{\rho + \tau'}{2} \in \tau' + \frac{5U}{8} \,; \; \frac{\rho + \tau'}{2} \in \rho + \frac{5U}{8} \,. \tag{18.80}$$

Thus, defining V as in (18.79), the second inclusion in this relation holds. Even though the point $(\rho + \tau')/2$ need not be in G, since the set S_ℓ is twice larger in every direction than an element of $\mathcal{P}_\ell$, it is obvious that V entirely contains an element A of $\mathcal{P}_\ell$. □

Lemma 18.5.8 *We have $y(A) \geq u/2$.*

Proof Since $u \geq N_5$, it follows from (18.16) that $\varphi(x) \leq u/4$ for $x \in 3U/4$. Consequently, if $\rho' \in G \cap V$ by (18.79), we have $\varphi(\rho' - \tau') \leq u/4$ and $\varphi(\rho' - \rho) \leq u/4$. Thus,

$$\widehat{h}(\rho') \geq h(\tau') - \varphi(\tau' - \rho') \geq h(\tau') - \frac{u}{4} \,.$$

Also, by (18.63), we have

$$h(\rho') \leq w(\rho) + \varphi(\rho' - \rho) \leq w(\rho) + \frac{u}{4} \,.$$

Thus, using (18.76) in the second line and (18.77) in the last line,

$$\begin{aligned}\min_{\rho' \in V \cap G} \widehat{h}(\rho') - \max_{\rho' \in V \cap G} h(\rho') &\geq h(\tau') - w(\rho) - \frac{u}{2} \\ &= h(\tau') - h(\tau) + \varphi(\tau - \rho) - \frac{u}{2}\end{aligned}$$

$$\geq \max(h(\tau') - h(\tau), \varphi(\tau - \rho)) - \frac{u}{2}$$

$$= \frac{u}{2} \,. \tag{18.81}$$

Since $A \subset V$, this implies the required result by the definition of $y(A)$. □

Proof of Proposition 18.5.6. We will prove that $u \leq z(C)$, from which (18.75) follows. Since $\tau' - \tau \in U/4$, using (18.79) in the second inclusion and (18.78) in the third one, we have

$$V \subset \tau' + \frac{3U}{4} \subset \tau + U \subset \tau + 32S_\ell \,. \tag{18.82}$$

Since $\ell \geq k$, and since $C \in \mathcal{P}_k$, there is a unique rectangle $D \in \mathcal{P}_\ell$ with $C \subset D$, and since $A \subset V$ and $\tau \in C$, (18.82) and Lemma 18.5.1 (b) imply that D and A are 32-adjacent. Moreover, by Lemma 18.5.8, we have $y(A) \geq u/2 \geq N_\ell/2$. By Proposition 18.5.4 (b), we have $\ell = k$. Thus, $D = C \in \mathcal{P}_k$ and $A \in \mathcal{P}_k$. Since A and C are 32-adjacent, it then follows from the definition (18.71) that $x(C) \geq y(A) \geq u/2$. Also, $u < N_{\ell+1} = N_{k+1}$, so that $u \leq \min(2x(C), N_{k+1}) = z(C)$, completing the proof of (18.75). □

Finally, it remains to prove that, with the notation of (18.62),

$$h(\tau) \leq h(\tau') + z(C) \,. \tag{18.83}$$

For this, we repeat the previous argument, exchanging the roles of τ and τ', up to (18.82), which we replace by

$$V \subset \tau + \frac{3U}{4} \subset \tau + U \subset \tau + 32S_\ell \,,$$

and we finish the proof in exactly the same manner. □

18.6 Probability, I

To prove a discrepancy bound involving the functions of the class $\mathcal{S}(\Delta)$ of Theorem 18.5.2, we must understand "how they oscillate". There are two sources for such oscillations.

- The function h oscillates within each rectangle $C \in \mathcal{C}_k$.
- The value of h may jump when one changes the rectangle C.

In the present section, we take care of the first type of oscillation. We show that we can reduce the proof of Theorem 18.4.1 to that of Theorem 18.6.9, that is, to the case where h is constant on each rectangle $C \in \mathcal{C}_k$. This is easier than the proof of

Theorem 18.6.9 itself, but already brings to light the use of the various conditions of Theorem 18.5.2. Throughout the rest of the proof, for $\tau \in G$, we consider the r.v.

$$Y_\tau = \operatorname{card}\{i \le N\,;\ U_i = \tau\} - N\mu(\{\tau\})\,, \tag{18.84}$$

where U_i are i.i.d. r.v.s on G with $\mathsf{P}(U_i = \tau) = \mu(\{\tau\})$. We recall the number m_0 of (18.30), and we note right away that

$$\mathsf{E}Y_\tau^2 \le N\mu(\{\tau\}) \le 2m_0\,,$$

so that we may think of $|Y_\tau|$ as being typically of size about $\sqrt{m_0}$. Before we start the real work, let us point out a simple principle.

Lemma 18.6.1 *Consider a sequence $(a_i)_{i\le M}$ of positive numbers. For a set $I \subset \{1,\dots,M\}$, define $S_I = \sum_{i\in I} a_i$. For an integer $r \le M$, define $A_r = \max\{S_I;\operatorname{card} I \le r\}$, and consider I with $\operatorname{card} I = r$ and $S_I = A_r$. Then for $i \notin I$, we have $a_i \le A_r/r$.*

Proof Assume without loss of generality that the sequence (a_i) is non-increasing. Then $I = \{1,\dots,r\}$, and $A_r \ge ra_r$, while for $i \notin I$, we have $i \ge r$ so that $a_i \le a_r \le A_r/r$. □

The main result of this section is as follows:

Proposition 18.6.2 *With probability $\ge 1 - L\exp(-100p)$, the following happens. Consider any $\Delta \ge 1$ and a partition $(B_k)_{k\ge 4}$ of G such that B_k is $\mathcal{P}_k$-measurable, and for each $C \in \mathcal{C}_k := \{C \in \mathcal{P}_k;\ C \subset B_k\}$, consider a number $z(C)$ such that the following properties hold:*

$$\sum_{k\ge 4}\sum_{C\in\mathcal{C}_k} 2^{s(k)} z(C) \le L2^{3p}\Delta\,, \tag{18.85}$$

$$k \ge 4\,,\ C \in \mathcal{C}_k \Rightarrow z(C) \le N_{k+1}\,. \tag{18.86}$$

Then

$$\sum_{k\ge 4}\sum_{C\in\mathcal{C}_k} z(C)\sum_{\tau\in C}|Y_\tau| \le L2^{3p}\sqrt{m_0}\Delta\,. \tag{18.87}$$

The basic problem to bound a sum as in (18.87) is that a few of the r.v.s $W_C := \sum_{\tau\in C}|Y_\tau|$ might be quite large due to random fluctuations. The strategy to take care of this is at kindergarten level. Given a subset I of $\mathcal{C}_k$, we write

$$\sum_{C\in\mathcal{C}_k} z(C)W_C \le \max_{C\in\mathcal{C}_k} z(C)\sum_{C\in I} W_C + \sum_{C\in\mathcal{C}_k} z(C)\max_{C\in\mathcal{C}_k\setminus I} W_C\,. \tag{18.88}$$

Next, consider an integer r and the quantity

$$A_{r,k} := \max_{I \subset \mathcal{C}_k, \operatorname{card} I = r} \sum_{C \in I} W_C . \tag{18.89}$$

In particular, $A_{1,k} = \max_{C \in \mathcal{C}_k} W_C$, and

$$\sum_{C \in \mathcal{C}_k} z(C) W_C \le A_{1,k} \sum_{C \in \mathcal{C}_k} z(C) . \tag{18.90}$$

If the set I is such that the maximum is attained in (18.89), then by Lemma 18.6.1, we have $W_C \le A_{r,k}/r$ for $C \notin I$, and (18.88) implies

$$\sum_{C \in \mathcal{C}_k} z(C) W_C \le A_{r,k} \max_{C \in \mathcal{C}_k} z(C) + \frac{A_{r,k}}{r} \sum_{C \in \mathcal{C}_k} z(C) . \tag{18.91}$$

We will prove (18.87) by application of these inequalities for a suitable value $r = r_k$. The probabilistic part of the proof is to control from above the random quantities $A_{r,k}$. This is done by application of the union bound in the most brutish manner. The wonder is that everything fits so well in the computations.

To carry out this program, we need first to understand the properties of this family (Y_τ) of r.v.s and other related entities. This is the motivation behind the following definition:

Definition 18.6.3 Consider a finite set V. We say that a family $(Y_v)_{v \in V}$ is of type $\mathcal{B}(N)$ if there exist a probability space (Ω, θ), i.i.d. r.v.s $(W_i)_{i \le N}$ valued in Ω, of law θ and functions ψ_v on Ω, $|\psi_v| \le 1$, with *disjoint* supports for different v, such that

$$\frac{1}{2 \operatorname{card} V} \le \theta(\{\psi_v \ne 0\}) \le \frac{2}{\operatorname{card} V} \tag{18.92}$$

and for which

$$Y_v = \sum_{i \le N} \Big(\psi_v(W_i) - \int \psi_v \mathrm{d}\theta \Big) . \tag{18.93}$$

The family $(Y_\tau)_{\tau \in G}$ is of type $\mathcal{B}(N)$, with $\psi_\tau = \mathbf{1}_{\{\tau\}}$ and $\Omega = G$ provided with the uniform probability. The following basic bound simply follows from Bernstein's inequality:

Lemma 18.6.4 *Consider any family* $(Y_v)_{v \in V}$ *of type* $\mathcal{B}(N)$ *and numbers* $(\eta_v)_{v \in V}$ *with* $|\eta_v| \le 1$. *Then for* $u > 0$, *we have*

$$\mathsf{P}\Big(\Big| \sum_{v \in V} \eta_v Y_v \Big| \ge u \Big) \le 2 \exp\Big(-\frac{1}{L} \min\Big(u, \frac{u^2 \operatorname{card} V}{N \sum_{v \in V} \eta_v^2} \Big) \Big) . \tag{18.94}$$

Proof Consider the r.v.s W_i and the functions ψ_v as in Definition 18.6.3. We define

$$S_i = \sum_{v \in V} \eta_v \psi_v(W_i) \ .$$

Since the functions ψ_v have disjoint supports, we have $|S_i| \le 1$ and also

$$\mathsf{E}(S_i - \mathsf{E}S_i)^2 \le \mathsf{E}S_i^2 \le \frac{2}{\operatorname{card} V} \sum_{v \in V} \eta_v^2 \ .$$

Since $\sum_{v \in V} \eta_v Y_v = \sum_{i \le N}(S_i - \mathsf{E}S_i)$, and since $|S_i| \le 1$, (18.94) follows from Bernstein's inequality (4.44). □

We are now ready to write our basic bound. We shall use the following well-known elementary fact: For $k \le n$, we have

$$\binom{n}{k} \le \Big(\frac{en}{k}\Big)^k \ . \qquad (18.95)$$

Lemma 18.6.5 *For each $u > 0$ and k, $r \le \operatorname{card} \mathcal{P}_k$, we have*

$$\mathsf{P}(A_{r,k} \ge u) \le 2^{r2^{s(k)}} \Big(\frac{e2^{3p}}{r}\Big)^r \exp\Big(-\frac{1}{L}\min\Big(u, \frac{u^2}{m_0 r 2^{s(k)}}\Big)\Big) \ . \qquad (18.96)$$

Proof By definition of $A_{r,k}$, when $A_{r,k} \ge u$, we can find a subset $I \subset \mathcal{C}_k$ with $\operatorname{card} I = r$ and for $\tau \in C \in I$ a number $\xi_\tau = \pm 1$ such that

$$\sum_{\tau \in C \in I} \xi_\tau Y_\tau \ge u \ . \qquad (18.97)$$

We bound the probability of this event by the union bound. Since $\operatorname{card} C = 2^{s(k)}$ for $C \in \mathcal{P}_k$ and $\operatorname{card} \mathcal{P}_k = 2^{3p-s(k)} \le 2^{3p}$, the factors in front of the exponential in (18.96) are a bound for the possible number of choices of I and the ξ_τ. To bound the probability that given I and the ξ_τ (18.97) occurs, we use the bound (18.94) for $V = G$ with $\eta_\tau = 0$ if $\tau \notin I$ and $\eta_\tau = \xi_\tau$ if $\tau \in I$, so that $\sum_{\tau \in G} \eta_\tau^2 = r2^{s(k)}$ and $\operatorname{card} V = 2^{3p}$. Finally, we use that by 18.31, we have $N \le 2m_0 3^p$ to reach the bound (18.96). □

The rest of the argument is hand-to-hand combat with the inequality (18.96) to choose the appropriate values of $r = r_k$ and u while maintaining the right-hand side at most $L\exp(-100p)$. It is convenient to distinguish the cases of "small k" and "large k". These are the cases $4 \le k < k_0$ and $k \ge k_0$ for an integer $k_0 \ge 4$ which we define now. If $2^{3p} < N_6$, we define $k_0 = 4$. Otherwise, we define k_0 as the largest integer with $N_{k_0+2} \le 2^{3p}$, so that $N_{k_0+3} \ge 2^{3p}$ and thus

$$\frac{p}{L} \le 2^{k_0} \le Lp \; . \tag{18.98}$$

Recalling (18.27), as N becomes large, so does p, and the condition $2^{k_0} \le Lp$ then forces that $k_0 \le p$. Recalling the quantity $s(k)$ of (18.54) and that $s(k) = k$ for $k \le p$ by (18.56), we then have $s(k) = k$ for $k \le k_0$.

We first take care of the values $k \ge k_0$. This is the easiest case because for these values, the r.v.s $W_C = \sum_{\tau \in C} |Y_\tau|$ are not really larger than their expectations, as the following shows:

Lemma 18.6.6 *With probability* $\ge 1 - L\exp(-100p)$ *for* $k_0 \le k \le p$, *we have*

$$A_{1,k} \le L2^{s(k)}\sqrt{m_0} \; . \tag{18.99}$$

Proof We recall that $2^{s(k)} = 2^k \ge 2^{k_0} \ge p/L$ by (18.98). Consider a parameter $B \ge 1$. Using (18.96) for $r = 1$ and $u = B\sqrt{m_0}2^{s(k)}$, we obtain

$$\mathsf{P}\Big(A_{1,k} \ge B\sqrt{m_0}2^{s(k)}\Big) \le 2^{2^{s(k)}} e 2^{3p} \exp\Big(-\frac{B2^{s(k)}}{L}\Big) \le \exp\Big(L2^{s(k)} - \frac{B2^{s(k)}}{L}\Big) \; .$$

If B is a large enough constant, with probability $\ge 1 - L\exp(-100p)$ for each $k_0 \le k \le p$, we have $A_{1,k} \le B2^{s(k)}\sqrt{m_0}$. □

We now take care of the small values of k.

Lemma 18.6.7 *With probability* $\ge 1 - L\exp(-100p)$ *for* $4 \le k \le k_0$, *we have*

$$A_{r_k,k} \le Lr_k2^k\sqrt{m_0} \tag{18.100}$$

where $r_k = \lfloor 2^{3p}/N_{k+2} \rfloor$.

Proof Since $4 \le k \le k_0$ by definition of k_0, we have $N_{k+2} \le 2^{3p}$, and thus $r_k \ge 1$. Since $\lfloor x \rfloor \ge 1 \Rightarrow \lfloor x \rfloor \ge x/2$, this yields $r_k \ge 2^{3p}/(2N_{k+2})$. Consequently,

$$\Big(\frac{e2^{3p}}{r_k}\Big)^{r_k} \le (2eN_{k+2})^{r_k} \le \exp(L2^k r_k) \; .$$

Thus, given a parameter $B \ge 1$, and choosing $u = B\sqrt{m_0}r_k2^k$ in (18.96), we obtain

$$\mathsf{P}\Big(A_{r_k,k} \ge B\sqrt{m_0}r_k2^k\Big) \le \exp\Big(\Big(L - \frac{B}{L}\Big)r_k2^k\Big) \; . \tag{18.101}$$

Now, since the sequence $(2^k/N_{k+2})$ decreases, using (18.98) in the last inequality,

$$2^k r_k \ge \frac{2^{3p+k}}{2N_{k+2}} \ge 2^{3p}\frac{2^{k_0}}{2N_{k_0+2}} \ge \frac{2^{k_0}}{2} \ge \frac{p}{L} . \tag{18.102}$$

Taking for B a large enough constant, (18.101) and (18.102) imply the result. □

Proof of Proposition 18.6.2. We assume that the events described in Lemmas 18.6.7 and 18.6.6 occur, and we prove (18.87). First, using (18.90),

$$\sum_{k\ge k_0}\sum_{C\in\mathcal{C}_k} z(C)W_C \le \sum_{k\ge k_0} A_{1,k}\sum_{C\in\mathcal{C}_k} z(C) ,$$

so that using (18.99), we have

$$\sum_{k\ge k_0}\sum_{C\in\mathcal{C}_k} z(C)W_C \le L\sqrt{m_0}\sum_{k\ge k_0}\sum_{C\in\mathcal{C}_k} 2^{s(k)}z(C) . \tag{18.103}$$

Next for $k \le k_0$, by (18.86), we have $z(C) \le N_{k+1}$ for $C \in \mathcal{C}_k$ so that (18.91) implies

$$\sum_{C\in\mathcal{C}_k} z(C)W_C \le A_{r,k}N_{k+1} + \frac{A_{r,k}}{r}\sum_{C\in\mathcal{C}_k} z(C) .$$

Using this with $r = r_k$, and using (18.100), we obtain (and since $r_k \le 2^{3p}/N_{k+2}$)

$$\sum_{C\in\mathcal{C}_k} z(C)\sum_{\tau\in C}|Y_\tau| \le L\sqrt{m_0}2^{3p}\frac{N_{k+1}}{N_{k+2}} + L2^k\sqrt{m_0}\sum_{C\in\mathcal{C}_k} z(C) . \tag{18.104}$$

Summation of these inequalities for $4 \le k \le k_0$ and use of (18.85) prove (18.87). □

Let us denote by L_0 the constant in (18.60).

Definition 18.6.8 Consider $\Delta > 1$. A function $h : G \to \mathbb{R}$ belongs to $\mathcal{S}^*(\Delta)$ if we can find a partition $(B_k)_{k\ge 4}$ of G where B_k is $\mathcal{P}_k$-measurable with the following properties, where we recall the notation $\mathcal{C}_k := \{C \in \mathcal{P}_k; C \subset B_k\}$ of (18.70). First

$$\text{for each } k \ge 4 ,\ h \text{ is constant on each rectangle } C \in \mathcal{C}_k . \tag{18.105}$$

Next, for each $C \in \mathcal{C}_k$ we can find a number $z(C)$ satisfying the following properties:

$$\sum_{k\ge 4}\sum_{C\in\mathcal{C}_k} 2^{s(k)}z(C) \le L_0 2^{3p}\Delta , \tag{18.106}$$

and

$$k \geq 4 \ , \ C \in \mathcal{C}_k \Rightarrow N_k \leq z(C) \leq N_{k+1} \ . \tag{18.107}$$

Moreover, if $C \in \mathcal{C}_k$ and if $C' \in \mathcal{P}_k$ is adjacent to C and such that for $k' > k$, we have $C' \not\subset B_{k'}$, then

$$\tau \in C \ , \ \tau' \in C' \Rightarrow |h(\tau) - h(\tau')| \leq z(C) \ . \tag{18.108}$$

Let us stress the difference between (18.108) and (18.62). In (18.108), C' is adjacent to C as in (18.62) but satisfies the further condition that $C' \not\subset B_{k'}$ for $k' \geq k$. Equivalently, $C' \subset \cup_{\ell \leq k} B_\ell$.

In the rest of this chapter, we shall prove the following:

Theorem 18.6.9 *Consider an i.i.d. sequence of r.v.s $(U_i)_{i \leq N}$ distributed like μ. Then with probability $\geq 1 - L\exp(-100p)$, the following occurs: For $\Delta \geq 1$, whenever $h \in \mathcal{S}^*(\Delta)$,*

$$\Big|\sum_{i \leq N} \Big(h(U_i) - \int h \mathrm{d}\mu\Big)\Big| \leq L\sqrt{m_0} 2^{3p} \Delta \ . \tag{18.109}$$

Let us observe right away the following fundamental identity, which is obvious from the definition of Y_τ:

$$\sum_{i \leq N} \Big(h(U_i) - \int h \mathrm{d}\mu\Big) = \sum_{\tau \in G} h(\tau) Y_\tau \ . \tag{18.110}$$

Proof of Theorem 18.4.1. Consider a function $h \in \mathcal{S}(\Delta)$, the sets B_k, and the numbers $z(C)$ as provided by Theorem 18.5.2. Let us define the function h^* on G as follows: if $C \in \mathcal{C}_k$, then h^* is constant on C and $\int_C h \mathrm{d}\mu = \int_C h^* \mathrm{d}\mu$. In other words, the constant value $h^*(C)$ of h^* on C is the average value of h on C:

$$h^*(C) = \frac{1}{\mu(C)} \int_C h \mathrm{d}\mu \ . \tag{18.111}$$

Using (18.62) for $C' = C$, averaging over $\tau' \in C$ and using Jensen's inequality yields that $|h - h^*| \leq z(C)$ on C, so that

$$\Big|\sum_{\tau \in C} h(\tau) Y_\tau - \sum_{\tau \in C} h^*(\tau) Y_\tau\Big| \leq z(C) \sum_{\tau \in C} |Y_\tau|$$

and by summation,

$$\Big|\sum_{\tau \in G} h(\tau) Y_\tau - \sum_{\tau \in G} h^*(\tau) Y_\tau\Big| \leq \sum_{k \geq 4} \sum_{C \in \mathcal{C}_k} z(C) \sum_{\tau \in C} |Y_\tau| \ .$$

According to Proposition 18.6.2, with probability $\geq 1 - L\exp(-100p)$, the last sum is $\leq L\sqrt{m_0}2^{3p}\Delta$, so that with the same probability for each function $h \in \mathcal{S}(\Delta)$, we have

$$\Big|\sum_{\tau\in G} h(\tau)Y_\tau\Big| \leq \Big|\sum_{\tau\in G} h^*(\tau)Y_\tau\Big| + L\sqrt{m_0}2^{3p}\Delta\,.$$

Using 18.110 for both h and h^*, we then have

$$\Big|\sum_{i\leq N}\Big(h(U_i) - \int h\mathrm{d}\mu\Big)\Big| \leq \Big|\sum_{i\leq N}\Big(h^*(U_i) - \int h^*\mathrm{d}\mu\Big)\Big| + L\sqrt{m_0}2^{3p}\,.$$

Therefore, using Theorem 18.6.9, it suffices to prove that $h^* \in \mathcal{S}^*(\Delta)$. Using the same sets B_k and the same values $z(C)$ for h^* as for h, it suffices to prove (18.108). Consider C and C' as in this condition and $\tau \in C, \tau' \in C'$. Consider k' such that $\tau' \in C'' \in \mathcal{C}_{k'}$. Then we have $k' \leq k$, for otherwise $C' \subset C'' \subset B_{k'}$, which contradicts the fact that we assume that $C' \not\subset B_{k'}$. Thus, $C'' \subset C'$, and consequently, by (18.62) for $\rho \in C$ and $\rho' \in C''$, we have $|h(\rho) - h(\rho')| \leq z(C)$. Recalling (18.111), and using Jensen's inequality, proves that $|h^*(C) - h^*(C')| \leq z(C)$ which concludes the proof since $|h^*(\tau) - h^*(\tau')| = |h^*(C) - h^*(C')|$. □

18.7 Haar Basis Expansion

The strategy to prove Theorem 18.6.9 is very simple. We write an expansion $h = \sum_v a_v(h)v$ along the Haar basis, where $a_v(h)$ is a number and v is a function belonging to the Haar basis. (See the details in (18.117).) We then write

$$\begin{aligned}\Big|\sum_{i\leq N}\Big(h(U_i) - \int h\mathrm{d}\mu\Big)\Big| &\leq \sum_v |a_v(h)|\Big|\sum_{i\leq N}\Big(v(U_i) - \int v\mathrm{d}\mu\Big)\Big| \\ &= \sum_v |a_v(h)||Y_v|\,, \qquad (18.112)\end{aligned}$$

where

$$Y_v = \sum_{i\leq N}\Big(v(U_i) - \int v\mathrm{d}\mu\Big)\,. \qquad (18.113)$$

The first task is to use the information $h \in \mathcal{S}^*(\Delta)$ to bound the numbers $|a_v(h)|$. This is done in Proposition 18.7.1, and the final work of controlling the sum in (18.112) is the purpose of the next and last section.

We first describe the Haar basis. For $1 \le r \le p+1$, we define the class $\mathcal{H}(r)$ of functions on $\{1, \dots, 2^p\}$ as follows:

$$\mathcal{H}(p+1) \text{ consists of the function that is constant equal to } 1\ . \tag{18.114}$$

For $1 \le r \le p$, $\mathcal{H}(r)$ consists of the 2^{p-r} functions $f_{i,r}$ for $0 \le i < 2^{p-r}$ that are defined as follows:

$$f_{i,r}(\sigma) = \begin{cases} 1 & \text{if } i2^r < \sigma \le i2^r + 2^{r-1} \\ -1 & \text{if } i2^r + 2^{r-1} < \sigma \le (i+1)2^r \\ 0 & \text{otherwise.} \end{cases} \tag{18.115}$$

In this manner, we define a total of 2^p functions. These functions are orthogonal in $L^2(\theta)$ where θ is the uniform probability on $\{1, \dots, 2^p\}$ and thus form a complete orthogonal basis of this space. Let us note that $|f_{i,r}| \in \{0, 1\}$. Also, if $f \in \mathcal{H}_{p+1}$, we have $\int f^2 \mathrm{d}\theta = 1$, while if $r \le p$, we have

$$\int f_{i,r}^2 \mathrm{d}\theta = 2^{-p+r}\ . \tag{18.116}$$

Given three functions f_1, f_2, f_3 on $\{1, \dots, 2^p\}$ denote by $f_1 \otimes f_2 \otimes f_3$ the function on $G = \{1, \dots, 2^p\}^3$ given by $f_1 \otimes f_2 \otimes f_3((\sigma_1, \sigma_2, \sigma_3)) = f_1(\sigma_1) f_2(\sigma_2) f_3(\sigma_3)$. For $1 \le q_1, q_2, q_3 \le p+1$, let us denote by $\mathcal{V}(q_1, q_2, q_3)$ the set of functions of the type $v = f_1 \otimes f_2 \otimes f_3$ where $f_j \in \mathcal{H}(q_j)$ for $j \le 3$. The functions $v \in \mathcal{V}(q_1, q_2, q_3)$ have disjoint supports, and for $v \in \mathcal{V}$, we have $|v|^2 \in \{0, 1\}$. As q_1, q_2, and q_3 take all possible values, these functions form a complete orthogonal system of $L^2(\nu)$, where ν *denotes the uniform probability on* G. Consequently, given any function h on G, we have the expansion

$$h = \sum_{1 \le q_1, q_2, q_3 \le p+1} \sum_{v \in \mathcal{V}(q_1, q_2, q_3)} a_v(h) v\ , \tag{18.117}$$

where

$$a_v(h) = \frac{\int h v \mathrm{d}\nu}{\int v^2 \mathrm{d}\nu}\ . \tag{18.118}$$

The decomposition (18.117) then implies

$$\Big|\sum_{\tau \in G} h(\tau) Y_\tau\Big| \le \sum_{1 \le q_1, q_2, q_3 \le p+1} \sum_{v \in \mathcal{V}(q_1, q_2, q_3)} |a_v(h)| \Big|\sum_{\tau \in G} v(\tau) Y_\tau\Big|\ . \tag{18.119}$$

This will be our basic tool to prove Theorem 18.6.9, keeping (18.110) in mind. Fixing q_1, q_2, and q_3, the main effort will be to find competent bounds for

$$\sum_{v \in \mathcal{V}(q_1,q_2,q_3)} |a_v(h)| \Big| \sum_{\tau \in G} v(\tau) Y_\tau \Big| . \tag{18.120}$$

Since we think of q_1, q_2, and q_3 as fixed, we lighten notation by writing

$$\mathcal{V} := \mathcal{V}(q_1, q_2, q_3) \ ; \ \forall v \in \mathcal{V} \ , \ Y_v = \sum_{\tau \in G} v(\tau) Y_\tau \ . \tag{18.121}$$

Let us set $q_j^* = \min(q_j, p)$. Since $q_j \le p + 1$, we have $q_j - 1 \le q_j^* \le q_j$, so that

$$2^{3p-q_1-q_2-q_3} \le \operatorname{card} \mathcal{V} = 2^{3p-q_1^*-q_2^*-q_3^*} \le L 2^{3p-q_1-q_2-q_3} \ . \tag{18.122}$$

Also, recalling (18.116), we obtain that for $v \in \mathcal{V}(q_1, q_2, q_3)$

$$\int v^2 \mathrm{d}\nu = 2^{q_1^*+q_2^*+q_3^*-3p} \ge \frac{1}{L} 2^{q_1+q_2+q_3-3p} \ . \tag{18.123}$$

Recall Definition 18.6.8 of the class $\mathcal{S}^*(\Delta)$. The next task is, given a function $h \in \mathcal{S}^*(\Delta)$, to gather information about the coefficients $a_v(h)$. This information depends on the information we have about h, that is, the sets B_k and the coefficients $z(C)$. We think of h as fixed, and for $k \ge 4$, we consider the function R_k on G defined as follows:

$$R_k = 0 \text{ outside } B_k \ . \tag{18.124}$$

$$\text{If } C \in \mathcal{C}_k, \text{ then } R_k \text{ is constant } = z(C) \text{ on } C \ . \tag{18.125}$$

These functions have disjoint supports. They will be essential for the rest of this chapter. We may think of them as the parameters which govern "the size of h". Since $\nu(C) = 2^{s(k)-3p}$ for $C \in \mathcal{P}_k$,

$$\sum_{C \in \mathcal{C}_k} 2^{s(k)} z(C) = 2^{3p} \int R_k \mathrm{d}\nu \ , \tag{18.126}$$

and thus from (18.106),

$$\sum_{k \ge 4} \int R_k \mathrm{d}\nu \le L\Delta \ . \tag{18.127}$$

Our basic bound is as follows:

Proposition 18.7.1 *Consider q_1, q_2, q_3, and $q = q_1 + q_2 + q_3$. Consider $j \le 3$ with $q_j \le p$. Define k_j as the largest k for which $n_j(k) < q_j$. Then for any $v \in \mathcal{V} = \mathcal{V}(q_1, q_2, q_3)$, we have*

$$|a_v(h)| \le L2^{3p-q+q_j} \sum_{4 \le \ell \le k_j} 2^{-n_j(\ell)} \int |v| R_\ell \mathrm{d}\nu \, . \tag{18.128}$$

Since $|v| \in \{0, 1\}$, $\int |v| R_\ell \mathrm{d}\nu$ is simply the integral of R_ℓ on the support of v. A crucial fact here is that these supports are disjoint, so that

$$\sum_{v \in \mathcal{V}} \int |v| R_\ell \mathrm{d}\nu \le \int R_\ell \mathrm{d}\nu \, .$$

The reason why only terms for $4 \le \ell \le k_j$ appear in (18.128) is closely related to the fact that $h \in \mathcal{S}(\Delta)$ is constant on the elements C of $\mathcal{C}_k$ and that $a_v(\mathbf{1}_C) = 0$ for $C \in \mathcal{P}_k$ as soon as $q_j \le n_j(k)$ and $q_j \le p$ for some $j \le 3$. Observe also that Proposition 18.7.1 offers different bounds, one for each value of j with $q_j \le p$. We will choose properly between these bounds.

In view of (18.118) and (18.123), to prove Proposition 18.7.1, it suffices to show that when $q_j \le p$, we have

$$\Big| \int vh \mathrm{d}\nu \Big| \le 2^{q_j} \sum_{4 \le \ell \le k_j} 2^{-n_j(\ell)} \int |v| R_\ell \mathrm{d}\nu \, . \tag{18.129}$$

The proof relies on a simple principle to which we turn now. We say that a subset of $\mathbb{N}^*$ is a *dyadic interval* if it is of the type $\{r2^q + 1, \ldots, (r+1)2^q\}$ for some integers $r, q \ge 0$. The support of a function $f_{i,r}$ of (18.115) is a dyadic interval. Given two dyadic intervals I and J with $\operatorname{card} J \ge \operatorname{card} I$, we have

$$I \cap J \ne \emptyset \Rightarrow I \subset J \, .$$

Lemma 18.7.2 *Consider one of the functions $f_{i,r}$ of (18.115), and call its support I. Consider a partition $\mathcal{Q}$ of I into dyadic intervals. Assume that to each $J \in \mathcal{Q}$ is associated a number $z(J)$. Consider a function $g : I \to \mathbb{R}$. Assume that whenever $J, J' \in \mathcal{Q}$ are adjacent and* $\operatorname{card} J \ge \operatorname{card} J'$,

$$\sigma \in J \, , \, \sigma' \in J' \Rightarrow |g(\sigma) - g(\sigma')| \le z(J) \, . \tag{18.130}$$

Then

$$\Big| \sum_{\sigma \in I} f_{i,r}(\sigma) g(\sigma) \Big| \le \operatorname{card} I \sum_{J \in \mathcal{Q}} z(J) \, . \tag{18.131}$$

Let us insist that (18.130) is required in particular if $J = J' \in \mathcal{Q}$.

Proof We first prove that for all σ, σ' in I, we have

$$|g(\sigma) - g(\sigma')| \le 2 \sum_{J \in \mathcal{Q}} z(J) . \tag{18.132}$$

Without loss of generality, we may assume that $\sigma \le \sigma'$. Let us enumerate $\mathcal{Q}$ as $J_1, J_2, \ldots$ in a way that $J_{\ell+1}$ is immediately to the right of J_ℓ. If for some $J \in \mathcal{Q}$ we have $\sigma, \sigma' \in J$, then (18.130) implies $|g(\sigma)-g(\sigma')| \le z(J)$ and hence (18.132). Otherwise, $\sigma \in J_{\ell_1}$ and $\sigma' \in J_{\ell_2}$ for some $\ell_1 < \ell_2$. For $\ell_1 < \ell < \ell_2$, consider a point $\sigma_\ell \in J_\ell$. Set $\sigma_{\ell_1} = \sigma$ and $\sigma_{\ell_2} = \sigma'$. Then

$$|g(\sigma') - g(\sigma)| = |g(\sigma_{\ell_2}) - g(\sigma_{\ell_1})| \le \sum_{\ell_1 \le \ell < \ell_2} |g(\sigma_{\ell+1}) - g(\sigma_\ell)| . \tag{18.133}$$

Moreover, it follows from (18.130) (distinguishing whether card $J_{\ell+1} \ge$ card J_ℓ or the other way around) that

$$|g(\sigma_{\ell+1}) - g(\sigma_\ell)| \le z(I_\ell) + z(I_{\ell+1}) , \tag{18.134}$$

and combining with (18.133) proves (18.132). Letting $I_1 = \{i2^r+1, \ldots, i2^r+2^{r-1}\}$ and $I_2 = \{i2^r + 2^{r-1} + 1, \ldots, (i+1)2^r\}$, we have $I = I_1 \cup I_2$. Recalling that $I = \{i2^r + 1, \ldots, (i+1)2^r\}$, we have

$$\Big|\sum_\sigma g(\sigma) f_{i,r}(\sigma)\Big| = \Big|\sum_{\sigma \in I_1} g(\sigma) - \sum_{\sigma \in I_2} g(\sigma)\Big| = \Big|\sum_{\sigma \in I_1} (g(\sigma) - g(\sigma + 2^{r-1}))\Big|$$

and using (18.132), this concludes the proof since card $I_1 =$ card $I/2$. □

Proof of (18.129). Without loss of generality, we assume that $j = 1$, so that $q_1 \le p$. By definition of the class $\mathcal{V}(q_1, q_2, q_3)$, v is of the type $f_1 \otimes f_2 \otimes f_3$ for $f_j \in \mathcal{H}(q_j)$. Also, $\nu = \nu_1 \otimes \nu_2 \otimes \nu_3$, where ν_j is the uniform probability on $\{1, \ldots, 2^p\}$. Therefore,

$$\Big|\int v h \mathrm{d}\nu\Big| \le \Big|\int \Big(\int f_1 h \mathrm{d}\nu_1\Big) f_2 f_3 \mathrm{d}\nu_2 \mathrm{d}\nu_3\Big| \le \int \Big|\int f_1 h \mathrm{d}\nu_1\Big| |f_2 f_3| \mathrm{d}\nu_2 \mathrm{d}\nu_3 . \tag{18.135}$$

Let us fix τ^2 and τ^3 in $\{1, \ldots, 2^p\}$. We shall prove that

$$\Big|\int f_1(\sigma) h(\sigma, \tau^2, \tau^3) \mathrm{d}\nu_1(\sigma)\Big| \le S := 2^{q_1} \sum_{4 \le \ell \le k_1} 2^{-n_1(\ell)} \int |f_1(\sigma)| R_\ell(\sigma, \tau^2, \tau^3) \mathrm{d}\nu_1(\sigma) . \tag{18.136}$$

Therefore, using Fubini's theorem, (18.136) yields

$$\int \Big| \int f_1 h \mathrm{d}\nu_1 \Big| |f_2 f_3| \mathrm{d}\nu_2 \mathrm{d}\nu_3 \le 2^{q_1} \sum_{4 \le \ell \le k_1} 2^{-n_1(\ell)} \int |v| R_\ell \mathrm{d}\nu \ ,$$

and combining with (18.135) yields the result.

We shall deduce (18.136) from Lemma 18.7.2. Recalling the definition of the class $\mathcal{V}(q_1, q_2, q_3)$, f_1 is of the type $f_{i,r}$ given by (18.115) for $r = q_1 \le p$ and a certain value of i. Let I be the support of f_1, and note that $\int_I f(\sigma)\mathrm{d}\nu_1(\sigma) = 0$. Consider the map $\psi : \{1, \ldots, 2^p\} \to G$ given by $\psi(\sigma) = (\sigma, \tau^2, \tau^3)$, and let $I^* = \psi(I)$. Assume first

$$\exists \ell \ , \ n_1(\ell) \ge r = q_1 \ , \ \exists C \in \mathcal{C}_\ell \ , \ C \cap I^* \ne \emptyset \ . \tag{18.137}$$

Since $C \in \mathcal{P}_\ell$, $J = \psi^{-1}(C)$ is a dyadic interval with $\operatorname{card} J = 2^{n_1(\ell)} \ge 2^{q_1} = \operatorname{card} I$, and since $I \cap J \ne \emptyset$ because $C \cap I^* \ne \emptyset$ by (18.137), we have $I \subset J$. Now h is constant on C, so that the function $\sigma \mapsto h(\sigma, \tau_1, \tau_2)$ is constant on I. Since $\int_I f_1 \mathrm{d}\nu_1 = 0$ and I is the support of f_1, the left-hand side of (18.136) is zero in this case. So we may assume that (18.137) fails, i.e.,

$$\forall \ell \ , \ \forall C \in \mathcal{C}_\ell \ , \ C \cap I^* \ne \emptyset \Rightarrow n_1(\ell) < q_1 \ . \tag{18.138}$$

We consider the partition $\mathcal{Q}$ of I that consists of the sets of the type $\psi^{-1}(C) \cap I$, where, for some $\ell \ge 1$, $C \in \mathcal{C}_\ell$ and $C \cap I^* \ne \emptyset$. When $J = \psi^{-1}(C) \in \mathcal{Q}$, we define

$$z(J) := z(C) \ .$$

We now prove that (18.130) follows from (18.108). Consider $J, J' \in \mathcal{Q}$ which are adjacent with $\operatorname{card} J \ge \operatorname{card} J'$. Then $J = \psi^{-1}(C)$ and $J' = \psi^{-1}(C')$ where C and C' are adjacent and $C \in \mathcal{C}_\ell$, $C' \in \mathcal{C}_{\ell'}$. We claim that we may assume that $\ell \ge \ell'$. Since $\operatorname{card} J = 2^{n_1(\ell)}$ and $\operatorname{card} J' = 2^{n_1(\ell')}$, this is automatically the case if $\operatorname{card} J \ge \operatorname{card} J'$. If $\operatorname{card} J = \operatorname{card} J'$, it suffices if necessary to exchange the names of J and J'. Thus, $C' \in \mathcal{C}_{\ell'}$ for some $\ell' \le \ell$ and in particular $C' \subset B_{\ell'}$ so that $C' \not\subset B_k$ for $k > \ell$, and then by (18.108), we have $|h(\tau) - h(\tau')| \le z(C) = z(J)$ for $\tau \in C$ and $\tau' \in C'$, and this proves (18.130).

We define

$$S^* := \sum_{J \in \mathcal{Q}} z(J) = \sum \Big\{ z(C) \ ; \ C \cap I^* \ne \emptyset \ , \ C \in \bigcup_{\ell \ge 1} \mathcal{C}_\ell \Big\} \ . \tag{18.139}$$

Using Lemma 18.7.2 in the inequality, we obtain

$$\Big|\int |f_1(\sigma)h(\sigma,\tau^1,\tau^2)\Big| = 2^{-p}\Big|\sum_{\sigma\in I} f_1(\sigma)h(\sigma,\tau^2,\tau^3)\Big|$$
$$\le (2^{-p}\operatorname{card} I)S^* = 2^{-p+q_1}S^* .$$

Recalling the quantity S of (18.136), we now prove that $2^{q_1}S^* = 2^p S$, finishing the proof of (18.136) and of the lemma. For this, we observe that if $C \in \mathcal{C}_\ell$ is such that $C \cap I^* \neq \emptyset$, then $J = \psi^{-1}(C)$ is a dyadic interval with $J \cap I \neq \emptyset$. Moreover, since (18.138) implies $\operatorname{card} J = 2^{n_1(\ell)} \le \operatorname{card} I = 2^{q_1}$, we have $J \subset I$, so that $\operatorname{card}(C \cap I^*) = 2^{n_1(\ell)}$. Consequently,

$$z(C) = 2^{-n_1(\ell)}\sum_{\sigma\in J}|f_1(\sigma)|R_\ell(\sigma,\tau^2,\tau^3) \tag{18.140}$$

because there are $2^{n_1(\ell)}$ non-zero terms in the summation, and for each of these terms, $|f_1(\sigma)| = 1$ and $R_\ell(\sigma,\tau^2,\tau^3) = z(C)$. We rewrite (18.140) as

$$z(C) = 2^{p-n_1(\ell)}\int_J |f(\sigma)|R_\ell(\sigma,\tau^2,\tau^3) .$$

Summation of the relations (18.140) over $C \in \bigcup_{\ell\ge 1}\mathcal{C}_\ell$ with $C\cap I^* \neq \emptyset$ then proves that $2^p S = 2^{q_1}S^*$. □

18.8 Probability, II

We go back to the problem of bounding the quantities (18.120)

$$\sum_{v\in\mathcal{V}(q_1,q_2,q_3)} |a_v(h)||Y_v| , \tag{18.141}$$

where the r.v.s Y_v are defined in (18.113). We think of q_1, q_2, and q_3 as fixed, and we write again $q = q_1 + q_2 + q_3$ and $\mathcal{V} = \mathcal{V}(q_1,q_2,q_3)$. We will then bound with high probability the sum (18.141) by a quantity $C(q_1,q_2,q_3)$ in such a way that the sum over q_1, q_2, and q_3 of these quantities is $\le L\sqrt{m_0}2^{3p}\Delta$. The plan is to combine the bound of Proposition 18.7.1 with probabilistic estimates. Computation of $\mathsf{E}Y_v^2$ shows that $|Y_v|$ is typically of size about $\sqrt{m_0}2^{q/2}$ but some of the quantities $|Y_v|$ might be much larger than their typical values.

We will use the same kindergarten technique as in Sect. 18.6, namely, if b_v are positive numbers and if

$$A_r = \max_{\mathrm{card}\, I = r} \sum_{v \in I} |Y_v| \,, \tag{18.142}$$

then

$$\sum_{v \in \mathcal{V}} b_v |Y_v| \le A_1 \sum_{v \in \mathcal{V}} b_v \tag{18.143}$$

and

$$\sum_{v \in \mathcal{V}} b_v |Y_v| \le \frac{A_r}{r} \sum_{v \in \mathcal{V}} b_v + A_r \max_{v \in \mathcal{V}} b_v \,. \tag{18.144}$$

The random quantity A_r depends also on q_1, q_2, and q_3, but this is kept implicit.

Lemma 18.8.1 *For each $u > 0$, and $r \le 2^{3p}$, we have*

$$\mathsf{P}(A_r \ge u) \le 2^r \Big(\frac{e 2^{3p}}{r}\Big)^r \exp\Big(-\frac{1}{L} \min\Big(u, \frac{u^2}{2^q m_0 r}\Big)\Big) \,. \tag{18.145}$$

Proof Since the functions $v \in \mathcal{V}$ have disjoint supports, it should be obvious from the definition (18.113) that the family of r.v.s $(Y_v)_{v \in \mathcal{V}}$ belongs to $\mathcal{B}(N)$. When $A_r > u$, we can find a set I with $\mathrm{card}\, I = r$ and numbers $\eta_v = \pm 1$, $v \in I$ with $\sum_{v \in I} \eta_v Y_v \ge u$. The result then follows from (18.94) and the union bound since $\mathrm{card}\, \mathcal{V} \le 2^{3p}$, $N \le 2^{3p+1} m_0$ and $\sum_{v \in I} \eta_v^2 = r$. □

Corollary 18.8.2 *With probability $\ge 1 - L\exp(-100p)$, the following occurs. Consider any $1 \le q_1, q_2, q_3 \le p + 1$ and $q = q_1 + q_2 + q_3$. Then for each $k \ge 4$, we have*

$$2^{3p} < N_{k+2} \,,\ k \le q \Rightarrow A_1 \le L\sqrt{m_0} 2^{k/2} 2^{q/2} \tag{18.146}$$

and

$$k \le q \le p, N_{k+2} \le 2^{3p}, r := \lfloor 2^{3p}/N_{k+2} \rfloor \Rightarrow A_r \le Lr\sqrt{m_0} 2^{k/2} 2^{q/2} \,. \tag{18.147}$$

Proof To prove (18.146), we observe that since $2^{3p} < N_{k+2}$, we have $2^k \ge p/L$. Considering a parameter $C \ge 1$, we take $u = C\sqrt{m_0} 2^{k/2} 2^{q/2} \ge 2^k$ in (18.145) so that $\min(u, u^2/(m_0 2^q)) \ge C 2^k$ and the result by taking C a large enough constant.

To prove (18.147), we note if $x \geq 1$ we have $\lfloor x \rfloor \geq 1$ so that $x \leq \lfloor x \rfloor + 1 \leq 2\lfloor x \rfloor$. Therefore, since $N_{k+2} \leq 2^{3p}$, we have $r = \lfloor 2^{3p}/N_{k+2} \rfloor \geq 1$ and $2^{3p}/N_{k+2} \leq 2r$ so that $2^{3p}/r \leq 2N_{k+2}$. We first prove that $r2^k \geq p/L$. If $N_{k+2} \leq 2^p$, then $2^{3p}/N_{k+2} \geq 2^{2p}$ and $r2^k \geq r \geq 2^{2p} \geq p/L$. If $N_{k+2} \geq 2^p$, this holds because then $2^k \geq p/L$.

Consider then a parameter $B \geq 1$. For $u = Br\sqrt{m_0}2^{k/2}2^{q/2}$, since $2^{q/2+k/2} \geq 2^k$, we have $\min(u, u^2/2^q m_0 r) \geq B2^k r$, and since $2^{3p}/r \leq \exp L2^k$, the bound in (18.145) is $\leq \exp(r2^k(L - B/L))$. Since $r2^k \geq p/L$, when B is a large enough constant, we obtain $\exp(r2^k(L - B/L)) \leq L\exp(-100p)$. □

Corollary 18.8.2 contains all the probabilistic estimates we need. From that point on, we assume that (18.146) and (18.147) hold, and we draw consequences. Thus, all these consequences will hold with probability $\geq 1 - L\exp(-100p)$. We first reformulate (18.143) and (18.144).

Proposition 18.8.3 *Consider numbers $(b_v)_{v\in\mathcal{V}}$, $b_v \geq 0$. Then for $k \leq q$, we have*

$$2^{3p} < N_{k+2} \Rightarrow \sum_{v\in\mathcal{V}} b_v|Y_v| \leq L\sqrt{m_0}2^{k/2}2^{q/2}\sum_{v\in\mathcal{V}} b_v \tag{18.148}$$

and

$$N_{k+2} \leq 2^{3p} \Rightarrow \sum_{v\in\mathcal{V}} b_v|Y_v| \leq L\sqrt{m_0}2^{k/2}2^{q/2}\Big(\sum_{v\in\mathcal{V}} b_v + \frac{2^{3p}}{N_{k+2}}\max_{v\in\mathcal{V}} b_v\Big) . \tag{18.149}$$

It is good to observe that (18.149) has the same form as (18.148) but with an extra term. Controlling this extra term requires controlling $\max_{v\in\mathcal{V}} b_v$. One problem in using Proposition 18.8.3 is to decide which value of k to use. If the value is too large, the term $2^{k/2}$ may become too large, but if the value of k is too small, then it is the extra term which creates issues.

Proof (18.148) is an immediate consequence of (18.143) and (18.146). To prove (18.149), we use (18.144) for the value $r = \lfloor 2^{3p}/N_{k+2} \rfloor$. In the second term of (18.144), we use the bound

$$A_r \leq Lr\sqrt{m_0}2^{k/2}2^{q/2} \leq L\frac{2^{3p}}{N_{k+2}}\sqrt{m_0}2^{k/2}2^{q/2} .$$

In the first term of (18.144), we use the bound $A_r/r \leq L\sqrt{m_0}2^{k/2}2^{q/2}$. □

We will typically use Proposition 18.8.3 with $b_v = |a_v(h)|$, which is why we systematically try to estimate $\sum_{v\in\mathcal{V}} |a_v(h)|$ and $\max_{v\in\mathcal{V}} |a_v(h)|$.

18.9 Final Effort

We turn to the task of bounding the quantities

$$S(q_1, q_2, q_3) := \sum_{v \in \mathcal{V}(q_1,q_2,q_3)} |a_v(h)||Y_v|$$

of (18.141). For each value of q_1, q_2, and q_3, our goal is to provide a suitable bound on $S(q_1, q_2, q_3)$ to imply that

$$\sum_{q} \sum_{q_1+q_2+q_3=q} S(q_1, q_2, q_3) \le L\sqrt{m_0}2^{3p}\Delta \,. \tag{18.150}$$

The proof relies on Proposition (18.8.3). The proof is not easy for a reason which is quite intrinsic. There are lower-order terms (many of them) for which there is plenty of room, but there are also "dangerous terms" for which there is no room whatsoever and for which every estimate has to be tight. This is to be expected when one proves an essentially optimal result.

Let us start by a simple result showing that the large values of q do not create problems.

Lemma 18.9.1 *We have*

$$S(q_1, q_2, q_3) \le Lp2^{3p-q/6}\sqrt{m_0}\Delta \,. \tag{18.151}$$

This bound is not useful for q small because of the extra factor p. But it is very useful for large values of q because of the factor $2^{-q/6}$.

Proof First, if $q_1 = q_2 = q_3 = p + 1$, then the unique element v of $\mathcal{V}(q_1, q_2, q_3)$ is the constant function equal to 1 and $Y_v = \sum_{\tau\in G} v(\tau)Y_\tau = \sum_{\tau\in G} Y_\tau = 0$ (as is obvious from the definition (18.84) of Y_τ). In all the other cases, one of the q_j is $\le p$ so that $q := q_1 + q_2 + q_3 < 3(p + 1)$. We then choose $j \le 3$ such that $q_j \le q/3 < p + 1$, so that we can use the bound (18.128). We use the trivial bound $n_j(\ell) \ge 0$, and we get[5]

$$|a_v(h)| \le L2^{3p-2q/3} \sum_{4\le\ell\le k_j} \int |\nu|R_\ell \mathrm{d}\nu \,.$$

[5] This was our first instance of choosing between the various bounds proposed by Proposition 18.7.1. There will be several others.

Thus, since the functions $v \in \mathcal{V}(q_1, q_2, q_3)$ have disjoint supports and satisfy $|v| \leq 1$, and using (18.127),

$$\sum_{v \in \mathcal{V}(q_1,q_2,q_3)} |a_v(h)| \leq L 2^{3p-2q/3} \Delta \,. \tag{18.152}$$

Consider the smallest k for which $2^{3p} \leq N_{k+2}$ so that $N_{k+1} \leq 2^{3p}$ and then $2^k \leq Lp$. Using (18.148) for $b_v = |a_v(h)|$, we obtain

$$\sum_{v \in \mathcal{V}(q_1,q_2,q_3)} |a_v(h)||Y_v| \leq L\sqrt{m_0} 2^{k/2} 2^{q/2} \sum_{v \in \mathcal{V}(q_1,q_2,q_3)} |a_v(h)| \,,$$

and combining with (18.152) and using that $2^k \leq p$ yields (18.151). □

To illustrate the use of (18.151), we note that there are at most q^3 possible choices of q_1, q_2, and q_3 for which $q_1 + q_2 + q_3 = q$, so that

$$\sum_{q_1+q_2+q_3=q} S(q_1, q_2, q_3) \leq Lpq^3 2^{3p-q/6} \sqrt{m_0} \Delta \,. \tag{18.153}$$

Summing these inequalities over $q \geq p$ and using that $p \sum_{q \geq p} q^3 2^{-q/6} \leq L$ shows that in the sum (18.151), we need only be concerned with the values of $q \leq p$. From now on, we always assume that $q \leq p$. We classify the elements $v \in \mathcal{V}$ by the integer

$$k(v) = \max \left\{ k \,;\, \int |v| R_k \mathrm{d}\nu \neq 0 \right\} , \tag{18.154}$$

the largest value of k for which the support of v meets B_k (so that $k(v) \geq 4$). Given q_1, q_2, and q_3 and an integer k, we define

$$S(q_1, q_2, q_3, k) = \sum_{v \in \mathcal{V}, k(v)=k} |a_v(h)||Y_v| \,, \tag{18.155}$$

so that $S(q_1, q_2, q_3) = \sum_{k \geq 4} S(q_1, q_2, q_3, k)$.

Lemma 18.9.2 *Set $q^* = \lfloor q/4 \rfloor$. Then for $k' \leq q^*$, we have*

$$S(q_1, q_2, q_3, k') \leq L\sqrt{m_0} 2^{3p} \Big(2^{-q/24} \Delta + \frac{2^{2q}}{N_{q^*+1}} \Big) \,. \tag{18.156}$$

Proof In one sentence, this bound follows from Proposition 18.8.3 by taking $k = q^*$. We choose j such that $q_j \le q/3$, so that using again the trivial bound $2^{-n_j(\ell)} \le 1$, the bound (18.128) implies that for $v \in \mathcal{V}$,

$$|a_v(h)| \le L2^{3p-2q/3} \sum_{4\le\ell\le k_j} \int |v|R_\ell \mathrm{d}\nu \; . \tag{18.157}$$

We recall that $\sum_{v\in\mathcal{V}} \int |v|R_\ell \mathrm{d}\nu \le \int R_\ell \mathrm{d}\nu$ because the functions $v \in \mathcal{V}$ have disjoint support and satisfy $|v| \le 1$. Using this in the first inequality and (18.127) in the second inequality,

$$\sum_{v\in\mathcal{V}} |a_v(h)| \le L2^{3p-2q/3} \sum_{\ell\ge 4} \int R_\ell \mathrm{d}\nu \le L2^{3p-2q/3}\Delta \; . \tag{18.158}$$

It is the factor $2^{-2q/3}$ which saves the day. We are going to apply Proposition 18.8.3 with $k = q^*$. The factor $2^{k/2+q/2}$ on the right of (18.148) and (18.149) is then $2^{q^*/2+q/2} \le 2^{5q/8}$, and $5q/8 - 2q/3 = -q/24$.

When $k(v) \le q^*$ by definition of $k(v)$, we have $\int |v|R_\ell \mathrm{d}\nu = 0$ for $\ell > q^*$, so that (18.157) implies

$$k(v) \le q^* \Rightarrow |a_v(h)| \le L2^{3p}2^{-2q/3} \sum_{4\le\ell\le q^*} \int |v|R_\ell \mathrm{d}\nu \; ,$$

Since $\int |v|\mathrm{d}\nu = 2^{q-3p}$ and $R_\ell \le N_{\ell+1}$ and $\int |v|\mathrm{d}\nu = 2^{q-3p}$, the supports of the functions R_ℓ being disjoint, we obtain $|a_v(h)| \le L2^q N_{q^*+1}$.

Let us now consider $k' \le q^*$ and set $b_v = |a_v(h)|$ if $k(v) = k'$ and $b_v = 0$ otherwise. Using (18.158) in the first inequality, we have proved that

$$\sum_{v\in\mathcal{V}} b_v \le L2^{3p-2q/3}\Delta \; ; \; \max_{v\in\mathcal{V}} b_v \le L2^q N_{q^*+1} \; . \tag{18.159}$$

If $2^{3p} < N_{q^*+2}$, we use (18.148) for $k = q^*$ and (18.159) to obtain

$$S(q_1, q_2, q_3, k') \le L\sqrt{m_0} 2^{q^*/2} 2^{q/2} 2^{3p-2q/3}\Delta$$

which implies (18.156) since $q^* \le q/4$ (and hence $q^*/2 + q/2 - 2q/3 \le -q/24$). If $N_{q^*+2} \le 2^{3p}$, we then use (18.149) and (18.159) to obtain

$$S(q_1, q_2, q_3, k') \le L\sqrt{m_0} 2^{q^*/2} 2^{q/2} \Big(2^{3p-2q/3}\Delta + \frac{2^{3p}}{N_{q^*+2}} 2^q N_{q^*+1} \Big) \; . \tag{18.160}$$

Using that $N_{q^*+2} = N^2_{q^*+1}$ and since $q^* \le q/4$, we obtain (18.156) again. □

The bound (18.156) sums well over q since N_{q^*} is so large.

Matters become more complicated in the case $k(v) \geq q^*$, because we no longer have a strong bound such as (18.159) for the values of $|a_v(h)|$. So we have to use a larger value of k to be able to control the last term in (18.149). But then the factor $2^{k/2+q/2}$ becomes large, and we have to be more sophisticated; we can no longer afford to use the crude bound $2^{-n_j(\ell)} \leq 1$ in (18.128). In fact, in that case, there are "dangerous terms" for which there is little room. How to use the factor $2^{-n_j(\ell)}$ is described in Lemma 18.9.4.

Let us set[6]

$$\beta_k = \int R_k \mathrm{d}\nu \;, \tag{18.161}$$

so that (18.127) implies

$$\sum_{k \geq 4} \beta_k \leq L\Delta \;, \tag{18.162}$$

and we also have

$$\sum_{v \in \mathcal{V}} \int |v| R_k \mathrm{d}\nu \leq \beta_k \;. \tag{18.163}$$

Let us also define

$$D(k, q_1, q_2, q_3) = \prod_{j \leq 3} \mathbf{1}_{\{n_j(k) \leq q_j\}} \;. \tag{18.164}$$

Proposition 18.9.3 *For $k > q^* = \lfloor q/4 \rfloor$, we have*

$$\begin{aligned} S(q_1, q_2, q_3, k) \leq{} & L D(k, q_1, q_2, q_3) \sqrt{m_0} 2^{3p} 2^{(k-q)/6} (\beta_k + \beta_{k-1}) \\ & + L\sqrt{m_0} 2^{3p} \Big(\Delta \frac{2^{2q}}{N_{k-1}} + \frac{2^{q/2+k/2}}{N_{k+1}} \Big) \;. \end{aligned} \tag{18.165}$$

The crucial term is the factor $2^{(k-q)/6}$, which will sum nicely over $q \geq k$. There is plenty of room to estimate the second-order quantities represented by the second term.

The proof of Proposition 18.9.3 requires two lemmas.

Lemma 18.9.4 *If $k(v) = k \geq q^*$, we have*

$$|a_v(h)| \leq b_v(h) + c_v(h) \;, \tag{18.166}$$

[6] β_k depends on h although this is not indicated in the notation.

where

$$c_v(h) := L2^{3p} \sum_{4\le\ell\le k-2} \int |v| R_\ell \mathrm{d}\nu \ , \tag{18.167}$$

and

$$b_v(h) := LD(k, q_1, q_1, q_3) 2^{3p} 2^{-2q/3-k/3} \int |v|(R_k + R_{k-1}) \mathrm{d}\nu \ . \tag{18.168}$$

As we will show in the next lemma, the terms $c_v(h)$ are a secondary nuisance, but the terms $b_v(h)$ will be harder to control, because to control them all the estimates have to be tight. The purpose of the decomposition is to identify this "dangerous part"$b_v(h)$ of $|a_v(h)|$ and, when using the bound (18.129), to choose the value of j with the goal of controlling this dangerous part as well as possible, in particular by creating the crucial factor $2^{-2q/3-k/3}$ in (18.168).

Proof We recall the bound (18.128): If k_j is the largest value of k for which $n_j(k) < q_j$, then for any $j \le 3$,

$$|a_v(h)| \le L2^{3p-q+q_j} \sum_{4\le\ell\le k_j} 2^{-n_j(\ell)} \int |v| R_\ell \mathrm{d}\nu \ .$$

We split the summation to obtain $|a_v(h)| \le b_{v,j}(h) + c_v(h)$ where

$$b_{v,j}(h) = L2^{3p-q+q_j} \sum_{\ell\in\{k,k-1\};4\le\ell\le k_j} 2^{-n_j(\ell)} \int |v| R_\ell \mathrm{d}\nu \tag{18.169}$$

and $c_v(h)$ is given by (18.167), using also there the crude bound $q_j - n_j(\ell) \le q$. Let us now choose j. If for some $j \le 3$ we have $k \ge k_j + 2$, we choose such a j. Then the term $b_{v,j}(h)$ is zero because there is no term in the summation and we are done. Otherwise, $k \le k_j + 1$ for all $j \le 3$, and since $n_j(k-1) \ge n_j(k) - 1$, we have

$$b_{v,j}(h) \le L2^{3p-q+q_j-n_j(k)} \int |v|(R_k + R_{k-1}) \mathrm{d}\nu \ .$$

Since $k \le k_j + 1$ for each $j \le 3$, we have $n_j(k) \le n_j(k_j+1) \le n_j(k_j) + 1 \le q_j$, and thus $D(k, q_1, q_2, q_3) = 1$. We choose $j \le 3$ such that

$$q_j - n_j(k) \le \frac{1}{3} \sum_{j'\le 3} (q_{j'} - n_{j'}(k)) = \frac{1}{3}(q - s(k)) = \frac{1}{3}(q-k) \ ,$$

so that $b_{v,j}(h) \le b_v(h)$ where $b_v(h)$ is given by (18.168). □

We lighten notation by writing $\sum_{v\in\mathcal{V},k(v)=k}$ as $\sum_{k(v)=k}$.

Lemma 18.9.5 *We have*

$$\sum_{k(v)=k} c_v(h) \le L2^{3p}\Delta\frac{2^q}{N_{k-1}} . \tag{18.170}$$

Here we see how useful it is to have controlled separately the small values of k and to assume $k \ge q^*$, which ensures that $2^q/N_{k-1}$ is very small.

Proof We bound the sum by the number of terms times the maximum of each term. Since $s(k) = k$, combining (18.107) and (18.106) shows that $\operatorname{card}\mathcal{C}_k \le L\Delta 2^{3p-s(k)}/N_k$. Each $C \in \mathcal{C}_k$ has cardinality $\le 2^{s(k)}$ and can meet at most $2^{s(k)}$ supports of different functions v so that

$$\operatorname{card}\{v \in \mathcal{V}\ ;\ k(v) = k\} \le L\Delta\frac{2^{3p}}{N_k} . \tag{18.171}$$

Since $R_\ell \le N_{\ell+1}$, we have, using that $\int |v|\mathrm{d}\nu = 2^{q-3p}$ for $v \in \mathcal{V}$,

$$\sum_{4\le\ell\le k-2}\int |v|R_\ell \mathrm{d}\nu \le LN_{k-1}\int |v|\mathrm{d}\nu \le L2^{q-3p}N_{k-1} , \tag{18.172}$$

so that $|c_v(h)| \le L2^q N_{k-1}$. Combining with (18.171) proves the result since $N_k = N_{k-1}^2$. □

Proof of Proposition 18.9.3. From Lemma 18.9.4 and (18.170), (18.163), we obtain

$$\sum_{k(v)=k} |a_v(h)| \le LD(k, q_1, q_2, q_3)2^{3p}2^{-2q/3-k/3}(\beta_k + \beta_{k-1}) + L2^{3p}\Delta\frac{2^q}{N_{k-1}} . \tag{18.173}$$

To bound $\max_{k(v)=k}|a_v(h)|$, we go back to (18.128). Since $\sum_{\ell\le k}\int |v|R_\ell \mathrm{d}\nu \le L2^{q-3p}N_{k+1}$, we obtain the bound

$$|a_v(h)| \le L2^q N_{k+1} . \tag{18.174}$$

The rest of the argument is nearly identical to the end of the proof of Lemma 18.9.2 using now (18.173) and (18.174) instead of (18.159) and (18.160). Let $b_v = |a_v(h)|$ for $k(v) = k$ and $b_v = 0$ otherwise. We use the bound (18.148) if $2^{3p} < N_{k+2}$. Otherwise, we use the bound (18.149). This concludes the proof, using also that $N_{k+2} = N_{k+1}^2$. □

Combining Lemmas 18.9.2 and Proposition 18.9.3, we obtain the following:

Proposition 18.9.6 *Recalling that $q^* = \lfloor q/4 \rfloor$ have*

$$S(q_1, q_2, q_3) \le L\sqrt{m_0}2^{3p}\Big(2^{-q/24}\Delta + \sum_{k \ge q^*} \big(A(k, q_1, q_2, q_3, h) + B(k, q)\big)\Big) , \tag{18.175}$$

where

$$A(k, q_1, q_2, q_3, h) := D(k, q_1, q_2, q_3)2^{(k-q)/6}(\beta_k + \beta_{k-1}) , \tag{18.176}$$

$$B(k, q) := \Delta \frac{2^{2q}}{N_{k-1}} + \frac{2^{q/2+k/2}}{N_{k+1}} . \tag{18.177}$$

Proof of Theorem 18.6.9. We have to prove that (18.109) holds. Combining (18.110) and (18.119), we have to prove that

$$\sum_{3 \le q \le 3p+3} \sum_{q_1+q_2+q_3=q} S(q_1, q_2, q_3) \le L\sqrt{m_0}2^{3p}\Delta . \tag{18.178}$$

Lemma 18.9.1 takes care of the summation over $q \ge p$. Control of the summation for $q \le p$ will be obtained by summing the inequalities (18.175) and interchanging the summation in k and q. Given q, there are at most q^3 possible values of (q_1, q_2, q_3) with $q = q_1 + q_2 + q_3$, and

$$\sum_{q \ge 1} q^3 2^{-q/24} \le L .$$

Also,

$$\sum_{q \ge 1} \sum_{k \ge q^*} q^3 B(k, q) \le L(1 + \Delta) \le L\Delta ,$$

because $q^* = \lfloor q/4 \rfloor$ and N_k is doubly exponential in k. It remains only to take care of the contribution of the term $A(k, q_1, q_2, q_3, h)$. We will prove that

$$\sum_{q_1,q_2,q_3} \sum_{k \ge 4} A(k, q_1, q_2, q_3, h) \le L\Delta , \tag{18.179}$$

and this will finish the proof. The first step is to exchange the order of summation

$$\sum_{q_1,q_2,q_3} \sum_{k \ge 4} A(k, q_1, q_2, q_3, h) = \sum_{k \ge 4}(\beta_k + \beta_{k-1}) \sum_{q_1,q_2,q_3} D(k, q_1, q_2, q_3)2^{(k-q)/6} . \tag{18.180}$$

Each term $D(k, q_1, q_2, q_3)$ is 0 or 1. When $D(k, q_1, q_2, q_3) = 1$, we have $n_j(k) \le q_j$ for each $j \le 3$. Since $\sum_{j\le 3} n_j(k) = k$, the non-negative integers $q_j - n_j(k)$ have a sum $\le q - k$. This can happen only for $q \ge k$, and then crudely there are at most $(q - k + 1)^3$ possible choices of q_1, q_2, and q_3 of a given sum q. That is,

$$\sum_{q_1,q_2,q_3} D(k, q_1, q_2, q_3) 2^{(k-q)/6} \le \sum_{q\ge k} (q - k + 1)^3 2^{(k-q)/6} \le L\,. \qquad (18.181)$$

The required inequality (18.179) then follows from (18.180), (18.181), and (18.162).

□

Chapter 19
Applications to Banach Space Theory

We concentrate on topics which make direct use of our previous results. Many more results in Banach space theory use probabilistic constructions, for which the methods of the book are relevant. Some of these results may be found in [132]. The reader should not miss the magnificent recent results of Gilles Pisier proved in Sect. 19.4.

As is customary, we use the same notation for the norm on a Banach space X and on its dual X^*. The norm on the dual is given by $\|x^*\| = \sup\{x^*(x); \|x\| \le 1\}$ so that in particular $|x^*(x)| \le \|x^*\|\|x\|$. The reader will keep in mind the duality formula

$$\|x\| = \sup\{x^*(x) \; ; \; \|x^*\| \le 1\} = \sup\{x^*(x) \; ; \; x^* \in X_1^*\} \, , \tag{19.1}$$

which is of constant use.

I am particularly grateful to Rafał Meller for his help with this chapter.

19.1 Cotype of Operators

The notion of cotype of a Banach space reflects a basic geometric property of this space, but we will study only very limited aspects related to our previous results.

19.1.1 Basic Definitions

We start by recalling some basic definitions. More background can be found in classical books such as [27] or [137].

M. Talagrand, *Upper and Lower Bounds for Stochastic Processes*, Ergebnisse der Mathematik und ihrer Grenzgebiete. 3. Folge / A Series of Modern Surveys in Mathematics 60, https://doi.org/10.1007/978-3-030-82595-9_19

Given an operator U (i.e., a continuous linear map) from a Banach space X to a Banach space Y and a number $q \geq 2$, we denote by $C_q^g(U)$ its Gaussian cotype-q constant, that is, the smallest number A (possibly infinite) for which, given any integer n, and any elements $x_1, \ldots, x_n$ of X, we have

$$\Big(\sum_{i \leq n} \|U(x_i)\|^q\Big)^{1/q} \leq A\mathsf{E}\Big\|\sum_{i \leq n} g_i x_i\Big\| . \tag{19.2}$$

Here, $(g_i)_{i \leq n}$ are i.i.d. standard Gaussian r.v.s, the norm of $U(x_i)$ is in Y and the norm of $\sum_{i \leq n} g_i x_i$ is in X.

The occurrence of the quantity

$$\mathsf{E}\Big\|\sum_{i \leq n} g_i x_i\Big\| = \mathsf{E} \sup_{x^* \in X_1^*} \sum_{i \leq n} g_i x^*(x_i) , \tag{19.3}$$

where $X_1^* = \{x^* \in X^*; \|x^*\| \leq 1\}$ suggests that results on Gaussian processes will bear on this notion. This is only true to a small extent. It is not really the understanding of the size of the quantity (19.3) at given $x_1, x_2, \ldots, x_n$ which matters but the fact that (19.2) has to hold for any elements $x_1, x_2, \ldots, x_n$.

Given a number $q \geq 2$, we define the Rademacher cotype-q constant $C_q^r(U)$ as the smallest number A (possibly infinite) such that, given any integer n, any elements $(x_i)_{i \leq n}$ of X, we have

$$\Big(\sum_{i \leq n} \|U(x_i)\|^q\Big)^{1/q} \leq A\mathsf{E}\Big\|\sum_{i \leq n} \varepsilon_i x_i\Big\| , \tag{19.4}$$

where $(\varepsilon_i)_{i \leq n}$ are i.i.d. Bernoulli r.v.s. The name "Rademacher cotype" stems from the fact that Bernoulli r.v.s are usually (but inappropriately) called Rademacher r.v.s in Banach space theory. Since Bernoulli processes are tricker than Gaussian processes, we expect that Rademacher cotype will be harder to understand than Gaussian cotype. This certainly seems to be the case.

Proposition 19.1.1 *We have*

$$C_q^g(U) \leq \sqrt{\frac{\pi}{2}}C_q^r(U) . \tag{19.5}$$

Proof Indeed (7.40) implies $\mathsf{E}\|\sum_{i \leq n} \varepsilon_i x_i\| \leq \sqrt{\pi/2}\mathsf{E}\|\sum_{i \leq n} g_i x_i\|$. □

Given $q \geq 1$, we define the $(q, 1)$-summing norm $\|U\|_{q,1}$ of U as the smallest number A (possibly infinite) such that, for any integer n, any vectors $x_1, \ldots, x_n$ of X we have

$$\Big(\sum_{i \leq n} \|U(x_i)\|^q\Big)^{1/q} \leq A \max_{\epsilon_i = \pm 1} \Big\|\sum_{i \leq n} \epsilon_i x_i\Big\| . \tag{19.6}$$

It should then be obvious that $\|U\|_{q,1} \le C_q^r(U)$. Consequently,

$$\sqrt{\frac{2}{\pi}} \max(C_q^g(U), \|U\|_{q,1}) \le C_q^r(U) . \tag{19.7}$$

Research Problem 19.1.2 Is it true that for some universal constant L and every operator U between Banach spaces we have

$$C_q^r(U) \le L \max(C_q^g(U), \|U\|_{q,1}) \ ? \tag{19.8}$$

A natural approach to this question would be a positive answer to the following far-reaching generalization of the Latała-Bednorz theorem:

Research Problem 19.1.3 (S. Kwapien) Does there exist a universal constant L with the following property: Given any Banach space X and elements $x_1, \dots, x_n$ of X, we can write $x_i = x_i' + x_i''$ where

$$\mathsf{E}\Big\| \sum_{i\le n} g_i x_i' \Big\| \le L\mathsf{E}\Big\| \sum_{i\le n} \varepsilon_i x_i \Big\| \ ; \ \max_{\epsilon_i=\pm1} \Big\| \sum_{i\le n} \epsilon_i x_i'' \Big\| \le L\mathsf{E}\Big\| \sum_{i\le n} \varepsilon_i x_i \Big\| \ .$$

Exercise 19.1.4 Prove that a positive answer to Problem 19.1.3 provides a positive answer to Problem 19.1.2. Hint: Study the proof of Theorem 19.1.5 below.

19.1.2 Operators from ℓ_N^∞

We now specialize to the case where X is the space ℓ_N^∞ of sequences $x = (x_j)_{j\le N}$ provided with the norm

$$\|x\| = \sup_{j\le N} |x_j| \ ,$$

and we give a positive answer to Problem 19.1.2.[1]

Theorem 19.1.5 *Given $q \ge 2$ and an operator U from ℓ_N^∞ to a Banach space Y, we have*

$$\sqrt{\frac{2}{\pi}} \max(C_q^g(U), \|U\|_{q,1}) \le C_q^r(U) \le L \max(C_q^g(U), \|U\|_{q,1}) . \tag{19.9}$$

[1] It is possible to show that similar results hold in the case where $X = C(W)$, the space of continuous functions over a compact topological space W. This is deduced from the case $X = \ell_N^\infty$ using a reduction technique unrelated to the methods of this book; see [60].

The reason we succeed is that in the case $X = \ell^\infty_N$ we can give a positive answer to Problem 19.1.3, as a simple consequence of the Latała-Bednorz Theorem.[2]

Proposition 19.1.6 *Consider n elements $x_1, \ldots, x_n$ of ℓ^∞_N. Then we can find a decomposition $x_i = x'_i + x''_i$ such that*

$$\mathsf{E}\Big\|\sum_{i\le n} g_i x'_i\Big\| \le L\mathsf{E}\Big\|\sum_{i\le n}\varepsilon_i x_i\Big\| \tag{19.10}$$

and

$$\max_{\epsilon_i=\pm 1}\Big\|\sum_{i\le n}\epsilon_i x''_i\Big\| \le L\mathsf{E}\Big\|\sum_{i\le n}\varepsilon_i x_i\Big\| . \tag{19.11}$$

Proof of Theorem 19.1.5. We shall prove the right-hand side inequality of (19.9):

$$C^r_q(U) \le L(C^g_q(U) + \|U\|_{q,1}) . \tag{19.12}$$

Let us consider a decomposition $x_i = x'_i + x''_i$ as in Proposition 19.1.6. Then

$$\Big(\sum_{i\le n}\|U(x'_i)\|^q\Big)^{1/q} \le LC^g_q(U)\mathsf{E}\Big\|\sum_{i\le n} g_i x'_i\Big\| \le LC^g_q(U)\mathsf{E}\Big\|\sum_{i\le n}\varepsilon_i x_i\Big\| \tag{19.13}$$

$$\Big(\sum_{i\le n}\|U(x''_i)\|^q\Big)^{1/q} \le L\|U\|_{q,1}\max_{\epsilon_i=\pm 1}\Big\|\sum_{i\le n}\epsilon_i x''_i\Big\| \le L\|U\|_{q,1}\mathsf{E}\Big\|\sum_{i\le n}\varepsilon_i x_i\Big\| . \tag{19.14}$$

Since $\|U(x_i)\| \le \|U(x'_i)\| + \|U(x''_i)\|$, the triangle inequality in ℓ^q_n implies

$$\Big(\sum_{i\le n}\|U(x_i)\|^q\Big)^{1/q} \le \Big(\sum_{i\le n}\|U(x'_i)\|^q\Big)^{1/q} + \Big(\sum_{i\le n}\|U(x''_i)\|^q\Big)^{1/q} ,$$

and combining with (19.13) and (19.14), this proves (19.12). □

Exercise 19.1.7 Before you read the next proof, make sure that you understand the following reformulation of the Latała-Bednorz theorem: Given any subset T of ℓ^2, we can write $T \subset T_1 + T_2$ with $\mathsf{E}\sup_{t\in T_1}\sum_{i\ge 1} g_i t_i \le Lb(T)$ and $\sup_{t\in T_2}\|t\|_1 \le Lb(T)$.

Proof of Proposition 19.1.6. Let us write $x_i = (x_{ij})_{1\le j\le N}$. For $1 \le j \le N$, consider $t_j \in \mathbb{R}^n$ given by $t_j = (x_{ij})_{i\le n}$. Let $t_0 = 0 \in \mathbb{R}^n$ and consider

[2] Weaker results suffice, and the author proved Theorem 19.1.5 long before the Bernoulli conjecture was solved.

$T = \{t_0, t_1, \ldots, t_N\}$ so that

$$b(T) = \mathsf{E}\max\Big(0, \sup_{1\le j\le N} \sum_{i\le n} \varepsilon_i x_{ij}\Big) \le \mathsf{E}\Big\| \sum_{i\le n} \varepsilon_i x_i \Big\| . \tag{19.15}$$

Theorem 6.2.8 (The Latała-Bednorz theorem) in the formulation of Exercise 19.1.7) provides for $0 \le j \le N$ a decomposition $t_j = t'_j + t''_j$, where $t'_j = (x'_{ij})_{i\le n}$, $t''_j = (x''_{ij})_{i\le n}$, and

$$\mathsf{E} \sup_{0\le j\le N} \sum_{i\le n} g_i x'_{ij} \le Lb(T) \tag{19.16}$$

$$\forall j \le N , \sum_{i\le n} |x''_{ij}| \le Lb(T) . \tag{19.17}$$

Since $t_0 = 0 = t'_0 + t''_0$, for each $0 \le j \le N$, we can replace t'_j by $t'_j - t'_0$ and t''_j by $t''_j - t''_0$, so that we may assume that $t'_0 = t''_0 = 0$. For $i \le n$, we consider the elements $x'_i = (x'_{ij})_{j\le N}$ and $x''_i = (x''_{ij})_{j\le N}$ of ℓ^∞_N. Thus $x_i = x'_i + x''_i$. Obviously (19.17) implies (19.11).

Let us now prove (19.10). When a process $(X_t)_{t\in T'}$ is symmetric and $X_s = 0$ for some $s \in T'$, then Lemma 2.2.1 implies

$$\mathsf{E} \sup_{t\in T} |X_t| \le \mathsf{E} \sup_{s,t\in T'} |X_s - X_t| = 2\mathsf{E} \sup_{t\in T'} X_t .$$

Using this for $X_t = \sum_{i\le n} g_i x_i$ when $t = (x_i)_{i\le n}$ and $T' = \{t'_0, t'_1, \ldots, t'_N\}$ yields (using that $X_{t'_0} = 0$ since $t'_0 = 0$), using also (19.16) in the last inequality,

$$\mathsf{E}\Big\| \sum_{i\le n} g_i x'_i \Big\| = \mathsf{E} \sup_{0\le j\le N} \Big| \sum_{i\le n} g_i x'_{ij} \Big| \le 2\mathsf{E} \sup_{0\le j\le N} \sum_{i\le n} g_i x'_{ij} \le Lb(T) . \qquad \square$$

19.1.3 Computing the Cotype-2 Constant with Few Vectors

The results of the present section are included not because they are very important but because the author cannot help feeling that they are part of an unfinished story and keeps hoping that someone will finish this story. The main result of the section is arguably a new comparison theorem between Gaussian and Rademacher averages (Theorem 19.1.11 below) which makes full use of Theorem 6.6.1.

When U is an operator between two finite dimensional Banach spaces X and Y, we recall the definition (19.4) of the Rademacher cotype-2 constant $C^r_2(U)$ of U.

Definition 19.1.8 Let us associate to each Banach space X an integer $M(X)$. We say that $M(X)$ vectors suffice to compute the Rademacher cotype-2 constant of an operator from X to any Banach space Y if for any such operator U one can find vectors $x_1, \ldots, x_{M(X)}$ in X with

$$\Big(\sum_{i\le M(X)} \|U(x_i)\|^2\Big)^{1/2} > \frac{1}{L} C_2^r(U) \mathsf{E}\Big\| \sum_{i\le M(X)} \varepsilon_i x_i \Big\| . \tag{19.18}$$

N. Tomczak-Jaegermann proved [137] that "N vectors suffice to compute the *Gaussian* cotype-2 constant of an operator from any Banach space X of dimension N". This motivated the previous definition.[3] Our main result is that $N \log N \log\log N$ vectors suffice to compute the Rademacher cotype-2 constant of an operator U from a Banach space X of dimension N. It does not appear to be known if N vectors suffice.

Consider a Banach space X of dimension $N \ge 3$, and its dual X^*. Consider elements $x_1, \ldots, x_n$ in X and assume without loss of generality that they span X (we will typically have $n \gg N$). We will now perform some constructions, and the reader should keep in mind that *they depend on this sequence* $(x_i)_{i\le n}$. We identify X^* with a subspace of ℓ_n^2 by the map $x^* \mapsto (x^*(x_i))_{i\le n}$, so that

$$\|x^*\|_2 = \Big(\sum_{i\le n} x^*(x_i)^2\Big)^{1/2} . \tag{19.19}$$

This norm arises from the dot product given by

$$(x^*, y^*) = \sum_{i\le n} x^*(x_i) y^*(x_i) .$$

Consider an orthonormal basis $(e_j^*)_{j\le N}$ of X^* for this dot product. Then

$$x^* = \sum_{j\le N} (x^*, e_j^*)\, e_j^* \ ; \ \|x^*\|_2^2 = \sum_{j\le N} (x^*, e_j^*)^2 ,$$

so that the elements x^* of X^* with $\|x\|_2 \le 1$ are exactly the elements $\sum_{j\le N} \beta_j e_j^*$ with $\sum_{j\le N} \beta_j^2 \le 1$. The dual norm $\|\cdot\|_2$ on X is then given by

$$\begin{aligned}\|x\|_2 &= \sup\{|x^*(x)| \ ; \ \|x^*\|_2 \le 1\} \\ &= \sup\Big\{\Big|\sum_{j\le N} \beta_j e_j^*(x)\Big| \ ; \ \sum_{j\le N} \beta_j^2 \le 1\Big\} = \Big(\sum_{j\le N} e_j^*(x)^2\Big)^{1/2} . \end{aligned} \tag{19.20}$$

[3] Similar questions in various settings are also investigated, for example, in [43].

Using (19.19) with $x^* = e_j^*$, we obtain $1 = \|e_j^*\|^2 = \sum_{i\le n} e_j^*(x_i)^2$ so that, using (19.20) to compute $\|x_i\|_2^2$, we get

$$\sum_{i\le n} \|x_i\|_2^2 = \sum_{i\le n}\sum_{j\le N} e_j^*(x_i)^2 = \sum_{j\le N}\sum_{i\le n} e_j^*(x_i)^2 = N\,. \tag{19.21}$$

Considering independent standard normal r.v.s $(\eta_j)_{j\le N}$, $G := \sum_{j\le N}\eta_j e_j^*$ is a standard Gaussian random vector valued in $(X^*, \|\cdot\|_2)$. For a subset T of X, we define

$$g(T) = \mathsf{E}\sup_{x\in T} G(x) = \mathsf{E}\sup_{x\in T}\sum_{j\le N}\eta_j e_j^*(x)\,. \tag{19.22}$$

The reason for the notation is that $g(T)$ is the usual quantity when we consider T as a subset of the Hilbert space $(X, \|\cdot\|_2)$, simply because by (19.20), the map $x \mapsto (e_j^*(x))_{j\le N}$ is an isometry from $(X, \|\cdot\|_2)$ to ℓ_N^2.[4] For further use, we spell now a consequence of (19.22) and of Sudakov's dual minoration (Lemma 15.2.7).[5]

Lemma 19.1.9 *Consider a subset T of X with $T = -T$ and the semi-norm $\|\cdot\|_T$ on X^* given by $\|x^*\|_T = \sup_{x\in T} x^*(x)$. Then the unit ball of $(X^*, \|\cdot\|_2)$ can be covered by N_n balls for the norm $\|\cdot\|_T$ of radius $Lg(T)2^{-n/2}$.*

Proof It follows from (19.22) that $g(T) = \mathsf{E}\|G\|_T$. The conclusion then follows from Lemma 15.2.7. □

We will not use anymore the formula (19.22) but only the general fact that in any Hilbert space, if $T = \{\pm t_k\,;\ k \ge 1\}$, then

$$g(T) \le L\sup_{k\ge 1}\left(\|t_k\|_2\sqrt{\log(k+1)}\right), \tag{19.23}$$

as shown in Proposition 2.11.6. Let us stress again that the quantity $g(T)$ depends on T *and on the whole sequence* $(x_i)_{i\le N}$.

Lemma 19.1.10 *If $T = \{\pm x_1, \ldots, \pm x_n\}$, then*

$$g(T) \le L\sqrt{\log(N+1)}\,. \tag{19.24}$$

When the sequence $(\|x_i\|_2)_{i\le n}$ is non-increasing, and if $M = \lfloor N\log N\rfloor$, the set $T' = \{\pm x_i\,;\ M \le i \le n\}$ satisfies

$$g(T') \le L\,. \tag{19.25}$$

[4] Please keep in mind, however, that this embedding of T in a Hilbert space depends on $x_1, \ldots, x_n$ and so does the quantity $g(T)$.

[5] Please note that the original norm of T plays no part in this result.

Proof Both results are based on (19.23). Since the norm $\|\cdot\|_2$ on X is the dual norm of the norm (19.19), it is obvious that $\|x_i\|_2 \le 1$. Assuming that the sequence $(\|x_i\|_2)_{i\ge 1}$ is non-increasing, we see from (19.21) that $\|x_i\|_2 \le \sqrt{N/i}$ and thus $\|x_i\| \le \min(1, \sqrt{N/i})$. Using (19.23) for the sequences $t_k = x_k$ $(1 \le k \le n)$, we obtain

$$g(T) \le L \sup_{k\ge 1}\Bigl(\min\Bigl(1, \sqrt{\frac{N}{k}}\Bigr)\sqrt{\log(k+1)}\Bigr) \le L\sqrt{\log N}\,.$$

Using again (19.23) for the sequences $t_k = x_{M+k}$ $(1 \le k \le n - M)$, we now obtain

$$g(T') \le L \sup_{k\ge 1}\Bigl(\sqrt{\frac{N}{M+k}}\sqrt{\log(k+1)}\Bigr) \le L\sqrt{\frac{N}{M}}\log M \le L\,. \qquad \square$$

In the next statement, we define $T = \{\pm x_1, \dots, \pm x_n\}$, and for a subset I of $\{1, \dots, n\}$, we define T^I as the collection of elements x_i for i *outside* I,

$$T^I = \{\pm x_i\ ;\ i \le n\,,\ i \notin I\}\,. \tag{19.26}$$

We are now ready to state our new comparison principle between Gaussian and Rademacher averages.

Theorem 19.1.11 *We have*

$$\mathsf{E}\Bigl\|\sum_{i\le n} g_i x_i\Bigr\| \le L\mathsf{E}\Bigl\|\sum_{i\le n} \varepsilon_i x_i\Bigr\|(1 + g(T))\,. \tag{19.27}$$

More generally, for any subset I of $\{1, \dots, n\}$, we have

$$\mathsf{E}\Bigl\|\sum_{i\notin I} g_i x_i\Bigr\| \le L\mathsf{E}\Bigl\|\sum_{i\notin I} \varepsilon_i x_i\Bigr\|\Bigl(1 + \frac{\mathsf{E}\|\sum_{i\le n} g_i x_i\|}{\mathsf{E}\|\sum_{i\notin I} g_i x_i\|}g(T^I)\Bigr)\,. \tag{19.28}$$

When $I = \emptyset$ (19.28) specializes into (19.27). Using (19.24), we see that (19.27) generalizes the classical inequality

$$\mathsf{E}\Bigl\|\sum_{i\le n} g_i x_i\Bigr\| \le L\sqrt{\log N}\mathsf{E}\Bigl\|\sum_{i\le n} \varepsilon_i x_i\Bigr\|\,. \tag{19.29}$$

Let us also stress that in (19.28), the quantity $g(T^I)$ is computed as in (19.22), that is, for the norm $\|\cdot\|_2$ on X involving the *whole sequence* $(x_i)_{i\le N}$ and *not only* the $(x_i)_{i\notin I}$.

We will prove Theorem 19.1.11 later. First, we draw some consequences.

Corollary 19.1.12 *There exists a subset I of $\{1,\ldots,n\}$ such that* $\operatorname{card} I \le N\log N$ *and that* ***either*** *of the following holds true:*

$$\mathsf{E}\Big\|\sum_{i\notin I} g_i x_i\Big\| \le \frac{1}{2}\mathsf{E}\Big\|\sum_{i\le n} g_i x_i\Big\| \tag{19.30}$$

or else

$$\mathsf{E}\Big\|\sum_{i\notin I} g_i x_i\Big\| \le L\mathsf{E}\Big\|\sum_{i\notin I} \varepsilon_i x_i\Big\| \,. \tag{19.31}$$

Proof The set $I = \{1,\ldots,M\}$ satisfies $g(T^I) \le L$ by (19.25) so that if (19.30) fails, (19.31) follows from (19.28). □

Corollary 19.1.13 *We can find elements $y_1,\ldots,y_M$ of X such that*

$$\frac{C_2^r(U)}{L}\mathsf{E}\Big\|\sum_{j\le M}\varepsilon_j y_j\Big\| < \Big(\sum_{j\le M}\|U(y_j)\|^2\Big)^{1/2} \tag{19.32}$$

and $M \le N\log N\log\log N$.

We have obtained (19.18) for $M \le N\log N\log\log N$, that is, "M vectors suffice to compute the Rademacher cotype-2 constant of U".

Proof We find elements $x_1,\ldots,x_n$ of X such that

$$\frac{C_2^r(U)}{2}\mathsf{E}\Big\|\sum_{i\le n}\varepsilon_i x_i\Big\| < \Big(\sum_{i\le n}\|U(x_i)\|^2\Big)^{1/2} \,. \tag{19.33}$$

The next step of the proof consists of showing that we can find a subset J of $\{1,\ldots,n\}$ with $\operatorname{card} J \le N\log N\log\log N$ such that

$$\mathsf{E}\Big\|\sum_{i\notin J} g_i x_i\Big\| \le L\mathsf{E}\Big\|\sum_{i\le n} \varepsilon_i x_i\Big\| \,. \tag{19.34}$$

To this aim, consider the largest integer k_0 with $2^{k_0} \le \sqrt{\log N}$ so that $k_0 \le \log\log N$. By induction over k, for $k \le k_0$, we construct subsets I_k of $\{1,\ldots,n\}$ with $\operatorname{card} I_k \le N\log N$ and either

$$\mathsf{E}\Big\|\sum_{i\notin I_1\cup\ldots\cup I_k} g_i x_i\Big\| \le \frac{1}{2}\mathsf{E}\Big\|\sum_{i\notin I_1\cup\ldots\cup I_{k-1}} g_i x_i\Big\| \tag{19.35}$$

or else

$$\mathsf{E}\Big\| \sum_{i \notin I_1 \cup \ldots \cup I_k} g_i x_i \Big\| \le L \mathsf{E}\Big\| \sum_{i \notin I_1 \cup \ldots \cup I_{k-1} \cup I_k} \varepsilon_i x_i \Big\| . \tag{19.36}$$

The induction step is performed by using Corollary 19.1.12 for the set of indices $\{i \le n; i \notin I_1 \cup \ldots \cup I_{k-1}\}$ rather than the set $\{1, \ldots, n\}$. If at the k-th step (19.36) holds, we then stop the construction, and we define $J = I_1 \cup \ldots \cup I_k$. Thus $M := \operatorname{card} J \le kN \log N \le k_0 N \log N$ and

$$\mathsf{E}\Big\| \sum_{i \notin J} g_i x_i \Big\| \le L \mathsf{E}\Big\| \sum_{i \notin J} \varepsilon_i x_i \Big\| \le L \mathsf{E}\Big\| \sum_{i \le n} \varepsilon_i x_i \Big\| ,$$

so that (19.34) holds. If instead (19.36) never occurs during the construction, we continue this construction until $k = k_0$, and we define now $J = I_1 \cup \ldots \cup I_{k_0}$. Thus $M := \operatorname{card} J \le k_0 N \log N$ and, iterating (19.35),

$$\mathsf{E}\Big\| \sum_{i \notin J} g_i x_i \Big\| \le 2^{-k_0} \mathsf{E}\Big\| \sum_{i \le n} g_i x_i \Big\| .$$

Combining with (19.29), this implies

$$\mathsf{E}\Big\| \sum_{i \notin J} g_i x_i \Big\| \le 2^{-k_0} L \sqrt{\log N} \mathsf{E}\Big\| \sum_{i \le n} \varepsilon_i x_i \Big\| ,$$

and this proves (19.34) by the choice of k_0.

Now that we have proved (19.34), we consider two cases.

Case 1. We have

$$\sum_{i \in J} \|U(x_i)\|^2 \ge \frac{1}{2} \sum_{i \le n} \|U(x_i)\|^2 . \tag{19.37}$$

Then, using (19.33) in the third inequality,

$$\begin{aligned} \frac{C_2^r(U)}{4} \mathsf{E}\Big\| \sum_{i \in J} \varepsilon_i x_i \Big\| &\le \frac{C_2^r(U)}{4} \mathsf{E}\Big\| \sum_{i \le n} \varepsilon_i x_i \Big\| \\ &< \frac{1}{2} \Big(\sum_{i \le n} \|U(x_i)\|^2 \Big)^{1/2} \le \Big(\sum_{i \in J} \|U(x_i)\|^2 \Big)^{1/2} , \end{aligned}$$

and this proves (19.32).[6]

[6] It is certainly disturbing at first that this case does not use at all any of the previous work. The point is that (19.37) is very unlikely to hold.

Case 2. (19.37) fails so that we have

$$\sum_{i \notin J} \|U(x_i)\|^2 \geq \frac{1}{2} \sum_{i \leq n} \|U(x_i)\|^2 .$$

Then (19.33) yields

$$\frac{C_2^r(U)}{4} \mathsf{E} \Big\| \sum_{i \leq n} \varepsilon_i x_i \Big\| < \Big(\sum_{i \notin J} \|U(x_i)\|^2 \Big)^{1/2}$$

and combining with (19.34), we obtain

$$\frac{C_2^r(U)}{L} \mathsf{E} \Big\| \sum_{i \notin J} g_i x_i \Big\| < \Big(\sum_{i \notin J} \|U(x_i)\|^2 \Big)^{1/2} , \tag{19.38}$$

which implies that the Gaussian cotype-2 constant $C_2^g(U)$ of U is $\geq C_2^r(U)/L$. It is proved in [137] that the Gaussian cotype-2 constant $C_2^g(U)$ of U "can be computed on N vectors", so that we can find N elements $y_1, \ldots, y_N$ of X such that

$$\frac{C_2^g(U)}{L} \mathsf{E} \Big\| \sum_{j \leq N} g_j y_j \Big\| \leq \Big(\sum_{j \leq N} \|U(y_j)\|^2 \Big)^{1/2} . \tag{19.39}$$

Using (6.6), we have $\mathsf{E}\| \sum_{j \leq N} \varepsilon_j y_j \| \leq L\mathsf{E}\| \sum_{j \leq N} g_j y_j \|$ so that (19.39) implies (19.32) since $C_2^g(U) \geq C_2^r(U)/L$. □

We turn to the proof of Theorem 19.1.11. We fix a set $I \subset \{1, \ldots, n\}$, and we recall the set T^I of (19.26). We consider the set

$$V_I = \{(x^*(x_i))_{i \notin I} \; ; \; x^* \in X_1^*\} \subset \mathbb{R}^{I^c} ,$$

where $I^c = \{1, \ldots, n\} \setminus I$ and X_1^* is the unit ball of X^* for the orginal norm. On V_I, we consider the distance d_∞ induced by the supremum norm on $\mathbb{R}^{I^c}$. The key step is the following:

Lemma 19.1.14 *We have*

$$\gamma_1(V_I, d_\infty) \leq Lg(T^I)\mathsf{E} \Big\| \sum_{i \leq n} g_i x_i \Big\| . \tag{19.40}$$

Proof of Theorem 19.1.11. By duality, we have

$$g(V_I) = \mathsf{E}\Big\| \sum_{i \notin I} g_i x_i \Big\| \; ; \; b(V_I) = \mathsf{E}\Big\| \sum_{i \notin I} \varepsilon_i x_i \Big\| \, . \tag{19.41}$$

We appeal to Theorems 6.6.1 and 2.7.11 to obtain

$$g(V_I) \le L\big(b(V_I) + \sqrt{b(V_I)\gamma_1(V_I, d_\infty)}\big) = L\big(b(V_I) + \sqrt{g(V_I)C}\big) \, ,$$

where

$$C = \frac{b(V_I)}{g(V_I)} \gamma_1(V_I, d_\infty) \le L b(V_I) g(T^I) \frac{\mathsf{E}\| \sum_{i \le n} g_i x_i \|}{\mathsf{E}\| \sum_{i \notin I} g_i x_i \|}$$

where we have used (19.40) and the first part of (19.41) in the inequality.

Using that for any $c > 0$ we have the inequality $\sqrt{xy} \le cx + y/c$, we conclude that

$$g(V_I) \le L b(V_I) + \frac{1}{2} g(V_I) + LC \, ,$$

so that $g(V_I) \le L b(V_I) + LC$, which, recalling (19.41), is the desired inequality (19.28). □

It remains to prove Lemma 19.1.14. The proof of this lemma involves several ingredients which have to be combined in an unusual way. One of them is the following general principle, where we recall that $N_0 = 1$ and that $N_n = 2^{2^n}$ for $n \ge 1$:

Lemma 19.1.15 *Consider a set W provided with two distances d_2 and d_1. Assume that for a certain number S and every integer $n \ge 0$, every number $a > 0$, every ball $B_{d_2}(t, a)$ of W can be covered by N_n sets of d_1-diameter at most $aS2^{-n/2}$. Then*

$$\gamma_1(W, d_1) \le LS\gamma_2(W, d_2) \, .$$

Proof Consider an admissible sequence $(\mathcal{B}_n)$ of W with

$$\forall t \in W \, , \sum_{n \ge 0} 2^{n/2} \Delta(B_n(t), d) \le 2\gamma_2(W, d_2) \, .$$

We construct by induction an increasing sequence of partitions $(\mathcal{C}_n)$ satisfying

$$\operatorname{card} \mathcal{C}_n \le N_{n+2} \tag{19.42}$$

$$\forall C \in \mathcal{C}_n \, , \exists B \in \mathcal{B}_n \, , \; C \subset B \, , \; \Delta(C, d_1) \le S2^{-n/2} \Delta(B, d_2) \, . \tag{19.43}$$

First, we set $\mathcal{C}_0 = \{W\}$. We note that using the hypothesis for $a = \Delta(W, d_2)$ and $n = 0$, we have

$$\Delta(W, d_1) \le S\Delta(W, d_2)\,. \tag{19.44}$$

Thus (19.43) is true for $n = 0$. Assuming that $\mathcal{C}_n$ has been constructed, we split each element C of $\mathcal{C}_n$ as follows: First, we split C into the sets $C \cap B$, $B \in \mathcal{B}_{n+1}$. Then we split each set $C \cap B$ into N_{n+1} pieces C' such that

$$\Delta(C', d_1) \le S2^{-(n+1)/2}\Delta(C \cap B, d_2)\,.$$

This is possible by hypothesis, and this completes the construction of $\mathcal{C}_{n+1}$. Clearly, $\mathcal{C}_{n+1}$ consists of at most $N_{n+2} \cdot N_{n+1}^2 = N_{n+3}$ sets, and it is obvious that (19.42) and (19.43) hold for $n + 1$. A consequence of (19.43) is that

$$\forall t\,,\ \Delta(C_n(t), d_1) \le S2^{-n/2}\Delta(B_n(t), d_2)$$

and thus

$$\begin{aligned}\sum_{n\ge 0} 2^n\Delta(C_n(t), d_1) &\le S\sum_{n\ge 0} 2^{n/2}\Delta(B_n(t), d_2)\\ &\le 2S\gamma_2(W, d_2)\,.\end{aligned}$$

Using (19.44) and Lemma 2.9.10 then yields the result. □

Proof of Lemma 19.1.14. Let us denote by X_1^* the unit ball of X^* and by d_2 the distance associated to the norm $\|\cdot\|_2$. By (19.1), the process given for x^* in X_1^* by $X_{x^*} = \sum_{i\le n} g_i x^*(x_i)$ satisfies $\sup_{x^*\in X_1^*}\sum_{i\le n} g_i x^*(x_i) = \|\sum_{i\le n} g_i x_i\|$. Theorem 2.10.1 yields

$$\gamma_2(X_1^*, d_2) \le L\mathsf{E}\Big\|\sum_{i\le n} g_i x_i\Big\|\,. \tag{19.45}$$

Consider the norm $\|\cdot\|_1$ on X^* given by

$$\|x^*\|_1 = \sup_{i\le n, i\notin I} |x^*(x_i)| = \sup_{x\in T^I} x^*(x)\,,$$

where $T^I = \{\pm x_i;\, i \le n, i \notin I\}$. Lemma 19.1.9 asserts that the unit ball of $(X^*, \|\cdot\|_2)$ can be covered by N_n balls for the norm $\|\cdot\|_1$ of radius $Lg(T^I)2^{-n/2}$. Denoting by d_1 the distance associated to the norm $\|\cdot\|_1$, Lemma 19.1.15 then implies

$$\gamma_1(X_1^*, d_1) \le Lg(T^I)\gamma_2(X_1^*, d_2)\,.$$

Combining with (19.45) completes the proof since obviously $\gamma_1(V_I, d_\infty) = \gamma_1(X_1^*, d_1)$. □

19.2 Unconditionality

19.2.1 Classifying the Elements of B_1

Consider a general σ-finite measure space (Ω, μ), and

$$B_1 = \left\{ f \in L^1(\mu) \; ; \; \int |f| \mathrm{d}\mu \le 1 \right\} .$$

Theorem 19.2.1 below provides a kind of classification of the elements of B_1. It is at the root of Proposition 10.14.3. It will be used a number of times in the following sections, allowing us to gain an excellent control of the subsets T of B_1 which are small in some other sense, for example, $\gamma_2(T, d_2) < \infty$. It has no content when μ is a probability and is of interest only in the case where the total mass of μ is large. The parameter τ below is of secondary importance, and one may assume $\tau = 0$ at first reading. We recall the notation $a \wedge b = \min(a, b)$.

Theorem 19.2.1 *For any integer $\tau \in \mathbb{Z}$, there exists an admissible sequence of partitions $(\mathcal{C}_n)$ of B_1, and for each $C \in \mathcal{C}_n$, an integer $\ell_n(C) \in \mathbb{Z}$, such that if we set*

$$\ell(f, n) = \ell_n(C_n(f)) \tag{19.46}$$

where as usual $C_n(f)$ denotes the element of $\mathcal{C}_n$ containing f, we have

$$\forall f \in B_1 , \; \int (2^{\ell(f,n)} f)^2 \wedge 1 \mathrm{d}\mu \le 2^{n+\tau} , \tag{19.47}$$

and

$$\forall f \in B_1 , \; \sum_{n \ge 0} 2^{n - \ell(f,n)} \le 18 \cdot 2^{-\tau} . \tag{19.48}$$

We note that (19.47) implies

$$\forall f \in B_1 , \; \mu(\{|f| > 2^{-\ell(f,n)}\}) \le 2^{n+\tau} . \tag{19.49}$$

A first (and partial) understanding of the meaning of this result is that it classifies the functions f of B_1 according to the values of the integers $\ell(f, n)$ for which

$$\mu(\{|f| > 2^{-\ell(f,n)}\}) \simeq 2^{n+\tau} .$$

The effectiveness of this result will be understood through multiple applications, the first of which is a proof of Proposition 10.14.3 at the end of the present subsection.

Lemma 19.2.2 *For any number $a \in \mathbb{R}$, we have*

$$\sum_{k \in \mathbb{Z}} (2^{k+2}a^2) \wedge 2^{-k} \le 8|a| \,. \tag{19.50}$$

Proof Without loss of generality, we assume that $a > 0$. Consider the largest integer k_0 such that $2^{k_0}a < 1$ so that $2^{k_0+1}a \ge 1$. Thus, $2^{k_0+2}a^2 \le 4a$, $2^{-k_0+1} \le 4a$ and

$$\sum_{k \in \mathbb{Z}} (2^{k+2}a^2) \wedge 2^{-k} \le \sum_{k<k_0} 2^{k+2}a^2 + \sum_{k \ge k_0} 2^{-k} = 2^{k_0+2}a^2 + 2^{-k_0+1} \le 8a \,. \quad \square$$

Lemma 19.2.3 *Given $f \in B_1$ and $n \ge 0$, we define $\ell(f,n)$ as the largest integer $\le 2n + \tau$ for which*

$$\int (2^{\ell(f,n)} f)^2 \wedge 1 \mathrm{d}\mu \le 2^{n+\tau} \,. \tag{19.51}$$

Then

$$\sum_{n \ge 0} 2^{n-\ell(f,n)+\tau} \le 18 \,. \tag{19.52}$$

Proof Let us consider the set $J(f) = \{n \,;\, \ell(f,n) < 2n + \tau\}$. Then, for $n \in J(f)$, we have

$$\int (2^{\ell(f,n)+1} f)^2 \wedge 1 \mathrm{d}\mu \ge 2^{n+\tau} \,,$$

and therefore

$$\int (2^{\ell(f,n)+2} f^2) \wedge 2^{-\ell(f,n)} \mathrm{d}\mu \ge 2^{n-\ell(f,n)+\tau} \,. \tag{19.53}$$

It is obvious by construction that the sequence $(\ell(f,n))_n$ is non-decreasing in n. For $k \in \mathbb{Z}$, we define $J_k(f) = \{n \in J(f); \ell(f,n) = k\}$. Let $I(f) = \{k \in \mathbb{Z}; J_k(f) \ne \emptyset\}$. It follows from (19.53) that when $k \in I(f)$, $J_k(f)$ has a largest element n_k, and then, using again (19.53) in the last inequality,

$$\sum_{n \in J_k(f)} 2^{n-\ell(f,n)+\tau} = \sum_{n \in J_k(f)} 2^{n-k+\tau} \le 2^{n_k+1-k+\tau} \le 2 \int (2^{k+2} f^2) \wedge 2^{-k} \mathrm{d}\mu \,.$$

Summing the previous equations over $k \in I(f)$ and using (19.50), we obtain

$$\sum_{n \in J(f)} 2^{n-\ell(f,n)+\tau} \le 2 \int \sum_{k \in I(f)} (2^{k+2} f^2) \wedge 2^{-k} \mathrm{d}\mu \le 16 \int |f| \mathrm{d}\mu \le 16 \,.$$

The result follows since $\sum_{n \notin J(f)} 2^{n-\ell(f,n)+\tau} \le \sum_{n \ge 0} 2^{-n} \le 2$. $\square$

Proof of Theorem 19.2.1. We define $\ell(f,n)$ as in Lemma 19.2.3. Since $h^2 \wedge 1 \le |h|$ and since $f \in B_1$

$$\int (2^{n+\tau} f)^2 \wedge 1 \mathrm{d}\mu \le 2^{n+\tau} \int |f| \mathrm{d}\mu \le 2^{n+\tau} ,$$

and the definition of $\ell(f,n)$ implies $\ell(f,n) \ge n+\tau$ and therefore $\tau+n \le \ell(f,n) \le \tau + 2n$ so that $\ell(f,n)$ can take at most $n+1$ different values. We define $\mathcal{C}_0 = \{B_1\}$, and $\ell_0(B_1) = \tau$. Consider the partition $\mathcal{C}_n$ of B_1 induced by the following equivalence relation: f and f' are equivalent if and only if $\ell(f,m) = \ell(f',m)$ for each $m \le n$. The sequence $(\mathcal{C}_n)$ is increasing. Moreover, since $\ell(f,m)$ can take at most $m+1$ values, and since the values of $\ell(f,m)$ for $m \le n$ determine to which element of $\mathcal{C}_n$ the function f belongs,

$$\operatorname{card} \mathcal{C}_n \le (n+1)! \le N_n , \tag{19.54}$$

so that the sequence $(\mathcal{C}_n)$ is admissible.

By construction of $\mathcal{C}_n$, for $f \in C \in \mathcal{C}_n$, $\ell(f,n)$ has a fixed value $\ell_n(C)$, that is, $f \in C$, we have $\ell_n(C_n(f)) = \ell_n(C) = \ell(f,n)$ so that (19.46) holds. Also (19.47) holds by construction. Finally (19.48) follows from (19.52). □

As the crude inequality (19.54) shows, the use of admissible sequences is not really canonical for a "classification result" such as Theorem 19.2.1 (one could consider sequences of partitions with a much smaller cardinality). This, however, suffices for the applications, and we have not yet found uses for sharper results.

Proof of Proposition 10.14.3. We have to produce an admissible sequence of partitions $(\mathcal{A}_n)$ of B_a and for $n \ge 0$ and $A \in \mathcal{A}_n$ an integer $j_n(A)$ satisfying the conditions of Definition 10.14.1, where the quantity S of (10.170) is $\le Lar$. Consider the admissible sequence of $(\mathcal{C}_n)$ of B_1 obtained by application of Theorem 19.2.1 with $\tau = -2$ (when μ is the counting measure on $\mathbb{N}$). Consider the bijection $f \mapsto af$ between B_1 and B_a. Define the admissible sequence $(\mathcal{A}_n)$ of partitions of B_a consisting of the sets aC for $C \in \mathcal{C}_n$. Thus, if $A \in \mathcal{A}_n$, then $A/a \in \mathcal{C}_n$. Define $j_0(B_a)$ as the largest integer with $2a \le r^{-j_0(B_a)}$ (so that (10.169) holds). For $n \ge 1$ and $A \in \mathcal{A}_n$, define $\ell'_n(A) = \ell_n(A/a)$ (recalling that $A/a \in \mathcal{C}_n$). Define $j_n(A)$ as the largest integer for which $ar^{j_n(A)} \le 2^{\ell'_n(A)}$, so that $r^{-j_n(A)} \le ar2^{-\ell'_n(A)}$. It then follows from (19.52) that for $f \in B_a$, we have $\sum_{n\ge0} 2^n r^{-j_n(A_n(f))} \le Lar$ so that as desired, the quantity S of (10.170) is $\le Lar$. Also, for $f \in A$, we have $f/a \in A/a$ so that, using (19.47) in the last inequality,

$$\varphi_{j_n(A)}(f,0) = \sum_{i\ge1} (r^{j_n(A)} f_i)^2 \wedge 1 \le \sum_{i\ge1} (2^{\ell_n(A/a)}(f_i/a))^2 \wedge 1 \le 2^{n-2} .$$

Since the function φ_j is the square of a distance, we have $\varphi_{j_n(A)}(f,g) \le 2^n$ for $f,g \in A_n$, and this proves (10.168). □

19.2.2 Subsets of B_1

To lighten notation, we write

$$a_n := \frac{1}{\sqrt{\log(n+1)}} . \tag{19.55}$$

To understand the main result of this section, Theorem 19.2.4 below, we have to keep in mind the following consequence of Theorem 2.11.9:

$$\text{A set } T \text{ with } 0 \in T \text{ and } \gamma_2(T,d) \le 1 \text{ is (basically) a subset of the convex hull of a sequence } (x_n) \text{ with } \|x_n\| \le a_n . \tag{19.56}$$

This is really a *structure theorem*, giving in a sense a complete description of the sets T with $\gamma_2(T,d) \le 1$. When furthermore $T \subset B_1 = \{y \in \ell^2; \sum_{i\ge 1} |y_i| \le 1\}$, we will obtain a much more precise description of T. Given a finite subset I of $\mathbb{N}^* = \mathbb{N} \setminus \{0\}$, and a number $a > 0$, we define $B_2(I,a)$ as the set of elements of ℓ^2 with support in I and with ℓ^2 norm $\le a$, that is,

$$B_2(I,a) = \Big\{x \in \mathbb{R}^{\mathbb{N}^*} \; ; \; i \notin I \Rightarrow x_i = 0 \; ; \; \sum_{i\in I} x_i^2 \le a^2\Big\} . \tag{19.57}$$

Theorem 19.2.4 *Consider a subset T of ℓ^2. Assume that for a certain number S, we have $\gamma_2(T,d_2) \le S$ and $T \subset SB_1$. Then there exist sets $I_n \subset \mathbb{N}^*$ such that* card $I_n \le \log(n+1)$ *with*

$$T \subset LS\,\overline{\mathrm{conv}} \bigcup_{n\ge 1} B_2(I_n, a_n) , \tag{19.58}$$

where $\overline{\mathrm{conv}}\, A$ denotes the closed convex hull of A.

We start by some simple observations.

Lemma 19.2.5 *Consider sets $I_n \subset \mathbb{N}^*$ such that*

$$\forall n \ge 1 \; , \; \operatorname{card} I_n \le \log(n+1) . \tag{19.59}$$

Consider independent standard Gaussian r.v.s $(g_i)_{i\ge 1}$. Then

$$\mathsf{E} \sup_{n\ge 1} a_n \Big(\sum_{i\in I_n} g_i^2\Big)^{1/2} \le L . \tag{19.60}$$

Proof For each i, we have $\mathsf{E}\exp(g_i^2/4) \le 2$ so that for any set I,

$$\mathsf{E}\exp\Big(\frac{1}{4}\sum_{i\in I} g_i^2\Big) \le 2^{\operatorname{card} I} ,$$

and, for $v \ge 8\operatorname{card} I$,

$$\mathsf{P}\Big(\sum_{i\in I} g_i^2 \ge v\Big) \le 2^{\operatorname{card} I}\exp\Big(-\frac{v}{4}\Big) \le \exp\Big(-\frac{v}{8}\Big) .$$

Now, (19.59) implies that for $w^2 \ge 8$, we have, using the value (19.55) of a_n,

$$\begin{aligned}\mathsf{P}\Big(\sup_{n\ge 1} a_n\Big(\sum_{i\in I_n} g_i^2\Big)^{1/2} \ge w\Big) &\le \sum_{n\ge 1}\mathsf{P}\Big(\sum_{i\in I_n} g_i^2 \ge w^2\log(n+1)\Big)\\ &\le \sum_{n\ge 1}\exp\Big(-\frac{w^2\log(n+1)}{8}\Big) ,\end{aligned}$$

and the last sum is $\le L\exp(-w^2/L)$ for w large enough. □

Exercise 19.2.6 If the sets I_n satisfy $\operatorname{card} I_n \ge \log(n+1)$, prove that

$$\mathsf{E}\sup_{n\ge 1}\Big(\frac{1}{\operatorname{card} I_n}\sum_{i\in I_n} g_i^2\Big)^{1/2} \le L .$$

Exercise 19.2.7 Find another proof of Lemma 19.2.5 by constructing a sequence (u_k) of ℓ^2 with $\|u_k\|_2 \le La_k$ and

$$\bigcup_{n\ge 1} B_2(I_n, a_n) \subset \operatorname{conv}\{u_k\ ;\ k\ge 1\} .$$

Hint: Recall Lemma 14.3.2. Use this for each ball $B_2(I,a)$ in the left-hand side above.

Exercise 19.2.8 We recall that for $T\subset \ell^2$, we write

$$g(T) = \mathsf{E}\sup_{t\in T} X_t = \mathsf{E}\sup_{t\in T}\sum_{i\ge 1} t_i g_i .$$

Consider subsets T_n of ℓ^2, and assume that for certain numbers b_n, we have $\|x\|_2 \le b_n$ for $x\in T_n$. Prove that

$$g\Big(\bigcup_{n\ge 1} T_n\Big) \le L\sup_n\Big(g(T_n) + b_n\sqrt{\log(n+1)}\Big) . \tag{19.61}$$

Use (19.61) with $T_n = B(a_n, I_n)$ and $b_n = a_n$ to give another proof of Lemma 19.2.5.

The following is a kind of converse to Theorem 19.2.4:

Proposition 19.2.9 *Consider sets* $I_n \subset \mathbb{N}^*$ *with* $\operatorname{card} I_n \le \log(n+1)$. *Then the set* $T_1 := \overline{\operatorname{conv}} \bigcup_{n \ge 1} B_2(I_n, a_n)$ *satisfies* $\gamma_2(T_1, d_2) \le L$ *and* $T_1 \subset B_1$.

Proof It follows from the Cauchy-Schwarz inequality that $B_2(I_n, a_n) \subset B_1$ so that $T_1 \subset B_1$. It follows from Lemma 19.2.5 that $g(T_1) \le L$ and from Theorem 2.10.1 that $\gamma_2(T_1, d_2) \le L$. □

Thus, Theorem 19.2.4 in a sense provides a complete description of the sets $T \subset B_1$ for which $\gamma_2(T, d_2) \le 1$.

Proof of Theorem 19.2.4. By homogeneity, we may assume that $S = 1$. We denote by $\Delta_2(A)$ the diameter of A for the distance d_2 induced by ℓ^2. Since $\gamma_2(T, d_2) \le 1$, we may consider an admissible sequence $(\mathcal{B}_n)_{n \ge 0}$ with

$$\sup_{t \in T} \sum_{n \ge 0} 2^{n/2} \Delta_2(B_n(t)) \le 2 . \tag{19.62}$$

Next, we take advantage of the fact that $T \subset B_1$. We consider the admissible sequence $(\mathcal{C}_n)$ provided by Theorem 19.2.1 when $\tau = 0$, $\Omega = \mathbb{N}^*$, and μ is the counting measure. We consider the increasing sequence of partitions $(\mathcal{A}_n)_{n \ge 0}$ where $\mathcal{A}_n$ is generated by $\mathcal{B}_n$ and $\mathcal{C}_n$, so $\operatorname{card} \mathcal{A}_n \le N_{n+1}$. The numbers $\ell(t, n)$ of (19.46) depend only on $A_n(t)$. Therefore,

$$s \in A_n(t) \Rightarrow \ell(s, n) = \ell(t, n) . \tag{19.63}$$

For every $A \in \mathcal{A}_n$, we pick an arbitrary element $v_n(A) = (v_{n,i}(A))_{i \ge 1}$ of A. We set

$$J_n(A) = \left\{ i \in \mathbb{N}^* \; ; \; |v_{n,i}(A)| > 2^{-\ell(v_n(A), n)} \right\} ,$$

so that $\operatorname{card} J_n(A) \le 2^n$ by (19.49) since μ is the counting measure, and we define

$$J_n'(A) = \bigcup \left\{ J_k(B) \; ; \; k < n \; , \; B \in \mathcal{A}_k \; , \; A \subset B \right\} .$$

For $n \ge 1$ and $A \in \mathcal{A}_n$, we set $I_n(A) = J_n(A) \backslash J_n'(A)$, so that $\operatorname{card} I_n(A) \le 2^n$. We define $I_0(T) = J_0(T)$ and $\mathcal{F}$ as the family of pairs $(I_n(A), 2^{-n/2})$ for $A \in \mathcal{A}_n$ and $n \ge 0$. The heart of the argument is to prove that

$$T \subset L \, \overline{\operatorname{conv}} \bigcup_{(I,a) \in \mathcal{F}} B_2(I, a) . \tag{19.64}$$

To prove this, let us fix $t \in T$ and for $n \geq 1$ define

$$I_n(t) = I_n(A_n(t)) = J_n(A_n(t)) \setminus \bigcup_{k<n} J_k(A_k(t)) \tag{19.65}$$

and $\pi_n(t) = v_n(A_n(t))$. Since $\pi_n(t) \in A_n(t)$, it follows from (19.63) that

$$\ell(\pi_n(t), n) = \ell(t, n) . \tag{19.66}$$

Thinking of a point $t \in \ell^2$ as a function from $\mathbb{N}^*$ to $\mathbb{R}$, for $I \subset \mathbb{N}^*$, we consider the element $t\mathbf{1}_I \in \ell^2$ given by

$$t\mathbf{1}_I = (t_i \mathbf{1}_I(i))_{i \geq 1} . \tag{19.67}$$

For $i \in I_n(t)$, we have $i \notin J_{n-1}(A_{n-1}(t))$ so that by definition of this set $|v_{n-1,i}(A_{n-1}(t))| \leq 2^{-\ell(t,n-1)}$. Since $\pi_{n-1}(t) = v_{n-1}(A_{n-1}(t))$, we have $|\pi_{n-1}(t)_i| = |v_{n-1,i}(A_{n-1}(t))| \leq 2^{-\ell(t,n-1)}$. We have proved that

$$\|\pi_{n-1}(t)\mathbf{1}_{I_n(t)}\|_\infty \leq 2^{-\ell(t,n-1)} , \tag{19.68}$$

so that $\|\pi_{n-1}(t)\mathbf{1}_{I_n(t)}\|_2 \leq 2^{n/2-\ell(t,n-1)}$ since card $I_n(t) =$ card $I_n(A_n(t)) \leq 2^n$. Since $t, \pi_{n-1}(t) \in A_{n-1}(t)$, we have $\|t\mathbf{1}_{I_n(t)} - \pi_{n-1}(t)\mathbf{1}_{I_n(t)}\|_2 \leq \|t - \pi_{n-1}(t)\|_2 \leq \Delta_2(A_{n-1}(t))$ and thus,

$$\|t\mathbf{1}_{I_n(t)}\|_2 \leq c(t,n) := \Delta_2(A_{n-1}(t)) + 2^{n/2-\ell(t,n-1)} . \tag{19.69}$$

Therefore,

$$t\mathbf{1}_{I_n(t)} \in 2^{n/2} c(t,n) B_2(I_n(t), 2^{-n/2}) . \tag{19.70}$$

For each $t \in T$, we define $c(t,0) = 1$. Since $t \in T \subset B_1$ and card $J_0(T) =$ card $I_0(t) \leq 2^0 = 1$, (19.70) also holds for $n = 0$. We claim now that

$$t = \sum_{n \geq 0} t\mathbf{1}_{I_n(t)} . \tag{19.71}$$

We first show that

$$|t_i| > 0 \Rightarrow i \in \bigcup_{n \geq 0} J_n(A_n(t)) . \tag{19.72}$$

To prove this, consider i with $|t_i| > 0$ and n large enough so that $\Delta_2(A_n(t)) < |t_i|/2$. Then for all $x \in A_n(t)$, we have $|x_i - t_i| \leq |t_i|/2$ and hence $|x_i| > |t_i|/2$. Recalling that (19.48) holds for $\tau = 0$, we have in particular $2^{n-\ell(x,n)} \leq 18 \leq 2^5$ so that $\ell(x,n) \geq n-5$, and for n large enough, for all $x \in A_n(t)$, we have $2^{-\ell(x,n)} < |x_i|$. This holds in particular for $x = \pi_n(t) = v_n(A_n(t))$. Thus, by definition of $J_n(A)$, this shows that $i \in J_n(A_n(t))$.

It follows from (19.72) that if $|t_i| > 0$, there is a smallest $n \geq 0$ such that $i \in J_n(A_n(t))$. If $n = 0$, then $i \in J_0(T) = I_0(T)$. If $n > 0$, then (19.65) implies that $i \in I_n(t)$ and that furthermore, the sets $I_n(t)$ are disjoint. We have proved (19.71).

Combining (19.71) and (19.70), we have

$$t = \sum_{n \geq 0} t\mathbf{1}_{I_n(t)} = \sum_{n \geq 0} 2^{n/2} c(t,n) u(n) , \tag{19.73}$$

where $u(n) \in B_2(I_n(t), 2^{-n/2})$. Furthermore, $\sum_{n \geq 0} 2^{n/2} c(t,n) \leq L$ by (19.62) and (19.48), so the relation (19.73) proves (19.64).

It remains to deduce (19.58) from (19.64). This tedious argument simply requires a cautious enumeration of the pairs $(I,a) \in \mathcal{F}$ as follows. Consider the set $\mathcal{I}_n$ consisting of all the sets of the type $I_n(A)$ for $A \in \mathcal{A}_n$ so that $\operatorname{card} \mathcal{I}_n \leq \operatorname{card} \mathcal{A}_n \leq N_{n+1}$. We then find a sequence $(I_k)_{k \geq 1}$ of sets with the following properties. First, $I_k = \emptyset$ if $k < N_2$. Next, for $n \geq 0$, $\mathcal{I}_n = \{I_k; N_{n+1} \leq k < N_{n+2}\}$.[7] This is possible because $\operatorname{card} \mathcal{I}_n \leq N_{n+1} \leq N_{n+2} - N_{n+1}$. Furthermore, $\operatorname{card} I_k \leq 2^n$ for $N_{n+1} \leq k < N_{n+2}$ since $\operatorname{card} I \leq 2^n$ for $I \in \mathcal{I}_n$.

Thus for $N_{n+1} \leq k < N_{n+2}$, we have

$$\operatorname{card} I_k \leq 2^n \leq 2^{n+1} \log 2 = \log N_{n+1} \leq \log(k+1) \leq 2^{n+2} . \tag{19.74}$$

This proves that for all k, we have $\operatorname{card} I_k \leq \log(k+1)$. Consider now $(I,a) \in \mathcal{F}$. We prove that for some k, we have $I = I_k$ and $a \leq La_k$, which obviously conclude the proof. By definition of $\mathcal{F}$, there exists $n \geq 0$ such that $I \in \mathcal{I}_n$ and $a = 2^{-n/2}$ so that by our construction $I = I_k$ for some k with $N_{n+1} \leq k < N_{n+2}$ and $a = 2^{-n/2}$ satisfies $a \leq 2/\sqrt{k+1} = 2a_k$ by the last inequality of (19.74). □

Numerous relations exist between the following properties of a set T: $\gamma_2(T,d_2) \leq 1$; $T \subset B_1$; $\gamma_1(T,d_\infty) \leq 1$ (where d_∞ denotes the distance associated with the supremum norm). We started exploring this theme in Chap. 6. For example, the essence of Theorem 19.2.10 below is that the conditions $T \subset B_1$ and $\gamma_1(T,d_\infty) \leq 1$ taken together are very restrictive. We pursue this direction in the rest of this section, a circle of ideas closely connected to the investigations of Sect. 19.3.1 below.

For $I \subset \mathbb{N}^*$ and $a > 0$, in the spirit of the definition 19.57 of $B_2(I,a)$, we define $B_\infty(I,a)$ as the set of elements of support in I and of ℓ^∞ norm $\leq a$, that is,

$$B_\infty(I,a) = \left\{x = (x_i)_{i \geq 1} \; ; \; i \notin I \Rightarrow x_i = 0 \; ; \; i \in I \Rightarrow |x_i| \leq a\right\} . \tag{19.75}$$

[7] The sets I_k are *not* required to be all different from each other.

We have

$$x \in B_\infty(I, a) \Rightarrow \sum_{i \geq 1} x_i^2 \leq a^2 \operatorname{card} I$$

and thus, recalling the sets $B_2(I, a)$ of (19.57), this implies

$$B_\infty(I, a) \subset B_2(I, a\sqrt{\operatorname{card} I}) . \tag{19.76}$$

Theorem 19.2.10 *Consider a set $T \subset SB_1$, and assume that $\gamma_1(T, d_\infty) \leq S$. Then we can find subsets I_n of $\mathbb{N}^*$ with* $\operatorname{card} I_n \leq \log(n+1)$, *for which*

$$T \subset LS\,\overline{\operatorname{conv}} \bigcup_{n \geq 1} B_\infty\Big(I_n, \frac{1}{\log(n+1)}\Big) . \tag{19.77}$$

Proof Replacing T by T/S, we may assume that $S = 1$. We proceed as in the proof of Theorem 19.2.4, but we may now assume

$$\forall t \in T , \sum_{n \geq 0} 2^n \Delta_\infty(A_n(t)) \leq 2 .$$

Using (19.68) rather than (19.69), we get

$$\|t\mathbf{1}_{I_n(t)}\|_\infty \leq c(t, n) := \Delta_\infty(A_{n-1}(t)) + 2^{-\ell(t,n-1)}$$

so that

$$t\mathbf{1}_{I_n(t)} \in 2^n c(t, n) B_\infty(I_n(t), 2^{-n})$$

and the proof is finished exactly as before. □

Corollary 19.2.11 *If $T \subset SB_1$ and $\gamma_1(T, d_\infty) \leq S$, then $\gamma_2(T, d_2) \leq LS$.*

Proof Indeed (19.76) and (19.77) imply that $T \subset LS\,\overline{\operatorname{conv}} \bigcup_{n \geq 1} B_2(I_n, a_n)$, and Lemma 19.2.5 shows that this implies that $\gamma_2(T, d_2) \leq LS$. □

The information provided by (19.77) is however very much stronger than the information $\gamma_2(T, d_2) \leq S$.

19.2.3 1-Unconditional Sequences and Gaussian Measures

Definition 19.2.12 A sequence $(e_i)_{i \leq N}$ of vectors of a Banach space X is 1-unconditional if for each numbers $(a_i)_{i \leq N}$ and signs $(\epsilon_i)_{i \leq N}$ we have

$$\Big\| \sum_{i \leq N} a_i e_i \Big\| = \Big\| \sum_{i \leq N} \epsilon_i a_i e_i \Big\| .$$

Exercise 19.2.13 Prove that a sequence $(e_i)_{i\le N}$ in $\mathbb{R}^n$ provided with the Euclidean norm is 1-unconditional if and only it is orthogonal.

There are many natural norms on $\mathbb{R}^N$ for which the canonical basis $(e_i)_{i\le N}$ is 1-unconditional, for example, the norms $\|\cdot\|_p$ for $p\ge 1$ given for $x=(x_i)_{i\le N}$ by $\|x\|_p^p=\sum_{i\le N}|x_i|^p$ or the norms given by the formula (19.82) below.

The main result of this section is Theorem 19.2.15 below. It is in a sense a dual version of Theorem 19.2.4. Our next result, which does not involve unconditionality, provides perspective on this result. It is in a sense a dual version of (19.56). We recall the notation $a_n:=1/\sqrt{\log(n+1)}$.

Theorem 19.2.14 *Consider elements $(e_i)_{i\le N}$ of a Banach space*[8] *X and $S=\mathsf{E}\|\sum_{i\le N}g_ie_i\|$. Then we can find a sequence $x_n^*\in X^*$ such that for each $n\ge 1$, we have*

$$\Big(\sum_{i\le N}x_n^*(e_i)^2\Big)^{1/2}\le a_n \tag{19.78}$$

and

$$\forall x\in X\ ,\ \|x\|\le LS\mathcal{N}(x)\ , \tag{19.79}$$

where $\mathcal{N}(x):=\sup_{n\ge 1}|x_n^*(x)|$.

The point of this result is that, using Proposition 2.11.6,

$$\mathsf{E}S\mathcal{N}\Big(\sum_{i\le N}g_ie_i\Big)\le LS=L\mathsf{E}\Big\|\sum_{i\le N}g_ie_i\Big\|\ .$$

In words, given a norm $\|\cdot\|$, if we are only interested in the quantity $S=\mathsf{E}\|\sum_{i\le N}g_ie_i\|$, our norm in a sense does not differ from a *larger* norm (see (19.79)) of the type $S\mathcal{N}$, where $\mathcal{N}(x)=\sup_n|x_n^*(x)|$ for a sequence (x_n^*) that satisfies (19.78). Thus, Theorem 19.2.14 is a *structure theorem*, just as its dual version (19.56). If $\mathsf{E}\|\sum_{i\le N}g_ie_i\|$ is the only characteristic of the norm in which we are interested, we basically have a complete understanding of this norm.

Proof of Theorem 19.2.14. Denoting by x^* an element of the dual X^*, consider the set of sequences

$$T=\{(x^*(e_i))_{i\le N}\ ;\ \|x^*\|\le 1\}\subset\mathbb{R}^N\ .$$

[8] The reason for the change of notation is the results of this chapter have a natural extension when the finite sequence $(e_i)_{i\le N}$ is replaced by an infinite sequence which is a basis of X. We refer to [132] for this.

As usual, for a sequence $t = (t_i)_{i\le N} \in \mathbb{R}^N$, we write $X_t = \sum_{i\le N} t_i g_i$. Thus,

$$S = \mathsf{E}\Big\|\sum_{i\le N} g_i e_i\Big\| = \mathsf{E}\sup_{\|x^*\|\le 1}\sum_{i\le N} x^*(e_i)g_i = \mathsf{E}\sup_{t\in T} X_t \ .$$

Consider then a countable subset T' of T such that

$$\sup\Big\{\sum_{i\le N} t_i x_i \ ; \ (t_i)\in T\Big\} = \sup\Big\{\sum_{i\le N} t_i x_i \ ; \ (t_i)\in T'\Big\} \ . \tag{19.80}$$

We apply Theorem 2.11.9 to T' to obtain a sequence $y_n = (y_{n,i})_{i\le N}$ with $\|y_n\|_2 \le a_n$ and $T' \subset LS\,\mathrm{conv}\{y_n; n\ge 1\}$ and where y_n is moreover a multiple of the difference of two elements of T'. Thus,

$$\sup\Big\{\sum_{i\le N} t_i x_i \ ; \ (t_i)\in T'\Big\} \le LS\sup\Big\{\sum_{i\le N} y_{n,i}x_i \ ; \ n\ge 1\Big\} \ .$$

Since y_n is a multiple of the difference of two elements of T, there exists x_n^* in X^* with $y_n = (x_n^*(e_i))_{i\le N}$, that is, $y_{n,i} = x_n^*(e_i)$. Thus (19.78) follows from the fact that $\|y_n\|_2 \le a_n$. Moreover, when $x = \sum_{i\le N} x_i e_i$, we obtain from (19.1) that

$$\begin{aligned}\|x\| &= \sup\Big\{\sum_{i\le N} x^*(e_i)x_i \ ; \ \|x^*\|\le 1\Big\} = \sup\Big\{\sum_{i\le N} t_i x_i \ ; \ (t_i)\in T\Big\}\\ &= \sup\Big\{\sum_{i\le N} t_i x_i \ ; \ (t_i)\in T'\Big\} \le LS\sup\Big\{\sum_{i\le N} y_{n,i}x_i \ ; \ n\ge 1\Big\}\\ &= LS\sup\Big\{\sum_{i\le N} x_n^*(e_i)x_i \ ; \ n\ge 1\Big\} = LS\sup_{n\ge 1} x_n^*(x) \le LS\sup_{n\ge 1}|x_n^*(x)| \ .\end{aligned}$$

This proves (19.79) and finishes the argument. □

Suppose now that the sequence $(e_i)_{i\le N}$ is 1-unconditional. Then Theorem 19.2.14 is not satisfactory because the sequence $(e_i)_{i\le N}$ is not 1-unconditional for the norm $\mathcal{N}$ produced by this theorem. We provide a version of Theorem 19.2.14 which is adapted to the case where the sequence $(e_i)_{i\le N}$ is 1-unconditional.

Theorem 19.2.15 *Consider a 1-unconditional sequence $(e_i)_{i\le N}$ in a Banach space X, and let $S = \mathsf{E}\|\sum_{i\le N} g_i e_i\|$. Then we can find a sequence (I_n) of subsets of $\{1,\dots,N\}$ satisfying (19.59) and*

$$\forall x\in X \ , \ x = \sum_{i\le N} x_i e_i \ , \ \|x\| \le LS\sup_{n\ge 1} a_n\Big(\sum_{i\in I_n} x_i^2\Big)^{1/2} \ . \tag{19.81}$$

To explain the meaning of this result, let us assume that the sequence $(e_i)_{i\le N}$ spans X, and when $x = \sum_{i\le N} x_i e_i$, let us define the new norm

$$\mathcal{N}(x) = \sup_{n\ge 1} a_n \Big(\sum_{i\in I_n} x_i^2\Big)^{1/2} . \tag{19.82}$$

This sequence $(e_i)_{i\le N}$ is 1-unconditional for this norm, and (19.81) implies $\|x\| \le LS\mathcal{N}(x)$. Moreover, Lemma 19.2.5 implies that $\mathsf{E}\mathcal{N}(\sum_{i\le N} g_i e_i) \le L$. In words, given the 1-unconditional sequence $(e_i)_{i\le N}$, if we are only interested in the quantity $S = \mathsf{E}\|\sum_{i\le N} g_i e_i\|$, we can replace our norm by a *larger* norm of the type $S\mathcal{N}$, for which the sequence $(e_i)_{i\le N}$ is still 1-unconditional. Again, this should be viewed as a structure theorem.

Exercise 19.2.16 In the statement of Theorem 19.2.15 prove that one may instead request card $I_n \ge \log(1+n)$ and replace (19.81) by

$$\|x\| \le LS \sup_{n\ge 1}\Big(\frac{1}{\operatorname{card} I_n}\sum_{i\in I_n} x_i^2\Big)^{1/2} .$$

We start the proof of Theorem 19.2.15 with a simple observation.

Lemma 19.2.17 *Consider a 1-unconditional sequence $(e_i)_{i\le N}$ and $S = \mathsf{E}\|\sum_{i\ge 1} g_i e_i\|$. Then the set*

$$T = \big\{(x^*(e_i))_{i\le N}\ ;\ x^*\in X^*,\ \|x^*\|\le 1\big\} \subset \mathbb{R}^N \tag{19.83}$$

satisfies

$$\forall\, y\in T,\ \sum_{i\le N}|y_i| \le 2S . \tag{19.84}$$

Proof Denote by η_i the sign of $g_i x^*(e_i)$ so that

$$\begin{aligned}\sum_{i\le N}|g_i||x^*(e_i)| &= \sum_{i\le N}|x^*(g_i e_i)| = \sum_{i\le N}\eta_i x^*(g_i e_i)\\ &= x^*\Big(\sum_{i\le N}\eta_i g_i e_i\Big) \le \Big\|\sum_{i\le N}\eta_i g_i e_i\Big\| = \Big\|\sum_{i\le N} g_i e_i\Big\| .\end{aligned}$$

Taking expectation completes the proof since $\mathsf{E}|g_i| = \sqrt{2/\pi} \ge 1/2$. □

Proof of Theorem 19.2.15. We recall the set T of (19.83). Lemma 19.2.17 implies that $T \subset 2SB_1$. (This is the only place where the fact that the sequence $(e_i)_{i\le N}$ is unconditional is used.) Moreover, Theorem 2.10.1 implies that $\gamma_2(T, d_2) \le Lg(T)$, whereas

$$g(T) = \mathsf{E} \sup_{\|x^*\| \le 1} x^*\Big(\sum_{i \le N} g_i e_i\Big) = \mathsf{E}\Big\| \sum_{i \le N} g_i e_i \Big\| = S \,.$$

Theorem 19.2.4 provides sets I_n that satisfy (19.59) and $T \subset LST_1$, where

$$T_1 = \overline{\operatorname{conv}} \bigcup_{n \ge 1} B_2(I_n, a_n) \,.$$

Thus, by duality, if $x = \sum_{i \le N} x_i e_i$, we have, using the Cauchy-Schwarz inequality in the last step,

$$\begin{aligned} \|x\| &= \sup_{t \in T} \sum_{i \le N} t_i x_i \le LS \sup_{t \in T_1} \sum_{i \le N} t_i x_i \le LS \sup_{n \ge 1} a_n \sup_{t \in B_2(I_n, a_n)} \sum_{i \le N} t_i x_i \\ &= LS \sup_{n \ge 1} a_n \sup_{t \in B_2(I_n, a_n)} \sum_{i \in I_n} t_i x_i \le LS \sup_{n \ge 1} a_n \Big(\sum_{i \in I_n} x_i^2\Big)^{1/2} \end{aligned}$$

and this proves (19.81). □

The following exercise is similar to Theorem 19.2.15 but for r.v.s with exponential tails rather than Gaussian:

Exercise 19.2.18 Assume that the r.v.s Y_i are independent and symmetric and satisfy $\mathsf{P}(|Y_i| \ge x) = \exp(-x)$. Consider a 1-unconditional sequence $(e_i)_{i \le N}$ in a Banach space E, and let $S = \mathsf{E}\| \sum_{i \ge 1} Y_i e_i \|$. Prove that we can find a sequence (I_n) of subsets of $\{1, \ldots, N\}$ with card $I_n \le \log(n+1)$ and

$$\forall x \in E \,,\; x = \sum_{i \le N} x_i e_i \,,\; \|x\| \le LS \sup_{n \ge 1} \frac{1}{\log(n+1)} \sum_{i \in I_n} |x_i| \,.$$

Hint: Use Theorem 8.3.3.

19.3 Probabilistic Constructions

To prove the existence of an object with given properties, the probabilistic method exhibits a random object for which one can prove through probabilistic estimates that it has the required properties with positive probability.[9]

[9] There are many situations where this method applies but where one does not know how to exhibit any explicit object with these properties.

19.3.1 Restriction of Operators

Consider $q > 1$, the space ℓ_N^q, and its canonical basis $(e_i)_{i\le N}$. Consider a Banach space X and an operator $U : \ell_N^q \to X$. We will use many times the trivial observation that such an operator is entirely determined by the elements $x_i = U(e_i)$ of X. Our goal is to give in Theorem 19.3.1 below (surprisingly mild) conditions under which there are large subsets J of $\{1,\dots,N\}$ such that the norm $\|U_J\|$ of the restriction U_J to the span of the elements $(e_i)_{i\in J}$ is much smaller than the norm of U. We first compute this norm. We denote by X_1^* the unit ball of the dual of X, by p the conjugate exponent of q. Setting $x_i = U(e_i)$, we have

$$\begin{aligned}\|U_J\| &= \sup\Big\{x^*\Big(\sum_{i\in J}\alpha_i x_i\Big)\ ;\ \sum_{i\in J}|\alpha_i|^q \le 1\,,\ x^*\in X_1^*\Big\} \\ &= \sup\Big\{\Big(\sum_{i\in J}|x^*(x_i)|^p\Big)^{1/p}\ ;\ x^*\in X_1^*\Big\}\,.\end{aligned} \tag{19.85}$$

The set J will be constructed by a random choice. Specifically, given a number $0<\delta<1$, we consider (as in Sect. 11.11) i.i.d. r.v.s $(\delta_i)_{i\le N}$ with

$$\mathsf{P}(\delta_i = 1) = \delta\ ;\ \mathsf{P}(\delta_i = 0) = 1-\delta\,, \tag{19.86}$$

and we set $J = \{i\le N;\, \delta_i = 1\}$. Thus (19.85) implies

$$\|U_J\|^p = \sup_{t\in T}\sum_{i\le N}\delta_i|t_i|^p\,, \tag{19.87}$$

where

$$T = \{(x^*(x_i))_{i\le N}\ ;\ x^*\in X_1^*\}\subset\mathbb{R}^N\,. \tag{19.88}$$

Setting

$$|T|^p := \{(|t_i|^p)_{i\le N}\ ;\ t\in T\}\subset\mathbb{R}^N\,,$$

we may rewrite (19.87) as

$$\|U_J\|^p = \sup_{t\in|T|^p}\sum_{i\le N}\delta_i t_i\,. \tag{19.89}$$

This brings forward the essential point: To control $\mathsf{E}\|U_J\|^p$, we need information on the set $|T|^p$. However, information we might gather from the properties of X as a Banach space is likely to bear on T rather than $|T|^p$. The link between the properties of T and $|T|^p$ is provided in Theorem 19.3.2 below, which transfers a certain "smallness" property of T (reflected by (19.96) below) into an appropriate smallness property of $|T|^p$ (witnessed by (19.99) below).

Let us start with an obvious observation: Interchanging the supremum and the expectation yields

$$\mathsf{E}\|U_J\|^p \geq \sup_{t\in T} \mathsf{E}\Big(\sum_{i\leq N} \delta_i |t_i|^p\Big) = \delta \sup_{t\in T} \sum_{i\leq N} |t_i|^p \,. \tag{19.90}$$

This demonstrates the relevance of the quantity

$$\sup_{t\in T} \sum_{i\leq N} |t_i|^p = \sup_{\|x^*\|\leq 1} \sum_{i\leq N} |x^*(x_i)|^p \,. \tag{19.91}$$

We can think of this quantity as an obstacle to making $\mathsf{E}\|U_J\|^p$ small. It might be sometimes to our advantage to change the norm (as little as we can) to decrease this obstacle (of a somewhat uninteresting nature). For this, given a number $C > 0$, we denote by $\|\cdot\|_C$ the norm on X such that the unit ball of the dual norm is (bearing in mind that $\|x^*\|$ is the dual norm of x^*)

$$X^*_{1,C} = \Big\{x^* \in X^* \; ; \; \|x^*\| \leq 1 \,, \sum_{i\leq N} |x^*(x_i)|^p \leq C\Big\} \,, \tag{19.92}$$

and we denote by $\|U\|_C$ the operator norm of U when X is provided with the norm $\|\cdot\|_C$. This definition is tailored so that for the norm $\|\cdot\|_C$, the quantity (19.91) is now $\leq C$. Another very nice feature is that the set

$$T_C = \{(x^*(x_i))_{i\leq N} \; ; \; x^* \in X^*_{1,C}\} \subset \mathbb{R}^N \tag{19.93}$$

of (19.88) corresponding to the new norm is a *subset* of the set $T = \{(x^*(x_i))_{i\leq N} \; ; \; x^* \in X^*_1\}$ corresponding to the original norm. We will then be able to prove that T_C is small in the sense of (19.96) below simply because T is already small in this sense. This will be done by using the geometric properties of the original norm, and we shall *not* have to be concerned with the geometric properties of the norm $\|\cdot\|_C$.

We are now ready to bound the operator norm of a random restriction U_J of U.

Theorem 19.3.1 *Consider $1 < q \leq 2$ and its conjugate exponent $p \geq 2$. Consider a Banach space X such that X^* is p-convex (see Definition 4.1.2). Then there exists a number $K(p,\eta)$ depending only on p and on the constant η in Definition 4.1.2 with the following property. Consider elements $x_1, \ldots, x_N$ of X, and $S = \max_{i\leq N} \|x_i\|$. Denote by U the operator $\ell^q_N \to X$ such that $U(e_i) = x_i$. Consider a number C and define $B = \max(K(p,\eta)S^p \log N, C)$. Assume that for some $\epsilon > 0$*

$$\delta \leq \frac{S^p}{B\epsilon N^\epsilon} \leq 1 \,. \tag{19.94}$$

Consider r.v.s $(\delta_i)_{i\le N}$ *as in (19.86) and* $J = \{i \le N; \delta_i = 1\}$*. Then*

$$\mathsf{E}\|U_J\|_C^p \le K(p,\eta)\frac{S^p}{\epsilon} . \tag{19.95}$$

It is remarkable that the right-hand side of (19.95) does not depend on $\|U\|_C$ but only on $S = \max_{i\le N}\|U(e_i)\|$. In the situations of interest, S will be much smaller than $\|U\|_C$ so that (19.95) brings information. The condition (19.94) is not very intuitive at first, but the reader will find in Lemma 19.3.9 below two specific examples of application.

There are three steps in the proof.

- We use geometry to show that the set T of (19.88) "is not too large". This is Theorem 19.3.4 below.
- We transfer this control of T to $|T|^p$. This is Theorem 19.3.2 below.
- The structure result obtained for $|T|^p$ in the previous step is perfectly adapted to obtain a statement of the same nature as Theorem 19.3.1 and Theorem 19.3.3 below.

We first perform the second of the previous steps, which is closely related to Theorem 19.2.10. We recall from (19.75) that for a subset I of $\{1,\dots,N\}$ and for $a > 0$, we write

$$B_\infty(I,a) = \big\{(t_i)_{i\le N}\ ;\ i \notin I \Rightarrow t_i = 0\,,\ \forall i \in I\,,\ |t_i| \le a\big\} \subset \mathbb{R}^N .$$

Theorem 19.3.2 *Consider a subset* T *of* $\mathbb{R}^N$ *with* $0 \in T$*. Assume that for a certain number* $A > 0$*, there exists an admissible sequence* $(\mathcal{B}_n)$ *of* T *such that*[10]

$$\forall t \in T\,,\ \sum_{n\ge 0} 2^n \Delta(B_n(t), d_\infty)^p \le A \tag{19.96}$$

and let

$$B = \max\Big(A, \sup_{t\in T}\sum_{i\le N}|t_i|^p\Big) . \tag{19.97}$$

Then we can find a sequence $(I_k)_{k\ge 1}$ *of subsets of* $\{1,\dots,N\}$ *with*

$$\operatorname{card} I_k \le \frac{LB}{A}\log(k+1) , \tag{19.98}$$

[10] In the language of the functionals $\gamma_{\alpha,\beta}$ of (4.5), the following condition basically states that $\gamma_{p,p}(T,d_\infty)^p \le A$.

and

$$|T|^p \subset K(p)A \ \mathrm{conv} \bigcup_{k\geq 1} B_\infty\Big(I_k, \frac{1}{\log(k+1)}\Big) . \tag{19.99}$$

Proof The proof is self-contained. However, your task will be much easier if you study first Theorems 19.2.4 and 19.2.10 which have nearly the same proof.

The set $|T|^p$ is a subset of the ball of $L^1(\mu)$ of center 0 and radius B, where μ is the counting measure on $\{1, \dots, N\}$. The first step of the proof is to take advantage of this through Theorem 19.2.1. Consider the largest integer τ for which $2^\tau \leq B/A$. Since $B \geq A$, we have $\tau \geq 0$, and $2^{-\tau} < 2A/B$. Recalling that for a subset I of $\{1, \dots, N\}$ we have $\mu(I) = \mathrm{card}\, I$, homogeneity and Theorem 19.2.1 provide us with an admissible sequence of partitions $(\mathcal{D}_n)$ of $|T|^p$ and for each $D \in \mathcal{D}_n$ an integer $\ell^*(D) \in \mathbb{Z}$, such that if for $t \in |T|^p$, we set[11]

$$\ell^*(t,n) = \ell^*(D_n(t)) \tag{19.100}$$

then (according to (19.49) and (19.48), respectively)

$$\forall t \in |T|^p \,,\ \mathrm{card}\{i \leq N \,;\ t_i \geq 2^{-\ell^*(t,n)}\} \leq 2^{n+\tau} \leq \frac{2^n B}{A} \tag{19.101}$$

$$\forall t \in |T|^p \,,\ \sum_{n\geq 0} 2^{n-\ell^*(t,n)} \leq 18 \cdot 2^{-\tau} B \leq LA \,. \tag{19.102}$$

The second step of the proof is to bring back to T the information we just gathered about $|T|^p$. This is done in the most straightforward manner. We consider the canonical map $\varphi : T \to |T|^p$ given by $\varphi((t_i)_{i\leq N}) = (|t_i|^p)_{i\leq N}$. We consider on T the admissible sequence of partitions $(\mathcal{C}_n)$ where $\mathcal{C}_n$ consists of the sets $\varphi^{-1}(D)$ where $D \in \mathcal{D}_n$. For $t \in T$, we define $\ell(t,n) = \ell^*(\varphi(t),n)$, and this number depends only on $C_n(t)$ because $\ell^*(\varphi(t),n)$ depends only on $D_n(\varphi(t))$. Moreover, we deduce from (19.101) and (19.102), respectively, that

$$\forall t \in T \,,\ \mathrm{card}\{i \leq N \,;\ |t_i|^p \geq 2^{-\ell(t,n)}\} \leq \frac{2^n B}{A} \tag{19.103}$$

$$\forall t \in T \,,\ \sum_{n\geq 0} 2^{n-\ell(t,n)} \leq LA \,. \tag{19.104}$$

The obvious move now is to combine this information with the information provided by (19.96). Denoting by $\mathcal{A}_n$ the partition generated by $\mathcal{B}_n$ and $\mathcal{C}_n$, the sequence $(\mathcal{A}_n)$

[11] As ususal, $D_n(t)$ is the element of $\mathcal{D}_n$ containing t.

is increasing and card $\mathcal{A}_n \le N_{n+1}$. Moreover since $A_n(t) \subset B_n(t)$, (19.96) implies

$$\forall t \in T\,,\ \sum_{n\ge 0} 2^n \Delta(A_n(t), d_\infty)^p \le A\,, \tag{19.105}$$

and furthermore, the integer $\ell(t,n)$ depends only on $A_n(t)$.

After these preparations, we start the main construction. For $D \in \mathcal{A}_n$, $n \ge 0$, let us choose in an arbitrary manner $v_n(D) \in D$, and set $\pi_n(t) = v_n(A_n(t))$. We write $\pi_n(t) = (\pi_{n,i}(t))_{i\le N}$, and we define

$$I_0(t) = \{i \le N\,;\ |\pi_{0,i}(t)|^p \ge 2^{-\ell(t,0)}\}\,. \tag{19.106}$$

For $n \ge 1$, we further define

$$I_n(t) = \left\{i \le N\,;\ |\pi_{n,i}(t)|^p \ge 2^{-\ell(t,n)}\,,\ 0 \le k < n \Rightarrow |\pi_{k,i}(t)|^p < 2^{-\ell(t,k)}\right\}.$$

It is important that $I_n(t)$ depends only on $A_n(t)$ so that there are at most card $\mathcal{A}_n \le N_{n+1}$ sets of this type. Next, since $|t_i - \pi_{n,i}(t)| \le \Delta(A_n(t), d_\infty) \le \Delta(B_n(t), d_\infty)$, we have $\lim_{n\to\infty} |t_i - \pi_{n,i}(t)| = 0$ and thus

$$\{i \le N\,;\ |t_i| \ne 0\} \subset \bigcup_{n\ge 0} I_n(t)\,. \tag{19.107}$$

Finally, we note from (19.103) that

$$\operatorname{card} I_n(t) \le \frac{2^n B}{A}\,. \tag{19.108}$$

The definition of $I_n(t)$ shows that for $n \ge 1$ and $i \in I_n(t)$, we have $|\pi_{n-1,i}(t)|^p < 2^{-\ell(t,n-1)}$ so that

$$|t_i| \le |t_i - \pi_{n-1,i}(t)| + |\pi_{n-1,i}(t)| \le \Delta(A_{n-1}(t), d_\infty) + 2^{-\ell(t,n-1)/p}$$

and hence

$$|t_i|^p \le K(p)(\Delta(A_{n-1}(t), d_\infty)^p + 2^{-\ell(t,n-1)}) := c(t,n)\,. \tag{19.109}$$

Let us define $c(t,0) = \Delta(T, d_\infty)^p$. Since $0 \in T$, (19.109) remains true for $n = 0$. We have then

$$n \ge 0\,,\ i \in I_n(t) \Rightarrow |t_i|^p \le c(t,n)\,. \tag{19.110}$$

Moreover (19.105) and (19.104) imply

$$\forall t \in T\,,\ \sum_{n\ge 0} 2^n c(t,n) \le K(p)A\,. \tag{19.111}$$

We consider the family $\mathcal{F}$ of all pairs $(I_n(t), 2^{-n})$ for $t \in T$ and $n \geq 0$, and we prove that

$$|T|^p \subset K(p)A \operatorname{conv} \bigcup_{(I,a)\in\mathcal{F}} B_\infty(I, a) . \tag{19.112}$$

We recall the notation (19.67). For $n \geq 0$, we define u_n by $2^n c(n,t) u_n = |t|^p \mathbf{1}_{I_n(t)}$ so that, using (19.110),

$$u_n = \frac{1}{2^n c(n,t)} |t|^p \mathbf{1}_{I_n(t)} \in B_\infty(I_n(t), 2^{-n}) . \tag{19.113}$$

We then have, using (19.107) in the first equality,

$$|t|^p = \sum_{n\geq 0} |t|^p \mathbf{1}_{I_n(t)} = \sum_{n \geq 0} 2^n c(n,t) u_n . \tag{19.114}$$

Together with (19.111) and (19.113), the relation 19.114 proves that $|t|^p \in K(p)A \operatorname{conv} \cup_{(I,a)\in\mathcal{F}} B_\infty(I,a)$ and (19.112). (The reason why we can take a convex hull rather than the closure of a convex hull is that there is only a finite number of possibilities for the sets $I_n(t)$.)

It remains now to deduce (19.99) from (19.112). This requires a careful enumeration of the pairs $(I, a) \in F$ for which we basically copy the argument given at the end of the proof of Theorem 19.2.4. Consider the set $\mathcal{I}_n$ consisting of all the sets of the type $I_n(t)$ for $t \in T$ so that $\operatorname{card} \mathcal{I}_n \leq N_{n+1}$. We find a sequence $(I_k)_{k\geq 1}$ of sets such that $I_k = \emptyset$ for $k < N_2$ and that for $n \geq 0$, $\mathcal{I}_n = \{I_k ; N_{k+1} \leq k < N_{k+2}\}$. This is possible because $\operatorname{card} \mathcal{I}_n \leq N_{n+1} \leq N_{n+2} - N_{n+1}$.

Then any $(I, a) \in \mathcal{F}$ is such that for some $n \geq 0$, we have $I \in \mathcal{I}_n$ and $a = 2^{-n}$. Thus, $I = I_k$ where $N_{n+1} \leq k < N_{n+2}$ so that $k + 1 \leq N_{n+2}$ and consequently $2^{-n} \leq 4/\log(k+1)$. Thus (19.99) follows from (19.112). Furthermore, since $k \geq N_{n+1}$, we have $2^n \leq L \log k$, and (19.108) implies (19.98). □

The smallness criterion provided by (19.99) is perfectly adapted to the control of $\mathsf{E}\|U_J\|^p$.

Theorem 19.3.3 *Consider the set* $T = \{(x^*(e_i))_{i\leq N} ; x^* \in X_1^*\}$ *of (19.88). Assume (19.96) and let* B *as in (19.97). Consider* $\epsilon > 0$ *and* $\delta \leq 1$ *such that*

$$\delta \leq \frac{A}{B\epsilon N^\epsilon \log N} . \tag{19.115}$$

Then if the r.v.s $(\delta_i)_{i\leq N}$ *are as in (19.86) and* $J = \{i \leq N ; \delta_i = 1\}$, *for* $v \geq 6$, *we have*

$$\mathsf{P}\Big(\|U_J\|^p \geq vK(p)\frac{A}{\epsilon \log N}\Big) \leq L \exp\Big(-\frac{v}{L}\Big)$$

and in particular

$$\mathsf{E}\|U_J\|^p \le K(p)\frac{A}{\epsilon \log N} \,. \tag{19.116}$$

Proof The magic is that

$$\sup_{t\in B_\infty(I,a)} \sum_{i\le N} \delta_i t_i = \sup_{t\in B_\infty(I,a)} \sum_{i\in I} \delta_i t_i \le a \sum_{i\in I} \delta_i \,,$$

so that (19.99) implies

$$\sup_{t\in |T|^p} \sum_{i\le N} \delta_i t_i \le K(p) A \sup_{k\ge 1} \frac{1}{\log(k+1)} \sum_{i\in I_k} \delta_i \,. \tag{19.117}$$

We will control the right-hand side using the union bound. For $k \ge 1$, we have $\operatorname{card} I_k \le L_0 \log(k+1) B/A$ by (19.98), so that

$$\delta \operatorname{card} I_k \le \frac{L_0 \log(k+1)}{\epsilon N^\epsilon \log N} \,.$$

We recall the inequality (11.70): If $u \ge 6\delta \operatorname{card} I$,

$$\mathsf{P}\Big(\sum_{i\in I} \delta_i \ge u\Big) \le \exp\Big(-\frac{u}{2}\log\frac{u}{2\delta \operatorname{card} I}\Big) \,.$$

Considering $v \ge 6$, we use this inequality for $u = L_0 v \log(k+1)/(\epsilon \log N) \ge 6\delta N^\epsilon \operatorname{card} I_k \ge 6\delta \operatorname{card} I_k$ to obtain

$$\begin{aligned}\mathsf{P}\Big(\sum_{i\in I_k} \delta_i \ge \frac{L_0 v \log(k+1)}{\epsilon \log N}\Big) &\le \exp\Big(-\frac{L_0 v \log(k+1)}{2\epsilon \log N}\log(N^\epsilon)\Big) \\ &= \exp\Big(-\frac{L_0 v \log(k+1)}{2}\Big) \,. \end{aligned} \tag{19.118}$$

Thus, if we define the event

$$\Omega(v) \;:\; \forall k \ge 1 \,,\; \sum_{i\in I_k} \delta_i \le \frac{L_0 v \log(k+1)}{\epsilon \log N} \,,$$

we obtain from (19.118) that $\mathsf{P}(\Omega(v)^c) \le L\exp(-v/L)$. When $\Omega(v)$ occurs, for $k \ge 1$, we have

$$\frac{1}{\log(k+1)} \sum_{i \in I_k} \delta_i \le \frac{L_0 v}{\epsilon \log N} \,.$$

Then (19.117) and (19.89) imply $\|U_J\|^p \le K(p)vA/(\epsilon \log N)$. □

We finally come to the control of T. We recall the functionals $\gamma_{\alpha,\beta}$ of (4.5).

Theorem 19.3.4 *Under the conditions of Theorem 19.3.1, the set T of (19.88) satisfies*

$$\gamma_{p,p}(T, d_\infty) \le K(p,\eta) S (\log N)^{1/p} \,. \tag{19.119}$$

Before the proof, we consider the (quasi) distance d_∞ on X_1^* defined by

$$d_\infty^*(x^*, y^*) = \max_{i \le N} |x^*(x_i) - y^*(x_i)| \,.$$

The map $\psi : X_1^* \to T$ given by $\psi(x^*) = (x^*(x_i))_{i \le N}$ satisfies

$$d_\infty(\psi(x^*), \psi(y^*)) = d_\infty^*(x^*, y^*) \,. \tag{19.120}$$

Lemma 19.3.5 *We have*

$$e_k(X_1^*, d_\infty^*) \le K(p,\eta) S 2^{-k/p} (\log N)^{1/p} \tag{19.121}$$

or, equivalently, for $\epsilon > 0$,

$$\log N(X_1^*, d_\infty^*, \epsilon) \le K(p,\eta) \Big(\frac{S}{\epsilon}\Big)^p \log N \,. \tag{19.122}$$

Here, X_1^* is the unit ball of X^* for the original dual norm, $N(X_1^*, d_\infty, \epsilon)$ is the smallest number of balls for d_∞ of radius ϵ needed to cover X_1^*, and e_k is defined in (2.36).

It would be nice to have a simple proof of this statement. The only proof we know is somewhat indirect. It involves geometric ideas. First, one proves a "duality" result, namely, that if W denotes the convex hull of the points $(\pm x_i)_{i \le N}$, to prove (19.122), it suffices to show that

$$\log N(W, \|\cdot\|, \epsilon) \le K(p,\eta) \Big(\frac{S}{\epsilon}\Big)^p \log N \,. \tag{19.123}$$

This duality result is proved in [24], Proposition 2, *(ii)*. We do not reproduce the simple and very nice argument, which is not related to the ideas of this work. The proof of (19.123) also involves geometrical ideas. Briefly, since X^* is p-convex, it is classical that "X is of type p, with a type p constant depending only on p and η" as proved in [57], and then the conclusion follows from a beautiful probabilistic argument of Maurey, which is reproduced, for example, in [120], Lemma 3.2. □

Exercise 19.3.6 Deduce (19.121) from (19.123) and Proposition 2, (ii) of [24].

Proof of Theorem 19.3.4. Recalling that we assume that the dual norm of X is p-convex, we combine (19.121) with Theorem 4.1.4 (used for $\alpha = p$). □

Proof of Theorem 19.3.1. We reformulate (19.119) as follows: There exists an admissible sequence $(\mathcal{B}_n)$ on X_1^* for which

$$\forall x^* \in X_1^* \,, \sum_{n \geq 0} 2^n \Delta(B_n(x^*), d_\infty^*)^p \leq K(p,\eta) S^p \log N := A \,. \tag{19.124}$$

It follows from (19.124) and (19.120) that the set T of (19.88) satisfies (19.96). Thus, this is also the case of the smaller set T_C of (19.93). Since $\sum_{i \leq N} |t_i|^p \leq C$ for $t \in T_C$, this set also satisfies (19.97) for $B = \max(A, C)$. We then conclude with Theorem 19.3.3. □

To conclude this section, we describe an example showing that Theorem 19.3.1 is very close to being optimal in certain situations. Consider two integers r, m and set $N = rm$. We divide $\{1, \ldots, N\}$ into m disjoint subsets $I_1, \ldots, I_m$ of cardinality r. Consider $1 < q \leq 2$ and the canonical bases $(e_i)_{i \leq N}, (e_j)_{j \leq m}$ of ℓ_N^q and ℓ_m^q, respectively. Consider the operator $U : \ell_N^q \to \ell_m^q = X$ such that $U(e_i) = e_j$ where j is such that $i \in I_j$. Thus, $S = 1$. It is classical [57] that $X^* = \ell_m^p$ is p-convex. Consider δ with $\delta^r = 1/m$. Then

$$\mathsf{P}(\exists j \leq m \,;\, \forall i \in I_j \,,\, \delta_i = 1) = 1 - \left(1 - \frac{1}{m}\right)^m \geq \frac{1}{L} \,,$$

and when this event occurs, we have $\|U_J\| \geq r^{1/p}$ since $\|\sum_{i \in I_j} e_i\| = r^{1/q}$ and $\|U_J(\sum_{i \in I_j} e_i)\| = r\|e_j\| = r$. Thus,

$$\mathsf{E}\|U_J\|^p \geq \frac{r}{L} \,. \tag{19.125}$$

Let us now try to apply Theorem 19.3.1 to this situation so that $x_i = e_j$ for $i \in I_j$. Then we must take C large enough such that $\|\cdot\|_C = \|\cdot\|$, that is, $C = \sup_{\|x^*\| \leq 1} \sum_{i \leq N} |x^*(x_i)|^p$. Since there are r values of i for which $s_i = e_j$, we get

$$\sum_{i \leq N} |x^*(x_i)|^p = r \sum_{j \leq m} |x^*(e_j)|^p \,.$$

This can be as large as r for $\|x^*\| \le 1$, so one has to take $C = r$. Then $B = r$ whenever $K(q)\log N \le r$. Let us choose $\epsilon = 1/(2r)$ so that for large m

$$\delta = \frac{1}{m^{1/r}} \le \frac{S^p}{B\epsilon N^\epsilon} = \frac{1}{r\epsilon N^\epsilon} = \frac{1}{r\epsilon m^\epsilon r^\epsilon} .$$

Thus (19.125) shows that (19.95) gives the exact order of $\|U_J\|$ in this case.

19.3.2 The $\Lambda(p)$-Problem

We denote by λ the uniform measure on $[0, 1]$. Consider functions $(x_i)_{i\le N}$ on $[0, 1]$ satisfying the following two conditions:

$$\forall i \le N\,,\ \|x_i\|_\infty \le 1\,, \tag{19.126}$$

$$\text{The sequence } (x_i)_{i\le N} \text{ is orthogonal in } L^2 = L^2(\lambda)\,. \tag{19.127}$$

For a number $p \ge 1$, we denote by $\|\cdot\|_p$ the norm in $L^p(\lambda)$. Thus, if $p \ge 2$ for all numbers $(\alpha_i)_{i\le N}$, we have

$$\Big\|\sum_{i\le N}\alpha_i x_i\Big\|_2 \le \Big\|\sum_{i\le N}\alpha_i x_i\Big\|_p .$$

J. Bourgain [22] proved the remarkable fact that we can find a set J, with card $J \ge N^{2/p}$, for which we have an estimate in the reverse direction:[12]

$$\forall(\alpha_i)_{i\in J}\,,\ \Big\|\sum_{i\in J}\alpha_i x_i\Big\|_p \le K(p)\Big(\sum_{i\in J}\alpha_i^2\Big)^{1/2} . \tag{19.128}$$

Bourgain's argument is probabilistic, showing in fact that a random choice of J works with positive probability. The most interesting application of this theorem is the case of the trigonometric system, say $x_i(t) = \cos 2\pi k_i t$ where the integers $(k_i)_{i\in I}$ are all different. Even in that case, no simpler proof is known.

We consider r.v.s δ_i as in (19.86) with $\delta = N^{2/p-1}$ and we set $J = \{i \le N\,;\ \delta_i = 1\}$.

Theorem 19.3.7 *There is a r.v. $W \ge 0$ with $\mathsf{E}W \le K$ such that for any numbers $(\alpha_i)_{i\in J}$, we have*

$$\forall(\alpha_i)_{i\in J}\,,\ \Big\|\sum_{i\in J}\alpha_i x_i\Big\|_p \le W\Big(\sum_{i\in J}\alpha_i^2\Big)^{1/2} . \tag{19.129}$$

[12] Proving this is (a version of) what was known as the Λ_p problem.

Here, as well as in the rest of this section, K denotes a number depending only on p, that need not be the same on each occurrence. Since $\mathsf{P}(\operatorname{card} J \ge N^{2/p}) \ge 1/K$,[13] recalling that $\mathsf{E}W \le K$ and using Markov's inequality, with positive probability, we have both $\operatorname{card} J \ge N^{2/p}$ and $W \le K$, and in this case, we obtain (19.128). It is possible with minimum additional effort to prove a slightly stronger statement than (19.129), where the L^p norm on the left is replaced by a stronger norm.[14] We refer the reader to [132] for this. In the present presentation, we have chosen the simplest possible result. I am grateful to Donggeun Ryou for a significant simplification of the argument.[15]

Theorem 19.3.7 is mostly a consequence of the following special case of Theorem 19.3.1, which we state again to avoid confusion of notation. We recall the Definition 4.1.2 of a 2-convex Banach space and the corresponding constant η. Given a number $C > 0$, we denote by $\|\cdot\|_C$ the norm on X such that the unit ball of the dual norm is the set

$$X^*_{1,C} = \{x^* \in X^*; \|x^*\|_C \le 1\} = \Big\{x^* \in X^* \; ; \; \|x^*\| \le 1 \, , \; \sum_{i \le N} x^*(x_i)^2 \le C\Big\} \, . \tag{19.130}$$

Theorem 19.3.8 *Consider a Banach space X such that X^* is 2-convex with corresponding constant η. Then there exists a number $K(\eta)$ depending only on η with the following property. Consider elements $x_1, \ldots, x_N$ of X with $\|x_i\| \le 1$. Denote by U the operator $\ell^2_N \to X$ such that $U(e_i) = x_i$. Consider a number $C > 0$ and define $B = \max(C, K(\eta)\log N)$. Consider a number $\delta > 0$ and assume that for some $\epsilon > 0$*

$$\delta \le \frac{1}{B\epsilon N^\epsilon} \le 1 \, . \tag{19.131}$$

Consider r.v.s $(\delta_i)_{i \le N}$ as in (19.86) and $J = \{i \le N; \delta_i = 1\}$. Then the restriction U_J of U to the span of the vectors $(e_i)_{i\in J}$ satisfies

$$\mathsf{E}\|U_J\|_C^2 \le \frac{K(\eta)}{\epsilon} \, , \tag{19.132}$$

where $\|U_J\|_C$ is the operator norm of U_J when X is provided with the norm $\|\cdot\|_C$.

Despite the fact that Bourgain's result is tight, there is some room in the proof of Theorem 19.3.7. Some of the choices we make are simply convenient and by no means canonical. We fix $p > 2$ once and for all, and we set $p_1 = 3p/2$. Denoting

---

[13] Note that this is obvious for large N by the Central Limit Theorem.

[14] More specifically, the so-called $L^{p,1}$ norm.

[15] And in particular for observing that there is no need of a special argument to control the large values of the function f in the proof below.

q_1 the conjugate exponent of p_1, the dual L^{q_1} of $X = L^{p_1}$ is 2-convex [57], and the corresponding constant η depends on p only.[16] For a number $C > 0$, we consider on $X = L^{p_1}$ the norm $\|\cdot\|_C$ constructed before the statement of Theorem 19.3.8. From now on, $\delta = N^{2/p-1}$.

Lemma 19.3.9 *Setting* $C_1 = N^{1/2-1/p}$ *and* $C_2 = e^{-1}N^{1-2/p}\log N$*, we have*

$$\mathsf{E}\|U_J\|_{C_1} \le K \; ; \; \mathsf{E}\|U_J\|_{C_2} \le K\sqrt{\log N} \; .$$

Proof We apply Theorem 19.3.8 to the case $X = L^{p_1}$, whose dual L^{q_1} is 2-convex, and we note that it suffices to prove the result for N large enough.

We choose first $C = C_1$ $(= N^{1/2-1/p})$ and $\epsilon = 1/2 - 1/p$. For N large enough $B = \max(C, K\log N) = N^{1/2-1/p}$ so that $BN^\epsilon = N^{1-2/p} = \delta^{-1}$. Since $\epsilon < 1$, (19.131) holds. Since $1/\epsilon \le K$, (19.132) then proves that $\mathsf{E}\|U_J\|^2_{C_1} \le K$.

Next, we choose $C = C_2$ $(= e^{-1}N^{1-2/p}\log N)$ and $\epsilon = 1/\log N$. Thus, $N^\epsilon = e$, and for N large enough $B = \max(C, K\log N) = C$, so that $B\epsilon = e^{-1}N^{1-2/p}$ and $B\epsilon N^\epsilon = N^{1-2/p} = \delta^{-1}$. Thus (19.131) holds again, and (19.132) proves now that $\mathsf{E}\|U_J\|^2_{C_2} \le K\log N$. □

Lemma 19.3.10 *Consider a measurable function* f*. Assume that* $\|f\|_{C_1} \le 1$ *and* $\|f\|_{C_2} \le \sqrt{\log N}$*. Then* $\|f\|_p \le K$*.*

Proof of Theorem 19.3.7. Let

$$V = \|U_J\|_{C_1} + \frac{1}{\sqrt{\log N}}\|U_J\|_{C_2} \; ,$$

so that $\mathsf{E}V \le K$ by Lemma 19.3.9. Consider numbers $(\alpha_i)_{i\in J}$ and $y := \sum_{i\in J}\alpha_i e_i$. For $j = 1, 2$, we have $\|U_J(y)\|_{C_j} \le \|U_J\|_{C_j}\|y\|_2$. The function $f = (V\|y\|_2)^{-1}U_J(y)$ satisfies the hypotheses of Lemma 19.3.10 so that $\|f\|_p \le K$. That is, $\|U_J(y)\|_p \le KV\|y\|_2$, that is, $\|\sum_{i\in J}\alpha_i x_i\|_p \le KV(\sum_{i\in J}\alpha_i^2)^{1/2}$. □

Before we prove Lemma 19.3.10, we need to learn to use information on $\|f\|_{C_j}$. This is through duality in the form of the following lemma, which is a consequence of the Hahn-Banach theorem:

Lemma 19.3.11 *If* $f \in L^{p_1}$ *satisfies* $\|f\|_C \le 1$*, then* $f \in \mathcal{C} := \mathrm{conv}(\mathcal{C}_1 \cup \mathcal{C}_2)$ *where* $\mathcal{C}_1 = \{g;\, \|g\|_{p_1} \le 1\}$ *and* $\mathcal{C}_2 = \{\sum_{i\le N}\beta_i x_i;\, \sum_{i\le N}\beta_i^2 \le C^{-1}\}$.

Proof The set $\mathcal{C}_2$ is closed and finite dimensional, so it is compact. The set $\mathcal{C}_1$ is closed so that the set $\mathcal{C}$ is closed and obviously convex. If $f \notin \mathcal{C}$, by the Hahn-Banach theorem, there exists $x^* \in L^{q_1}$ such that $x^*(f) > 1$ but $x^*(g) \le 1$ for $g \in \mathcal{C}$. That $x^*(g) \le 1$ for $g \in \mathcal{C}_1$ implies that $\|x^*\|_{q_1} \le 1$. That $x^*(g) \le 1$ for

[16] So that when we will apply Theorem 19.3.8 to X, the corresponding constant $K(\eta)$ depends only on p.

$\in \mathcal{C}_2$ implies that $\sum_{i\le N} x^*(x_i)^2 \le C$. Thus, $\|x^*\|_C \le 1$ by definition of the norm $\|x^*\|_C$. Since $x^*(f) > 1$ from (19.1), we have $\|f\|_C > 1$. □

Proof of Lemma 19.3.10. The proof is based on the formula

$$\|f\|_p^p = p\int_0^\infty t^{p-1}\lambda(\{|f|\ge t\})\mathrm{d}t\ , \tag{19.133}$$

where λ is Lebesgue's measure on $[0,1]$ and suitable bounds for the integrand. The method to bound $\lambda(\{|f|\ge t\})$ differs depending on the value of t. For a certain quantity D (depending on N), we will distinguish three cases. For $t \le 1$, we will use that $\lambda(\{|f|\ge t\}) \le 1$ so that $\int_0^1 t^{p-1}\lambda(\{|f|\ge t\})\mathrm{d}t \le K$. For $1 \le t \le D$, we will use the hypothesis that $\|f\|_{C_1} \le 1$. For $t \ge D$, we will use the hypothesis that $\|f\|_{C_2} \le \sqrt{\log N}$.

Since $\|f\|_{C_1} \le 1$, by Lemma 19.3.11 (used for $C = C_1 = N^{1/2-1/p}$), we may write f as a convex combination[17] $f = \tau_1 u_1 + \tau_2 u_2$ where $\|u_1\|_{p_1} \le 1$ and $u_2 = \sum_{i\le N}\beta_i x_i$ with $\sum_{i\le N}\beta_i^2 \le N^{1/p-1/2}$. By (19.126) and (19.127), we have $\|u_2\|_2^2 \le N^{1/p-1/2}$. Markov's inequality implies that for each $s \ge 1$

$$\lambda(\{|u|\ge t\}) \le \frac{\|u\|_s^s}{t^s}\ , \tag{19.134}$$

and we combine (19.134) with the obvious inequality

$$\lambda(\{|f|\ge t\}) \le \lambda(\{|u_1|\ge t\}) + \lambda(\{|u_2|\ge t\}) \tag{19.135}$$

to obtain

$$\lambda(\{|f|\ge t\}) \le t^{-p_1} + t^{-2}N^{1/p-1/2}\ . \tag{19.136}$$

Recalling that $p > 2$, let us define $\alpha > 0$ by the relation $\alpha(p_1-2) = 1/2 - 1/p$, and let us set $D = N^\alpha$. Then for $t \le D$, we have $t^{-p_1} + t^{-2}N^{1/p-1/2} = t^{-p_1} + t^{-2}N^{-\alpha(p_1-2)} \le 2t^{-p_1}$. Since $p_1 = 3p/2$, we get $\int_1^D t^{p-1}\lambda(\{|f|\ge t\})\mathrm{d}t \le 2\int_1^\infty t^{-1-p/2}\mathrm{d}t \le K$. It remains only to control $\int_D^\infty t^{p-1}\lambda(\{|f|\ge t\})\mathrm{d}t$.

Since $\|f\|_{C_2} \le \sqrt{\log N}$, by Lemma 19.3.11 again (used now for $C = C_2 = e^{-1}N^{1-2/p}\log N$), we may write f as a convex combination $f = \tau_1 v_1 + \tau_2 v_2$ where $\|v_1\|_{p_1} \le \sqrt{\log N}$ and $v_2 = \sum_{i\le N}\beta_i x_i$, with $\sum_{i\le N}\beta_i^2 \le \log N C_2^{-1} = eN^{2/p-1}$. Thus, $\|v_2\|_2^2 \le eN^{2/p-1}$, and since $\|x_i\|_\infty \le 1$, we also have $\|v_2\|_\infty \le \sum_{i\le N}\beta_i \le \sqrt{N}\sqrt{\sum_{i\le N}\beta_i^2} \le 2N^{1/p}$. Just as in (19.136), we then obtain

$$\lambda(\{|f|\ge t\}) \le \lambda(\{|v_1|\ge t\}) + \lambda(\{|v_2|\ge t\}) \le (\log N)^{p_1/2}t^{-p_1} + \lambda(\{|v_2|\ge t\})\ ,$$

[17] That is, $\tau_1, \tau_2 \ge 0$, $\tau_1 + \tau_2 = 1$.

so that $\int_D^\infty t^{p-1}\lambda(\{|f| \ge t\})\mathrm{d}t \le (\log N)^{p_1/2} I_1 + I_2$ where $I_1 = \int_D^\infty t^{p-p_1-1}\mathrm{d}t$ and $I_2 = \int_D^\infty t^{p-1}\lambda(\{|v_2| \ge t\})\mathrm{d}t$. Now $I_1 \le KD^{p-p_1} = KD^{-p/2} = KN^{-\alpha p/2}$. Since $\|v_2\|_2^2 \le eN^{2/p-1}$, we have $\lambda(\{|v_2| \ge t\}) \le eN^{2/p-1}t^{-2}$ by (19.134), and also $\lambda(\{|v_2| \ge t\}) = 0$ for $t \ge 2N^{1/p} \ge \|v_2\|_\infty$, so that

$$I_2 \le \int_0^{2N^{1/p}} t^{p-1}\lambda(\{|v_2| \ge t\})\mathrm{d}t \le eN^{2/p-1}\int_0^{2N^{1/p}} t^{p-3}\mathrm{d}t = K ,$$

and we have proved as desired that $\int_D^\infty t^{p-1}\lambda(\{|f| \ge t\})\mathrm{d}t \le K$. □

19.4 Sidon Sets

Let us recall that if T is a compact abelian group, a character χ is a continuous map from T to $\mathbb{C}$ with $|\chi(t)| = 1$ and $\chi(s+t) = \chi(s)\chi(t)$. Thus, $\chi(0) = 1$ and $\chi(-s) = \bar{\chi}(s)$. Throughout this section, we denote by μ the Haar probability measure on T. We recall from Lemma 7.3.6 that two different characters are orthogonal in $L^2(T, \mathrm{d}\mu)$. Given a set Γ of characters, we define its Sidon constant Γ_{si} (possibly infinite) as the smallest constant such that for each sequence of complex numbers $(\alpha_\chi)_{\chi\in\Gamma}$, only finitely many of them nonzero, we have

$$\sum_{\chi\in\Gamma} |\alpha_\chi| \le \Gamma_{\mathrm{si}} \sup_t \Big| \sum_{\chi\in\Gamma} \alpha_\chi \chi(t) \Big| . \tag{19.137}$$

We say that Γ is a Sidon set if $\Gamma_{\mathrm{si}} < \infty$. To understand this definition and the next one, it is very instructive to consider the case where $T = \{-1, 1\}^N$ and $\Gamma = \{\varepsilon_i, i \le N\}$ where $\varepsilon_i(t) = t_i$.[18] Sidon sets are standard fare in harmonic analysis. In a certain sense, the characters in a Sidon set are "independent".[19] Another measure of independence of the elements of Γ is given by the smallest constant Γ_{sg} such that for all $p \ge 1$, we have, for all families, α_χ as above

$$\Big(\int \Big|\sum_{\chi\in\Gamma} \alpha_\chi \chi\Big|^p \mathrm{d}\mu\Big)^{1/p} \le \sqrt{p}\,\Gamma_{\mathrm{sg}} \Big(\sum_{\chi\in\Gamma} |\alpha_\chi|^2\Big)^{1/2} . \tag{19.138}$$

This should be compared with Khinchin's inequality (6.3), a comparison which supports the idea that controlling Γ_{sg} is indeed a measure of independence. The subscript "sg" stands for "subgaussian", as in the subgaussian inequality (6.1.1). This is because, as is shown in Exercise 2.3.8 (which we advise the reader to

---

[18] Here of course, the group structure on $\{-1, 1\}$ is given by ordinary multiplication.

[19] In a much stronger sense than just linear independence.

carefully review now), the constant $\Gamma_{\rm sg}$ is within a universal constant[20] of the constant $\Gamma'_{\rm sg}$ defined as the smallest number such that

$$\Big\|\sum_{\chi\in\Gamma}\alpha_\chi\chi\Big\|_{\psi_2}\le\Gamma'_{\rm sg}\Big(\sum_{\chi\in\Gamma}|\alpha_\chi|^2\Big)^{1/2}\,, \tag{19.139}$$

where the norm $\|\cdot\|_{\psi_2}$ defined in (14.14) is for the measure μ. The reader should also review (14.18) which explains how (19.139) is related to the subgaussian inequality (6.1.1).

One of the main results of this section is the following classical result. It relates the two "measures of independence" which we just considered.

Theorem 19.4.1 (W. Rudin, G. Pisier) *We have*

$$\Gamma_{\rm sg}\le L\Gamma_{\rm si}\,. \tag{19.140}$$

There exists a function $\varphi:\mathbb{R}^+\to\mathbb{R}^+$ *such that*[21]

$$\Gamma_{\rm si}\le\varphi(\Gamma_{\rm sg})\,. \tag{19.141}$$

Furthermore, we will prove a considerable generalization of (19.141), due also to Gilles Pisier [89] (after contributions by J. Bourgain and M. Lewko [20]). The most important consequence of this theorem is that $\Gamma_{\rm sg}<\infty$ if and only if $\Gamma_{\rm si}<\infty$.

As this is a book of probability rather than analysis, we will simplify some analytical details by *assuming that* T *is finite*. We start with the rather easy part, the proof of (19.140).

Lemma 19.4.2 *Consider complex numbers* β_χ *for* $\chi\in\Gamma$. *Then we can find a function* f *on* T *with*

$$\|f\|_1\le\Gamma_{\rm si}\sup_{\chi\in\Gamma}|\beta_\chi| \tag{19.142}$$

and

$$\forall\chi\in\Gamma\,;\ \int f\chi\,{\rm d}\mu=\beta_\chi\,. \tag{19.143}$$

Here, the norm $\|f\|_1$ is the norm in $L^1({\rm d}\mu)$.

---

[20] That is, $\Gamma_{\rm sg}\le L\Gamma'_{\rm sg}\le L\Gamma_{\rm sg}$.

[21] The proof we give shows that $\Gamma_{\rm si}\le L(1+\Gamma_{\rm sg})^4$. It is known that $\Gamma_{\rm si}\le L\Gamma_{\rm sg}^2(1+\log\Gamma_{\rm sg})$ but that it is not true that $\Gamma_{\rm si}\le L\Gamma_{\rm sg}$; see [88].

Proof Let us denote by V the complex linear span of the functions χ for $\chi \in \Gamma$. Recalling that the characters are linearly independent because they form an orthogonal set, consider the linear functional ϕ on V given by $\phi(\sum_{\chi\in\Gamma}\alpha_\chi\chi) = \sum_{\chi\in\Gamma}\alpha_\chi\beta_\chi$. Thus, if $h = \sum_{\chi\in\Gamma}\alpha_\chi\chi$ then, using (19.137)

$$|\phi(h)| \le \sup_{\chi\in\Gamma}|\beta_\chi|\sum_{\chi\in\Gamma}|\alpha_\chi| \le \Gamma_{\mathrm{si}}\sup_{\chi\in\Gamma}|\beta_\chi|\|h\|_\infty\,. \tag{19.144}$$

Let us provide the space of functions on T and its subspaces with the supremum norm. The content of (19.144) is that ϕ is of norm $\le \Gamma_{\mathrm{si}}\sup_{\chi\in\Gamma}|\beta_\chi|$ on V. By the Hahn-Banach theorem, we can extend ϕ to a linear functional $\bar\phi$ of the same norm on the space of all functions on T. Since T is finite, there exists a function f on T such that $\bar\phi(h) = \int fh\mathrm{d}\mu$ for all functions h on T, and $\|f\|_1 = \|\bar\varphi\| \le \Gamma_{\mathrm{si}}\sup_{\chi\in\Gamma}|\beta_\chi|$. In particular, $\int fh\mathrm{d}\mu = \phi(h)$ whenever $h \in V$. Taking $h = \chi$ implies (19.143). □

Lemma 19.4.3 *Consider for $\chi \in \Gamma$ numbers $\varepsilon_\chi = \pm 1$. Then for each p and each sequence of numbers $(\alpha_\chi)_{\chi\in\Gamma}$, we have*

$$\int\Big|\sum_{\chi\in\Gamma}\alpha_\chi\chi\Big|^p\mathrm{d}\mu \le \Gamma_{\mathrm{si}}^p\int\Big|\sum_{\chi\in\Gamma}\varepsilon_\chi\alpha_\chi\chi\Big|^p\mathrm{d}\mu\,. \tag{19.145}$$

Proof According to Lemma 19.4.2, we can find a function f on T such that $\int f\chi\mathrm{d}\mu = \varepsilon_\chi$ for each $\chi \in \Gamma$, whereas $\|f\|_1 \le \Gamma_{\mathrm{si}}$. Let us define $g(t) = \sum_{\chi\in\Gamma}\varepsilon_\chi\alpha_\chi\chi(t)$. Thus, for $t \in T$, we have

$$\begin{aligned}\sum_{\chi\in\Gamma}\alpha_\chi\chi(t) &= \sum_{\chi\in\Gamma}\alpha_\chi\varepsilon_\chi\chi(t)\varepsilon_\chi\\ &= \sum_{\chi\in\Gamma}\alpha_\chi\varepsilon_\chi\chi(t)\int f(x)\chi(x)\mathrm{d}\mu(x)\\ &= \sum_{\chi\in\Gamma}\alpha_\chi\varepsilon_\chi\int f(x)\chi(x+t)\mathrm{d}\mu(x)\\ &= \int f(x)g(x+t)\mathrm{d}\mu(x)\,. \end{aligned} \tag{19.146}$$

If a function h satisfies $\int|h(x)|\mathrm{d}\mu(x) = 1$, we have

$$\begin{aligned}\Big|\int h(x)g(x+t)\mathrm{d}\mu(x)\Big|^p &\le \Big(\int|h(x)||g(x+t)|\mathrm{d}\mu(x)\Big)^p\\ &\le \int|h(x)||g(x+t)|^p\mathrm{d}\mu(x)\,,\end{aligned}$$

where in the second line, we use the convexity of the map $x \to |x|^p$. Using this for the function $h(x) = f(x)/\|f\|_1$, we obtain

$$\Big|\int f(x)g(x+t)\mathrm{d}\mu(x)\Big|^p \le \|f\|_1^{p-1}\int |f(x)||g(x+t)|^p\mathrm{d}\mu(x)\,,$$

and combining with (19.146), we obtain

$$\Big|\sum_{\chi\in\Gamma}\alpha_\chi\chi(t)\Big|^p \le \|f\|_1^{p-1}\int |f(x)||g(x+t)|^p\mathrm{d}\mu(x)\,.$$

Integrating in t and using the translation invariance of μ yields

$$\begin{aligned}\int\Big|\sum_{\chi\in\Gamma}\alpha_\chi\chi(t)\Big|^p\mathrm{d}\mu(t) &\le \|f\|_1^{p-1}\iint |f(x)||g(x+t)|^p\mathrm{d}\mu(x)\mathrm{d}\mu(t)\\ &= \|f\|_1^p\int|g(t)|^p\mathrm{d}\mu(t)\,.\end{aligned}$$

□

Proof of (19.140). We average (19.145) over all choices of $\varepsilon_\chi = \pm 1$, using Khintchin's inequality (6.3). □

Next, we turn to the proof of (19.141). We will deduce it from the considerably more general recent result of [89].

Theorem 19.4.4 *For $j = 1, 2$, consider a sequence $(\varphi_{j,n})_{n\le N}$ of functions on a probability space $(\Omega, \mathrm{d}\nu)$. Assume that each of these sequences is an orthonormal system.*[22] *Assume that for certain numbers A, B, for $j = 1, 2$, and each numbers $(\alpha_n)_{n\le N}$, we have*

$$\forall n\le N,\ \|\varphi_{j,n}\|_\infty \le A\,, \tag{19.147}$$

$$\Big\|\sum_{n\le N}\alpha_n\varphi_{j,n}\Big\|_{\psi_2} \le B\Big(\sum_{n\le N}|\alpha_n|^2\Big)^{1/2}\,. \tag{19.148}$$

Then for each complex numbers $(\alpha_n)_{n\le N}$, we have

$$\sum_{n\le N}|\alpha_n| \le LA^2(A+B)^4\Big\|\sum_{n\le N}\alpha_n\varphi_{1,n}\otimes\varphi_{2,n}\Big\|_\infty\,, \tag{19.149}$$

where $\varphi_{1,n}\otimes\varphi_{2,n}$ is the function on $\Omega\times\Omega$ defined by $\varphi_{1,n}\otimes\varphi_{2,n}(\omega_1,\omega_2) = \varphi_{1,n}(\omega_1)\varphi_{2,n}(\omega_2)$.

[22] That is, each vector is of norm 1 and they are orthogonal.

Proof of (19.141). Consider a set $\Gamma = (\chi_n)_{n\le N}$ of characters. Let $\varphi_{1,n} = \varphi_{2,n} = \chi_n$ so that (19.147) holds for $A = 1$ and (19.148) holds for $B = \Gamma'_{\rm sg} \le L\Gamma_{\rm sg}$. Taking $(\Omega,\nu) = (T,\mu)$, since $\chi_n(s)\chi_n(t) = \chi_n(s+t)$, (19.149) becomes $\sum_{n\le N}|\alpha_n| \le L(1+\Gamma_{\rm sg})^4\|\sum_{n\le N}\alpha_n\chi_n\|_\infty$, and this proves that $\Gamma_{\rm si} \le L(1+\Gamma_{\rm sg})^4$. □

Throughout the rest of this section, we write $T = \{-1,1\}^N$ and μ denotes the Haar measure. We denote by ε_n the n-th coordinate function on T. The first ingredient of the proof of Theorem 19.4.4 is the following:

Proposition 19.4.5 *Under the conditions of Theorem 19.4.4, there exist operators $U_j : L^1(T,\mu) \to L^1(\Omega,\nu)$ such that $U_j(\varepsilon_n) = \varphi_{j,n}$ and $\|U_j\| \le L(A+B)$.*

To understand the issue there, let us first prove a simple fact about operators from $\ell^1 = \ell^1(\mathbb{N})$ to $L^1(\Omega,\nu)$. Given elements $u_1,\dots,u_m$ of $L^1(\Omega,\nu)$, we define $\max_{k\le m}|u_k|$ pointwise, $(\max_{k\le m}|u_k|)(\omega) = \max_{k\le m}|u_k(\omega)|$. Similarly, elements of ℓ^1 are seen as functions on $\mathbb{N}$.

Lemma 19.4.6 *Consider a bounded operator $U : \ell^1 \to L^1(\Omega,\nu)$. Then given elements $f_1,\dots,f_m$ of ℓ^1, we have*

$$\Big\|\max_{k\le m}|U(f_k)|\Big\|_1 \le \|U\|\Big\|\max_{k\le m}|f_k|\Big\|_1\,. \tag{19.150}$$

Proof Let (e_n) be the canonical basis of ℓ^1 and $h_n = U(e_n)$ so that $\|h_n\|_1 \le \|U\|$. Consider elements $f_1,\dots,f_m$ of ℓ^1 with $f_k = \sum_{n\ge1}a_{k,n}e_n$. Then

$$\max_{k\le m}|U(f_k)| = \max_{k\le m}\Big|\sum_{n\ge1}a_{k,n}h_n\Big| \le \sum_{n\ge1}\max_{k\le m}|a_{k,n}||h_n|$$

and thus

$$\Big\|\max_{k\le m}|U(f_k)|\Big\|_1 \le \sum_{n\ge1}\max_{k\le m}|a_{k,n}|\|h_n\|_1 \le \|U\|\sum_{n\ge1}\max_{k\le m}|a_{k,n}|\,.$$

Finally, $\sum_{n\ge1}\max_{k\le m}|a_{k,n}| = \|\max_{k\le m}|f_k|\|_1$. □

Mireille Lévy proved the following converse to Lemma 19.4.6. The proof does not use probabilistic ideas but relies on the Hahn-Banach theorem, and we refer the reader to [56] for it.

Lemma 19.4.7 *Consider a subspace E of ℓ^1 and an operator V from E to $L^1(\Omega,\nu)$. Assume that for a certain number C and any elements $f_1,\dots,f_m$ of E, we have*

$$\Big\|\max_{k\le m}|V(f_k)|\Big\|_1 \le C\Big\|\max_{k\le m}|f_k|\Big\|_1\,. \tag{19.151}$$

Then there exists an operator $U : \ell^1 \to L^1(\Omega, \nu)$ such that $\|U\| \le C$ and that the restriction of U to E coincides with V.

Proof of Proposition 19.4.5. Since T is finite for each point $t \in T$, we have $\mu(\{t\}) = 1/\operatorname{card} T$. Thus, the space $L^1(T, \mu)$ is isomorphic to a space ℓ^1. We consider the span E of the elements ε_n in $L^1(T, \mu)$, and fixing $j \in \{1, 2\}$ on E, we consider the operator V defined by $V(\varepsilon_n) = \varphi_{j,n}$. The plan is to prove (19.151) for $C = L(A + B)$ and to use Lemma 19.4.7.

Let $f_k = \sum_{n \le N} a_{k,n}\varepsilon_n$. Thinking of (T, μ) as a probability space, and denoting accordingly by E integration with respect to μ, we then have

$$S := \big\| \max_{k \le m} |f_k| \big\|_1 = \mathsf{E} \max_{k \le m} \Big| \sum_{n \le N} a_{k,n}\varepsilon_n \Big| .$$

We recognize the supremum of a Bernoulli process. Setting $a_{0,k} = 0$, for $0 \le k \le m$, consider the sequence $a_k = (a_{k,n})_{n \le N}$. We can then use Theorem 6.2.8 to find a set $W \subset \ell^2(N)$ with $\gamma_2(W) \le LS$ such that each sequence a_k can be decomposed as $a_k^1 + a_k^2$ where $a_k^1 \in W$ and where $\sum_{n \le N} |a_{k,n}^2| \le LS$. Since $a_0 = 0$, we have $a_0^1 + a_0^2 = 0$. We may then replace a_k^1 by $a_k^1 - a_0^1$ and a_k^2 by $a_k^2 - a_0^k$. This replaces W by $W - a_k^1$ so that now $0 \in W$ and consequently $\|a\|_2 \le LS$ for $a \in W$. Since $V(\varepsilon_n) = \varphi_{j,n}$, we then have

$$\max_{k \le m} |V(f_k)| = \max_{k \le m} \Big| V\Big(\sum_{n \le N} a_{k,n}\varepsilon_n \Big) \Big| \le \mathrm{I} + \mathrm{II} , \tag{19.152}$$

where

$$\mathrm{I} = \max_{k \le m} \Big| \sum_{n \le N} a_{k,n}^1 \varphi_{j,n} \Big| \; ; \; \mathrm{II} = \max_{k \le m} \Big| \sum_{n \le N} a_{k,n}^2 \varphi_{j,n} \Big| .$$

We will prove that

$$\|\mathrm{I}\|_1 \le LBS \; ; \; \|\mathrm{II}\|_1 \le LAS . \tag{19.153}$$

Combining with (19.152), this proves as desired that $\| \max_{k \le m} |V(f_k)| \|_1 \le L(A + B) \| \sup_{k \le m} |f_k| \|_1$ and concludes the proof.

Since $\|\varphi_{j,n}\|_\infty \le A$ by (19.147) and since $\sum_{n \le N} |a_{k,n}^2| \le LS$, it follows that $\|\mathrm{II}\|_\infty \le LAS$ so that $\|\mathrm{II}\|_1 \le LAS$. To control the term I, let us consider the process $Y_a := \sum a_n \varphi_{j,n}$ so that $\|\mathrm{I}\|_1 \le \mathsf{E} \sup_{a \in W} |Y_a|$. Then (19.148) means that the process $X_a = Y_a/(LB)$ satisfies the increment condition (2.4). Using (2.60) together with the fact that $\mathsf{E}|Y_a| \le LS$ because $\|a\|_2 \le LS$ for $a \in W$, this implies that $\|\mathrm{I}\|_1 \le LBS$ and finishes the proof of (19.153). □

The next ingredient to Theorem 19.4.4 is a nearly magical observation. Recalling that $T = \{-1, 1\}^N$, for $n \le N$ and $j = 1, 2$, we define on $T \times T$ the functions ε_n^j by $\varepsilon_n^j(t^1, t^2) = t_n^j$ where $t^j = (t_n^j)_{n \le N} \in T$.

Proposition 19.4.8 *Given $0 < \delta < 1$, we can decompose the function $F = \sum_{1 \le n \le N} \varepsilon_n^1 \varepsilon_n^2$ as $F = F_1 + F_2$ in a way that*

$$\|F_1\|_1 = \iint |F_1| \mathrm{d}\mu \otimes \mathrm{d}\mu \le 2/\delta \tag{19.154}$$

and such that for any two functions g_1 and g_2 on T, we have

$$\Big| \iint F_2 g_1 \otimes g_2 \mathrm{d}\mu \otimes \mathrm{d}\mu \Big| \le \delta \|g_1\|_2 \|g_2\|_2 \,. \tag{19.155}$$

Given a set $I \subset \{1, \ldots, N\}$, we define the function ε_I on T by $\varepsilon_\emptyset = 1$ and if $I \ne \emptyset$ by $\varepsilon_I(t) = \prod_{n \in I} t_n$. As I varies, the functions ε_I form an orthonormal basis of $L^2(T, \mathrm{d}\mu)$. For $j = 1, 2$ and a set $I \subset \{1, \ldots, N\}$, we define $\varepsilon_I^j = \prod_{n \in I} \varepsilon_n^j$ so that $\varepsilon_I \otimes \varepsilon_J = \varepsilon_I^1 \varepsilon_J^2$. As I and J vary, the functions $\varepsilon_I^1 \varepsilon_J^2$ form an orthonormal basis of $L^2(T \times T, \mathrm{d}\mu \otimes \mathrm{d}\mu)$.

Lemma 19.4.9 *The function $\tau := \sum_{I \subset \{1,\ldots,N\}} \varepsilon_I^1 \varepsilon_I^2 \delta^{\operatorname{card} I}$ satisfies $\|\tau\|_1 = 1$.*

Proof Indeed, $\tau = \prod_{n \le N}(1 + \delta \varepsilon_n^1 \varepsilon_n^2)$ is ≥ 0 and of integral 1. □

Proof of Proposition 19.4.8. Define $F_1 = (\tau - 1)/\delta = \sum_{\operatorname{card} I \ge 1} \varepsilon_I^1 \varepsilon_I^2 \delta^{\operatorname{card} I - 1}$ so that $\|F_1\|_1 \le 2/\delta$ by the previous lemma. Then $F_2 := F - F_1 = -\sum_{\operatorname{card} I \ge 2} \delta^{\operatorname{card} I - 1} \varepsilon_I^1 \varepsilon_I^2$. For $j = 1, 2$, let us decompose the function g_j on T in the basis (ε_I): $g_j = \sum_I g_{j,I} \varepsilon_I$. Then, $g_1 \otimes g_2 = \sum_{I,J} g_{1,I} g_{2,J} \varepsilon_I^1 \varepsilon_J^2$ and thus, using the Cauchy-Schwarz inequality,

$$\begin{aligned}\Big| \iint F_2 g_1 \otimes g_2 \mathrm{d}\mu \otimes \mathrm{d}\mu \Big| &= \Big| \sum_{\operatorname{card} I \ge 2} \delta^{\operatorname{card} I - 1} g_{1,I} g_{2,I} \Big| \le \delta \sum_{\operatorname{card} I \ge 2} |g_{1,I} g_{2,I}| \\ &\le \delta \Big(\sum_I |g_{1,I}|^2 \Big)^{1/2} \Big(\sum_I |g_{2,I}|^2 \Big)^{1/2} = \delta \|g_1\|_2 \|g_2\|_2 \,. \end{aligned}$$ □

Lemma 19.4.10 *Under the hypotheses of Theorem 19.4.4, consider numbers $(\theta_n)_{n \le N}$ with $|\theta_n| = 1$. Then given any $\delta > 0$, the function $\Phi := \sum_{n \le N} \theta_n \varphi_{1,n} \otimes \varphi_{2,n}$ can be written as $\Phi_1 + \Phi_2$ where $\|\Phi_1\|_1 \le L(A + B)^2/\delta$ and where for any functions g_1, g_2 on Ω,*

$$\Big| \iint \Phi_2 g_1 \otimes g_2 \mathrm{d}\nu \otimes \mathrm{d}\nu \Big| \le L\delta (A + B)^2 \|g_1\|_\infty \|g_2\|_\infty \,. \tag{19.156}$$

Proof We may assume that $\theta_n = 1$ by replacing $\varphi_{1,n}$ by $\theta_n\varphi_{1,n}$. Consider the operators U_j as provided by Proposition 19.4.5. There exists an operator $U_1 \otimes U_2 : L^1(T \times T, \mu\otimes\mu) \to L^1(\Omega \times \Omega, \nu\otimes\nu)$ with the property that $U_1 \otimes U_2(f_1 \otimes f_2) = U_1(f_1) \otimes U_2(f_2)$. This is obvious by considering first the functions f_1, f_2 which are supported by a single point. This construction also shows that it has a norm $\le \|U_1\|\|U_2\|$. Furthermore, $\Phi = \sum_{n\le N} \varphi_{1,n} \otimes \varphi_{2,n} = U_1 \otimes U_2(\sum_{n\le N} \varepsilon_n^1 \otimes \varepsilon_n^2)$.

Considering the functions F_1 and F_2 of Proposition 19.4.8, we set $\Phi_1 = U_1 \otimes U_2(F_1)$ and $\Phi_2 = U_1 \otimes U_2(F_2)$ so that $\Phi = \Phi_1 + \Phi_2$. Now, $\|\Phi_1\|_1 \le L\|U_1\|\|U_2\|\|F_1\|_1 \le L(A+B)^2/\delta$. It remains only to prove (19.156). Let $U_j^* : L^\infty(\Omega, \nu) \to L^\infty(T, \mu)$ denote the adjoint of U_j. Given a function $H \in L^1(T\times T)$, the identity

$$\iint U_1 \otimes U_2(H) g_1 \otimes g_2 \mathrm{d}\nu \otimes \mathrm{d}\nu = \iint H U_1^*(g_1) \otimes U_2^*(g_2) \mathrm{d}\mu \otimes \mathrm{d}\mu \tag{19.157}$$

holds because it holds when H is of the type $H_1 \otimes H_2$ and because the elements of that type span $L^1(T \times T)$. Also, $\|U_j^*(g_j)\|_2 \le \|U_j^*(g_j)\|_\infty \le L(A+B)\|g_j\|_\infty$. Using (19.157) for $H = F_2$ and (19.155) yields (19.156). □

Proof of Theorem 19.4.4. Let us fix the numbers α_n and consider θ_n with $|\theta_n| = 1$ and $\bar{\alpha}_n\theta_n = |\alpha_n|$. Since the systems $(\varphi_{1,n})$ and $(\varphi_{2,n})$ are orthonormal, we have

$$\sum_{n\le N} |\alpha_n| = \iint \bar{\Psi}\Phi \mathrm{d}\mu \otimes \mathrm{d}\mu \,, \tag{19.158}$$

where $\bar{\Psi} = \sum_{n\le N} \bar{\alpha}_n\varphi_{1,n} \otimes \varphi_{2,n}$ and $\Phi = \sum_{n\le N} \theta_n\varphi_{1,n} \otimes \varphi_{2,n}$. Let us then use the decomposition $\Phi = \Phi_1 + \Phi_2$ provided by Lemma 19.4.10. First,

$$\Big|\iint \bar{\Psi}\Phi_1 \mathrm{d}\mu \otimes \mathrm{d}\mu\Big| \le \|\Phi_1\|_1\|\Psi\|_\infty \le L\delta^{-1}(A+B)^2\|\Psi\|_\infty \,.$$

Also, using (19.156), we have

$$\Big|\iint \bar{\Psi}\Phi_2 \mathrm{d}\mu \otimes \mathrm{d}\mu\Big| \le \sum_{n\le N} |\alpha_n|\Big|\iint \Phi_2\varphi_{1,n} \otimes \varphi_{2,n}\Big| \le L\delta A^2(A+B)^2 \sum_{n\le N} |\alpha_n| \,.$$

Then (19.158) yields

$$\sum_{n\le N} |\alpha_n| \le L\delta^{-1}(A+B)^2\|\Psi\|_\infty + L\delta A^2(A+B)^2 \sum_{n\le N} |\alpha_n| \,,$$

from which (19.149) follows by taking $\delta = 1/(2LA^2(A+B)^2)$. □

Appendix A
Discrepancy for Convex Sets

A.1 Introduction

The purpose of this appendix is to bring forward the following (equally beautiful) close cousin of the Leighton-Shor theorem. We denote by λ_3 the usual volume measure and by $(X_i)_{i\le N}$ independent uniformly distributed points in $[0,1]^3$.

Theorem A.1.1 *Consider the class $\mathcal{C}$ of convex sets in $\mathbb{R}^3$. Then*

$$\frac{1}{L}\sqrt{N}(\log N)^{3/4} \le \mathsf{E}\sup_{C\in\mathcal{C}}\Big|\sum_{i\le N}(\mathbf{1}_C(X_i)-\lambda_3(C))\Big| \le L\sqrt{N}(\log N)^{3/4}\,. \tag{A.1}$$

The upper bound is proved in [113]. In this appendix, we sketch how to adapt the lower bound machinery of Chap. 4 to the present case. The following exercise highlights the parallels between Theorem A.1.1 and the Leighton-Shor theorem:

Exercise A.1.2 Convince yourself that as a consequence of the Leighton-Shor theorem (A.1) holds when $(X_i)_{i\le N}$ are uniformly distributed points in $[0,1]^2$ and $\mathcal{C}$ is the class of sets which are the interior of a closed curve of length ≤ 1.

Consider independent uniformly distributed points $(X_i)_{i\le N}$ in $[0,1]^k$, $\mathcal{C}$ a class of subsets of $[0,1]^k$ and define $S_N := \mathsf{E}\sup_{C\in\mathcal{C}}|\sum_{i\le N}(\mathbf{1}_C(X_i)-\lambda_k(C))|$. This quantity is of interest when there is some constraint on the size of $\mathcal{C}$. Interestingly, the constraint in dimension $k=3$ that the elements of $\mathcal{C}$ are convex or in dimension $k=2$ that they are the interiors of curves of length ≤ 1 yield the same rate of growth for S_N.

M. Talagrand, *Upper and Lower Bounds for Stochastic Processes*, Ergebnisse der Mathematik und ihrer Grenzgebiete. 3. Folge / A Series of Modern Surveys in Mathematics 60, https://doi.org/10.1007/978-3-030-82595-9

A.2 Elements of Proof of the Upper Bound

Unfortunately, a complete proof of the upper bound in (A.1) requires considerable work, not all of which is very exciting, so we will give only a very short outline of it. The understanding of subgraphs[1] of convex functions is closely related to the understanding of convex sets. In fact, it is rather elementary to show that the boundary of a convex set in $\mathbb{R}^3$ can be broken into six pieces which are graphs of convex 1-Lipschitz functions (defined on a subset of $\mathbb{R}^2$.)

It is time to recall that a twice differentiable function g on $[0,1]^2$ is convex if and only if at each point we have $\partial^2 g/\partial x^2 \geq 0$ and

$$\left(\frac{\partial^2 g}{\partial x \partial y}\right)^2 \leq \frac{\partial^2 g}{\partial x^2}\frac{\partial^2 g}{\partial y^2} . \tag{A.2}$$

We denote by λ_2 the two-dimensional Lebesgue measure on $[0,1]^2$.

Lemma A.2.1 *A convex differentiable 1-Lipschitz function g on $[0,1]^2$ satisfies*

$$\int \frac{\partial^2 g}{\partial x^2} \mathrm{d}\lambda_2 \leq 2 \,; \; \int \frac{\partial^2 g}{\partial y^2} \mathrm{d}\lambda_2 \leq 2 \,; \; \int \left|\frac{\partial^2 g}{\partial x \partial y}\right| \mathrm{d}\lambda_2 \leq 2 . \tag{A.3}$$

Proof We write $\int_0^1 \partial^2 g/\partial x^2(x,y)\mathrm{d}x = \partial g/\partial x(1,y) - \partial g/\partial x(0,y) \leq 2$, and we integrate over $y \in [0,1]$ to obtain the first part of (A.3). The second part is similar, and the third follows from the first two parts using (A.2) and the Cauchy-Schwarz inequality. □

An important step in the proof of Theorem A.1.1 is as follows:

Theorem A.2.2 *The class $\mathcal{C}$ for functions $[0,1]^2 \to [0,1]$ which satisfy (A.3) is such that $\gamma_{1,2}(\mathcal{C}) < \infty$.*

To understand why this theorem can be true, consider first the smaller class $\mathcal{C}^*$ consisting of functions which are zero on the boundary of $[0,1]^2$ and which satisfy the stronger conditions $\|\partial^2 g/\partial x^2\|_2 \leq L$ and $\|\partial^2 g/\partial y^2\|_2 \leq L$. Then the use of Fourier transform shows that $\mathcal{C}^*$ is isometric to a subset of the ellipsoid $\sum_{n,m}(1 + n^4 + m^4)|x_{nm}|^2 \leq L$, and Corollary 4.1.7 proves that $\gamma_{1,2}(\mathcal{C}^*) < \infty$. Surely, the assumption that the function is zero on the boundary of $[0,1]^2$ brings only lower-order effects. The fact that for a function in $\mathcal{C}$ we require only integrability of the partial derivatives as is (A.3) (rather than square integrability for a function of $\mathcal{C}^*$) is a far more serious problem. It is solved in [113] following the same approach as in Sect. 17.3: One shows that a function $f \in \mathcal{C}$ can be written as a sum $\sum_{k \geq 0} f_k$ where $f_k \in \mathcal{C}_k$, the class of functions g which satisfy $\|\partial^2 g/\partial x^2\|_\infty \leq 2^k$, $\|\partial^2 g/\partial y^2\|_\infty \leq 2^k$ and moreover $\lambda_2(\{g \neq 0\}) \leq L2^{-k}$. One main step of the

[1] Look at (4.118) if you forgot what is the subgraph of a function.

proof is to show that the sequence $(\gamma_{1,2}(\mathcal{C}_k))$ decreases geometrically (in fact, $\gamma_{1,2}(\mathcal{C}_k) \le Lk2^{-k/2}$).

A.3 The Lower Bound

Our strategy to construct convex functions is based on the following elementary lemma:

Lemma A.3.1 *Consider a function* $h : [0,1]^2 \to \mathbb{R}$ *and assume the following:*

$$h(0,0) = 0 = \frac{\partial h}{\partial x}(0,0) = \frac{\partial h}{\partial y}(0,0) , \tag{A.4}$$

$$\left\| \frac{\partial^2 h}{\partial x^2} \right\|_\infty \le \frac{1}{16} ; \left\| \frac{\partial^2 h}{\partial y^2} \right\|_\infty \le \frac{1}{16} ; \left\| \frac{\partial^2 h}{\partial x \partial y} \right\|_\infty \le \frac{1}{16} . \tag{A.5}$$

Then the function g given by

$$g(x,y) = \frac{1}{2} + \frac{1}{8}(x^2 + y^2) + h(x,y) \tag{A.6}$$

is valued in $[0,1]$*, and is convex.*

Proof It is elementary to prove that $|h(x,y)| \le 1/4$ so that g is valued in $[0,1]$. Moreover for each $x, y \in [0,1]$, we have

$$\frac{\partial^2 g}{\partial x^2}(x,y) = \frac{1}{4} + \frac{\partial^2 h}{\partial x^2} \ge \frac{1}{8}$$

and similarly $\partial^2 g/\partial y^2(x,y) \ge 1/8$, $|\partial^2 g/\partial x \partial y(x,y)| \le 1/16$ so that (A.2) is satisfied and thus g is convex. □

Thus, to construct large families of convex functions, it will suffice to construct large families of functions satisfying (A.4) and (A.5). We will do this using a variation on the method of Sects. 4.6 and 4.8. The control of the mixed partial derivatives requires a more clever choice of our basic function, to which we turn now.

Let us consider the function f on $\mathbb{R}$ given by $f(x) = 0$ unless $0 \le x \le 1$, $f(0) = f'(0) = 0$, $f''(x) \in \{-1, 1\}$, $f''(x) = 1$ if $1/8 \le x < 3/8$, $4/8 \le x < 5/8$ or $7/8 \le x \le 1$ and $f''(x) = -1$ otherwise (Fig. A.1).

We observe that $f''(1-x) = -f''(x)$, $f'(1-x) = f'(x)$, and $f(1-x) = -f(x)$ and also that

$$\int f'' \mathrm{d}\lambda = \int f' \mathrm{d}\lambda = \int f \mathrm{d}\lambda = 0 . \tag{A.7}$$

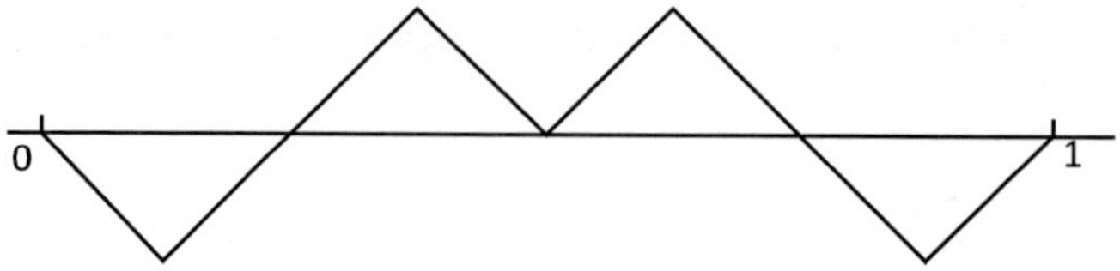

Fig. A.1 The graph of f'

For $q \geq 1$ and $1 \leq \ell \leq 2^q$, let us define $f_{q,\ell}$ by $f_{q,\ell}(x) = 2^{-2q} f(2^q(x - (\ell - 1)2^{-q}))$. We note right away that

$$f_{q,\ell}(x) \neq 0 \Rightarrow (\ell - 1)2^{-q} \leq x \leq \ell 2^{-q} , \tag{A.8}$$

so that at a given q, the functions $f_{q,\ell}$ have disjoint supports. Let us list some elementary properties of these functions. We denote by λ the Lebesgue measure on $[0, 1]$. The following lemma resembles Lemma 4.6.2, but the nontrivial new piece of information is (A.17).

Lemma A.3.2 *We have the following:*

$$\|f''_{q,\ell}\|_\infty = 1 . \tag{A.9}$$

$$\|f''_{q,\ell}\|_2^2 = 2^{-q} . \tag{A.10}$$

$$\|f'_{q,\ell}\|_\infty \leq 2^{-q-3} . \tag{A.11}$$

$$\|f'_{q,\ell}\|_2^2 = 2^{-3q-6}/3 . \tag{A.12}$$

$$\|f_{q,\ell}\|_\infty \leq 2^{-2q-6} . \tag{A.13}$$

$$\|f_{q,\ell}\|_2^2 \leq 2^{-5q-12} . \tag{A.14}$$

$$\|f_{q,\ell}\|_1 \geq 2^{-q}/L . \tag{A.15}$$

$$q' \geq q + 3, \ell \leq 2^q, \ell' \leq 2^{q'} \Rightarrow \int f''_{q,\ell} f''_{q',\ell'} \mathrm{d}\lambda = 0 . \tag{A.16}$$

$$q' \geq q + 3, \ell \leq 2^q, \ell' \leq 2^{q'} \Rightarrow \int f'_{q,\ell} f'_{q',\ell'} \mathrm{d}\lambda = 0 . \tag{A.17}$$

Proof Only (A.16) and (A.17) are not obvious. To prove (A.16), we observe that on the support of $f''_{q',\ell'}$ the function $f''_{q,\ell}$ is constant, so the result follows from the obvious fact that $\int f'' \mathrm{d}\lambda = 0$. To prove (A.17), we observe that on the support of

$f'_{q',\ell'}$, the function $f'_{q,\ell}$ is affine, of the type $a+bx$. Since $\int f'\mathrm{d}\lambda = 0$, it suffices to check that $\int f'(x)(x-1/2)\mathrm{d}\lambda(x) = 0$ which follows by making the change of variables $x \to 1-x$ and since $f'(x) = f'(1-x)$. □

We consider a number r with $r \simeq \log N/100$. Given two functions f, g on $[0,1]$, we write a usual $f \otimes g$ the function on $[0,1]^2$ given by $f \otimes g(x,y) = f(x)g(y)$. We consider an integer $c \geq 3$ which is designed to give us some room (just as the integer c of Sect. 4.8). It is a universal constant which will be determined later. We will be interested in functions of the type[2]

$$f_k = \frac{2^{2ck-5}}{\sqrt{r}} \sum_{\ell,\ell' \leq 2^{ck}} z_{k,\ell,\ell'} f_{ck,\ell} \otimes f_{ck,\ell'} , \tag{A.18}$$

where $z_{k,\ell,\ell'} \in \{0,1\}$.

Lemma A.3.3 *Given functions f_k as above for $ck \leq r$ and setting $f = \sum_{ck \leq r} f_k$, we have*

$$\Big\| \frac{\partial^2 f}{\partial x^2} \Big\|_2^2 \leq 2^{-22} \,; \quad \Big\| \frac{\partial^2 f}{\partial y^2} \Big\|_2^2 \leq 2^{-22} \,. \tag{A.19}$$

$$\Big\| \frac{\partial^2 f}{\partial x \partial y} \Big\|_2^2 \leq 2^{-22} \,. \tag{A.20}$$

Proof According to (A.16), and since at a given k the functions $f''_{ck,\ell}$ have disjoint supports, the functions $f''_{ck,\ell}$ form an orthogonal system. Thus, given any y, we have

$$\begin{aligned}
&\int \Big(\sum_{ck \leq r} \sum_{\ell,\ell' \leq 2^{ck}} z_{k,\ell,\ell'} \frac{2^{2ck}}{\sqrt{r}} f''_{ck,\ell}(x) f_{ck,\ell'}(y) \Big)^2 \mathrm{d}x \\
&= \sum_{ck \leq r} \sum_{\ell,\ell' \leq 2^{ck}} z^2_{k,\ell,\ell'} \frac{2^{4ck}}{r} \| f''_{ck,\ell} \|_2^2 f_{ck,\ell'}(y)^2 \leq \sum_{ck \leq r} \sum_{\ell' \leq 2^{ck}} \frac{2^{4ck}}{r} f_{ck,\ell'}(y)^2 .
\end{aligned}$$

where we have used that $(\sum_{\ell' \leq 2^{ck}} z_{k,\ell,\ell'} f_{ck,\ell'}(y))^2 = \sum_{\ell' \leq 2^{ck}} z^2_{k,\ell,\ell'} f_{ck,\ell'}(y)^2$ because the functions $(f_{ck,\ell'}(y))_{\ell' \leq 2^{ck}}$ have disjoint supports, that $z^2_{k,\ell,\ell'} \leq 1$, and that $\|f''_{ck,\ell}\|_2^2 = 2^{-ck}$. Integrating in y and using (A.14) proves the first part of (A.19), and the second part is identical. The proof of (A.20) is similar, using now (A.17) and (A.12). □

To prove the lower bound, we construct numbers $z_{q,\ell,\ell'} \in \{0,1\}$ by induction over q, for $cq \leq r$. Defining f_k by (A.18), our goal is that the function $h = \sum_{cq \leq r} f_q$

[2] The coefficient 2^{-5} is just there to ensure there is plenty of room.

satisfies (A.4) and (A.5) while for the corresponding (convex!) function (A.6), there is an excess of points under the graph of this function. After we construct f_q, we define a dangerous square as a $c(q+1)$-square which contains a point where one of the second-order partial derivatives of f has an absolute value $\geq 1/32$. Using a bit of technical work in the spirit of Lemma 4.6.7 (which does not use any tight estimate), it automatically follows from Lemma A.3.3 that at most, $1/2$ of the $c(q+1)$-squares are dangerous. It is also crucial to observe that the family of dangerous squares is *entirely determined by* f_q. We ensure that the function $h_q = \sum_{k\leq q} f_k$ satisfies (A.4) and (A.5) as follows: When choosing $z_{q+1,\ell,\ell'}$, we take this quantity to be 0 if the corresponding $c(q+1)$-square is dangerous. If it is not, we choose $z_{q+1,\ell,\ell'} = 1$ if and only if by doing so we increase the number of points X_i in the subgraph of the function $g_q := 1/2 + (x^2+y^2)/8 + h_q$, and otherwise, we choose $z_{q+1,\ell,\ell'} = 0$. Let us estimate this increase. Denoting by $S(g)$ the subgraph of a function g, consider the regions $A_+ = S(g_q + \tilde{h}) \setminus S(g_q)$ and $A_- = S(g_q) \setminus S(g_q + \tilde{h})$ where $\tilde{h} = af_{c(q+1),\ell} \otimes f_{c(q+1),\ell'}$ for $a = 2^{2c(q+1)-5}/\sqrt{r}$. Let $N_\pm = \operatorname{card}\{i \leq N;\, X_i \in A_\pm\}$. Thus, by choosing $z_{q+1,\ell,\ell'} = 1$, we increase the number of points in the subgraph by $N_+ - N_-$. Now since $\tilde{h}$ is of average zero, we have $\lambda_3(A_+) = \lambda_3(A_-)$, and this volume is $\geq V := 2^{-4c(q+1)}/(L\sqrt{r})$. We can then expect that with probability $\geq 1/4$, we will have $N_+ - N_- \geq \sqrt{NV}/L \geq \sqrt{N}2^{-2c(q+1)}/(Lr^{1/4})$. The key point is to show that this will happen for at least a fixed proportion of the possible choices of ℓ and ℓ' because if this is the case at each step of the construction, we increase the number of points X_i in the subgraph of g_q by at least $\sqrt{N}/(Lr^{1/4})$, and in r/c steps, we reach the required excess number $r^{3/4}\sqrt{N}/L$ of points X_i in this subgraph.

Let us detail the crucial step, showing that with high probability, a fixed proportion of the possible choices works. Let us say that a $c(q+1)$-square is *safe* if it is not dangerous and is *favorable* if (with the notation above) we have $N_+ - N_- \geq \sqrt{NV}/L \geq \sqrt{N}2^{-2c(q+1)}/(Lr^{1/4})$. Our goal is to show that with probability close to 1, a proportion at least $1/16$ of the $c(q+1)$-squares are both safe and favorable.[3] The argument for doing this is a refinement of the arguments given at the end of Sect. 4.8. For each $k \leq q$, there are 2^{2ck} numbers $z_{k,\ell,\ell'}$ to choose, for a number $2^{2^{2ck}}$ of possible choices. As $k \leq q$ varies, this gives a total number of at most $2^{\sum_{k\leq q} 2^{2ck}}$ choices for the function f_q and therefore for the family of safe $c(q+1)$ squares. Using poissonization, given any family of $c(q+1)$-squares of cardinality M, the probability that less than $M/8$ are favorable is $\leq \exp(-\beta M)$, where β is a universal constant (see (4.98)). The family of safe $c(q+1)$-squares has cardinality $M \geq 2^{2(q+1)c-1}$ so that for this family, this probability is $\leq \exp(-\beta M) \leq \exp(-\beta 2^{2(q+1)c-1})$. The probability that this happens for at least one of the at most $2^{L2^{cq}}$ possible families of safe $c(q+1)$-squares is then at most $2^{2^{cq}(L-\beta 2^{2c-1})}$. Choosing the constant c such that $\beta 2^{2c-1}$ is large enough, it is almost certain that for each of the possible families of safe $c(q+1)$-squares, at least a proportion of $1/8$ of its $c(q+1)$-squares will be favorable.

[3] One difficulty being that the previous steps of the construction as well as the set of dangerous $c(q+1)$-squares depend on the X_i.

Appendix B
Some Deterministic Arguments

B.1 Hall's Matching Theorem

Proof of Proposition 4.3.2. Let us denote by a the quantity $\sup \sum_{i\le N}(w_i + w_i')$, where the supremum is taken over the families $(w_i)_{i\le N}$, $(w_i')_{i\le N}$ which satisfy (4.28), that is, $w_i + w_j' \le c_{ij}$ for all $i, j \le N$. For families $(w_i)_{i\le N}$, $(w_i')_{i\le N}$ satisfying (4.28), then for any permutation π of $\{1, \dots, N\}$, we have

$$\sum_{i\le N} c_{i\pi(i)} \ge \sum_{i\le N}(w_i + w_i')$$

and taking the supremum over the values of w_i and w_i', we get

$$\sum_{i\le N} c_{i\pi(i)} \ge a \,,$$

so that $M(C) \ge a$.

The converse relies on the Hahn-Banach theorem. Consider the subset $\mathcal{C}$ of $\mathbb{R}^{N\times N}$ that consists of the vectors $(x_{ij})_{i,j\le N}$ for which there exists numbers $(w_i)_{i\le N}$ and $(w_i')_{i\le N}$ such that

$$\sum_{i\le N}(w_i + w_i') > a \tag{B.1}$$

$$\forall i, j \le N \,,\; x_{ij} \ge w_i + w_j' \,. \tag{B.2}$$

M. Talagrand, *Upper and Lower Bounds for Stochastic Processes*, Ergebnisse der Mathematik und ihrer Grenzgebiete. 3. Folge / A Series of Modern Surveys in Mathematics 60, https://doi.org/10.1007/978-3-030-82595-9

Then by definition of a, we have $(c_{ij})_{i,j\le N} \notin \mathcal{C}$. It is obvious that $\mathcal{C}$ is an open convex subset of $\mathbb{R}^{N\times N}$. Thus, we can separate the point $(c_{ij})_{i,j\le N}$ from $\mathcal{C}$ by a linear functional, that is, we can find numbers $(p_{ij})_{i,j\le N}$ such that

$$\forall (x_{ij}) \in \mathcal{C}\,, \sum_{i,j\le N} p_{ij}c_{ij} < \sum_{i,j\le N} p_{ij}x_{ij}\,. \tag{B.3}$$

By definition of $\mathcal{C}$, it is obvious that if $(x_{ij}) \in \mathcal{C}$ and $y_{ij} \ge 0$, then $(x_{ij}+y_{ij}) \in \mathcal{C}$. In particular (B.2) remains true when one replaces x_{ij} by $x_{ij} + y_{ij}$. This implies that $p_{ij} \ge 0$ for each i, j. Furthermore, because of the strict inequality in (B.3), not all the numbers p_{ij} are 0. Thus, there is no loss of generality to assume that $\sum_{i,j\le N} p_{ij} = N$. Consider families $(w_i)_{i\le N}$, $(w'_i)_{i\le N}$ that satisfy (B.1). Then if $x_{ij} = w_i + w'_j$, the point $(x_{ij})_{i,j\le N}$ belongs to $\mathcal{C}$, and using (B.3) for this point, we obtain

$$\sum_{i,j\le N} p_{ij}c_{ij} < \sum_{i,j\le N} p_{ij}(w_i + w'_j)\,. \tag{B.4}$$

This holds whenever the numbers (w_i) and (w'_i) satisfy (B.1). Considering numbers $(y_i)_{i\le N}$ with $\sum_{i\le N} y_i = 0$, the numbers $(w_i + y_i)$ and (w'_i) satisfy (B.1), and from (B.4), we have

$$\begin{aligned}\sum_{i,j\le N} p_{ij}c_{ij} &< \sum_{i,j\le N} p_{ij}(w_i + y_i + w'_j)\\ &= \sum_{i,j\le N} p_{ij}(w_i + w'_j) + \sum_{i\le N} y_i\Big(\sum_{j\le N} p_{ij}\Big)\,.\end{aligned} \tag{B.5}$$

This inequality holds whenever $\sum_{i\le N} y_i = 0$ so that replacing y_i by λy_i for $\lambda \in \mathbb{R}$, the previous inequality remains true. Therefore, the last term in (B.5) must be 0. We have shown that

$$\sum_{i\le N} y_i = 0 \Rightarrow \sum_{i\le N} y_i\Big(\sum_{j\le N} p_{ij}\Big) = 0\,,$$

and this forces in turn all the sums $\sum_{j\le N} p_{ij}$ to be equal. Since $\sum_{i,j\le N} p_{ij} = N$, we have $\sum_{j\le N} p_{ij} = 1$, for all i. Similarly, we have $\sum_{i\le N} p_{ij} = 1$ for all j, that is, the matrix $(p_{ij})_{i,j\le N}$ is bistochastic. Thus (B.4) becomes

$$\sum_{i,j\le N} p_{ij}c_{ij} \le \sum_{i\le N}(w_i + w'_i)$$

so that $\sum_{i,j\le N} p_{ij}c_{ij} \le a$. The set of bistochastic matrices is a convex set, so the infimum of $\sum_{i,j\le N} p_{ij}c_{ij}$ over this convex set is obtained at an extreme point. The extreme points are of the type $p_{ij} = \mathbf{1}_{\{\pi(i)=j\}}$ for a permutation π of $\{1,\dots,N\}$ (a classical result known as Birkhoff's theorem) so that we can find such a permutation with $\sum_{i \le N} c_{i\pi(i)} \le a$. □

Proof of Hall's Marriage Lemma We set $c_{ij} = 0$ if $j \in A(i)$ and $c_{ij} = 1$ otherwise. Using the notations of Proposition 4.3.2, we aim to prove that $M(C) = 0$. Using (4.27), it suffices to show that given numbers $u_i (= -w_i)$, $v_i (= w_i')$, we have

$$\forall i, \forall j \in A(i) \,,\; v_j \le u_i \Rightarrow \sum_{i \le N} v_i \le \sum_{i \le N} u_i \,. \tag{B.6}$$

Adding a suitable constant, we may assume $v_i \ge 0$ and $u_i \ge 0$ for all i, and thus

$$\sum_{i \le N} u_i = \int_0^\infty \operatorname{card}\{i \le N \,;\; u_i \ge t\} dt \tag{B.7}$$

$$\sum_{i \le N} v_i = \int_0^\infty \operatorname{card}\{i \le N \,;\; v_i \ge t\} dt \,. \tag{B.8}$$

Given t, using (4.33) for $I = \{i \le N \,;\; u_i < t\}$ and since $v_j \le u_i$ if $j \in A(i)$, we obtain

$$\operatorname{card}\{j \le N \,;\; v_j < t\} \ge \operatorname{card}\{i \le N \,;\; u_i < t\}$$

and thus

$$\operatorname{card}\{i \le N \,;\; u_i \ge t\} \le \operatorname{card}\{i \le N \,;\; v_i \ge t\} \,.$$

Combining with (B.7) and (B.8), this proves (B.6). □

B.2 Proof of Lemma 4.7.11

Consider the subset $\mathcal{L}^*$ of $\mathcal{L}$ consisting of the functions f for which $f(1/2) = 0$. To $f \in \mathcal{L}^*$, we associate the curve $W(f)$ traced out by the map

$$u \mapsto \Big(\tau^1 + 2^{k+1} f\big(\frac{u}{2}\big), \tau^2 + 2^{k+1} f\big(\frac{u+1}{2}\big)\Big) \,,$$

where $(\tau^1, \tau^2) = \tau$. A curve in $\mathcal{C}(\tau, k)$ can be parameterized, starting at τ and moving at speed 1 along each successive edges so that it is the range of a map of the type $t \mapsto \tau + \varphi(t)$ where φ is a 1-Lipschitz map from $[0, 2^k]$ to $\mathbb{R}^2$ with $\varphi(0) = 0 = \varphi(2^k)$. Denoting by g_0 and g_1 the components of φ, the curve is therefore the range of a map of the type $t \mapsto (\tau^1 + g_0(t), \tau^2 + g_1(t))$ where g_0 and g_1 are 1-Lipschitz maps from $[0, 2^k]$ to $\mathbb{R}$ with $g_0(0) = g_1(0) = g_0(2^k) =$

$g_1(2^k) = 0$. Considering the function f on $[0, 1]$ given by $f(u) = 2^{-k-1}g_0(2^{k+1}u)$ for $u \le 1/2$ and $f(u) = 2^{-k-1}g_1(2^{k+1}(u - 1/2))$ for $1/2 \le u \le 1$ proves that $\mathcal{C}(\tau, k) \subset W(\mathcal{L}^*)$. We set $T = W^{-1}(\mathcal{C}(\tau, k))$. Consider f_0 and f_1 in T and the map $h : [0, 1]^2 \to [0, 1]^2$ given by

$$h(u, v) = \Big(\tau^1 + 2^{k+1}\Big(vf_0\big(\frac{u}{2}\big) + (1 - v)f_1\big(\frac{u}{2}\big)\Big), \\ \tau^2 + 2^{k+1}\Big(vf_0\big(\frac{1+u}{2}\big) + (1 - v)f_1\big(\frac{1+u}{2}\big)\Big)\Big) .$$

The area of $h([0, 1]^2)$ is at most $\iint_{[0,1]^2} |Jh(u, v)|\mathrm{d}u\mathrm{d}v$, where Jh is the Jacobian of h, and a straightforward computation gives

$$Jh(u, v) = 2^{2k+1}\bigg(\Big(vf_0'\big(\frac{u}{2}\big) + (1 - v)f_1'\big(\frac{u}{2}\big)\Big)\Big(f_0\big(\frac{1+u}{2}\big) - f_1\big(\frac{1+u}{2}\big)\Big) \\ - \Big(vf_0'\big(\frac{1+u}{2}\big) + (1 - v)f_1'\big(\frac{1+u}{2}\big)\Big)\Big(f_0\big(\frac{u}{2}\big) - f_1\big(\frac{u}{2}\big)\Big)\bigg) ,$$

so that since $|f_0'| \le 1$, $|f_1'| \le 1$,

$$|Jh(u, v)| \le 2^{2k+1}\Big(\Big|f_0\big(\frac{u}{2}\big) - f_1\big(\frac{u}{2}\big)\Big| + \Big|f_0\big(\frac{1+u}{2}\big) - f_1\big(\frac{1+u}{2}\big)\Big|\Big) .$$

The Cauchy-Schwarz inequality implies

$$\iint |Jh(u, v)|\mathrm{d}u\mathrm{d}v \le L2^{2k}\|f_0 - f_1\|_2 . \tag{B.9}$$

If x does not belong to the range of h, both curves $W(f_0)$ and $W(f_1)$ "turn the same number of times around x". This is because "the number of times the closed curve $u \mapsto h(u, v)$ turns around x" is then a continuous function of v, and since it is integer valued, it is constant. In particular, it takes the same value for $v = 0$ and $v = 1$. Consequently, either $x \in \overset{o}{W}(f_0) \cap \overset{o}{W}(f_1)$ or $x \notin \overset{o}{W}(f_0) \cup \overset{o}{W}(f_1)$. Thus, the range of h contains the symmetric difference $\overset{o}{W}(f_0) \,\triangle\, \overset{o}{W}(f_1)$, and (B.9) implies (4.115). □

B.3 The Shor-Leighton Grid Matching Theorem

Let us say that a simple curve C traced on G is a *chord* if it is the range of $[0, 1]$ by a continuous map φ where $\varphi(0)$ and $\varphi(1)$ belong to the boundary of $[0, 1]^2$. If C is a chord, $]0, 1[^2 \setminus C$ is the union of two regions R_1 and R_2, and (assuming without loss of generality that no point X_i belongs to G),

$$\sum_{i\le N}\big(\mathbf{1}_{R_1}(X_i)-\lambda(R_1)\big)=-\sum_{i\le N}\big(\mathbf{1}_{R_2}(X_i)-\lambda(R_2)\big)\,.$$

We define

$$\mathcal{D}(C)=\Big|\sum_{i\le N}(\mathbf{1}_{R_1}(X_i)-\lambda(R_1))\Big|=\Big|\sum_{i\le N}(\mathbf{1}_{R_2}(X_i)-\lambda(R_2))\Big|\,.$$

If C is a chord, "completing C by following the boundary of $[0,1]^2$" produces a closed simple curve C' on G such that either $R_1=\overset{o}{C'}$ or $R_2=\overset{o}{C'}$. The length we add along each side of the boundary is less or equal than the length of the chord itself so that $\ell(C')\le 3\ell(C)$. Thus, the following is a consequence of Theorem 4.7.2:

Theorem B.3.1 *With probability at least* $1-L\exp(-(\log N)^{3/2}/L)$, *for each chord* C, *we have*

$$\mathcal{D}(C)\le L\ell(C)\sqrt{N}(\log N)^{3/4}\,. \tag{B.10}$$

Proof of Theorem 4.7.1. Consider a number $\ell_2<\ell_1$ to be determined later, and the grid $G'\subset G$ of mesh width $2^{-\ell_2}$. (This is the slightly coarser grid we mentioned on page 153.)

A union of squares of G' is called a *domain*. Given a domain R, we denote by R' the union of the squares of G' such that at least one of the four edges that form their boundary is entirely contained in R (recall that squares include their boundaries). The main argument is to establish that if (4.103) and (B.10) hold, and provided ℓ_2 has been chosen appropriately, then for any choice of R, we have

$$N\lambda(R')\ge \operatorname{card}\{i\le N\,;\,X_i\in R\}\,. \tag{B.11}$$

We will then conclude with Hall's Marriage Lemma. The basic idea to prove (B.11) is to reduce to the case where R is the closure of the interior of a simple closed curve minus a number of "holes" which are themselves the interiors of simple closed curves.

Let us say that a domain R is *decomposable* if $R=R_1\cup R_2$ where R_1 and R_2 are non-empty unions of squares of G' and when every square of G' included in R_1 has at most one vertex belonging to R_2. (Equivalently, $R_1\cap R_2$ is finite.) We can write $R=R_1\cup\ldots\cup R_k$ where each R_j is undecomposable (i.e., not decomposable) and where any two of these sets have a finite intersection. This is obvious by writing R as the union of as many domains as possible, under the condition that the intersection of any of two of these domains is finite. Then each of them must be undecomposable.

We claim that

$$\frac{1}{4}\sum_{\ell \le k} \lambda(R'_\ell \backslash R_\ell) \le \lambda(R' \backslash R) . \tag{B.12}$$

To see this, let us set $\mathcal{S}_\ell = R'_\ell \backslash R_\ell$ so that by definition of R'_ℓ, $\mathcal{S}_\ell$ is the union of the squares $\mathcal{D}$ of G' that have at least one of the edges that form their boundary contained in R_ℓ but are not themselves contained in R_ℓ. Obviously, we have $\mathcal{S}_\ell \subset R'$. When $\ell \ne \ell'$, the sets R_ℓ and $R_{\ell'}$ have a finite intersection so that a square $\mathcal{D}$ contained in $\mathcal{S}_\ell$ cannot be contained in $R_{\ell'}$, since it has an entire edge contained in R_ℓ. Since $\mathcal{D}$ is not contained in R_ℓ either, it is not contained in R. Thus, the interior of $\mathcal{D}$ is contained in $R' \backslash R$, and since this is true for any square $\mathcal{D}$ of $\mathcal{S}_\ell$ and any $\ell \le k$, we have

$$\lambda\Big(\bigcup_{\ell \le k} \mathcal{S}_\ell\Big) \le \lambda(R' \backslash R) .$$

Moreover, a given square $\mathcal{D}$ of G' can be contained in a set $\mathcal{S}_\ell$ for at most four values of ℓ (one for each of the edges of $\mathcal{D}$) so that

$$\sum_{\ell \le k} \lambda(R'_\ell \setminus R_\ell) = \sum_{\ell \le k} \lambda(\mathcal{S}_\ell) \le 4\lambda\Big(\bigcup_{\ell \le k} \mathcal{S}_\ell\Big) .$$

This proves (B.12).

To prove that (B.11) holds for any domain R, it suffices to prove that when R is an undecomposable domain, we have (pessimistically)

$$\frac{N}{4}\lambda(R' \backslash R) \ge \operatorname{card}\{i \le N \,;\, X_i \in R\} - N\lambda(R) . \tag{B.13}$$

Indeed, writing (B.13) for $R = R_\ell$, summing over $\ell \le k$ and using (B.12) implies (B.11).

We turn to the proof of (B.13) when R is an undecomposable domain. The boundary S of R is a subset of G'. Inspection of the cases shows that:

$$\text{If a vertex } \tau \text{ of } G' \text{ belongs to } S, \text{ either 2 or 4 of the edges of } G' \text{ incident to } \tau \text{ are contained in } S. \tag{B.14}$$

Next, we show that any subset S of G' that satisfies (B.14) is a union of closed simple curves, any two of them intersecting only at vertices of G'. (This is simply the decomposition into cycles of Eulerian graphs.) To see this, it suffices to construct a closed simple curve C contained in S, to remove C from S and to iterate, since $S \backslash C$ still satisfies (B.14). The construction goes as follows. Starting with an edge $\tau_1\tau_2$ in S, we find successively edges $\tau_2\tau_3, \tau_3\tau_4, \ldots$ with $\tau_k \ne \tau_{k-2}$, and we continue the construction until the first time $\tau_k = \tau_\ell$ for some $\ell \le k-2$ (in fact $\ell \le k-3$). Then the edges $\tau_\ell\tau_{\ell+1}, \tau_{\ell+1}\tau_{\ell+2}, \ldots, \tau_{k-1}\tau_k$ define a closed simple curve contained in S.

Thus, the boundary of an undecomposable domain R is a union of closed simple curves $C_1, \ldots, C_K$, any two of them having at most a finite intersection.

We next show that for each ℓ, the set R is either contained in the closure C^*_ℓ of $\overset{o}{C}_\ell$ (so that C_ℓ is then the "outer boundary" of R) or else $\overset{o}{C}_\ell \cap R = \emptyset$ (in which case $\overset{o}{C}_\ell$ is "a hole" in R). Let us fix ℓ and assume otherwise that $\overset{o}{C}_\ell \cap R \ne \emptyset$ and $R \not\subset C^*_\ell$. We will show that this contradicts the fact that R is undecomposable. Consider the domain R_1 which is the union of the squares of G' that are contained in R but not in C^*_ℓ so that R_1 is not empty by hypothesis. Consider also the domain R_2 that is the union of the squares of G' contained in R whose interiors are contained in $\overset{o}{C}_\ell$. Then R_2 is not empty either. Given a square of G', and since $\overset{o}{C}_\ell$ is the interior of C^*_ℓ, either its interior is contained in $\overset{o}{C}_\ell$ or else the square is not contained in C^*_ℓ. This proves that $R = R_1 \cup R_2$. Next, we show that the domains R_1 and R_2 cannot have an edge of the grid G' in common. Assuming for contradiction that such an edge exists, it is an edge of exactly two squares A and B of G'. One of these squares is a subset of R_1, and the other is a subset of R_2. Thus, the edge must belong to C_ℓ for otherwise A and B would be "on the same side of C_ℓ", and they would both be subsets of R_1 or both subsets of R_2. Next, we observe that this edge cannot be on the boundary of R because both A and B are subsets of R. This contradicts the fact that C_ℓ is contained in the boundary of R, therefore proving that R_1 and R_2 cannot have an edge in common. Since $R = R_1 \cup R_2$, this in turn would imply that R is decomposable, contradicting our assumption.

If C_ℓ is an outer boundary of R, then $R \subset C^*_\ell$, and consequently for each ℓ', we have $C^*_{\ell'} \subset C^*_\ell$. Thus, $C_{\ell'}$ is an outer boundary of R, then $C^*_\ell = C^*_\ell$, so that $C_\ell = C_{\ell'}$, contradicting the fact that these two curves have a finite intersection.

Thus, without loss of generality, we may assume that C_1 is the only outer boundary of R, and that for $2 \le \ell \le K$, we have $R \cap \overset{o}{C}_\ell = \emptyset$. The goal now is to prove that

$$R = C^*_1 \setminus \bigcup_{2 \le \ell \le k} \overset{o}{C}_\ell \,. \tag{B.15}$$

It is obvious that $R \subset C^*_1 \setminus \bigcup_{2\le\ell\le k} \overset{o}{C}_\ell$ so that we have to show that $D := (C^*_1 \setminus \bigcup_{2\le\ell\le k} \overset{o}{C}_\ell) \setminus R$ is empty. We assume for contradiction that D is not empty. Consider a square A of G' which is contained in D, and a square A' of G' which has an edge in common with A. First, we claim that $A' \subset C^*_1$. Otherwise, A and A' would have to be on different sides of C_1, which means that their common edge has to belong to C_1 and hence to the boundary of R. This is impossible because neither A nor A' is then a subset of R. Indeed in the case of A', this is because we assume that $A' \not\subset C^*_1$, and in the case of A, this is because we assume that $A \subset D$.

Exactly, the same argument shows that the interior of A' cannot be contained in $\overset{o}{C_\ell}$ for $2 \le \ell \le k$. Indeed then, A and A' would be on different sides of C_ℓ so that their common edge would belong to C_ℓ and hence to the boundary of R, which is impossible since neither A nor A' is a subset of R. We have now shown that A and A' lie on the same side of each curve C_ℓ so that their common edge cannot belong to the boundary of R, and since A is not contained in R, this is not the case of A' either. Consequently, the definition of D shows that $A' \subset D$, but since A was an arbitrary square contained in D, this is absurd and completes the proof that $D = \emptyset$ and of (B.15).

Let $R_\ell^\sim$ be the union of the squares of G' that have at least one edge contained in C_ℓ. Thus, as in (B.12), we have

$$\sum_{\ell \le k} \lambda(R_\ell^\sim \backslash R) \le 4\lambda(R' \backslash R)$$

and to prove (B.13), it suffices (recalling that we assume that no point X_i belongs to G) to show that for each $1 \le \ell \le k$, we have

$$\left| \operatorname{card}\left\{i \le N \; ; \; X_i \in \overset{o}{C_\ell}\right\} - \lambda(\overset{o}{C_\ell})\right| \le N2^{-4}\lambda(R_\ell^\sim \backslash R) \,. \tag{B.16}$$

For $\ell \ge 2$, C_ℓ does not intersect the boundary of $[0,1]^2$. Each edge contained in C_ℓ is in the boundary of R. One of the two squares of G' that contain this edge is included in $R_\ell^\sim \backslash R$ and the other in R. Since a given square contained in $R_\ell^\sim \backslash R$ must arise in this manner from one of its four edges, we have

$$\lambda(R_\ell^\sim \backslash R) \ge \frac{1}{4} 2^{-\ell_2} \ell(C_\ell) \,. \tag{B.17}$$

On the other hand, (4.103) implies

$$\left| \operatorname{card}\left\{i \le N \; ; \; X_i \in \overset{o}{C_\ell}\right\} - \lambda(\overset{o}{C_\ell})\right| \le L\ell(C_\ell)\sqrt{N}(\log N)^{3/4} \,. \tag{B.18}$$

Assuming

$$2^{-\ell_2} \ge \frac{2^6 L}{\sqrt{N}} (\log N)^{3/4} \,, \tag{B.19}$$

where L is the constant of (B.18), we have, using (B.17) in the last inequality,

$$L\ell(C_\ell)\sqrt{N}(\log N)^{3/4} \le 2^{-\ell_2 - 6} N \ell(C_\ell) \le N2^{-4}\lambda(R_\ell^\sim \backslash R)$$

and (B.16) follows.

When $\ell = 1$, (B.17) need not be true because parts of C_1 might be traced on the boundary of $[0, 1]^2$. In that case, we simply decompose C_1 into a disjoint union of chords and of parts of the boundary of $[0, 1]^2$ to deduce (B.16) from (B.10).

Thus, we have proved that (4.103) and (B.10) imply (B.11) provided that (B.19) holds. Next, for a domain R, we denote by R^* the set of points which are within distance $2^{-\ell_2}$ of R', and we show that, provided

$$2^{-\ell_2} \geq \frac{20}{\sqrt{N}} \tag{B.20}$$

we have

$$\mathrm{card}\{i \leq N \ ; \ Y_i \in R^*\} \geq N\lambda(R') \ . \tag{B.21}$$

This is simply because since the sequence Y_i is evenly spread, the points Y_i are centers of disjoint rectangles of area $1/N$ and diameter $\leq 20/\sqrt{N}$. There are at least $N\lambda(R')$ points Y_i such that the corresponding rectangle intersects R' (because the union of these rectangles cover R') and (B.20) implies that these little rectangles are entirely contained in R^*. Therefore (B.11) and (B.21) imply

$$\mathrm{card}\{i \leq N \ ; \ Y_i \in R^*\} \geq \mathrm{card}\{i \leq N \ ; \ X_i \in R\} \ . \tag{B.22}$$

Next, consider a subset I of $\{1, \dots, N\}$ and let R be the domain that is the union of the squares of G' that contain at least a point X_i , $i \in I$. Then, using (B.22),

$$\mathrm{card}\, I \leq \mathrm{card}\{i \leq N \ ; \ X_i \in R\} \leq \mathrm{card}\{i \leq N \ ; \ Y_i \in R^*\} \ . \tag{B.23}$$

A point of R' is within distance $2^{-\ell_2}$ of a point of R. A point of R^* is within distance $2^{-\ell_2+1}$ of a point of R. A point of R is within distance $\sqrt{2} \cdot 2^{-\ell_2} \leq 2^{-\ell_2+1}$ of a point X_i with $i \in I$. Consequently, each point of R^* is within distance $\leq 2^{-\ell_2+2}$ of a point X_i with $i \in I$. Therefore, if we define

$$A(i) = \left\{j \leq N \ ; \ d(X_i, Y_j) \leq 2^{-\ell_2+2}\right\} ,$$

we have proved that $\{j \leq N; Y_j \in R^*\} \subset \bigcup_{i \in I} A(i)$, and combining with (B.23) that

$$\mathrm{card} \bigcup_{i \in I} A(i) \geq \mathrm{card}\, I \ .$$

Hall's Marriage Lemma (Corollary 4.3.5) then shows that we can find a matching π for which $\pi(i) \in A(i)$ for any $i \leq N$ so that by definition of $A(i)$

$$\sup_{i \leq N} d(X_i, Y_{\pi(i)}) \leq 2^{-\ell_2+2} \leq \frac{L}{\sqrt{N}} (\log N)^{3/4} \ ,$$

by taking for ℓ_2 the largest integer that satisfies (B.19) and (B.20). Since this is true whenever (4.103) and (B.10) occur, the proof of (4.101) is complete. □

B.4 End of Proof of Theorem 17.2.1

The most difficult point is to ensure that the functions $h\mathbf{1}_R$ satisfy (17.47). In fact, rather than (17.47), we shall prove that

$$\forall (k, \ell) \in R \ , \ |h(k, \ell)| \le L(k_2 - k_1) \ , \tag{B.24}$$

which suffices by homogeneity.

For $j \le q \le p$, we consider the partition $\mathcal{D}(q)$ of G consisting of all the sets of the type

$$\{a2^q + 1, \dots, (a+1)2^q\} \times \{b2^{q-j} + 1, \dots, (b+1)2^{q-j}\} \ , \tag{B.25}$$

where a and b are integers with $0 \le a < 2^{p-q}$ and $0 \le b < 2^{p-q+j}$. For $3 \le q \le j$, we define $\mathcal{D}(q)$ as the partition consisting of all the sets of the type

$$\{a2^q + 1, \dots, (a+1)2^q\} \times \{b\} \tag{B.26}$$

where $0 \le a < 2^{p-q}$ and $1 \le b \le 2^p$.

We observe that if $q' > q$, $R' \in \mathcal{D}(q')$ and $R \in \mathcal{D}(q)$, then either $R \subset R'$ or $R \cap R' = \emptyset$.

Fixing a function $h \in \mathcal{H}_j(2^{2p-j})$, we consider the set $C = \{(k, \ell); h(k, \ell) \ne 0\}$ so $\operatorname{card} C \le 2^{2p-j}$. We proceed to the following construction. Keeping in mind that the sequence $(\mathcal{D}(q))$ of partitions increases so that $\mathcal{D}(p)$ consists of the largest rectangles, we first consider the set $U(p)$ that is the union of all rectangles $R \in \mathcal{D}(p)$ such that

$$\operatorname{card}(R \cap C) \ge \frac{1}{8} \operatorname{card} R \, . \tag{B.27}$$

Then we consider the union $U(p-1)$ of all the rectangles $R \in \mathcal{D}(p-1)$ that are not contained in $U(p)$ and that satisfy (B.27), and we continue in this manner until we construct $U(3)$. Since the sets $U(p)$, … , $U(3)$ are disjoint and each is a union of disjoint sets satisfying (B.27), we get

$$\sum_{3 \le q \le p} \operatorname{card} U(q) \le 8 \operatorname{card} C \le 2^{2p-j+3} \, . \tag{B.28}$$

Moreover

$$C \subset \bigcup_{1 \le q \le p} U(q) \, . \tag{B.29}$$

This is simply because if $(k,\ell) \in C$ and $(k,\ell) \in R \in \mathcal{D}(3)$, then if $(k,\ell) \notin \bigcup_{q\ge 4} U(q)$, we have $R \subset U(3)$ since (B.27) holds because $\operatorname{card} R = 8$. We also note that

$$R \in \mathcal{D}(q)\, ,\; q \le p-1\, ,\; R \subset U(q) \Rightarrow \operatorname{card}(R \cap C) \le \frac{1}{2} \operatorname{card} R \, . \tag{B.30}$$

Indeed, if $R' \supset R$ and $R' \in \mathcal{D}(q+1)$, then $\operatorname{card} R' \le 4 \operatorname{card} R$. Since $R \subset U(q)$, we have $R' \not\subset U(q+1)$ so that

$$\operatorname{card}(R \cap C) \le \operatorname{card}(R' \cap C) \le \frac{1}{8} \operatorname{card}\ R' \le \frac{1}{2} \operatorname{card} R \, .$$

Now (B.29) implies

$$h = \sum h \mathbf{1}_R \, , \tag{B.31}$$

where the summation is over $3 \le q \le p\, ,\; R \in \mathcal{D}(q)$ and $R \subset U(q)$. Writing $R = \{k_1, \ldots, k_2\} \times \{\ell_1, \ldots, \ell_2\}$ as in Proposition 17.3.2, we observe by construction that, first, (17.43) holds for $q \ge j$ and that the function $h\mathbf{1}_R$ satisfies (17.44) to (17.46); second, that (17.49) holds for $3 \le q \le j$, and the function $h\mathbf{1}_R$ then satisfies (17.50) to (17.52).

We turn to the proof of (B.24). We start with the typical case, $R \in \mathcal{D}(q)\, ,\; 3 \le q < p$. Then (B.30) implies that there exists $(k_0, \ell_0) \in R$ with $h(k_0, \ell_0) = 0$, and (17.39) implies, using also (17.43),

$$\begin{aligned} |h(k,\ell)| = |h(k,\ell) - h(k_0,\ell_0)| &\le |h(k,\ell) - h(k,\ell_0)| + |h(k,\ell_0) - h(k_0,\ell_0)| \\ &\le 2^j |\ell - \ell_0| + |k - k_0| \le 2(k_2 - k_1) \, , \end{aligned}$$

and this proves (B.24).

Next, we consider the case $q = p$ so that $R \in \mathcal{D}(p)$ and $R = \{1, \ldots, 2^p\} \times \{b2^{p-j}+1, \ldots, (b+1)2^{p-j}\}$. Given an integer r, define

$$\tilde{R} = G \cap (\{1, \ldots, 2^p\} \times \{b2^{p-j} + 1 - r, \ldots, (b+1)2^{p-j} + r\}) \, .$$

Then, for $r \le 2^p$, we have

$$\operatorname{card}(\{b2^{p-j} + 1 - r, \ldots, (b+1)2^{p-j} + r\} \cap \{1, \ldots, 2^p\}) \ge r/2 \, .$$

Thus card $\tilde{R} \geq 2^p r/2$ so that if $2^p r > 2^{2p-j+1}$, then card $\tilde{R} > 2^{2p-j}$, and therefore, $\tilde{R}$ contains a point (k, ℓ') with $h(k, \ell') = 0$. Then R contains a point (k, ℓ) with $|\ell - \ell'| \leq r$ so that the second part of (17.39) implies

$$|h(k, \ell)| \leq r 2^j .$$

Assuming that we choose r as small as possible with $2^p r > 2^{2p-j+1}$, we then have

$$|h(k, \ell)| \leq L 2^{2p-j} 2^{-p} 2^j \leq L 2^p ,$$

and (17.39) shows that this remains true for each point (k, ℓ) of R, completing the proof of (B.24).

Consequently for $R \in \mathcal{D}(q)$, $R \subset U(q)$ and $j \leq q \leq p$, we can use (17.48), which implies

$$\Big|\sum_{i \leq N}(h\mathbf{1}_R(U_i) - \int h\mathbf{1}_R \mathrm{d}\mu)\Big| \leq L\sqrt{pm_0}\, 2^{j/2} \operatorname{card} R . \tag{B.32}$$

Moreover for $3 \leq q \leq j$, this inequality remains true from (17.53). Recalling (B.31), summation of these inequalities over $R \in \mathcal{D}(q)$, $R \subset U(q)$ yields (17.55) and completes the proof. □

B.5 Proof of Proposition 17.3.1

In this section, we prove Proposition 17.3.1. We will denote by I an interval of $\{1, \ldots, 2^p\}$, that is a set of the type

$$I = \{k \; ; \; k_1 \leq k \leq k_2\} .$$

Lemma B.5.1 *Consider a map $f : \{1, \ldots, 2^p\} \to \mathbb{R}^+$, a number $a > 0$ and*

$$A = \Big\{\ell \; ; \; \exists I \, , \, \ell \in I \, , \, \sum_{\ell' \in I} f(\ell') \geq a \operatorname{card} I\Big\} .$$

Then

$$\operatorname{card} A \leq \frac{L}{a} \sum_{\ell \in A} f(\ell) .$$

Proof This uses a discrete version of the classical Vitali covering theorem (with the same proof). Namely, a family $\mathcal{I}$ of intervals contains a disjoint family $\mathcal{I}'$ such that

$$\operatorname{card} \bigcup_{I \in \mathcal{I}} I \le L \operatorname{card} \bigcup_{I \in \mathcal{I}'} I = L \sum_{I \in \mathcal{I}'} \operatorname{card} I \;.$$

We use this for $\mathcal{I} = \{I; \sum_{\ell' \in I} f(\ell') \ge a \operatorname{card} I\}$ so that $A = \bigcup_{I \in \mathcal{I}} I$ and $\operatorname{card} A \le L \sum_{I \in \mathcal{I}'} \operatorname{card} I$. Since $\sum_{\ell' \in I} f(\ell') \ge a \operatorname{card} I$ for $I \in \mathcal{I}'$, and since the intervals of $\mathcal{I}'$ are disjoint and contained in A, we have $a \sum_{I \in \mathcal{I}'} \operatorname{card} I \le \sum_{\ell' \in A} f(\ell')$. □

Proof of Proposition 17.3.1. We consider $h \in \mathcal{H}$, and for $j \ge 2$, we define

$$B(j) = \Big\{(k, \ell) \in G \; ; \; \exists I \, , \, \ell \in I \, , \sum_{\ell' \in I} |h(k, \ell' + 1) - h(k, \ell')| \ge 2^j \operatorname{card} I\Big\} \, .$$

We claim that when $r, s, \ell \le 2^p$, then

$$(r, s) \notin B(j) \Rightarrow |h(r, \ell) - h(r, s)| \le 2^j |\ell - s| \, . \tag{B.33}$$

To see this, assuming for specificity that $s < \ell$, we note that

$$|h(r, \ell) - h(r, s)| \le \sum_{\ell' \in I} |h(r, \ell' + 1) - h(r, \ell')| < 2^j \operatorname{card} I$$

where $I = \{s, s+1, \dots, \ell - 1\}$, and where the last inequality follows from the fact that $s \in I$ and $(r, s) \notin B(j)$.

Now we use Lemma B.5.1 for each k, for the function $f_k(\ell) = |h(k, \ell + 1) - h(k, \ell)|$ and for $a = 2^j$. Summing over k, we obtain

$$\operatorname{card} B(j) \le \frac{L}{2^j} \sum_{(k,\ell) \in B(j)} |h(k, \ell + 1) - h(k, \ell)| \, .$$

Now (17.14) implies $\sum_{k,\ell} |h(k, \ell + 1) - h(k, \ell)| \le 2^{2p}$, and therefore, we get

$$\operatorname{card} B(j) \le L_1 2^{2p-j} \, . \tag{B.34}$$

We consider the smallest integer j_0 such that $L_1 2^{-j_0} < 1/4$ so that $L_1 \le 2^{j_0-2}$, and hence for $j \ge j_0$, we have

$$\operatorname{card} B(j) \le 2^{2p-j+j_0-2} \, , \tag{B.35}$$

and in particular $B(j) \ne G$. For $j \ge j_0$, we define

$$g_j(k, \ell) = \min \big\{h(r, s) + |k - r| + 2^j |\ell - s| \; ; \; (r, s) \notin B(j)\big\} \, .$$

The idea here is that g_j is a regularization of h. The larger the j, the better the g_j approximates h, but this comes at the price that the larger the j, the less regular the g_j is. We will simply use these approximations to write

$$h = g_{j_0} + (g_{j_0+1} - g_{j_0}) + \cdots$$

to obtain the desired decomposition (17.41).

It is obvious that for $(k,\ell) \notin B(j)$, we have $g_j(k,\ell) \le h(k,\ell)$ and that

$$|g_j(k+1,\ell) - g_j(k,\ell)| \le 1 \tag{B.36}$$

$$|g_j(k,\ell+1) - g_j(k,\ell)| \le 2^j , \tag{B.37}$$

since g_j is the minimum over $(s,t) \notin B(j)$ of the functions $(k,\ell) \mapsto h(r,s) + |k - r| + 2^j|\ell - s|$ that satisfy the same properties. Consider $(r,s) \notin B(j)$. Then (B.33) yields

$$|h(r,\ell) - h(r,s)| \le 2^j|\ell - s| ,$$

while the first part of (17.39) yields

$$|h(r,\ell) - h(k,\ell)| \le |k - r| ,$$

and thus, we have proved that

$$(r,s) \notin B(j) \Rightarrow |h(k,\ell) - h(r,s)| \le |k - r| + 2^j|\ell - s| . \tag{B.38}$$

This implies that $g_j(k,\ell) \ge h(k,\ell)$ for all $(k,\ell) \in G$. Consequently, since we already observed that $g_j(k,\ell) \le h(k,\ell)$ for $(k,\ell) \notin B(j)$, we have proved that

$$(k,\ell) \notin B(j) \Rightarrow g_j(k,\ell) = h(k,\ell) . \tag{B.39}$$

We define $h_1 = g_{j_0}$ so that $h_1 \in L\mathcal{H}_1$ by (B.36) and (B.37). For $j > 1$, we define $h_j = g_{j+j_0-2} - g_{j+j_0-1}$. By (B.39), and since $B(j+j_0-1) \subset B(j+j_0-2)$, for $(k,\ell) \notin B(j+j_0-2)$, we have $g_{j+j_0-2}(k,\ell) = h(k,\ell) = g_{j+j_0-1}(k,\ell)$ so that $h_j(k,\ell) = 0$. Consequently,

$$h_j(k,\ell) \ne 0 \Rightarrow (k,\ell) \in B(j+j_0-2) ,$$

and thus, from (B.35) that

$$\operatorname{card}\{(k,\ell)\ ;\ h_j(k,\ell) \ne 0\} \le 2^{2p-j} .$$

Combining with (B.36) and (B.37), we obtain $h_j \in L\mathcal{H}_j(2^{2p-j})$.

Now for $j > 2p$, we have $B(j) = \emptyset$ (since for each k and ℓ, we have $|h(k, \ell + 1) - h(k, \ell)| \le 2^{2p}$ by (17.14)) so that then $g_j = h$ from (B.39). Consequently, $h_j = 0$ for large j and thus $h = \sum_{j \ge 1} h_j$. □

B.6 Proof of Proposition 17.2.4

The next lemmas prepare for the proof of Proposition 17.2.4.

Lemma B.6.1 *Consider numbers* $(v_k)_{k \le 2^p}$ *and* $(v'_k)_{k \le 2^p}$. *We define*

$$g(k) = \inf\left\{v_r + |k - r| \; ; \; 1 \le r \le 2^p\right\} \; ; \; g'(k) = \inf\left\{v'_r + |k - r| \; ; \; 1 \le r \le 2^p\right\} . \tag{B.40}$$

Then

$$\sum_{k \le 2^p} |g(k) - g'(k)| \le \sum_{k \le 2^p} \left(v_k + v'_k - g(k) - g'(k) + |v_k - v'_k|\right) . \tag{B.41}$$

Proof Obviously, $g(k) \le v_k$ and $g'(k) \le v'_k$. If $g'(k) \ge g(k)$, then

$$\begin{aligned} g'(k) - g(k) \le v'_k - g(k) &= v'_k - v_k + v_k - g(k) \\ &\le |v'_k - v_k| + v_k - g(k) + v'_k - g'(k) . \end{aligned}$$

A similar argument when $g(k) \ge g'(k)$ and summation finish the proof. □

We consider numbers $u(k, \ell)$ for $(k, \ell) \in G$, and $h(k, \ell)$ as in (17.31). We set

$$v(k, \ell) = \min\{u(k, s) \; ; \; |\ell - s| \le 1\} , \tag{B.42}$$

so that

$$h(k, \ell) = \inf\left\{v(r, \ell) + |k - r| \; ; \; 1 \le r \le 2^p\right\} . \tag{B.43}$$

We observe that $v(k, \ell) \le u(k, \ell)$. We lighten notation by writing $n(k, \ell)$ for $n(\tau)$ when $\tau = (k, \ell)$.

Lemma B.6.2 *We have*

$$m_0 \sum_{k \le 2^p, \ell < 2^p} |v(k, \ell + 1) - v(k, \ell)| \le 10 \sum_{k, \ell \le 2^p} n(k, \ell)(u(k, \ell) - v(k, \ell)) . \tag{B.44}$$

Proof We observe that $|a - b| = a + b - 2\min(a, b)$ and that

$$v(k,\ell) \le \min(u(k,\ell+1), u(k,\ell))$$
$$v(k,\ell+1) \le \min(u(k,\ell+1), u(k,\ell)) .$$

Thus

$$\begin{aligned}|u(k,\ell+1) - u(k,\ell)| &= u(k,\ell) + u(k,\ell+1) - 2\min(u(k,\ell+1)\,,\, u(k,\ell))\\ &\le u(k,\ell) - v(k,\ell) + u(k,\ell+1) - v(k,\ell+1) .\end{aligned}$$

By summation, we get

$$\sum_{k \le 2^p, \ell < 2^p} |u(k,\ell+1) - u(k,\ell)| \le 2 \sum_{k,\ell \le 2^p} (u(k,\ell) - v(k,\ell))$$

and since $m_0 \le n(k,\ell)$ for $(k,\ell) \in G$ by (17.9)

$$m_0 \sum_{k \le 2^p, \ell < 2^p} |u(k,\ell+1) - u(k,\ell)| \le 2 \sum_{k,\ell \le 2^p} n(k,\ell)(u(k,\ell) - v(k,\ell)) . \tag{B.45}$$

Now

$$|v(k,\ell) - u(k,\ell)| \le |u(k,\ell+1) - u(k,\ell)| + |u(k,\ell-1) - u(k,\ell)|$$

so that

$$\begin{aligned}|v(k,\ell+1) - v(k,\ell)| &\le |v(k,\ell+1) - u(k,\ell+1)| + |u(k,\ell+1) - u(k,\ell)|\\ &\quad + |u(k,\ell) - v(k,\ell)|\\ &\le |u(k,\ell-1) - u(k,\ell)| + 3|u(k,\ell+1) - u(k,\ell)|\\ &\quad + |u(k,\ell+2) - u(k,\ell+1)| .\end{aligned} \tag{B.46}$$

Plugging (B.46) in the left-hand side of (B.44) and using (B.45) prove (B.44), the factor 10 being 2(1+3+1). □

Proof of Proposition 17.2.4. Given $1 \le \ell < 2^p$, we use Lemma B.6.1 for $v_k = v(k,\ell)$, and $v'_k = v(k,\ell+1)$, where $v(k,\ell)$ is given by (B.42). Thus, $g(k) = h(k,\ell)$ and $g'(k) = h(k,\ell+1)$. Summing the inequalities (B.41) for $1 \le k \le 2^p$, we get

$$\begin{aligned}\sum_{k \le 2^p, \ell < 2^p} |h(k,\ell+1) - h(k,\ell)| &\le 2 \sum_{k,\ell} (v(k,\ell) - h(k,\ell))\\ &\quad + \sum_{k,\ell} |v(k,\ell) - v(k,\ell+1)| .\end{aligned}$$

Using (B.44) and since $m_0 \le n(k, \ell)$, we get

$$
\begin{aligned}
m_0 \sum_{k \le 2^p, \ell < 2^p} |h(k, \ell+1) - h(k, \ell)| &\le 2 \sum_{k,\ell} n(k, \ell)(v(k, \ell) - h(k, \ell)) \\
&\quad + 10 \sum_{k,\ell} n(k, \ell)(u(k, \ell) - v(k, \ell)) \\
&\le 10 \sum_{k,\ell} n(k, \ell)(u(k, \ell) - h(k, \ell)) ,
\end{aligned}
$$

using that $h(k, \ell) \le v(k, \ell) \le u(k, \ell)$ in the last line. This proves (17.33), and (17.32) is obvious. □

Appendix C
Classical View of Infinitely Divisible Processes

In this appendix, we explain the classical view of infinitely divisible processes and why it coincides with our direct definition of Sect. 12.2.

C.1 Infinitely Divisible Random Variables

Definition C.1.1 We say that a r.v. X is *positive infinitely divisible* if there exists a positive measure ν on $\mathbb{R}^+$ such that

$$\int (\beta \wedge 1) \mathrm{d}\nu(\beta) < \infty \,, \tag{C.1}$$

$$\forall \alpha \in \mathbb{R} \,, \; \mathsf{E} \exp i\alpha X = \exp\Big(- \int (1 - \exp(i\alpha\beta)) \mathrm{d}\nu(\beta)\Big) \,. \tag{C.2}$$

The use of (C.1) is to ensure that the integral in the right-hand side of (C.2) makes sense. To motivate this definition, let us recall the definition (12.1) of Poisson r.v.s. Consider finitely many independent Poisson r.v.s X_k with $\mathsf{E}X_k = a_k$ and numbers $\beta_k \geq 0$. Then, by independence, (12.3) implies

$$\begin{aligned} \mathsf{E} \exp\Big(i\alpha \sum_k \beta_k X_k\Big) &= \exp\Big(- \sum_k a_k (1 - \exp(i\alpha\beta_k))\Big) \\ &= \exp\Big(- \int (1 - \exp(i\alpha\beta)) \mathrm{d}\nu(\beta)\Big) \,, \end{aligned} \tag{C.3}$$

M. Talagrand, *Upper and Lower Bounds for Stochastic Processes*, Ergebnisse der Mathematik und ihrer Grenzgebiete. 3. Folge / A Series of Modern Surveys in Mathematics 60, https://doi.org/10.1007/978-3-030-82595-9

where ν is the discrete positive measure on $\mathbb{R}^+$ such that for each $\beta \in \mathbb{R}^+$, we have $\nu(\{\beta\}) = \sum\{a_k; \beta_k = \beta\}$. Let us observe the formula

$$\mathsf{E}\sum_k \beta_k X_k = \sum_k \beta_k a_k = \int \beta \mathrm{d}\nu(\beta) .$$

It is appropriate to think of a positive infinitely divisible r.v. X as a (continuous) sum of independent r.v.s of the type βY where Y is a Poisson r.v. and $\beta \geq 0$. This is a sum of quantities that are ≥ 0, and *there is no cancellation* in this sum. The r.v. X has an expectation if and only if $\int \beta d\nu(\beta) < \infty$ (and the value of this expectation is then $\int \beta d\nu(\beta)$).

Definition C.1.2 We say that a r.v. X is *infinitely divisible* (real, symmetric, without Gaussian component) if there exists a positive measure ν on $\mathbb{R}^+$ such that

$$\int (\beta^2 \wedge 1)\mathrm{d}\nu(\beta) < \infty , \tag{C.4}$$

$$\forall \alpha \in \mathbb{R} , \ \mathsf{E}\exp i\alpha X = \exp\Big(- \int (1 - \cos(\alpha\beta))\mathrm{d}\nu(\beta)\Big) . \tag{C.5}$$

The use of (C.4) is to ensure the existence of the integral in the right-hand side of (C.5). We shall prove the existence of X in Sect. C.3. To motivate this definition, consider again a Poisson r.v. Y of expectation a and an independent copy Y' of Y. Then (12.3) implies

$$\mathsf{E}\exp i\alpha(Y - Y') = \exp(-2a(1 - \cos(\alpha))) . \tag{C.6}$$

Thus, when a r.v. X is a sum of independent terms $\beta_k(Y_k - Y'_k)$ where Y_k and Y'_k are independent Poisson r.v.s of expectation a_k and $\beta_k \geq 0$, it satisfies (C.5), where now ν is the discrete positive measure on $\mathbb{R}^+$ such that $\nu(\{\beta\}) = 2\sum\{a_k; \beta_k = \beta\}$ for each $\beta \in \mathbb{R}^+$.

It is appropriate to think of an infinitely divisible r.v. X as a continuous sum of independent r.v.s of the type $\beta(Y - Y')$ where Y and Y' are independent Poisson r.v.s with the same expectation. These r.v.s are symmetric rather than positive, and there is *a lot of cancellation* when one adds them. This is why the formula (C.5) makes sense under the condition (C.4) rather than the much stronger condition (C.1).[1]

[1] This dichotomy no cancellation versus cancellation is a leitmotiv of the book.

C.2 Infinitely Divisible Processes

Consider a finite set T and let us denote by $\beta = (\beta(t))_{t\in T}$ a generic point of $\mathbb{R}^T$. A stochastic process $(X_t)_{t\in T}$ is called (real, symmetric, without Gaussian component) infinitely divisible if there exists a positive measure ν on $\mathbb{R}^T$ such that $\int_{\mathbb{R}^T}(\beta(t)^2 \wedge 1)\mathrm{d}\nu(\beta) < \infty$ for all t in T, and such that for all families $(\alpha_t)_{t\in T}$ of real numbers, we have

$$\mathsf{E}\exp i\sum_{t\in T}\alpha_t X_t = \exp\Big(-\int_{\mathbb{R}^T}\Big(1-\cos\Big(\sum_{t\in T}\alpha_t\beta(t)\Big)\Big)\mathrm{d}\nu(\beta)\Big)\,. \tag{C.7}$$

The positive measure ν is called the *Lévy measure* of the process.[2] Each of the linear combinations $\sum_{t\in T}\alpha_t X_t$ is an infinitely divisible r.v.

Exercise C.2.1 Assume that for $t \in T$, the r.v.s X_t is infinitely divisible and that these r.v.s are independent. Prove that $(X_t)_{t\in T}$ is an infinitely divisible process.

As an example of infinitely divisible process, assume that ν consists of a mass a at a point $\beta \in \mathbb{R}^T$. Then, in distribution, $(X_t)_{t\in T} = (\beta(t)(Y - Y'))_{t\in T}$ where Y and Y' are independent Poisson r.v.s of expectation $a/2$. One can view the formula (C.7) as saying that the general case is obtained by taking a (kind of continuous) sum of independent processes of the previous type. Lots of cancellations occur when taking such sums.

For the purpose of studying the supremum of a process, our definition of infinitely divisible processes is the most general one: It is essentially not a restriction to consider only the symmetric case (using the familiar symmetrization procedure which replaces the process (X_t) by the process $(X_t - X_t')$ where (X_t') is an independent copy of the process (X_t)), and it is not a real restriction to exclude Gaussian components which are very well understood.

When T is infinite, we still say that the process $(X_t)_{t\in T}$ is infinitely divisible if (C.7) holds for each family $(\alpha_t)_{t\in T}$ such that only finitely many coefficients are not 0.[3] An infinitely divisible process indexed by T is thus parameterized by a σ-finite measure on $\mathbb{R}^T$ (with the sole restriction that $\int(\beta(t)^2 \wedge 1)\mathrm{d}\nu(\beta) < \infty$ for each $t \in T$). Only some extremely special subclasses have yet been studied in any detail. The best known such subclass is that of infinitely divisible processes with *stationary increments*. Then $T = \mathbb{R}^+$ and ν is the image of $\mu \otimes \lambda$ under the map $(x, u) \mapsto (x\mathbf{1}_{\{t\geq u\}})_{t\in\mathbb{R}^+}$, where μ is a positive measure on $\mathbb{R}$ such that $\int(x^2 \wedge 1)\mathrm{d}\mu(x) < \infty$ and where λ is Lebesgue measure.

---

[2] Assuming without loss of generality that $\nu(\{0\}) = 0$, it is unique.

[3] One then considers the Lévy measure as a "cylindrical measure" that is known through its projections on $\mathbb{R}^S$ for S a finite subset of T, projections that are positive measures and satisfy the obvious compatibility conditions on how these projections relate to each other. It is sometimes necessary to go beyond this naive point of view. The final word on how to define the Lévy measure seems to be [93], but we are not really concerned with these matters here.

C.3 Representation

We show that a process as in Definition 12.2.1 is indeed "an infinitely divisible process of Lévy measure ν" in the classical sense of (C.7). We simply have to show that if $X_t = \sum_{j\geq 1} \varepsilon_j Z_j(t)$, then

$$\mathsf{E}\exp i\sum_{t\in T}\alpha_t X_t = \exp\Big(-\int_{\mathbb{R}^T}\Big(1-\cos\Big(\sum_{t\in T}\alpha_t\beta(t)\Big)\Big)\mathrm{d}\nu(\beta)\Big)\,. \tag{C.8}$$

Leaving some convergence details to the reader, we first take expectation E_ε in the r.v.s ε_j given the Z_j to obtain, setting $u_j = \sum_{t\in T}\alpha_t Z_j(t)$,

$$\mathsf{E}_\varepsilon \exp i\sum_{j\geq 1}\varepsilon_j u_j = \prod_{j\geq 1}\cos u_j = \exp\sum_{j\geq 1}\log\cos u_j\,, \tag{C.9}$$

and we simply take expectation using the formula (12.8) to obtain (C.8). Conversely, an infinitely divisible process of Lévy measure ν has the representation $X_t = \sum_{j\geq 1}\varepsilon_j Z_j(t)$ where $(Z_j)_{j\geq 1}$ is a realization of a Poisson point process of intensity measure ν, the Lévy measure of the process.

C.4 p-Stable Processes

Finally, we prove the claim that p-stable processes are infinitely divisible and conditionally Gaussian. It is proved in [53], Theorem 5.2, that a p-stable process has a spectral measure and that there exists a finite positive measure m on $\mathbb{R}^T$ such that for any family $(\alpha_t)_{t\in T}$ we have

$$\mathsf{E}\exp i\sum_{t\in T}\alpha_t X_t = \exp\Big(-\frac{1}{2}\int_{\mathbb{R}^T}\Big|\sum_{t\in T}\alpha_t\beta(t)\Big|^p\mathrm{d}m(\beta)\Big)\,. \tag{C.10}$$

We observe the formula

$$\int_{\mathbb{R}^+}(1-\cos(ax^{-1/p}))\mathrm{d}\lambda(x) = C(p)|a|^p/2\,,$$

which is obvious through change of variable. Let us denote by λ Lebesgue's measure on $\mathbb{R}^+$. Consider the probability measure ν on $\mathbb{R}^T$ such that

$$C(p)\nu \text{ is the image of } \lambda\otimes m \text{ under the map } (x,\gamma)\mapsto x^{-1/p}\gamma\,. \tag{C.11}$$

Then

$$\begin{aligned}
&\int_{\mathbb{R}^T} \Big(1 - \cos\big(\sum_{t\in T} \alpha_t \beta(t)\big)\Big)\mathrm{d}\nu(\beta) \\
&= \frac{1}{C(p)} \int_{\mathbb{R}^T} \int_{\mathbb{R}^+} \Big(1 - \cos\big(x^{-1/p} \sum_{t\in T} \alpha_t \gamma(t)\big)\Big)\mathrm{d}\lambda(x)\mathrm{d}m(\gamma) \\
&= \frac{1}{2} \int_{\mathbb{R}^T} \big|\sum_{t\in T} \alpha_t \gamma(t)\big|^p \mathrm{d}m(\gamma) \,, \qquad \text{(C.12)}
\end{aligned}$$

and combining with (C.10), this shows that ν is a Lévy measure for the process $(X_t)_{t\in T}$.

To prove that a p-stable process is conditionally Gaussian, consider then a finite positive measure m on $\mathbb{R}^T$ and the measure ν as in (C.11). Consider a Poisson point process $(Z_j)_{j\geq 1}$ of intensity measure ν and independent Gaussian r.v.s (g_j). Then one checks as previously that for $\alpha > 0$, the process $X_t = \alpha \sum_{j\geq 1} g_j Z_j(t)$ is p-stable with spectral measure $\alpha K(p) m$. Consequently, if the p-stable process $(X_t)_{t\in T}$ has spectral measure m, it has the same distribution as the process $(K(p)^{-1} \sum_{j\geq 1} g_j Z_j(t))_{t\in T}$.

Appendix D
Reading Suggestions

It has been a deliberate choice not to include results of other authors which were proved later than the first edition of this book, as we try to present only results in their final form. The single exception to this policy concerns the recent results of G. Pisier in Sect. 19.4. In this appendix, we point out some directions connected to the main ideas of this book.

D.1 Partition Schemes

The work of R. van Handel [140, 141] attempts to provide a new view on partition schemes. The formulation of Theorem 2.9.8 is directly inspired by this work although in a sense, this theorem is a hybrid between the methods of van Handel and the original methods of the author.

D.2 Geometry of Metric Spaces

Let us recall that a distance δ on a T is called ultrametric if

$$\forall s, t, v \in T \ , \ \delta(s,t) \leq \max(\delta(s,v), \delta(t,v)) \ .$$

Given $A > 1$, let us say that a subset S of a metric space (T, d) is an A distortion of an ultrametric space if for some ultrametric distance δ on S we have

$$\forall s, t \in S \ ; \ \delta(s,t) \leq d(s,t) \leq A\delta(s,t) \ .$$

M. Talagrand, *Upper and Lower Bounds for Stochastic Processes*, Ergebnisse der Mathematik und ihrer Grenzgebiete. 3. Folge / A Series of Modern Surveys in Mathematics 60, https://doi.org/10.1007/978-3-030-82595-9

One may then investigate, given a metric space, whether it contains subsets which are A distortion of ultrametric spaces, and which are in a sense large. This question is investigated in great detail in the papers [65] and [67]. The equivalence of the quantity (3.13) with $\gamma_2(T, d)$ means that for any metric space, one may find a subset S which is an L distortion of an ultrametric space and for which $\gamma_2(S, d) \geq \gamma_2(T, d)/L$, and the results of [65] and [67] can be seen as far-reaching extensions of this result. We quote here the main result of [67]:

Theorem D.2.1 *Given $A > 1$, there exists a constant $K = K(A)$ with the following property. Consider a metric space (T, d) and a probability measure on μ on T. Then there is a subset S of T which is an A distortion of an ultrametric space, and a probability measure ν on S such that for each $x \in X$ and each $r > 0$, one has*

$$\nu(B(x, r)) \leq \mu(B(x, Kr))^{1-1/A} .$$

D.3 Cover Times

Consider a connected graph with set of edges E and set of vertices V. Starting with a given vertex v, we consider the following random walk: At each step, if the walker is at a given vertex w, he chooses at random with equal probability to move to one of the vertices connected to w by an edge. The cover time τ_v is defined as the first time all vertices have been visited. On the other hand, let us think to the graph as an electrical network, each edge being given conductance 1. Then the effective resistance $R(u, v)$ of the network between edges u and v defines a distance R on V. The main result of [28] is very clean:

$$\frac{1}{L}\gamma_2(V, \sqrt{R})^2 \leq \max_{v \in V} \mathsf{E}\tau_v \leq L\gamma_2(V, \sqrt{R})^2 .$$

D.4 Matchings

Under the lead in particular of Giorgio Parisi, physicists are bringing new ideas to the theory of matchings (see, e.g., https://arxiv.org/pdf/1402.6993.pdf), and some of these ideas have been made rigorous; see [4]. Many other possible nontrivial directions in the theory of matchings have not been fully explored; see [133].

D.5 Super-concentration in the Sense of S. Chatterjee

Given a Gaussian process $(X_t)_{t\in T}$, we have learned in (2.10.6) that the fluctuations of $\sup_{t\in T} X_t$ are typically not larger than $\sigma := \sup_{t\in T}(\mathsf{E}X_t^2)^{1/2}$. However, it turns out that in many cases, these fluctuations are of a lower order. Probably the simplest case is that if $(g_i)_{i\le N}$ are standard independent Gaussian r.v.s, then $\max_{i\le N} g_i$ has fluctuations of order $1/\sqrt{\log N}$ (an easy exercise). A more elaborate example is as follows. Denoting by T the unit sphere of $\mathbb{R}^n$, let us set $Y_t = \sum_{i\le n} t_i g_i$ and $X_t = \sum_{i,j\le n} t_i t_j g_{i,j}$ where $g_i, g_{i,j}$ are i.i.d. standard Gaussian r.v.s. The fluctuations of $\sup_{t\in T} Y_t$ are of order 1, but the fluctuations of $\sup_{t\in T} X_t$ are known to be of order $n^{-1/6}$. In such cases, one says that one has super-concentration. Sourav Chatterjee [25] has discovered (among many other things) that super-concentration is related to the fact that the function $t \mapsto X_t$ has typically many near maxima (which is not the case of the function $t \mapsto Y_t$).

D.6 High-Dimensional Statistics

An important and rapidly expanding area of research is High-Dimensional Statistics, where the goal is to study large amounts of data, living in a high-dimensional space. Because high-dimensional data appears in important and diverse applications (e.g., DNA sequencing, image and speech recognition, autonomous car systems, Internet search engines, and many more), a well-established theoretical understanding of the area is of the utmost importance.

Chaining plays a key role in High-Dimensional Statistics: The analysis of statistical recovery procedures often calls for the study of the supremum (or infimum) of certain random processes, and that is where chaining methods take center stage.

To give a flavor of the problems one encounters, consider the incredibly important area of compressed sensing (used, e.g., in MRI imaging and remote sensing (radar)). One wishes to recover an unknown signal (vector or function), living in a very high-dimensional space. The signal is sparse: Its expansion with respect to some natural basis has a few nonzero coefficients.[1] One receives, as data, relatively few linear measurements of the signal, and those may be further corrupted by noise. The goal is to use the given data to construct a good approximation of the unknown signal. A general introduction to this topic can be found in [34], and a recent example the way chaining is used in [74].

The novelty in compressed sensing was the realization that under rather minimal assumptions, a rather small number of measurements were required to generate a good approximation of the signal—essentially scaling linearly in the degree of

[1] The *degree of sparsity* is then the number of nonzero components.

sparsity and only logarithmically in the dimension of the space in which the signal lives) . Moreover, recovery can be performed in an efficient way computationally.[2]

A simplified way of viewing recovery procedures is as follows: If t_0 is the unknown vector, and $\langle t_0, X_1\rangle, \ldots, \langle t_0, X_N\rangle$ are the given linear measurements (noise-free), a reasonable guess of an approximating vector would be some t, that is also sparse and satisfies that $(\langle t, X_i\rangle)_{i=1}^N$ is close to the given measurements. The success of recovery is based on the fact that if x is any sparse vector that is far away from t_0, that fact will be exhibited in the value of $N^{-1}\sum_{i=1}^N \langle x - t_0, X_i\rangle^2$; to that end, one has to study the behavior of the quadratic empirical process $u \to N^{-1}\sum_{i=1}^N \langle u, X_i\rangle^2 - \mathsf{E}\langle u, X_i\rangle^2$ on the set of sufficiently sparse vectors [73]. Related topics may be found in Sects. 14.2 and 14.3.

Empirical processes often occur when dealing with high-dimensional data. Two generic examples of empirical processes are the centered product empirical process indexed by two classes of functions, $\mathcal{F}$ and $\mathcal{H}$, that is, $(f, h) \to N^{-1}\sum_{i=1}^N f(X_i)h(X_i) - \mathsf{E}fh$, and the centered multiplier empirical process indexed by the class $\mathcal{F}$, that is, $f \to N^{-1}\sum_{i=1}^N \xi_i f(X_i) - \mathsf{E}\xi f$, with $(\xi_i)_{i=1}^N$ being a random vector and $(X_1, \ldots, X_N)$ selected randomly according to some procedure (e.g., independent sampling). The study of these two processes appears in diverse applications besides sparse recovery. General references are [55, 138, 142] and [71] are a few more interesting examples of random processes one encounters in High-Dimensional Statistics and the chaining arguments that are used in their analysis.

[2] Sparse recovery procedures do not require preliminary information on the degree of sparsity of t_0. In the very basic setup, the procedure selects t that agrees with t_0 on the sample points and has the smallest possible ℓ_1 norm. And if t_0 happens to be sparse, one can show that so will be the vector selected by this procedure (see, e.g., [139] and [91]).

Appendix E
Research Directions

It seems worthwhile to recapitulate some of the research problems we stressed in this book. It is very risky to attempt to evaluate the potential of a research problem, but we will try.[1]

E.1 The Latała-Bednorz Theorem

Find a proof that you can explain to your grandmother. It is hard to understand why the current proof works. This is a pity. The core material of this book, the theory presented in Chaps. 2 to 13, is rather beautiful, with the exception of that proof. Ideally, one wishes for a new conceptual idea.

E.2 The Ultimate Matching Conjecture

It is stated in Problem 17.1.2. Possibly, it is only a hard combinatorial problem to crack, to understand the geometry of certain classes of functions, and that the solution would not open new horizons.

[1] This author's most notable contribution to mathematics, the discovery of new directions for concentration inequalities, started by studying a problem of secondary importance.

M. Talagrand, *Upper and Lower Bounds for Stochastic Processes*, Ergebnisse der Mathematik und ihrer Grenzgebiete. 3. Folge / A Series of Modern Surveys in Mathematics 60, https://doi.org/10.1007/978-3-030-82595-9

E.3 My Favorite Lifetime Problem

It is explained in Sect. 13.3. There is no telling how difficult this is and what new horizons a positive solution would open. It is reserved for the very ambitious.

E.4 From a Set to Its Convex Hull

The general problem is to understand geometrically how the smallness of a set, in the sense of certain γ functionals, transfers to its convex hull. The first and most important occurrence of this problem[2] is Problem 2.11.2. Further occurrences are Problem 8.3.5 and Problem 8.3.12. But it is not certain that this is possible. Also related are Problems 2.11.13 and 2.11.14.

[2] It is while discussing this problem with Keith Ball that I invented the generic chaining.

Appendix F
Solutions of Selected Exercises

As the purpose of the exercises is to have the reader (rather than the author) work, the solutions are sketchy and have not been worked out with the same dedication as the rest of this book. Therefore, expect much lousiness and some plain nonsense.

Exercise 1.3.1 Considering for each (s, t), the largest k with $d(s, t) \le 2^{-k}$ yields

$$\sup_{s,t\in G} \frac{|X_s - X_t|^p}{d(s,t)^\beta} \le L \sum_{k\ge 0} \sup_{s,t\in G; d(s,t)\le 2^{-k}} 2^{k\beta} |X_s - X_t|^p .$$

Since $\mathsf{E} \sup_{s,t\in G; d(s,t)\le 2^{-k}} |X_s - X_t|^p \le K(m, p, \alpha) 2^{k(m-\alpha)}$ by (1.10), taking expectation yields (1.12) since $\beta + m - \alpha < 0$.

Exercise 1.3.2 By Jensen's inequality, we have $\varphi(\mathsf{E} \max_i V_i) \le \mathsf{E}\varphi(\max_i V_i)$. Furthermore, $\varphi(\max_i V_i) \le \sum_i \varphi(V_i)$ so that $\mathsf{E}\varphi(\max_i V_i) \le \sum_i \mathsf{E}\varphi(V_i)$.

Exercise 1.3.3 It follows from (1.13) and (1.14) that the r.v. Y_n of (1.5) satisfies $\mathsf{E}Y_n/c_n \le \varphi^{-1}(K(m)2^{nm} d_n)$, and (1.15) follows by combining with (1.7).

Exercise 1.4.3 The distance d associated to Brownian motion is given by $d(s, t) = \sqrt{|s - t|}$ and $N([0, 1], d, \epsilon) \le L\epsilon^{-2}$. The condition $|s - t| \le \delta$ implies $d(s, t) \le \sqrt{\delta}$. Dudley's bound is then $L \int_0^{\sqrt{\delta}} \sqrt{\log (L/\epsilon^2)} d\epsilon \le L\sqrt{\delta \log(2/\delta)}$.

Exercise 2.2.2 Just use that $|X_t| \le |X_t - X_{t_0}| + |X_{t_0}| \le \sup_{s,t} |X_s - X_t| + |X_{t_0}|$.

Exercise 2.3.1 Because $\mathsf{P}(Y \ge a\mathsf{E}Y) \le 1/a$ by Markov's inequality.

Exercise 2.3.3 (a) This means that given $L_1 > 0$, there exists L_2 such that $\sup_x xy - L_1 x^3 \le y^{3/2}/L_2$, which is proved by computing this supremum. (b) Let us then assume that $p(u) \le L_1 \exp(-u^2/L_1)$ for $u \ge L_1$. Given a parameter A, for $Au \ge L_1$, we have $p(Au) \le L_1 \exp(-A^2u^2/L_1)$. Also, we have $p(Au) \le 1$ so that $p(Au) \le 2\exp(-u^2)$ for $u \le \sqrt{\log 2}$. Assuming that $A\sqrt{\log 2} \ge L_1$, it suffices that $L_1 \exp(-A^2u^2/L_1) \le 2\exp(-u^2)$ for $u \ge \sqrt{\log 2}$. This is true as soon as $L_1 \le 2\exp(u^2(A/L_1 - 1))$ for $u \ge \sqrt{\log 2}$ and in particular as soon as A is large enough that $L_1 \le 2\exp(\log 2(A^2/L_1 - 1))$. (c) Taking logarithms, it suffices to prove that for $x \ge 0$ and a constant L_1, one has $L_1 x - x^{3/2}/L_1 \le L_2 - x^{3/2}/L_2$

M. Talagrand, *Upper and Lower Bounds for Stochastic Processes*, Ergebnisse der Mathematik und ihrer Grenzgebiete. 3. Folge / A Series of Modern Surveys in Mathematics 60, https://doi.org/10.1007/978-3-030-82595-9

for a certain constant L_2. Assuming first $L_2 \geq 2L_1$, it suffices to find L_2 for which $L_1 - x^{3/2}/(2L_1) \leq L_2$ which is obvious.

Exercise 2.3.5 (a) Since the relation $\mathsf{P}(\sup_{k\leq N} |g_k| \geq u) \leq 2N\exp(-u^2/2)$ does not require the r.v.s to be Gaussian. (c) Consider sets Ω_k with $\mathsf{P}(\Omega_k) = 1/N$, which we split into two sets $\Omega_{k,1}$ and $\Omega_{k,2}$ of probability $1/(2N)$. The centered r.v.s $g_k = \sqrt{\log N}(\mathbf{1}_{\Omega_{k,1}} - \mathbf{1}_{\Omega_{k,2}})$ satisfy $\mathsf{P}(|g_k| \geq u) \leq \exp(-u^2)$ because the left-hand side is 0 for $u > \sqrt{\log N}$ and $1/N$ for $u \leq \sqrt{\log N}$. When the sets Ω_k are disjoint, $\sup_{k\leq N} g_k = \sqrt{\log N}$ on $\bigcup_{k\leq N} \Omega_{k,1}$ and is zero elsewhere. Thus, $\mathsf{E}(\sup_{k\leq N} g_k) = \sqrt{\log N}\mathsf{P}(\bigcup_{k\leq N} \Omega_k) = \sqrt{\log N}/2$. (c) When the sets $\Omega_{k,1}$ are independent, it is still true that $\sup_{k\leq N} g_k = \sqrt{\log N}$ on $\bigcup_{k\leq N} \Omega_{k,1}$, and by (2.18), this union has probability $\geq 1/L$. Next, $\sup_{k\leq N} g_k \geq 0$ except on the set $\bigcap_{k\leq N} \Omega_{k,2}$ where this supremum is $-\sqrt{\log N}$, and this set has probability $\leq 1/(2N)$. Thus, $\mathsf{E}\sup_{k\leq N} g_k \geq \sqrt{\log N}(1/L - 1/(2N))$.

Exercise 2.3.6 $\mathsf{P}(\cup_{k\leq N} A_k) = 1 - \prod_{k\leq N}(1 - \mathsf{P}(A_k))$ and $1 - x \leq \exp(-x)$ for $x > 0$.

Exercise 2.3.7 Use (b) for u such that $\mathsf{P}(g_1 \geq u) = 1/N$ so that u is about $\sqrt{\log N}$.

Exercise 2.3.8 (a) We have $\mathsf{E}\exp(Y/(2B)) = \int_0^\infty \mathsf{P}(\exp(Y/2B) \geq u)\mathrm{d}u = 1 + \int_1^\infty \mathsf{P}(Y \geq 2B\log u)\mathrm{d}u \leq 1 + 2\int_1^\infty u^{-2}\mathrm{d}u = 3$. Calculus shows that $(x/a)^a$ takes its maximum at $a = x/e$ so $(x/p)^p \leq \exp(x/e)$. Using this for $x = Y/B$ and taking expectations yield the result. The rest is obvious. (b) follow by using (a) for the variable Y^2. (c) If $\mathsf{E}Y^p \leq p^pB^p$, Markov's inequality yields $P(Y \geq u) \leq (Bp/u)^p$, and for $u \geq Be$, one takes $p = u/(Be)$ to get a bound $\exp(-u/(Be))$.

Exercise 2.3.9 Given any value of $x \geq 0$, we have $(\mathsf{E}|g|^p)^{1/p} \geq x\mathsf{P}(|g| \geq x)^{1/p}$, and for $x = \sqrt{p}$, one has $\mathsf{P}(|g| \geq p)^{1/p} \geq 1/L$.

Exercise 2.4.1 We have $d(t_1, T_n) = 0$ for each n. For $k \geq 2$, let $n(k)$ be the smallest integer with $N_{n(k)} \geq k$ so that $N_{n(k)-1} < k$ and thus $2^{n(k)} \leq L\log k$. Furthermore, $d(t_k, T_n) \leq d(t_k, t_1)$ for $n < n(k)$ and $d(t_k, T_n) = 0$ for $n \geq n(k)$ since then $t_k \in T_n$. Thus, $\sum_{n\geq 0} 2^{n/2}d(t_k, T_n) \leq L2^{n(k)/2}d(t_k, t_1) \leq L2^{n(k)/2}/\sqrt{\log k} \leq L$.

Exercise 2.5.3 Take $T = \{-1, 0, 1\}$, $U = \{-1, 1\}$, $d(x, y) = |x-y|$, so $e_0(T) = 1$, $e_0(U) = 2$.

Exercise 2.5.4 For $\epsilon > e_0(T)$, we have $N(T, d, \epsilon) = 1$. For $\epsilon > e_n(T)$, we have $N(T, d, \epsilon) \leq N_n$. Consequently,

$$\int_0^\infty \sqrt{\log N(T, d, \epsilon)}\,\mathrm{d}\epsilon = \sum_{n\geq 0} \int_{e_{n+1}(T)}^{e_n(T)} \sqrt{\log N(T, d, \epsilon)}\,\mathrm{d}\epsilon$$

$$\leq \sum_{n\geq 0} e_n(T)\sqrt{\log N_{n+1}}$$

and the result holds since $\log N_{n+1} \leq 2^{n+1}$.

Exercise 2.5.5 If $(A/\epsilon_n)^\alpha = \log N_n$, then $e_n(T) \leq \epsilon_n$.

Exercise 2.5.6 Indeed if n_0 is the smallest integer with $\operatorname{card} T \le N_{n_0}$, then $\sum_{n\ge 0} 2^{n/2} e_n(T) = \sum_{n\le n_0} 2^{n/2} e_n(T)$, and there are about $\log\log \operatorname{card} T$ terms in this sum. Furthermore, we control each term of the sum: For each k, we have $\sup_{t\in T} \sum_{n\ge 0} 2^{n/2} d(t, T_n) \ge 2^{k/2} \sup_{t\in T} d(t, T_k) \ge 2^{k/2} e_k(T)$.

Exercise 2.5.8 Consider a subset W of T maximal with respect to the property that $s, t \in W \Rightarrow d(s,t) > 2\epsilon$. Since the balls of radius ϵ centered at the points of W are disjoint, each of measure $\ge a$, we have $\operatorname{card} W \le 1/a$, and the balls centered at the points of W and of radius 2ϵ cover U by Lemma 2.5.7.

Exercise 2.5.9 (a) Consider a subset W of T which is maximum with respect to the property that any two of its points are at a distance $\ge \epsilon$. Then the balls centered on W of radius ϵ cover T. Furthermore, two points within a ball of radius ϵ are at mutual distance $\le 2\epsilon$. (b) If A is covered by N balls of radius ϵ, these balls have the same volume $\operatorname{Vol}(\epsilon B)$ so $\operatorname{Vol}(A) \le N\operatorname{Vol}(\epsilon B)$. Consider $W \subset A$ as in (a) but for 2ϵ rather than ϵ. The open balls of radius ϵ centered at the points of W are disjoint and entirely contained in $A + \epsilon B$ so that $\operatorname{card} W \operatorname{Vol}(\epsilon B) \le \operatorname{Vol}(A + \varepsilon B)$. This proves (2.45). (c) Since $\operatorname{Vol}(\epsilon B) = \epsilon^k \operatorname{Vol}(B)$. (d) If ϵ_n is defined by $(1/\epsilon_n)^k = N_n$ and ϵ'_n by $(1 + 2/\epsilon'_n)^k = N_n$, then by (c) $\epsilon_n \le e_n(B) \le \epsilon'_n$. Then $e_n(B)$ is of order 1 for $2^n \le k$ and of order $2^{-2^n/k}$ for $2^n \ge k$. As a result $\sum_{n\ge 1} 2^{n/2} e_n(B) = \sum_{2^n \le k} 2^{n/2} e_n(B) + \sum_{2^n > k} 2^{n/2} e_n(B) \le L\sqrt{k}$. (e) We cover T with N_{n_0} balls of radius $2e_{n_0}(T)$ and each of these by N_n balls of radius $\le L2^{-2^n/k} e_{n_0}(T)$. There are at most $N_{n_0} N_n \le N_{n+1}$ of these balls, and this proves the claim.

Exercise 2.5.10 (a) We have $\binom{m}{\ell} = (\ell+1)/(m-\ell)\binom{m}{\ell+1}$, and for $\ell \le k-1$, we have $(\ell+1)/(m-\ell) \le 2k/m$. Iteration of this relation shows that for $\ell \le k$, we have $\binom{m}{\ell} \le (2k/m)^{k-\ell}\binom{m}{k}$. When $k \le m/4$, we have $2k/m \le 1/2$, and thus, $\sum_{0\le \ell \le k/2} (2k/m)^{k-\ell} \le 2(2k/m)^{k/2}$. This proves (2.48). (b) We have $\operatorname{card} \mathcal{I} = \binom{m}{k}$. On $\mathcal{I}$, consider the distance $d(I,J) := \operatorname{card}(I \bigtriangleup J)$ where $I \bigtriangleup J$ is the symmetric difference $(I \setminus J) \cup (J \setminus I)$. We bound from above the cardinality of a ball of $\mathcal{I}$ of radius $k/2$. Given $I \in \mathcal{I}$, a set J is entirely determined by $I \bigtriangleup J$. Thus, the number of sets in $\mathcal{I}$ for which $\operatorname{card}(I \bigtriangleup J) \le k/2$ is bounded by $\sum_{0\le\ell\le k/2} \binom{m}{\ell}$. Thus (2.48) shows that $\mu(B(I, k/2)) \le 2(2k/m)^{k/2}\mu(\mathcal{I})$ and $N(\mathcal{I}, d, k/2) \ge (m/(2k))^{k/2}/2$ are the desired result.

Exercise 2.5.11 Consider an integer k with $k/m \le 1/4$. It follows from Exercise 2.5.10 that $N(T_k, d, 1/\sqrt{2k}) \ge (m/(2k))^{k/2}/2$. In particular for $1 \le k \le \sqrt{m}$, we have $\log N(T, d, 1/L\sqrt{k}) \ge (k/L)\log m$ so that for $m^{-1/4}/L \le \epsilon \le 1/L$, we have $\sqrt{\log N(T, d, \epsilon)} \ge (1/L\epsilon)\sqrt{\log m}$, from which it readily follows that the right-hand side of (2.41) is $\ge (\log m)^{3/2}/L$.

Exercise 2.7.5 Consider the smallest integer n_0 with $\operatorname{card} T \le N_{n_0}$ so that $2^{n_0/\alpha} \le K(\log \operatorname{card} T)^{1/\alpha}$. Take $\mathcal{A}_n = \{T\}$ for $n \le n_0$ and $\mathcal{A}_n$ consisting of the sets $\{t\}$ for $t \in T$ when $n \ge n_0$.

Exercise 2.7.6 (a) We enumerate $\mathcal{B}_n$ as $(B_\ell)_{\ell \le N_n}$. The sets $C_j = B_j \setminus \cup_{\ell<j} B_\ell$ then provide a partition $\mathcal{C}_n$ of T of cardinality $\le N_n$, and $C_j \subset B_j \in \mathcal{B}_n$. We define $\mathcal{A}_0 = \{T\}$ and/or $n \ge 1$ define $\mathcal{A}_n$ as the partition generated by $\mathcal{A}_{n-1}$ and $\mathcal{C}_{n-1}$ so that $\operatorname{card} \mathcal{C}_n \le N_{n-1}^2 = N_n$, and for $n \ge 1$, each element A of $\mathcal{A}_n$ is contained is an element of $\mathcal{C}_{n-1}$ and thus in an element of $\mathcal{B}_n$. (b) We can cover T by N_n balls of

radius $\leq 2e_n(T)$. We use these balls as covering $\mathcal{B}_n$. Then each element of $\mathcal{A}_n$ has diameter $\leq 2e_{n-1}(T)$.

Exercise 2.7.7 We set $\mathcal{A}_0 = \{T\}$ and for $n \geq 1$ $\mathcal{A}_n$ consists of the sets of the type $B \cap C$ for $B \in \mathcal{B}_{n-1}$ and $C \in \mathcal{C}_{n-1}$. There are at most $N_{n-1}^2 \leq N_n$ such sets.

Exercise 2.7.8 (a) Consider an admissible sequence $(\mathcal{A}_n)$ with

$$\sup_{t \in T} \sum_{k \geq 0} 2^{k/2} \Delta(A_k(t)) \leq 2\gamma_2(T, d) .$$

In particular for each $A \in \mathcal{A}_n$, we have $\Delta(A) \leq 2^{-n/2+1}\gamma_2(T, d)$. Thus, A is contained in a ball of radius $\leq 2^{-n/2+1}\gamma_2(T, d)$. As these sets A for $A \in \mathcal{A}_n$ cover T, we have $e_n(T) \leq 2^{-n/2+1}\gamma_2(T, d)$. (b) Consider $\epsilon > 0$ with $N(T, d, \epsilon) > 1$, and the smallest n with $N(T, d, \epsilon) \leq N_n$ so that $\sqrt{\log N(T, d, \epsilon)} \leq L2^{n/2}$. By definition of n, we have $N_{n-1} < N(T, d, \epsilon)$ so that by definition of $e_{n-1}(T)$, we have $\epsilon \leq e_{n-1}(T)$, and $\epsilon\sqrt{\log N(T, d, \epsilon)} \leq L2^{n/2}e_{n-1}(T) \leq L\gamma_2(T, d)$.

Exercise 2.7.9 Consider the smallest integer n_0 such that $2^{n_0} \geq m$. By Exercise 2.5.9 (e) for $n \geq n_0$, we have $e_{n+1}(T) \leq L2^{-2^n/m}e_{n_0}(T)$ so that $\sum_{n \geq n_0} 2^{n/2}e_n(T) \leq L2^{n_0/2}e_{n_0}(T)$. Then

$$\sum_{n \geq 0} 2^{n/2}e_n(T) \leq L \sum_{n \leq n_0} 2^{n/2}e_n(T) \leq L \log(m+1)\gamma_2(T, d)$$

because each term of the middle summation is at most $L\gamma_2(T, d)$, and there are $n_0 + 1 \leq L \log(m+1)$ terms.

Exercise 2.7.10 It was already shown in Exercise 2.5.11 that the estimate (2.58) is optimal, but the present construction is easier to visualize. According to (2.45) in $\mathbb{R}^M$, there exists a set of cardinality 2^M in the unit ball consisting of points at mutual distance $\geq 1/2$. Denote by n_0 the largest integer with $2^{n_0} \leq M$. Thus, for $n \leq n_0$, there exists a set T_n of cardinality N_n consisting of points within distance $2^{-n/2+1}$ of the origin but at mutual distances $\geq 2^{-n/2}$. Set $T = \cup_{n \leq n_0} T_n$ so that for $n \leq n_0$, we have $e_n(T) \geq e_n(T_n) \geq 2^{-n/2}$. Consequently, $\sum_{n \geq 0} 2^{n/2}e_n(T) \geq n_0 \geq (\log M)/L$. One can prove that $\gamma_2(T, d) \leq L$ by proceeding as in Exercise 2.4.1.

Exercise 2.7.12 The inequality (2.33) never used that the process is centered, so it remains true, and combining it with (2.6) implies $\mathsf{E} \sup_{t \in T} |X_t - X_{t_0}| \leq LS$, and the result.

Exercise 2.8.2 For example, T consists of a sequence of real numbers which converges fast to 0, for example, $T = \{1/n!; n \geq 1\}$. For each value of r, the largest value of m such that (2.76) holds is finite.

Exercise 2.9.2 The growth condition is satisfied because $m = N_n \geq N_1 = 4$, and there exist no separated family. In this case, the inequality $\gamma_2(T, d) \leq LrF(T)/c^*$ is false because the left-hand side is positive and the right-hand side is zero.

Exercise 2.10.4 Let $a = \min_{p \leq m}(\mathsf{E}X_p^2)^{1/2} > 0$ where the r.v.s $(X_p)_{p \leq m}$ are independent Gaussian. (Observe that it is not required that the numbers $\mathsf{E}X_p^2$ be all equal.) Consider a number c_p such that $\mathsf{P}(X_p \geq c_p) = 1/m$ so that

$c_p \geq a\sqrt{\log m}/L$. The m sets $\Omega_p = \{X_p \geq c_p\}$ are independent of probability $1/m$ so that their union Ω is of probability $1 - (1 - 1/m)^m \geq 1/L$. Given $\omega \in \Omega$, there is $p \leq m$ with $\omega \in \Omega_p$ so that $X_p \geq c_p$ and thus $\mathbf{1}_\Omega \max_{p \leq m} X_p \geq \mathbf{1}_\Omega \min_{p \leq m} c_p$ and consequently by taking expectation $\mathsf{E}\mathbf{1}_\Omega \max_{p \leq m} X_p \geq (a/L)\sqrt{\log m}$. On the other hand, $\max_{p \leq m} X_p \geq -|X_1|$, and since we may assume without loss of generality that $\mathsf{E}X_1^2 = a^2$, we get $\mathsf{E}\mathbf{1}_{\Omega^c} \max_p X_p \geq -\mathsf{E}|X_1| \geq -La$. Consequently, $\mathsf{E}\max_{p \leq m} X_p \geq a((1/L)\sqrt{\log m} - L)$, the desired result.

Exercise 2.10.7 Then $\sigma = 1$ and $Y = \sup_{t \in T} X_t = \sqrt{\sum_{i \leq n} g_i^2}$ is about $\sqrt{n}$. Here, you can visualize the fluctuations of Y: $\mathsf{E}Y^4 - (\mathsf{E}Y^2)^2 \leq Ln$ so that the fluctuations of Y^2 are of order $\sqrt{n}$, and since Y^2 is of order n, the fluctuations of Y are of order 1. (Remember that $\sqrt{A+a} - \sqrt{A}$ is of order $a/\sqrt{A}$.)

Exercise 2.11.7 Enumerate $T \cup \{0\}$ as a sequence $(x_i)_{i \geq 1}$ with $x_1 = 0$ and $d(x_1, x_i) \leq L/\log i$.

Exercise 2.11.10 Look at the case where T consists of one single point.

Exercise 2.11.11 The first part is obvious. For the second part, simply replace (2.134) by $t = \sum_{n \geq 1} \pi_n(t) - \pi_{n-1}(t) = \sum_{n \geq 1} a_n(t) u_n(t)$.

Exercise 2.11.12 By homogeneity, we may assume that $S = \mathsf{E}\sup_t X_t = 1$. Consider an integer n_0 with $2^{-n_0} \simeq \delta$. Starting with an admissible sequence $(\mathcal{B}_n)$ of T such that $\sup_t \sum_{n \geq 0} 2^{n/2} \Delta(B_n(t)) \leq LS$, we consider the admissible sequence $(\mathcal{A}_n)$ given by $\mathcal{A}_n = \mathcal{B}_n$ if $n > n_0$ and $\mathcal{A}_n = \{T\}$ for $n \leq n_0$. Then $\sup_t \sum_{n \geq 0} 2^{n/2} \Delta(A_n(t)) \leq LS$. We set $T_n = \{0\}$ for $n \leq n_0$ and $U_n = \emptyset$ for $n \leq n_0$, and we follow the proof of Theorem 2.11.9. For $n \geq n_0$ and $u \in U_n$, we have $\|u\| \leq L2^{-n_0} \leq L\delta$.

Exercise 2.11.15 To prove (a), we observe that

$$\|U_q\| \geq \|U_1\| \geq \sup\Big\{\sum_{i \leq n} \varepsilon_{i,1} x_i \,;\, \sum_{i \leq n} x_i^2\Big\} = \sqrt{\sum_{i \leq n} \varepsilon_{i,1}^2} = \sqrt{n}\,.$$

If $\|x\| \leq 1$, then (2.140) implies $\mathsf{E}\exp(\|U_q(x)\|^2/4) \leq L^q$ so that $\mathsf{P}(\|U_q(x)\| \geq a\sqrt{q}) \leq \exp((L - a^2/4)q)$ by Markov's inequality and (2.141) by homogeneity. Consider now a sequence $t_k \in \mathbb{R}^n$ such that $\|t_k\| \leq \min(\delta, LA/\sqrt{\log(k+1)})$. Then for $w \geq 1$,

$$\mathsf{P}(\|U_q(t_k)\| \geq Lw\delta\sqrt{q}) = \mathsf{P}(\|U_q(t_k)\| \geq Lv\|t_k\|\sqrt{q}) \leq \exp(-v^2 q)$$

where $v = w\delta/\|t_k\| \geq w$, and thus,

$$\sum_{k \geq 1} \mathsf{P}(\|U_q(t_k)\| \geq Lw\delta\sqrt{q}) \leq \sum_{k \geq 1} \min(\exp(-w^2 q), \exp(-w^2\delta^2 q \log(k+1)/A^2)),$$

and when $\delta^2 q/A^2 \geq 1$, this is $\leq \exp(-w^2 q/L)$. The result follows since we can find such a sequence (t_k) for which $T \subset L\overline{\mathrm{conv}}\{t_k, k \geq 1\}$ and $A = \mathsf{E}\sup_{t \in T} X_t \leq \delta\sqrt{q}$.

Exercise 2.11.17 (a) The subsets of T obtained by fixing the coordinates of t in each T_k for $k \in I^c$ have diameter $\le \epsilon$, and there are $\prod_{k\in I^c} M_k$ such sets. (b) Take $\eta_k = \sqrt{\log M_k}$ and use (a) to see that $\sqrt{\log N(T,d,\epsilon)} \le S(\epsilon)$. (c) We reduce to the case $\sum_{k\le N} \epsilon_k \eta_k = 1$ by replacing ϵ_k by $\lambda\epsilon_k$ for a suitable λ. We set $\alpha_k = \epsilon_k\eta_k$, and we consider the probability μ on $\{1,\dots,N\}$ such that $\mu(\{k\}) = \alpha_k$. Define $h(k) = \epsilon_k/\eta_k$. Then $\int_I h d\mu = \sum_{k\in I} \epsilon_k^2$ and $\int_{I^c} 1/h d\mu = \sum_{k\in I^c} \eta_k^2$. (d) For any non-increasing function $\tilde{S}$ on $\mathbb{R}^+$, we have $\int_{\mathbb{R}^+} \tilde{S}(\epsilon)d\epsilon \le \sum_{\ell\in\mathbb{Z}} 2^{-\ell} S(2^{-\ell-1})$. (e) Thus, $\tilde{S}(2^{-\ell})^2 \le \int_{A_\ell^c}(1/h)d\mu$. Setting $B_\ell = A_\ell \setminus A_{\ell+1}$, we have $\tilde{S}(2^{-\ell})^2 \le \sum_{m<\ell} c_m$ where $c_m = \int_{B_m}(1/h)d\mu$. Thus, $\tilde{S}(2^{-\ell}) \le \sum_{m<\ell} \sqrt{c_m}$ and $\sum_\ell 2^{-\ell}\tilde{S}(2^{-\ell}) \le \sum_\ell 2^{-\ell}\sum_{m<\ell}\sqrt{c_m} \le \sum_m 2^{-m}\sqrt{c_m}$. Recalling that the set A_m is of the type $A_m = \{h \le t_m\}$ for a certain t_m, we have $B_m = \{t_{m+1} < h \le t_m\}$ and in particular $c_m \le \mu(B_m)/t_{m+1}$. Now, $\int_{B_m} h d\mu = 2^{-2m} - 2^{-2(m+1)} \ge 2^{-2(m+1)}$ so that $2^{-2(m+1)} \le \int_{B_m} h d\mu \le t_m \mu(B_m)$, and changing m into $m+1$, we have $2^{-m} \le L\sqrt{t_{m+1}\mu(B_{m+1})}$. Thus, $2^{-m}\sqrt{c_m} \le L\sqrt{\mu(B_m)\mu(B_{m+1})}$. The Cauchy-Schwarz inequality then shows that $\sum_m 2^{-m}\sqrt{c_m} \le L$ as desired.

Exercise 2.11.18 (a) First note that if $\epsilon^2 = \sum_{k\le N}\epsilon_k^2$, then $N(T,d,\epsilon) \le \prod_{k\le N} N(T_k,d_k,\epsilon_k)$, and then use (2.146) for $f_k(\epsilon) = \sqrt{\log N(Tk,d_k,\epsilon)}$. (b) Consider a set I with $\sum_{k\in I}\theta_k^2 \le \epsilon^2$. Taking $\epsilon_k = \theta_k$ if $k \in I$ and $\epsilon_k = 0$ otherwise shows that $V(\epsilon)^2 \le \sum_{k\in I^c}\eta_k^2$. Thus, $V(\epsilon) \le S(\epsilon)$. (d) First note that $\sum_\ell 2^{-\ell}\theta_{k,\ell} \le L\int_0^\infty f_k(\epsilon)d\epsilon$. Denoting by $\bar{V}(\epsilon)$ the quantity corresponding to $V(\epsilon)$ for the family $f_{k,\ell}$, it then suffices to prove that $V(\epsilon) \le L\bar{V}(\epsilon)$. Consider a family $\epsilon_{k,\ell}$ of numbers with $\sum_{k,\ell}\epsilon_{k,\ell}^2 \le \epsilon^2$. For $k \le N$, define $\bar{\epsilon}_k$ by $\bar{\epsilon}_k^2 = \sum_\ell \epsilon_{k,\ell}^2$ so that $\sum_k \bar{\epsilon}_k^2 \le \epsilon^2$. It suffices to prove that $f_k(\bar{\epsilon}_k)^2 \le L\sum_\ell f_{k,\ell}(\epsilon_{k,\ell})^2$. Consider $\bar{\ell} = \inf\{\ell;\ f_{k,\ell}(\epsilon_{k,\ell}) \ne 0\}$ so that $f_{k,\bar{\ell}}(\epsilon_{k,\ell}) = 2^{-\bar{\ell}}$, and it suffices to prove that $f_k(\bar{\epsilon}_k^2) \le L2^{-\bar{\ell}}$. But since $\bar{\epsilon}_k \ge \epsilon_{k,\ell}$ for each ℓ, for $\ell < \bar{\ell}$, we have $f_{k,\ell}(\epsilon_k) = 0$. Thus, $f_k(\epsilon_k) \le \sum_{\ell\ge\bar{\ell}} f_{k,\ell}(\epsilon_k) \le \sum_{\ell\ge\bar{\ell}} 2^{-\ell} \le L2^{-\bar{\ell}}$.

Exercise 2.13.3 In $\mathbb{R}^m$, the sum $\sum_{n\ge} 2^{n/2}e_n(\mathcal{E})$ has to be replaced by $\sum_{n\le n_0}$ where n_0 is the smallest integers with $2^{n_0} \ge m$ (because as shown in Exercise 2.5.9 (e) $e_n(\mathcal{E})$ decreases very fast for larger values of n), so n_0 is of order $\log(m+1)$. The result is then a consequence of the Cauchy-Schwarz inequality, $\sum_{0\le n\le n_0} b_n \le \sqrt{n_0+1}(\sum_{0\le n_0} b_n^2)^{1/2}$ for $b_n = 2^{n/2}a_n$. This estimate is optimal in the case where b_n is independent of n. Thus, for ellipsoids, Dudley's entropy integral is off by a factor at most $\sqrt{\log m}$ for ellipsoids, compared with a factor $\log m$ for general sets.

Exercise 2.15.2 For each $n \ge 0$ can find a set T_n such that card $T_n \le N_n$ and $d(t,T_n) \le 2e_n(T)$ for each t. Then (2.172) holds for $B = 2\sum_{n\ge m} 2^{n/2}e_n(T)$. Given $\delta > 0$, consider the smallest m with $e_m(T) \le \delta$. Then, as we have already seen, $\sum_{n\ge m} 2^{n/2}e_n(T) \le LI(\delta)$ where $I(\delta) = \int_0^\delta \sqrt{\log N(T,d,\epsilon)}d\epsilon$. On the other hand, since $e_{m-1}(T) \ge \delta$ for $\epsilon \le \delta$, we have $\sqrt{\log N(T,d,\epsilon)} \ge 2^{(m-1)/2}$ so that $2^{m/2}\delta \le LI(\delta)$. It then follows from (2.173) that with probability $\ge 1-\exp(-u^2 2^m)$, we have $\sup_{d(s,t)\le\delta}|X_s - X_t| \le LuI(\delta)$, and in particular that $\mathsf{E}\sup_{d(s,t)\le\delta}|X_s - X_t| \le LI(\delta)$.

Exercise 3.1.3 We write $\int_0^\Delta f(\epsilon)\mathrm{d}\epsilon = \sum_{n\geq 0}\int_{\epsilon_{n+1}}^{\epsilon_n} f(\epsilon)\mathrm{d}\mu(\epsilon)$. For $n \geq 1$, we have $2^n < f(\epsilon) \leq 2^{n+1}$ for $\epsilon_{n+1} < \epsilon < \epsilon_n$ so that $2^n(\epsilon_n - \epsilon_{n+1}) \leq \int_{\epsilon_{n+1}}^{\epsilon_n} f(\epsilon)\mathrm{d}\mu(\epsilon) \leq 2^{n+1}(\epsilon_n - \epsilon_{n+1}) \leq 2^{n+1}\epsilon_n$. Also, $f(\epsilon) \leq 2$ for $\epsilon > \epsilon_1$ so that the previous upper bound remains true for $n = 0$ and $\int_0^\Delta f(\epsilon)\mathrm{d}\epsilon \leq 2\sum_{n\geq 0} 2^n\epsilon_n$. For the lower bound, we observe that $\sum_{n\geq 1} 2^n\epsilon_{n+1} = \sum_{n\geq 2} 2^{n-1}\epsilon_n$ so that $\sum_{n\geq 1} 2^n(\epsilon_n - \epsilon_{n+1}) \geq (1/2)\sum_{n\geq 1} 2^n\epsilon_n$.

Exercise 3.1.6 (a) It is obvious that if $\epsilon^2 = \sum_{k\leq n}\epsilon_k^2$, then $B_d(t,\epsilon) \supset \prod_{k\leq N} B_{d_k}(t_k,\epsilon_k)$ so that $\mu(B_d(t,\epsilon)) \geq \prod_{k\leq N}\mu_k(B_{d_k}(t,\epsilon_k))$, and consequently, $\log(1/\mu(B_d(t,\epsilon))) \leq \sum_{k\leq N}\log(1/\mu_k(B_{d_k}(t_k,\epsilon_k)))$ so that the desired result is a consequence of (2.147) used for the functions $f_k(\epsilon) = \sqrt{\log(1/\mu_k(B_k(t_k,\epsilon)))}$. (b) should be obvious from (a) and the equivalence of (3.20) and $\gamma_2(T,d)$ for any metric space.

Exercise 3.4.2 Just by considering the term $n = 1$ of the sum, it is obvious that $\chi_2(T,d) \geq \Delta(T)$, so it suffices to prove the following growth condition. Consider $n \geq 1$, $m = N_n$ a number $a > 0$, points $t_\ell \in T$ with $d(t_\ell, t_{\ell'}) \geq 6a$ for $\ell \neq \ell'$ and sets $H_\ell \subset B(t_\ell, a)$. Then $\chi_2(\cup_{\ell\leq m} H_\ell, d) \geq a2^{n/2}/L + \min_\ell \chi_2(H_\ell, d)$. Given $\epsilon > 0$, consider for $\ell \leq m$ a probability measure μ_ℓ on H_ℓ such that for each admissible sequence $(\mathcal{B}_k)$ of partitions of H_ℓ we have $\int_{H_\ell}\sum_{k\geq 0} 2^{k/2}\Delta(B_k(t))\mathrm{d}\mu_\ell(t) \geq \chi_2(H_\ell,d) - \epsilon$. Consider the probability measure $\mu = m^{-1}\sum_{\ell\leq m}\mu_\ell$. Our goal is to prove that for any admissible sequence $(\mathcal{A}_k)$ of partitions of $H = \cup_{\ell\leq m} H_\ell$, we have

$$\int_T \sum_{k\geq 0} 2^{k/2}\Delta(A_k(t))\mathrm{d}\mu(t) \geq a2^{n/2}/L + \min_{\ell\leq m}\chi_2(H_\ell,d) - \epsilon\ . \tag{F.1}$$

For $t \in H$, let $\ell(t)$ be the unique integer $\ell \leq m$ such that $t \in H_{\ell(t)}$. For $t \in H$, let us define $f(t) = \Delta(A_{n-1}(t)) - \Delta(A_{n-1}(t) \cap H_{\ell(t)})$ so that

$$\int_T \sum_{k\geq 0} 2^{k/2}\Delta(A_k(t))\mathrm{d}\mu(t) \geq 2^{(n-1)/2}\int_T f(t)\mathrm{d}\mu(t) + m^{-1}\sum_{\ell\leq m} U_\ell$$

where $U_\ell = \int_T \sum_{k\geq 0} 2^{k/2}\Delta(A_k(t) \cap H_\ell)\mathrm{d}\mu_\ell(t) \geq \chi_2(H_\ell,d) - \epsilon$ because if $\mathcal{A}_k^\ell = \{A \cap H_\ell;\ A \in \mathcal{A}_k\}$, then the sequence $(\mathcal{A}_k^\ell)$ is an admissible sequence of partitions of H_ℓ. Thus, it suffices to prove that $\int_T f(t)\mathrm{d}\mu(t) \geq a/L$. Let $B = \cup\{A \in \mathcal{A}_{n-1};\ \exists \ell \leq m,\ A \subset H_\ell\}$. Then $\mu(B) \leq N_{n-1}/m = N_{n-1}/N_n \leq 1/2$ because B is the union of $\leq N_{n-1}$ sets each of measure $\leq 1/m$. Now for $t \notin B$, the set $A_n(t)$ meets at least two different sets H_ℓ so that its diameter is $\geq 3a$ while $A_n(t) \cap H_{\ell(t)}$ had diameter $\leq \Delta(H_{\ell(t)}) \leq 2a$ so that $f(t) \geq a$.

Exercise 4.1.5 (a) Find a partition $\mathcal{B}_n$ with $\operatorname{card}\mathcal{B}_n \leq N_n$ and $\Delta(B) \leq 3e_n(T)$ for $B \in \mathcal{B}_n$. Define $\mathcal{A}_0 = \{T\}$ and $\mathcal{A}_n$ as the partition generated by $\mathcal{A}_{n-1}$ and $\mathcal{B}_{n-1}$. Thus, the sequence $(\mathcal{A}_n)$ is admissible, and for $A \in \mathcal{A}_n$, we have $\Delta(A) \leq \Delta(T)$ for $n = 0$ and $\Delta(A) \leq e_{n-1}(T)$ for $n \geq 1$. Writing $e_{-1}(T) = \Delta(T)$ for each t, we have $\sum_{n\geq 0}(2^{n/\alpha}\Delta(A_n(t)))^p \leq \sum_{n\geq 0}(2^{n/\alpha}e_{n-1}(T))^p \leq K\sum_{n\geq 0}(2^{n/\alpha}e_n(T))^p$. It is then straightforward that this sequence of partitions witnesses (4.9). (b) The key idea is that entropy bounds "are optimal when the space is homogeneous" so one should try to see what happens on such spaces. A good class of spaces are those of the form $T = \prod_{i\leq m} U_i$ where U_i is a finite set of cardinality N_i. Given a decreasing sequence $b_i \geq 0$, we define a distance on T as follows. For $t = (t_i)_{i\leq m}$, then $d(s,t) = b_j$ where $j = \min\{i \leq m; s_i \neq t_i\}$, and then one has $b_n \leq e_n(T) \leq b_{n-1}$ from which one may estimate the right-hand side of (4.9). To get lower bounds on $\gamma_{\alpha,p}(T,d)$, an efficient method is to use the uniform probability μ on T and the appropriate version of (3.32).

Exercise 4.1.9 Even though the space T shares with the unit interval the property that $N(T,d,\epsilon)$ is about $1/\epsilon$, there are far more 1-Lipschitz functions on this space than on $[0,1]$. The claim (a) is obvious from the hint. To prove (b), for $\ell \geq 0$, let us denote by $\mathcal{B}_\ell$ the partition of T into the 2^ℓ sets of T determined by fixing the values of $t_1,\ldots,t_\ell$. These sets are exactly the balls of radius $2^{-\ell}$. Consider the class $\mathcal{F}_\ell$ of functions f_ℓ on T such that $|f_\ell(t)| = 2^{-\ell-1}$ for each t and that f_ℓ is constant on each set of $\mathcal{B}_\ell$. Then each sum $f = \sum_{\ell\geq 0} f_\ell$ where $f_\ell \in \mathcal{F}_\ell$ is 1-Lipschitz. Denoting by $\mathcal{F}$ the class of all such functions f, it is already true that $\gamma_{1,2}(\mathcal{F},d_2) \geq \sqrt{k}/L$. It is not so difficult to prove that $e_n(\mathcal{F},d_2) \geq 2^{-n}/L$, and to bound below $\gamma_{1,2}(\mathcal{F},d_2)$ itself, again the appropriate version of (3.32) is recommended, or an explicit construction of a tree as in Sect. 3.2.

Exercise 4.2.2 Try the functional

$$F(A) = 1 - \inf\{\|u\| \; ; \; u \in A\} \tag{F.2}$$

for $A \subset T$, and the growth condition

$$F(\cup_{\ell\leq m} H_\ell) \geq c^* a^p 2^{np/\alpha} + \min_{\ell\leq m} F(H_\ell) \,,$$

when the sets H_ℓ are (a,r)-separated as in Definition 2.8.1.

Exercise 4.3.1 In dimension d, an hypercube of side 2^{-k} has an excess (or deficit) of points of about $\sqrt{N2^{-dk}}$, and you (heuristically) expect to have to move these points about 2^{-k} to match them. Summing over the hypercubes of that side gives a contribution of $\sqrt{N}2^{k(d/2-1)}$. For $d = 3$, it is the large values of k which matter (i.e., the small cubes), while for $d = 1$, it is the small values of k.

Exercise 4.3.4 For an optimal matching, we have $\sum_i d(X_i, Y_{\pi(i)}) = \sum_i f(X_i) - f(Y_{\pi(i)})$, and since $f(X_i) - f(Y_{\pi(i)}) \leq d(X_i, Y_{\pi(i)})$ for each i, we have $f(X_i) - f(Y_{\pi(i)}) = d(X_i, Y_{\pi(i)})$.

Exercise 4.5.2 For $k \leq n_0$ denote by M_k the number of points X_i which have not been matched after we perform k-th step of the process. Thus, there are $M_{k-1} - M_k$ points which are matched in the k-step, and this are matched to points within distance $L2^{k-n_0}$. Thus, the cost of the matching is $\leq L\sum_{k\leq n_0} 2^{k-n_0}(M_{k-1}-M_k) \leq L\sum_{k\leq n_0} 2^{k-n_0}M_k$. At the k-th step of the process, we match points in a given square A of $\mathcal{H}_{n_0-k+1}$. It should be obvious that the number of points of that square A which are not matched after this step is completed equal the excess of the number of X_i over the number of Y_i in that square, that is, $\max(\mathrm{card}\{i \le N, X_i \in A\} - \mathrm{card}\{i \le N; Y_i \in A\}, 0)$, so that

$$M_k = \sum_{A \in \mathcal{H}_{n_0-k+1}} \max\left(\mathrm{card}\{i \le N, X_i \in A\} - \mathrm{card}\{i \le N; Y_i \in A\}, 0)\right) .$$

Let $a = 2^{-n_0+k-1}$ the length of the side of A so that the area of A is a^2, and the expected number of points X_i in A is Na^2. Skipping some details about the Y_i, the excess number of points X_i in A is then of expected value $\le L\sqrt{Na^2} = La\sqrt{N}$. As there are $1/a^2$ elements of $\mathcal{H}_{n_0-k+1}$, the expected value of aM_k is then $\le L\sqrt{N}$, and since in the summation $\sum_{k \le n_0}$ there are about $\log N$ terms, the result follows.

Exercise 4.5.8 (a) Let us denote by $\mathcal{C}$ the family of 2^{2n} little squares into which we divide $[0, 1]^2$. For $\epsilon = (\epsilon_C)_{C \in \mathcal{C}} \in \{-1, 1\}^{\mathcal{C}}$, consider the function f_ϵ such that for $x \in C \in \mathcal{C}$, we have $f_\epsilon(x) = \epsilon_c d(x, B)$ where B is the boundary of C. It should be pretty obvious that $\|f_\epsilon - f_{\epsilon'}\|_2^2 \ge 2^{-4n}\,\mathrm{card}\{C \in \mathcal{C}; \epsilon_C \ne \epsilon'_C\}$. Then Exercise 2.5.11 (a) used for $m = 2^{2n}$, and $k = m/4$ proves that $N(\widehat{\mathcal{C}}, d_2, 2^{-n}/L) \ge 2^{2n}/L$.

Exercise 4.5.14 (a) Making the change of variable $u = t^p$ in (2.6), we obtain

$$\mathsf{E}Y^p = \int_0^\infty pt^{p-1}\mathsf{P}(Y \ge t)\mathrm{d}t \le 2\int_0^\infty pt^{p-1}(\exp(-t^2/A^2) + \exp(-t/B))\mathrm{d}t .$$

One may calculate these integrals by integration by parts, but it is simpler to use bounds such as $u^{p-1}\exp(-u^2) \le Lp^{(p-1)/2}\exp(-u^2/2)$ for $u = t/A$ for the first term and $u^{p-1}\exp(-u) \le (Lp)^{p-1}\exp(-u/2)$ for $u = t/B$ for the second term. (b) We denote by $\Delta_j(A)$ the diameter of the set A for d_j. We consider an admissible sequence $(\mathcal{B}_n)_{n \ge 0}$ which satisfies 4.52 and an admissible sequence $(\mathcal{C}_n)_{n \ge 0}$ which satisfies (4.53). We define the admissible sequence $(\mathcal{A}_n)$ as in the proof of Theorem 4.5.13. It then follows from (a) that $D_n(A_n(t)) \le L(2^{n/2}\Delta_2(C_{n-1}(t)) + 2^n\Delta_1(B_{n-1}(t)))$ and (4.52) and (4.53) and summation imply (4.54).

Exercise 4.5.17 Of course, (4.56) holds in any probability space, not only on $[0, 1]^2$! Then $\gamma_1(\mathcal{F}, d_\infty) \le L\log N$ by Exercise 2.7.5. Also, $e_n(\mathcal{F}, d_2) \le N_n^{-1/2}$ so that $\sum_{n \ge 0} 2^{n/2}e_n(\mathcal{F}, d_2) < \infty$. Then (4.56) shows that $\mathsf{E}S \le L\sqrt{N}$ where $S = \sup|\mathrm{card}\{i \le N\ ;\ X_i \le t\} - Nt|$ for a supremum taken over all t of the type k/M. Finally, $\sup_{0 \le t \le 1}|\mathrm{card}\{i \le N\ ;\ X_i \le t\} - Nt| \le 3S + 1/N$, using the fact that an interval of length $\le 1/N$ contains at most $2S + 1/N$ points X_i.

Exercise 4.5.20 Using Dudley's bound in the form (2.38), in the right-hand side, there are about $\log N$ terms of order 1.

Exercise 4.5.22 Denote by m the largest integer such that $2^{-m} \ge 1/N$. We find a subset T of $\widehat{\mathcal{C}}$ such that $\mathrm{card}\,T \le N_m$ and such that for $f \in \widehat{\mathcal{C}}$, we have $d_\infty(f, T) \le L2^{-m/3} \le LN^{-1/3}$. We then have $e_n(T, d_\infty) \le L2^{-n/3}$ for $n \le m$ and $e_n(T, d_\infty) = 0$ for $n > m$. Thus, $\gamma_1(T, d_\infty) \le \sum_{n \ge 0} 2^n e_n(T, d_\infty) \le \sum_{n \le m} L2^{2n/3} \le L2^{2m/3} \le LN^{2/3}$ and $\gamma_2(T, d_2) \le$

$\gamma_2(T, d_\infty) \le \sum_{n\le m} 2^{n/2} e_n(T, d_\infty) \le L\sum_{n\le m} 2^{n/6} \le LN^{1/6}$ and then $\mathsf{E}\sup_{f\in T} |\sum_{i\le N}(f(X_i) - \int f d\mu)| \le LN^{2/3}$ by (4.56).

Exercise 4.5.23 Given $n \ge 1$, consider the set $\mathcal{C}$ of 1-Lipschitz functions on T which are 0 at the point $(0, 0, \ldots,)$ and which depend only on the first n coordinates. The estimate (4.59) used for $\alpha = 2$ shows that $\log N(\mathcal{C}, d_2, \epsilon) \le \log N(\mathcal{C}, d_\infty, \epsilon) \le L/\epsilon^2$, and thus, $\gamma_1(\mathcal{C}, d_\infty) \le L2^{n/2}$ and $\gamma_2(\mathcal{C}, d_2) \le Ln$. We may then appeal to Theorem 4.5.16 to obtain that $\mathsf{E}\sup_{f\in\mathcal{C}} |\sum_{i\le N}(f(X_i) - \int f d\mu)| \le L(n\sqrt{N} + 2^{n/2})$ and the result when n is chosen with $2^n \simeq N$.

Exercise 4.6.8 Consider the partition $\mathcal{B}_\ell$ of T obtained by fixing the values of $t_1, \ldots, t_\ell$ so that $\operatorname{card}\mathcal{B}_\ell = 2^\ell$, and each set B in $\mathcal{B}_{\ell-1}$ contains two sets in $\mathcal{B}_\ell$. Consider the class $\mathcal{F}_\ell$ of functions f on T such that $|f(t)| = 2^{-\ell/2}$ for each t and f is constant on each set $B \in \mathcal{B}_\ell$. Thus, as in the case of Exercise 4.1.9, the functions of the type $\sum_{\ell\ge 0} f_\ell$ are 1-Lipschitz. Assuming for simplicity $N = N_n$ for a certain n to obtain an evenly spread family (Y_i), we simply put one point Y_i in each set of $\mathcal{B}_{2^n}$. For $\ell < 2^n$, we construct a function $f_\ell \in \mathcal{F}_\ell$ as follows. Recalling that each set B in $\mathcal{B}_{\ell-1}$ contains two sets in $\mathcal{B}_\ell$, the function f_ℓ equals $2^{-\ell/2}$ on one of these sets and $-2^{-\ell/2}$ on the other. Subject to this rule, we choose f_ℓ so that $\sum_{i\le N} f_\ell(X_i)$ is as large as possible. In each set, $B \in \mathcal{B}_{\ell-1}$ are about $2^{-\ell+1}N$ points X_i, which typically gives rise to fluctuations of order $\sqrt{2^{-\ell}N}$ between the two halves of B, and thus we expect $\sum_{X_i\in B} f_\ell(X_i)$ to be of order $2^{-\ell/2}\sqrt{2^{-\ell}N} = 2^{-\ell}\sqrt{N}$ and $\sum_{i\le N} f_\ell(X_i)$ to be of order $\sqrt{N}$. The function $f = \sum_{\ell<2^n} f_\ell$ is a 1-Lipschitz function with $\sum_{i\le N} f(X_i)$ of order $2^n\sqrt{N}$ (which is about $\log N\sqrt{N}$) whereas $\sum_{i\le N} f(Y_i) = 0$.

Exercise 4.7.4 See Exercise 2.3.3(c).

Exercise 4.9.4 (a) Integration by parts as in the case of AKT. (b) $(a_{p,C})_{p\in\mathbb{Z}}$ are the Fourier coefficients of the function $f_C(x) = \int h_C(t) f(x, t) d\mu_m(t)$ so that $\sum_p |a_{p,C}|^2 = \int f_C(x)^2 dx$. Now, fixing a point t_0 in C, we have $f_C(x) = \int h_C(t) f(x, t) d\mu_m(t) = \int h_C(t)(f(x, t) - f(x, t_0)) d\mu_m(t)$. We have $|f(x, t) - f(x, t_0)| \le 2^{-n}$ because f is 1-Lipschitz in t. Also, $|h_C(t)| \le 2^{n/2}$, and the integration is restricted to C so that $|f_C(x)| \le 2^{-n} \times 2^{n/2} \times 2^{-n} = 2^{-3n/2}$ and $\int f_C(x)^2 dx \le 2^{-3n}$. The case $C = \emptyset$ is easier.

Exercise 4.9.5 (a) for each $C \in \mathcal{C}_n$, we have $\sum_p |a_{p,C}|^2 \le 2^{-3n}$. Summing over $C \in \mathcal{C}_n$ and then over $n \le m$ yields $\sum_{p\in\mathbb{Z}} \sum_{0\le n\le m} \sum_{C\in\mathcal{C}_n} 2^{2n} |a_{p,C}|^2 \le m$ (recalling that $\operatorname{card}\mathcal{C}_n = 2^n$), and the desired result follows when combining with (4.129). (b) The ellipsoid $\mathcal{E}$ given by $\sum_{p,C} \alpha_{p,C}^2 |a_{p,C}|^2 \le 1$ satisfies $\gamma_2(\mathcal{E}) \le L\sqrt{\sum_{p,C} \alpha_{p,C}^{-2}}$. We compute $\sum_{p,C} \alpha_{p,C}^{-2}$ by distinguishing three sets of indices and recalling that $\alpha_{p,C}^{-2} \le L\min(p^{-2}, m2^{-2n})$. First, the set I_0 consisting of the $(p, \emptyset)$. Then the set I_1 of indices (p, C) with $C \in \mathcal{C}_n$, $0 \le n \le m$, and $|p| \le 2^n/\sqrt{m}$, in which case $\alpha_{p,C}^{-2} \le Lm2^{-2n}$. Finally, the set I_2 of indices (p, C) with $C \in \mathcal{C}_n$, $0 \le n \le m$ for which $|p| \ge 2^n/\sqrt{m}$, in which case $\alpha_{p,C}^{-2} \le Lp^{-2}$. Then $\sum_{I_0} \alpha_{p,C}^{-2} \le L$ and

$$\sum_{I_2} p^{-2} = \sum_{n \le m} \sum_{C \in \mathcal{C}_n} \sum_{|p| \ge 2^n/\sqrt{m}} p^{-2}$$

$$\le L \sum_{n \le m} \sum_{C \in \mathcal{C}_n} \sqrt{m}/2^n \le L \sum_{n \le m} \sqrt{m} \le Lm^{3/2},$$

and we proceed in a similar way to prove that the other sum is also $\le Lm^{3/2}$.

Exercise 4.9.6 Since this section is for the experts, this solution will be suitably concise. A function on U is a function of $(x, t) \in [0, 1] \times T$. A first observation is that a function $\sum_{k \le q} f_k$ is always 1/2-Lipschitz in the t variable. Problems arise in the x variable, and the way this is solved is to set $z_{q+1,\ell,C} = 0$ whenever the little "square" $[(\ell - 1)2^{-(q+1)}, \ell 2^{-(q+1)}[\times C$ is "dangerous", that is, the function $f = \sum_{k \le q} f_k$ is such that at one point of this "square", we have $|\partial f/\partial x| \ge 1/2$. Matters are set up in a way that at most, 1/2 of the little squares are dangerous, exactly by the same type of arguments as we used for the AKT theorem. Next, defining $D_{\ell,C} = \sum_{i \le N}(f_{q+1,\ell,C}(X_i) - \int f_{q+1,\ell,C} d\theta)$, the point is that $|D_{\ell,C}|$ is often of order $\sqrt{N}2^{-q-p}\sqrt{2^{-2q-p}}$ because if $h = f_{q+1,\ell,C} - \int f_{q+1,\ell,C} d\theta$, then $(\int h^2 d\theta)^{1/2}$ is about 2^{-q-p} (the typical order of the value of h on the support of $f_{q+1,\ell,C}$) times $\sqrt{2^{-2q-p}}$ (the square root of the measure of this support). Now, $\sqrt{N}2^{-q-p}\sqrt{2^{-2q-p}} = \sqrt{N}2^{-2q-3p/2}$, and since there are 2^{2q+p} terms to be summed, this gives a total contribution of order $\sqrt{N}2^{-p/2} = \sqrt{N}/r^{1/4}$ as desired.

Exercise 4.9.8 The magic relation is $(N_n)^{2^k} = 2^{2^{n+k}} = N_{n+k}$. By hypothesis for $n \ge 0$, there is a partition $\mathcal{A}_n$ of T such that $\operatorname{card} \mathcal{A}_n \le N_n$ and each set of $\mathcal{A}_n$ has diameter $L2^{-n}$. Consider a subset B_n of T_m such that each element of T_m is within distance 2^{-n} of B_n. We can take $\operatorname{card} B_n = 2^n$ for $n \le m$ and $\operatorname{card} B_n = 2^m$ for $n \ge m$. We classify the elements f of U by looking to which set of $\mathcal{A}_n$ the value of each $s \in B_n$ belongs. In this manner, we break U into $(\operatorname{card} \mathcal{A}_n)^{\operatorname{card} B_n} = N_{2n}$ sets (or N_{n+m} sets for $n \ge m$). Since f is assumed to be 1-Lipschitz, the diameter of each such set in U is $\le L2^{-n}$. This implies that $e_{2n}(U, D) \le L2^{-n}$ for $n \le m$ and $e_{n+m} \le L2^{-n}$ for $n \ge m$, from which the last assertion readily follows.

Exercise 5.2.2 (a) is a consequence of (5.2). (b) In that case, $\sup_{t \in T} X_t = \sum_{i \le N} |Y_i|$ has expectation $N\mathsf{E}|Y_1|$. To show that $\gamma_q(T, d)$ is of the same order, we use the bound $\gamma_q(T, d) \ge 2^{n/q} e_n(T)$. According to Exercise 2.5.10, for $2^n = N/L$, $e_n(T)$ is about the diameter $N^{1/p}$ of T. (c) The metric space (T, d) consists of N points within distance at most two of each other, and the left-hand side of (5.7) is of order $(\log N)^{1/q}$. However, $\mathsf{E} \sup_{t \in T} X_t = \mathsf{E} \max_{i \le N} Y_i$ is $\ge N^{1/p}/K$. This is because from (5.3) for each $i \le N$, there exists a set Ω_i of probability $1/N$ on which $Y_i \ge N^{1/p}/K$, where K depends on p only, so that since the sets Ω_i are independent, $\max_{i \le N} Y_i$ is at least $N^{1/p}/K$ on a set of probability at least

$1-(1-1/N)^N \geq 1/L$. Thus, $\mathsf{E}\max_{i\leq N} Y_i^+ \geq N^{1/p}/K$, and it is then a simple matter to conclude.

Exercise 5.5.1 This is just as simple as it sounds. Considering an increasing sequence $(\mathcal{B}_n)$ of partitions with card $\mathcal{B}_n \leq M_n$, the sequence $(\mathcal{A}_m)$ of partitions such that $\mathcal{A}_m = \mathcal{B}_n$ for $2^n \leq m < 2^{n+1}$ satisfies card $\mathcal{A}_m \leq M_n = N_{2^n} \leq N_m$ and $\sum_{m\geq 0} \Delta(A_m(t)) \leq L\sum_{n\geq 0} 2^n \Delta(B_n(t))$. Conversely, given an admissible sequence of partitions $(\mathcal{A}_m)$, the sequence $\mathcal{B}_n$ given by $\mathcal{B}_m = \mathcal{A}_{2^m}$ satisfies $\sum_{n\geq 0} 2^n \Delta(B_n(t)) \leq \sum_{m\geq 0} \Delta(A_m(t))$.

Exercise 6.2.1 (a)

$$\mathsf{E}\Phi(X) = \mathsf{E}\sup_{f\in\mathcal{C}} f(X) \geq \sup_{f\in\mathcal{C}} \mathsf{E}f(X) = \sup_{f\in\mathcal{C}} f(\mathsf{E}X) = \Phi(\mathsf{E}X) .$$

(b) Use Jensen's inequality for the function $\Phi(y) = |x+y|^p$. (c) Use Jensen's inequality conditionally on θ and X.

Exercise 6.3.5 Write $Y^2 = Y^{1/2}Y^{3/2}$, and use the Cauchy-Schwarz inequality. Take $Y = |\sum_{i\geq 1} t_i\varepsilon_i|$, and use that $\mathsf{E}Y^3 \leq L\|t\|_2^3$ by Khintchin's inequality (6.3).

Exercise 6.4.2 If $t_{\ell,i} \in \{0,b\}$, then $\sum_i |t_{\ell,i}| \leq a^2/b$ so that the left-hand side of (6.21) is bounded by a^2/b.

Exercise 6.6.3 An instructive example is given by $T = \{t_1,\dots,t_N\}$ where $t_n = (t_{n,i})_{i\geq 1}$, $t_{n,i} = 0$ if $i \neq n$ and $t_{n,n} = 1$. Then $\gamma_2(T)$ is about $\sqrt{\log N}$ and $\gamma_1(T,d_\infty)$ is about $\log N$.

Exercise 7.1.5 Given $t \in T$, the sets B and $t+B$ are not disjoint because $\mu(B\cap(B+t)) = 2\mu(B) - \mu(B\cup(B+t)) \geq 2\mu(B) - 1 > 0$, and if $s\in B$ and $s \in t+B$, that is, $s = t+u$ with $u\in B$, then $t = s-u \in B-B$.

Exercise 7.1.8 This follows from Theorem 7.1.1 since

$$\inf\left\{\epsilon > 0\,;\; \mu(B_d(0,\epsilon)) \geq 2^{-2^n} = N_n^{-1}\right\} \leq \epsilon_n .$$

Exercise 7.3.2 (a) Given ω, the set $A_\omega = \{t \in T; d_\omega(0,t) \leq \Delta(T,d_\omega)/4\}$ satisfies $\mu(A_\omega) \leq 1/2$. Indeed, if this is not the case, then for each $s \in T$, we have $(s+A_\omega)\cap A_\omega \neq \emptyset$, that is, $T = A_\omega - A_\omega$. Then given $a,b\in T$, we have $a-b = s-t$ for $s,t\in A_\omega$ and

$$d_\omega(a,b) = d_\omega(a-b,0) = d_\omega(s-t,0) = d_\omega(s,t) \leq d_\omega(s,0) + d_\omega(0,t) .$$

Since $d_\omega(s,0) \leq \Delta(T,d_\omega)/4$ and $d_\omega(t,0) \leq \Delta(T,d_\omega)/4$, by the definition of A_ω, we obtain $d_\omega(a,b) \leq \Delta(T,d_\omega)/2$ which is absurd. (b) Setting $B_\omega = \{s; d_\omega(0,s) \geq \Delta(T,d_\omega)/4\}$, then $\mu(B_\omega) \geq 1/2$. Consequently

$$\int \mathsf{E}\Delta(T,d_\omega)\mathbf{1}_{\{s\in B_\omega\}}\mathrm{d}\mu(s) \geq (1/2)\mathsf{E}\Delta(T,d_\omega)$$

and thus there exists $s \in T$ such that $\mathsf{E}\Delta(T, d_\omega)\mathbf{1}_{\{s\in B_\omega\}} \ge (1/2)\mathsf{E}\Delta(T, d_\omega)$. Now

$$\mathsf{E}\Delta(T, d_\omega)\mathbf{1}_{\{s\in B_\omega\}} \le L\mathsf{E}d_\omega(s, 0) = L\bar{d}(s, 0) \le L\Delta(T, \bar{d}) .$$

Thus $\mathsf{E}\Delta(T, d_\omega) \le L\Delta(T, d) \le L\gamma_2(T, d)$.

Exercise 7.3.3 Choose, for example, $T = 1, \ldots, N$. For a subset I of T, define the distance d_I on T by $d_I(s, t) = 0$ if $s, t \notin I$ and $d_I(s, t) = 1$ otherwise ($s \ne t$). If $\operatorname{card} I = k$, then $\gamma_2(T, d_I)$ is about $\sqrt{\log k}$. On the other hand, the distance d on T such that d^2 is the average of the distances d_I^2 over all possible choices of I with $\operatorname{card} I = k$ satisfies $d(s,t)^2 = \binom{N-2}{k-2}/\binom{N}{k} = k(k-1)/N(N-1) \le k/N$ for $s, t \in T$ so that $\gamma_2(T, d) \le L\sqrt{k/N}\sqrt{\log N}$.

Exercise 7.3.8 Observe that $\mathsf{E}\|\xi_{i_0}\chi_{i_0}\| = \mathsf{E}|\xi_{i_0}|$. Use Jensen's inequality for the lower bound and the triangle inequality $\mathsf{E}\|\sum_i \xi_i\chi_i\| \le \mathsf{E}\|\xi_{i_0}\chi_{i_0}\| + \mathsf{E}\|\sum_{i\ne i_0}\xi_i\chi_i\|$ and (7.33) for the upper bound.

Exercise 7.3.9 The Haar measure μ is normalized so that $\mu(T) = 1$. If $|t - 1| \le 2\pi 2^{-2^n}$, then $d(1, t) \le \eta_n$ where $\eta_n^2 = L\sum_i c_i \min(1, |i|^2 2^{-2^{n+1}})$ and so that $\mu(B(1, \eta_n)) \ge N_n^{-1}$ and thus $\epsilon_n \le \eta_n$. It remains only to show that $\sum_{n\ge 0} 2^{n/2}\eta_n \le L\sum_{n\ge 0} 2^{n/2}\sqrt{b_n}$. For this, we write $\eta_n^2 \le L\sum_{k<n} b_k N_k^2 N_n^{-2} + L\sum_{k\ge n} b_k$ so that $\sqrt{\eta_n} \le L\sum_{k<n}\sqrt{b_k}N_k N_n^{-1} + L\sum_{k\ge n}\sqrt{b_k}$, from which the result follows.

Exercise 7.3.10 (a) Use that $A^s(u) = \sum_i a_i\chi_i(s)\chi_i(u)$ and that the χ_i are orthonormal. (b) $a_{i_0} = \int A(s)\mathrm{d}\mu(s)$. (c) From (7.36). (d) With obvious notation, the hint implies that $d_{AB}(s,t) \le \|A\|d_B(s,t) + \|B\|d_A(s,t)$ so that (4.55) and Exercise 2.7.4 imply that $\gamma_2(T, d_{AB}) \le L\|B\|\gamma_2(T, d_A) + L\|A\|\gamma_2(T, d_B)$ and the result by (c).

Exercise 7.3.11 (a) Note that $\chi^t = \chi(t)\chi$, so that $U(\chi)(t) = U(\chi)^t(0) = U(\chi^t)(0) = \chi(t)U(\chi)(0) = u_\chi\chi(t)$ where $u_\chi = U(\chi)(0)$. Thus, $U(\chi) = u_\chi\chi$. (b) $X_s - X_t = U(A)^s - U(A)^t = U(A^s) - U(A^t) = U(A^s - A^t)$ so that $\|X_s - X_t\|_{\psi_2} \le \|U\|_{2,\psi_2}\|A^s - A^t\|_2 = \|U\|_{2,\psi_2}d(s,t)$. Then (2.60) states that $\int_T \sup_{s,t\in T}|U(A)^s(u) - U(A)^t(u)|\mathrm{d}\mu(u) \le L\|U\|_{2,\psi_2}\gamma_2(T, d)$, but the integrand is independent of u and equals $\sup_{s,t\in T}|U(A)(s) - U(A)(t)|$. Furthermore, $\gamma_2(T, d) \le L\mathcal{N}(A)$. (c)

$$|U(A)(0) - \int U(A)(t)\mathrm{d}\mu(t)| \le \sup_{s,t\in T}|U(A)(s) - U(A)(t)| .$$

Now, $\int U(A)(t)\mathrm{d}\mu(t) = a_{i_0}u_{i_0}$. We have seen that $|a_{i_0}| \le L\mathcal{N}(A)$ and $|u_{i_0}| = \|U(\mathbf{1})\|_2 \le \|U\|_{2,\psi_2}$ since $\|\mathbf{1}\|_2 = 1$.

Exercise 7.4.1 The arch-typical example is when x_i is the canonical basis of ℓ_n^∞, and the example boils down to $\mathsf{E}\sup_{i\le n}|\varepsilon_i| = 1$ and $\mathsf{E}\sup_{i\le n}|g_i|$ of order $\sqrt{\log n}$, as follows from Exercise 2.3.7.

Exercise 7.4.3 Consider the set T of sequences $(t_i)_{i\le N}$ with $t_i = \pm 1$ and $\operatorname{card}\{i \le N; t_i = -1\} \le \sqrt{N}$. Consider the function χ_i on T given by $\chi_i(t) = t_i$. Then for $t \in T$, we have $|\sum_i \varepsilon_i\chi_i(t) - \sum_i \varepsilon_i| = 2\operatorname{card}\{i \le N; t_i = -1\} \le 2\sqrt{N}$ so that $\mathsf{E}\sup_t|\sum_{i\le N}\varepsilon_i\chi_i(t)| \le 2\sqrt{N}$. On the other hand, $\sum_{i\le N} g_i\chi_i(t) =$

$\sum_{i\le N} g_i - 2\sum_{i;t_i=-1} g_i$ so that $\mathsf{E}\sup_t \sum_{i\le N} g_i\chi_i(t) = \mathsf{E}\sup_I \sum_{i\in I} -g_i$ where the supremum is taken over all the sets I of cardinality $\le \sqrt{N}$. Consider α such that $\mathsf{P}(-g \ge \alpha) = 1/\sqrt{N}$, where g is standard Gaussian. The events $A_i = \{-g_i \le \alpha\}$ are independent, each of probability $1/\sqrt{N}$, so that about $N/\sqrt{N} = \sqrt{N}$ of them occurs. This shows that $\sup_I \sum_{i\in I} -g_i$ is typically of order $\sqrt{N}\alpha$, and α is about $\sqrt{\log N}$, so that $\mathsf{E}\sup_I \sum_{i\in I} -g_i$ is of order $\sqrt{N\log N}$.

Exercise 7.4.4 (a) is an application of Theorem 6.2.8 to the set $T = \{(a_i\chi(t))_{i\ge 1}\}$. We write $a_i\chi(t) = u_i(t) + v_i(t)$ where $\sup_i \sum_i |v_i(t)| \le LS$ and $\mathsf{E}\sum_t |\sum_i u_i(t)g_i| \le LS$. We then have $a_i\chi_i(s+t) = u_i(s+t) + v_i(s+t)$ so that $a_i\chi_i(t) = \chi_i(-s)u_i(s+t) + \chi_i(-s)v_i(s+t)$. Averaging over s for the Haar measure $\mathrm{d}\mu$, we obtain $a_i\chi_i(t) = \bar{u}_i(t) + \bar{v}_i(t)$ where $\bar{u}_i(t) = \int \chi_i(-s)u_i(s+t)\mathrm{d}\mu(s) = \int \chi_i(t-s)u_i(s)\mathrm{d}\mu(s) = \chi_i(t)\bar{u}_i(0)$. Similarly, $\bar{v}_i(t) = \chi_i(t)\bar{v}_i(0)$ so that $a_i = \bar{u}_i(0) + \bar{v}_i(0)$. Moreover, $\sum_i |\bar{v}_i(0)| \le LS$ so that $\mathsf{E}\sup_t |\sum_i g_i\bar{v}_i(0)\chi_i(t)| \le \mathsf{E}\sum_i |g_i||\bar{v}_i(0)| \le LS$. Finally (easily) for each $s \in T$, we have $\mathsf{E}\sup_t |\sum_i g_iu_i(t)\chi_i(-s)| \le LS$ so that $\mathsf{E}\sup_t |\sum_i g_iu_i(t+s)\chi_i(-s)| \le LS$ and by averaging over s, $\mathsf{E}\sup_t |\sum_i g_i\bar{u}_i(t)| \le LS$. Finally, since $a_i = \bar{u}_i(0) + \bar{v}_i(0)$, we have $\mathsf{E}\sup_t |\sum_i g_ia_i(t)| \le LS$.

Exercise 7.4.9 We will do only the easy part. Assume without loss of generality that the sequence $(|a_i|)$ is non-increasing. Then $V_n = \{t \in T; \forall i \le 2^n, t_i = 1\}$ is a neighborhood of the identity 1 of Haar measure $1/N_n$, and for $t \in V_n$, we have

$$d(t,1)^2 \le 4\sum_{i>2^n} |a_i|^2 \le 4\sum_{m\ge n} 2^n|a_{2^n}|^2$$

so that $\epsilon_n \le 2\sum_{m\ge n} 2^{m/2}|a_{2^m}|$ and $\sum_{n\ge 0} 2^n\epsilon_n \le L\sum_{m\ge 0} 2^m|a_{2^m}| \le L\sum_i |a_i|$.

Exercise 7.5.3 Since $\mathsf{E}(|a(Z_i(s)-Z_i(t)|\wedge 1) \le \mathsf{P}(|Z_i| \ne 0)$, we have $\varphi_j(s,t) \le 1$ for all $s,t \in G$, and the claim is obvious.

Exercise 7.5.4 (a) Write the triangle inequality for the distance $\sqrt{\varphi_j}$ and raise to the square, using that $(a+b)^2 \le 2(a^2+b^2)$. (b) It suffices to consider that case of $D-D$. We prove first that for $s \in D-D$, we have $\varphi_j(s,0) \le 2d$. For this, we write $s = a-b$ where $a,b \in D$ so that using (a) and translation invariance $\varphi_j(s,0) = \varphi_j(a-b,0) = \varphi_j(a,b) \le 2(\varphi_j(a,0)+\varphi_j(0,b)) \le 2d$. We then use (a) again to conclude. (c) By (7.67) for $n \ge 1$, the set $D_n = \{s \in T; \varphi_{j_n}(s,0) \le 2^n\}$ satisfies $\mu(D_n) \ge 1/N_n$. According to Lemma 7.1.3, we can cover T by at most N_n translates of $D_n - D_n$. According to Exercise 2.7.6, there exists an admissible sequence of partitions $(\mathcal{A}_n)$ such that for $\mathcal{A}_0 = \{T\}$ and that for $n \ge 1$, each element $A \in \mathcal{A}$ is included in a translate of $D_{n-1} - D_{n-1}$ so that by (b) for $s,t \in A$, we have $\varphi_{j_{n-1}}(s,t) \le 4\cdot 2^{n-1}$.

Exercise 7.5.7 When $\varphi_j(s,t) = \sum_i |r^{2j}a_i(\chi_i(s) - \chi_i(t))|^2 \wedge 1 < 1$, we have $\sum_i |r^{2j}a_i(\chi_i(s)-\chi_i(t))|^2 < 1$. We then integrate in s with respect to μ using (7.34).

Exercise 7.5.12 Using symmetry and independence, we have $\mathsf{E}|\sum_i a_i\theta_i| \le L\mathsf{E}\sqrt{\sum_i a_i^2\theta_i^2}$. Given $u > 0$, let Ω_u the event defined by $|a_i\theta_i| \le u$ for each i. Then $\mathsf{P}(\Omega_u^c) \le \sum_i \mathsf{P}(|\theta_i| \ge u/|a_i|) \le KS/u^p$ where $S = \sum_i |a_i|^p$. The trick is then to compute $\mathsf{E}a_i^2\theta_i^2\mathbf{1}_{\Omega_u}$. Using (2.6) as in (7.88), one obtains that this is $\le Ku^{2-p}|a_i|^p$. Thus, $\mathsf{E}\sum_i a_i^2\theta_i^2\mathbf{1}_{\Omega_u} \le KSu^{2-p}$, and Markov's inequality proves that $\mathsf{P}(\Omega_u \cap \{\sum_i a_i^2\theta_i^2 \ge u^2\}) \le KSu^{-p}$. Finally, we have proved that $\mathsf{P}(\sqrt{\sum_i a_i^2\theta_i^2} \ge u) \le KSu^{-p}$, from which follows that $\mathsf{E}\sqrt{\sum_i a_i^2\theta_i^2} \le KS^{1/p}$. To prove that $S \le K\Delta(T, d_p)^p$, we consider the inequality $\sum_i |a_i|^p|\chi_i(s) - 1|^p \le \Delta(T, d_p)^p$, and we integrate in $s \in T$, using that $\int |\chi_i(s) - 1|^p \mathrm{d}\mu(s) \ge 1/K$ since $\int |\chi_i(s) - 1|^2 \mathrm{d}\mu(s) = 2$ and since $2^{2-p}|x|^p \ge |x|^2$ for $|x| \le 2$.

Exercise 7.6.5 This is a consequence of Lemma 7.6.4, setting $T_\omega = \{t \in T; \omega \in \Xi_t\}$, and using Lemma 7.6.1 to obtain the $\mathsf{E}\mu(T_\omega) \ge c$ since $p(t) = \mathsf{P}(t \in T_\omega) = \mathsf{P}(\Xi_t) \ge c$ for each s.

Exercise 7.8.5 (b). For $s, t \in U, s \ne t$ since $s - t \notin B$, there exists $i \le N$ with $|\chi_i(s - t) - 1| > \alpha/2$ so that $|\chi_i(s) - \chi_i(t)| > \alpha/2$ and $|\chi_i(5s) - \chi_i(5t)| > 2\alpha$ by (a). Now for $u \in A$, we have $|\chi_i(5s + u) - \chi_i(5s)| = |\chi_i(u) - 1| \le \alpha$. Thus, for $u, v \in A$, we have $\chi_i(5s + u) \ne \chi_i(5t + v)$. (c) We can find U as in (b) with $\operatorname{card} U \ge \mu(A)/\mu(B)$ because for the largest possible U, the sets $s + B$ for $s \in U$ cover A. On the other hand, since the sets $s + A$ for $s \in U$ are disjoint, we have $\mu(A)\operatorname{card} U \le 1$. Thus, $\mu(A)^2/\mu(B) \le 1$.

Exercise 7.8.17 This is Fubini's theorem. Informally, we take expectation in the equality $\mu(D_n \backslash B_{n,u}) = \int_{D_n} \mathbf{1}_{\{s \notin B_{n,u}\}} \mathrm{d}\mu(s)$.

Exercise 7.8.20 Instead of integrating over all T, integrate over $D_0 = \{s \in T; \varphi_{j_0}(s, 0) \le 1\}$ and use that $\int_{D_0} |Z_i(s) - Z_i(0)|^1 \ge |Z_i(0)|^2$ (as we have seen several times).

Exercise 7.9.8 So for $s \in D_n$, we have $\sum_{i \in I} |a_i|^2|\chi_i(s) - 1|^2 \le \epsilon_n^2$, and by integration over D_n, we get $\sum_{i \in U_n} |a_i|^2 \le 2\epsilon_n^2$. Assuming without loss of generality that $|a_i| > 0$ for $i \in I$, this shows that $\cap_n U_n = \emptyset$ and consequently that $I = \cup_n I_n$. Using Theorem 7.8.1, this shows also that $\mathsf{E}\|\sum_{i \in I_n} a_i g_i \chi\| \le L2^{n/2}\epsilon_n$.

Exercise 7.9.9 The canonical distance $d(s, t) = \sum_{i \le N} |\chi_i(s) - \chi_i(t)|^2$ satisfies $\int_D d(s, 0)^2 \mathrm{d}\mu(s) = 2N\mu(D) - 2\mathrm{Re}\int_D \sum_{i \le N} \chi_i(s)\mathrm{d}\mu(s)$. By (7.125), we have

$$\int_D \Big|\sum_{i \le N} \chi_i(s)\Big|\mathrm{d}\mu(s) \le LC\sqrt{N}\mu(D)\sqrt{\log(2/\mu(D))}\,,$$

so that if $\log(2/\mu(D)) \le N/(LC^2)$, we have $\int_D d(s, 0)^2 \mathrm{d}\mu(s) \ge N$ and $\sup_D d(s, 0)^2 \ge N$. This prove that $\mu(\{s; d(s, 0) \le \sqrt{N}\}) \le 2\exp(-N/LC^2)$, from which the result follows by (7.4).

Exercise 7.10.2 (a) The series $\sum_{i \ge 1} \mathsf{P}(|W_i| \ge a)$ converges since $a^2\mathsf{P}(|W_i| \ge a) \le \mathsf{E}(W_i^2 \wedge a^2)$, and so does the series $\sum_{i \ge 1} W_i \mathbf{1}_{\{|W_i| > a\}}$ because a.s. it has only finitely many nonzero terms. Thus, it suffices to prove the convergence of the series $\sum_{i \ge 1} W_i \mathbf{1}_{\{|W_i| \le a\}}$, but symmetry and (7.197) imply that this series converges in L^2 (using Cauchy's criterion) and hence in probability. The conclusion then follows from Lemma 7.10.1. (b) If the series $\sum_i W_i$ converges a.s. it has finitely many terms which are ≥ 1 so the series $\sum_i W_i \mathbf{1}_{\{|W_i| \le 1\}}$ also converges, and we may assume

$|W_i| \le 1$ without loss of generality. Consider a finite set I of indices and $X = (\sum_{i\in I} W_i)^2$. Then $\mathsf{E}X = \sum_{i\in I}\mathsf{E}W_i^2$ and $\mathsf{E}X^2 = \sum_{i_1,i_2,i_3,i_4\in I}\mathsf{E}W_{i_1}W_{i_2}W_{i_3}W_{i_4} = 2\sum_{i,j\in I}\mathsf{E}W_i^2\mathsf{E}W_j^2 + \sum_{i\in I}\mathsf{E}W_i^4$, and since $\mathsf{E}W_i^4 \le \mathsf{E}W_i^2$, we obtain that when $\mathsf{E}X \ge 1$, then $\mathsf{E}X^2 \le 3(\mathsf{E}X)^2$ so that by the Paley-Zygmund inequality (6.15), we obtain $\mathsf{P}(X \ge \mathsf{E}X/2) \ge 1/L$, and this should make the result obvious.

Exercise 7.10.3 If $\mathsf{P}(f \ge \delta) \le 1-\delta$, then $\int_\Omega |f|\mathrm{dP} \le \delta$ where $\Omega = \{|f| \le \delta\}$ satisfies $\mathsf{P}(\Omega) \ge 1-\delta$. Conversely, if $\int_\Omega |f|\mathrm{dP} = \alpha$ is small, then by Markov's inequality, $\mathsf{P}(\Omega \cap \{|f| \ge \sqrt{a}\}) \le \sqrt{\alpha}$ so that $\mathsf{P}(|f| \ge \sqrt{a}) \le \sqrt{\alpha} + P(\Omega^c)$. Finally, if $\mathsf{E}|f|^p = \alpha$ is small, then $\mathsf{P}(|f| \ge \alpha^{1/(2p)}) \le \sqrt{\alpha}$.

Exercise 7.12.3 Here in (7.38), one has $c_i = |a_i|^2$, so the condition

$$\sum_{n\ge 0} 2^{n/2}\Big(\sum_{i\ge N_n} |a_i|^2\Big)^{1/2} < \infty$$

is identical to the condition $\sum_{n\ge 0} 2^{n/2} b_n < \infty$.

Exercise 8.2.2 You prove as usual that for $|\lambda| \le 1/2$, we have $\mathsf{E}\exp\lambda Y \le \exp L\lambda^2$ and using (2.6) that for $|\lambda| \ge 1/2$ you have $\mathsf{E}\exp|\lambda Y| \le \exp(K|\lambda|^q)$ where $1/p + 1/q = 1$ so that $\mathsf{E}\exp\lambda Y \le \exp(K\max(\lambda^2, |\lambda|^q))$ for $p < 2$ and $\mathsf{E}\exp\lambda Y \le \exp(K\min(\lambda^2, |\lambda|^q))$ for $p > 2$. The rest is straightforward.

Exercise 8.2.10 We prove first that if $t \in B(u)$, then the sequence $t' = (t_i')_{i\ge 1}$ where $t_i' = t_i\mathbf{1}_{\{|t_i|\le 4\}}$ satisfies $\|t'\| \le 4\sqrt{u}$. Let us assume for contradiction that $S := \sum_{i\ge 1}(t_i')^2 \ge 16u$. Then $a_i := t_i'\sqrt{u/S}$ satisfies $|a_i| \le 1$ so that $\hat{U}(a_i) = a_i^2$ and $\sum_{i\ge 1} a_i^2 = u$. This, by definition of $B(u)$, we have $\sum_{i\ge 1} a_i t_i \le u$. However, this is impossible because $\sum_{i\ge 1} a_i t_i' = \sqrt{Su} \ge 4u$.

Next, consider the sequence $t'' = (t_i'')_{i\ge 1}$ where $t_i'' = t_i\mathbf{1}_{\{|t_i|>4\}}$. Our goal is to prove that $\sum_{i\ge 1}|t_i''|^q \le 2^{2q-1}u$. Assuming that this is not the case, we may by decreasing some t_i if necessary assume that $S := \sum_{i\ge 1}|t_i''|^q = 2^{2q-1}u$. Let $a_i = c|t_i''|^{q/p}$ where $c = (u/(2S))^{1/p} = 4^{-q/4p}$ so that $|a_i| \ge 1$ if $a_i \ne 0$. Thus, $\hat{U}(a_i) \le 2|a_i|^p$ and $\sum_{i\ge 1}\hat{U}(a_i) \le 2\sum_{i\ge 1}|a_i|^p = 2Sc^p = u$. But then, $\sum_{i\ge 1} a_i t_i = c\sum_{i\ge 1}|t_i''|^{q/p+1} = cS = (u/2)^{1/p}S^{1/q} = 2u$.

Exercise 8.3.6 Assume without loss of generality that $N = N_\tau$ for a certain integer τ. For each integer p, consider the set H_p of sequences $t = (t_i)$ with the following properties: Each $0 \le t_i \le 1$ is a multiple of $1/N$, at most p coordinates t_i are not 0, and the corresponding i are $\le N$. Then, since there are at most N^p ways to choose the indices i where $t_i \ne 0$, by trivial bounds, one has $\operatorname{card} H_p \le N^{2p} = 2^{p2^{\tau+1}}$. Let us define $T_n = \{0\}$ if $n \le \tau$ and $T_n = H_{p(n)}$ where $p(n) = 2^{n-\tau-1}$ otherwise. Thus, $\operatorname{card} T_n \le N_n$. Fix $t \in \operatorname{conv} T$, and assume without loss of generality that the sequence (t_i) is non-increasing. Now $\sum_{\tau\le n\le 2\tau+1} t_{p(n)}(p(n) - p(n-1)) \le \sum t_i = 1$ so that $\sum_{\tau\le n\le 2\tau+1} 2^{n-\tau-1}t_{p(n)} \le 2$. Obviously, $d(t, T_n) \le t_{p(n)} + 1/N$ so that we have shown that $\sum_{\tau\le n\le 2\tau+1} 2^n d(t, T_n) \le L2^\tau = L\log N$. Now $\sum_{n\le\tau} 2^n d(t, T_n) \le L2^\tau$ because $d(t, T_n) \le 1$ and $\sum_{n>2\tau+1} 2^n d(t, T_n)$ can be bounded by the usual dimensionality arguments.

Exercise 8.3.7 First observe that $\gamma_1(T, d_\infty)$ is of order $\log k$. On the other hand, if x_i denotes the i-th coordinate function, $\|\sum_i \alpha_i x_i\|_\infty = \sum_i |\alpha_i|$, from which it follows (by considering the case where $\alpha_i \in \{0, 2/k\}$ and $\sum_i \alpha_i = 1$) that $\log N(\operatorname{conv} T, d_\infty, 1/4)$ is of order k, so that $\gamma_1(\operatorname{conv} T, d_\infty)$ is at least of order k. Note that in that case, $\gamma_2(T, d_2)$ is of order $2^{k/2}\sqrt{\log k}$.

Exercise 9.2.3 For $n \geq 0, s, t \in A \in \mathcal{A}_n$, we have $\int |2^{j_n'(A)}(s(\omega) - t(\omega))|^2 \wedge 1 d\nu(\omega) \leq u2^n$, and since $\sum_{n\geq 0} 2^n 2^{-j_n'(A)} \leq 2\sum_{n\geq 0} 2^n r^{-j_n(A)}$, Theorem 9.2.1 indeed follows from the special case $r = 2$.

Exercise 9.4.5 Obviously, ψ_j is translation invariant, $\psi_j(t+s, t'+s) = \psi_j(t, t')$. And since it is the square of a distance, it satisfies the inequality $\psi_j(t, t'') \leq 2(\psi_j(t, t') + \psi_j(t', t''))$. As simple consequence is that for a set $C = \{s \in T\ ;\ \psi_j(s, 0) \leq u\}$ and any $s \in T$, then for $t, t' \in s + C - C$, we have $\psi_j(t, t') \leq 2^4 u$. By Lemma 7.1.3, T can be covered by $\leq N_n$ translates $s + C_n - C_n$ of $C_n - C_n$. So for each n, we have a partition $\mathcal{B}_n$ of T in N_n appropriately small sets, and we construct the required admissible sequence of partitions in the usual manner, $\mathcal{A}_n$ being the partition generated by $\mathcal{A}_{n-1}$ and $\mathcal{B}_{n-1}$.

Exercise 9.4.6 The idea is to apply Theorem 9.4.1 using the sequence of partitions built in the previous exercise. A problem is that the sequence of partitions lives on T, whereas Theorem 9.4.1 applies to sets of sequences, so some translation is necessary. For this, it helps to write our finite sums $\sum_i$ as infinite sums $\sum_{i\geq 1}$ with the understanding that the terms of the sum are eventually zero. We may assume that the sequence $(j_n)_{n\geq 0}$ is non-decreasing, simply by replacing j_n by $\max_{k\leq n} j_k$. For $s \in T$ and $i \geq 1$, let us define $\theta_i(s) = a_i(\chi_i(s) - \chi_i(0))$ and $\theta(s) = (\theta_i(s))_{i\geq 1} \in \ell_2$. Thus,

$$\sup_{s\in T} \Big|\sum_{i\geq 1} \varepsilon_i a_i(\chi_i(s) - \chi_i(0))\Big| = \sup_{s\in T} \Big|\sum_{i\geq 1} \varepsilon_i \theta_i(s)\Big| = \sup_{x\in\theta(T)} \Big|\sum_{i\geq 1} \varepsilon_i x_i\Big|$$

and $\sum_{i\geq 1} |r^j(\theta_i(s) - \theta_i(t))|^2 \wedge 1 = \psi_j(s, t)$. We transport the admissible sequence $(\mathcal{A}_n)$ constructed in Exercise 7.223 to $T^* := \theta(T)$ to obtain an admissible sequence $(\mathcal{A}_n^*)$. For $A \in \mathcal{A}_n^*$, we set $j_n(A) = j_n'$. It is then a simple matter to deduce (9.54) from the application of Theorem 9.4.1 to the set T^*, using also that for $x = (a_i(\chi_i(s) - \chi_i(0)))_{i\geq 1} \in T^*$, we have $|x_i| \leq 2|a_i|$ so that $\sum_{i\geq 1} |x_i| \mathbf{1}_{\{2|x_i|\geq r^{-j_0}\}} \leq 2\sum_{i\geq 1} |a_i| \mathbf{1}_{\{4|a_i|\geq r^{-j_0}\}}$.

Exercise 10.3.5 Let us fix $x > y$ and write $a_\ell = \varphi_{c\ell, c(\ell+1)}(x) - \varphi_{c\ell, c(\ell+1)}(y)$. Thus, $0 \leq a_\ell \leq c$ and $\sum_\ell a_\ell = x - y$ so that $\sum_\ell a_\ell^2 \leq c\sum_\ell a_\ell = c(x - y)$ and $\sum_\ell a_\ell^2 \leq (\sum_\ell a_\ell)^2 = (x - y)^2$, proving (10.39). Next, let k be the number of a_ℓ which are $\neq 0$. If $k = 2$, then $(\sum_\ell a_\ell)^2 \leq 2\sum_\ell a_\ell^2$. This is particularly the case if $x < y + c$. Furthermore, $c|x - y| = \sum_\ell ca_\ell \leq kc^2$ and $\sum_\ell a_\ell^2 \geq (k - 2)c^2$ because all the a_ℓ which are not zero but 2 are equal to c. Since $k \leq 3(k - 2)$ for $k > 2$ (10.38) follows.

Exercise 10.3.7 If $b < 1$, it is not possible to cover B_1 by finitely many translates of the set bB_2 because given such a translate A, for n large enough, the basis unit vector e_n does not belong to A. Also, $\epsilon B_2 + aB_1 \subset (\epsilon + a)B_2$.

Exercise 10.3.8 According to Theorem 6.2.8, we can write $T \subset T_1 + T_2$ with $\gamma_2(T_1) \le Lb(T)$ and $T_2 \subset Lb(T)B_1$. Then $N(T_1 + T_2, \epsilon B_2 + Lb(T)B_1) \le N(T_1, \epsilon B_2)$ and we use Exercise 2.7.8 (c) to bound the last term.

Exercise 10.14.6 The main point is the inequality $\varphi_j(s,t) \le r^{2j} d(s,t)^2$, where d is the ℓ^2 distance. Given an admissible sequence $(\mathcal{A}_n)$ of partitions, for $A \in \mathcal{A}_n$, define $j_n(A)$ as the largest integer $j \in \mathbb{Z}$ such that $r^j \Delta(A)^2 \le 2^{n/2}$, so that $r^{-j_n(A)} \le r 2^{n/2} D(A)$ and $\sum_{n\ge 0} 2^n r^{-j_n(A_n(t))} \le r \sum_{n \ge 0} 2^{n/2} \Delta(A_n(t))$ while $\varphi_{j_n(A)}(s,t) \le 2^n$ for $s, t \in A \in \mathcal{A}_n$.

Exercise 10.15.3 Given the $(\varepsilon_i)_{i\in I}$, we construct recursively a sequence $\sigma_1, \ldots, \sigma_n, \ldots$ with $\sigma_n \in \{0,1\}$ as follows. Assuming that $\sigma_1, \ldots, \sigma_n$ have been constructed, let $i = (\sigma_1, \ldots \sigma_n) \in I_n$, $i_0 = (\sigma_1, \ldots \sigma_n, 0) \in I_{n+1}$ and $i_1 = (\sigma_1, \ldots \sigma_n, 1) \in I_{n+1}$. We chose $\sigma_{n+1} = 0$ if $\varepsilon_{i_0} = 1$, and otherwise, we take $\sigma_{n+1} = 1$. It is straightforward to show by induction that $\mathsf{E} \sum_{k \le n} \sum_{i \in I_k} \varepsilon_i t_i = \sum_{k \le n} \alpha_k / 2$ where for $i \in I_k$, we have $t_i = \alpha_k$ if $i = (\sigma_1, \ldots, \sigma_k)$ and $t_i = 0$ otherwise.

Exercise 11.6.2 The argument is given at the beginning of the proof of Theorem 11.7.1.

Exercise 11.12.2 (a) Since $\delta(T) = \delta(\operatorname{conv} T)$ and by Theorem 11.12.1. (b) Taking $A = \gamma_1(T, d_\infty)$ the inequality $\gamma_2(T, d_2) \le A/\sqrt{\delta}$ is satisfied for δ small enough, but $\gamma_1(\operatorname{conv} T, d_\infty)$ may be much larger than $\gamma_1(T, d_\infty)$.

Exercise 11.12.4 The use of (11.56) gives a bound of $-\log \mathsf{P}(\sum_{i\in I} \delta_i \ge u)$ of order $\min(u^2/\delta \operatorname{card} I, u)$ which can match the bound (11.70) only when u is not much larger than $\delta \operatorname{card} I$.

Exercise 12.1.2 Given $M \ge k$, the probability of finding k points in A is $p_M := \binom{n}{k} P(A)^k (1 - \mathsf{P}(A)^{M-k}$, so the probability that $\operatorname{card}(A \cap \Pi) = k$ is $\sum_{m \ge k} \mathsf{P}(M = m) p_m$ which you compute to be $\exp(-\mathsf{P}(A))\mathsf{P}(A)^k/k!$. As for the property of Lemma 12.1.1, it is simply because given M and $I = \{i \le M, Y_i \in A\}$, the points $(Y_i)_{i\in I}$ are uniform i.i.d. in A, distributed according to the probability of this lemma. When Ω is σ-finite but not finite, you break Ω into a countable disjoint union of sets Ω_n of finite measure, define independent random sets Π_n in each of them according to the previous procedure, set $\Pi = \cup_n \Pi_n$, and check that this works by similar arguments.

Exercise 12.3.6 For the Lévy measure ν of a p-stable process, it never happens that $\int |\beta(t)| \wedge 1 d\nu(\beta) < \infty$ for all $t \in T$ unless this Lévy measure is concentrated at zero, because for $a \ne 0$, the integral $\int |ax^{-1/p}| \wedge 1 dx$ is divergent. In fact, it is not difficult to show that when $\int |\beta(t)| \wedge 1 d\nu(\beta) = \infty$ if the sequence Z_i is generated by a Poisson point process of intensity ν, then $\sum_i |Z_i(t)| = \infty$ a.s.

Exercise 12.3.14 Really straightforward from the hint.

Exercise 12.3.15 It is better to use the functional $\gamma^*(T,d)$ of (5.21), which in the homogeneous case is just $\sum_{n\ge 0} e_n(T)$ and the result by (7.5).

Exercise 12.3.16 By hypothesis, ν is the image of the measure $\mu \otimes m$ on $\mathbb{R}^+ \times \mathcal{C}$ under the map $(x, \beta) \mapsto x\beta$ where m is supported by G, and μ has density x^{-p-1} with respect to Lebesgue's measure. The result follows since $\beta(0) = 1$ for $\beta \in G$.

Exercise 12.3.17 Combining Exercise 12.3.14 with Theorem 12.3.13, the only point which is not obvious is that for $p > 1$, we have $\mathsf{E}\sup_{t\in T}|X_t| < \infty$. Reducing to the case where the Lévy measure is the measure ν^1 of Theorem 12.3.11, this follows from the fact that $\mathsf{E}|X_t| \le \mathsf{E}\sum_i |Z_i(0)| = \int |\beta(0)| d\nu^1(\beta) < \infty$ by Exercise 12.3.16.

Exercise 12.3.18 As in the previous exercise, we reduce to the case where the Lévy measure is the measure ν^1 of Theorem 12.3.11, and we have to prove an estimate $\mathsf{P}(\sum_i |Z_i(0)| \ge u) \le C/u$. Consider the probability P_N, the conditional probability given the event that there are N numbers Z_i, we bound crudely $\mathsf{P}_N(\sum_i |Z_i(0)| \ge u) \le N\mathsf{P}_N(\exists i, Z_i \ge u/N) \le CN^2/u$, using both Exercises 12.1.2 and 12.3.16, and the result follows by summation over N.

Exercise 13.2.2 Consider the subset A' of A defined as follows: A sequence $(\delta_i)_{i\le M}$ is in A' if and only if $\delta_i = 0$ when i does not belong to any one of the sets I_ℓ and $\sum_{i\in I_\ell}\delta_i = 1$ for each $\ell \le r$. The set A' obviously identifies with $\prod_{\ell\le r} I_\ell$. Consider the uniform probability measure ν on A'. Consider a set $I \subset \{1,\dots,M\}$. Then $A' \cap H_I = \emptyset$ unless $I \subset \cup_{\ell\le r} I_\ell$ and $\operatorname{card} I \cap I_\ell \le 1$ for each $\ell \le r$, and in that case, $\nu(H_I) = k^{-\operatorname{card} I}$ (as belonging to H_I amounts to fixing $\operatorname{card} I$ coordinates in the product $\prod_{\ell\le r} I_\ell$). Thus, $\nu(H_I) \le k^{-\operatorname{card} I}$ for each I. Hence if $A \subset \cup_{I\in\mathcal{G}} H_I$, then $\sum_{I\in\mathcal{G}} k^{-\operatorname{card} I} \ge 1$.

Exercise 13.4.2 The class $\mathcal{J}$ consisting of M disjoint sets of cardinality k satisfies $\delta(\mathcal{J}) \le k$. Fixing δ and k, $\mathcal{S}_\delta(\mathcal{J})$ is such that $M(\delta k/\mathcal{S}_\delta(\mathcal{J}))^{\mathcal{S}_\delta(\mathcal{J})} \le 1$ so that it goes to ∞ as $M \to \infty$.

Exercise 13.4.3 Consider integers $2 \ll N_2 \ll N_1$ and a number $\delta > 0$ such that $S := 4\delta N_1 = 2$. Consider the class $\mathcal{I}_0$ of sets consisting of one single set D of cardinality N_1 and of M disjoint sets $(B_i)_{i\le M}$, each of cardinality N_2, where M is the largest integer with $M(N_2/N_1)^S \le 1/2$ (so that $M \ge (N_1/N_2)^S/4$). It is straightforward that $S_\delta(\mathcal{I}_0) \le S$. Consider the class $\mathcal{I}$ consisting of sets which are union of two sets of $\mathcal{I}_0$. Then $\delta(\mathcal{I}) \le 2\delta(\mathcal{I}_0) \le LS$. Consider now a class $\mathcal{J}$ of sets such that $\mathcal{I} \subset \mathcal{J}(1,m)$. The goal is to prove that given $A > 0$, for suitable choices of N_1 and N_2, we have $\mathcal{S}_\delta(\mathcal{J}) + m \ge AS$. Assume for contradiction that $\mathcal{S}_\delta(\mathcal{J}) + m \le AS$ so that $m \le AS$ and

$$\sum_{J\in\mathcal{J}} \Big(\frac{\delta \operatorname{card} J}{AS}\Big)^{AS} \le 1\,. \tag{F.3}$$

In particular, $\operatorname{card} J \le AS/\delta = 4AN_1$ for $J \in \mathcal{J}$ so that

$$\operatorname{card}\{i \le M;\, J \cap B_i \ne \emptyset\} \le 4AN_1\,. \tag{F.4}$$

Since $\mathcal{I} \subset \mathcal{J}(1,m)$, given $I \in \mathcal{I}$, there exists $J \in \mathcal{J}$ with $\operatorname{card} I \setminus J \le m$. In particular, given $i \le M$, there exists $J_i \in \mathcal{J}$ with $\operatorname{card}((D \cup B_i) \setminus J_i) \le m$. Assume now $N_2 > AS \ge m$. When $\operatorname{card}((D \cup B_i) \setminus J_i) \le m$, since $\operatorname{card} B_i = N_2$, we must have $J_i \cap B_i \ne \emptyset$. Combining with (F.3) shows that there are at least $M/(4AN_1)$ different sets J_i. For each of these sets J_i, we have $\operatorname{card}(D \setminus J_i) \le m$ so that card $J_i \geq N_1/2$, and since $S = 4\delta N_1$, this implies δ card $J_i/(AS) \geq 1/8A$. The sum (F.4) is then at least

$$\frac{M}{4AN_1}\Big(\frac{1}{8A}\Big)^{AS} \geq \frac{1}{16AN_1}\Big(\frac{N_1}{N_2}\Big)^{S}\Big(\frac{1}{8A}\Big)^{AS},$$

and since $S = 2$, this cannot be ≤ 1 when $N_1 \gg N_2$. In conclusion, given a number A, we choose $N_2 > 2A$, N_1 large enough, and then δ so that $S = 4\delta N_2 = 2$, and the previous construction provides a class $\mathcal{I}$ such that $\mathcal{S}(\mathcal{J}) + m \geq AS$ whenever $\mathcal{I} \subset \mathcal{J}(1, m)$.

Exercise 14.1.3 Defining $b_n(\mathcal{F}) = \inf\{\epsilon > 0, N_{[\,]}(\mathcal{F}, \epsilon) \leq N_n\}$, the right-hand side of (14.8) is at least of order $\sum_{n\geq 0} 2^{n/2} b_n(\mathcal{F})$, by the same argument as was used to prove (2.40). One then proceeds exactly as in Exercise 2.7.6 to prove that the previous quantity dominates the quantity (14.6).

Exercise 14.2.1 (a) $\int \exp(|f|/k)d\mu \leq (\int \exp|f|d\mu)^{1/k} \leq 2$. (b) should be obvious to you at this stage. (c) Since for $x \geq 0$ we have $\exp x \leq \exp(1/4)\exp x^2 \leq 2\exp x^2$, when $\|f\|_{\psi_2} \leq 1$, we have $\int \exp(|f|)d\mu \leq 4$ and (14.16) by (14.15). To prove (14.17), assume $\|f_1\|_{\psi_2} \leq 1$, $\|f_2\|_{\psi_2} \leq 2$, and use that $|f_1 f_2| \leq f_1^2 + f_2^2$ and the Cauchy-Schwarz inequality. (d) It is elementary that $\int \exp(g^2/A^2) = A/\sqrt{A^2 - 2}$ for $0 < A < 2$, so this norm is the positive solution of the equation $A^2 = 4(A^2 - 2)$, that is, $A = \sqrt{8/3}$. (e) Combine the subgaussian inequality and (b). For the rest, see the next exercise.

Exercise 14.2.3 The reader should review the proof of Theorem 6.7.2. We consider the same chaining as in the proof a Theorem 14.2.2. Along the chains, for we decompose each function, $\pi(f) - \pi_{n_1}(f)$ as $f_{n_1} + f_{n,2}$ where $f_{n,1} = (\pi_n(f) - \pi_{n_1}(f))\mathbf{1}_{|\pi(f)-\pi_{n_1}(f)|\leq 2^{-n/2}\sqrt{N}\Delta(A_{n-1}(f),\psi_2)}$. We define $\mathcal{F}_1$ as the set of sums $\sum_{1\leq n\leq n_1} f_{n,1}$. By the method of Theorem 14.2.2, we then have $\gamma_2(\mathcal{F}_1, \psi_2) \leq L\gamma_2(\mathcal{F}, \psi_2)$ and $\gamma_1(\mathcal{F}_1, d_\infty) \leq \sqrt{N}\gamma_2(\mathcal{F}, \psi_2)$. Let $\mathcal{F}_2'$ be the sets of sums $\sum_{1\leq n\leq n_1} f_{n,2}$. We will show that $\mathsf{E}\sup_{f\in\mathcal{F}_2'}\sum_{i\leq N}|f(X_i)|$. This will finish the proof since the chaining beyond n_1 involves no cancellations as is shown in the proof of Theorem 6.7.2. This follows from Lemma 14.2.4 and the following observation: If $\|f\|_{\psi_2} \leq 1$ and $a \geq 1$, then $\|f\mathbf{1}_{|f|\geq a}\|_{\psi_1} \leq L/a$. Thus, $\mathsf{P}(\sum_{i\leq N}|f_{2,n}(X_i)| \leq u2^{n/2}\sqrt{N}\Delta(A_{n-1}(f), \psi_2)) \leq \exp(-LNu)$, etc.

Exercise 14.2.10 As always, we start with the relation $\exp x \leq 1 + x + x^2 \exp|x|$ which is obvious on power series expansions. Thus, using Hölder's inequality, we get

$$\mathsf{E}\exp(\lambda Y) \leq \lambda^2 \mathsf{E}Y^2 \exp|\lambda Y| \leq \lambda^2(\mathsf{E}Y^6)^{1/3}(\mathsf{E}\exp 3|\lambda Y|/2)^{2/3}.$$

It then remains to prove that $\mathsf{E}Y^6 \leq LA^6$ and $\mathsf{E}\exp 3|\lambda Y|/2 \leq L\exp\lambda^2 A^2$ which follows from a routine use of (2.6).

Exercise 15.1.9 (a) Consider an admissible sequence $(\mathcal{A}_n)$ of partitions of T such that

$$\forall t \in T\ , \ \sum_{n\geq 0} 2^{n/2}\Delta(A_n(t), d_\omega) \leq 2\gamma_2(T, d_\omega)\ . \tag{F.5}$$

For $n \geq 1$ set $T_n' = \cup\{A \cap D\ ;\ A \in \mathcal{A}_n;\ D \in \mathcal{D}_n, \mu(A \cap D) \leq 2N_{n++2}^{-1}\}$. Since T_n' is the union of $\leq N_n^2 = N_{n+1}$ sets of measure $\leq 2N_{n+2}^{-1} = 2N_{n+1}^{-2}$, we have $\mu(T_n') \leq 2N_{n+1}^{-1}$. Also, for $t \in T_n := T \setminus T_n'$, we have $\mu(A_n(t) \cap D_n(t)) \geq 2N_{n+2}^{-1}$ so that $\eta_{n,\omega}(t) \leq \Delta(A_n(t), d_\omega)$. Thus,

$$\int_T 2^{n/2}\eta_{n,\omega}(t) \leq 2^{n/2}N_{n+1}^{-1}\Delta(T, d_\omega) + \int_T 2^{n/2}\Delta(A_n(t), d_\omega)\mathrm{d}\mu(t)\ .$$

Summing over $n \geq 0$ and since $\Delta(T, d_\omega) \leq L\gamma_2(T, d_\omega)$ we obtain the result, using also (F.5). (b) It suffices to assume that $\epsilon_n(t) > K2^{n/2}\Delta(D_n(t), d_1)$. By definition of $\epsilon_n(t)$,

$$\mu(\{s \in D_n(t)\ ;\ d(s,t) \leq \theta_n(t)/2\}) \leq \mu(\{s; d(s,t) \leq \theta_n(t)/2\}) \leq N_{n+2}^{-1}\ .$$

If $s \in D_n(t)$ and $d(s,t) \geq \epsilon_n(t)/2$, then $d(s,t) \geq K_1 2^{n/2}\Delta(D_n(t), d_1)/2 \geq K_1 2^{n/2} d_1(s,t)/2$. Thus, the right-hand side of (15.23) is $\leq \exp(-2^n)$ for a suitable choice of K_1. As usual, this implies that with large probability, we have

$$\mu(\{s \in D_n(t); d(s,t) \leq \alpha\theta_n(t)/2)\}) \leq \alpha^{-1}\exp(-2^n),$$

and then by Fubini theorem that with large probability $\mu(\{s \in D_n(t) \leq \alpha\epsilon_n(t)/2\}) \leq 2N_{n+2}^{-1}$ so that $\eta_{n,\omega}(t) \geq \alpha\theta_n(t)/2$ and (15.28) is proved. Combining the preceding, we have

$$\int_T \sum_{n\geq 0} 2^{n/2}\epsilon_n(t)\mathrm{d}\mu(t) \leq K \sup_{t\in T}\sum_{n\geq 0} 2^n\Delta(D_n(t), d_1) + K\mathsf{E}\gamma_2(T, d_\omega)\ .$$

By an appropriate choice of the $\mathcal{D}_n$, we obtain $\int_T \sum_{n\geq 0} 2^{n/2}\epsilon_n(t)\mathrm{d}\mu(t) \leq K\gamma_1(T, d_1) + K\mathsf{E}\gamma_2(T, d_\omega)$ and consequently $\int_T I_\mu(t)\mathrm{d}\mu(t) \leq K\gamma_1(T, d_1) + K\mathsf{E}\gamma_2(T, d_\omega)$. The result follows since μ is arbitrary.

Exercise 15.1.14 First, we show that $\epsilon\Delta(T,d) \leq KM$. Indeed, for $s, t \in T$, by (15.22), we have $\mathsf{P}(d_\omega(s,t) \geq \alpha d(s,t)) \geq \alpha$ so that according to (15.24), with positive probability, we have at the same time $d_\omega(s,t) \geq \alpha d(s,t)$ and $\gamma_2(T, d_\omega) \leq M$, and since $d_\omega(s,t) \leq L\gamma_2(T, d_\omega)$, we have $d(s,t) \leq KM$. Thus, $\Delta(T,d) \leq K_1 M$. Consequently, $N(T,d,\epsilon) = 1$ for $\epsilon \geq K_1 M$ and thus $\epsilon\sqrt{\log N(T,d,\epsilon)} = 0$.

Consider $0 < \epsilon \leq K_1 M$ and assume that we can find points $t_1, \ldots, t_N$ of T with $d(t_i, t_j) \geq \epsilon > 0$ for $i \neq j$. Then by (15.23) for $i \neq j$, we have $\mathsf{P}(d_\omega(t_i, t_j) \leq \alpha\epsilon) \leq (1/\alpha)\exp(-\alpha\epsilon^2/\Delta(T, d_1))$. For $\epsilon \geq K\sqrt{\log N}\Delta(T, d_1)$, the right-hand side is $\leq \alpha/2N^2$. According to (15.24), with positive probability, we have at the same time $d_\omega(t_i, t_j) \geq \alpha\epsilon$ for all $i \neq j$ and $\gamma_2(T, d_\omega) \leq M$. Since $\epsilon\alpha\sqrt{\log N} \leq L\gamma_2(T, d_\omega)$, this proves that $\epsilon\sqrt{\log N} \leq KM$. That is, we have proved that

$$\sqrt{\log N} \leq \epsilon/(K_2\Delta(T, d_1)) \Rightarrow \sqrt{\log N} \leq K_2M/\epsilon\ , \tag{F.6}$$

and without loss of generality, we may assume that $K_2 \geq K_1$. Let us assume now that $\epsilon/(K_2\Delta(T, d_1)) \geq 2K_2M/\epsilon$ (or, equivalently, $\epsilon \geq K_2\sqrt{2\Delta(T, d_1)M}$). Observe that then since $\epsilon < K_1M$, we have $2K_2M/\epsilon \geq 2$. Let us prove that we must have $\sqrt{\log N} \leq 2K_2M/\epsilon$. Otherwise, we may replace N by the largest integer N' for which $\sqrt{\log N'} < 2K_2M/\epsilon$. Then $2K_2M/\epsilon \leq \sqrt{\log(N'+1)} < 2\sqrt{\log N'}$, and thus, we obtain the relations $K_2M/\epsilon < \sqrt{\log N'} \leq 2K_2M/\epsilon \leq \epsilon/(K_2\Delta(T, d_1))$ which contradict (F.6). Thus, $\sqrt{\log N(T, d, \epsilon)} \leq KM/\epsilon$ which concludes the proof. The application to Proposition 15.1.13 is straightforward as (15.24) holds for $M = LS(T)$.

Exercise 15.1.17 Indeed (15.44) implies $\log N(T, d_\infty, \alpha) \leq LS(T)^2/\alpha^2$. Now, if B is a ball $B_\infty(t, \alpha)$ of T for $\alpha = \epsilon^2/L'S(T)$, since $\Delta(B_\infty(t, \alpha), d_\infty) \leq 2\alpha$, for L' large enough, the right-hand side of (15.40) holds and this inequality implies $\log N(B, d_2, \epsilon) \leq LS(T)^2/\epsilon^2$. Since $N(T, d_2, \epsilon) \leq N(T, d_\infty, \alpha) \times \max_{t\in T} N(B_\infty(t, \alpha), d_2, \epsilon)$, combining these yields

$$\log N(T, d_2, \epsilon) \leq L\Big(\frac{S(T)^4}{\epsilon^4} + \frac{S(T)^2}{\epsilon^2}\Big)\ .$$

Thus, $N(T, d_2, \epsilon) = 1$ for $\epsilon \geq LS(T)$. For $\epsilon \leq LS(T)$, the term $S(T)^4/\epsilon^4$ dominates, and this implies (15.43).

Exercise 15.2.2 To each tensor A, we associate the r.v. X_A given by (15.55). We deduce from (15.69) that $\|X_A\|_p \leq K(d)\sum_{1\leq k\leq d} p^{k/2}\|A\|_{(k)}$ where $\|A\|_{(k)} = \sum_{\operatorname{card}\mathcal{P}=k} \|A\|_{\mathcal{P}}$. To turn this inequality in a tail estimate, we use that $t\mathsf{P}(|X_A| \geq t)^{1/p} \leq \|X_A\|_p$ so that

$$\log \mathsf{P}(|X_A| \geq t) \leq p\log\Big(\frac{1}{t}\sum_{1\leq k\leq d} p^{k/2}\|A\|_{(k)}\Big)\ .$$

We then take $p = (1/K(d))\min_{1\leq k\leq d}(t/\|A\|_{(k)})^{2/k}$ to obtain a bound

$$\mathsf{P}(|X_A| \geq t) \leq K(d)\exp\Big(-\frac{1}{K(d)}\min_{1\leq k\leq d}\Big(\frac{t}{\|A\|_{(k)}}\Big)^{2/k}\Big)\ .$$

Considering a set T of tensors A and d_k the distance on T induced by the norm $\|\cdot\|_{(k)}$, we then have the bound $\mathsf{E}\sup_{A\in T} X_A \leq K(d)\sum_{1\leq k\leq d}\gamma_{2/k}(T, d_k)$.

Exercise 15.2.8 Since A is symmetric, we have $\alpha(y) = \sup\{\langle x, y\rangle; x \in A\}$ so that $\mathsf{E}\alpha(G) = \mathsf{E}\sup_{x\in A}\langle x, G\rangle = g(A)$ by definition of that quantity.

Exercise 15.2.16 We have $\|\langle A, G^1\rangle\|_{\{2,3\}} = (\sum_{j,k}(\sum_i a_{i,j,k} g_i^1)^2)^{1/2}$ so that by the Cauchy-Schwarz inequality, $\mathsf{E}\|\langle A, G^1\rangle\|_{\{2,3\}} \le (\sum_{i,j,k} a_{i,j,k}^2)^{1/2} = \|A\|_{\{1,2,3\}}$.

Exercise 16.2.2 It is obvious that $N(T, d, 2^{-n/2}) \le M_n := \operatorname{card}\{I \in \mathcal{I}_n \; ; \; I \cap T \neq \emptyset\}$. On the other hand, one has $M_n \le 2N(T, d, 2^{-n/2})$. Indeed, if the intervals $I \in \mathcal{I}_n$ which meet T are numbered as $I_1, I_2 \ldots$ from left to right, then any points in $I_1, I_3, I_5, \ldots$ are at mutual distances $> 2^{-n/2}$. We have already proved (d) as a consequence of 16.5. To prove (c), one simply uses the Cauchy-Schwarz inequality to obtain $\sum_{I\in\mathcal{I}_n} \sqrt{2^{-n}\mu(I)} \le \sqrt{2^{-n}\operatorname{card}\{I \in \mathcal{I}_n \; ; \; I \cap T \neq \emptyset\}}$.

Exercise 16.3.4 We observe that the function $x(\log(2/x))^2$ increases for x small. Distinguishing whether $a_n \ge 1/n^2$ or not, we get $a_n(\log(2/a_n))^2 \le La_n(\log n)^2 + Ln^{-2}(\log n)^2$, and by summation, this proves that (16.16) implies (16.21)[1] Assuming now that the sequence (a_n) decreases, (16.21) implies that $\sum_k 2^k a_{2^k}(\log(2/a_{2^k}))^2 < \infty$ so that $a_{2^k} \le C2^{-k}$ and for k large enough $\log(2/a_{2^k}) \ge k/L$ and thus $\sum_k 2^k a_{2^k}(\log k)^2 < \infty$ which implies (16.16).

Exercise 16.3.5 The recursion formula is simply

$$M(T) = \frac{1}{\sqrt{2^{n(T)}\mu(I_{n(T)}(T))}} + \max_{i=1,2} M(T \cap I_i) \, .$$

Exercise 16.5.3 Take $\epsilon_n = 2^{-n}\Delta(T, d)$ and T with $\operatorname{card} T_n = N(T, d, \epsilon_n)$ and $d(t, T_n) \le \epsilon_n$ for $t \in T$.

Exercise 16.8.17 Let us follow the hint. Then given s, t, we have $|X_s - X_t| = \varphi^{-1}(N/(2\epsilon))$ on the set $\Omega_s \cup \Omega_t$ of probability $2\epsilon/N$ and zero elsewhere, and thus, $\mathsf{E}\varphi(X_s - X_t) \le 1$. The r.v. $\sup_{s,t}|X_s - X_t|$ equals $\varphi^{-1}(N/(2\epsilon))$ on the set $\bigcup_{t\in T}\Omega_t$ of probability ϵ and zero elsewhere so that for $A > 0$, we have $\|\sup_{s,t}|X_s - X_t|\|_\psi \le A \Rightarrow \epsilon\psi(\varphi^{-1}(N/(2\epsilon))/A) \le 1$. Since in our case the integral in the right-hand side of (16.177) is $\varphi^{-1}(N)$, this inequality implies that for all choices of N and ϵ, one must have $\epsilon\psi(\varphi^{-1}(N/(2\epsilon))/(L\varphi^{-1}(N))) \le 1$. Setting $x = L\varphi^{-1}(N)$ and $y = \varphi^{-1}(N/(2\epsilon))/x$ and eliminating ϵ and N yields the relation $\varphi(x/L)\psi(y) \le 2\varphi(xy)$ which is basically (16.170).

Exercise 16.9.2 It follows from the Cauchy-Schwarz inequality that for any probability μ, we have $1 \le \int_0^1 \mu(B(t,\epsilon))\mathrm{d}t \int_0^1 \mathrm{d}t/\mu(B(t,\epsilon))$. Now, by Fubini theorem, denoting by λ Lebesgue's measure on $[0, 1]$, $\int_0^1 \mu(B(t,\epsilon))\mathrm{d}t = \int \mathrm{d}\mu(u)\lambda(B(t,\epsilon)) \le 2\epsilon$, and thus $\int_0^1 \mathrm{d}t/\mu(B(t,\epsilon)) \ge 1/(2\epsilon)$. Integrating in ϵ yields the result.

Exercise 16.9.3 For $k \ge 1$ and $t \in T$, define the function $Y_{k,t}$ by $Y_{k,t}(s) = 1$ if $d(s,t) \le 2^{-k}$ and $Y_{k,t}(s) = 0$ otherwise. Consider $Y_{k,t}$ as a r.v. on the basic probability space (T, μ) where μ is the uniform measure on T. Consider the r.v. $X_t = \sum_{k\ge1} Y_{k,t}$. It should be obvious that $\sup_t X_t = \infty$. On the other hand,

[1] This is also a consequence of Corollaries 16.3.1 and 16.3.2, but the direct proof is much clearer.

consider $s, t \in T$ with $d(s,t) = 2^{-\ell}$. Then $Y_{k,t} = X_{k,t}$ for $k \leq \ell$. Thus, $\mathsf{E}|X_s - X_t| \leq \sum_{k>\ell} \mathsf{E}(Y_{k,s} + Y_{k,t}) = \sum_{k>\ell} 2^{-k+1} = 2^{-\ell+1} = 2\delta(s,t)$.

Exercise 16.9.4 Simply replace the inequality (16.129) by the simpler $|X_s - X_{\theta_n(s)}| \leq d(s, \theta_n(s))\big(|X_s - X_{\theta_n(s)}|/d(s, \theta_n(s))\big)$, and follow the same proof.

Exercise 18.1.4 According to (18.11) for each $u \geq 1$, the convex set $C(u) := \{\psi \leq u\}$ is invariant by the symmetries $(x_1, x_2, x_3) \mapsto (\pm x_1, \pm x_2, \pm x_3)$. Consider the smallest box $B = [-a, a] \times [-b, b] \times [-c, c]$ containing C. Then the points $(\pm a, 0, 0)$, $(0, \pm b, 0)$, $(0, 0, \pm c)$ belong to $C(u)$, and thus $B/3 \subset C(u)$. Given $k \geq 0$, let us define $(m_j(k))_{j\leq 3}$ as the smallest integers such that $C(N_k) \subset B(k) := [-2^{m_1(k)}, 2^{m_1(k)}] \times [-2^{m_2(k)}, 2^{m_2(k)}] \times [-2^{m_3(k)}, 2^{m_3(k)}]$. Observe the fundamental fact that $B(k)/6 \subset C(N_k)$. According to (18.10), we have $2^{m_1(k)+m_2(k)+m_3(k)+3} \geq \log N_k \geq 2^{k-1}$ so that $m_1(k) + m_2(k) + m_3(k) \geq k - 4$. Also, $m_j(k) \geq 0$ from (18.12). It should be clear that we can find sequences $(n_j(k))_{j\leq 3}$ such that $n_j(k) \leq m_j(k+4)$ which satisfy (18.2). Observe then that $S_k \subset B_{k+4}$. Recalling (18.3), we consider the function φ given by (18.5), and we proceed to prove that $\psi(x) \leq \varphi(Lx)$. Since $\varphi \geq 1$, this is always the case when $\psi(x) \leq 1$. It follows from (18.12) that the set $\{\psi \leq 1\}$ contains the set $D = [-1/3, 1/3]^3$ so that if L is large enough $S_5/L \subset D$ so that $\psi(x) \geq 1 \Rightarrow \varphi(x) \geq N_5$, and it suffices to consider the case $\psi(x) \geq N_5$. Consider then the largest k such that $\psi(x) > N_k$, so $k \geq 5$. Then $S_{k-4}/6 \subset B_k/6 \subset C(N_k)$ and then $S_{k+1}/L \subset C(N_k)$. Since $x \notin C(N_k)$, this proves that $\varphi(x/L) \geq N_{k+1}$ so that $\varphi(x/L) \geq \psi(x)$ by definition of k.

Exercise 19.2.6 Small variation on the proof of Lemma 19.2.5.

Exercise 19.2.7 So there exists a set $U_n \subset \ell^2$ with $\operatorname{card} U_n \leq 5^{\operatorname{card} I_n}$, $\|u\| \leq 2a_n$ and $B_2(I_n, a_n) \subset \operatorname{conv} U_n$. Let $U = \bigcup_{n\geq 1} U_n$ so that given $a > 0$, we have $N_a := \operatorname{card}\{u \in U \,;\, \|u\| \geq a\} \leq \sum\{5^{\operatorname{card} I_n} \,;\, 2a_n \geq a\}$. Now for $2a_n \geq a$, we have $n + 1 \leq \exp(4/a^2)$, and since $5^{\operatorname{card} I_n} \leq (n+1)^{\log 5}$, we have $N_a \leq \exp(L/a^2)$. Thus, if we enumerate U as a sequence (u_k) such that the sequence $(\|u_k\|)_{k\geq 1}$ is non-increasing, we have $\|u_k\| \leq L/\sqrt{\log(k+1)}$. On the other hand, $\sup_{n\geq 1} a_n(\sum_{i\in I_n} g_i^2)^{1/2} \leq \sup_k \sum_{i\geq 1} u_{k,i} g_i$.

Exercise 19.2.8 We use the following form of (2.61): $\mathsf{P}(\sup_{s,t\in T_n} |X_s - X_t| \geq L(g(T_n) + u b_n)) \leq L \exp(-u^2)$, and by the union bound as usual

$$\mathsf{P}\Big(\forall n \geq 1,\ \sup_{s,t\in T_n} |X_s - X_t| \geq Lu(g_2(T_n) + b_n\sqrt{\log n})\Big) \leq L \exp(-u^2)$$

and hence $\mathsf{E} \sup_{n,s,t\in T_n} |X_s - X_t| \leq L \sup_n L(g(T_n) + b_n\sqrt{\log(n+1)})$ which implies (19.61). We then apply this inequality to the case $T_n = B_2(I_n, a_n)$.

Exercise 19.2.16 Consider a set $J_n \supset I_n$ with $\log(n+1) \leq \operatorname{card} J_n \leq 2\log(n+1)$. Then

$$\|x\| \leq LS \sup_{n\geq 1} \left(\frac{1}{\log(n+1)} \sum_{i\in I_n} x_i^2\right)^{1/2} \leq LS \sup_{n\geq 1} \left(\frac{1}{\operatorname{card} J_n} \sum_{i\in J_n} x_i^2\right)^{1/2}.$$

Exercise 19.2.18 The set T of (19.83) satisfies $T \subset LB_1 S$ by a simple adaptation of Lemma 19.2.17, and it satisfies $\gamma_1(T, d_\infty) \le LS$ by Theorem 8.3.3. The result follows from Theorem 19.2.10 as in the proof of Theorem 8.3.3.

Exercise 19.3.6 The result of Proposition 2, (ii) of [24] states (changing ε into $\varepsilon/5$ and taking $\theta = 2\varepsilon$) that

$$N(X_1^*, d_\infty, \varepsilon) \le N(X_1^*, \delta_\infty, 2\varepsilon) N(W, \|\cdot\|, \varepsilon/K) .$$

Using (19.123) to bound the last term, we obtain

$$\log N(X_1^*, d_\infty, \varepsilon) \le \log N(X_1^*, \delta_\infty, 2\varepsilon) + K(p, \eta)\Big(\frac{S}{\epsilon}\Big)^p \log N ,$$

through which the desired result holds using iteration.

Exercise A.1.2 When there is a matching π between the points X_i and evenly spread points Y_i with $d(X_i, Y_{\pi(i)}) \le A$, then for any set C, we have $\mathrm{card}\{i \le N; X_i \in C\} \le \mathrm{card}\{i \le N, d(Y_i, C) \le A\}$. Any point Y_i with $d(Y_i, C) \le A$ is such that every point of the corresponding little rectangle is within distance $\le A + L/\sqrt{N}$ of A. When C is the interior of a curve of length ≤ 1, the set of points within distance ϵ of C has area $\le \lambda(C) + L\epsilon$. For $A \ge 1/\sqrt{N}$, this proves that $\mathrm{card}\{i \le N; d(Y_i, C) \le A\} \le N\lambda(C) + LNA$. This provides an upper bound $\mathrm{card}\{i \le N; X_i \in C\} \le N\lambda(C) + LNA$. The lower bound is similar.

Exercise C.2.1 This is because (C.7) is satisfied for $\nu = \sum_{u \in T} \mu_u$, where μ_u is the image of ν_u under the map $\varphi_u : \mathbb{R} \to \mathbb{R}^T$ given by $\varphi_u(x) = (\beta(t))_{t \in T}$ where $\beta(u) = x$ and $\beta(t) = 0$ for $t \neq u$ and where ν_u is the measure which satisfies (C.5) for X_u.

Appendix G
Comparison with the First Edition

This section will try to answer two questions:

- If you have some knowledge of the first edition (hereafter referred to as OE), what can you find of interest in the present edition?
- If you bought the present edition (hereafter referred to as NE), may you find anything of interest in OE?

The short answer to the first question is that yes, there have been some dramatic improvements in some key mathematics (both in the proofs and the results themselves), and to the second question, it is no, unless you are a specialist of Banach space theory.

Generally speaking, the entire text has been revised and polished, so at every place, NE should be better than OE. Greater attention has been paid to pedagogy, by breaking long proofs into smaller pieces which are made to stand out on their own. Some points, however, are a matter of taste. More variations on the theory of "functionals" are presented in OE, although nothing essential is omitted here. Another technical choice which is a matter of taste is as follows. In the basic constructions of partitions in metric spaces, in OE, the size of the pieces is controlled by the radius of these pieces, whereas in NE, it is controlled by their diameter. This leads to slightly simpler proofs at the expense of some worse numerical constants (whose value is anyway irrelevant).

At the level of global organization, a major decision was to present the proof of Theorem 6.2.8 (the Latała-Bednorz theorem) at a later stage of the book, in the tenth chapter rather than in the fifth. The author started working on this problem as soon as he identified it, around 1989, and a significant part of the results of this book here discovered during this effort. Whereas, strictly speaking, some of these results are not needed to understand the proof of the Latała-Bednorz result, the underlying ideas are part of that proof. It might require suprahuman dedication to understand the proof of the Latała-Bednorz theorem as it is given in OE, but now we try to prepare the reader by studying random Fourier series and families of distances first.

M. Talagrand, *Upper and Lower Bounds for Stochastic Processes*, Ergebnisse der Mathematik und ihrer Grenzgebiete. 3. Folge / A Series of Modern Surveys in Mathematics 60, https://doi.org/10.1007/978-3-030-82595-9

Although we could not simplify the proof of the Latała-Bednorz result itself, we tried to prepare the reader by working backward and rewriting the proofs of the other "partition schemes" of the same general nature exactly in the same form as will be done for this theorem. Thus, if OE frustrated your attempts to understand this proof, you may try again, reading in order Chaps. 2, 6, 7, 9, and 10.

A major difference between OE and NE is that a number of proofs are now quite shorter. At some places, like Chap. 18, this was simply achieved by reworking the arguments, but at many other places, the mathematics are just better. Most of the improvements can be traced back to a simple new idea: The method which we use to provide lower bounds for random Fourier (the key step of which is Lemma 7.7.3) series can be generalized to conditionally Bernoulli processes by appealing to Theorem 10.15.1 (which makes full use of the Latała-Bednorz theorem), leading to Lemma 11.4.1. The author then combined this idea with the idea of witnessing measures (as in Sect. 3.4) which replace the use of the Haar measure on groups. Then Witold Bednorz and Rafał Martynek [18] observed that in the case of infinitely divisible processes, this method could be combined with Fernique's convexity argument to remove a technical condition the author was assuming. As the case of infinitely divisible processes had been showcased by the author precisely because it looked like an entry door to more general situations (such as those of empirical processes), this shortly lead to a positive solution of three of the main conjectures of OE, which are presented, respectively, in Theorems 6.8.3, 11.12.1, and 12.3.5.

We have considerably shortened Chap. 19 for the simple reason that the field of Banach Spaces attracts much less attention than it used to do. We have kept only the topics which are very directly related to other material in the book. This is the one single area where the specialist may like to look at OE. We have also deleted results and arguments which are too tedious or too specialized compared to what they achieve. For example, we have deleted parts of the proof of Shor's matching theorem (Theorem 17.1.3), as the method we follow there cannot yield an optimal result. It serves no purpose to make an exhaustive list of the deleted results which are mentioned at the relevant places in the present text for the sake of the (purely hypothetical) reader who really wants to master all details and go fetch them in OE. The single simple result which we have not reproduced and which is not too specialized is the abstract version of the Burkholder-Davis-Gundy of the appendix A.6 of OE (while the other material of this appendix has now found its way elsewhere).

Bibliography

1. Adamczak, R., Litvak, A.E., Pajor, A., Tomczak-Jaegermann, N.: Quantitative estimates of the convergence of the empirical covariance matrix in log-concave ensembles. J. Amer. Math. Soc. **23**(2), 535–561 (2010)
2. Adamczak, R., Litvak, A.E., Pajor, A., Tomczak-Jaegermann, N.: Sharp bounds on the rate of convergence of the empirical covariance matrix. C. R. Math. Acad. Sci. Paris **349**(3–4), 195–200 (2011)
3. Ajtai, M., Komlós, J., Tusnády, G.: On optimal matchings. Combinatorica **4**(4), 259–264 (1984)
4. Ambrosio, L., Federico Stra, F., Trevisan, D.: A PDE approach to a 2-dimensional matching problem (2017). https://arxiv.org/abs/1611.04960
5. Artstein, S.: The change in the diameter of a convex body under a random sign-projection. In: Geometric Aspects of Functional Analysis. Springer's Lecture Notes in Mathematics, vol. 1850, pp. 31–39 (2004)
6. Batson, J., Spielman, D., Srivastava, N.: Twice-Ramanujan sparsifiers. In: Proceedings of the 41st Annual Symposium on the Theory of Computing, pp. 47–55 (2009)
7. Bednorz, W.: On a Sobolev type inequality and its applications. Studia Math. **176**(2), 113–137 (2006)
8. Bednorz, W.: Majorizing measures on metric spaces. C. R. Math. Acad. Sci. Paris **348**(1–2), 75–78 (2006)
9. Bednorz, W.: A theorem on majorizing measures. Ann. Probab. **34**(5), 1771–1781 (2006)
10. Bednorz, W.: Hölder continuity of random processes. J. Theoret. Probab. **20**(4), 917–934 (2007)
11. Bednorz, W.: On Talagrand's admissible net approach to majorizing measures and boundedness of stochastic processes. Bull. Pol. Acad. Sci. Math. **56**(1), 83–91 (2008)
12. Bednorz, W.: Majorizing measures and ultrametric spaces. Bull. Pol. Acad. Sci. Math. **60**(1), 91–100 (2012)
13. Bednorz, W.: The complete characterization of a.s. convergence of orthogonal series. Ann. Probab. **41**(2), 1055–1071 (2013)
14. Bednorz, W.: The majorizing measure approach to sample boundedness. Colloq. Math. **139**(2), 205–227 (2015)
15. Bednorz, W.: Bounds for stochastic processes on product index spaces. In: High Dimensional Probability VII, pp. 327–357, vol. 71. Progress in Probability. Springer, Berlin (2016)
16. Bednorz, W., Latała, R.: On the suprema of Bernoulli processes. C. R. Math. Acad. Sci. Paris **351**(3–4), 131–134 (2013)

M. Talagrand, *Upper and Lower Bounds for Stochastic Processes*, Ergebnisse der Mathematik und ihrer Grenzgebiete. 3. Folge / A Series of Modern Surveys in Mathematics 60, https://doi.org/10.1007/978-3-030-82595-9

17. Bednorz, W., Latała, R.: On the boundedness of Bernoulli processes. Ann. Math. (2) **180**(3), 1167–1203 (2014)
18. Bednorz, W., Martynek, R.: A note on infinitely divisible processes. Manuscript (2020)
19. Bobkov, S., Ledoux, M.: A simple analytic proof of the AKT optimal matching theorem. Manuscript 2019 (2020). https://arxiv.org/abs/1909.06193
20. Bourgain, J., Lewko, M.: Sidonicity and variants of Kaczmarz's problem. Ann. Inst. Fourier **67**(3), 1321–1352 (2017)
21. Bollobas, B.: Combinatorics. In: Set Systems, Hypergraphs, Families of Vectors and Combinatorial Probability. Cambridge University Press, Cambridge (1986)
22. Bourgain, J.: Bounded orthogonal sets and the Λ_p-problem. Acta. Math. **162**, 227–246 (1989)
23. Bourgain, J.: Random points in isotropic convex sets. In: Convex Geometric Analysis (Berkeley, 1996). Mathematical Sciences Research Institute Publications, vol. 34, pp. 53–58. Cambridge University Press, Cambridge (1999)
24. Bourgain, J., Pajor, A., Szarek, S., Tomczak-Jaegerman, N.: On the duality problem for entropy numbers. In: Geometric Aspects of Functional Analysis. Springer's Lecture Notes in Math, vol. 1376, pp. 53–63 (1989)
25. Chatterjee, S.: Superconcentration and related topics. Springer Monographs in Mathematics, x+156 pp. Springer, Cham (2014). ISBN: 978-3-319-03885-8; 978-3-319-03886-5
26. Coffman, E., Shor, P.: A simple proof of the $O(\sqrt{n}\log^{3/4} n)$ upright matching bound. SIAM J. Discret. Math. **4**(1), 48–57 (1991)
27. Diestel, J., Jarchow, H., Tonge, A.: Absolutely summing operators. In: Cambridge Studies in Advanced Mathematics, vol. 43. Cambridge University Press, Cambridge (1975)
28. Ding, J., Lee, J., Peres, Y.: Cover times, blanket times, and majorizing measures. Ann. Math. **175**(3), 1409–1471 (2012)
29. Dudley, R.: The sizes of compact subsets of Hilbert space and continuity of Gaussian processes. J. Funct. Anal. **1**, 290–330 (1967)
30. Dudley, R.: A course on empirical processes. École d'été de probabilités de Saint-Flour XII, 1982, pp. 1–142, vol. 1097. Spinger's Lecture Notes in Mathematics. Springer, Berlin (1984)
31. Eames, C., Eames, R.: Powers of Ten. https://en.wikipedia.org/wiki/PowersofTen(film)
32. Fernique, X.: Regularité des trajectoires des fonctions aléatoires gaussiennes. École d'été de Probabilités de Saint-Flour, IV-1974. Springer's Lecture Notes in Mathematics, Berlin, vol. 480, pp. 1–96 (1975)
33. Fernique, X.: Fonctions aléatoires gaussiennes, vecteurs aléatoires gaussiens. Université de Montréal, Centre de Recherches Mathématiques, Montreal, iv+217 pp. (1977)
34. Foucart, S., Rauhut, H.: A Mathematical Introduction to Compressive Sensing, pp. xviii+625. Birkhäuser/Springer, New York (2013)
35. Giné, E., Zinn, J.: Some limit theorems for empirical processes. Ann. Probab. **12**, 929–989 (1984)
36. Guédon, O., Rudelson, M.: L_p-moments of random vectors via majorizing measures. Adv. Math. **208**(2), 798–823 (2007)
37. Guédon, O., Mendelson, S., Pajor, A., Tomczak-Jaegermann, N.: Majorizing measures and proportional subsets of bounded orthonormal systems. Rev. Mat. Iberoam. **24**(3), 1071–1095 (2008)
38. Hanson, D.L., Wright, F.T.: A bound on tail probabilities for quadratic forms of independent random variables. Ann. Math. Stat. **42**(6), 1079–1083 (1971)
39. Jain, N., Marcus, M.: Integrability of infinite sums of independent vector valued random variables. Trans. Amer. Math. Soc. **212**, 1–36 (1975)
40. Kahane, J.-P.: Some Random Series of Functions, 2nd edn. Cambridge Studies in Advanced Mathematics, vol. 5, xiv+305 pp. Cambridge University Press, Cambridge (1985)
41. Klartag, B., Mendelson, S.: Empirical processes and random projections. J. Funct. Anal. **225**(1), 229–245 (2005)
42. Kôno, N.: Sample path properties of stochastic processes. J. Math. Kyoto Univ. **20**, 295–313 (1980)

43. König, H., Tzafriri, L.: Some estimates for type and cotype constants. Math. Ann. **256**(1), 85–94 (1981)
44. Kramher, F., Mendelson, S., Rauhut, H.: Suprema of chaos processes and the restricted isometry property. Comm. Pure Appl. Math. **67**(11), 1877–1904 (2014)
45. Kwapien, S., Rosinski, J.: Sample Hölder continuity of stochastic processes and majorizing measures. Seminar on Stochastic Analysis, Random Fields and Applications IV. Progress in Probability, vol. 58, pp. 155–163. Birkhäuser, Basel (2006)
46. Last, G., Penrose, M.: Lectures on the Poisson process. Institute of Mathematical Statistics Textbooks, vol. 7, xx+293 pp. Cambridge University Press, Cambridge (2018)
47. Latała, R.: Sudakov minoration principle and supremum of some processes. Geom. Funct. Anal. **7**, 936–953 (1997)
48. Latała, R.: Estimates of moments and tails of Gaussian chaoses. Ann. Probab. **34**(6), 2315–2331 (2006)
49. Latała, R.: On the boundedness of Bernoulli processes on thin sets. Elec. Comm. Probab. **13**, 175–186 (2008)
50. Latała, R.: On weak tail domination of random vectors. Bull. Pol. Acad. Sci. Math. **57**(1), 75–80 (2009)
51. Latała, R.: Weak and strong moments of random vectors. In: Marcinkiewicz Centenary Volume, Banach Center Publications, vol. 95, pp. 115–121. Polish Academy of Sciences Institue of Mathematics, Warsaw (2011)
52. Ledoux, M.: The concentration of measure phenomenon. In: Mathematical Surveys and Monographs, vol. 89, x+181 pp. American Mathematical Society, Providence (2001) ISBN: 0-8218-2864-9
53. Ledoux, M., Talagrand, M.: Probability in Banach spaces: isoperimetry and processes. In: Ergebnisse der Mathematik und ihrer Grenzgebiete (3), vol. 23, xii+480 pp. Springer, Berlin (1991). ISBN: 3-540-52013-9
54. Lehec, J.: Moments of the Gaussian Chaos. In: Séminaire de Probabilités XLIII. Springer's Lecture Notes in Mathematics, Berlin, vol. 2006, pp. 327–340 (2011)
55. Leighton, T., Shor, P.: Tight bounds for minimax grid matching with applications to the average case analysis of algorithms. Combinatorica **9**(2), 161–187 (1989)
56. Lévy, M.: Prolongement d'un opérateur d'un sous-espace de $L^1(\mu)$ dans $L^1(\nu)$. Séminaire d'Analyse Functionelle, 1979–1980, Exp. No. 5, 5 pp. École Polytech, Palaiseau (1980). www.numdam.org
57. Lindenstrauss, J., Tzafriri, L.: Classical Banach spaces, II. Function spaces. In: Ergebnisse der Mathematik und ihrer Grenzgebiete, vol. 97, x+243 pp. Springer, Berlin (1979). ISBN: 3-540-08888-1
58. Lugosi, G., Mendelson, S.: Risk minimization by median-of-means tournaments. J. Eur. Math. Soc. (JEMS) **20**(3), 925–965 (2020)
59. Marcus, M.B.: ξ-radial processes and random Fourier series. Mem. Amer. Math. Soc. **68**(368), viii+181 (1987)
60. Maurey, B.: Type et cotype dans les espaces munis de structures locales inconditionelles. In: Séminaire Maurey-Schwartz 1973–74, Exposé 12, Paris, Ecole Polytechnique (1974)
61. Marcus, M.B., Pisier, G.: Random Fourier series with applications to harmonic analysis. In: Annals of Mathematics Studies, vol. 101, v+151 pp. Princeton University Press, Princeton; University of Tokyo Press, Tokyo (1981). ISBN: 0-691-08289-8; 0-691-08292-8
62. Marcus, M.B., Pisier, G.: Characterizations of almost surely continuous p-stable random Fourier series and strongly stationary processes. Acta Math. **152**(3–4), 245–301 (1984)
63. Marcus, M.B., Pisier, G.: Continuity and boundedness of infinitely divisible processes: a Poisson point process approach. J. Theoret. Probab. **18**, 109–160 (2005)
64. Marcus, M., Talagrand, M.: Continuity conditions for a class of Gaussian chaoses related to Wick squares. J. Theor. Probab. **11**, 157–179 (1998)
65. Mendel, M., Naor, A.: Ultrametric subsets with large Hausdorff dimension. Invent. Math. **192**(1), 1–54 (2013)
66. Menchov, D.: Sur les séries de fonctions orthogonales. Fund. Math. **4**, 82–105 (1923)

67. Mendel, M., Naor, A.: Ultrametric skeletons. Proc. Natl. Acad. Sci. USA **110**(48), 19256–19262 (2013)
68. Mendelson, S.: On weakly bounded empirical processes. Math. Ann. **340**(2), 293–314 (2008)
69. Mendelson, S.: Empirical processes with a bounded ψ_1-diameter. Geom. Funct. Anal. **20**(4), 988–1027 (2010)
70. Mendelson, S.: Upper bounds on product and multiplier empirical processes. Stochastic Process. Appl. **126**, 3652–3680 (2016)
71. Mendelson, S.: An unrestricted learning procedure. J. ACM **66**(6), 1–42 (2020)
72. Mendelson, S., Paouris, G.: On generic chaining and the smallest singular value of random matrices with heavy tails. J. Funct. Anal. **262**(9), 3775–3811 (2012)
73. Mendelson, S., Pajor, A., Tomczak-Jaegermann, N.: Reconstruction and subgaussian operators in asymptotic geometric analysis. Geom. Funct. Anal. **17**(4), 1248–1282 (2007)
74. Mendelson, S., Rauhut, H., Ward, R.: Improved bounds for sparse recovery from subsampled random convolutions. Ann. Appl. Probab. **28**(6), 3491–3527 (2018)
75. Olejnik, J.: On Paszkiewicz-type criterion for a.e. continuity of processes in L^p-spaces. In: Stability in Probability. Banach Center Publications, vol. 90, pp. 103–110. Polish Academy of Sciences Institue of Mathematics, Warsaw (2010)
76. Olejnik, J.: On a construction of majorizing measures on subsets of $\mathbb{R}^n$ with special metrics. Studia Math. **197**(1), 1–12 (2010)
77. Ossiander, M.: A central limit theorem under metric entropy with L_2 bracketing. Ann. Probab. **15**(3), 897–919 (1987)
78. Paley, R., Zygmund, A.: On some series of functions (1). Proc. Camb. Phil. Soc. **26**, 337–357 (1930)
79. Paley, R., Zygmund, A.: On some series of functions (2). Proc. Camb. Phil. Soc. **26**, 458–474 (1930)
80. Paley, R., Zygmund, A.: On some series of functions (3). Proc. Camb. Phil. Soc. **28**, 190–205 (1932)
81. Paszkiewicz, A.: A complete characterization of coefficients of a.e. convergent orthogonal series and majorizing measures. Invent. Math. **180**(1), 50–110 (2010)
82. Pisier, G.: Ensembles de Sidon et processus gaussiens. C. R. Acad. Sci. Paris Sér. A-B **286**(15), A671–A674 (1978)
83. Pisier, G.: Sur l'espace de Banach des séries de Fourier aléatoires presque sûrement continues. In: Séminaire sur la Géométrie des Espaces de Banach 1977–1978, Exp. No. 17–18, 33 pp. École Polytech. Palaiseau (1978)
84. Pisier, G.: A remarkable homogeneous Banach algebra. Israel J. Math. **34**(1–2), 38–44 (1980)
85. Pisier, G.: Conditions d'entropie assurant la continuité de certains processus et applications à l'analyse harmonique. In: Séminaire d'analyse fonctionelle 1979–1980, Exp. No. 13–14, 43 pp., École Polytechnique, Palaiseau (1980)
86. Pisier, G.: De nouvelles caractérisations des ensembles de Sidon. In: Mathematical Analysis and Applications, Part B, pp. 685–726. Advances in Mathematical Suppl. Studies, 7b, Academic Press, New York-London (1981)
87. Pisier, G.: Probabilistic methods in the geometry of Banach spaces. In: Probability and Analysis (Varenna, 1985). Springer's Lecture Notes in Mathematics, Berlin, vol. 1206, pp. 167–241 (1986)
88. Pisier, G.: Subgaussian sequences in probability and Fourier analysis (1998). https://arxiv.org/abs/1607.01053v3
89. Pisier, G.: On uniformly bounded orthonormal Sidon systems. Math. Res. Lett. **24**(3), 893–932 (2017)
90. Radmacher, H.: Einige Sätze über Reihen von allgemeinen Orthogonalfunktionen. Math. Ann. **87**(1–2), 112–138 (1922)
91. Rauhut, H.: Compressive sensing and structured random matrices. In: Theoretical Foundations and Numerical Methods for Sparse Recovery. Radon Series on Computational and Applied Mathematics, vol. 9, pp. 1–92. Walter de Gruyter, Berlin (2010)

92. Rhee, W., Talagrand, M.: Exact bounds for the stochastic upward matching problem. Trans. Amer. Math. Soc. **109**, 109–126 (1988)
93. Rosinski, J.: Representations and isomorphism identities for infinitely divisible processes. Ann. Probab. **46**, 3229–3274 (2018)
94. Salem, R., Zygmund, A.: Some properties of trigonometric series whose terms have random signs. Acta Math. **91**, 245–301 (1954)
95. Shor, P.: The average-case analysis of some on-line algorithms for bin packing. Theory of computing (Singer Island, Fla., 1984). Combinatorica **6**(2), 179–200 (1986)
96. Shor, P.: How to pack better than Best Fit: tight bounds for average-case on-line bin packing. In: Proceedings of the 32nd Annual Symposium on Foundations of Computer Science, pp. 752–759 (1991)
97. Shor, P., Yukich, J.: Minimax grid matching and empirical measures. Ann. Probab. **19**(3), 1338–1348 (1991)
98. Talagrand, M.: Regularity of Gaussian processes. Acta Math. **159**, 99–149 (1987)
99. Talagrand, M.: The Glivenko-Cantelli problem. Ann. Probab. **15**, 837–870 (1987)
100. Talagrand, M.: The structure of sign-invariant GB-sets, and of certain Gaussian measures. Ann. Probab. **16**, 172–179 (1988)
101. Talagrand, M.: Necessary conditions for sample boundedness of p-stable processes. Ann. Probab. **16**, 1584–1595 (1988)
102. Talagrand, M.: Donsker classes of sets. Probab. Theo. Relat. Fields **78**, 179–191 (1988)
103. Talagrand, M.: On subsets of L^p and p-stable processes. Ann. Inst. H. Poincaré **25**, 153–166 (1989)
104. Talagrand, M.: Sample boundedness of stochastic processes under increment conditions. Ann. Probab. **18**, 1–49 (1990)
105. Talagrand, M.: Characterization of almost surely continuous 1-stable random Fourier series and strongly stationary processes. Ann. Probab. **18**, 85–91 (1990)
106. Talagrand, M.: A new isoperimetric inequality for product measure and the concentration of measure phenomenon. In: Israel Seminar (GAFA), Springer's Lecture Notes in Mathematics, vol. 1469, pp. 94–124 (1991)
107. Talagrand, M.: Cotype of operators from $C(K)$. Invent. Math. **107**, 1–40 (1992)
108. Talagrand, M.: Necessary and sufficient conditions for sample continuity of Random Fourier series and of harmonic infinitely divisible processes. Ann. Probab. **20**, 1–28 (1992)
109. Talagrand, M.: Sudakov-type minoration for Gaussian chaos. Israel J. Math **79**, 207–224 (1992)
110. Talagrand, M.: The Ajtai-Komlos-Tusnady matching theorem for general measures. In: Probability in a Banach Space, vol. 8, pp. 39–54. Progress in Probability. Birkhäusser, Basel (1992)
111. Talagrand, M.: A simple proof of the majorizing measure theorem. Geom. Funct. Anal. **2**, 119–125 (1992)
112. Talagrand, M.: Regularity of infinitely divisible processes. Ann. Probab. **21**, 362–432 (1993)
113. Talagrand, M.: The supremum of certain canonical processes. Amer. J. Math. **116**, 283–325 (1994)
114. Talagrand, M.: Matching theorems and discrepancy computations using majorizing measures. J. Amer. Math. Soc. **7**, 455–537 (1994)
115. Talagrand, M.: Constructions of majorizing measures, Bernoulli processes and cotype. Geom. Funct. Anal. **4**, 660–717 (1994)
116. Talagrand, M.: The transportation cost from the uniform measure to the empirical measure in dimension ≥ 3. Ann. Probab. **22**, 919–959 (1994)
117. Talagrand, M.: A remark on Sudakov minoration for chaos. In: Probability in Banach Spaces, vol. 9, pp. 341–346. Birkhäuser, Basel (1994)
118. Talagrand, M.: A borderline random Fourier series. Ann. Probab. **23**, 776–785 (1995)
119. Talagrand, M.: Embedding subspaces of L_p into ℓ_p^n. In: Operator Theory: Advances and Applications, vol. 77, pp. 311–326. Birkhäuser, Basel (1995)

120. Talagrand, M.: Sections of smooth convex bodies via majorizing measures. Acta. Math **175**, 273–306 (1995)
121. Talagrand, M.: Concentration of measure and isoperimetric inequalities in product spaces. Publ. Math. I.H.E.S. **81**, 73–205 (1995)
122. Talagrand, M.: Majorizing measures: the generic chaining. Ann. Probab. **24**, 1049–1103 (1996)
123. Talagrand, M.: The Glivenko-Cantelli problem, ten years later. J. Th. Probab. **9**, 371–384 (1996)
124. Talagrand, M.: A general lower bound for Gaussian chaos. J. Th. Probab. **11**, 157–179 (1998)
125. Talagrand, M.: Selecting a proportion of characters. Israel J. Math. **108**, 275–284 (1999)
126. Talagrand, M.: Majorizing measures without measures. Ann. Probab. **29**, 411–417 (2001)
127. Talagrand, M.: Gaussian averages, Bernoulli averages, and Gibbs' measures. Random Struct. Algorithm. **21**(3–4), 197–204 (2002)
128. Talagrand, M.: Vapnik-Chervonenkis type conditions and uniform Donsker classes of functions. Ann. Probab. **31**(2), 1565–1582 (2003)
129. Talagrand, M.: The Generic Chaining. Springer Monographs in Mathematics, viii+222 pp. Springer, Berlin (2005). ISBN: 3-540-24518-9
130. Talagrand, M.: Selector processes on classes of sets. Probab. Theory Relat. Fields **135**(4), 471–486 (2006)
131. Talagrand, M.: Are many small sets explicitly small? In: STOC'10–Proceedings of the 2010 ACM International Symposium on Theory of Computing, pp. 13–35. ACM, New York (2010). http://research.microsoft.com/apps/video/dl.aspx?id=137091
132. Talagrand, M.: Upper and Lower Bounds for Stochastic Processes, 1st edn. Springer, Berlin (2014)
133. Talagrand, M.: Scaling and non-standard matching theorems. C. R. Math. Acad. Sci. Paris **356**(6), 692–695 (2018)
134. Talagrand, M., Yukich, J.: The integrability of the square exponential transportation cost. Ann. Applied Probab. **3**, 1100–1111 (1993)
135. Tandori, K.: Über die divergenz der Orthogonalreihen. Publ. Math. Drecen **8**, 291–307 (1961)
136. Tandori, K.: Úber die Konvergenz der Orthogonalreihen. Acta Sci. Math. (Szeged) **24**, 131–151 (1963)
137. Tomczak-Jaegermann, N.: Banach-Mazur distances and finite-dimensional operator ideals. In: Pitman Monographs and Surveys in Pure and Applied Mathematics, vol. 38, xii+395 pp. J. Wiley, New York (1989). ISBN: 0-582-01374-7
138. Tsybakov, A.: Introduction to nonparametric estimation. In: Revised and Extended from the 2004 French Original. Translated by Vladimir Zaiats. Springer Series in Statistics, xii+214 pp. Springer, New York. ISBN: 978-0-387-79051-0 62-01. Ann. Probab. **46**(3), 1764–1805 (2009)
139. van de Geer, S.: Estimation and testing under sparsity. Lecture Notes in Mathematics, vol. 2159. École d'Été de Probabilités de Saint-Flour 2015 [Saint-Flour Probability Summer School], xiii+274 pp. Springer, Berlin (2016). ISBN 978-3-319-32773-0; 978-3-319-32774-7
140. van Handel, R.: Chaining, interpolation, and convexity. J. Eur. Math. Soc. **20**(10), 2413–2435 (2018)
141. van Handel, R.: Chaining, interpolation, and convexity II. Ann. Probab. **46**(3), 176–1805 (2018)
142. van der Vaart, A., Wellner, J.: Weak convergence and empirical processes. In: With Applications to Statistics. Springer Series in Statistics, xvi+508 pp. Springer, New York (1996). ISBN: 0-387-94640-3
143. Weber, M.: Analyse infinitésimale de fonctions aléatoires. Eleventh Saint Flour Probability summer school (Saint Flour, 1981). Lecture Notes in Mathematics, vol. 976, pp. 383–465, Springer, Berlin (1981)
144. Weber, M.: Estimating random polynomials by means of metric entropy methods. Math. Inequal. Appl. **3**(3), 443–457 (2000)

145. Weber, M.: Uniform bounds under increment conditions. Trans. Amer. Math. Soc. **358**(2), 911–936 (2006)
146. Yukich, J.: Some generalizations of the Euclidean two-sample matching problem. In: Probability in Banach Spaces, 8 (Brunswick, ME, 1991). Progress in Probability, vol. 30, pp. 55–66. Birkhäuser, Boston (1992)

Index

M. Talagrand, *Upper and Lower Bounds for Stochastic Processes*, Ergebnisse der Mathematik und ihrer Grenzgebiete. 3. Folge / A Series of Modern Surveys in Mathematics 60, https://doi.org/10.1007/978-3-030-82595-9